Ken Ring's

UNITED STATES OF AMERICA WEATHER ALMANAC

2025

BY THE SAME AUTHOR

Predict Weather Almanacs for New Zealand 1999-Present)
Predict Weather Almanacs for Australia (2006-Present)
Predict Weather Almanac for Ireland (2010-Present)
How to Get Your Kid to Like Maths
Predicting Weather by The Moon
New Meteorological Techniques
We Used to Call It Summer
The Romany's Apprentice
Moon and WeatherLore
Secrets of the Moon
Notes From the Pit
The Lunar Code
Fingermaths
Supertramp
A Fiery Tale

Paperback ISBN: 978-1-73862-309-9
Hardback ISBN: 978-1-0670445-0-3
eBook ISBN: 978-1-0670445-1-0

Email: enquiries@predictweather.com
Web: www.predictweather.com

CONTENTS

INTRODUCTION

DISCLAIMER

The forecasts in this work are best-endeavor opinions of the authors and associates. No guarantees of 100% accuracy is claimed. Neither can the author and/or publisher be held accountable for decisions made on the basis of information in these pages.

HOW TO USE THIS BOOK

Turn to the date required. The maps have been carefully prepared to describe the weather over the whole country. Next, turn to the Seasonal and Weekly Summaries in the Appendix. Look up the date in question. Finally, check the graphs at the end of the book that pertain to your town or state.

Regional Map of the USA

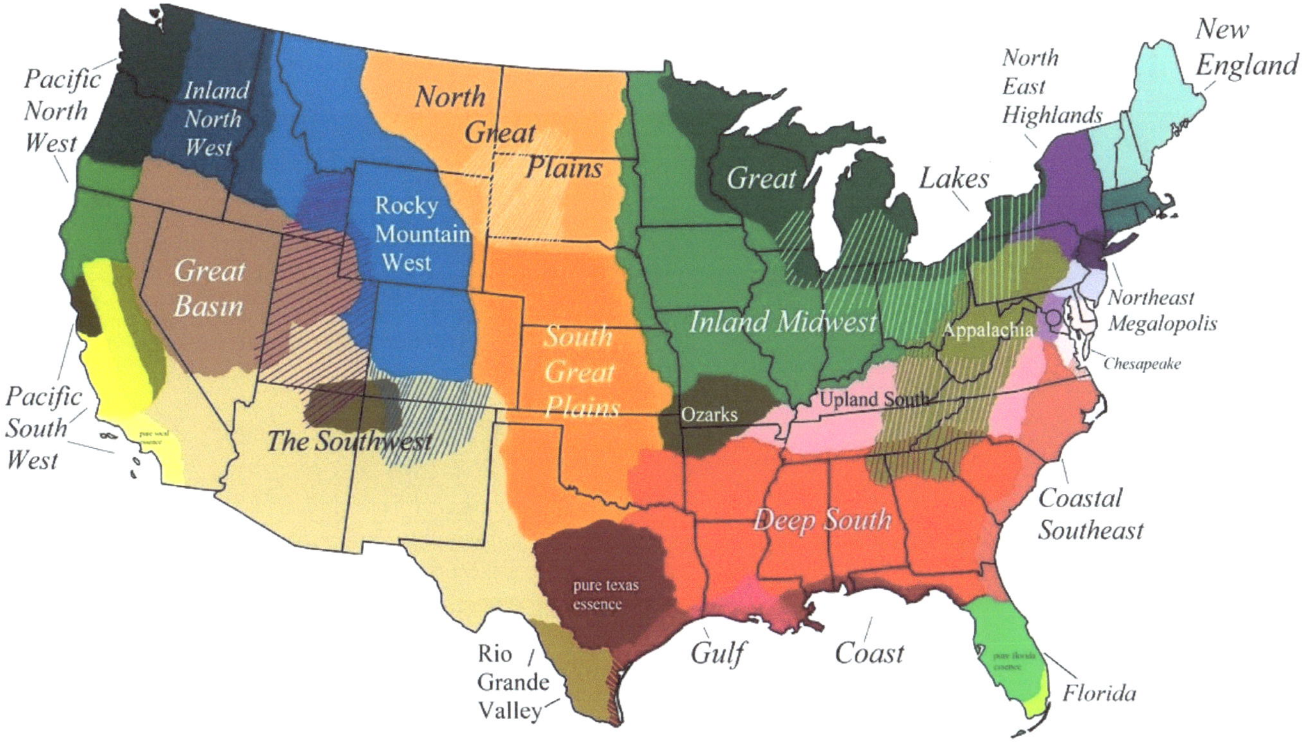

KEYS TO READING MAPS AND PHASES

MOON PHASES

Full Moon

Third Quarter

New Moon

First Quarter

READING AN ISOBARIC MAP

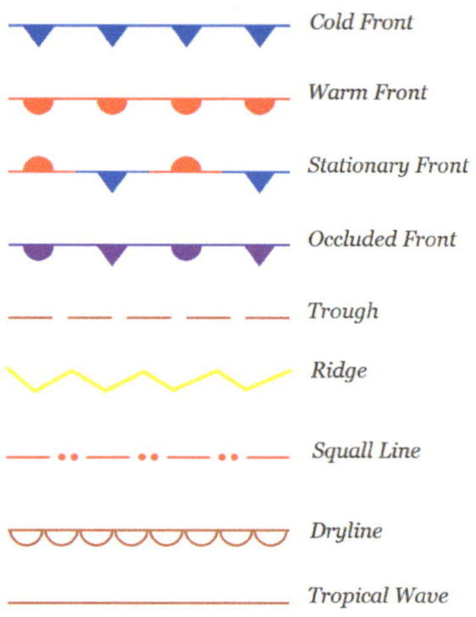

FRONTS

A leading edge of cold air that is displacing a warm air mass. Cold fronts nearly always prevail south and west of the center of a low pressure area and never from high-pressure systems.

A leading edge of warm air that is replacing a colder air mass. Warm fronts extend from the center of low-pressure areas, mostly from their east side.

A stationary front is one that is not moving (i.e. the two air masses on either side are not moving and one is not displacing the other). It can either be a cold or warm front by nature. Generally, there is little variation in temperature or wind on either side of a stationary front.

There are two types of occluded fronts – warm and cold, with cold occlusions being the most common. An occluded front usually indicate that the originating weather system has reached its maturity and is now moving into its decaying phase.

Cold occlusions are when the cold front overtakes the warm front and undercuts the cool air mass ahead of it.

Warm occlusions are when the air behind the cold front is in fact, not as cold as the cool air ahead of the warm front. As such, the warm air is forced up, but the colder air remains at the surface level.

OTHER BOUNDARY LINES

Dryline

A dry line can indicate a boundary for a strong storm system moving eastwards into the Mississippi Valley or beyond, regardless of the time of day. It can sometimes indicate the development of severe tornadic thunderstorms along its edge or just to the east of it.

It usually brings a sharp drop in humidity, a rise in temperature, a wind shift from south or southeasterly to west or southwesterly, and clearing skies. It may result in blowing dust, particularly if it occurs during the daytime.

A dry line marks the boundary between a moist air and dry air masses. It typically lies north-south across the central and southern high Plains states during the spring and early summer, where it separates moist air from the Gulf of Mexico to the east and dry desert air from the southwestern states.

A typical dry line moves in an easterly trajectory usually in the afternoons, before retreating back in a westerly trajectory during the night.

Squall Line

A squall line usually forms along a front and moves ahead of it. It generally brings a line of thunderstorms in its path. In some cases, it can develop into straight-line wind storms often associated with severe thunderstorms and hurricane-force winds, heavy rain, flash floods, wind shears and even tornadoes in summer months, or blizzards in winter months.

OTHER SYMBOLS

Trough

Troughs can be areas where showers and thunderstorms can form.

It is an area of lower air pressure extending from the center of a low pressure system. Changes in wind direction usually occur within a trough, but there is no change in air mass. It usually reflects a change in atmospheric conditions in the upper atmosphere.

Ridge

A ridge usually indicates an area of sunny weather, sometimes where fog or haze occurs.

It is an area of high pressure extending from the center of a high pressure system.

Tropical Wave

These formations can lead to the development of tropical cyclones.
It is an inverted trough of low pressure by nature, usually moving from east to west across the tropics. It can also be referred to as an 'easterly wave'.

WEATHER GLOSSARY

To get the most out of the forecasts, it helps to be able to read a weather map and to understand the language used.

A high air pressure – a fair-weather sign, is when the isobaric values increase inwards towards the center.

A low-pressure zone or depression, indicates worsening conditions, and when isobaric values decrease towards the center.

Closer isobars = stronger winds.

TERM	MEANING	WEATHER
Apogee	Moon furthest from Earth for month	Winds generally light. Calm conditions.
Perigee	Moon closest to Earth for month	Spells trouble Earthquake risk in some countries.
Cold front	Cold air advancing, pushing away warm air	Changeable
Depression	Area of low barometric pressure	Unpredictable, bad weather, strong winds
Ridge of high pressure	Edge of anticyclone	Good weather approaching, cooler nights in winter
Stationary front	Slow-moving weather	No change for 2-3 days
Northern declination	Moon at northernmost point for month. Moon rises in the north.	Slow moving weather system, northerlies, rain in north, colder conditions
Southern declination	Moon at southernmost point for month	Slow moving system, southerlies, precipitation in south, generally milder.
Moon crossing equator	Faster moving weather system, moonrise is due east.	Turbulent, changeable, often electrical storms, gales or winds expected.

SEASONAL SUMMARIES 2025

WINTER (JANUARY 1 – MARCH 19)

Winter storms proliferate more frequently in the second half of this winter, generally following a pattern of starting in the southern Rocky Mountains to the South Great Plains before spreading in a north-eastward trajectory into the Great Lakes and Northeast.

Many wintry storms bring heavy snow and ice, and very bitterly cold winds, from the Rocky Mountains, across the mid-section of the country to the Great Lakes and Northeast while below average precipitation prevails over the southern half of the country.

Unusually mild over most of the north and east of the country from the Upper Midwest eastwards, particularly over the first half of the season, then a cool change comes in the second half expanding colder than normal temperatures from the Great Plains to the East Coast.

Cooler than normal in the south-central parts of the country, including from California to the central and southern Great Plains over the first half of the season, before a warmer change moves into the West during the latter stages, particularly in February.

SPRING (MARCH 20 – JUNE 20)

Precipitation below average in many parts west of the Rocky Mountains and in the Southeast, particularly during March and May, when the Mississippi Delta to the Tennessee Valley and east to the Atlantic coast may receive less than half the normal amount of rainfall resulting in drought conditions developing.

Unusually wetter than normal in the central regions, where late season snow and abundant rains will prevail.

Temperatures will be unusually mild at the start and end of spring compared to normal, with an intense cold snap in the middle, when snow events return.

SUMMER (JUNE 21 – SEPTEMBER 22)

Precipitation will be frequent and intense in the south-central parts of the country, including in the Midwest. The Southeast will be very dry with drought conditions, particularly from North Carolina to Tennessee.

The Northwest, including Montana, will be unusually hot and dry, increasing the risk of wildfires, while regular monsoon showers in the Southwest will ease the risk of seasonal fires.

Heat wave temperatures through the middle of summer, especially in the north Rocky Mountains and North Great Plains, then shifting to the Southeast in the last month of summer. Cooler than average in Texas.

AUTUMN (SEPTEMBER 23 – DECEMBER 21)

A late start to autumn almost everywhere, except in a line from Oregon to the southern Rocky Mountains.

Precipitation slow to build during the season, with the first significant rains to bring some drought relief in the southeast regions arriving at the end of October into early November, and in late November to early December in the Pacific Northwest and California to the Four Corner States, when storms bring high winds and flooding rains to the west.

Overall, autumn temperatures well above normal almost everywhere, despite the occasional cold snap, except for near normal temperatures in the Deep South, and below average temperatures in the Far West.

Autumn freezes, particularly in the southeast, arrive late October, while the central and southern Great Plains and southern Atlantic States season will be warm and dry.

WEEKLY SUMMARIES 2025

JANUARY 1 - 10

- Winter storms may prevail across the central High Plains and into the adjacent Rockies, causing disruptions at times to travel corridors and livestock.
- One storm in particular may be severe through the mid-section of the country.
- The Northwest may start out wet at first, before becoming mostly dry mid-week.
- Rain and showers expected in Southern California and possibly be accompanied by very high winds in the Four Corners States.
- Rain may be heavy in the coastal Pacific Northwest, with lighter rain further inland.
- Snow and freezing rain expected regularly on the Plains to the upper Midwest, as well as in much of the Corn Belt, (an area that may remain almost untouched by ice and snow this week with temperatures possibly up to 20°F warmer than is typical for this time of year).
- Towards the end of this week, a cold front may bring localized thunderstorms across the eastern parts of Texas and into the Southern Atlantic states, with the strong thunderstorms bringing funnel formation winds in some parts of Florida.
- The eastern Texas region to the lower Mississippi Valley may receive enough rain to cause localized flooding during this week.
- Only isolated to light showers expected in the Southwest, North Great Plains, northern parts of the Corn Belt, and around the mid and northern Atlantic Coastal States, with some of these districts being mostly dry throughout the week.

JANUARY 10 - 16

- The west may continue to be mostly wet, while the east may be drier.
- The Pacific Northwest to the eastern parts of the Midwest including in the Corn Belt may continue to be wet with a mix of wintry precipitation of rain or snow, and although it may be wintry, snow may be scarce in most parts of the Corn Belt.
- There may be some rain relief sprinkling the Southern Atlantic districts, with moderate levels of rainfall expected over much of the Gulf Coast, Atlantic coast states and down through the Ohio River Valley.
- Snow will be a major feature of the Central Plains to the upper Midwest.
- The southern half of the West, including on the Great Plains, may be mostly dry.
- Despite temperatures being above average in most parts of the country this week, it may still be very cold with heavy snow in eastern parts of Colorado and into the western districts of Kansas.
- The Midwestern states may be unusually warmer than is usual for this time of year, with some parts possibly up to 20°F warmer than normal in the upper Midwest.
- Bands of severe thunderstorms cross the southern states, with some possibly generating a tornado watch, especially in parts of Louisiana to South Carolina.
- An intense storm moving through the Great Plains is expected to drop heavy snow from New Mexico to North Dakota.

JANUARY 17 - 23

- Very cold air and a significant snow storm moving up through the middle of the country, with the cold even reaching into parts of the Californian agricultural valleys and agricultural regions of Arizona, possibly seeing freezing temperatures and frosts threatening damage to crops.
- Widespread sleet and snow across the Great Plains and Midwest states, possibly causing disruptions to infrastructure and power supplies, but with lighter snowfalls in parts of Montana and western South Dakota areas.
- Rain activity expected to ease in the Northwest for a time, followed by a few days of light snow across the mountainous West to the Southwest.
- Snow may be lighter on the central High Plains, while particularly heavy around the northeast of Texas to the Great Lakes region.
- Stormy conditions bring some further snow to the west of the Corn Belt, as well as wintry rain into the Mississippi Valley and heavy rain in the Ohio Valley.
- It may be mostly dry and warm in Florida.
- Overall, this week may see a storm system crossing into the West Coast states at first, while milder conditions prevail in the east, followed by cold snap starting in the Northwest and spreading quickly southwards and snow and almost blizzard conditions envelope the Great Lakes to the Northeastern states.
- Showers daily along the Pacific Northwest coast. Further inland may be mostly dry.

JANUARY 24 - 31

- Freezing conditions down through the Southern California regions to the southwest of Arizona, possibly still threatening local crops in the valleys, with these areas temperatures around 10°F below normal.
- The areas from the east Rocky Mountains, down the Cascade Ranges and into the Sierra Nevada regions may be up to 20°F colder than is normal for this time of year.
- Southern Texas may be milder, with temperatures closer to frost to slightly above the freezing level over this period, and the Northern High Plains and the eastern States nearer to normal temperatures.
- During this outlook period, another couple of snow and ice storms can be expected crossing the Southern Rockies and into the central and south Great Plains up into the Northwest, with the chance of further disruptions to travel and power possible, with the first arriving around midway through the week.
- A second storm is likely to bring heavy rain into the western and central Gulf Coast states, and move northeasterly into the Ohio Valley, possibly resulting in some localized flooding along the lowlands.
- Further north however, it may be mostly dry and mild mostly in Montana to South Dakota areas, despite still be covered with patches of shallow snow.
- Elsewhere, some showers of rain or snow prevail in the Northwest.
- Florida is likely to remain mostly warm and dry.

WEEKLY SUMMARIES 2025

FEBRUARY 1 – 6

- From the southern half of Texas, through the Gulf Coast areas and into the Southeast, including most of Florida, can all expect productive rainfall, while mild and dry conditions may prevail for most of the time in the states of Montana and South Dakota, with much of the snow cover melting away.
- It may also be mostly dry in the Southwest, but rainfall may be significant at times in the Midwest just south of the Great Lakes, including with the chance of wintry rainfall falling as snow in the region.
- Temperatures may return to close to normal for this time of year in many parts of the country, with the California to the Southwest regions gradually warning to close to norm as the week progresses.
- It is likely to also be much warmer than average by up to 10-15°F above normal across the Great Plains.
- Many eastern and southern regions may be cooler than what is normal for this time of year with regions from the Rocky Mountains to the central and southern High Plains possibly down to 5-10°F below norm.

FEBRUARY 7 - 13

- Colder than normal almost everywhere, with particularly deep snow covering the central High Plains as cold air sweeps across the northern and central Plains into the Northeast regions.
- During this cold snap, temperatures may average around 10-20°F below what is typical for this time of year in some states, particularly those in the central Great Plains to the west of the Mississippi Valley, and including across most of the states of Wyoming and Colorado.
- This cold spell may bring wintry precipitation over much of the country, with much of it falling as snow with significant accumulations in places, particularly over the central High Plains and severe freezing temperatures in the Great North Plains, while significant rainfall is likely in the Southeast and in eastern parts of the Gulf Coast states.
- Milder temperatures return to California and rainfall amounts low.
- Storm activity producing tornados midway through this week, in central parts of Florida.

FEBRUARY 14 - 20

- Good rainfall prevails down through the Pacific Northwest into Northern California and the northern Rockies, but dry conditions are likely to persist in Southern California to the Southwest including Arizona.
- In the east, most rainfall prevails in the southern parts of Texas and Florida and in a narrow band of wintry precipitation from the Upper Midwest to the Mid-Atlantic States.
- Snow may occur almost daily in the west of New York over this week.
- Temperatures may generally be above normal in the West, but below normal across the Great Plains, with some parts possibly cooler by 25°F than normal for this time of year.
- It is also likely to be cooler than average through the Mississippi and Ohio Valleys and up around the Great Lakes, to the middle Atlantic Coast states and Northeast.
- The Northern Plains and eastward to the Ohio Valley can expect further light snow, as well as further extreme cold and snow across the northern to middle states, while temperatures may become milder in the Southeast region.
- This coming week may be mostly dry and with milder temperatures across Alaska, with some parts possibly up to 15°F above normal, particularly in the west of the State, and only some showers mostly in southern parts.
- Hawaii may be cooler and drier than normal.

FEBRUARY 21 - 28

- Intense storm activity over much of the Midwest and Northeast, bringing heavy snow, sleet, and wintry rain to the regions, and bands of thunderstorms crossing the southern States.
- Possible blizzard conditions may develop over this week across the Great Plains to the Midwest, possibly causing disruptions to travel corridors and infrastructure.
- Ahead of the storms, a very cold airflow can be expected to cross southward through the Plains and Midwest, and later into most parts of the Rockies.
- Many areas in the northern and central Great Plains may see temperatures of around 15°F below normal, while further west temperatures may be generally up to 10°F above normal.
- A warmer change arrives in the Southwest to California, while a cooler change is likely in northern Florida.
- Alaska may be mostly dry and warmer than normal for this time of year, while Hawaii may be showery.

WEEKLY SUMMARIES 2025

MARCH 1 - 6

- A strong storm system will cross the southern to central Plains in a Northeasterly track, which may bring further heavy wintry precipitation, sleet, and snow across the upper Midwest, including around parts of the central Corn Belt.
- Heavy snow expected in the northwest of Kansas.
- In the west, significant rain prevails in the Pacific Northwest to northern and central California, and with snow showers in the Rockies.
- Frequent shower and rain activity may also occur regularly in the Gulf Coast and Southeast regions, possibly bringing pockets of localized flooding.
- This week may see mild and mostly dry conditions with strong winds stirring up a dust storm in the Texas Panhandle region, and mostly dry, warmer, and sunny conditions in Florida.
- A strong southerly flow is expected over much of this week bringing temperatures of around 5-10°F above the norm in many central parts of the country from the Rocky Mountains to the Mississippi River and upper Great Lakes regions as well as in most parts of the Southeast.
- The Pacific Coast to Nevada and Arizona, as well as Florida, the Northeast, and the Ohio Valley to the Mid-Atlantic may be as much as 5-10°F below normal.

MARCH 7 – 13

- Another strong storm system crosses the eastern half of the country, with the generation of further snowstorms developing around the Great Lakes, North Great Plains and Inland Midwest to the Northeast regions, with such strong winds that blizzard conditions may develop for a time.
- Outbreaks of heavy rain expected along both the Pacific Coast and the Atlantic Coast regions and further stormy weather in the Southeast region may possibly generate thunderstorms, some with hail, and the possible formation of tornadoes.
- Much of the Southwest and the central to southern Great Plains may have a respite from the worst of wintry weather and be mostly dry, as well as in southern California to Arizona.
- Temperatures may be colder than normal in the western half of the country, while despite the unsettled stormy conditions, it is likely to be warmer than normal for this time of year in the east. Very cold in Alaska with temperatures possibly as much as 25-30°F below normal.

MARCH 13 - 20

- The Pacific Northwest receives significant rainfall, with snow in the higher reaches, while warmer and dry conditions are likely to prevail over much of the Great Plains.
- The built up snow in the Inland Midwest may begin melt, possibly resulting in some flooding in the lowlands around Illinois, Indiana and Ohio.
- The Deep South can anticipate drier conditions over most of this coming week.
- The California region to the Southwest including Arizona and across into Texas can expect mostly dry and warmer conditions to prevail, similarly over most of Florida and up through the west of the Appalachians.
- Many parts of the country may enjoy a respite from the recent stormy activity before the arrival of another brief cold snap.
- Temperatures over the coming week may average around 5-15°F warmer than normal throughout the Rocky Mountains, High Plains, in Kansas and Oklahoma, while parts of the Great Lakes to the Mid-Atlantic and Northeast regions are likely to be colder than normal by a similar variant.
- A cold wave may affect much of Alaska during this week with many parts of the state seeing temperatures around 10°F below normal for this time of year. Heavy snow in the southeast of the state; but it is likely to be mostly dry elsewhere.

MARCH 21 – 27

- Warm air over the Gulf of Mexico may lead to scattered thunderstorms and widespread heavy showers through Texas, and outbreaks of severe weather with the chance of flash flooding pockets in southern Texas and into the Gulf regions.
- Significant rainfall expected in parts of the Pacific Northwest, in the lower parts of the Mississippi Valley, Ohio Valley, and through the Mid-Atlantic regions into the Northeast, where sleet and snow may once again cover parts of the Northeast.
- Snowmelts may continue to cause some pockets of lowland flooding, particularly in the Corn Belt areas.
- Dry weather at first in the Great Plains and Southeast, before the arrival of showers towards the end of the week. Dry conditions are likely to continue down through the Florida Peninsula as well as in some areas west of the southern Appalachians.
- The last few days of March may see brief outbursts of precipitation across the Pacific Northwest and Southwest regions.
- Cold and dry conditions are likely to persist most of this week in Alaska, except for heavy snow likely for a time in the southeast, while it may be showery and warm in Hawaii.
- Warmer than normal almost everywhere by as much as 10-15°F above the average in the west from the Pacific coast to the Rocky Mountains and into the Great Plains.
- A colder change arrives in the Northeast over the second half of this week.

WEEKLY SUMMARIES 2025

MARCH 28 – APRIL 3

- Brief outbursts of precipitation can be expected across the Pacific Northwest and Southwest regions, while further east rainfall events are expected to be heavier and more widespread especially across the Great Plains and in particular towards the end of the week.
- Melting snow in the Inland Midwest and the Mississippi Valley may aggravate the localized lowland flooding that may have already occurred in these regions.
- Thunderstorms and tornado development towards the end of this week.
- This week may see mostly warm and dry conditions prevailing across the South.
- Above-average temperatures may swiftly re-establish themselves early in this week, with many parts of the country once more up to 10-15°F higher than normal for this time of year, including over much of the Central Plains.
- Only parts of the Northeast are likely to be colder than average but only by around 5°F compare to their norm.
- These warmer than normal temperatures may accelerate the snowmelt process, particularly in the West, leading to heightened wildfire risk later in the year.
- Once again, Alaska can expect this week to be mostly dry and unusually colder than average, but still with falls of heavy snow mostly in the southeast.
- Hawaii is likely to enjoy a warm week with only the occasional light showers.

APRIL 4 – 10

- Very severe storms over the Great Plains to the Inland Midwest, with the risk of large hail, strong winds, tornadoes, heavy rain and flooding in the lowlands.
- This stormy event is likely to spread over a couple of days into central parts of Texas, Oklahoma, and Kansas, with the risk of flash flooding.
- Parts of the north Rocky Mountains to Wyoming may receive falls of heavy snow.
- This week also brings some good rains into the Mississippi Delta region after having been mostly dry over the recent past, while moderate to heavy rainfall may prevail throughout the Great Plains, the Mississippi Valley, and Ohio Valley regions.
- It may generally warm and dry in the Southeast, and mostly dry and windy in Southern California to New Mexico.
- Temperatures are likely to be close to average for most of this week in the West and Coastal Atlantic regions, while being warmer than average over most of the Midwest and South.
- Temperatures up to 10°F above normal across the east of the Great Plains to the Appalachians, and up to 20°F above average over much of the Inland Midwest.
- The storm system expected over the first few days of this week may bring a slightly cooler change in the West to the Great Plains.
- Some shower activity prevails in Hawaii.
- In Alaska, while the bitterly cold conditions of recent may begin to ease, temperatures remain around 10°F below the norm in the interior of the state.

APRIL 11 - 17

- Rainfall may be variable across much of the North Great Plains, as well as in the west and central Gulf Coast regions and further north into the Great Lakes and into the Northeast, with some good rainfall likely around the central Gulf Coast and the Northeast regions, and lesser elsewhere.
- Possible snowfall in and around the Great Lakes and Northeast states this week.
- The Southwest to California, including in Nevada, Utah and Arizona states can all expect to be mostly dry and with temperatures of up to 5-15°F above their norms.
- There is likely to be a return to winter-like conditions in many regions east of the Rocky Mountains, with a very cold wave spreading across the Great Plains and into the Inland Midwest, Mississippi Valley, Ohio Valley, and Great Lakes region bringing a sharp end the warmer spell of the previous week.
- This cold outbreak is likely to be due to the passage of a storm crossing the north-central parts of the country towards the end of the week.
- This blast of Artic air may reach down into many parts as far north as Kansas and Missouri reaching freezing levels once more, including into parts of Texas and the Deep South states as up through into the Southeast.
- Only Florida and the west and central Gulf Coast regions may avoid the cold wave.
- Many districts may see average temperatures dropping by as much as 10-15°F below normal, except in the Mississippi Delta to the coastal Southeast, where maximum temperatures may be closer to their average for this time of year despite overnight minimums still dropping between 5-10°F below their norms.
- In Hawaii, unusually cooler than normal and mostly dry weather prevails.
- In Alaska, the cold spell is expected to break with temps 10-15°F warmer than normal. Wet conditions persist in the southeast, and drier elsewhere.

APRIL 18 - 24

- A mid-latitude cyclone develops and crosses into the Northeast early in the week, producing widespread rain over much of the Northeast region, and possibly resulting in flooding along a number of the New Jersey river systems.
- Some of this precipitation may be very wintry. A blizzard situation may develop for a time with heavy snow dumps and strong winds in this region.
- In other parts of the country, moderate to heavy rain can be expected in the Pacific Northwest region, as well as from the central Great Plains eastward into southern parts of the Inland Midwest to the northern parts of the Deep South and into the Southeast region and up along the Atlantic coastal states.
- There is a further chance of late seasonal snowfall as far south as Kansas.
- Rainfall late in the week may be very heavy for a time in the Mississippi Delta.
- Towards the end of this week, the Pacific Northwest may fall under the influence of a couple of major storms, bringing frequent showers to the region, as well as in parts of the northern and central Rocky Mountains.

WEEKLY SUMMARIES 2025

APRIL 18 – 24

- At the same time, a couple of storms may maintain wintry weather across the Great Plains and into the Midwest.
- It may be cooler than normal almost everywhere, except in the far south in Florida.
- Overall, most places east of the Rocky Mountains may see temperatures averaging around 5-15°F below normal for this time of year.
- Slightly cooler than normal to the west of the Rocky Mountains, but only by around 5°F below normal, with only the Southwest possibly slightly warmer than average.
- Alaska will be mostly dry and mild with temps 10°F above normal in some places.
- Warm and mostly dry in Hawaii, with occasional heavy showers in the west.

APRIL 25 – MAY 1

- A subtle shift into more favorable spring-like weather, although many parts around the Southwest of the Inland Midwest may remain more wintry, cold and wet.
- Mostly dry conditions prevail around southern Georgia, heightening the risk of fires from the development of tinder-like conditions in the environment.
- Other central and southern parts of the Mississippi Delta can expect scattered showers, while northern parts of the country can expect more significant rain.
- A powerful Atlantic storm is likely to cross the Northeast, bringing heavy rain and strong winds, as well as the chance of further snow in the higher reaches.
- This event may bring disruptions to travel and infrastructure and with a warmer change not arriving until near the end of the week. This storm is also likely to bring a cooler air across the south and east of the country for several days.
- Around midweek, a wet spell will bring rain across the Great Plains and northern Rocky Mountains to the Pacific Northwest and down into northern California, as well as the chance of further snow mostly in Montana to Nevada and Utah.
- Towards the end of this coming week, southern California can expect some heavy rain. Severe thunderstorms likely across the Great Plains and the Inland Midwest.
- It is likely to be a mostly dry week throughout the Southwest.
- There is a chance of storms feeding tornadoes Texas at the end of this week.
- Temperatures may be below average over much of the West this week, from Oregon and down through the Pacific coastal regions and into Arizona to the south and up into the Rocky Mountains, but only by around 3-5°F below the norm.
- Temperatures average to slightly below average in the Mississippi Delta, Atlantic Coast and Gulf of Mexico, into the Southeast and as far north as the Appalachians.
- Many parts of the Great Plains and the Inland Midwest are likely to be warmer than average, but also only by a small variant of around 3-8°F above the norm.
- It is likely to be a warm week with local heavy showers in Hawaii.
- Alaska will be mild and mostly dry in Alaska for the third week in a row, with temperatures up to 10°F above the norm for this time of year in the interior and northern parts of the State, while near normal elsewhere.

MAY 2 - 8

- Warm and mostly dry except for some isolated showery spells in the West to the North Great Plains and into the upper Midwest.
- Mostly dry in the southeast, including in South Carolina, Georgia, and Florida.
- Heavy rain in parts of Nebraska, Iowa, and surrounding regions. Localized flooding.
- Early in this week ahead, there is a chance of scattered thunderstorms mostly in the south-central parts of the country.
- Later, some storms arrive midway through this week across the Central Plains and into the Corn Belt regions of the Inland Midwest.
- Moderate amounts of rainfall expected from the Delta region and spreading across the country in a northeastward direction, although some parts of the Mississippi Delta region may not receive enough to alleviate parched areas.
- Temperatures likely to be average to slightly warmer in the Mississippi Delta, while most other parts of the country will have a warmer than normal week almost everywhere, bringing a respite from the recent cooler than normal spell.
- Temperatures will be close to average in central and southern parts of the Rocky Mountains and in the adjacent Great Plains, while most other parts of the country will be around 5°F or more above normal.
- Mostly dry in Hawaii except for some showery spells midway through the week.
- In Alaska, most rainfall will be confined mostly to the southern parts, while mostly dry elsewhere, and temperatures generally milder than the norm.

MAY 9 - 15

- Some productive rainfall prevails around eastern parts of the Delta and parts of the Tennessee Valley to the Carolinas and central parts of the country between the Mississippi Delta and Tennessee Valley regions can expect above average rainfall.
- Very heavy rain in some parts of Texas, Kansas, Nebraska, Iowa, and the Dakotas.
- Average to heavy rainfall expected over the Great Plains, particularly towards the end of this week and into the next week, with a high risk of flooding developing, especially in the states of South Dakota and Kansas.
- River levels to fall unusually lower than normal in Florida around this time as drought conditions establish in the southeast of the country.
- Thunderstorms likely on the Great Plains towards the end of this week, with bursts of hail, wind, and flooding that may cause some localized damage. Heightened risk of tornadoes once more in the Oklahoma to Kansas region.
- Increased risk of fire across the Deep South, particularly in southern parts of Georgia where there is a chance of wildfires in the second half of this week.
- Light to moderate rainfall in the west, possibly including some late-season snow.
- Chance of late season frosts in some parts of the Pacific Inland Northwest.
- Drought conditions intensify as the week progresses in parts of southern California and into Arizona.

WEEKLY SUMMARIES 2025

MAY 9 - 15

- Almost all parts of the country can expect temperatures to be above the average.
- Most regions can expect temperatures of around 2-10°F above the norm, with the only exceptions of slightly cooler being down through the Pacific Coast to the Pacific Northwest, and in the New York to northern New England regions, where temperatures are more likely to average around 1-5°F below normal.
- Warm and mostly dry weather prevails in Hawaii.
- Cooler weather may return once more to Alaska.

MAY 16 - 22

- Heavy rain continues, with the chance of causing some pockets of lowland flooding in parts of the central and Southern Great Plains into the central parts of the Mississippi Valley.
- Also, very heavy rain may occur in parts of Kansas, Iowa, Missouri, Oklahoma, and Texas during this week, particularly early in the week and possibly resulting in some flooding along local river systems in the region by the end of the week.
- A storm situated in the west of the Atlantic Ocean may bring very strong winds up through the eastern coastal regions, particularly affecting Virginia.
- This storm may be named by mid-week before degrading into a low-pressure system as it drifts further southward alongside the Florida peninsula.
- It is likely to be a mostly dry and warm week in the West of the country, with most precipitation out of the Northwest and Rocky Mountain regions early on.
- Despite this warmer trend in the West, there is a chance of some lingering light snow in the Colorado Rockies.
- A hot week in the Mississippi Delta region, with a wet spell about mid-week bringing an uneven mix of light showers in some places, while heavier outbreaks prevail in other parts, particularly in the central to northern parts.
- More favorable weather conditions through the northern and eastern Corn Belt.
- Drought conditions may continue to develop in the Southeast of the country with the fire risks intensified by strong winds blowing inland from the Atlantic Ocean throughout the week.
- Temperatures may rise to be as much as 10°F above normal by the end of this week in the middle states, and in general, most parts of the country can expect above average temperatures on most days.
- It is likely to be very warm mid-week in California, particularly in the south.
- It will be another mostly dry and warm week in Hawaii.
- It will be cooler than normal in most of Alaska, with many areas possibly around 5°F below normal.

MAY 23 - 29

- Warm and mostly dry conditions in the West continues, except for some occasional scattered showers south of the Rocky Mountains, with most of the precipitation confined to the southern Rockies.
- Snow-packs in the West will continue to melt.
- The Great Plains continues to be cooler than normal with bands of showers, especially in the south, while the north is likely to be warm and dry.
- Some shower activity continues in parts of western Texas to the eastern Dakotas.
- Further to east, more rainfall is likely, possibly adding to pockets of flooding in low-lying areas in the southwest of the Inland Midwest Corn Belt.
- A slow-moving storm brings heavy rain once more into the Northeast, especially in the New York to New England regions.
- It will continue to be mostly dry and warm in the Southeast, with the increasing threat of drought possibly beginning to cause concern.
- There is a chance for a cold front crossing the Mississippi Delta region towards the end of the week, bringing a cooler change.
- Cooler than normal temperatures will prevail in the east and south, and warmer than normal temperatures in the west and north with a range of 5-10°F variance from the norm either above or below throughout.
- There is a risk of some late season frosts in some parts of the Inland Midwest region at the end of this week, with the northwest of Michigan possibly at higher risk.
- Warm and mostly dry conditions will prevail over much of this week in Hawaii.
- Mostly dry weather with near normal temperatures prevails in most parts of Alaska, except in the southern parts where further wintry precipitation is expected, some of it as snow, particularly over the first half of this week.

MAY 30 – JUNE 5

- On the Atlantic coastal side of the country, this week will be mostly dry and warm around much of the Inland Midwest, with increasing dry conditions starting to develop in the central and eastern Corn Belt regions.
- Drought conditions in the Southeast regions continue to be a concern, especially around the Mississippi Delta, as the warm and mostly dry weather continues.
- It will be mostly dry and warm this week in the Southwest.
- There may be some significant storms, particularly midway through the week.
- Bands of thunderstorms are likely mid-week through central regions from the Mississippi Valley and down into northern Texas, possibly bringing tornadoes, as well as very large showers of hail and pockets of flooding in low-lying areas.
- Parts of the central and southern Great Plains to the Missouri Valley can also expect thunderstorms and showers, possibly resulting in some localized damage.

WEEKLY SUMMARIES 2025

MAY 30 – JUNE 5

- Scattered outbreaks of rain expected over the North Great Plains.
- In the Northwest, there is a chance for further rain and snow showers mostly over the Rocky Mountains, while areas west of the Rockies may remain mostly dry.
- Temperatures will steadily rise in the Mississippi Delta region during this week, while it is likely to be slightly cooler in some parts of the Great North Plains but only by around 5°F of the norm for this time of year.
- There is also the chance of some pockets of frost this week, mostly around the Great Lakes region but also for a time up into the Northeast, as well as around the North Great Plains and northern parts of the Inland Midwest.
- Hawaii may continue to be mostly dry with only a brief wet spell.
- It will also continue to be mostly dry and mild in Alaska this week, except for further outbreaks of heavy precipitation prevailing once more in some southern parts, and pockets of unusually warmer than normal temperatures in inland parts.

JUNE 6 - 12

- Significant heavy rainfall in Texas for a couple of days.
- Some showers and scattered thunderstorms expected across the Great Plains from Texas to the Dakotas, where some lowland flooding pockets may persist.
- Midway through this week, there is a chance of a tropical storm forming in the Gulf of Mexico crossing Florida by the end of the week.
- Heavy rain is likely with this storm, and should bring relief to parts of the Deep South suffering from a prolonged dry spell.
- Some wind is also likely.
- Hot and mostly dry in the Mississippi Delta region, where only some showers, occasionally accompanying thunderstorm activity, will be more regionally specific, leaving some other parts of the region still remaining dry.
- Much of the Southeast will continue to be mostly dry with drought conditions.
- Mostly dry and warm in the West, and most of the eastern Inland Midwest areas and the Ohio Valley region may not receive any rainfall at all.
- Mostly dry and hot west of the Rocky Mountains to the Pacific coasts.
- This week may see average temperatures of at least 10°F above normal in eastern parts of the Corn Belt to the Northeast region, while closer to average temperatures may prevail through the country's mid-section states.
- In the West however, the cooler temperatures of recent may take a swift change to heat wave temperatures during the week ahead.
- Very warm and dry conditions in Hawaii continue throughout this week.
- Alaska can expect a week with average temperatures for this time of year, as well as the passage of occasional bands of showers, with heavier falls at times in southern parts, particularly around mid-week.

JUNE 13 – 19

- The Deep South region, in particular around the Mississippi Delta, moves into a more stable summer weather pattern around this time, with hot and more humid conditions becoming more stable.
- Most of the Inland Midwest, including in eastern parts of the Corn Belt, will be mostly hot and dry during this week, while the western regions of the Corn Belt and into the Great Plains may enjoy some productive precipitation during the week, possibly accompanying bands of supercell thunderstorms towards the end of this week, possibly most severe in Wisconsin.
- Similarly, this divide between wet areas and dry areas may also prevail in the Deep South to the Southeast regions, because despite the passage of a tropical storm through the Gulf of Mexico and across Florida early in the week, only the southern Atlantic regions are likely to benefit from the rain the storm brings.
- A cold front is likely to cross the southern parts of the Mississippi Delta towards the end of this week, accompanied by strong winds and occasional shower activity, but no significant rainfall expected.
- Generally, the Pacific Northwest region and up into the northern Rocky Mountains may be cooler with the occasional showery spell during this week.
- Southern California and into the Southwest may remain mostly dry.
- It will be hot over the Southeast, while average to cooler temperatures will prevail almost everywhere else.
- Overall, temperatures may be around 5°F below normal in the Rocky Mountains, while warmer than normal by a similar variant in the Southeast, where temperatures may near 100°F towards the end of this week.
- Chance of frosts on the central Great Plains near the end of this week.
- It is likely to be unusually warm and dry in Hawaii.
- Unusually warm week in Alaska, despite continuing wet conditions prevailing in southern parts of the state, and lighter shower activity elsewhere.

JUNE 20 – 26

- Hot and dry weather pattern continues to prevail around the Mississippi Delta region, as well as in the central and eastern parts of the Corn Belt.
- Similarly, the interior regions of the Southeast may also be mostly dry and hot.
- The southern Atlantic Coast may receive some good precipitation during this week.
- Early in this week, heavy rain may affect many parts of the Great Plains.
- Severe thunderstorms with hail and strong winds cross the Inland Northwest to the North Great Plains at the end of this week, particularly through Montana.
- Occasional heavy showers and thunderstorms prevail from Texas in the south to the Dakotas in the north around the same time.
- There is the possibility of further pockets of low-lying flooding lingering in some Great Plains regions, particularly in central and southern parts.

WEEKLY SUMMARIES 2025

JUNE 20 – 26

- Very little rainfall from the Mississippi River to the Appalachians. Drought conditions in Alabama, Mississippi, Tennessee, Kentucky, and east Arkansas.
- The coastal regions along the Mid-Atlantic to southern Atlantic states can expect some shower activity, especially from the Carolinas to Florida.
- It is likely to be mostly hot and dry throughout the Pacific Southwest, throughout the Great Basin and into the Southwest.
- Temperatures may reach into the 90°F range on most days in the Mississippi Delta region as well as in the Inland Midwest.
- Much of this week will be cooler than last week in the Pacific Northwest region and in southern parts of the Rocky Mountains, while hot and dry conditions may prevail elsewhere west of the Rockies to the Pacific coasts, including through California.
- Temperatures in the Great Basin may be up to 10° F above average.
- Most of Hawaii can expect to be dry, except for some showers mostly on the Big Island around mid-week.
- In Alaska, some occasional showers prevail with temperatures around normal.

JUNE 27 – JULY 3

- Some very disturbed conditions in many parts of the country early in the week.
- A cold front over the North Great Plains, Midwest and Northeast early in the week.
- During this time, there is a chance of pockets of snow falling in the northern Rocky Mountains to Montana, while heavy rain may fall in some parts of the Northeast.
- Some showers also prevail in the Mississippi Delta, Southeast, and Tennessee Valley regions in the first couple of days this week, before hot and drier weather returns to these regions by week's end.
- Mid-week, the arrival of a cluster of severe thunderstorms with hail and strong winds is likely to spread across the South Great Plains, possibly spreading through the states of Oklahoma, New Mexico, Texas and Arkansas.
- The dry spell in the Mississippi Delta may break this week, with pockets of heavy rain, which may also result in some brief flooding in a few low-lying areas.
- Outbreaks of rain can also be expected occasionally in the Ohio Valley and central and eastern parts of the Inland Midwest Corn Belt region.
- It is likely to be mostly dry and warm in most areas west of the Rocky Mountains, before a cooler change at the end of the week.
- Temperatures warmer than normal for this time of year, with particularly hot weather likely through the mid-section of the country throughout the Great Plains.
- A very warm spell early in the week turns into a cold change mid-way in areas such as Virginia and Nebraska. Very cold midweek in Idaho and Oregon.
- Mostly dry in Hawaii, except for some spotty showers mostly on the Big Island.
- In Alaska, it may be unusually warmer, with some parts up to 80°F or more possible despite the occasional showers prevailing, which will be mostly light and scattered.

JULY 4 – 10

- This week sees showers almost daily in the South Great Plains.
- During this period, rainfall will expand as it intensifies, with spells of heavy rain and thunderstorms in the central and southern Great Plains, with areas from Texas to Kansas seeing heavy rainfall at times, and possible pockets of localized flooding.
- The Mississippi Delta region and the Inland Midwest Corn Belt can also expect several spells of light rainfall during this week, including in the Ohio Valley and down into the Tennessee Valleys and the Mid-Atlantic States.
- Another band of thunderstorms and heavy showers arrive towards the end of this coming week, particularly in Oklahoma, which may have already had almost daily showers for the past couple of weeks.
- In the West, northern and central parts of the Pacific Coast can expect some light showers over this week, while the rest of the West regions are likely to remain dry, with the risk of wildfires remaining strong; this includes the northwestern half of the Great Plains, which may be mostly dry and warm.
- Temperatures are likely to be warmer than average over the Great Plains, while further east, it will be much warmer than average in the Corn Belt regions.
- It is likely to be cooler than of recent in the Northwest at the start of this week.
- Moderate seasonal temperatures are likely to prevail in the Ohio Valley and Mid-Atlantic states, and many areas in the Southeast region including in the Mississippi Delta area can anticipate temperatures within the 95-100°F range on most days.
- Most regions in the West can expect temperatures to be hot almost everywhere, which increases the fire hazard risks in many western areas.
- A cooler airflow arrives over the Northeast at the end of this week after temperatures may have peaked around the 100°F plus mid-week, while high temperatures continue to expand in the West.
- This week is likely to see the return of some light showers over Hawaii, with some good falls on the Big Island early in the week.
- In Alaska, near average temperatures prevail, with some showers in most districts, including in Anchorage.

JULY 11 - 17

- Dryness in the Great Plains may be increasing around this time, although during this week, some rainfall occurs in the eastern parts of the central and southern Great Plains, which may result in some pockets of localized flooding.
- Rains in the western Gulf Coast may also produce some localized flooding.
- Rainfall drifts eastward, spreading into the lower Mississippi Valley and the Southeast, although showers will be more scattered further eastward.
- Any significant rainfall over this coming week may be mostly confined to the Ohio Valley and southern parts of the Northeast.
-

WEEKLY SUMMARIES 2025

JULY 11 - 17

- Rainfall in the Mississippi Delta may be mostly moderate to heavy during this week.
- Rain gradually eases across the central and southern Plains. Some heavy showers may bring pockets of flooding on the Southeastern Plains.
- Very heavy rain in the western Gulf Coast regions for a time, possibly resulting in some flooding in the lowlands, with the heavy rain reaching into the Delta to the Southern Atlantic States by the end of the week.
- The Inland Midwest region, including the Ohio Valley and parts of the Upper Midwest, are likely to receive some productive showers.
- This week may also bring the chance of flooding in several river basins from the eastern parts of Texas to the western regions of Missouri.
- Thunderstorms and heavy rain may prevail for a time on the Great Plains, from the south of Kansas into Texas for a time, possibly bringing some localized flooding.
- Some good rains fall in the central and eastern Inland Midwest regions.
- Hot and dry conditions in the west may see the threat of wildfires grow, particularly in parts of Utah, Nevada, and California.
- Temperatures may be average to slightly cooler than normal during this week east of the Mississippi River, in many parts of the central and southern Great Plains, as well as the Atlantic coastal plains, and along the Pacific coastal fringes.
- Heat-wave temperatures may build in the west of the country, including in South Dakota, Nevada and California with heat and possible drought conditions adding to the threat of wildfire activity as the week progresses.
- Temperatures may exceed 100°F in many western regions, including in Las Vegas.
- Heat is also likely to spread into the North Great Plains during this week.
- Extremely low humidity may accompany the heat wave temperatures in the west.
- Rainfall will be mostly light in Alaska with some thunderstorms around mid-week and temperatures may be unusually mild with many areas at least 5°F above normal, particularly across the central parts, and towards the end of the week.
- Hawaii can expect to be warmer and drier than normal with just an isolated shower on most days.

JULY 18 – 24

- It is likely to be mostly mild and dry across the northern parts of the Great Plains during this week, with most the precipitation and humidity, sometimes accompanying thunderstorm activity, on the central and southern Plains, with some outbreaks of heavy rain possibly causing some localized flooding.
- This wet spell over the southern half of the Great Plains may bring temperature relief across the region.
- Mostly dry in the Midwest over the western and eastern Inland Midwest regions.
- Hot and mostly dry weather in the west with continuing risk of wildfires.
- This week may see very heavy rain for a time in the Mississippi Delta regions.
- Thunderstorms and lightning strikes in the Southwest possibly triggering wildfires.
- Downpours possible on the South Great Plains, in direct contrast to hot and dry conditions over the North Great Plains.
- There may be some significant rainfall for a time in the Southeast.
- Mostly dry conditions may see the drought regions expanding further northwards around now, up into parts of the Mid-Atlantic States.
- Heat wave temperatures may continue to prevail over the west of the country maintaining a high wildfire risk.
- Many parts of the Northwest may see temperatures of at least 10°F above the norm for this time of year.
- Some relief to heat wave temperatures across the west of the Inland Midwest during towards the end of this coming week, as a cooler spell moves into the Midwest and Northeast regions, possibly as a result strong gusty winds at times.
- It will be mostly dry and very warm in Hawaii, with talk of drought development possible concerning the Big Island.
- Alaska may be mostly mild and dry especially in the west and north of the state, with rain mostly in the Southeast regions.

JULY 25 – 31

- Scattered rainy spells throughout most of this week in the Mississippi Delta region, easing temperatures and lowering humidity.
- Showers in the Pacific Northwest coastal region bring some relief to the wildfire risks, but high fire risks will remain in the Rocky Mountains.
- There may be some heavy rain for a time in the Pacific Northwest to northern California midway through the week.
- Chance of localized flooding in central and southern parts of Texas.
- Much of the northern Inland Midwest areas will remain mostly dry, spreading southward as the week progresses.
- Scattered showers may prevail for a time in the Southeast, while it may remain mostly dry in the Mid-Atlantic areas.
- There may be an increase in shower activity in the Four Corner States during this week, bringing some relief to drought and wildfire threats in these affected states.
- It may be mostly dry in the South Great Plains.
- Showers and thunderstorms are likely for a time in the Midwest, but there may still be some pockets remaining dry in the east and west of the Inland Midwest.
- Some spotty showers may sprinkle the Southeast.
- Some showers are likely from the Great Plains to the Atlantic Coast, followed by cooler and drier conditions.
- Rain may be particularly heavy midweek in Iowa to Wisconsin states in the Inland Midwest, while other parts in the Upper Midwest may remain mostly dry.
- Heatwave conditions may continue to prevail throughout the northern and central Rocky Mountains and into the Great Plains.

WEEKLY SUMMARIES 2025

JULY 25 – 31

- Increased wildfire risk in the Great Basin to the northern Rocky Mountains during this week, as mostly hot, breezy and dry conditions are likely to prevail.
- Despite heatwave temperatures in many parts of the country, close to average temperatures prevail in the south-central parts, including in Texas.
- A tropical depression south of Hawaii may bring heavy showers and gusty winds, with particularly heavy rain on the Big Island near the end of the week.
- Widespread showers and average temperatures prevail in Alaska, although temperatures may be slightly above average in some northern parts.

AUGUST 1 – 7

- During this week ahead, most of the rain in the Mississippi Delta region will be confined to the south and mostly fall as light showers.
- Some showers are likely in the Northwest region, as well as down through the central and southern parts of the Rocky Mountains.
- It will be mostly dry in the South Great Plains.
- Central and southern parts of Texas can anticipate some more heavy rain, again possibly causing some localized flooding.
- Some good rains are likely in the Inland Midwest in the southwest, central and eastern parts, while it may remain mostly dry in the western areas.
- Some rain expected in Florida and the Coastal Southeast during the week.
- Heavy showers may prevail for a time in the Four Corner States, bringing some relief to the wildfire and drought risks.
- Some light rain prevails on the Plains, with heavier bursts of rain towards the end of the week in the southern Nebraska region to the northern Kansas areas.
- Outbreaks of rain break a dry spell in the Coastal Southeast and Mid-Atlantic regions, with the chance of heavy rain in the western Gulf Coast causing flooding.
- Temperatures may be unusually low in the northern parts of the Mississippi Delta.
- Hot weather continues through the Great Basin and western Rocky Mountains.
- Similarly, hot and dry conditions are likely to persist in Montana to the Dakotas.
- Hot and dry in California, the Northwest regions, and the South Great Plains.
- The Great North Plains may endure persistent heat with temperatures averaging up to 10°F higher than normal for this time of year.
- Showers and gusty winds may prevail over the northern Pacific coastal fringes, easing the threat of wildfires.
- Rain in central and southern Texas likely.
- Calmer conditions return across Hawaii this week as the tropical storm of last week dissipates, with warmer and drier conditions throughout.
- In Alaska, temperatures may be above normal in many parts and showers are likely to be frequent, with some heavy rain expected for a time in the south-central parts of the state around mid-week.

AUGUST 8 – 14

- This coming week will be showery, hot, and humid in the Mississippi Delta regions, particularly in the first half of the week.
- Drier conditions will settle over the Mississippi Delta areas towards the end of the week, but high temperatures will persist, often about 90°F in many districts.
- Generally, much of this week will be hot and dry in the West, with only some outbreaks of isolated showers in some areas.
- Occasional spells of rain prevail over the Great Plains, with some heavier falls in the south and central parts.
- Some showers along the Gulf Coast regions may bring some relief during the week ahead, while it is likely to be mostly dry in the Southeast.
- Hot and dry conditions are likely to continue in the Inland Midwest regions, especially in northern parts.
- Some rain expected in southern parts of the Inland Midwest through Minnesota, Wisconsin and Iowa, with only spotty shower activity elsewhere in the Midwest.
- Light scattered showers prevail in the Mid-Atlantic regions and in the west; some shower activity prevails in the Four Corner States and into Wyoming.
- Mostly dry from California and up into the northern Rocky Mountains.
- Some showers are likely to cross the Plains south of South Dakota with some heavy falls at times, while North Dakota to Montana may be mostly dry.
- Occasional showers expected in some pockets in the Midwest during this week, while other parts of the region may be completely dry.
- Heavy showers in South Dakota and southern Wisconsin, with only lighter rain activity crossing into the drought affected parts of the Inland Midwest.
- Drier conditions return to central and southern parts of Texas.
- Temperatures are likely to be slightly above normal in the Inland Midwest.
- Unusually hot over a third of the country to the east of the Rocky Mountains, with the heat slowly shifting eastward across the country.
- A cooler flow is likely to spread over the areas west of the Rocky Mountains.
- Although much of this coming week will be warm and dry, Hawaii can still expect some heavy showers, mostly around mid-week.
- Occasional showers in Alaska throughout the week, with temperatures average to slightly above normal, with the southwest area receiving the most precipitation.

AUGUST 15 – 21

- Very hot temperatures prevail throughout the Mississippi Delta during this week with many days seeing temperatures of 100°F or more.
- It may be mostly dry in the Mississippi Delta regions, except for some isolated shower activity towards the end of the week.
- Heavy rain from Arizona and New Mexico, and northeast into the Inland Midwest.
- Light showers in the southern Atlantic region including the Western Carolinas.

WEEKLY SUMMARIES 2025

AUGUST 15 – 21

- Many parts of the South Great Plains, the Mississippi Delta, and the Coastal Southeast may only see light precipitation to mostly dry days, resulting in above normal temperatures throughout.
- A band of thunderstorms and heavy showers may cross the central Great Plains and over into the Northeast.
- There is a chance of frequent showers in the center of the Midwest.
- The monsoonal weather pattern may have established in the southwest around this time, with occasional showers prevailing in the Four Corner States.
- It is likely to be mostly dry during this week in the Northwest and West Coast states.
- During this week, almost one third of the western regions can expect a cooler and dry spell period, particularly towards the end of the week in the Northwest.
- California may be as much as 5°F below normal for this time of year.
- Temperatures 5-10°F higher than typical for this time of year in the Coastal Southeast, Inland Midwest and lower Northeast regions. High temperatures accompanying mostly dry weather add to the drought stressed areas.
- At first, this week will be pleasant and calm over Hawaii, before a tropical storm nears towards the end of the week increasing shower activity at first, before becoming more intense with strong winds and heavy rain across the Big Island as it passes just to the south early next week.
- Despite this, it may continue to be mostly dry in other islands in the state.
- In Alaska, it will be a warmer than normal week in western parts, while temperatures will be closer to average in the southeast region.

AUGUST 22 – 28

- Chance of a hurricane forming in the Atlantic Ocean near Mexico during the coming week, possibly making a category 5 force towards the end of the week.
- Some isolated heavy showers likely towards the end of this week in the Mississippi Delta, heaviest downpours in northern parts.
- Another tropical storm may also form around mid-week in the Gulf of Mexico, near the Texas coast, bringing heavy rain, including the chance of localized flooding in some parts of Texas to Oklahoma.
- This storm may swiftly downgrade, despite bringing flooding rains along its path.
- There may also be some heavy rain and possible pockets of lowland flooding in the northern Corn Belt region, particularly in some parts of Minnesota, Iowa, and Wisconsin, possibly because of passing thunderstorm bands that cross the Great Lakes to Northeast regions during this week.
- Lighter showers expected elsewhere in the Midwest.
- Florida can expect some rain during this week. Mostly dry elsewhere throughout much of the south of the country will intensify the heat wave conditions.
- It is also likely to be very dry in California to the northern Rockies.
- Continuing dry conditions continue in some parts of Alabama and Arkansas, with possibly no rainfall received in these states since the start of the month.
- Heat wave temperatures are likely throughout this week in the Mississippi Delta regions, with daily maxima often about 95°F.
- Temperatures will be warmer than average over most of the country, with pockets of extreme heat possible in the central Great Plains, the Mississippi Delta, including the Corn Belt regions, the Southeast, the Four Corner States, Tennessee Valley, and the Rocky Mountains to the Great Basin regions as well as in California.
- Very high temperatures, above 100°F, likely on most days in many southeastern areas from Texas to Tennessee and Alabama.
- Cooler than average along the Pacific coastal fringes to the Pacific Northwest, and in the northeastern regions from North Dakota, through to the Great Lakes and up into the Northeast.
- This week may see a hurricane pass near to the south of Hawaii around mid-week, with only some light scattered showers reaching the islands, with heavier showers on the Big Island, and fresh to strong winds only as the storm passes.
- In Alaska, it may be unusually warmer than average, with many parts of the State up to 10°F above normal, with brief scattered showers now and then.

AUGUST 29 – SEPTEMBER 4

- Another hurricane may form in the Gulf of Mexico during this week, possibly making landfall midweek in Mexico, but it may not exceed a Cat 2 strength.
- Most days will be dry in the Mississippi Delta region, as a high-pressure system dominates the week, with no significant rainfall expected and only the occasional isolated showers now and then.
- This high-pressure system may ridge over the southeast of the country, keeping fronts a bay throughout the week, locking most precipitation to areas over the north of the Midwest. Chance of widespread flooding in the northern Corn Belt.
- Heavy rain is likely from the central Great Plains, across the Inland Midwest and over the Appalachians to the Mid-Atlantic Coast, with some pockets of flooding in the lowlands possible, particularly in central parts of the Corn Belt.
- Heavy rain also prevails for a time in Iowa, Wisconsin, Michigan, and in northern parts of Indiana, Ohio, and Illinois.
- Hot and mostly dry in Southern Corn Belt areas, including the lower Ohio Valley.
- Some brief scattered showers may occur this week in the Southeast and despite the showers, temperatures will remain very high, by up to 10°F above the average.
- Overall, the western half of the country will be mostly dry.
- Wetter conditions occur on the Great Plains, from Oklahoma to South Dakota, while dry conditions will prevail in the northern and southern edges of the region.
- Scattered showers and cooler temperatures cross the Northwest may bring some relief to the wildfire risks.
- Hot and mostly dry conditions will prevail from California to the Four Corner States.

WEEKLY SUMMARIES 2025

AUGUST 29 – SEPTEMBER 4

- Remnants of a tropical storm may bring heavy rain across Oklahoma to the Ozarks.
- Drought conditions may strengthening in the Southeast, where no rainfall is expected in the state of Arkansas and Alabama.
- Much of the country will endure above-average temperatures during this week.
- Temperatures are likely to be very hot once more in the Mississippi Delta states, with many places seeing temperatures near 100°F on most days.
- A ridge of high pressure is likely to be keep the southeastern regions of the country enduring persistent high temperatures.
- Many parts of the Southwest, Southeast, Tennessee Valley and southern parts of the Corn Belt may be as much as 5-10°F above average for this time of year.
- However, the Pacific Northwest to the North Great Plains, and parts of the Mid-Atlantic Coast to the Northeast, may be slightly cooler than average.
- A cooler airflow may finally arrive in the south of the country by the end of this week, as well as scattered showers bringing some drought relief.
- The risk of wildfires may be extreme now in the Great Basin to the Northwest and northern Rocky Mountains.
- It will be mostly warm and dry in Hawaii, except for a burst of isolated heavy showers towards the end of the week, due to an upper-level disturbance.
- Rainfall will be sparse in Alaska and temperatures will be near average in southern parts, while warmer than average in central parts.

SEPTEMBER 5 - 11

- A couple more hurricanes may manifest in the Atlantic Ocean during this week, but are unlikely to make landfall in the US. They are more likely to land in Mexico, and further south in Nicaragua.
- The Mississippi Delta region can expect showery spells, occasionally heavy.
- Mostly dry in the west, with most rainfall confined to the Four Corner States.
- The spotty showers that may occur in the West will give little relief from the heat.
- Moderate rain in the Corn Belt region mostly in the areas of Nebraska to Wisconsin.
- Texas can also expect moderate rainfall, moving into the Southeast at times.
- Some showers are likely to cross Georgia and into southern South Carolina.
- This week may see a change in the weather patterns with it becoming drier in the Midwest, while wetter in parts of the Southeast.
- However, despite the local heavy showers at times in the Southeast region, many parts continue to endure drought conditions, especially in southern parts of the Corn Belt, including in the lower Ohio Valley.
- The heaviest rainfall expected this week may concentrate around Georgia, and heavy showers may prevail throughout the week in the western Gulf Coast regions.
- Shower activity on the Great Plains will decrease as the week progresses.
- The remnants of a hurricane bring showers into California and Nevada.
- Texas may see shower activity almost daily this week.
- The shower activity in the Mississippi Delta region will bring a slight relief to the high temperatures of recent.
- Above average temperatures may still persist in some parts of Georgia, Arkansas, South Carolina, Kentucky, Mississippi, Tennessee, and Alabama.
- Temperatures in the West will be above average on most days while temperatures will ease slightly in the Southeast.
- Overall, many parts of the Great Plains and the West may be considerably warmer than average, by up to 5-10°F above normal for this time of year.
- These above average temperatures in the West of the country may continue to aggravate the wildfire risks.
- The heatwave temperatures may shift westward as this week progresses.
- Hawaii may be mostly dry with only some brief showers.
- In Alaska, the south is expected to be mostly dry and mild, with near-normal temperatures and the occasional shower elsewhere.

SEPTEMBER 12 – 18

- Early in the week ahead, there is a chance of a hurricane forming in the Gulf of Mexico, and then strengthening quickly before possibly making landfall in Texas, to then track in a northeasterly direction through the southeast of the state and up into Louisiana, bringing floods and wind damage in its path.
- The Mississippi Delta states can expect very dull skies and outbreaks of heavy rain, but still with high temperatures prevailing, often over 90°F in a number of places.
- Many parts of the upper Great Lakes regions can expect some heavy rainfall just after midweek, including in Minnesota.
- The Pacific Coast regions to the Rockies can expect long dry spells broken by only a passing shower now and then, with the heaviest falls expected only in a narrow band down through the Rocky Mountains.
- Heavier rain from the Inland Midwest Corn Belt to the northern parts of Texas.
- The Coastal Southeast and Atlantic Coastal states may be mostly dry and warm, except for some isolated showers spotting into the southern parts of Florida and in the coastal fringes of Georgia and South Carolina.
- Some heavy rain at times on the South Great Plains, with showers elsewhere through the mid-section states of the Great Plains.
- One significant rain event may occur this week in the east of the country, possibly because of a decaying hurricane bringing heavy rain into the Ozarks to Missouri when the cold front associated with it crosses the region.
- Overall, this week will be warmer than normal for this time of year, almost everywhere, for the second week in a row.
- Some parts of the Midwest may be up to 5-10°F above normal.
- Much of the Southeast may be dry and hot, contrasting with the more showery parts in southern Florida and the western and central Gulf States regions.

WEEKLY SUMMARIES 2025

SEPTEMBER 12 – 18

- Mostly dry and unseasonably warmer than normal in the west, including California, the Great Basin, Colorado Plateau, and central and northern parts of the Rocky Mountains, with temperatures of about 100°F or more on a number of days.
- Later in the week, the heat is likely to shift into the Northeast region of the country.
- A cooler flow of air moves into Alabama, Tennessee, and North Carolina at the end of this week, bringing a respite to the heat wave temperatures of recent.
- Mostly warm and dry Hawaii, except for occasional brief showers.
- In Alaska, it will be milder and showery across the mainland, with some heavy rain expected in the south of the state around the end of the week.

SEPTEMBER 19 – 25

- A number of serious tropical storms may affect the Southeast of the country during this coming week, as well as the remnants of last week's hurricane in the south of the country that affected Texas and the Deep South.
- Remnants of last week's hurricane in the Texas to the Deep South regions may enhance the shower activity and bring strong gusty winds across the Southeast and Southern Delta.
- This precipitation will also spread up through the Mississippi Delta region where outburst of heavier precipitation is likely, including the risk of some flash flooding developing in the Delta River basins, with only the far northern parts of this region likely to avoid the heavy downpours.
- Frequent bands of showers are also likely throughout the central and southern Great Plains, resulting in cooler temperatures.
- The Gulf Coast regions and northeastwards into the Mid-Atlantic States, can also expect frequent showers, mostly because of nearby tropical systems.
- In the West however, showers will be lighter and more isolated, except in the southern parts of the Rocky Mountains.
- Early in this week, there is the chance of a tropical storm developing near the Coastal Southeast region, possibly making landfall near North Carolina.
- Another tropical storm may also build in the Gulf of Mexico about midway during this week ahead, near the Texan coast, swiftly building to hurricane level as it makes landfall and begins to weaken.
- As the former hurricane degrades while passing over the southeast of the country, the severe winds associated with it may develop into tornadoes in the Carolinas.
- The effect of the heavy rains in the Mississippi Delta region may also bring a cooler change across the region as well as including the chance of some overnight frost pockets developing, scarce at first, but becoming more frequent by the end of the week in the upper Inland Midwest regions in particular.
- Cooler temperatures spread across the Great Plains with many places seeing temperatures of up to 10°F below average, particularly in the Upper Midwest.
- The end of this week sees heat wave temperatures fade in many parts of the country east of the Rocky Mountains, including in the Dakotas, Montana, Texas, Tennessee, Missouri, Iowa, and Kansas with the arrival of frosts overnight in many regions especially on the North Great Plains to the Upper Inland Midwest states.
- However, the West may continue to have above normal temperatures throughout most of this week before a cooler change in the last couple of days, and most days this week cooler than of recent down through the Pacific coastal fringes.
- Warm temperatures may also persist over the Southeast and Southwest this week, but only warmer by around 5°F of their average for this time of year.
- Mostly dry and warm in Hawaii except for some isolated heavy showers.
- In Alaska, mild and showery conditions will prevail, with temperatures possibly up to 5°F above normal in southern parts of the State, but up to 10°F warmer along the Artic Coast fringes, particularly around midweek.

SEPTEMBER 26 – OCTOBER 2

- Another loose sub-tropical depression is likely to form around midweek near the Northeast of Florida, before drifting westward across the State and into the Gulf of Mexico by the end of this week producing showers and gusty winds over the state and showers throughout the lower Southeast.
- This storm system may produce some tornadoes towards the end of the week in southeastern parts from Florida to Georgia.
- The storm system that crossed the Mississippi Delta in the previous week should fade to a tropical depression, producing only occasional showers, with most of them contracting to the southern parts of the Delta by the end of this week.
- Further west, patches of light showers as well as the occasional heavier pockets with rain activity becoming more frequent and intensified by the end of the week.
- Chance of snow towards the end of this week in Montana and Idaho.
- Most rainfall will be confined to the Great Plains and the Northwest of the Inland Midwest, into the Corn Belt, with only isolated showers reaching into the Dakotas and Minnesota states.
- Occasional showers expected on the central and southern Great Plains.
- Moderate to heavy showers expected around the eastern Gulf Coast and southern Atlantic coastal regions, primarily in Florida, and southern Alabama and Georgia.
- Chance of some heavy rain in the Southwest over this week, possibly resulting in some flash flooding in the region.
- Mostly dry and average temperatures along the rest of the Atlantic Coast.
- Early signs of autumn appear in the Delta region, with temperatures dropping.
- It will also be cooler this week further west from the Rockies to the Pacific coast, with the passage of storm systems bringing cooler temperatures including into California, where temperatures may be as much as 10°F below normal.
- Most days will be warm in the Midwest, including in Illinois, with daily highs of near 100°F still possible, particularly around mid-week.

WEEKLY SUMMARIES 2025

SEPTEMBER 26 – OCTOBER 2

- The late-summer warmth drifts eastwards across the mid-section regions from the Rocky Mountains eastward to the Appalachians where many places can expect above average temperatures once more by around 5-10°F above the norm for this time of years, mostly from the southern Great Plains to the central parts of the Corn Belt districts.
- Most parts of the Inland Midwest, including the Corn Belt states and up into the Mid-Atlantic regions will be warm and dry, while cooler temperatures settle in across the Great Lakes and Northeast regions.
- By the end of the week, generally temperatures will stay low in the West, but return to warmer once more on the Great Plains.
- It will be another mostly dry and warm week in Hawaii, except for the chance of some heavy showers mostly on the Big Island early in the week.
- Alaska may start the week with storms, and temperatures near normal in eastern parts to slightly above normal temperatures in the west before a cold turn comes nearer to the end of the week, which may produce the first freeze of the season.

OCTOBER 3 - 9

- This week may see the threat of yet another hurricane developing near the Mexican coast, starting as a tropical depression early in the week, before increasing to a low-grade hurricane by midweek.
- Regular showers prevail through the Mississippi Delta states, with a slow moving cold front bringing some heavier rainfall with its passage around midweek, although the shower activity may completely miss some areas.
- Many parts of the Missouri and the Mississippi Delta regions, the central Rocky Mountains, and the Great Plains to the Northeast can all expect some rainfall.
- Heavy rain expected in the southern coastal parts of Florida as well as in the Northwest into Oregon, and southeast into New Mexico, as well as in southern parts of Wisconsin, Minnesota, and the Ohio Valley.
- For a while, rainfall on the Great Plains may linger longer in southern parts.
- Precipitation in the Pacific Northwest to the Rockies and into the North Great Plains including in Montana, Wyoming and Utah may fall as snow this week.
- Heavy rain in the North Great Plains to the Rocky Mountains, Montana and Colorado, tracking into the Upper Midwest and Michigan by the next day.
- Warm temperatures expected to continue throughout the mid-section regions of the country, from the eastern slopes of the Rocky Mountains to the East Coast, expanding for a time into the south and east.
- Up to 10°F above normal in some places , particularly in the Ohio Valley and interior regions of the Southeast.
- Cooler than average to the west by as much as 10°F, including in high elevations areas of the Rocky Mountains to the Pacific Northwest and into the Great Basin.
- Even California will be chillier than normal this week, particularly towards the end of the week.
- Hawaii may again be mostly dry with mild temperatures.
- In Alaska, unsettled showery conditions prevail over southern parts of the state, with less to the north, while temperatures will be average in the southeast, to as much as 5°F above normal in western parts of the state.

OCTOBER 10 - 16

- Most western parts of the country from the Pacific coast to the Rocky Mountains will be cooler this week, and with most precipitation confined to northern areas, including snow once more in isolated higher reaches of the Rocky Mountains.
- California, the Southwest and Mississippi Delta regions may all be mostly dry.
- Significant rainfall may fall in the west of the Inland Midwest region adjacent to the Great Plains as well as in parts of the South Great Plains.
- Around midweek, some heavy rain and thunderstorms may prevail across the east-central Great Plains regions moving into the North Great Plains by the end of the week, where much of it will fall as a mix of rain and snow.
- The Florida peninsula may also receive some significant rainfall this week, particularly in central and southern parts and in the coastal fringes.
- Heavy showers in the Inland Midwest, particularly in the Corn Belt regions.
- Heavy showers may persist once more in the southern Atlantic region, while lighter rain prevails in the Coastal Southeast.
- West of the Deep South, some rainfall occurs for a time, mostly as the result of another low-pressure system traversing the Gulf of Mexico.
- This coming week will be cooler than normal in the West with parts of the Pacific Northwest up to 10°F cooler than what is normal for this time of year.
- Overall, most parts west of the Rocky Mountains will be cool including in California, Utah, and Idaho, especially at the start of the week before a milder change midweek.
- In contrast, east of the Rockies, temperatures may be warmer than average from the Great Plains to the Atlantic Ocean, except when interrupted by shower activity.
- Even the Mississippi Delta regions and the Great Plains may still have temperatures up to 10°F or more warmer than is typical for this time of year.
- Like in the West, mid-week may also see a warmer turn in most eastern parts from the Great Plains to the Atlantic Coast with a return to almost summer-like temperatures, particularly by the end of the week.
- Hawaii may see an increasing in shower activity, although some places remain mostly dry.
- In Alaska, dry conditions are expected to start the week with temperatures at or slightly below average. Towards the end of this week, some parts of Alaska, mostly inland, may see their first significant snowfalls of the season, while frequent showers may prevail in the Southeast.

WEEKLY SUMMARIES 2025

OCTOBER 17 - 23

- The most significant rain activity expected during this week in the West of the country will prevail mostly through the northern California areas to the Pacific Northwest, and further east, a band of heavy rain may cross the central Great Plains and into the upper Inland Midwest, with showers affecting many areas of Kansas, Missouri, Nebraska and Iowa.
- Heavy snow is likely in the Rocky Mountains early in this week, as well as in Utah, where snowfalls may accompany very strong storm winds.
- Snowfall is likely around midweek in the Great Lakes region including in Minnesota.
- Most other parts of the Midwest may be dry, with some regions possibly still experience drought conditions, including in parts of the Southeast.
- Most shower activity in the Mississippi Delta region will be confined to the north, while most other parts remain almost completely dry during this week.
- Rain in the Northeast will ease throughout the week ahead.
- Mostly dry conditions prevail west of the Rockies, including in the Northwest and Southwest regions except towards the end of the week when a band of rain may spread from California to the central Great Plains and later into the west of the Corn Belt region of the Inland Midwest.
- Unusually warm for this time of year in the Inland Midwest, and south and east of the country at first, before temperatures return to their seasonal levels.
- During this spell, numerous places may record maximum temperatures of up 10°F more than is typical for this time of year.
- There will be a few days in the Mississippi Delta region when daily maximums may reach above 90°F, even with minimum temperatures dropping to below 50°F.
- The Pacific coastal regions to the Rocky Mountains may be cooler than normal this week, as well as in Minnesota and North Dakota.
- Temperatures in California 5-10°F cooler than is normal for this time of year.
- A wet period in the Northeast towards the end of this week will bring temperatures back to normal in that region.
- Many parts of the southern half of the Great Plains will be warmer than normal this week, while most northern parts will be colder than normal.
- Quite pleasant in Hawaii, with mostly calm and dry with light showers, growing unusually cold towards the end of the week.
- In Alaska, a cold air moves over the state, bringing temperatures to around 5°F below what is normal for this time of year in the mainland region. Rain and snow likely on the colder days, particularly in the first half of this coming week.

OCTOBER 24 - 30

- Some regular outbreaks of showers expected in the Mississippi Delta region during this coming week.
- Mostly dry in the Southwest and the North Great Plains.
- Heavy rain is likely for a time in the Pacific Northwest, particularly the Cascades, as well as storms towards the end of the week from the Northwest to Montana.
- Other wet areas expected further inland over much of Idaho and Wyoming, with lighter showers into the Rocky Mountains.
- The western parts of the Inland Midwest and Corn Belt will receive some good rainfall as well as the adjacent regions of the Great Plains.
- The Texas to Michigan regions, up through the eastern Gulf Coast areas will also receive some heavy rainfall this week. Chance of tornadoes around midweek.
- Even the Northeast region and the Gulf Coast regions can expect some heavy showers during this week, particularly at the end of the week.
- Overall, rain will occur throughout the eastern third of the country this week.
- Only the areas from the Mid-Atlantic region to parts of the Coastal Southeast will see very little rainfall, but still even some light showers may prevail now and then.
- This week may produce some heavy rain accompanying severe thunderstorms at times, particularly around midweek in the eastern Great Plains regions and later across the south and Midwest in subsequent days.
- Some of these thunderstorms may produce multiple tornadoes.
- Heavy precipitation in the Northwest may fall as snow in the higher reaches.
- The Pacific Southwest is expected to be mostly dry with very strong and gusty winds. Lingering wildfires may still pose a threat in California due to this wind.
- The mostly dry spell in Washington DC may finally break at the end of this week.
- The eastern half of the country from the Great Plains to the Atlantic coasts, can expect to be warmer by around 5-10°F more than is normal for this time of year, particularly ahead of any rain.
- Some regions in the east of the country may still see maximums of up to 90°F at times, especially in the Deep South states of Alabama and Tennessee, while the West will be generally cooler than normal.
- It may be mostly dry in Hawaii, with only some spotty showers now and then.
- Alaska can expect significant precipitation this week, mostly confined however, to the southern parts of the State, although some snow may extend into the interior.
- Overall, temperatures in Alaska may be milder than normal in the northern half of the state, while average to below average over the southern half of the state.

OCTOBER 31 – NOVEMBER 6

- If any wildfires are burning in the West, from the Rocky Mountains to the Great Basin and southern California, then the combination of strong winds, high temperatures, and low humidity are likely to pose a threat, particularly at the start of this coming week. Wind should subside by midweek.
- Many parts of the West and eastern North Great Plains will be mild and dry, with temperatures well above normal for this time of year.
- The eastern half of the Great Plains to the Inland Midwest including the Corn Belt will be cooler and have some productive rainfall, particularly early in the week.

WEEKLY SUMMARIES 2025

OCTOBER 31 – NOVEMBER 6

- Any regions in the east that may have been experiencing drought conditions of recent may enjoy some rain around this time, including in parts of the Tennessee and Ohio Valleys, in the Mississippi Delta and through the Mid-Atlantic regions.
- Some rain activity expected on at least half of the days of this week over the Mississippi Delta states.
- Some regions in the east and south may also receive some very heavy rain events, particularly around the central Gulf Coast region and in the Mid-Atlantic States, including in Kentucky, Ohio, North Carolina, and Washington DC.
- It will be unusually warmer than normal again across the eastern half of the country, with some areas seeing maximum temperatures of up to 10°F above their norm for this time of year.
- While some parts of the Northeast region may still have maximum temperatures up into the 80's this week, there will be some parts of the Southeast Great Plains to the Mississippi Delta regions that will be around 5°F cooler than normal.
- Overnight frost level temperatures may occur in some southern parts, including on the South Great Plains and into Texas, particularly towards the end of the week.
- Despite much of the country being warmer than normal, particularly early in this coming week, cooler temperatures may prevail throughout most of the Rocky Mountains including in Utah and Wyoming.
- Hawaii continues to be mostly dry with temperatures closer to normal, although some days may see a very wide variation between maximum and minimum.
- In Alaska, most days will have temperatures within 5°F of normal throughout the State, except in the very far northern regions, where temperatures may be up to 10°F more than is normal for this time of year, and it will be mostly dry, except for some precipitation towards the end of this week.

NOVEMBER 7 - 13

- The threat of another tropical storm builds off the Atlantic coast around this time and lingers near Cuba early in this week, before possibly drifting northwards.
- This storm may pass near the Northeast coasts nearer the end of this week to early into the start of next week with the greatest impact being breezier than normal weather along the Atlantic coastal fringes, where heavy surf may be generated.
- Almost completely dry and mostly tranquil with little to no rain expected almost throughout the country, except for some isolated to light showers along the Atlantic coastal fringes from Florida to the Northeast, and a brief showery spell in some parts of the Inland Midwest to the Great Lakes, including in the Corn Belt.
- Some snow briefly expected in the Rocky Mountains.
- Most regions in the Mississippi Delta can anticipate a dry week.
- Some southern regions of the Mississippi Delta may still see daily maxima's of around 80°F during this week.
- Frosts in many regions as far south as the Mid-Atlantic states to the southern Appalachians, as well as the odd patches in the Mississippi Delta regions, where temperatures may be very low during the night but still rise quickly during the day.
- Further west, temperatures will be cooler than average, but only by around 3-5°F below the norm for this time of year. Daytime temperatures from the North Great Plains to the northern Inland Midwest may rise as much as 5°F more than average.
- Overall, generally the southeast and northeast parts of the country will see temperatures during this week close to the average for this time of year, while the east-central regions may be slightly cooler than average.
- Temperatures may be around 5°F warmer in the Southwest region and over the Great North Plains, while parts of the Pacific Northwest and the Ohio Valley to the Mid-Atlantic regions may be cooler by a similar variant.
- Mostly warm and dry this week in California.
- Some heavy outbreaks of rain expected in Hawaii towards the end of this week, with significant falls seeing an end to what was an unusually long dry spell over the past three months.
- Stormy weather across the southeast of Alaska, bringing some good snowfalls. Milder and mostly dry conditions elsewhere, particularly in central parts, and temperatures in the state up to 10°F above normal for this time of year.

NOVEMBER 14 – 20

- Most parts of the country can expect to be mostly dry for the second week in a row, except for some isolated precipitation in the Pacific Northwest, northern Rocky Mountains, and around the Great Lakes region.
- A band of heavier rain may also briefly affect the east of the country from the Tennessee Valley to the Northeast.
- By the end of this week, a passing storm system may bring some rain and snow showers into the Northwest.
- This week ahead may be warmer than is normal for this time of year over the western half of the country, with a number of areas possibly up to 10°F or more above their average, particularly in some parts of Wyoming.
- Temperatures will average below normal by a similar variant in many regions east of the Great Plains, including in Florida.
- Frosts will be becoming more frequent in the Mississippi Delta regions, enhanced by the mostly dry conditions, and other parts of the Southeast also experience temperatures dropping below freezing.
- Possibly more than 10°F warmer than average in the West, including in a large part of the Rocky Mountains. Southern Atlantic states cooler by a similar variant.
- Hawaii can expect heavy showers early in the week, mostly dry conditions later.
- This week ahead brings milder weather across Alaska with some areas possibly up to 10°F above normal for this time of year, as well as a series of storms to cross the state, with accompanying snowfalls each time.

WEEKLY SUMMARIES 2025

NOVEMBER 21 - 27

- Storm and rain activity return with some gusto this week, with storms bringing high winds and hail through the Mississippi Delta around mid-week, and a cold front bringing thunderstorms and widespread showers in the regions from Ohio to the Tennessee Valley, as well as up into the Northeast regions.
- A stormy end to the week also prevails in Texas, where heavy rain may cause some localized flooding.
- Bands of thunderstorms may cross the states of Alabama, Kentucky and Tennessee around mid-week, possibly generating tornadoes.
- The end of this week may see a mix of showers and snowfall over the Northeast.
- Rain also prevails in the Pacific Northwest region to the northern Rocky Mountains including in Oregon and Washington, with heavy falls likely in many places.
- Elsewhere, the Pacific Southwest to the central and southern Rocky Mountains may be mostly dry this week with only the brief isolated showers possible.
- Similarly, the Great Plains will also be unusually warm and dry, except for the rain expected in the Texas gulf areas.
- Following a similar pattern to the Great Plains, the northern and western regions of the Corn Belt may also have a mild and dry week.
- The breezy conditions ahead of each front will bring freezing temperatures.
- Overall, it will still be a warmer than normal week almost everywhere, except in the Northeast and Southeast.
- In some parts of the North Great Plains and the Southwest regions may see temperatures up to 10°F higher than is normal for this time of year.
- Elsewhere, the Northeast region can expect normal temperatures, while the southeast of the country may be average to around 5°F cooler than normal.
- Hawaii may be mostly dry and warm, with some showers near the end of the week.
- In Alaska, significant snow expected across southern parts, mostly around mid-week, with temperatures generally mild in northern and southeastern parts of Alaska, while colder conditions will prevail elsewhere.

NOVEMBER 28 – DECEMBER 4

- A strong storm system originating from Mexico will to bring snow into the southwest of Texas and into the Southeast of New Mexico towards the end of this week and early into next week.
- Other storms bring scattered showers on a regular basis across the Great Plains to the Inland Midwest, and with heavier bursts of rain spreading into the southern Delta at times, particularly around early to mid-week.
- In the west of the country, most precipitation is likely to occur in the Pacific Northwest to the northern Rocky Mountains, including along the Pacific coastal areas of Oregon, as well as extending into Idaho, Montana, and Wyoming.
- Most other regions in the west of the country will be mostly dry.
- There may be bands of heavy rain crossing the central Corn Belt, Tennessee Valley and the Northeast.
- A band of snow crosses the central Plains to the Northeast early during this week, as well as some heavy snow in the Pacific Northwest to the North Great Plains.
- Some significant snowfalls expected towards the week's end in southern parts from the southern Rocky Mountains into the west of Texas.
- Chance of heavy rain for a time in the west and central Gulf Coast regions.
- At first, most parts of the country can expect warmer temperatures before the arrival of storms which will bring a shift to colder temperatures in many areas.
- Temperatures will be below normal in many western parts of the country from through the Northwest and spreading eastward over the Rocky Mountains and into the North and central Great Plains, with many parts possibly up to 5-10°F below average, particularly in Wyoming, Montana, and Colorado.
- Other parts of the Great Plains and east through the Inland Midwest, temperatures will be fairly normal to slightly warmer within 5°F of the average, while in contrast, the Northeast may be slightly cooler but also within 5°F of the norm.
- During this week, a very cold airflow follows a storm system, affecting not just the North Great Plains and in the Rocky Mountains, but also into the west of Texas.
- Hawaii can expect showers over the first half of the week before mostly dry weather returns.
- In Alaska, temperatures may be warmer than average for this time of year in the southwest of the state, and some central areas possibly as much as 20°F above the norm and snow prevails early in the week in the southeast of Alaska, and later in southern parts.

DECEMBER 5 – 11

- Most parts of the Mississippi Delta regions can expect some level of precipitation this week, although in many parts the amounts will be only very small, with the overall perception of this week being cooler, and mostly dry.
- Some rainfall expected in the Southwest to the Four Corners region, and in southern California, while heavy precipitation prevails in the Northwest.
- Heavy rain may also affect central parts of Arizona, with lighter precipitation in New Mexico to Colorado.
- Overall, most regions on the Great Plains, the Great Basin, and northern California can anticipate a mostly dry week.
- Further east, the Inland Midwest can expect some showers from eastern Nebraska to the Great Lakes, as well as through the Corn Belt districts.
- Moderate to heavy rain is likely early in the week in a line stretching from the Mississippi Delta to the Tennessee and Ohio Valleys and into the Northeast.
- Snow expected in the Northwest of the country in Montana, South Dakota, Oregon and Washington states.
- Texas can expect some passing showers this week, mostly in the panhandle region.

WEEKLY SUMMARIES 2025

- A major winter storm comes this week crossing through the Southwest to the Great Lakes regions, bringing some rain to the drier parts of southern California and the Four Corner States.
- Another storm later in the week will bring high winds and heavy rain into the Pacific Northwest.
- Widespread wintry conditions can be expected throughout the Inland Midwest at the end of this week, particularly in the northern Corn Belt regions, when wintry precipitation of sleet, snow, and freezing rain is likely, followed by very cold and gusty winds at times, with these winter forces swiftly spreading into the Northeast.
- Blizzard conditions may develop in the Inland Midwest to the Great Lakes towards the end of the week.
- Temperatures may be average to slightly above average over most of the southern half of the country, from Arizona to Florida, as well as along the Atlantic coastal fringes into the Northeast, while cooler in Texas and California.
- Further west, the Mississippi Valley to the central parts of the Great Plains can also expect near-average temperatures this week.
- Most other parts of the country can expect typical seasonal temperatures.
- It will be very cold across the North Great Plains, where covers of snow will prevail.
- Hawaii can expect some heavy showers around midweek, which may cause some flash flooding in some locations, particularly on the Big Island, with the other days mostly dry and warm.
- In Alaska, some heavy precipitation will prevail around the end of the week, mostly in the southwestern quarter of the state, while most other days will be mostly dry.
- Milder temperatures may prevail over the Alaskan mainland with many parts of the region up to 10-20°F warmer than normal for this time of year.

DECEMBER 12 – 18

- Early in this week, a series of storms move in from the Pacific Ocean into the Northwest, bringing heavy rain, outbreaks of flooding, and further east, snow into the northern Rocky Mountains before calmer and drier conditions settle in over the second half of the week.
- However, as conditions ease over the Northwest, rain moves into the Pacific Southwest, and across into the Four Corner States, and snow may spread across into the Central Great Plains by the end of the week, including in Wyoming.
- Mostly dry on the Great Plains, except for some snow in the Nebraska region towards the end of this week and only the isolated showery pockets.
- Unusually mild in the Great Plains, in contrast to a very cold airflow over the Inland Midwest to the Northeast, where severe low temperatures may prevail at times.
- Further south, with the exception of some showers in the Ohio Valley, the remaining regions of the Gulf and Atlantic Coast states may be mostly dry.
- Rainfall will be infrequent in the Mississippi Delta regions, with quite a variation in temperatures in the region, with some days very cold and others very warm.
- Very cold in the northeast quarter of the country where many places may see temperatures as low as -20°F at times, which is about 15°F below what is normal for this time of year, particularly in the upper Midwest and Northeast.
- The southern Rocky Mountains may be as much as 15°F above average, and the southeast will warm towards the end of this week after a cold start earlier on.
- Temperatures west of the Great Plains may be average to warmer by around 5-10°F more than average, with the warmer variation expected in the southernmost areas of the Plains.
- Warmer than normal in Arizona and Texas until towards the end of the week the passage of a cold front brings temperatures back to nearer to normal.
- A very cold week in the North Great Plains, the Midwest, and the Northeast ahead of snow events.
- Snow likely in the Great Lakes to the Megalopolis over the first half of this week.
- Hawaii may have to endure a storm around midway through week associated with a strong cold front, which is likely to bring heavy rain and high winds across the region, as well as the chance of a blizzard warning issued for the highest peaks on the Big Island.
- Very unsettled in Alaska this week, as a storm early-to-midway through the week may bring high winds and then further snow in the state, but temperatures may remain mild in most parts except in the southeast, where it will be very cold.

DECEMBER 19 – 25

- A late sub-tropical storm may build in the Atlantic Ocean to the Southeast of Florida early in the week, but not affect any part of the US until the end of this week, when remnants of it may develop into a winter storm as it moves through the east of the country close to Christmas.
- The Mississippi Delta regions can expect some icy rainfall this week, particularly towards the end of the week.
- Many parts in the West of the country may be mostly dry, except for some light to moderate precipitation in the Northwest, around Washington, and in some parts of the Four Corners region.
- Further east, central and southern parts of the Great Plains and up into the Northeast can expect some precipitation, mostly in the form of freezing rain, sleet or snow, although the southwest of Texas may be mostly dry.
- Most Atlantic coastal regions can also expect outbreaks of heavy rain this week, from Florida northwards.
- Stormy conditions carrying a widespread mix of snow, sleet, wintry rain and rain expected in many parts of the Southwest region, central and South Great Plains, in the Midwest and the Northeast.
- Snow packs may be deepened in many parts, possibly resulting in disruptions along the travel corridors and damage to infrastructure.
- Dry conditions are likely to stabilize over California and in the Great Basin.

WEEKLY SUMMARIES 2025

DECEMBER 19 – 25

- There is likely to be a stark difference between the cold, dry conditions in Montana to the Dakotas, with the wintry, icy conditions further south, particularly in Kansas and Oklahoma towards the end of this week.
- Further east, the Midwest may be cold and dry in the Upper Mississippi Valley to wet and milder in the Ohio Valley, with a band of snow in between the two.
- The Northeast may be mostly cold and dry over much of this week, while the northern Mid-Atlantic regions wetter and wintry.
- Rain expected from Arkansas to Kentucky. Mostly warm and dry in the southeast for much of the week, except for some showers at the end of the week.
- A storm may cross the mid-section of the country towards the end of the week, bringing very wintry conditions into Kansas and Illinois.
- This week may start with a milder airflow over the Mississippi Delta with many places up to 10-15°F warmer than average, with a colder change later.
- Unusually warm at the start of the week in Louisiana, Georgia and North Carolina.
- Texas may be average to warmer than normal in the southeast of the state as well as areas in the southern Atlantic Coast region, where temperatures may be up to 15°F warmer than is normal for this time of year, particularly over the southeast.
- It may be colder than average elsewhere, particularly over the central Rocky Mountains where temperatures may be as much as 10-15°F below normal.
- Snow may cover much of the states of Kansas, Nebraska, Colorado, Missouri, Iowa, Illinois and Texas by now, as well as in some parts of the Northwest, including Washington, Oregon, and Idaho.
- More spotty snow packs are likely to prevail in Montana, the South Great Plains and the Southeast.
- Hawaii may be mostly dry this week except for some heavy showers at times, mostly on the Big Island.
- In Alaska, it may be generally mild, despite daily snowfalls likely, particularly early in the week, with temperatures up to 10°F above normal almost everywhere except in the south.

DECEMBER 26 – JANUARY 1 2026

- This week sees a series of storms off the Pacific Ocean cross the western coasts bringing rain into the Northwest and California, wettest along the coastal fringes.
- Some heavy rain is also likely through Oregon and further inland in the Northwest up into the northern Rocky Mountains, than elsewhere in the west.
- Further east, showers can also be expected in the southeast of the Mississippi Delta region and up into the Inland Midwest Corn Belt areas.
- Most other parts of the country can expect only trace amounts of precipitation.
- There may be a lack of snow coverage in the North Great Plains to the northern Rocky Mountains due to a milder turn.
- Some light rain expected in Florida this week.
- Despite a snowmelt beginning in the Inland Midwest during this week, a storm towards the end of the week may see the refreshing of snow accumulations once more, particularly around the west of the Corn Belt.
- Elsewhere, the Northeast can expect a mix of precipitation between rain, freezing rain, and occasional snow.
- Some local heavy showers are likely around the Deep South, where even the colder temperatures may see the appearance of some pockets of frost.
- Severe thunderstorms at the start of this week in the south and east, possibly generating tornadoes in the Deep South, including in Georgia and down to Florida.
- Around mid-week, another spell of thunderstorms may accompany a warmer turn in the south, particularly affecting Louisiana, Texas, Alabama and Georgia.
- A storm at the very end of this week, at the turn of the New Year, may prevail across the central and southern Great Plains, bringing heavy snowfall into the regions, especially in Kansas and Texas.
- Overall, many parts of the country can expect temperatures within 5°F of what is typical for this time of year, with the only exception being in some pockets in Montana, Minnesota and Wisconsin, where temperatures may be much colder.
- Temperatures may be slightly above normal in the northwest quarter, and below average in most southwest regions, including parts of Arizona, Colorado, and Utah.
- The North Great Plains, eastward over the Great Lakes, and into the Ohio Valley and from Texas to the Tennessee Valley regions may all be slightly warmer than is normal for this time of year.
- The Atlantic Coast states may be also mostly close to average temperatures but slightly on the cooler side.
- The cool airflows behind the storms over the west of the country this week may see the appearances of frost in some parts of southern California.
- Meanwhile, the milder spell on the Great Plains may also see the snow cover dissipate in many parts, especially around central parts.
- Milder weather may also see a snow melt in the Inland Midwest early in the week, before a colder turn brings the return of further snowfalls later in the week.
- Midway through the week ahead, a spell of warmth may spread across the south, which may see the generation of thunderstorms and showers.
- This week may be warm and showery over Hawaii.
- In contrast, Alaska can expect very cold temperatures, but with only some light wintry precipitation in the western and southern parts of the State.

JANUARY

PHASES OF THE MOON

6th	First Quarter
7th	Perigee #13
7th	Crossing Equator
12th	Northern Declination
13th	Full Moon
20th	Crossing Equator
21st	Third Quarter
21st	Apogee
26th	Southern Declination
29th	New Moon

1 Wed	2 Thur	3 Fri	4 Sat	5 Sun	6 Mon	7 Tue	8 Wed	9 Thur	10 Fri	11 Sat	12 Sun	13 Mon	14 Tue	15 Wed	16 Thur	17 Fri	18 Sat	19 Sun	20 Mon	21 Tue	22 Wed	23 Thur	24 Fri	25 Sat	26 Sun	27 Mon	28 Tue	29 Wed	30 Thur	31 Fri
					1Q	P13 XhN					^	F							XhS	3Q A					V			N		

JANUARY 1 WEDNESDAY

ISOBARIC

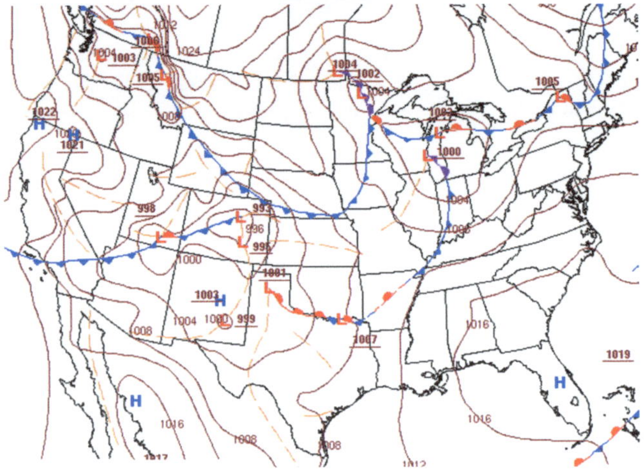

RAIN DISTRIBUTION (Inches)

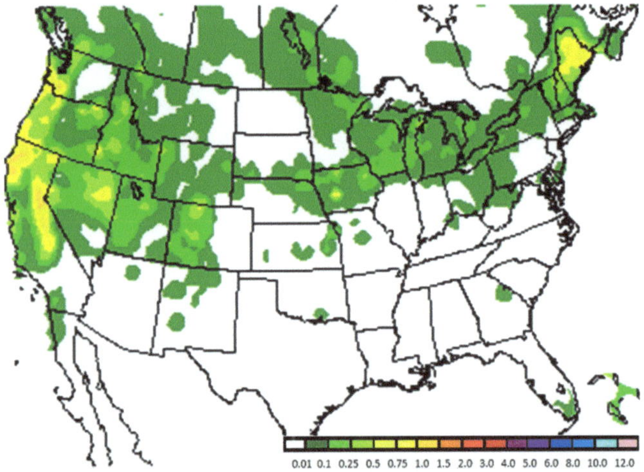

MAXIMUM TEMPERATURE / MINIMUM TEMPERATURE

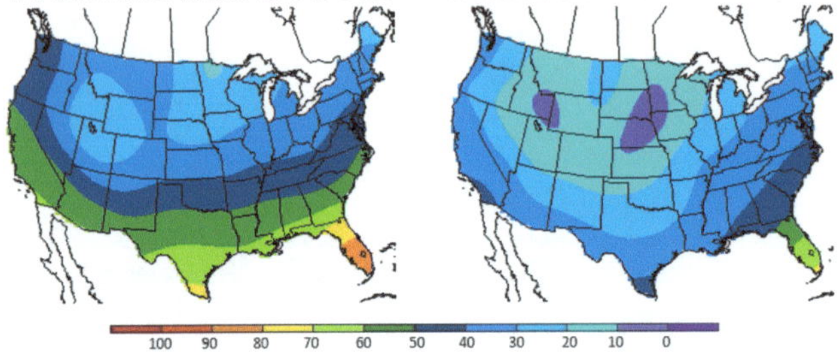

01/01/2025

Wet fronts prevail over the northern half of the country, with rain, wintry showers and snow from the Pacific Northwest to the Northeast.

Heavy snow in the Inland Midwest, especially in Wisconsin.

Blizzard conditions ease central High Plains to the Rocky Mountains.

Heavy rain northern California to Pacific Northwest and east to the Great Basin.

Mostly dry and cool in the southern parts from the Southwest to the Coastal Southeast.

Seasonal temperatures in Florida and the Gulf.

Isolated showers and mild in Hawaii.

Snow prevails in Alaska.

01/02/2025

A large depression prevails centered over the South Great Plains, while another intensifies over the Rocky Mountains.

Wintry precipitation prevails Pacific and Inland Northwest to the northern Rocky Mountains, clearing in California.

Rain in the Southwest to the southern Rockies.

Wintry precipitation, from the Great Lakes to the Deep South, but dry east of the Appalachians.

Snow likely in the Rocky Mountains and around the Great Lakes to the Inland Midwest.

Heavy rain in the Ozarks to the Deep South expected, particularly in Arkansas, spreading eastwards later.

Chance of localized flooding in the lower Mississippi Valley over the next couple of days.

JANUARY 2 THURSDAY

ISOBARIC

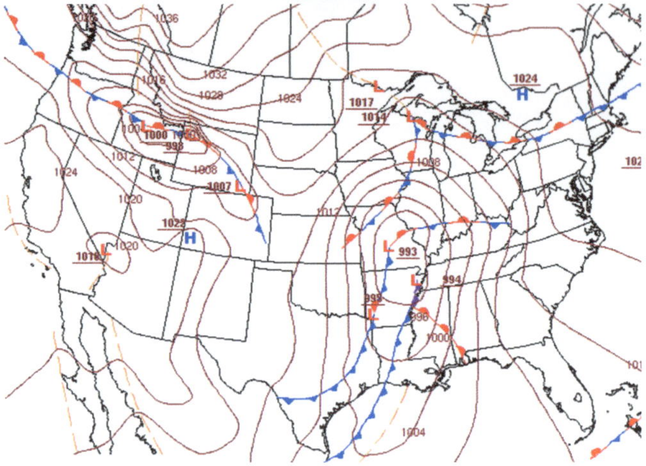

RAIN DISTRIBUTION (Inches)

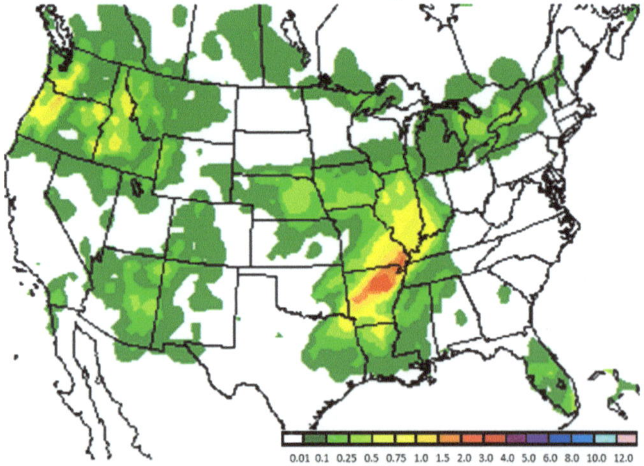

MAXIMUM TEMPERATURE / MINIMUM TEMPERATURE

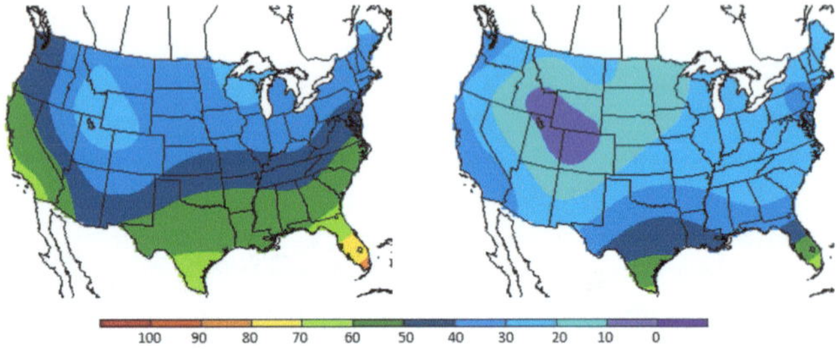

JANUARY 3 FRIDAY

ISOBARIC

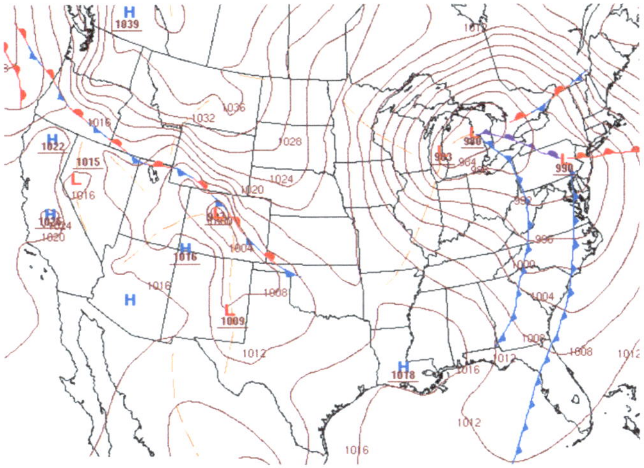

RAIN DISTRIBUTION (Inches)

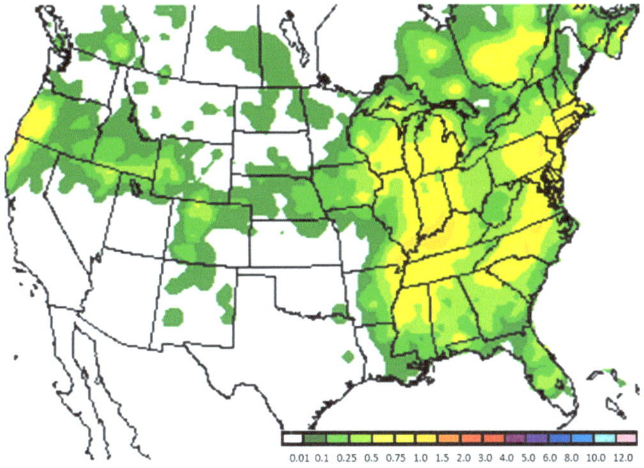

MAXIMUM TEMPERATURE / MINIMUM TEMPERATURE

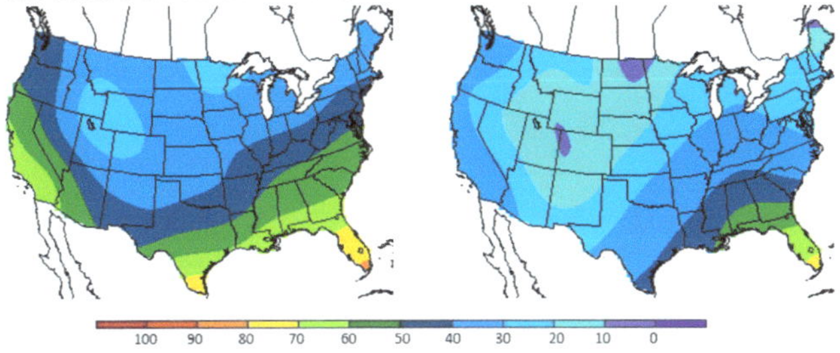

01/03/2025

A large storm system centered over the Great Lakes spreads widespread precipitation throughout the eastern half of the country.

Severe storms and blizzard conditions expected over the Great Lakes, Inland Midwest and in the Northeast with wintry rain in the Megalopolis.

Chance of flooding in the Mississippi Valley.

Wintry precipitation also prevails in Oregon to the Great Basin, over the Rocky Mountains and Colorado Plateau onto the central Great Plains.

Rain also in northern California.

Mostly dry elsewhere.

It may be unusually warm in Florida.

It may be very cold in Alaska.

01/04/2025

The intense depression that covered the eastern half of the country yesterday drifts further northwards into Canada, clearing precipitation over the southeast of the country while maintaining wintry wet conditions, including snow, over the northeast quarter from the Great Lakes, Inland Midwest, and the Northeast.

Chance of unstable winds forming tornadoes in northern Florida with rain in the south of the Peninsula.

Precipitation intensifies in the west, from Oregon to northern California, across the Great Basin and into the southern half of the Rocky Mountains.

Precipitation expected to be heavy in Nevada and Utah, and in the Sierra Nevada Ranges.

Mostly dry elsewhere.

It remains very cold in Alaska.

JANUARY 4 SATURDAY

ISOBARIC

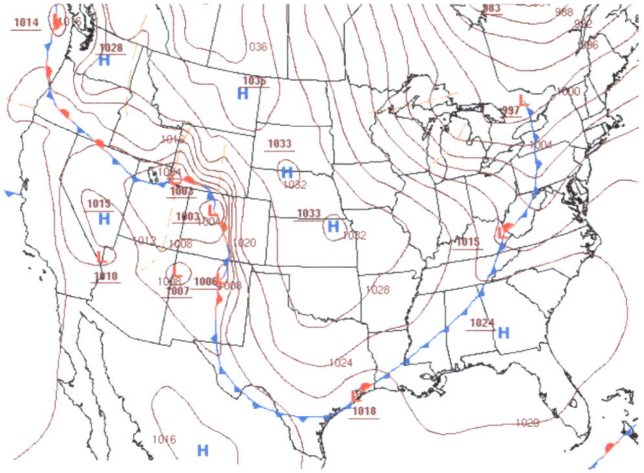

RAIN DISTRIBUTION (Inches)

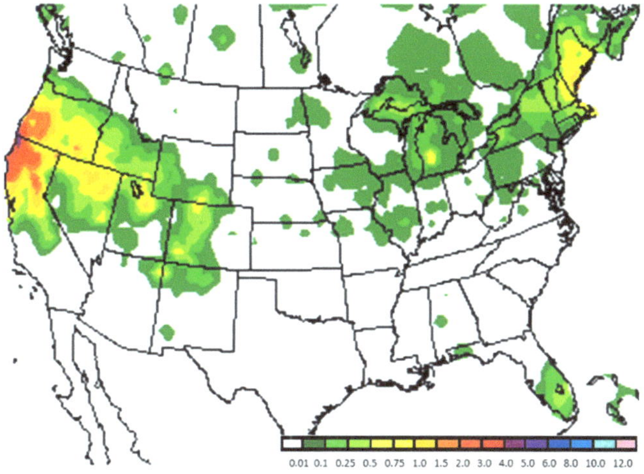

MAXIMUM TEMPERATURE / MINIMUM TEMPERATURE

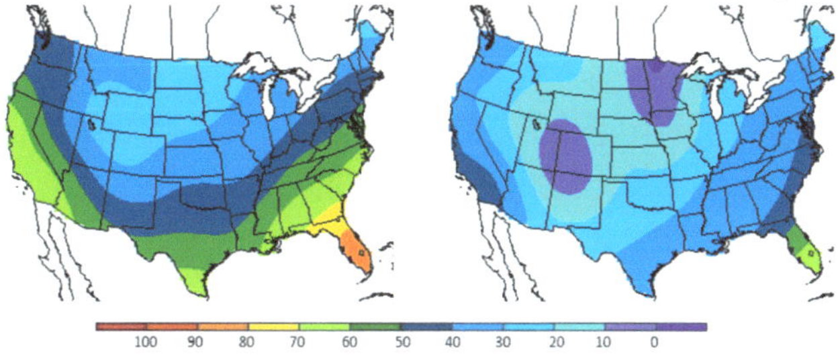

JANUARY 5 SUNDAY

ISOBARIC

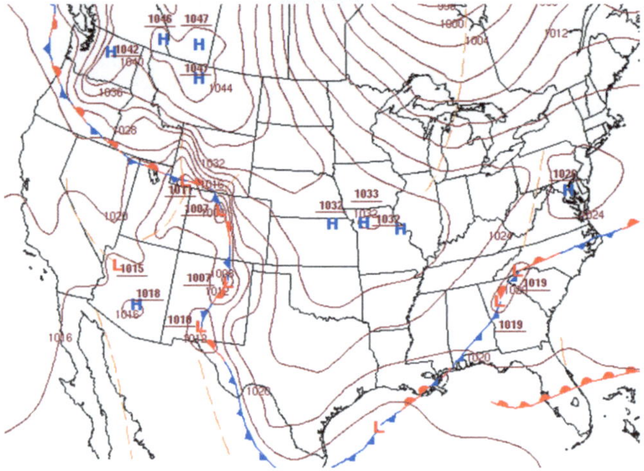

RAIN DISTRIBUTION (Inches)

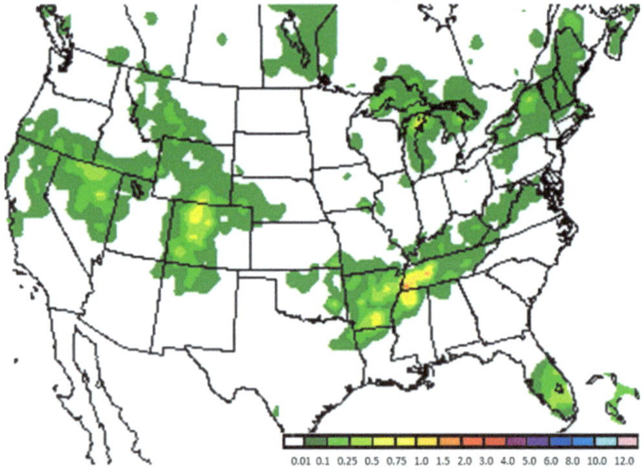

MAXIMUM TEMPERATURE

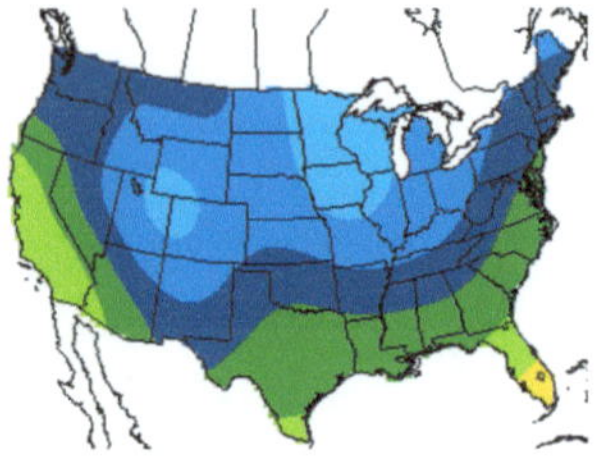

MINIMUM TEMPERATURE

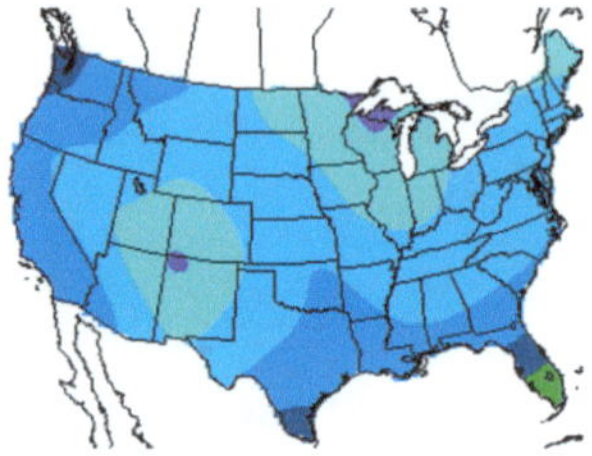

01/05/2025

High pressure expands from the central Plains towards the northeast bringing a milder flow into the northeast.

Precipitation eases over the Great Lakes into the Northeast and clearing through the Inland Midwest.

Very little snowfall expected over the Great Lakes, Ohio Valley and Northeast.

A cold front brings rain over the Appalachians to the northwest of the Deep South.

Showers in Florida.

Possibly no snow yet in the New York region.

In the west, scattered wintry rain also eases, with isolated showers in Oregon to northern California, occasional showers in the Great Basin, and some heavy rain over the Rocky Mountains with precipitation still expected in Nevada, Montana, and Utah.

Milder change in Alaska with plenty of snowfall.

01/06/2025

High pressure prevails through the midsection of the country with low pressure through the Rocky Mountains and around the Great Lakes.

Widespread precipitation expected in the east of the country from the Appalachians to the Northeast, as well as in a band through the central Great Plains to the southern Rocky Mountains.

Heavy rain expected in Florida and the Deep South.

Mostly dry west of the Rocky Mountains, North Great Plains and in Texas.

Very cold in the North Great Plains where recent heavy snow may have caused significant problems with ice, including disruptions to travel corridors and power grids.

JANUARY 6 MONDAY

ISOBARIC

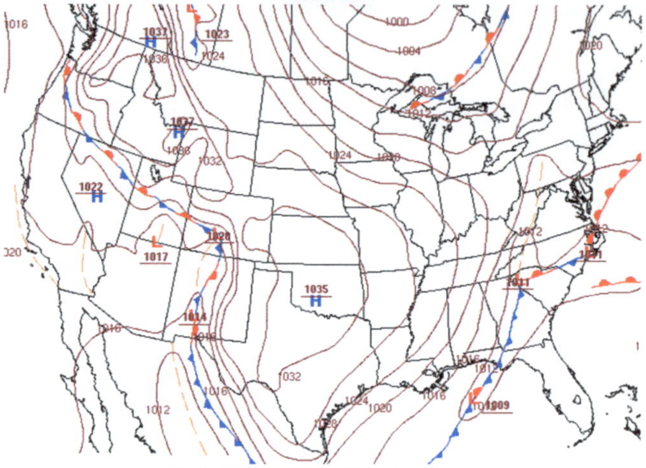

RAIN DISTRIBUTION (Inches)

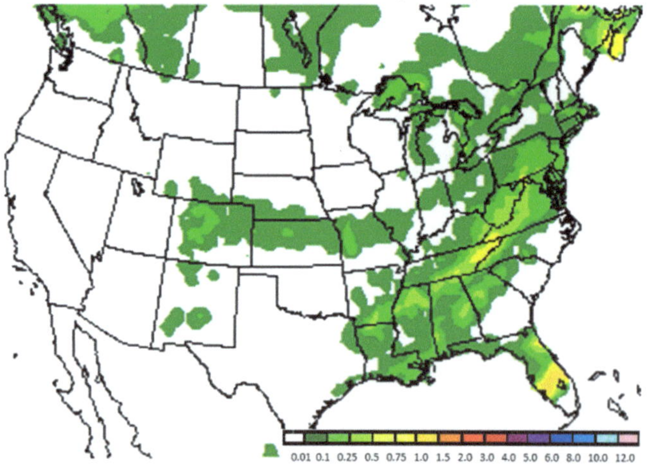

MAXIMUM TEMPERATURE

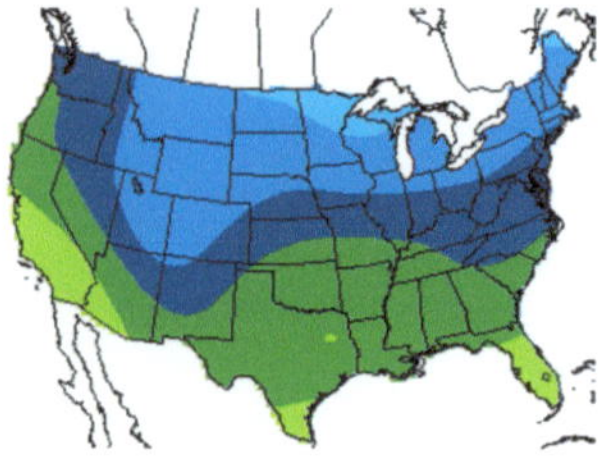

MINIMUM TEMPERATURE

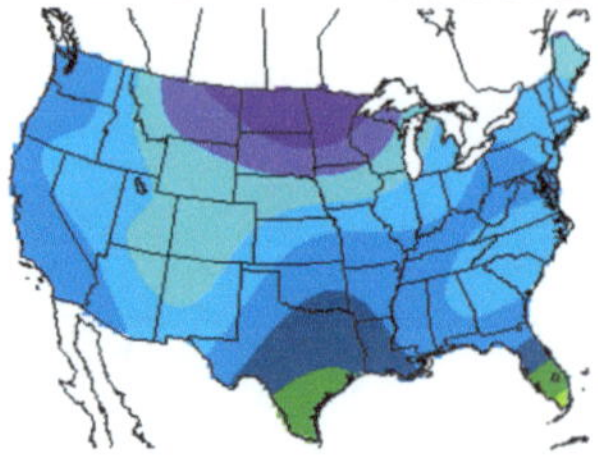

ISOBARIC

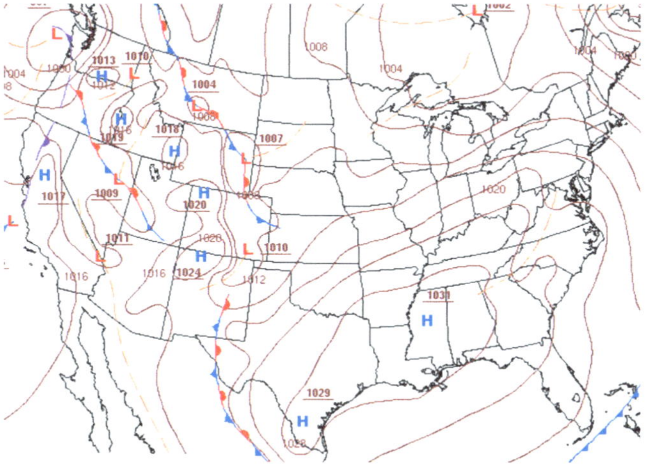

RAIN DISTRIBUTION (Inches)

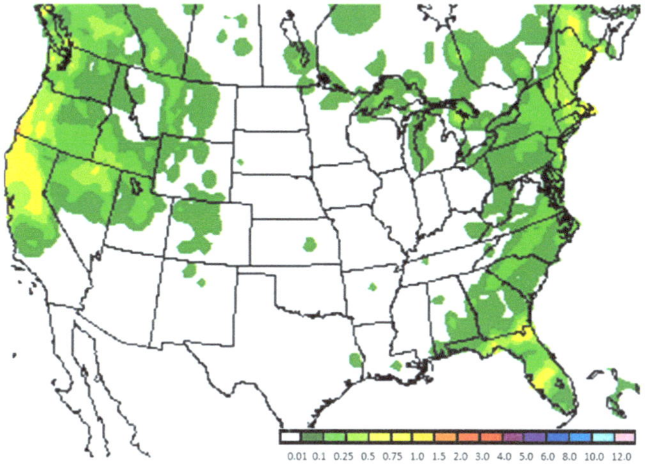

MAXIMUM TEMPERATURE

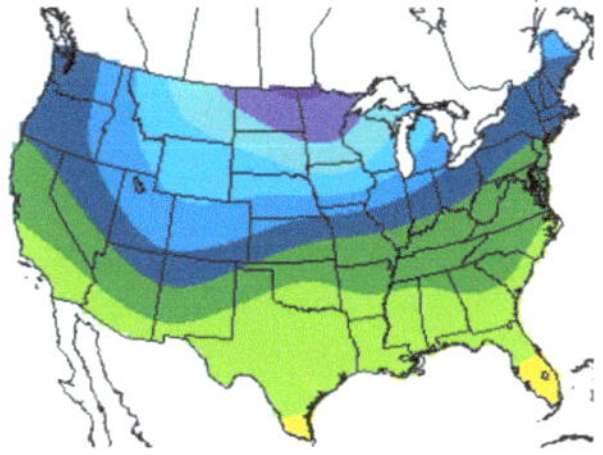

MINIMUM TEMPERATURE

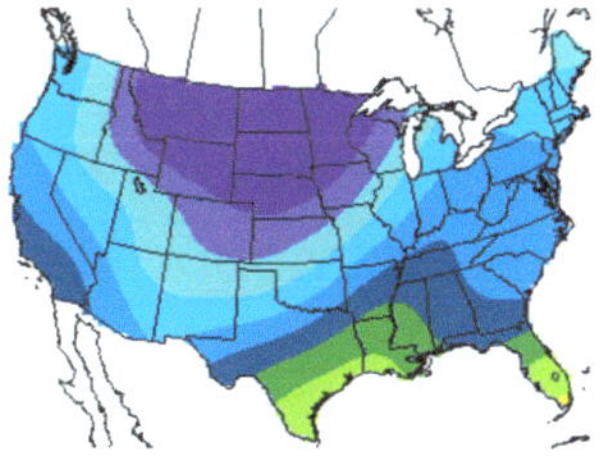

01/07/2025

High pressure dominates the eastern half of the country from the Great Plains to the Atlantic coast, with an unsettled complex of high and low pressures to the west of the Rocky Mountains.

Mostly dry through the midsection of the country into the Inland Midwest, and the Southwest to Texas, with precipitation on both coasts.

Storms in the west from the Pacific and Inland Northwest to the northern Rocky Mountains and into the north of California and adjacent Great Basin.

Heavy snow in Rocky Mountains, including Montana, Wyoming, with chance of further disruptions to power grids, travel corridors and possible stock losses.

Some wintry showers over the Great Lakes, and widespread precipitation along the Atlantic coasts from the Northeast to Florida.

Chance of heavy showers in Hawaii.

01/08/2025

Stormy from the Pacific Northwest over the Rocky Mountains and into the north and central Great Plains, with wintry precipitation prevailing, including snow.

Rain in northern California with showers in the south, and storms extending into the Four Corner states.

Freezing rain and snow expected in the upper Midwest, with isolated wintry showers in the Great Lakes to the Northeast.

Heavy showers in Hawaii. Mostly dry elsewhere.

A warmer southerly flow expected through the Inland Midwest, the Mississippi Valley and Deep South.

Chance of thunderstorms in eastern Texas and the tornado risk may be high for a time in the State.

Chance of snow in New Mexico.

ISOBARIC

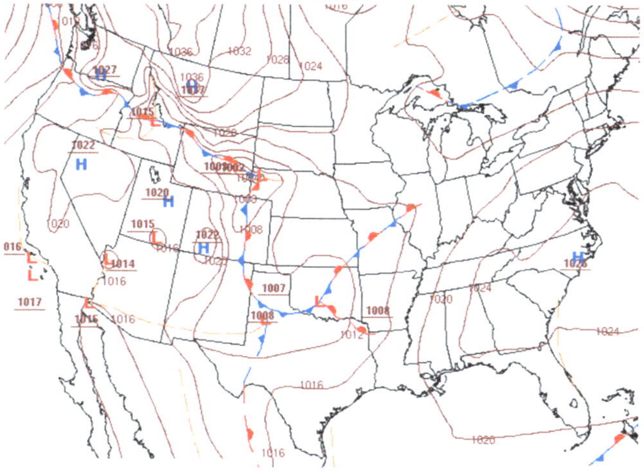

RAIN DISTRIBUTION (Inches)

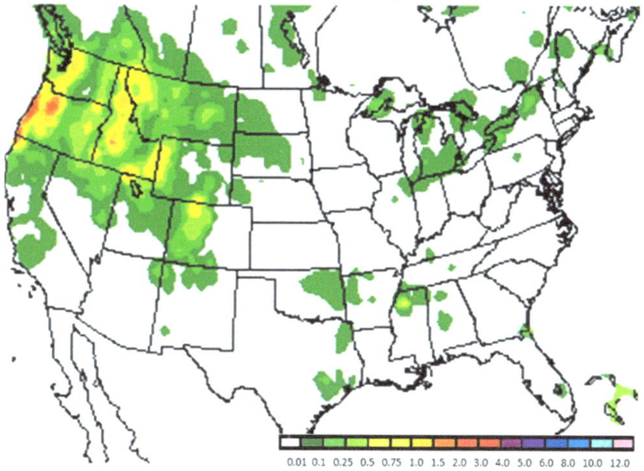

MAXIMUM TEMPERATURE

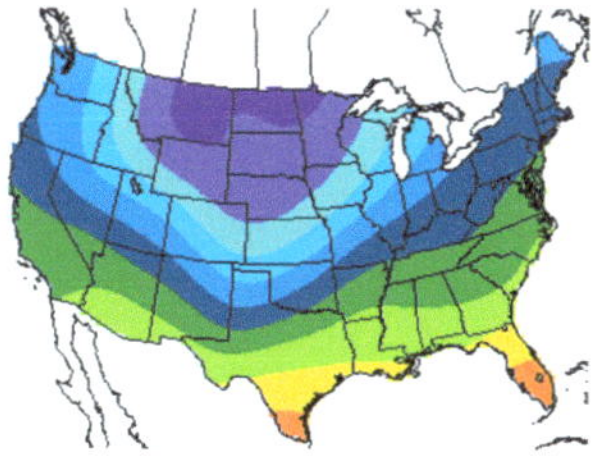

MINIMUM TEMPERATURE

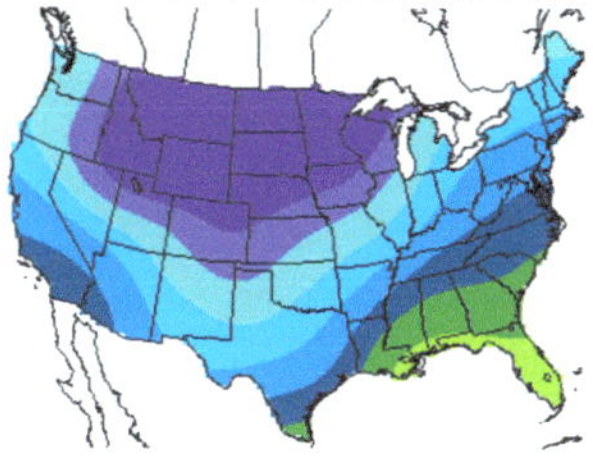

ISOBARIC

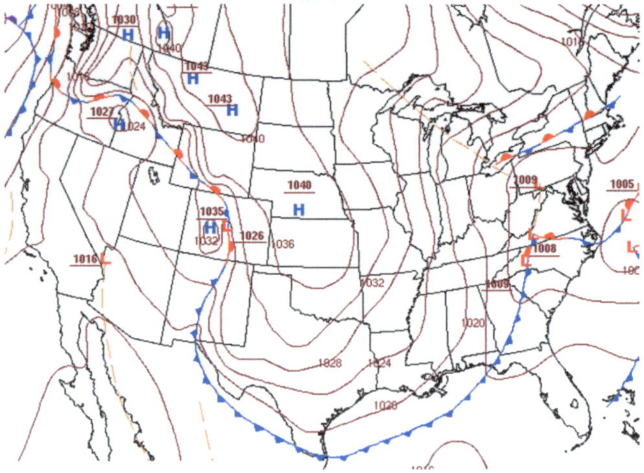

RAIN DISTRIBUTION (Inches)

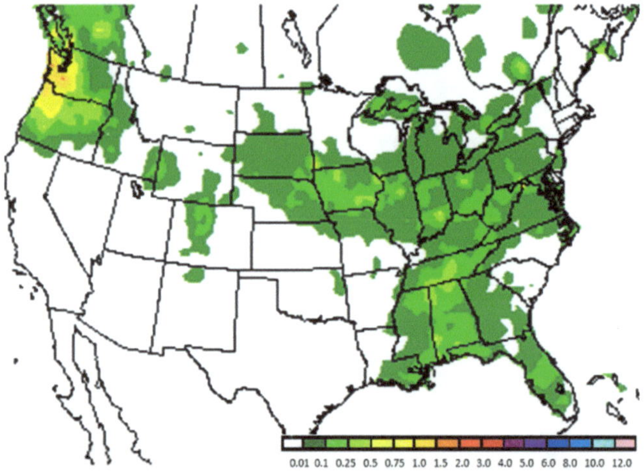

MAXIMUM TEMPERATURE

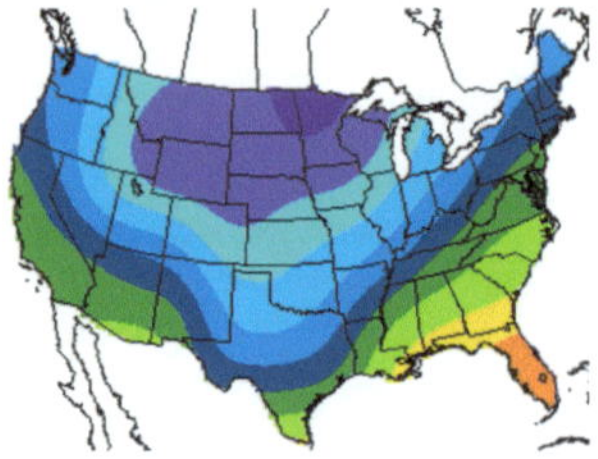

MINIMUM TEMPERATURE

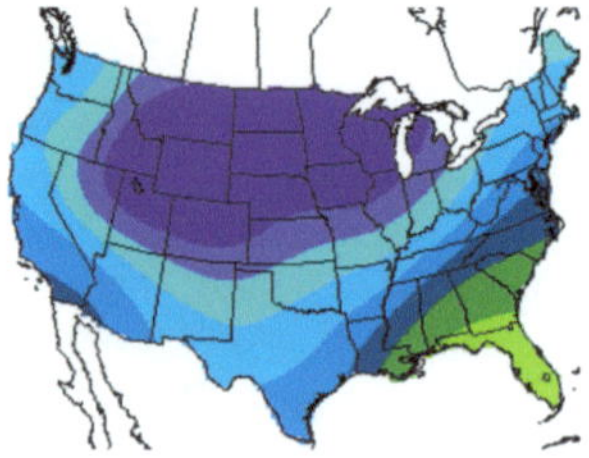

100 90 80 70 60 50 40 30 20 10 0

01/09/2025

High pressure builds over the midsection of the country with precipitation clearing in the Rocky Mountains to California.

Wintry rain continues in the Pacific Northwest.

Widespread precipitation prevails from the southern Northeast to Florida and west through the Deep South, over the Appalachians, throughout the Midwest and Great Lakes and over the North Great Plains.

Snow from the Great Lakes and Inland Midwest to the Northeast and central Great Plains.

Snow may be heavy in Nebraska.

Thunderstorms may occur from eastern Texas to the southern Atlantic areas, with some outbreaks of heavy rain especially around the Deep South States to Florida.

Mostly dry elsewhere.

01/10/2025

Wintry precipitation, much of it as snow, continues over the Great Lakes to the southern Northeast and through the north Appalachians to the Metropolitan.

Wet in the eastern Corn Belt and Pacific Northwest, with heavy rain in northern California.

Chance of lingering snow in Minnesota and Kansas.

Scattered showers in Florida.

Mostly dry through the midsection from the Northern Great Plains to Texas, as well as to the west in the Southwest to southern California.

Unusually milder temperatures in parts of the upper Midwest.

Heavy showers in Hawaii and heavy snow in Alaska.

ISOBARIC

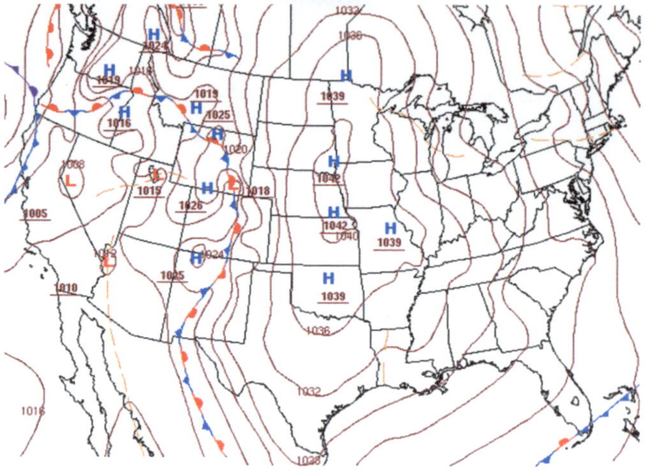

RAIN DISTRIBUTION (Inches)

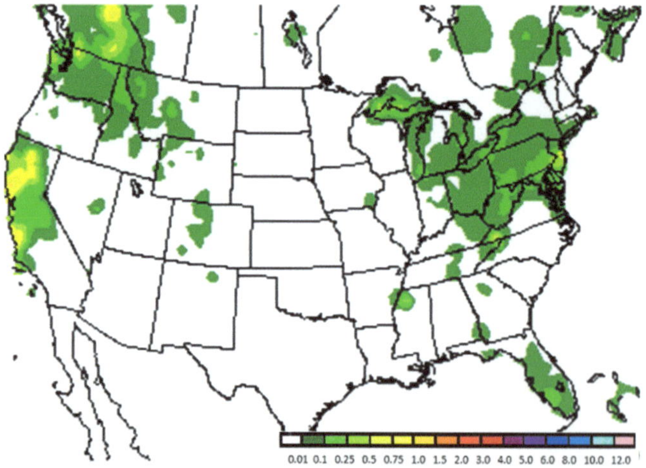

MAXIMUM TEMPERATURE

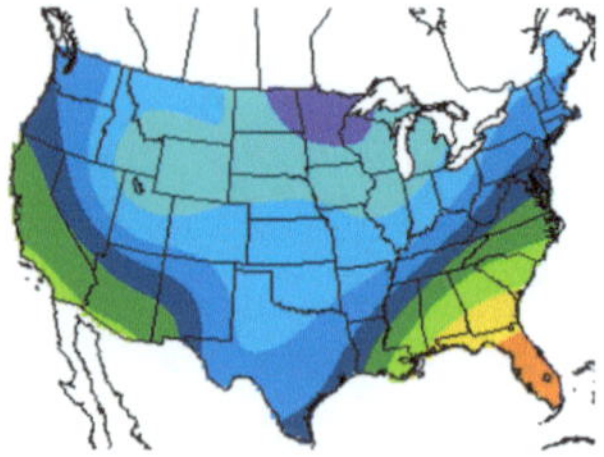

MINIMUM TEMPERATURE

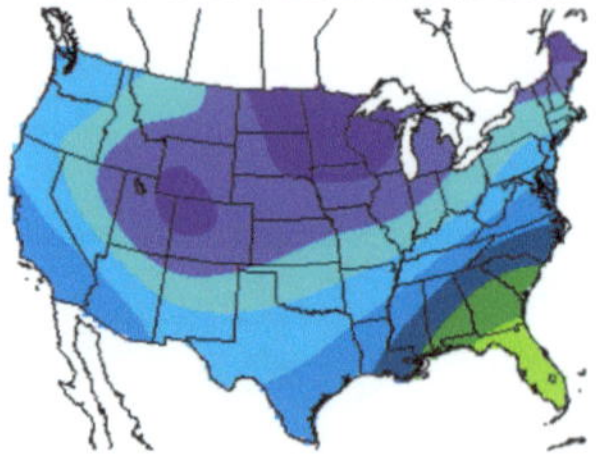

100 90 80 70 60 50 40 30 20 10 0

JANUARY 11 SATURDAY

ISOBARIC

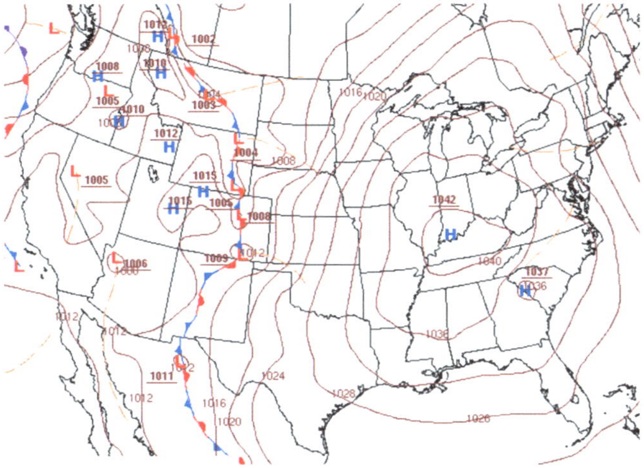

RAIN DISTRIBUTION (Inches)

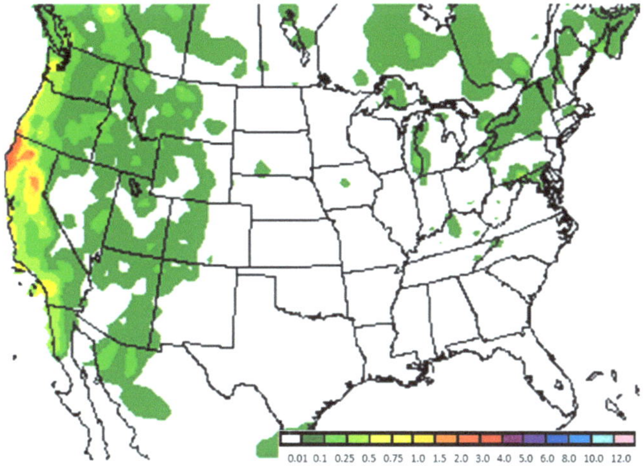

MAXIMUM TEMPERATURE

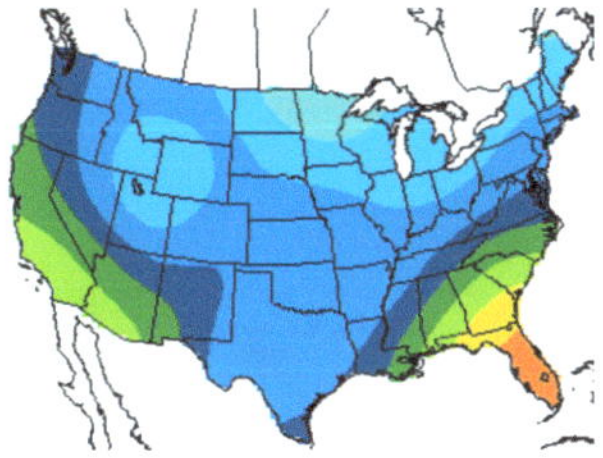

MINIMUM TEMPERATURE

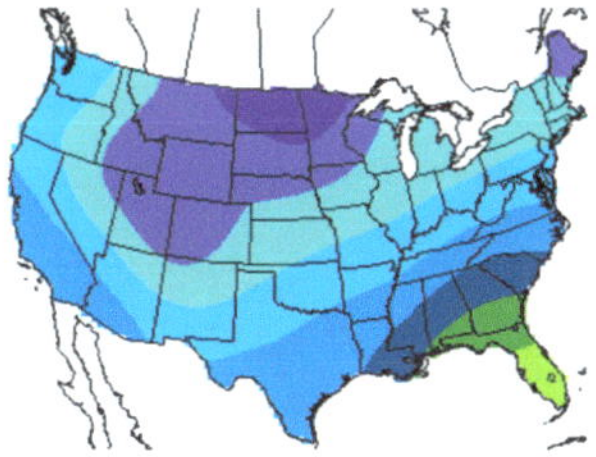

01/11/2025

High pressure strengthens over the eastern half of the country, easing precipitation to isolated outbreaks mostly around the Great Lakes to the Northeast.

Mostly dry through the midsection from the Great Plains to Texas, and throughout the southeast quarter of the country except for the chance of some showers around the far southern tip of Texas and lingering in the Appalachians.

Widespread precipitation prevails over the west of the country from the Rocky Mountains and Southwest to the Pacific coasts.

Heavy precipitation likely in northern California, while snow likely in the higher reaches of the Rocky Mountains, Idaho and Montana, to the Four Corners.

Winds may be 'chinook' in nature on the Montana High Plains for a time over the next few days.

01/12/2025

High pressure and mostly dry from the Atlantic coast from the Northeast to Florida, over the Appalachians, and west to the North Great Plains and the Deep South with just some spotty showers around the Great Lakes.

Widespread rain from Texas to the Ozarks, heavier around the Gulf coasts.

Widespread precipitation continues in the west from the Pacific coast to the Rocky Mountains, with rain particularly heavy along the Pacific coasts from the Northwest to northern California, as well as in parts of the Southwest.

JANUARY 12 SUNDAY

ISOBARIC

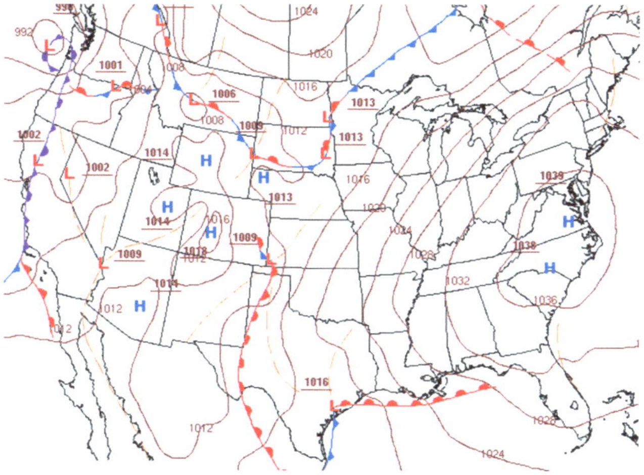

RAIN DISTRIBUTION (Inches)

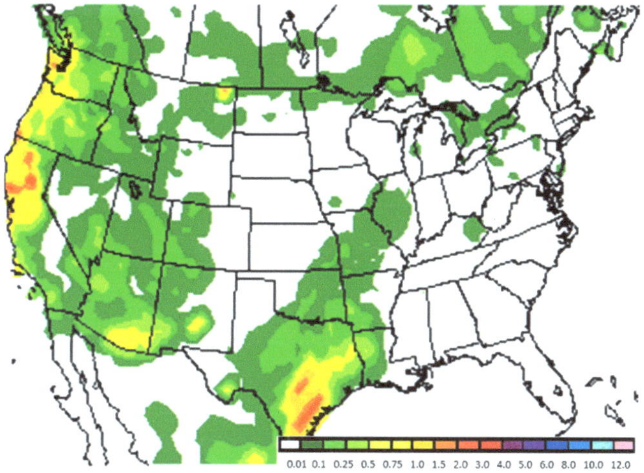

MAXIMUM TEMPERATURE

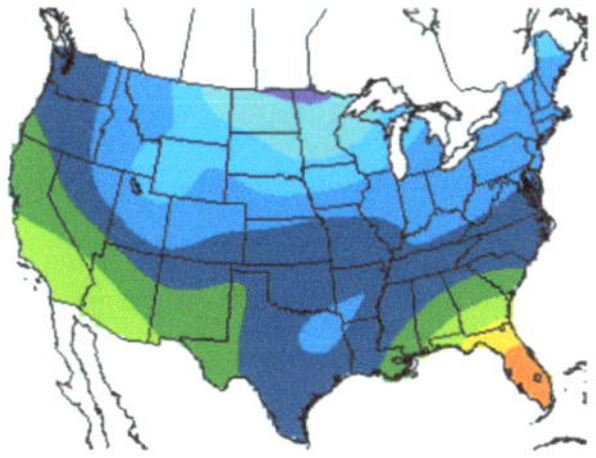

MINIMUM TEMPERATURE

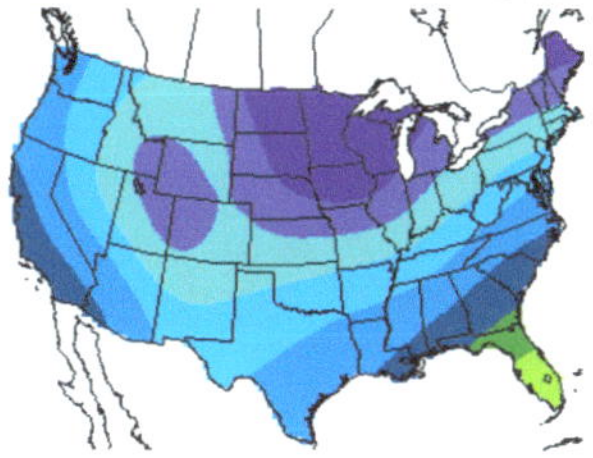

JANUARY 13 MONDAY

ISOBARIC

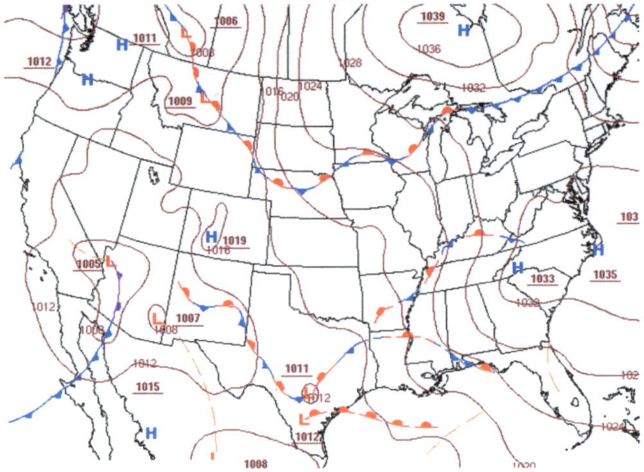

RAIN DISTRIBUTION (Inches)

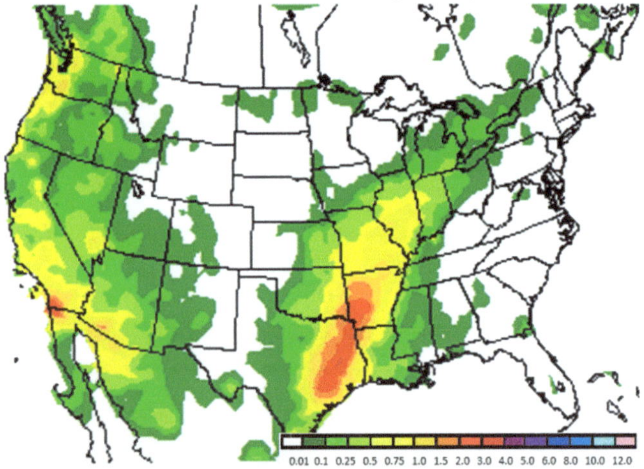

MAXIMUM TEMPERATURE

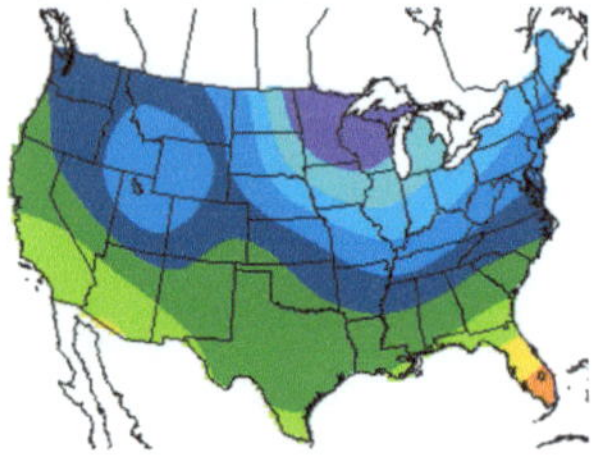

MINIMUM TEMPERATURE

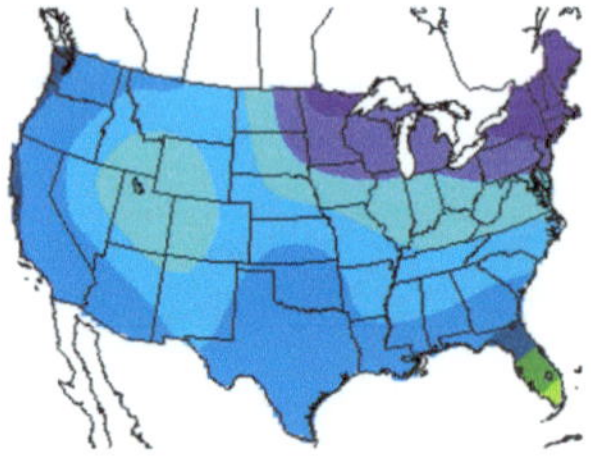

01/13/2025

A cold front spreads precipitation up through the midsection of the country from Texas and the western Deep South, into the South Great Plains, across the Inland Midwest and up to the Great Lakes.

Rainfall may be particularly heavy in the east of Texas to the Ozarks.

Widespread precipitation also dominates the west of the country from the Pacific Ocean to the Rocky Mountains and from the Pacific Northwest to the Southwest.

Precipitation may be particularly wintry with heavy snow possibly in Montana over the next few days.

Mostly dry over the North Great Plains and along the Atlantic coasts from the Northeast to Florida and up to the eastern Appalachian foothills.

Heavy snow also in the south of Alaska.

01/14/2025

Mostly dry over the next few days in Colorado and surrounding States to the east, bringing a brief respite to the recent snow, as well as mostly dry on the North Great Plains to the western Great Lakes, in the far north of the Northeast, and in Florida.

Some precipitation almost everywhere else, with particularly heavy rain in the east of Texas to the Deep South, particularly around the Mississippi River Valley, which may see the development of localized flooding over the next few days.

JANUARY 14 TUESDAY

ISOBARIC

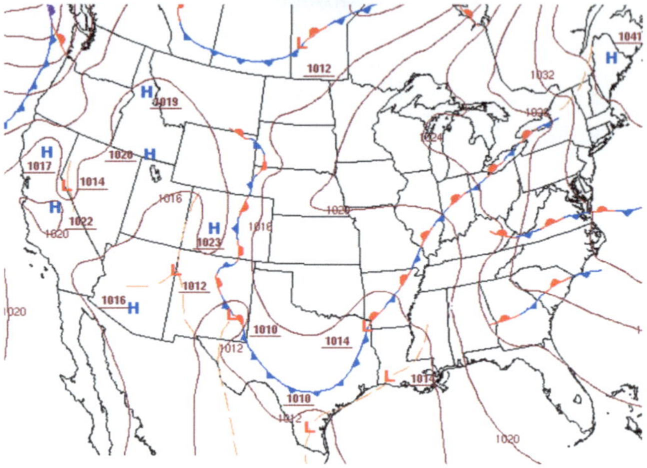

RAIN DISTRIBUTION (Inches)

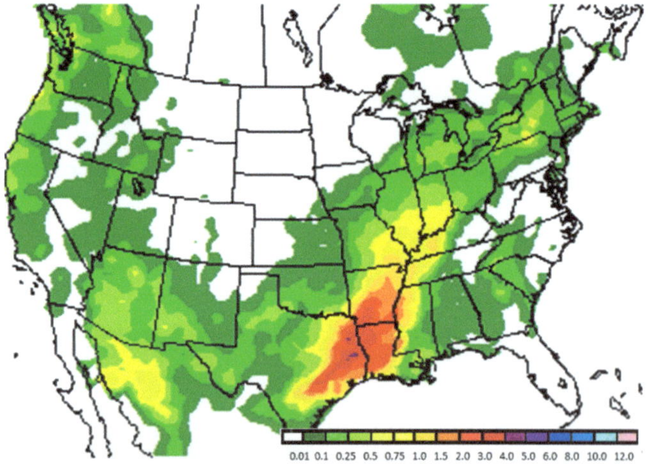

MAXIMUM TEMPERATURE

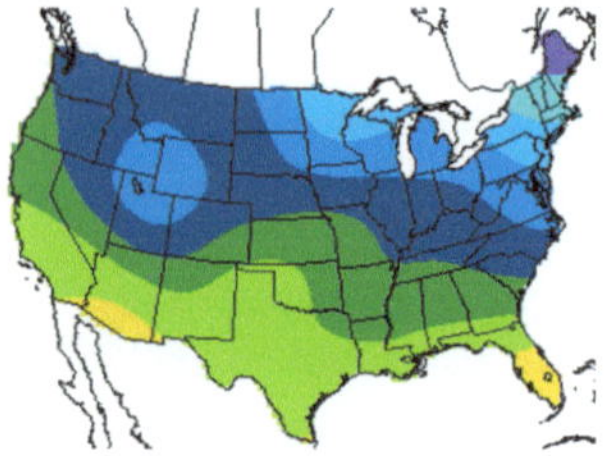

MINIMUM TEMPERATURE

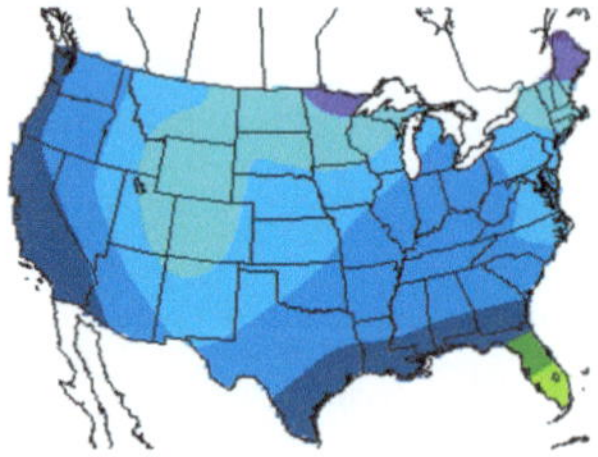

JANUARY 15 WEDNESDAY

ISOBARIC

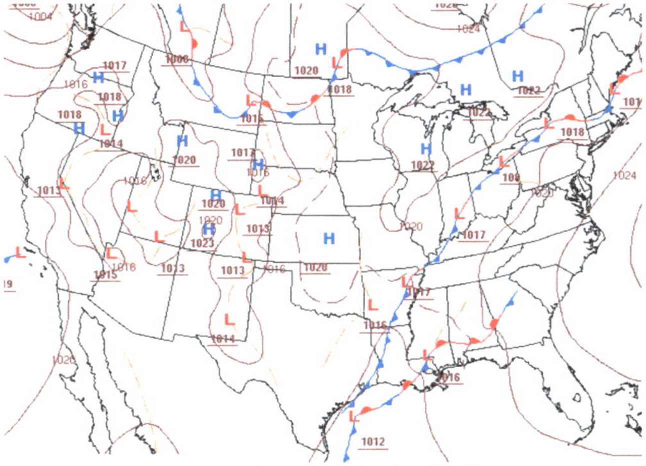

RAIN DISTRIBUTION (Inches)

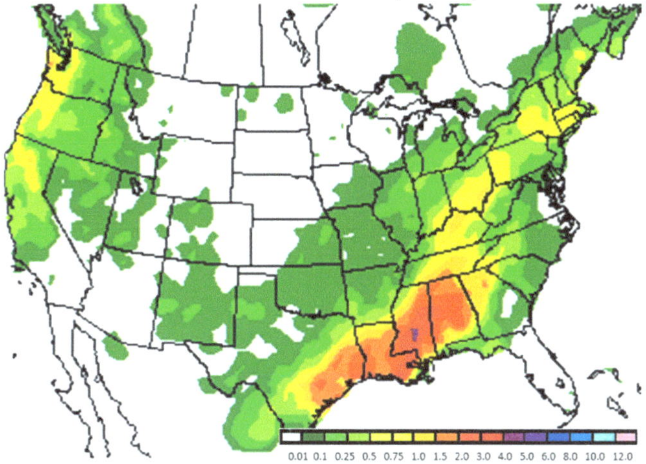

MAXIMUM TEMPERATURE

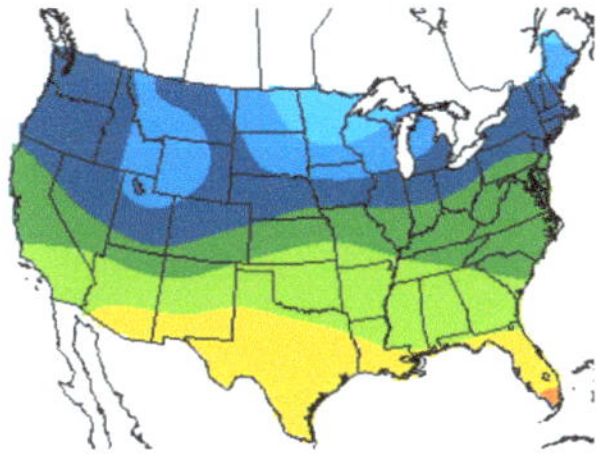

MINIMUM TEMPERATURE

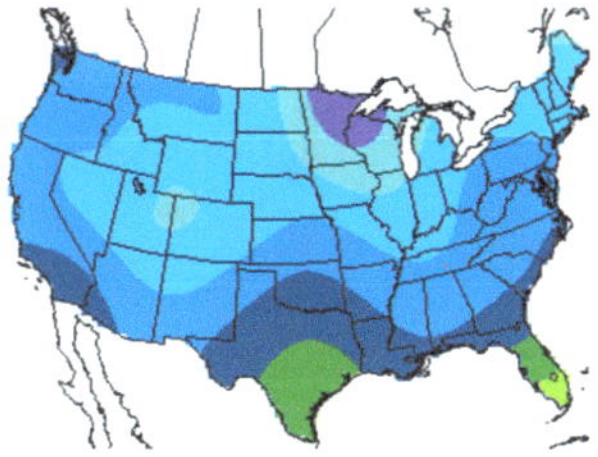

01/15/2025

A wet front spreads precipitation from the Gulf of Mexico, northwards either side of the Appalachians from the Atlantic coast to the Great Plains, across the Inland Midwest and throughout the Northeast, with showers into the Southwest and around the Great Lakes.

Further snow likely in the Great Lakes to the Northeast, and in the higher reaches of the Rocky Mountains.

Precipitation will be particularly heavy from the northeast of Texas and through the Deep South, where severe thunderstorms may generate tornadoes from Louisiana to the Carolinas over the next couple of days.

Widespread precipitation also prevails from Washington to northern California with showers into the Great Basin and southern Rocky Mountains.

Mostly dry elsewhere, including in Florida.

Winds may be unusually strong in southern California.

01/16/2025

High pressure maintains mostly dry conditions over the North Great Plains.

Rain clears California and eases over Texas.

Precipitation prevails elsewhere.

Widespread wintry precipitation persists over the eastern half of the country, from the South Great Plains to the Atlantic coasts and from the Northeast to the Deep South, and into the north of Florida, with heavy snow possible in the northeastern states.

Showers prevail from the Pacific Northwest, in the Great Basin and from the southern Rocky Mountains into the Southwest.

It may be unusually mild in Wisconsin.

JANUARY 16 THURSDAY

ISOBARIC

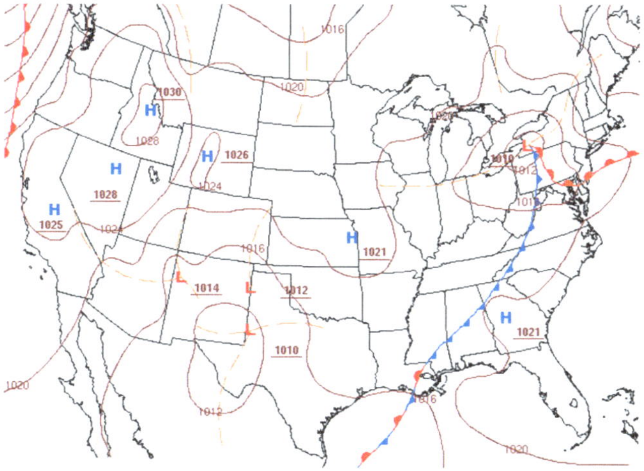

RAIN DISTRIBUTION (Inches)

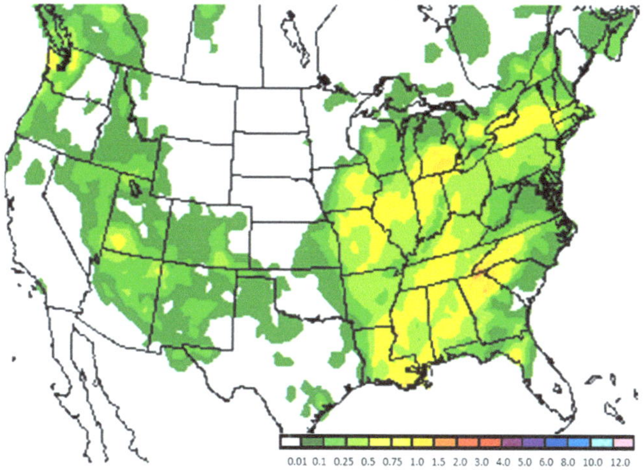

MAXIMUM TEMPERATURE

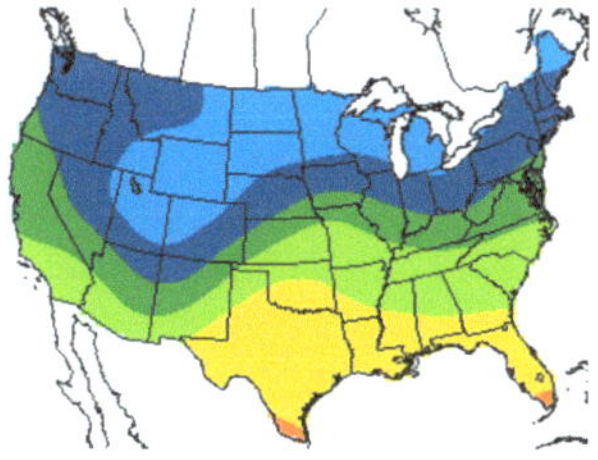

MINIMUM TEMPERATURE

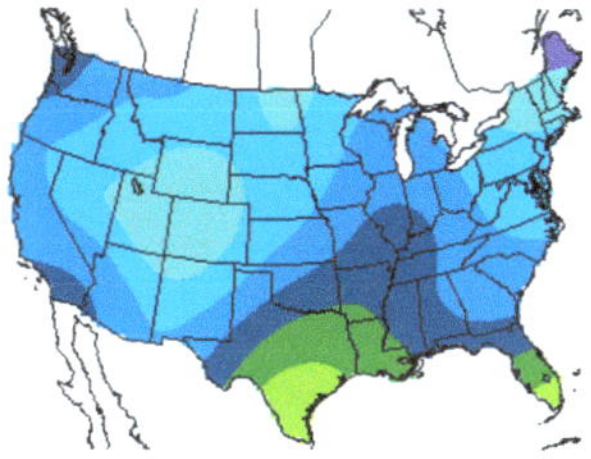

ISOBARIC

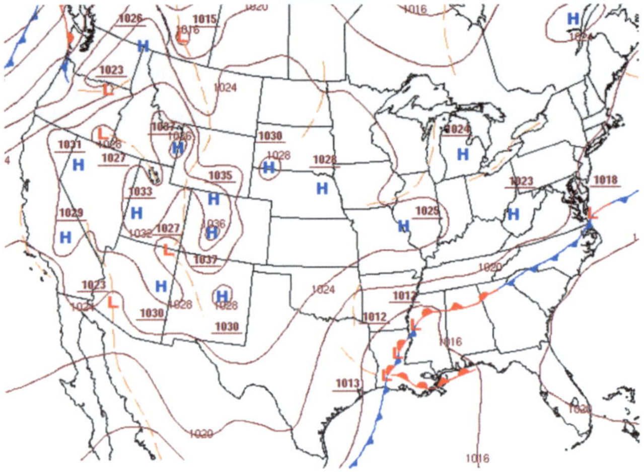

RAIN DISTRIBUTION (Inches)

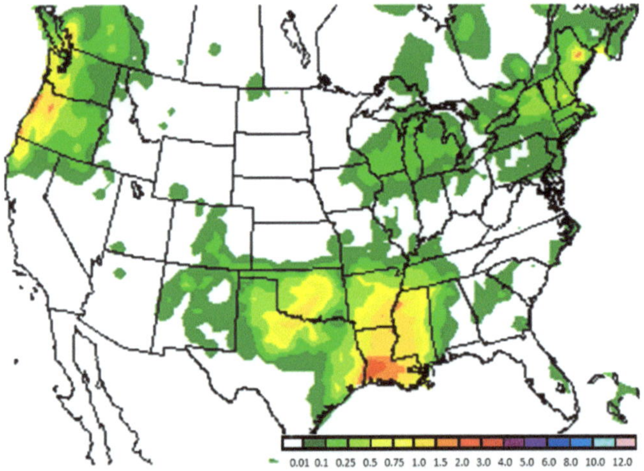

MAXIMUM TEMPERATURE

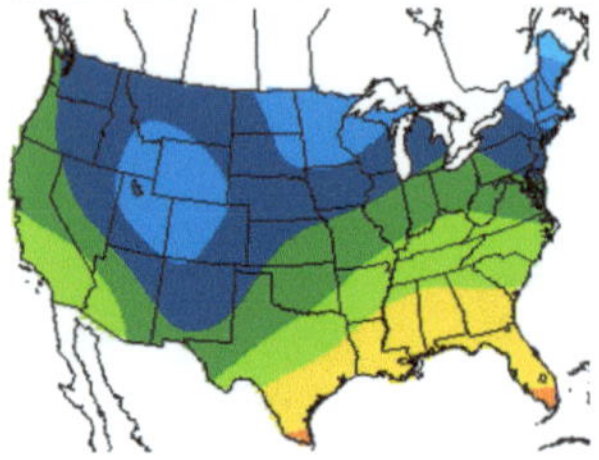

MINIMUM TEMPERATURE

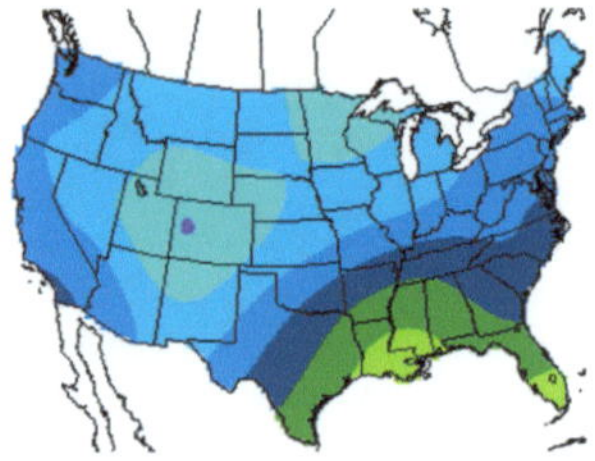

01/17/2025

Rain from Washington to northern California while clearing in the Great Basin to the Southwest.

Mostly dry over the northern and central Great Plains and northern Rocky Mountains.

Precipitation eases from the Inland Midwest to the Coastal Southeast and along the coast to Florida.

Further wintry precipitation, mostly of snow, prevails over the Great Lakes to the Northeast.

Showers in the eastern half of the Southwest.

Widespread rain and thunderstorms over the northern half of Texas to the western half of the Deep South, with the threat of tornado formation winds in the regions.

Very heavy precipitation likely around the Gulf of Mexico coastal fringes of the Deep South.

Milder temperatures in the east.

Milder temperatures in Alaska with the arrival of further snow.

01/18/2025

High pressure prevails from the west coast to the midsection, depression in the east.

A cooler change in the east brings widespread precipitation from the Great Lakes to Florida, with showers into the Northeast.

Heavy falls in the Southeast, east of the Appalachians.

Widespread rain over the Pacific Northwest to northern California, with heavy falls along the coasts, especially in Washington.

Mostly dry elsewhere.

Heavy showers likely in Hawaii over the next couple of days.

ISOBARIC

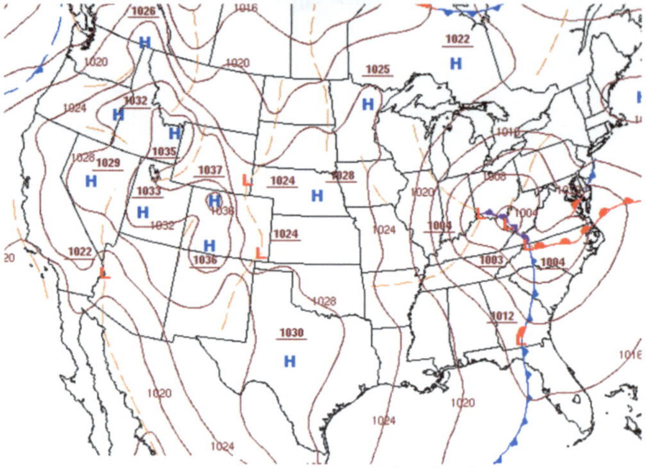

RAIN DISTRIBUTION (Inches)

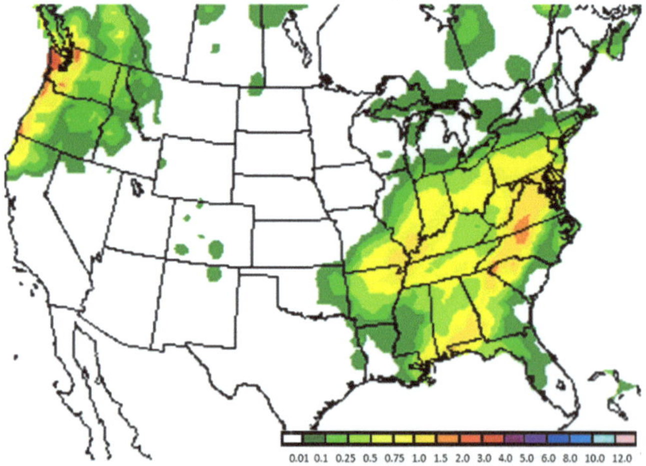

MAXIMUM TEMPERATURE

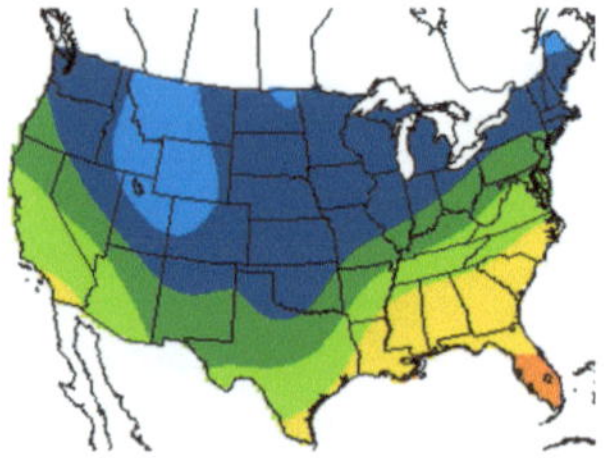

MINIMUM TEMPERATURE

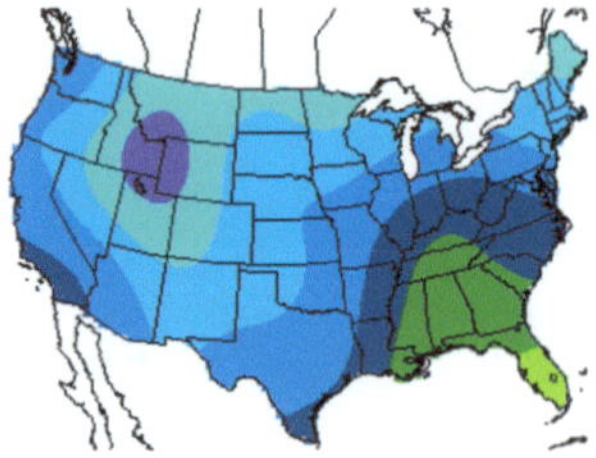

JANUARY 19 SUNDAY

ISOBARIC

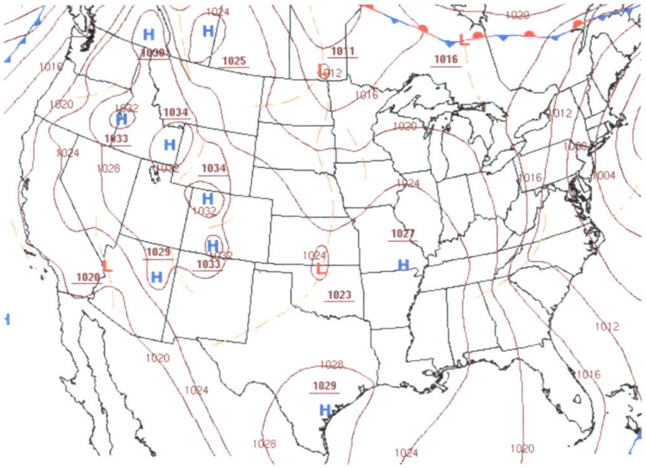

RAIN DISTRIBUTION (Inches)

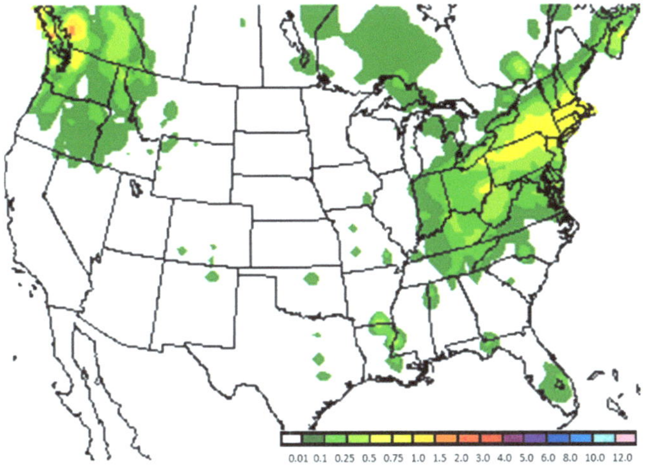

MAXIMUM TEMPERATURE

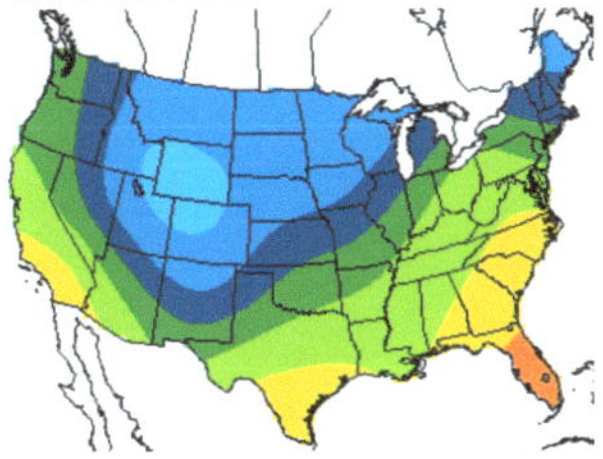

MINIMUM TEMPERATURE

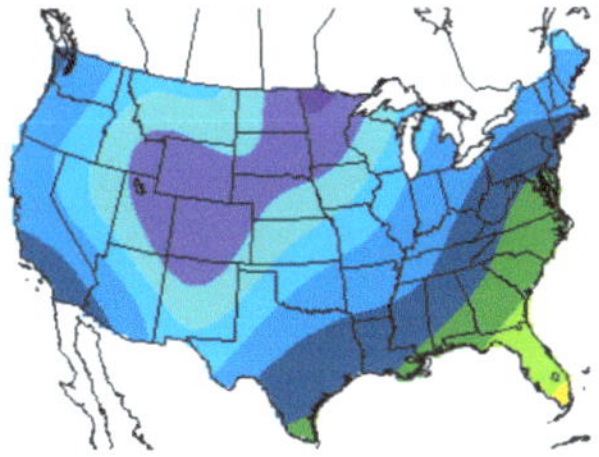

01/19/2025

Widespread wintry precipitation from the Northeast to the northern Appalachians persists, with showers in the Great Lakes, Coastal Southeast, Deep South and Florida.

Scattered precipitation persists in the Pacific to Inland Northwest.

Chance of light snow in the mountainous parts of the West and Southwest.

Mostly dry elsewhere.

Severe snowstorm expected in western Alaska.

JANUARY 20 MONDAY

01/20/2025

An occluded front brings wintry precipitation, including snow, over the Great Lakes, with wintry showers also extending over the Inland Midwest, and isolated snow flurries in the Northeast.

Chance of the first snowfalls in New York.

Some isolated showers remaining in the coastal parts of Washington.

Mostly dry elsewhere.

It may be colder in the Northwest to the Great Plains with the band of cold air spreading southward and some snow possible ahead of the cold snap.

ISOBARIC

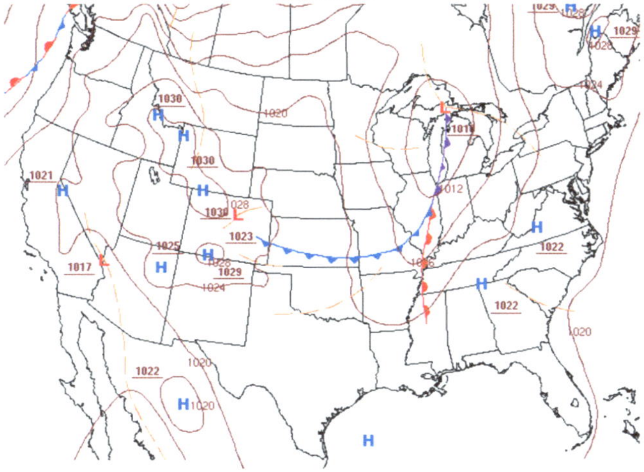

RAIN DISTRIBUTION (Inches)

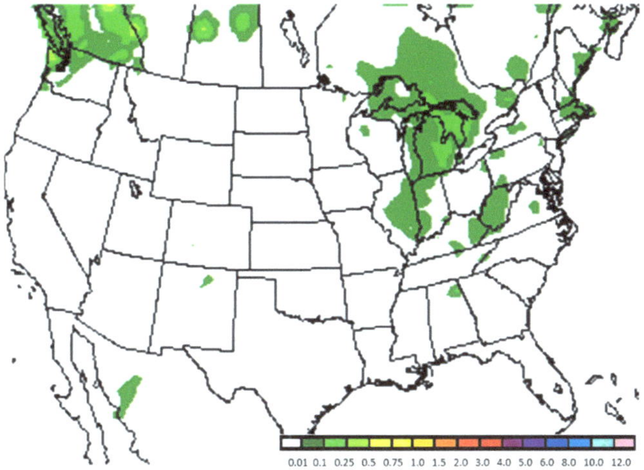

MAXIMUM TEMPERATURE

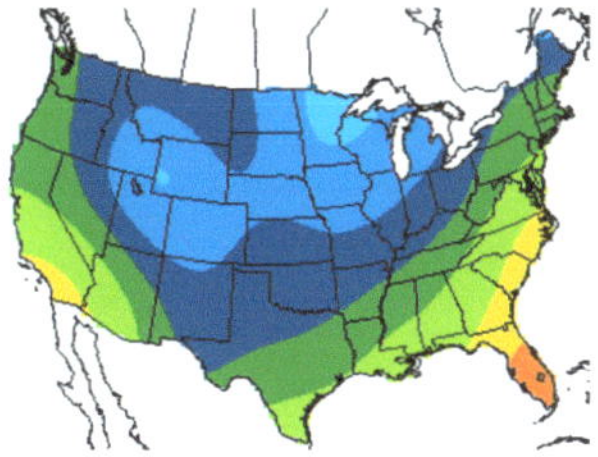

MINIMUM TEMPERATURE

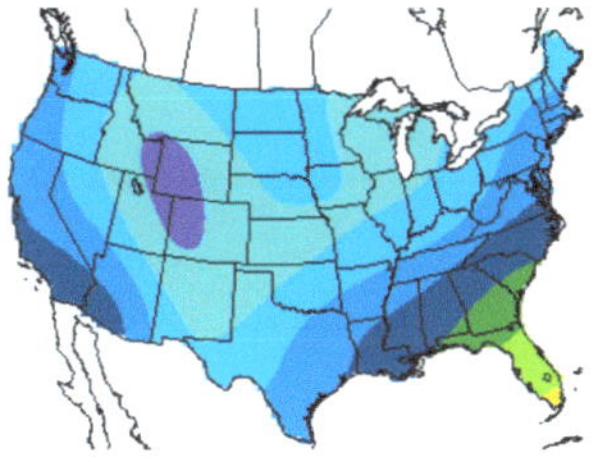

JANUARY 21 TUESDAY

ISOBARIC

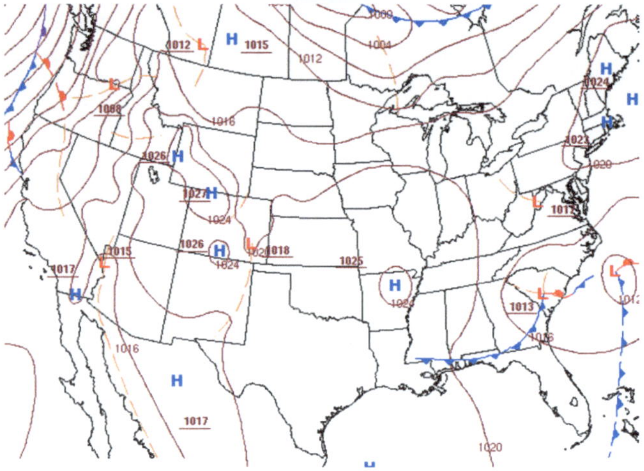

RAIN DISTRIBUTION (Inches)

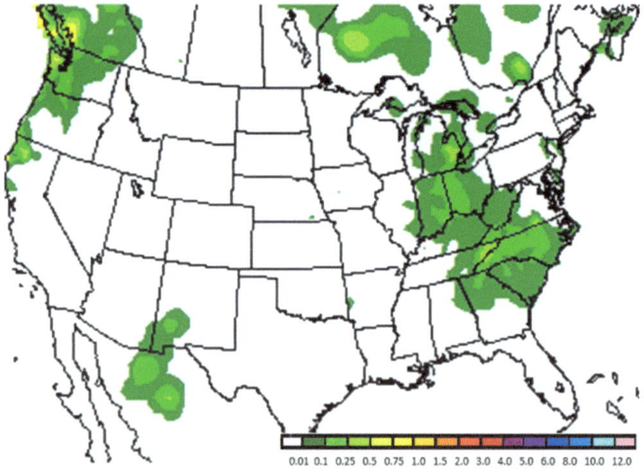

MAXIMUM TEMPERATURE | MINIMUM TEMPERATURE

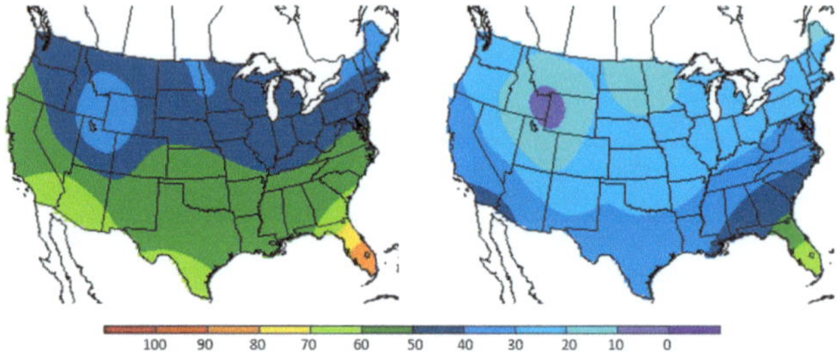

01/21/2025

High pressure prevails almost everywhere, with most of the country under a dry cold snap.

The only precipitation of any significance expected may be in the Pacific Northwest, mostly along the coastal areas, in southern New Mexico, and from the Great Lakes, across the Appalachians and into the southeast areas from the Carolinas to Georgia.

Chance of showers in the Sierra Nevada and Cascade Mountain Ranges.

Deep snow in Alaska.

01/22/2025

Widespread precipitation in the Pacific Northwest and south into California, with particularly heavy falls in coastal California to the Sierra Nevada Ranges.

Isolated wintry precipitation in the Northeast to the Great Lakes and around the Coastal Southeast.

Mostly dry elsewhere.

The cold snap may also see frosts in California's southern agricultural valleys and into the southwest of Arizona possibly causing some damage to crops over the next few days.

It may also be very cold in Idaho.

JANUARY 22 WEDNESDAY

ISOBARIC

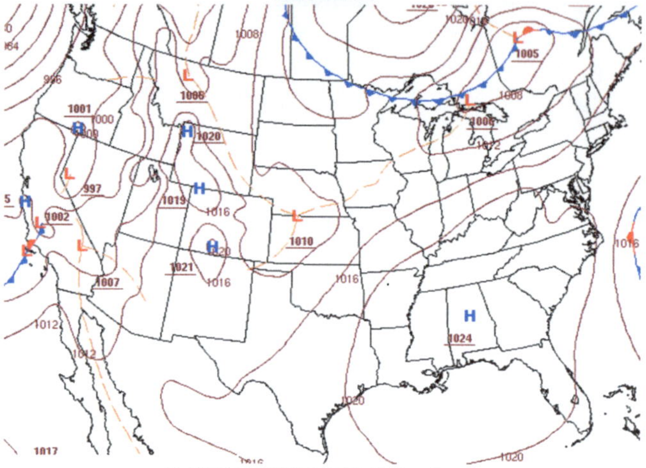

RAIN DISTRIBUTION (Inches)

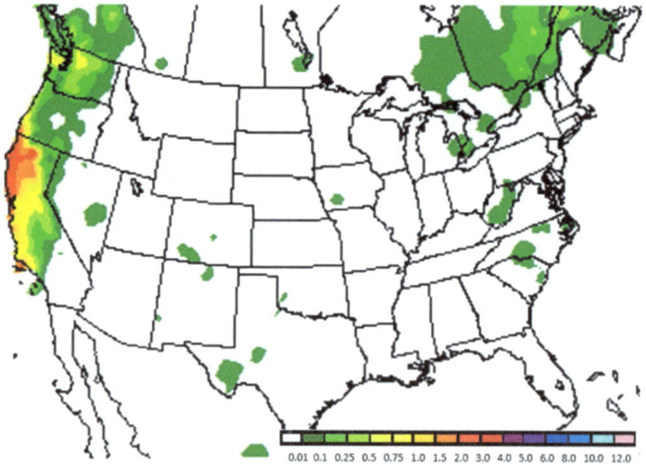

MAXIMUM TEMPERATURE | MINIMUM TEMPERATURE

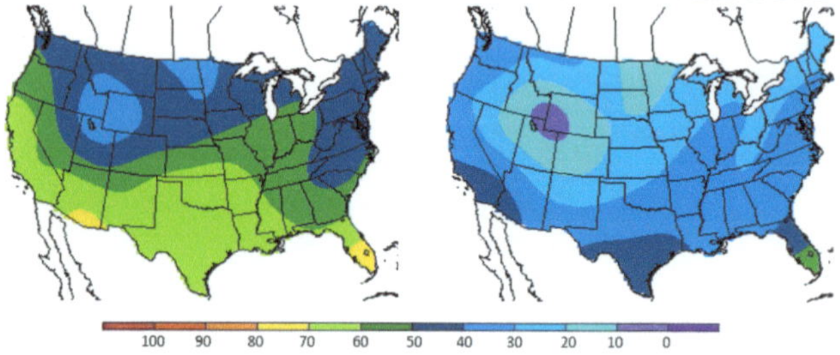

JANUARY 23 THURSDAY

ISOBARIC

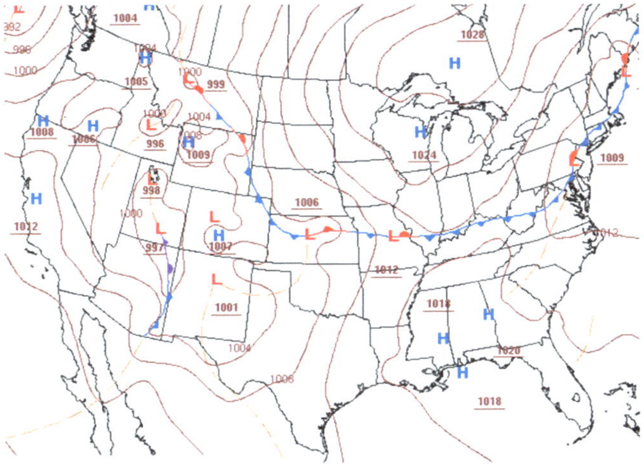

RAIN DISTRIBUTION (Inches)

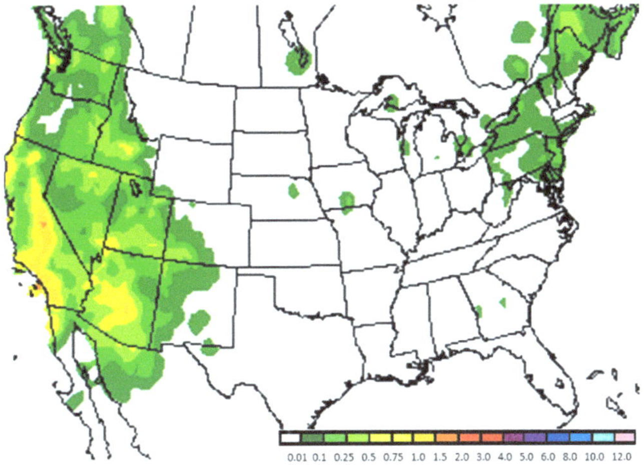

MAXIMUM TEMPERATURE | MINIMUM TEMPERATURE

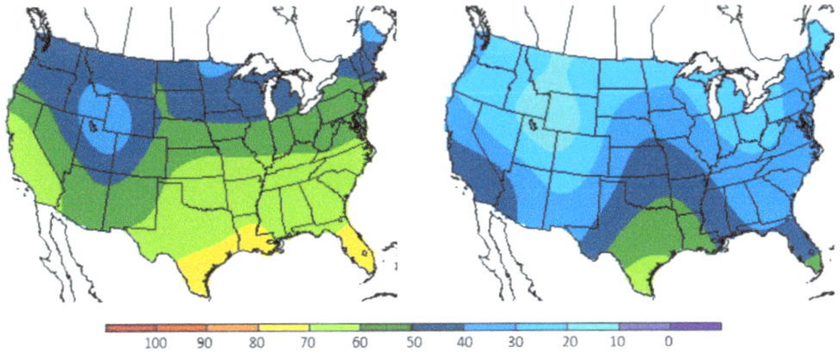

01/23/2025

Widespread precipitation over the west of the country from the Pacific coast, across the Great Basin, down through the Rocky Mountains and into the Southwest.

Similarly, wintry precipitation prevails in the Northeast States.

Mostly dry elsewhere.

The risk of frosts still prevails in the southern agricultural valleys of California to the southwest of Arizona.

Milder in the Deep South to Texas, with these regions possibly avoiding the current freeze being experienced elsewhere.

01/24/2025

A cold air mass over the west of the country brings widespread precipitation from the Pacific coast to Texas.

A winter snowstorm crosses the midsection, including throughout the South Great Plains, Oklahoma, Arkansas and possibly Nebraska with a mix of snow in the higher reaches and heavy rain elsewhere.

Chance of further frosts in California's southern agricultural valleys across to the southwest of Arizona.

Mostly dry elsewhere, with temperatures closer to seasonal norms on the North Great Plains, and milder than average in the east.

Mostly dry throughout the coming week in Hawaii.

JANUARY 24 FRIDAY

ISOBARIC

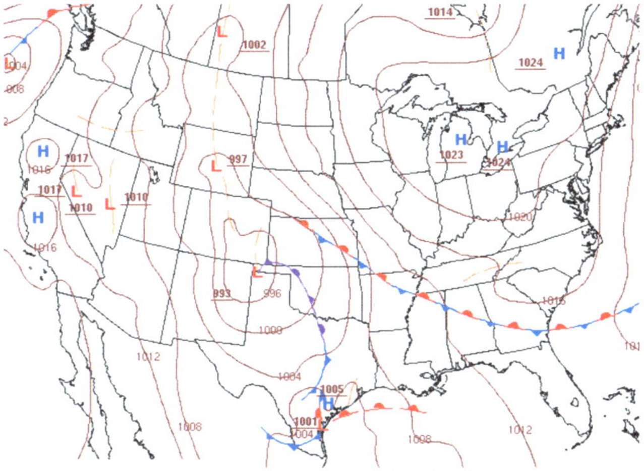

RAIN DISTRIBUTION (Inches)

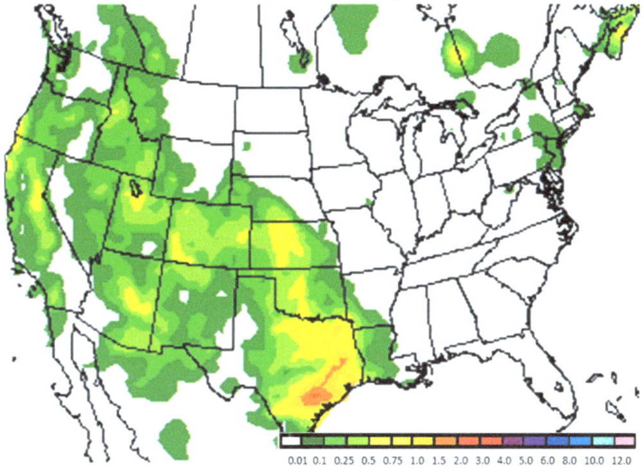

MAXIMUM TEMPERATURE | MINIMUM TEMPERATURE

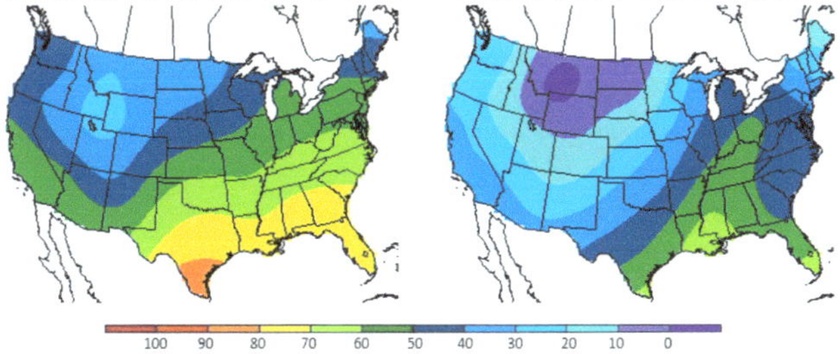

JANUARY 25 SATURDAY

ISOBARIC

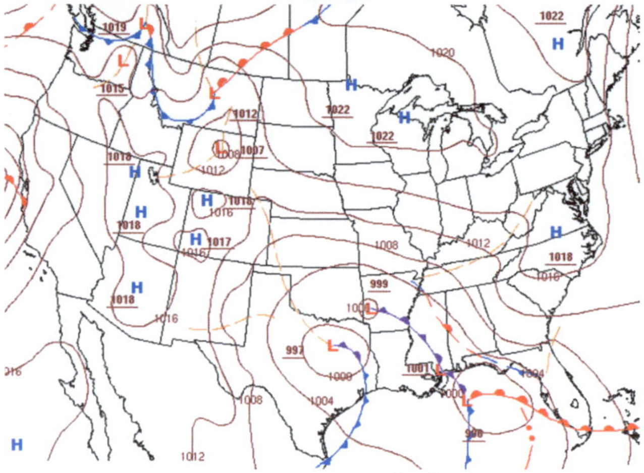

RAIN DISTRIBUTION (Inches)

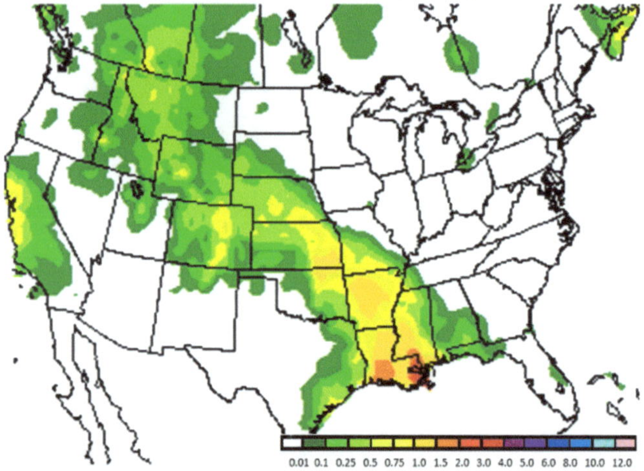

MAXIMUM TEMPERATURE

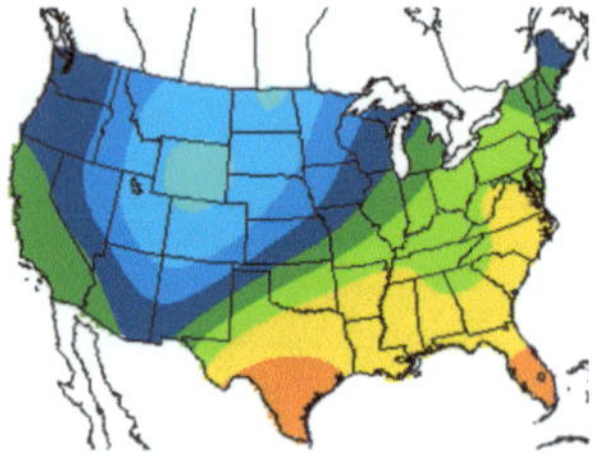

MINIMUM TEMPERATURE

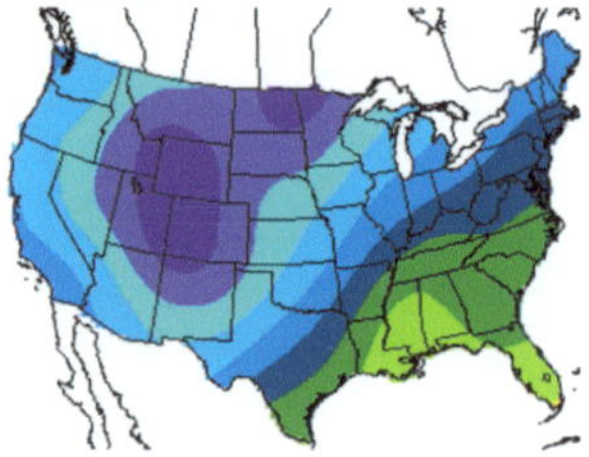

01/25/2025

Rain eases to mostly dry in the Pacific Northwest, but persists in California.

A large band of rain spreads through central US, from Inland Northwest and the Rocky Mountains, down through the Great Plains and into the Deep South.

Rainfall amounts may be particularly heavy in the Mississippi Basin to the Ozarks.

Chance of a snowstorm bringing heavy falls in the Michigan region and some lingering precipitation in New England.

Mostly dry elsewhere, including in the Southwest to the western half of Texas, throughout the Inland Midwest and the northeast quarter of the country to Florida in the south.

01/26/2025

Precipitation spreads from the Pacific Coast to the Rocky Mountains, and in a band through the central and southern Great Plains, into the Deep South to the coastal Southeast and Florida, with showers into the east of Texas.

Heavy rain in the Mississippi River Basin to the Ozarks may bring further risk of flooding, spreading east into Georgia, as well as chance of flooding in California and the Sierra Nevada over the next couple of days.

Mostly dry elsewhere, although some wintry precipitation may persist in New England and isolated outbreaks around the Great Lakes.

Frost temperatures may persist from California and Arizona, and possibly into Texas.

JANUARY 26 SUNDAY

ISOBARIC

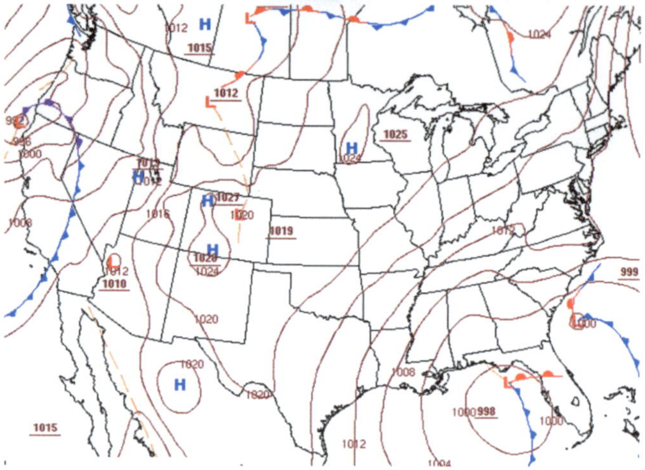

RAIN DISTRIBUTION (Inches)

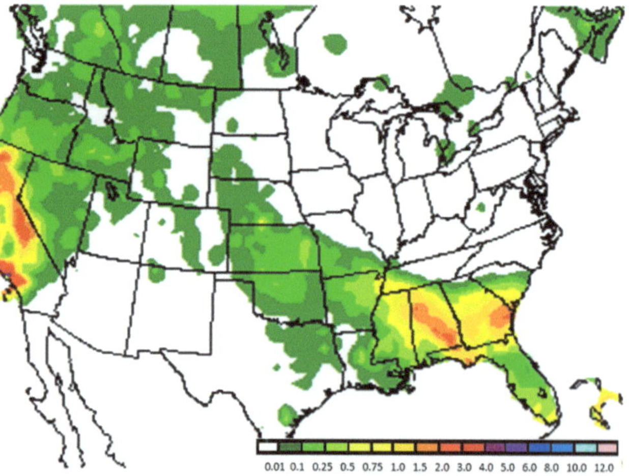

MAXIMUM TEMPERATURE

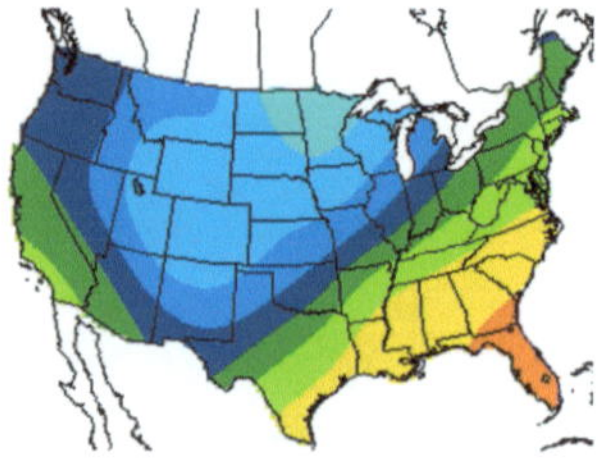

MINIMUM TEMPERATURE

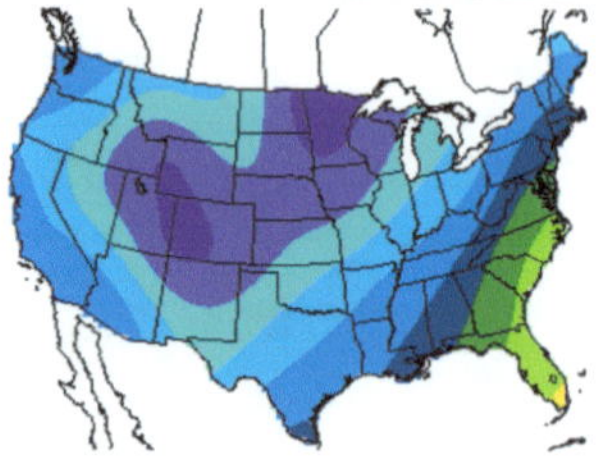

JANUARY **27** **MONDAY**

ISOBARIC

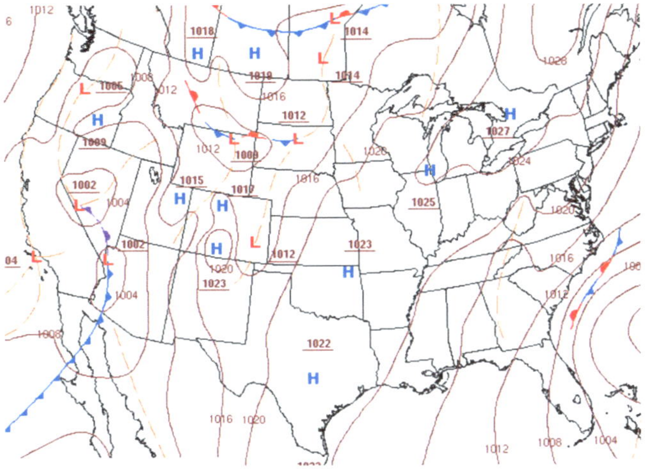

RAIN DISTRIBUTION (Inches)

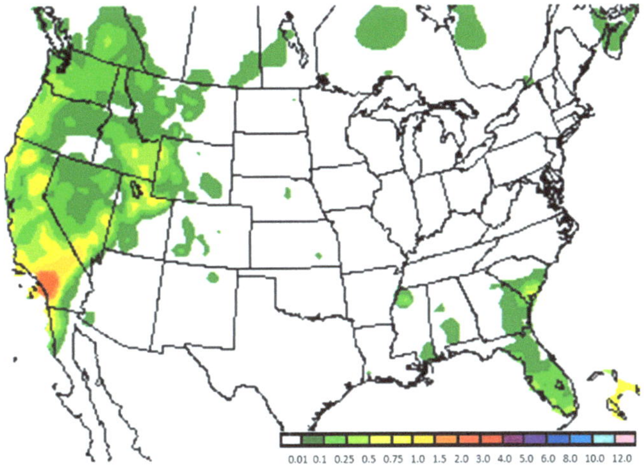

MAXIMUM TEMPERATURE

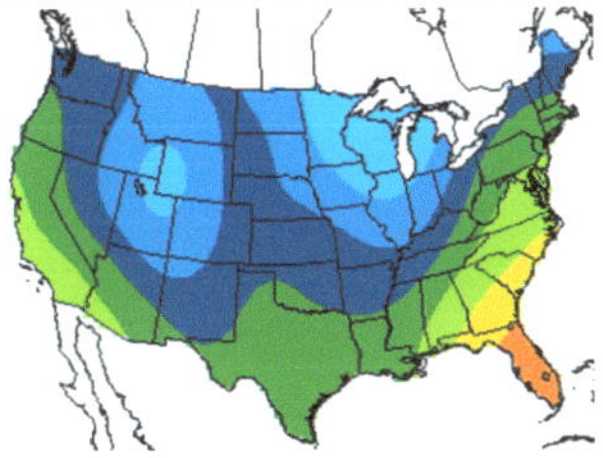

MINIMUM TEMPERATURE

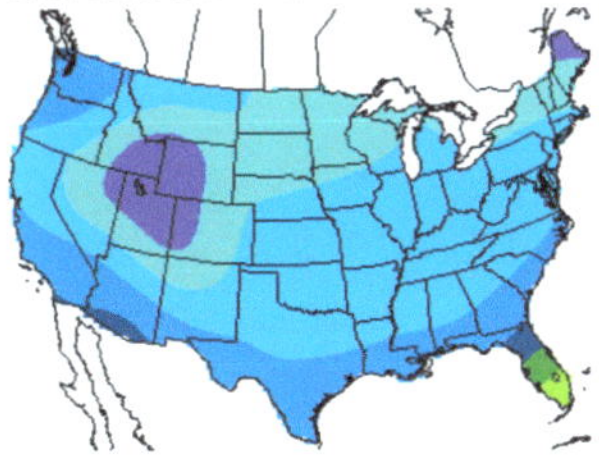

01/27/2025

Precipitation persists in the west from the Pacific coasts to the Rocky Mountains, including in the Inland Northwest and Great Basin, and down throughout California, where the risk of localized flooding may still remain.

Rain also continues in Florida.

All other parts of the country can expect to clear to mostly dry and remain dry in the eastern half of the country over the next few days.

Storms prevail in Alaska over the next couple of days, especially in the west.

01/28/2025

Mostly dry and cold throughout the eastern half of the country from the Great Plains to Texas in the west and eastward to the Atlantic coastal States.

Also, mostly dry in the Inland Northwest to northern California.

Widespread precipitation prevails elsewhere, from the Pacific coasts to the Rocky Mountains, including throughout the Great Basin and the Southwest States.

JANUARY **28** **TUESDAY**

ISOBARIC

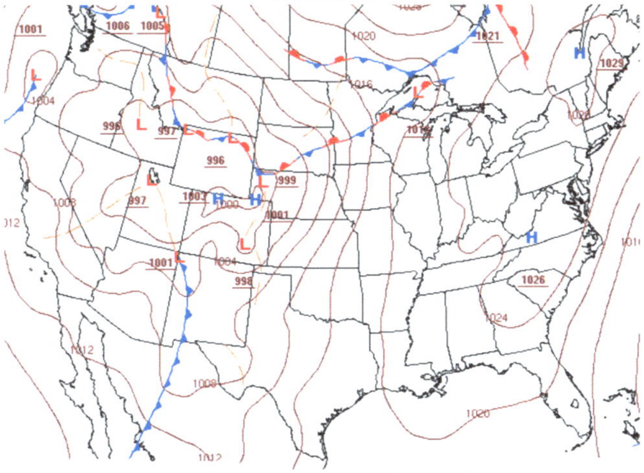

RAIN DISTRIBUTION (Inches)

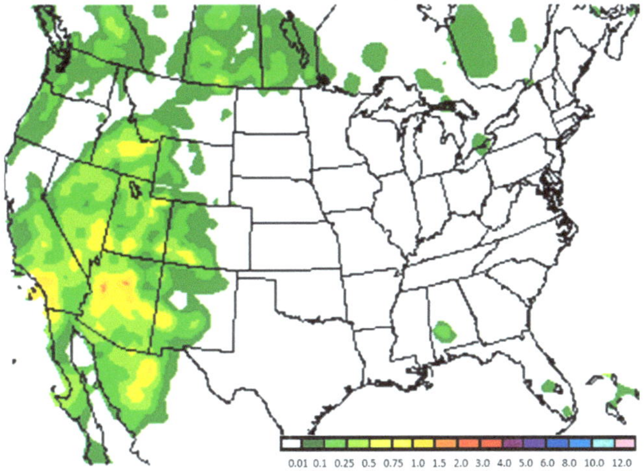

MAXIMUM TEMPERATURE

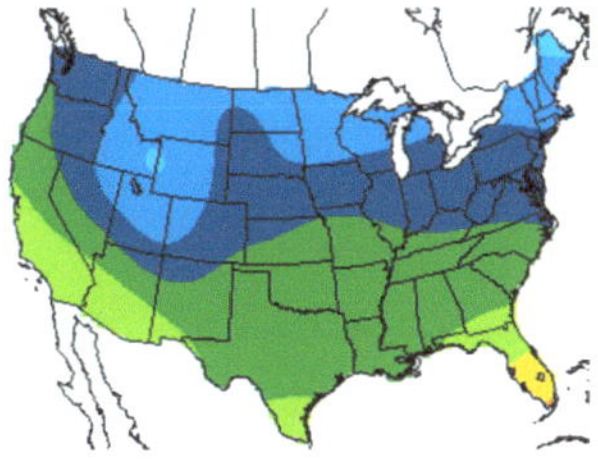

MINIMUM TEMPERATURE

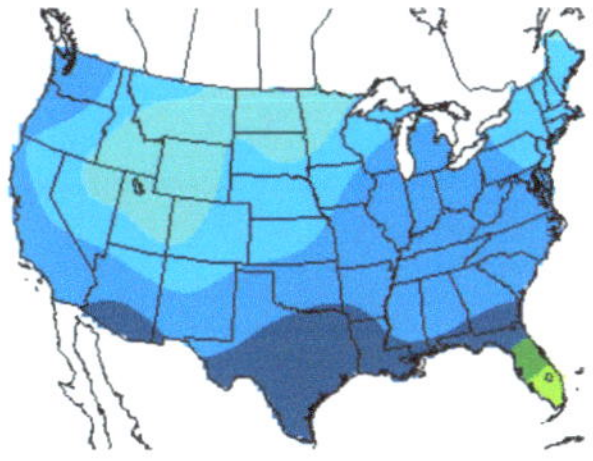

ISOBARIC

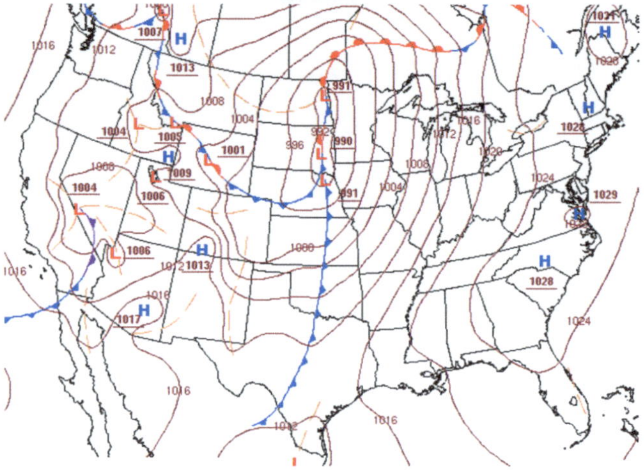

RAIN DISTRIBUTION (Inches)

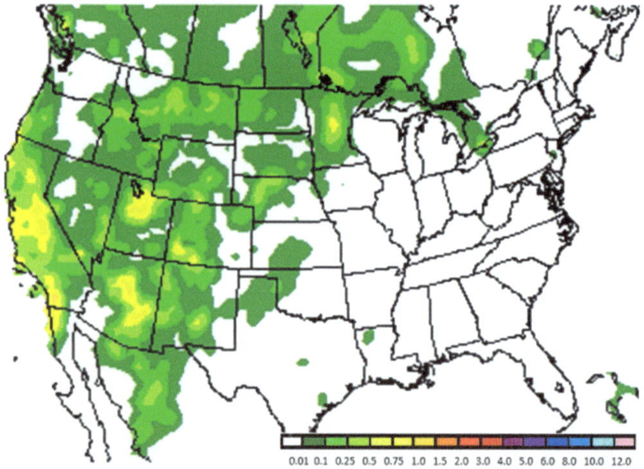

MAXIMUM TEMPERATURE

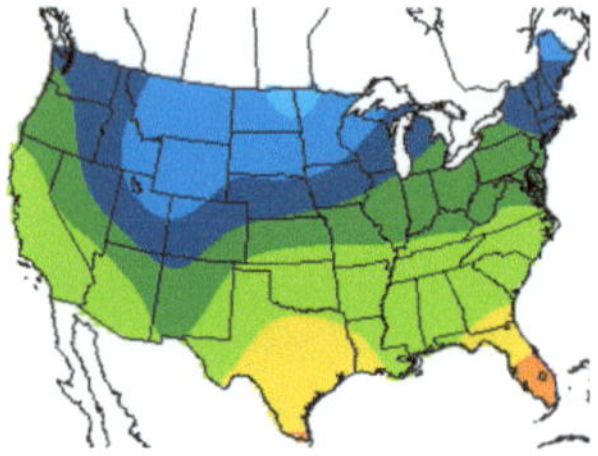

MINIMUM TEMPERATURE

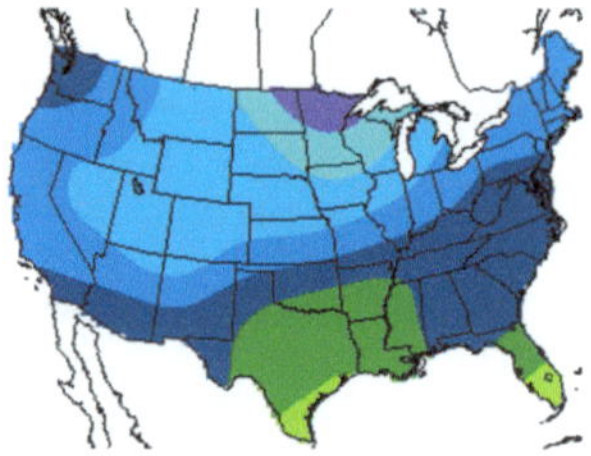

01/29/2025

Widespread wintry precipitation spreads eastward from the Pacific coast into the Great Plains east of the Rocky Mountains, with snowfalls likely in many parts of the Rockies, and North Great Plains to west of the Great Lakes, including Minnesota.

Mostly dry elsewhere, from Texas through the Ozarks and into the Inland Midwest, and east to the Atlantic coast.

Chance of the first real snowfalls occurring in Central Park in New York.

Snow may also occur in New Mexico over the next couple of days.

01/30/2025

A depression with an occluded front brings widespread wintry precipitation over the North Great Plains to the Great Lakes and Upper Midwest.

Another front brings some showers over the Appalachians into the west of the Deep South.

In the west, outbreaks of wintry showers prevail from the Pacific Northwest to the North Great Plains, with chance of snow in the Central Plains, including in Colorado, Kansas, and Nebraska.

Isolated showers in California with more frequent precipitation in the Great Basin to the Rocky Mountains.

Mostly dry elsewhere.

Chance of heavy rain developing in Texas, the Ohio Valley, and Indiana over the next couple of days, possibly causing localized flooding along river courses.

Showers in Hawaii.

ISOBARIC

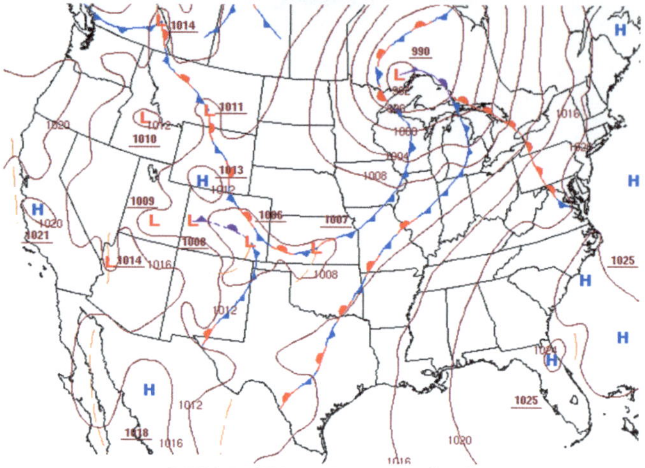

RAIN DISTRIBUTION (Inches)

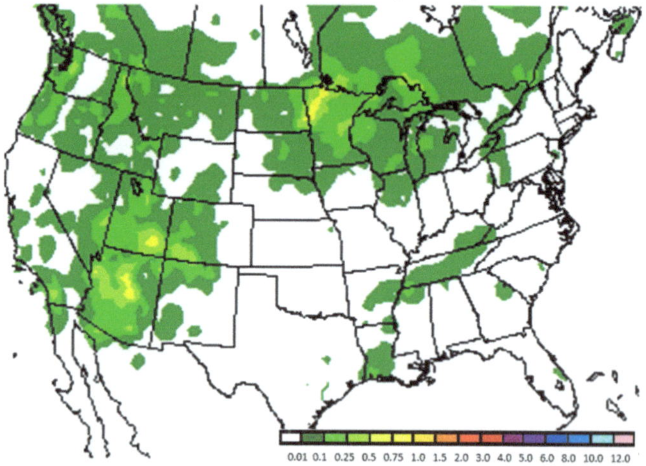

MAXIMUM TEMPERATURE

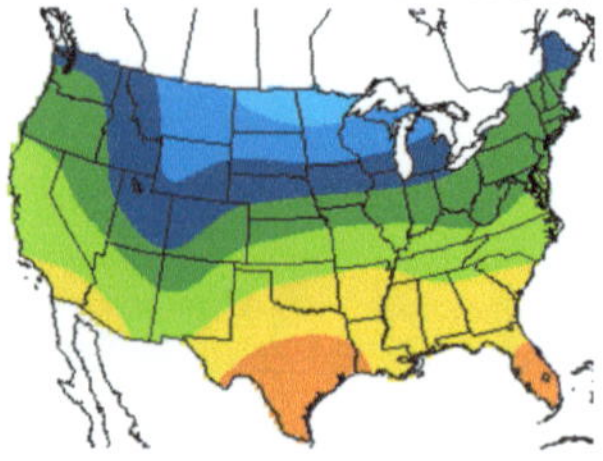

MINIMUM TEMPERATURE

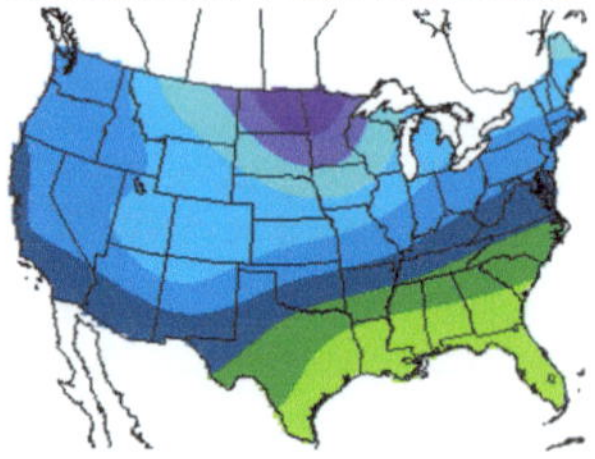

ISOBARIC

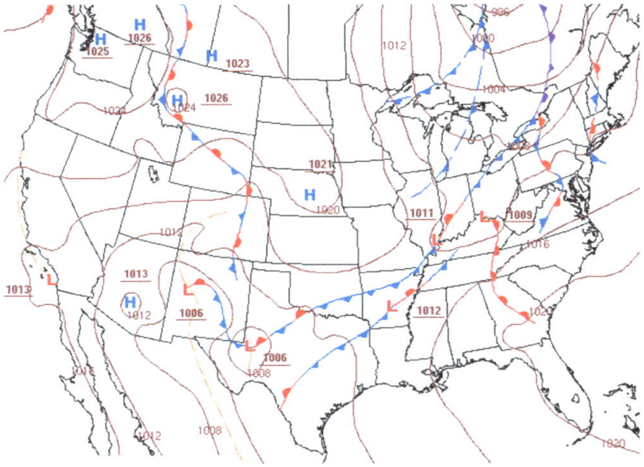

RAIN DISTRIBUTION (Inches)

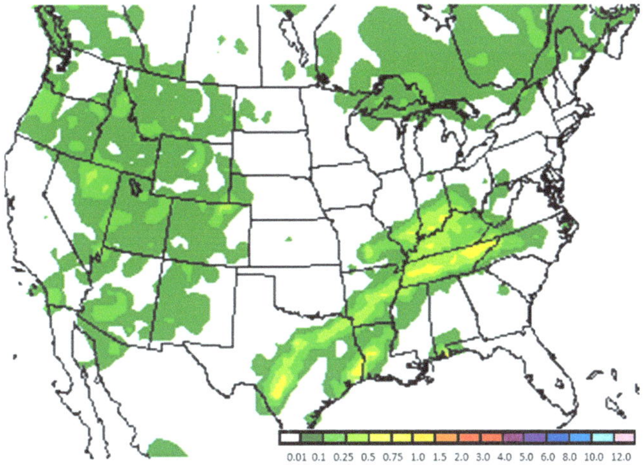

MAXIMUM TEMPERATURE MINIMUM TEMPERATURE

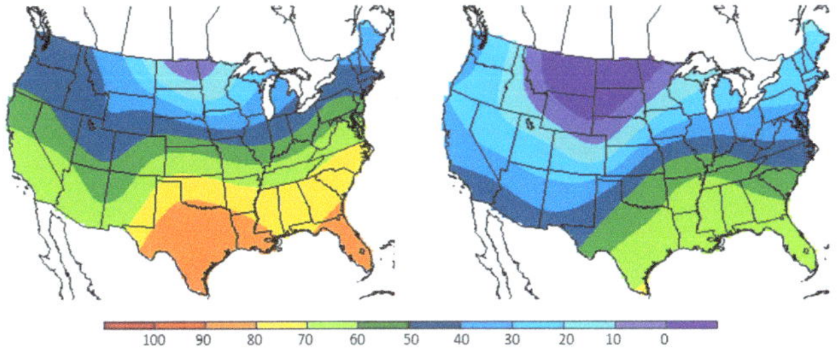

01/31/2025

An occluded front continues wintry precipitation over the northern half of the Great Lakes to the Northeast.

Cold fronts bring heavy precipitation from Texas across the Deep South to the southern half of the Appalachians.

Chance of localized flooding along the White River and Indiana Wabash River Basins expected over the next few days.

In the west, precipitation prevails over the Northwest to the western half of the Great Plains including into the Southwest, although most of California may be mostly dry.

Chance of snow in the central Great Plains and Arizona, spreading to southern parts of New Mexico and Texas over the next couple of days.

Mostly dry elsewhere.

Mostly dry in Hawaii.

Storms prevail in Alaska, with heavy snow in southern parts of the State expected over the next couple of days.

FEBRUARY

PHASES OF THE MOON

2nd	Perigee #10
3rd	Crossing Equator
5th	First Quarter
8th	Northern Declination
12th	Full Moon
16th	Crossing Equator
18th	Apogee
20th	Third Quarter
22nd	Southern Declination
28th	New Moon

1 Sat	2 Sun	3 Mon	4 Tue	5 Wed	6 Thur	7 Fri	8 Sat	9 Sun	10 Mon	11 Tue	12 Wed	13 Thur	14 Fri	15 Sat	16 Sun	17 Mon	18 Tue	19 Wed	20 Thur	21 Fri	22 Sat	23 Sun	24 Mon	25 Tue	26 Wed	27 Thur	28 Fri
	P10	XhN		1Q			^				F				XhS		A		3Q		V						N

FEBRUARY 1 SATURDAY

ISOBARIC

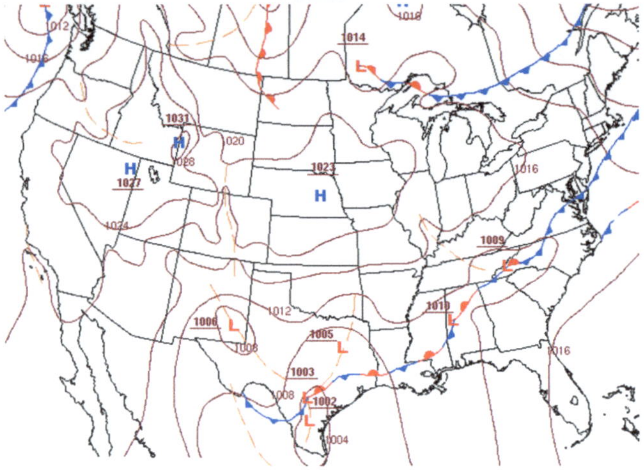

RAIN DISTRIBUTION (Inches)

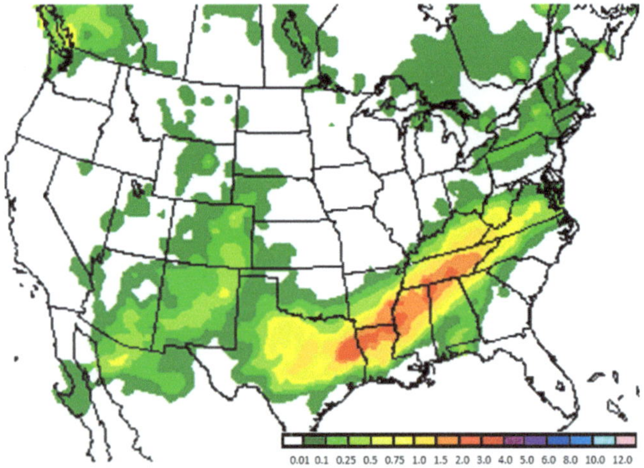

MAXIMUM TEMPERATURE

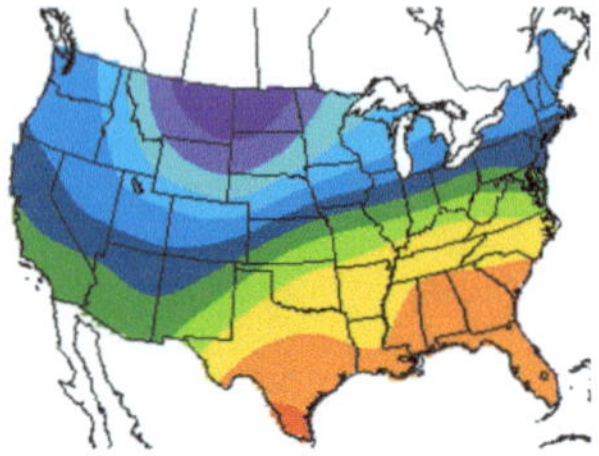

MINIMUM TEMPERATURE

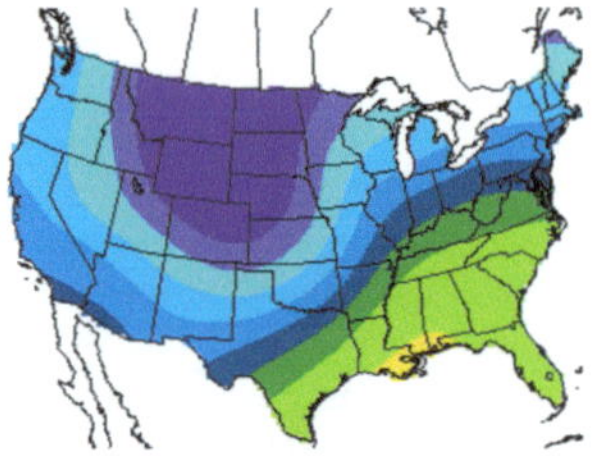

02/01/2025

Mostly dry in the Pacific coastal states to the Great Basin, over the North Great Plains to the eastern half of the South Great Plains and into the Inland Midwest, as well as in Florida to the coastal Carolinas.

Scattered precipitation elsewhere.

Widespread rain over the southern half of the country from the Southwest to the Atlantic coasts.

Heavy outbursts of rain from Texas, across the Deep South and Appalachians to the Coastal Southeast.

Light to moderate snow in the Northeast to the Megalopolis and northern side of the Great Lakes.

Further snow expected over the next few days spreading from the Midwest to the Mid-Atlantic states, including into Iowa and Ohio.

Some isolated showers in Hawaii.

02/02/2025

Scattered precipitation slowly spreads into the Pacific Northwest and northern Rockies over the next few days.

Heavy rain in the south from the northern half of Texas, across the Deep South to the Coastal Southeast Carolinas with flooding risks in local river systems.

Some showers in New Mexico and isolated wintry precipitation around the Great Lakes to the Northeast.

Warm and dry in Florida, milder on the North Great Plains, and mostly dry but chilly elsewhere.

Temperatures may be unusually chilly and cool along the west coast, with chance of frosts in the valleys over the next couple of days, as well as in Arizona.

FEBRUARY 2 SUNDAY

ISOBARIC

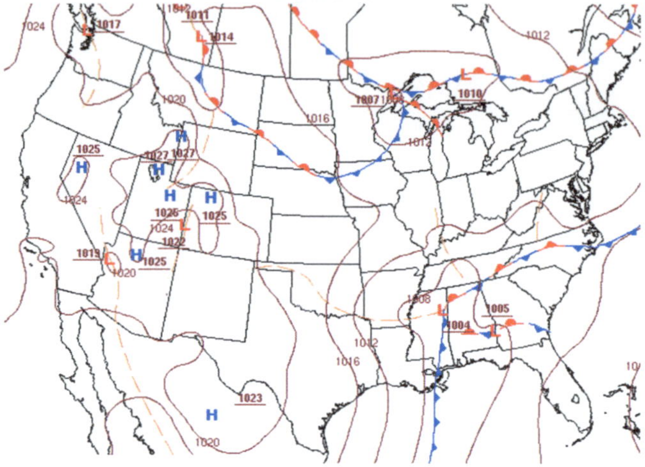

RAIN DISTRIBUTION (Inches)

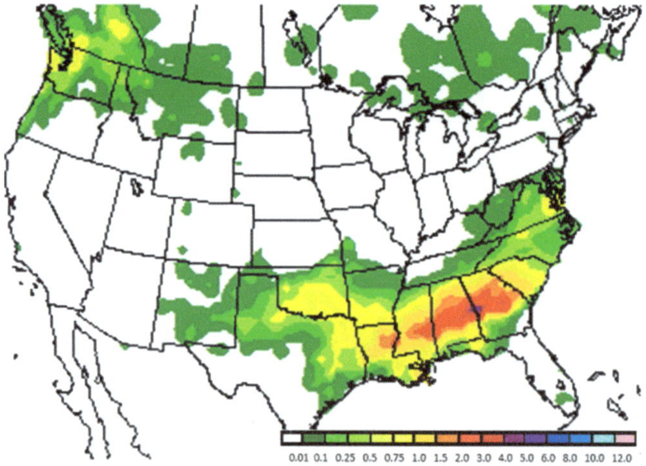

MAXIMUM TEMPERATURE

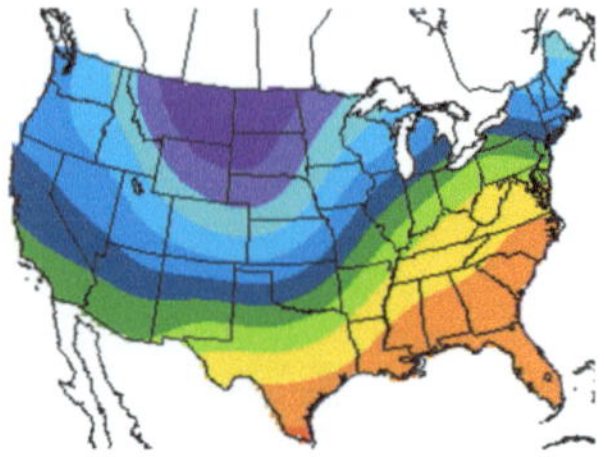

MINIMUM TEMPERATURE

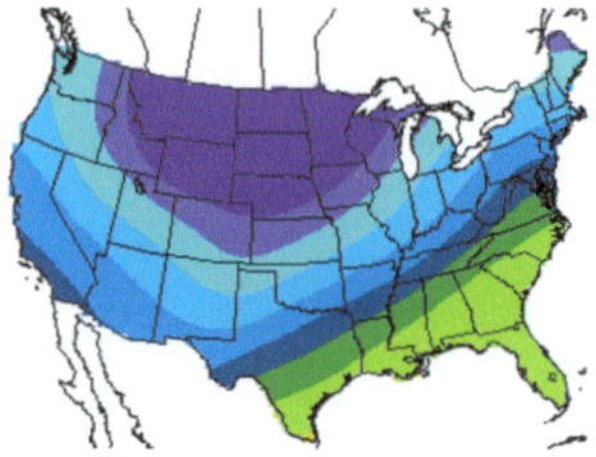

FEBRUARY **3** **MONDAY**

ISOBARIC

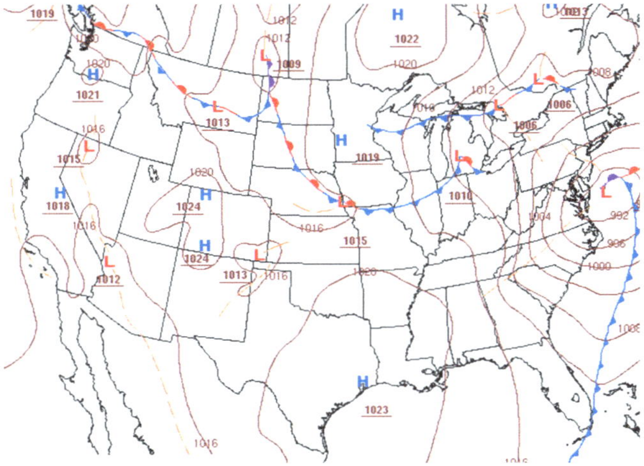

RAIN DISTRIBUTION (Inches)

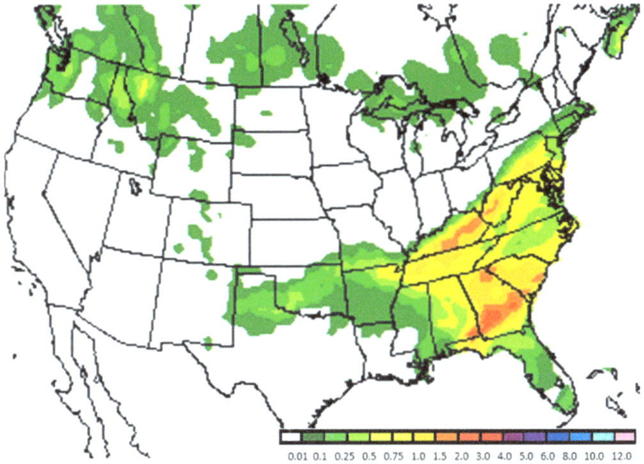

MAXIMUM TEMPERATURE

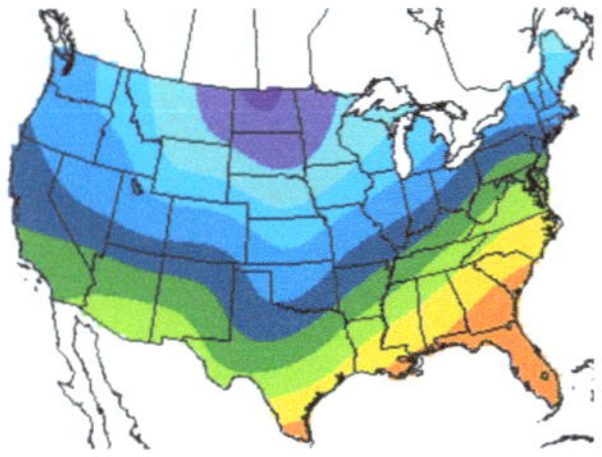

MINIMUM TEMPERATURE

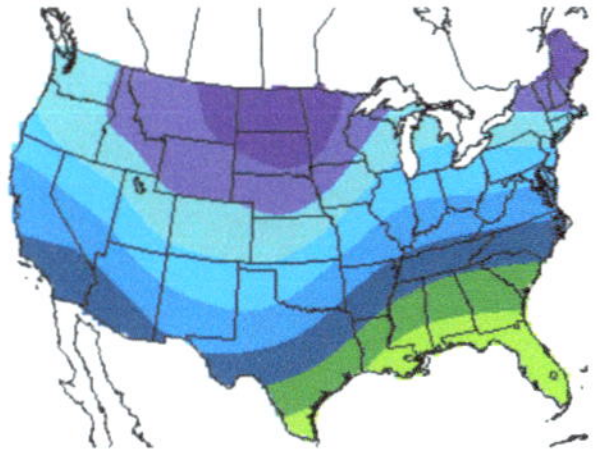

02/03/2025

An Atlantic depression forms off the Northeast coast and intensifies briefly as it slowly drifts northwards over the next couple of days.

Widespread, often heavy rain continues throughout the southeast of the country from the Appalachian Ranges to the Atlantic coast and west across the Ozarks and into the northern region of Texas, clearing in the Southwest while spreading into Florida.

Brief wintry precipitation over the northern Great Lakes and some showers in the Pacific Northwest to Montana and the over the northern Rocky Mountains.

Mostly dry and cool elsewhere.

02/04/2025

The Atlantic depression over the northeast continues to track deeper into the Atlantic Ocean with only some wintry precipitation linger over the Northeast to the Metropolitan.

Rain clears throughout the southern half of the country with just the isolated shower lingering.

Some outbreaks of showers, including snow in the Inland Northwest region and down through the northern Rocky Mountains to the North Great Plains.

Mostly dry elsewhere.

Isolated pockets of frost are still possible in the Californian valleys while milder temperatures prevail in Montana to North Dakota.

Very warm in Hawaii.

A storm crosses Alaska over the next few days bringing some heavy snowfalls.

FEBRUARY **4** **TUESDAY**

ISOBARIC

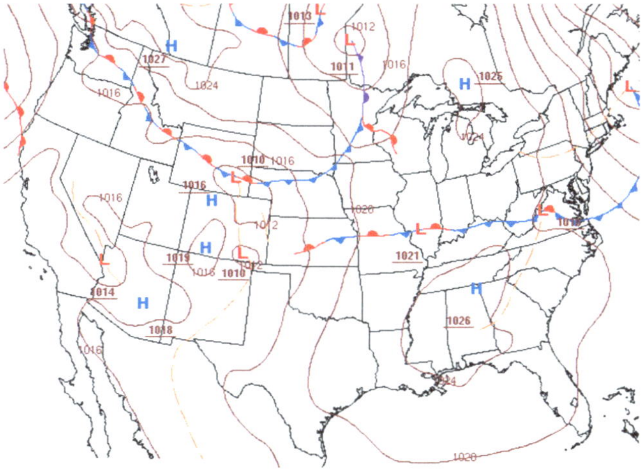

RAIN DISTRIBUTION (Inches)

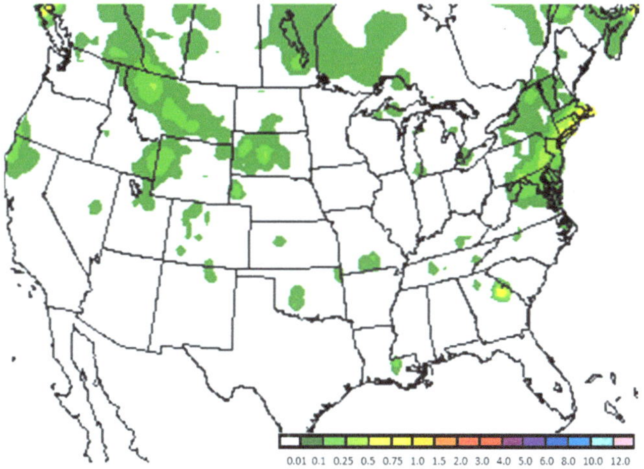

MAXIMUM TEMPERATURE

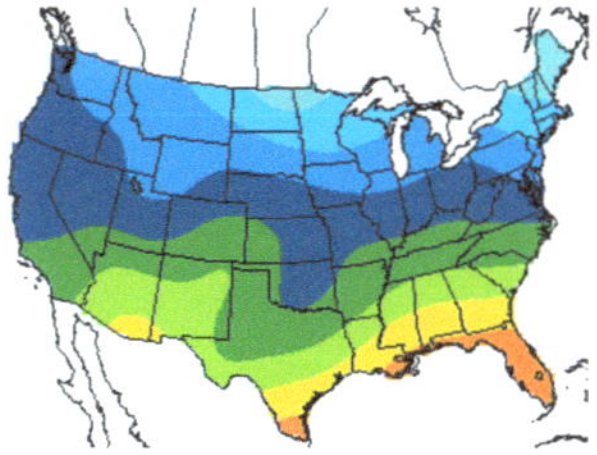

MINIMUM TEMPERATURE

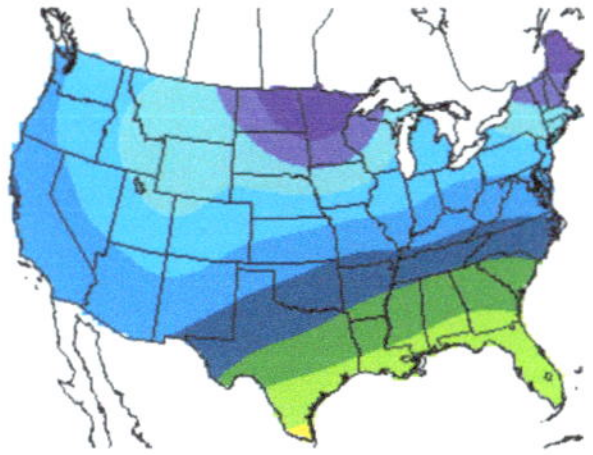

FEBRUARY 5 WEDNESDAY

ISOBARIC

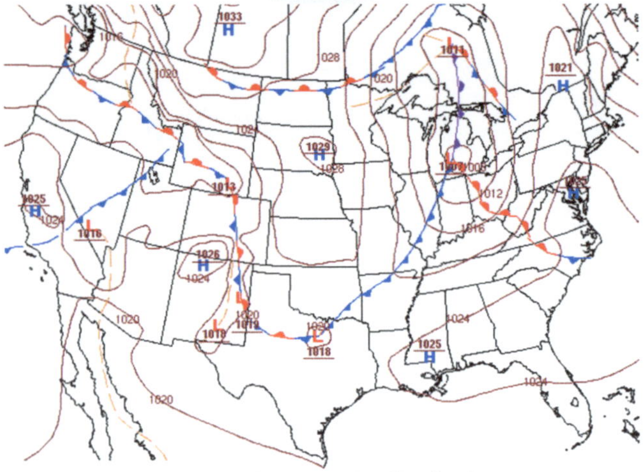

RAIN DISTRIBUTION (Inches)

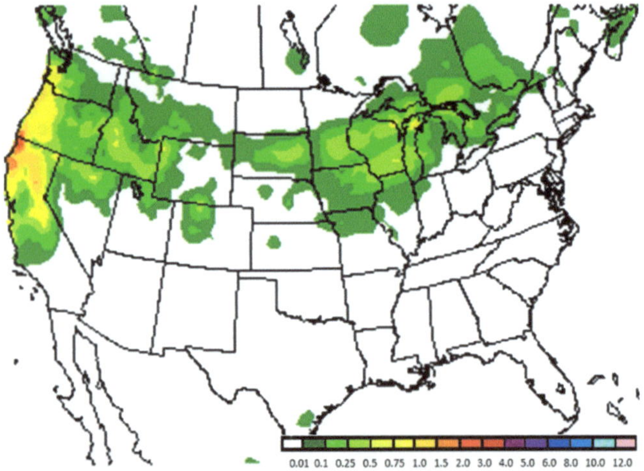

MAXIMUM TEMPERATURE

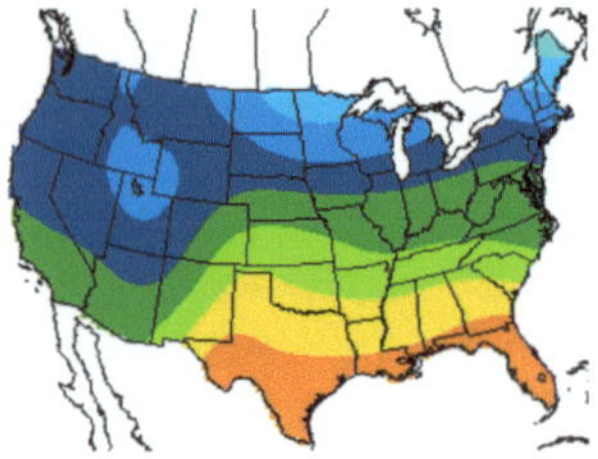

MINIMUM TEMPERATURE

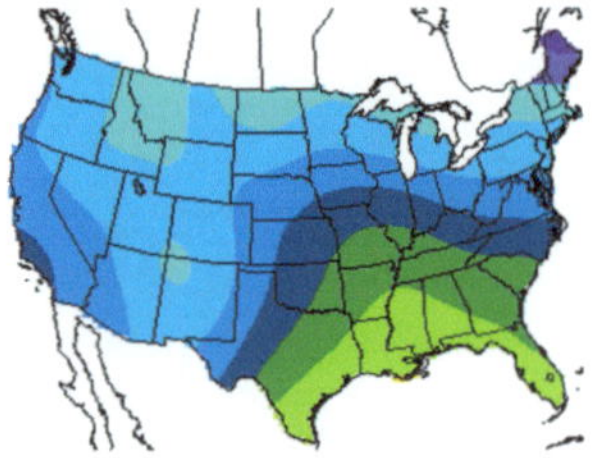

02/05/2025

A depression forms over the Great Lakes and wet fronts embedded with wintry precipitation cross the country from the Pacific Northwest to northern California and east across the Rocky Mountains and North Great Plains and into the Midwest and Great Lakes.

Heavy rain in some parts of northern California, through the Sierra Nevada ranges, and chance of heavy snow in the Inland Midwest, Great Lakes and North Great Plains.

Mostly dry elsewhere.

Isolated pockets of frost are still possible in the Californian valleys, and it may very cold over much of Vermont, Connecticut and New York, where temperatures may not rise above freezing.

02/06/2025

A weak Atlantic depression continues over the northeast of the country from the Great Lakes to the Northeast and across to the Megalopolis.

Widespread precipitation also continues in the west, from northern California and the Pacific Northwest, over the Rocky Mountains and across the central Great Plains.

Snow likely in many parts of the Rocky Mountains and the Great Lakes to the Northeast, including in the Upper Midwest to the South Great Plains.

Outbreaks of rain in southern Texas, with heavy falls likely in the Rio Grande Valley.

Mostly dry elsewhere.

Temperatures may be unusually cool in Hawaii over the next few days.

FEBRUARY 6 THURSDAY

ISOBARIC

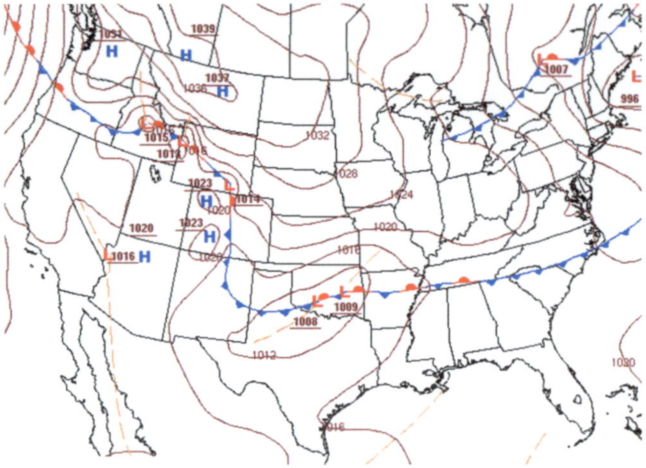

RAIN DISTRIBUTION (Inches)

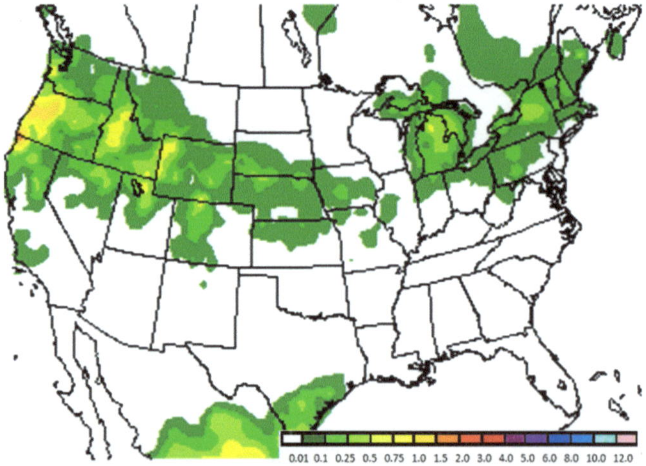

MAXIMUM TEMPERATURE

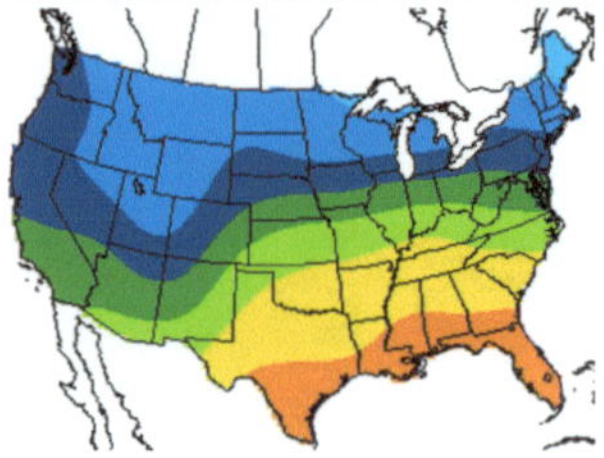

MINIMUM TEMPERATURE

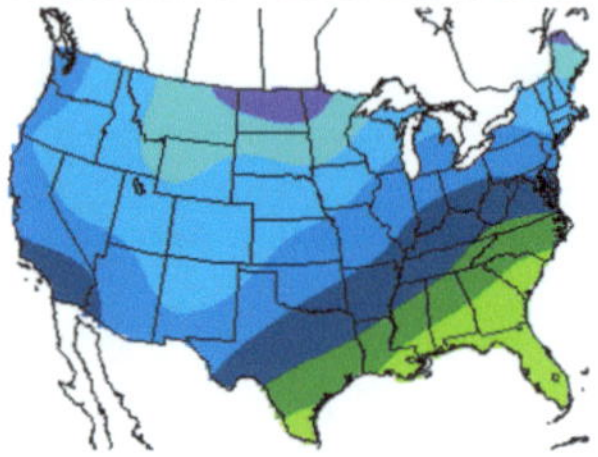

FEBRUARY 7 FRIDAY

ISOBARIC

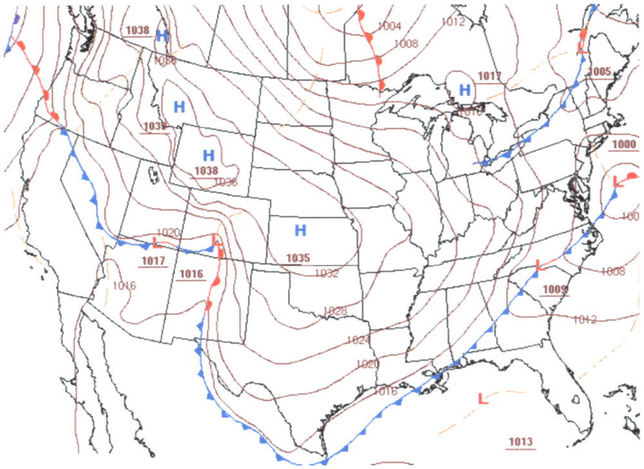

RAIN DISTRIBUTION (Inches)

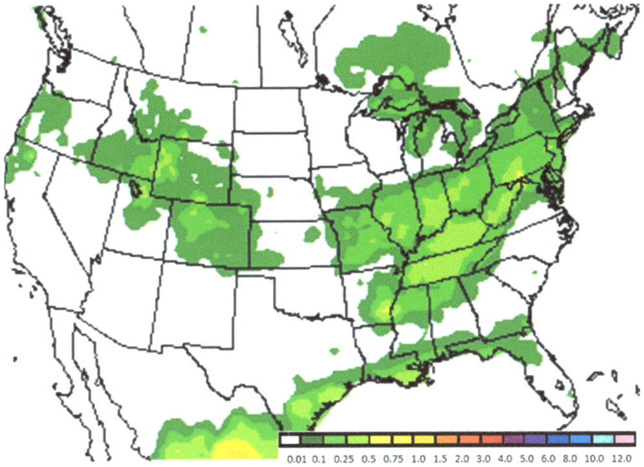

MAXIMUM TEMPERATURE / MINIMUM TEMPERATURE

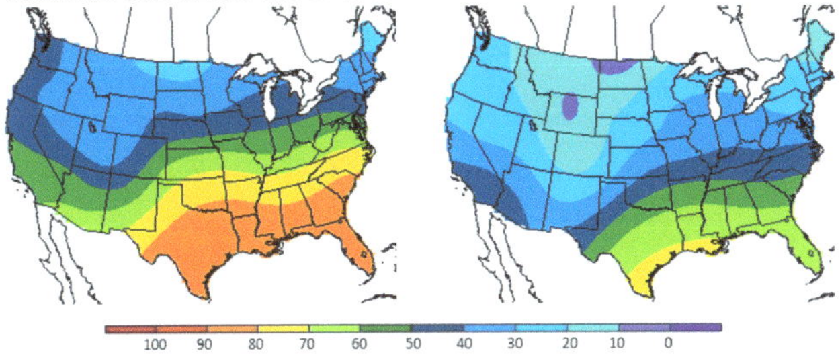

02/07/2025

High pressure through the midsection of the country.

Wintry precipitation follows a wet front from the Pacific Northwest, across the Rocky Mountains as snow, and through the South Great Plains, with some isolated showers possible in northern California.

Further frontal rain also prevails over the Gulf of Mexico regions from Texas to the north of Florida.

Cold fronts also brings widespread precipitation from the Deep South and up through the Appalachians to the Metropolitan and widespread over the Inland Midwest, over the Great lakes to the Northeast.

Snow likely on the central Great Plains to the Inland Midwest and in the higher reaches of the Appalachians to the Northeast and Great Lakes.

Mostly dry elsewhere, including in Washington.

Unusually low temperatures in Hawaii.

02/08/2025

A depression brings scattered wintry precipitation over the Great Lakes to the Northeast.

A very cold and wet in Florida.

Precipitation returns in the northwest from Oregon to California, with heavy falls likely in the Sierra Nevada.

Showers in the Great Basin and north Rocky Mountains and isolated showers in the South Great Plains.

Some rain around the Rio Grande Valley.

Mostly dry elsewhere and very cold especially in the North Great Plains to the Upper Midwest.

Storm activity expected in Hawaii, bringing very strong blustery winds and heavy rain around the State.

Southern parts of Alaska can also expect very heavy wintry precipitation over the next couple of days.

FEBRUARY 8 SATURDAY

ISOBARIC

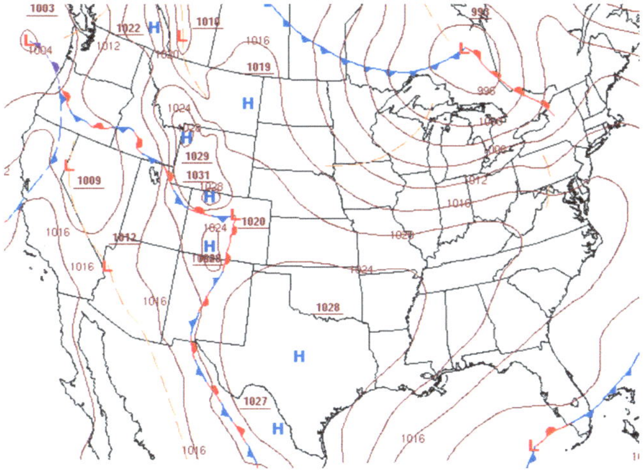

RAIN DISTRIBUTION (Inches)

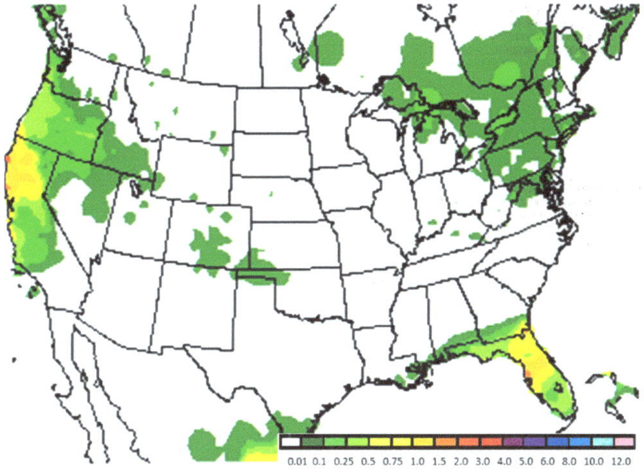

MAXIMUM TEMPERATURE / MINIMUM TEMPERATURE

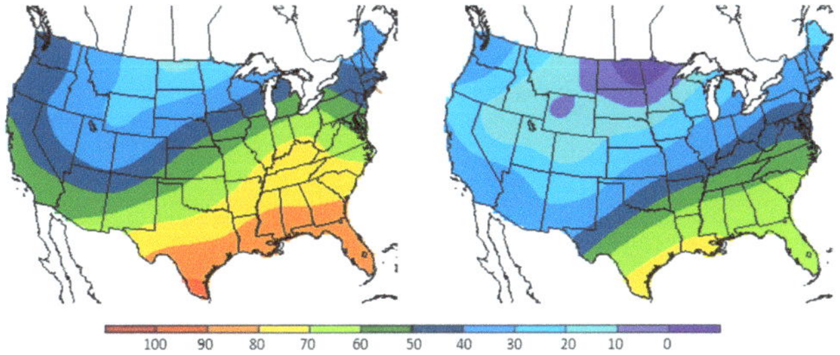

FEBRUARY 9 SUNDAY

ISOBARIC

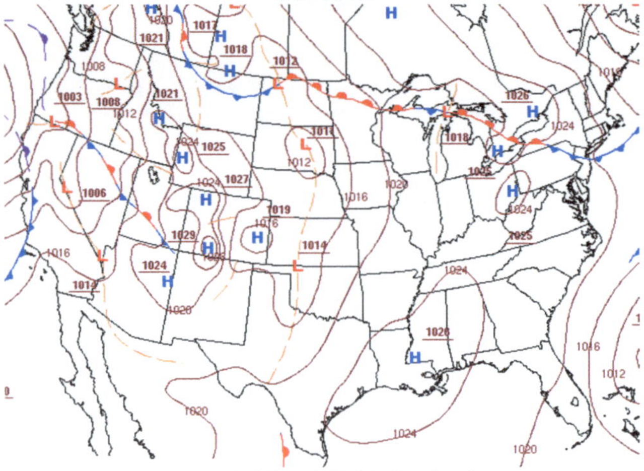

RAIN DISTRIBUTION (Inches)

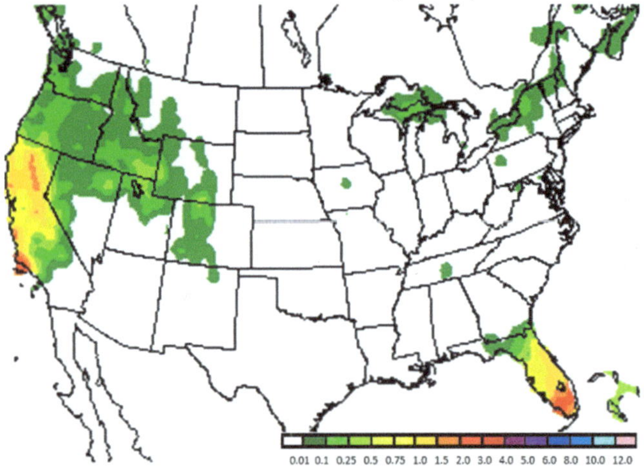

MAXIMUM TEMPERATURE

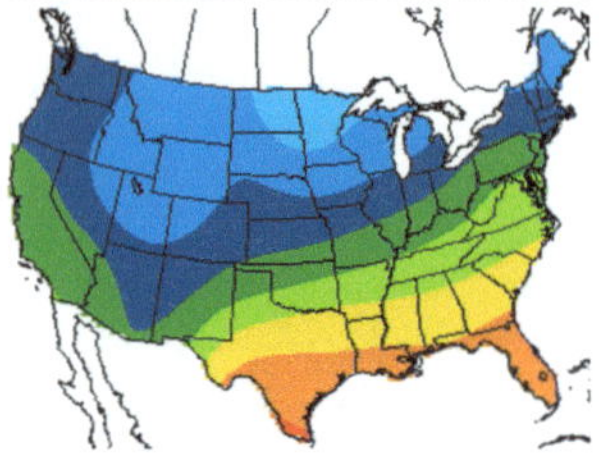

MINIMUM TEMPERATURE

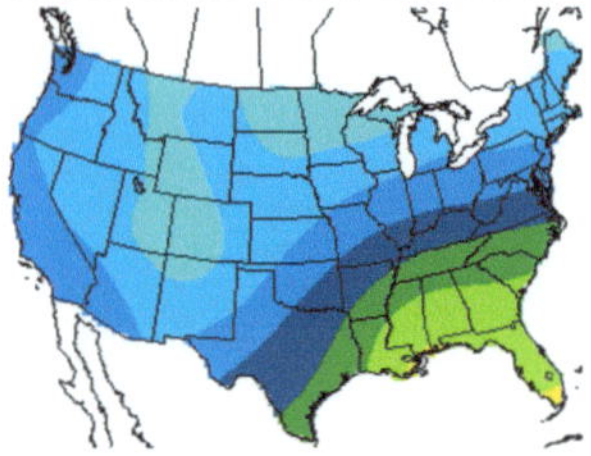

02/09/2025

A complex of high-pressure centers, troughs and fronts over the western half of the country brings widespread precipitation into the Pacific Northwest to the northern Rocky Mountains.

An occluded front brings heavy rain into California, particularly in the Sierra Nevada ranges.

In the east, an Atlantic depression off the coast of Florida brings widespread and rain and heavy downpours into the State.

Mostly dry elsewhere, except for some pockets of lingering sleety showers that may still prevail over the Great Lakes to the Northeast.

A cold snap builds over the eastern coastal regions to the far Northeast over the next few days, which may also bring some unseasonably cold temperatures in Florida.

02/10/2025

An occluded front over the northwest of the country maintains widespread precipitation in the Pacific Northwest to California and into the Great Basin and central Rocky Mountains, with heavy falls likely in northern California and the Sierra Nevada ranges.

Some isolated showers around the Great Lakes and Florida.

Mostly dry elsewhere.

Chance of tornadoes in the next couple of days in the Mississippi Basin including in northern Louisiana, and southern Arkansas.

Strong winds and heavy rain in Hawaii.

A storm system crossing Alaska brings very strong wind gusts that may cause localized damage.

FEBRUARY 10 MONDAY

ISOBARIC

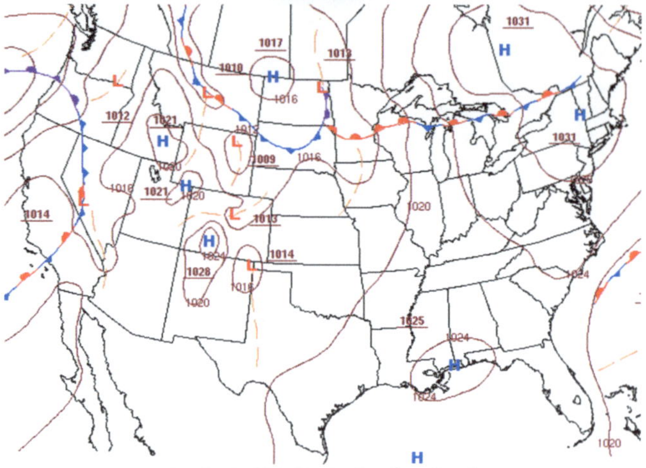

RAIN DISTRIBUTION (Inches)

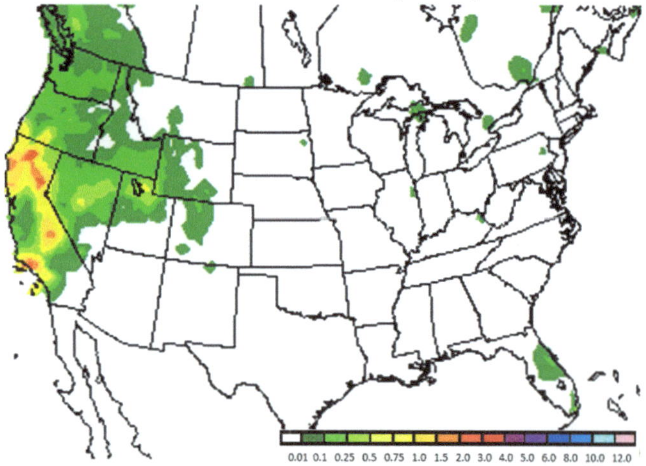

MAXIMUM TEMPERATURE

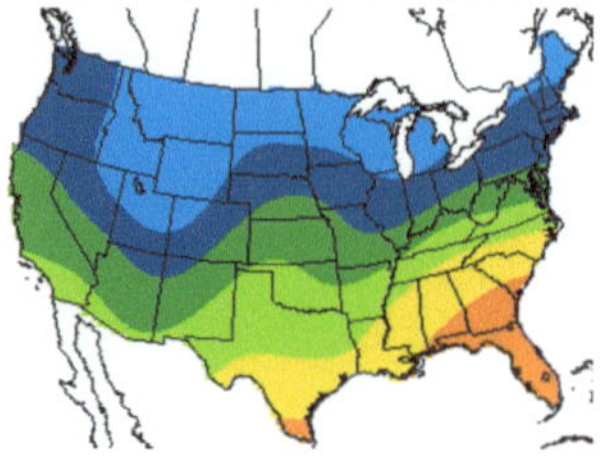

MINIMUM TEMPERATURE

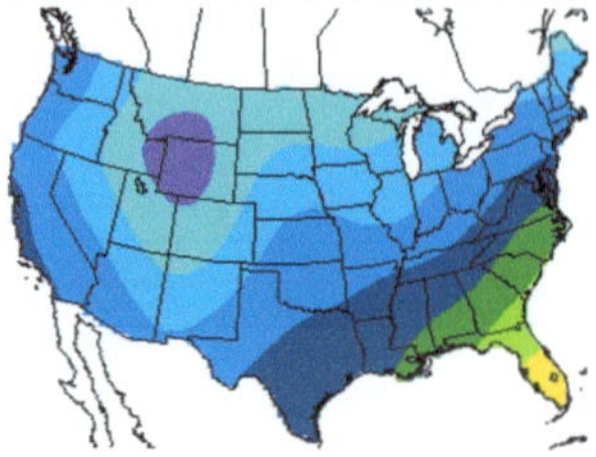

FEBRUARY 11 TUESDAY

ISOBARIC

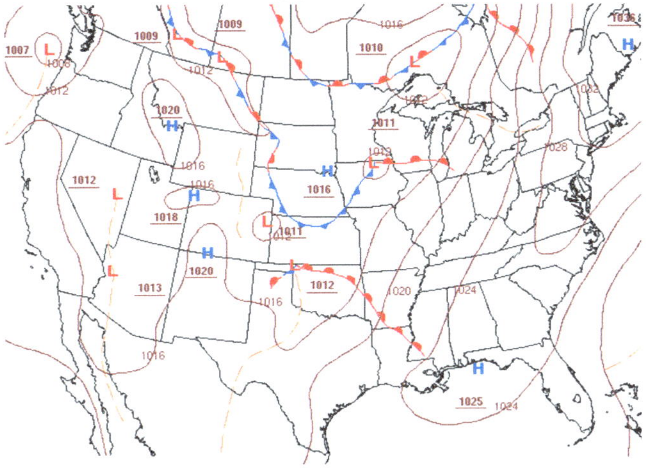

RAIN DISTRIBUTION (Inches)

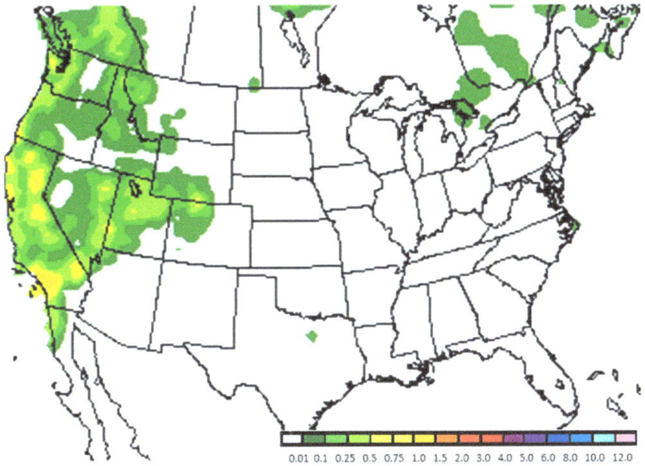

MAXIMUM TEMPERATURE MINIMUM TEMPERATURE

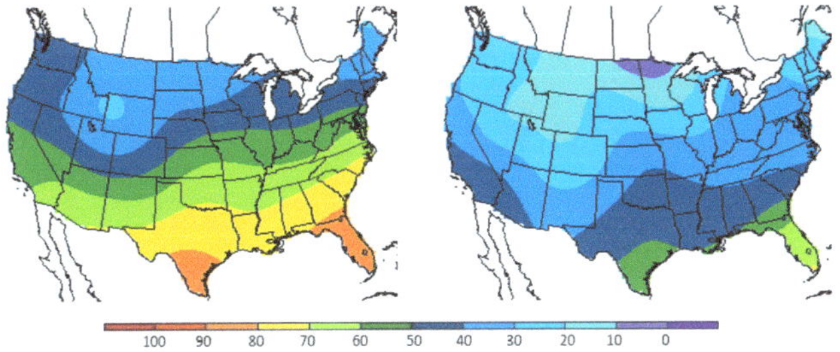

02/11/2025

Widespread rain continues over the west of the country from the Pacific Northwest down through California and across the Great Basin and into the Rocky Mountains.

Mostly dry elsewhere.

There is a chance of heavy squally snowfall south of the Great Lakes developing the next few days.

A tornado risk may still prevail in the Mississippi Basin including in northern Louisiana, and southern Arkansas.

Heavy rain likely in Hawaii.

02/12/2025

The west of the country can still expect further precipitation in the Pacific Northwest to the Great Basin and sporadically through California, the Rocky Mountains and the South Great Plains.

Some wintry precipitation also returns in the Upper Midwest, mostly adjacent to the Great Lakes.

Mostly dry elsewhere.

Rain threatens in the Mexican Gulf states and a warmer change in Florida may see an increased risk of tornadoes in the State.

A very cold air mass may move over the Northern Plains to the Midwest over the next few days.

Heavy rain may still prevail in Hawaii.

FEBRUARY 12 WEDNESDAY

ISOBARIC

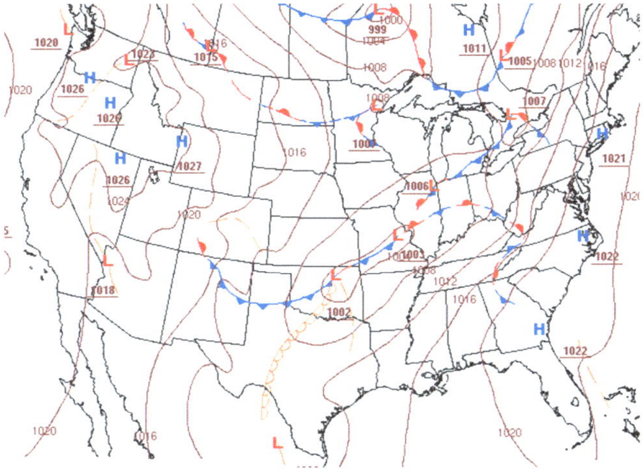

RAIN DISTRIBUTION (Inches)

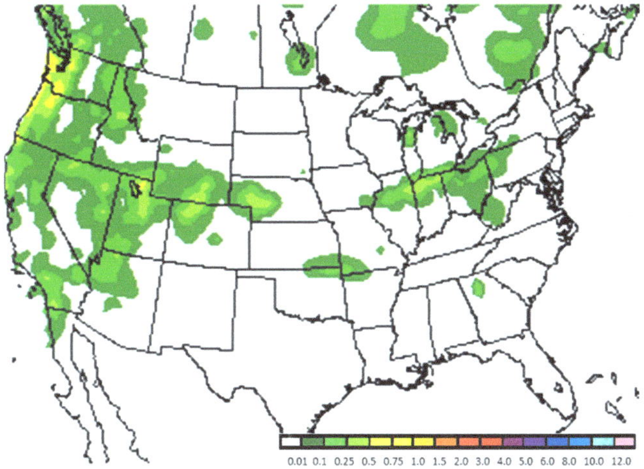

MAXIMUM TEMPERATURE MINIMUM TEMPERATURE

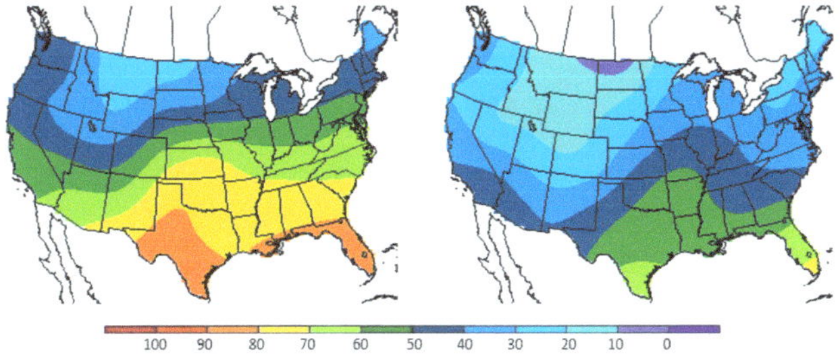

FEBRUARY 13 THURSDAY

ISOBARIC

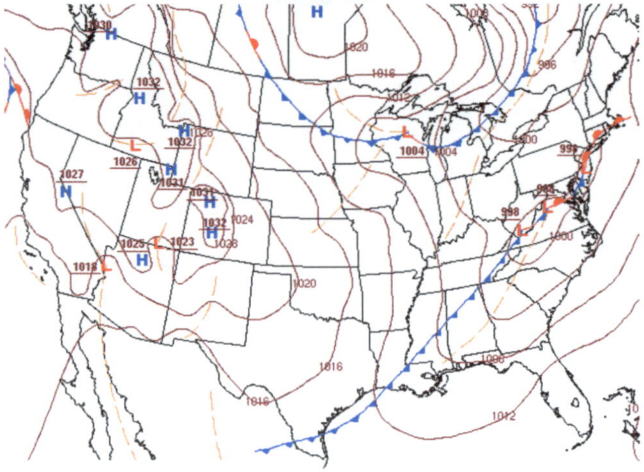

RAIN DISTRIBUTION (Inches)

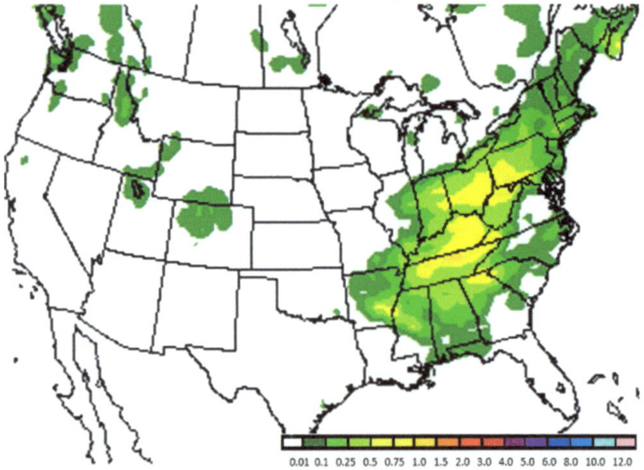

MAXIMUM TEMPERATURE

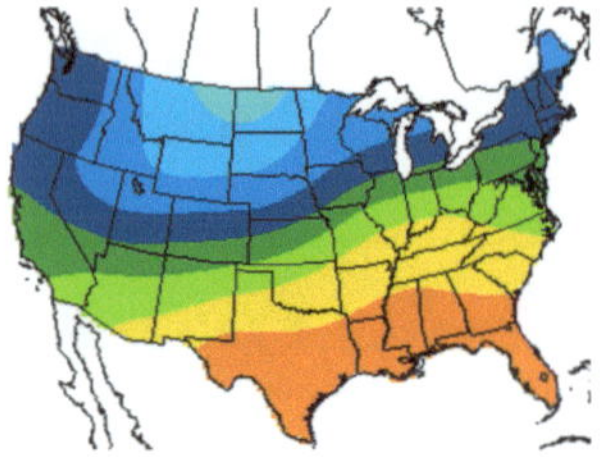

MINIMUM TEMPERATURE

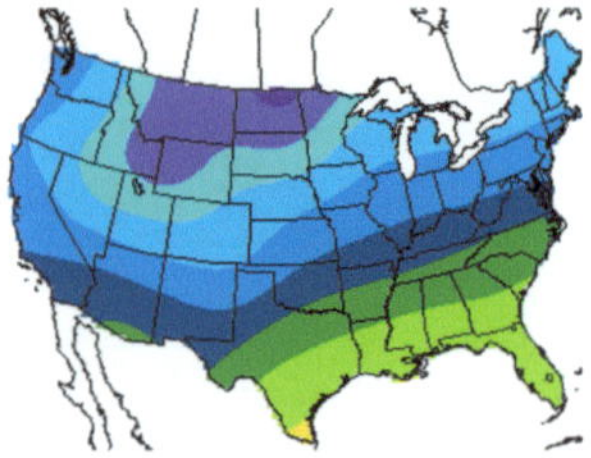

02/13/2025

Most rain activity clears throughout the western half of the country with just isolated showers lingering in the Pacific Northwest to the Rocky Mountains and mostly dry through the midsection over the Great Plains to Texas, and into the western fringes of the Upper Midwest.

Dry in Florida.

Widespread wintry rain, sleet, and snow returns to the northeast of the country including an intense snowstorm with very high wind gusts in the lower Midwest and heavy snow in the northern Great Lakes to Wisconsin, with the freezing rain and snow possibly causing damage to infrastructure.

Heavy snow expected in Oklahoma to Indiana and Michigan over the next couple of days.

Heavy precipitation also through the Deep South, the Appalachians, and showers into the Carolinas.

Heavy rain likely in Hawaii.

02/14/2025

Widespread precipitation prevails from the Upper Inland Midwest to the Megalopolis and southward from the eastern foothills of the Appalachians into Georgia and Florida.

Snow prevails over the east of the Northeast region with heavy snow likely in the far north.

Some isolated wintry showers likely in the southern Rocky Mountains.

Mostly dry elsewhere.

Warmer than normal temperatures expected in both the Great Basin as well as in Alaska.

Hawaii may still be cooler than average.

FEBRUARY 14 FRIDAY

ISOBARIC

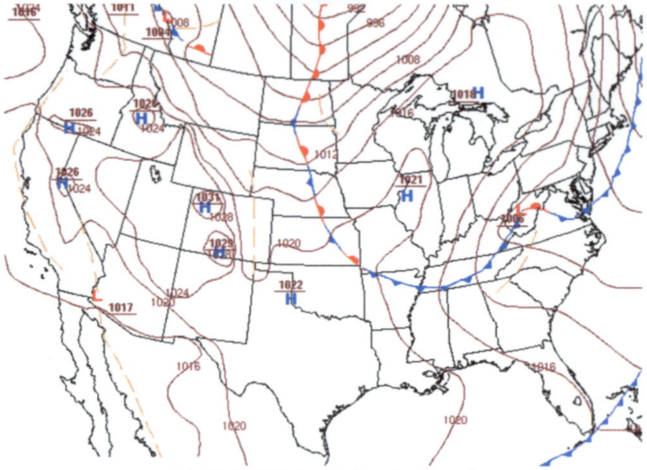

RAIN DISTRIBUTION (Inches)

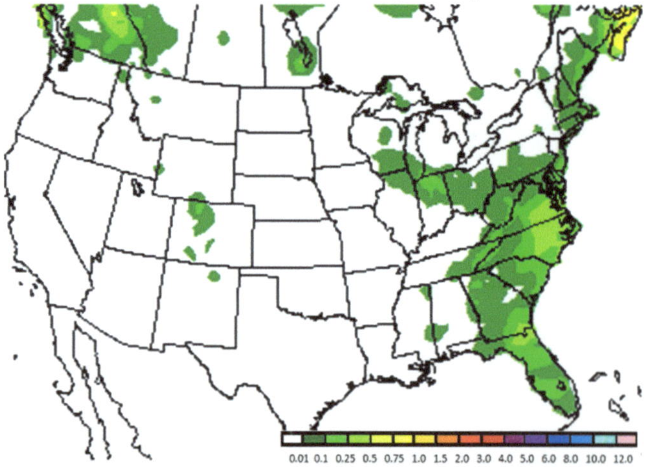

MAXIMUM TEMPERATURE

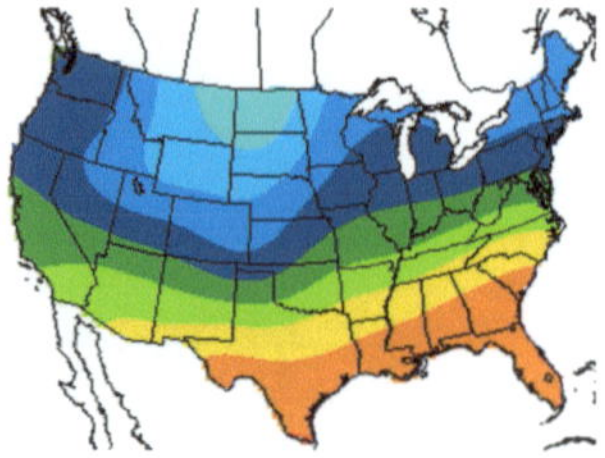

MINIMUM TEMPERATURE

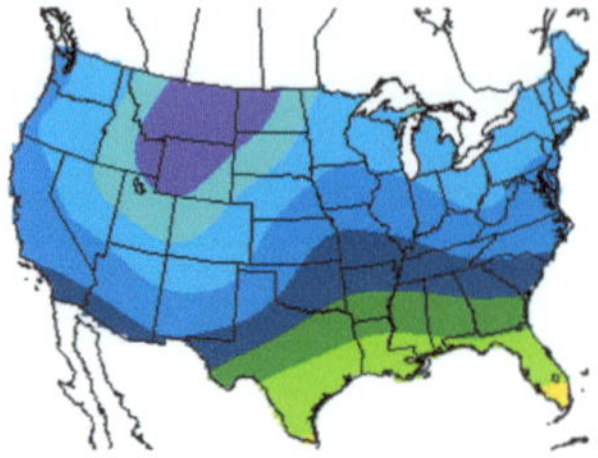

FEBRUARY 15 SATURDAY

ISOBARIC

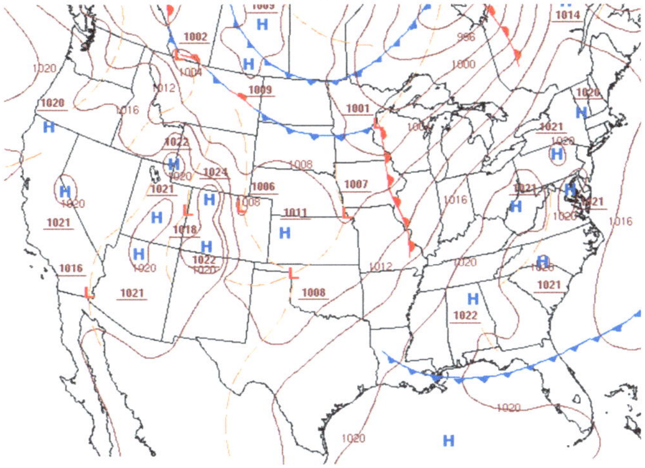

RAIN DISTRIBUTION (Inches)

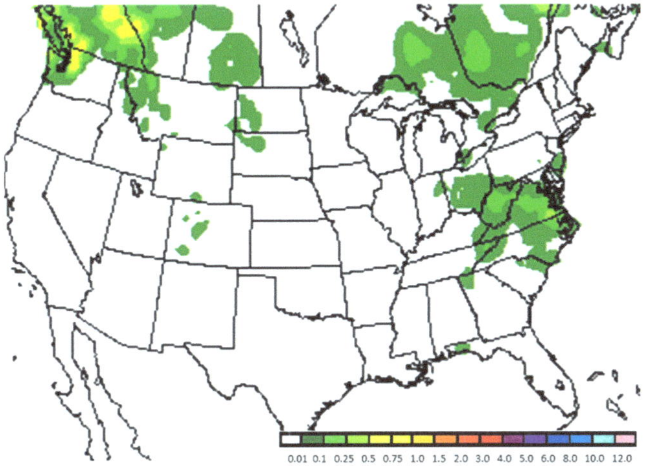

MAXIMUM TEMPERATURE

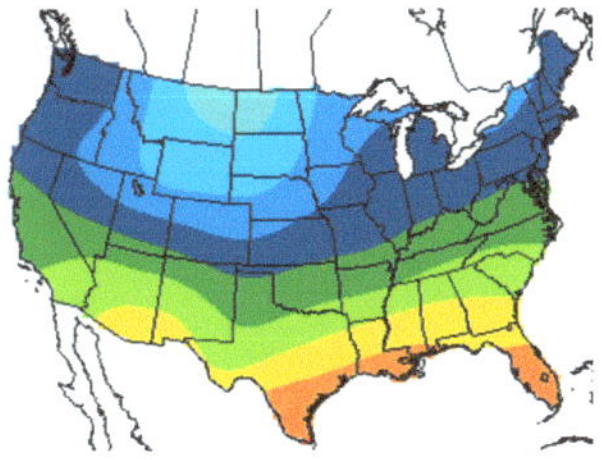

MINIMUM TEMPERATURE

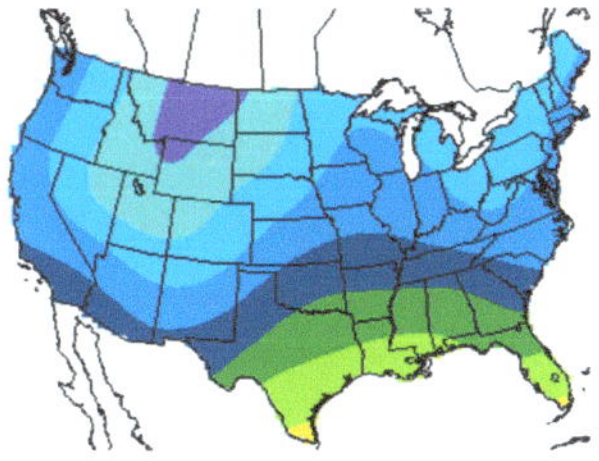

02/15/2025

Isolated wintry precipitation expected to linger around the Great Lakes, and showers in the Coastal Southeast, mostly in the Carolinas.

Some showers in the US/Canadian border areas on the Pacific to Inland Northwest and isolated wintry showers in the North Great Plains and southern Rocky Mountains.

Mostly dry elsewhere.

Very cold temperatures expected in the Ohio Valley and into the central parts of the Appalachians over the next couple of days.

Elsewhere, most of the country may have warmer temperatures spreading from the southern half of the country into the Western Plains over the next couple of days.

02/16/2025

A wet front extends rain activity along the US/Canada border regions from the Pacific Northwest to the Inland Northwest, eastward along the northern parts of the North Great Plains to the north of the Great Lakes and Northeast.

Isolated showers are also likely to linger around the Appalachians to the Coastal Southeast and briefly around the Four Corners Southwest and southern Rockies.

Mostly dry elsewhere.

There is a chance of some heavy snow in the Lake Ontario region.

FEBRUARY 16 SUNDAY

ISOBARIC

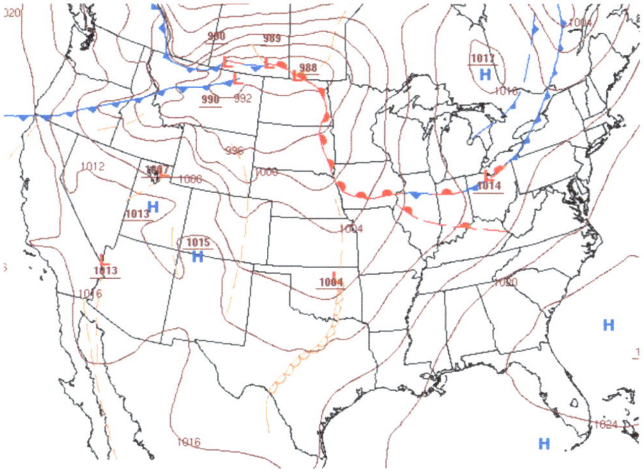

RAIN DISTRIBUTION (Inches)

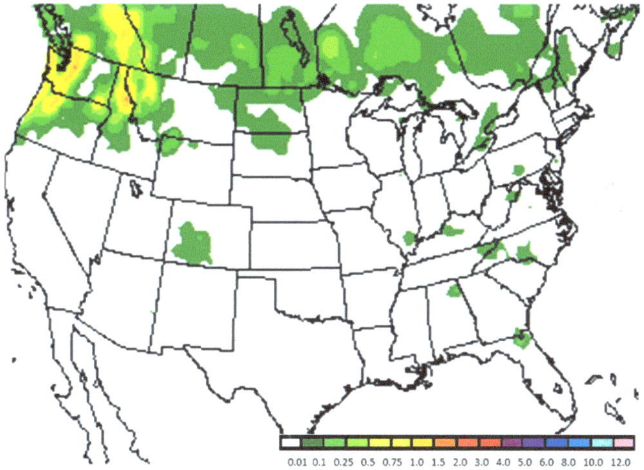

MAXIMUM TEMPERATURE

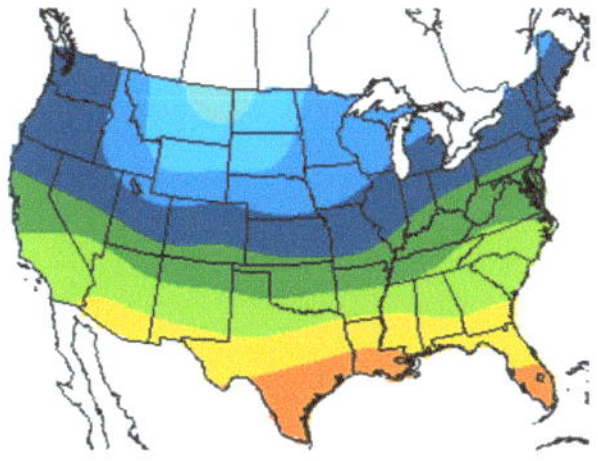

MINIMUM TEMPERATURE

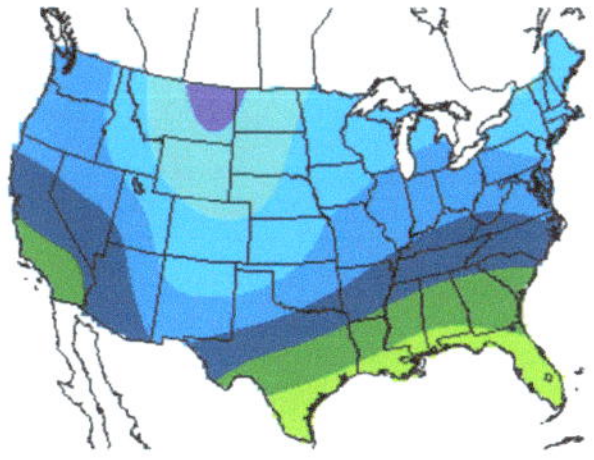

FEBRUARY 17 MONDAY

ISOBARIC

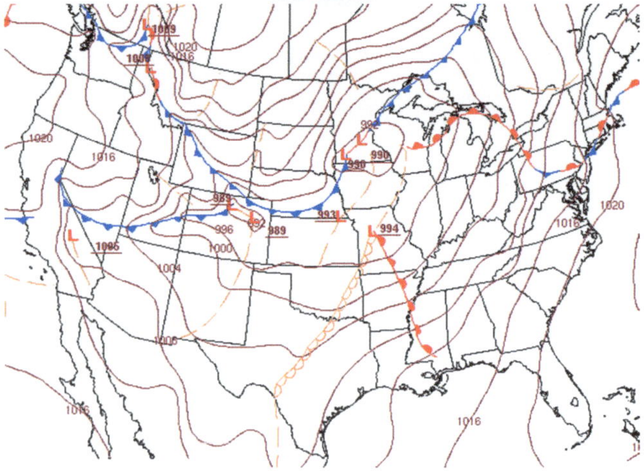

RAIN DISTRIBUTION (Inches)

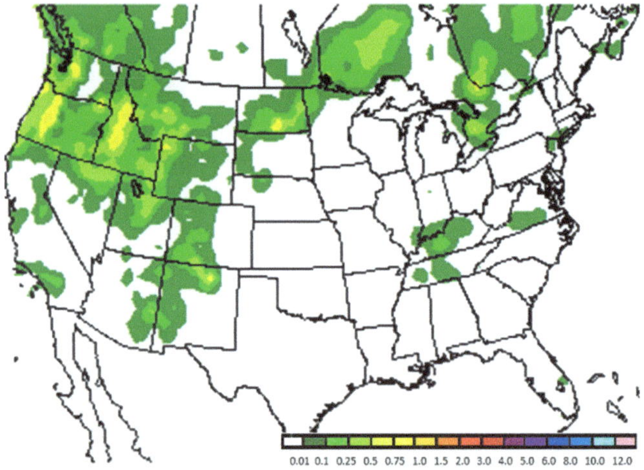

MAXIMUM TEMPERATURE

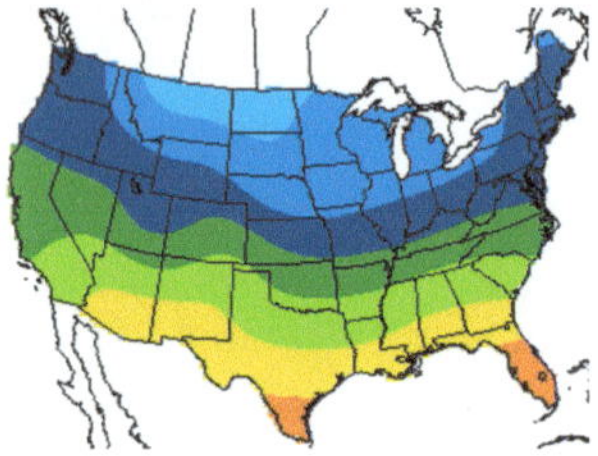

MINIMUM TEMPERATURE

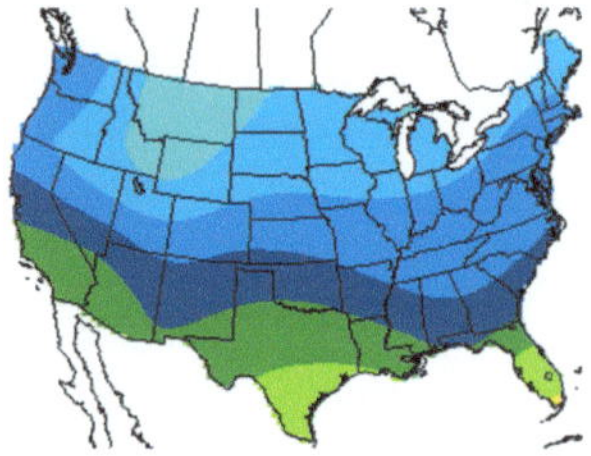

02/17/2025

Wet fronts bring widespread rain activity throughout the northwest of the country from the Pacific Coast to the Rocky Mountains, into northern California, and down along the Rocky Mountain ranges into the Great Basin and the western half of the Southwest.

Wintry precipitation also covers much of the North Great Plains, and around the Great Lakes.

Chance of blizzard conditions from the Rocky Mountains, over the Great Plains and into the Great Lakes in the next few days.

Isolated showers in the Appalachians.

Mostly dry elsewhere.

Strong southerly winds prevail throughout eastern regions of the country, while a strong and blustery northerly flow prevails across the North Great Plains.

02/18/2025

A large and intense stormy depression prevails from the Great Lakes to the Atlantic Coast, and brings widespread storms, rain or snow throughout the eastern half of the country.

Blizzard conditions in the northeast, including heavy snowfalls in New York over the next couple of days.

Widespread rain also expected in the west from the Pacific Northwest, down through the Rocky Mountains to the northern fringes of the Southwest, and with patchy rain into the Great Basin and in the Great Plains.

Heavy snow possible in Montana and Nevada over the next couple of days.

Mostly dry through the southern states of the country.

Winds are likely to be very strong in the Northeast of the country around this time, enhancing the severity of storm activity, particularly in the Midwest, Great Lakes and Northeast regions.

FEBRUARY 18 TUESDAY

ISOBARIC

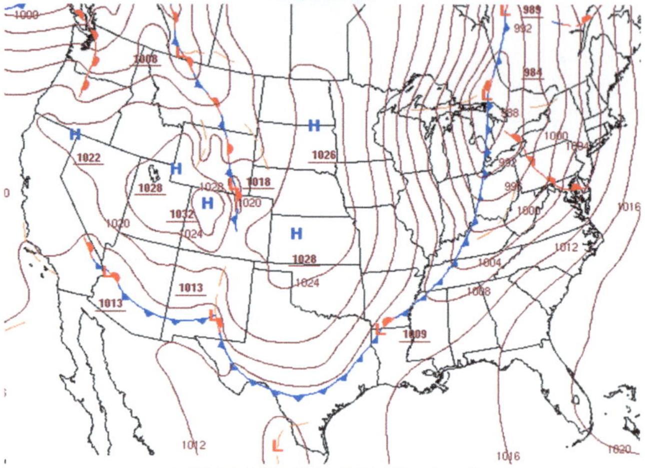

RAIN DISTRIBUTION (Inches)

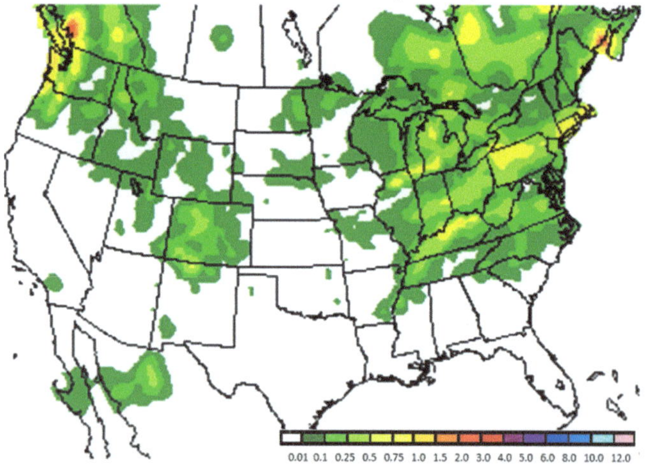

MAXIMUM TEMPERATURE

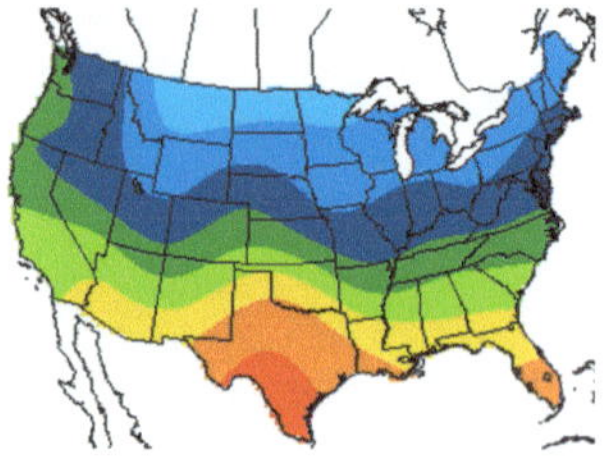

MINIMUM TEMPERATURE

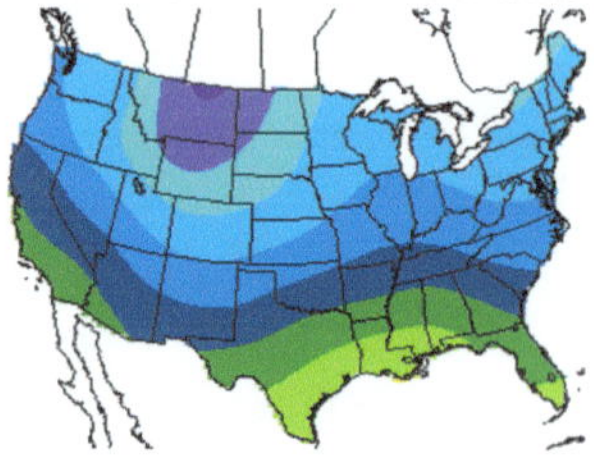

FEBRUARY 19 WEDNESDAY

ISOBARIC

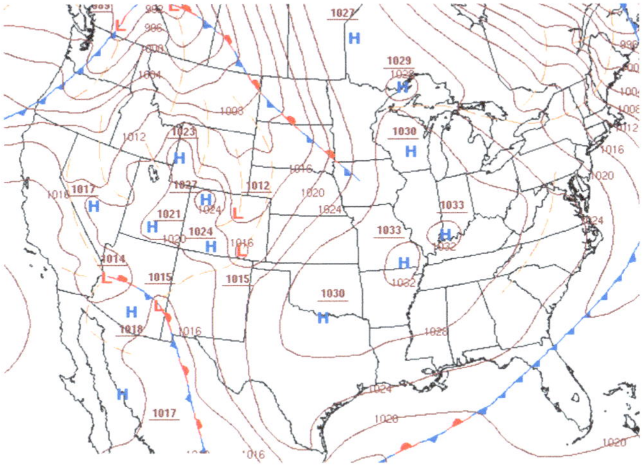

RAIN DISTRIBUTION (Inches)

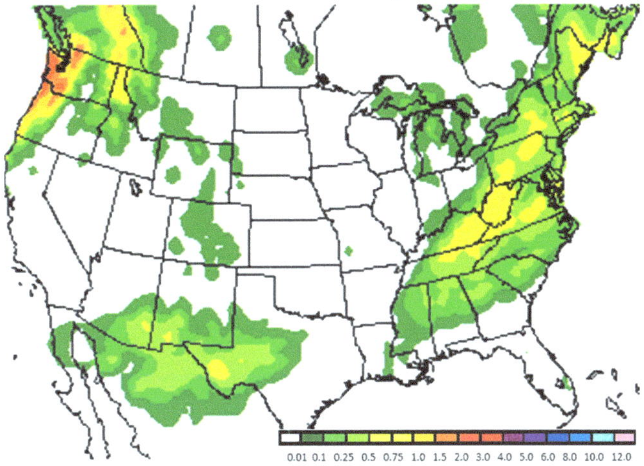

MAXIMUM TEMPERATURE / MINIMUM TEMPERATURE

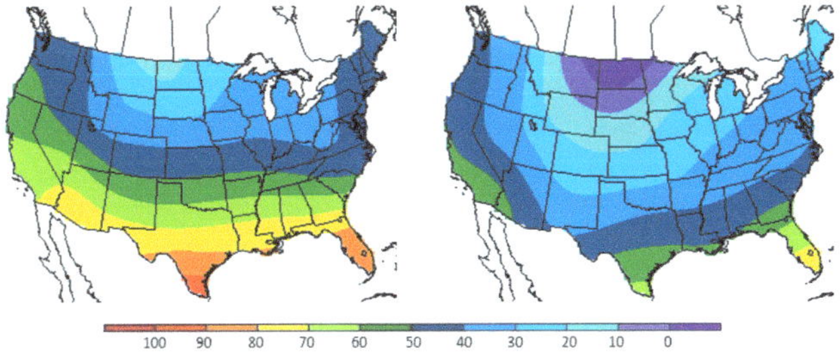

02/19/2025

The intense depression that covered much of the northeast quarter of the country yesterday drifts away to the north with high pressure filling in behind.

Heavy rain and storms continue in the Pacific Northwest to the west of the North Great Plains, with patches of rain down through the Rocky Mountains.

A wet front also brings widespread rain into the southern half of the Southwest to western Texas.

Wintry precipitation eases over the Great Lakes, while storms and widespread precipitation prevail in the east from the Northeast, throughout the Appalachians to the Coastal Southeast and into central and eastern parts of the Deep South.

Heavy snow may occur around the Great Lakes to the Northeast and into the Megalopolis.

Mostly dry elsewhere.

02/20/2025

Wet fronts cross the Pacific west coasts up into the Rocky Mountains continuing widespread precipitation throughout the Pacific and Inland Northwest to the adjacent Rocky Mountains, and heavy rain into northern California and along the Sierra Nevada and Cascade ranges.

Another wet front brings rain from New Mexico, through the northern half of Texas, into the Deep South to Georgia and into the southern reaches of the Great Plains to the Ozarks, but clears in Arizona as it slips deeper into the Gulf of Mexico.

Mostly dry elsewhere except for some lingering showers around the Great Lakes.

FEBRUARY 20 THURSDAY

ISOBARIC

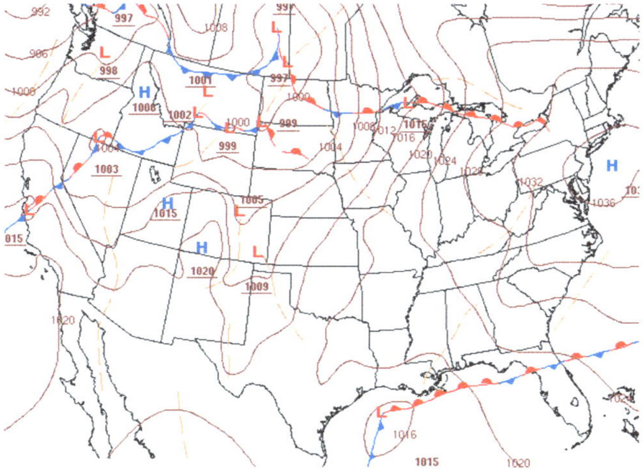

RAIN DISTRIBUTION (Inches)

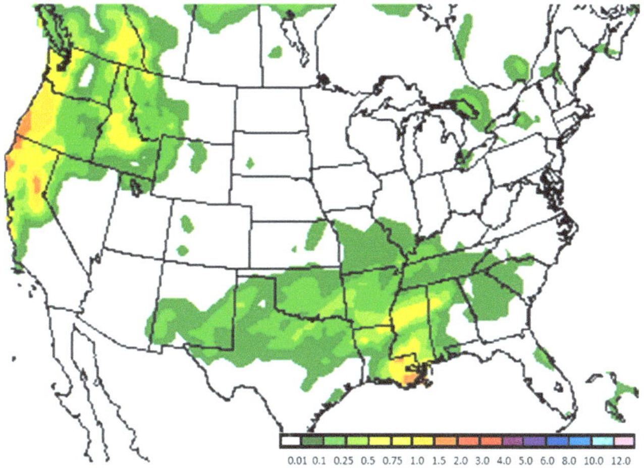

MAXIMUM TEMPERATURE / MINIMUM TEMPERATURE

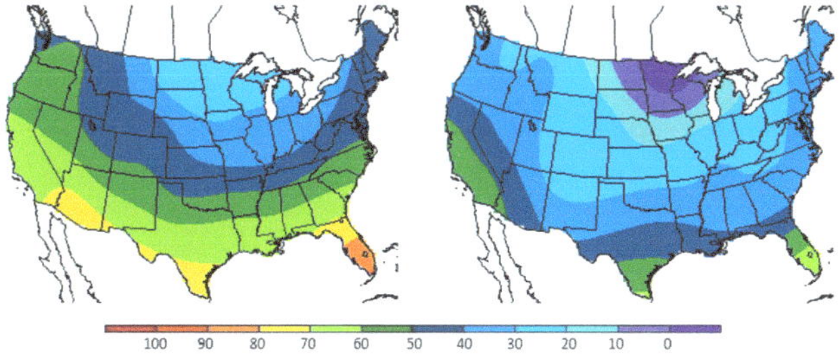

FEBRUARY 21 FRIDAY

ISOBARIC

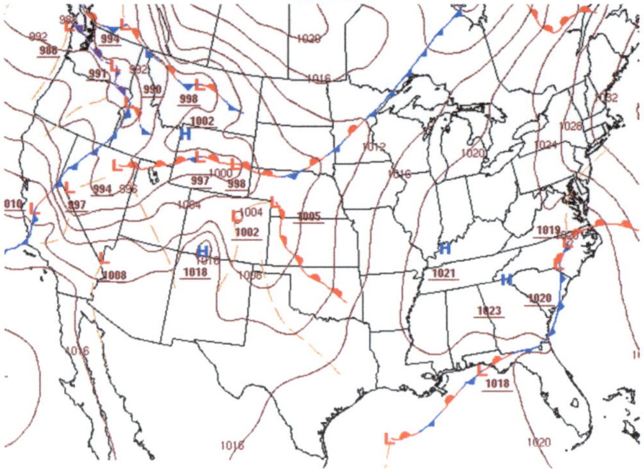

RAIN DISTRIBUTION (Inches)

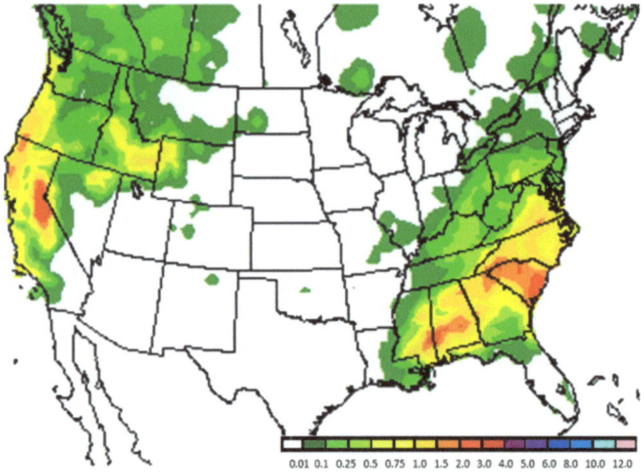

MAXIMUM TEMPERATURE

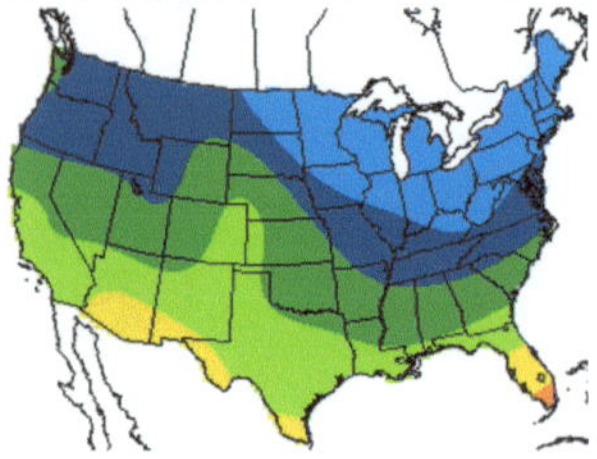

MINIMUM TEMPERATURE

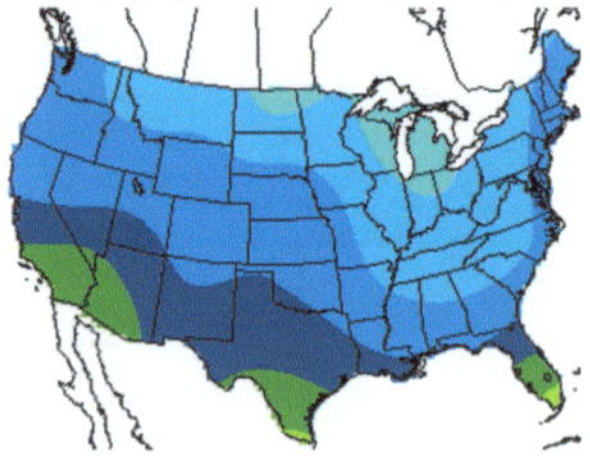

02/21/2025

Widespread precipitation continues in the Pacific Northwest to the northern Rockies, with wintry showers onto the North Great Plains.

Heavy precipitation also prevails down throughout most of California and into Nevada, and especially through the Sierra Nevada and Cascade ranges.

In the east, a wet front continues to maintain rain activity throughout most of the Deep South and spreading into the Atlantic coastal states from northern Florida to the Megalopolis.

Rain may be particularly heavy at times from the Coastal Southeast region to east of the Appalachians, and into the Deep South, with chance of flash flooding in the Carolinas.

Mostly dry elsewhere.

Snow expected to move eastward into Montana and Nevada over the next few days.

02/22/2025

A large depression moves onto the North Great Plains, low pressure to the west and high pressure to the east.

Rain activity clears in the southeast, except for some showers lingering from Florida to Georgia.

A depression over the northeast however, brings widespread wintry precipitation throughout the Northeast to the Coastal Southeast regions.

Widespread rain over the west of the country and spreads eastwards from the Pacific Coast through the Great Basin and into the Rocky Mountains the North Great Plains, with much of the precipitation falling as snow in the higher reaches and in the Northern Plains.

Heavy rain still possible in the Sierra Nevada ranges.

Mostly dry elsewhere, with strong winds on the Great Plains to the Great Lakes.

FEBRUARY 22 SATURDAY

ISOBARIC

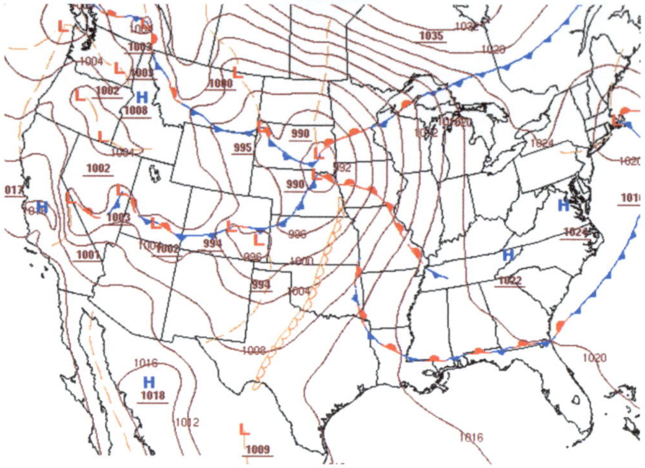

RAIN DISTRIBUTION (Inches)

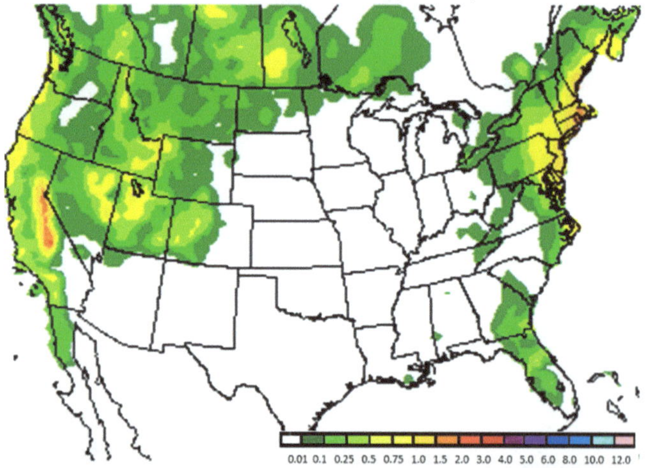

MAXIMUM TEMPERATURE

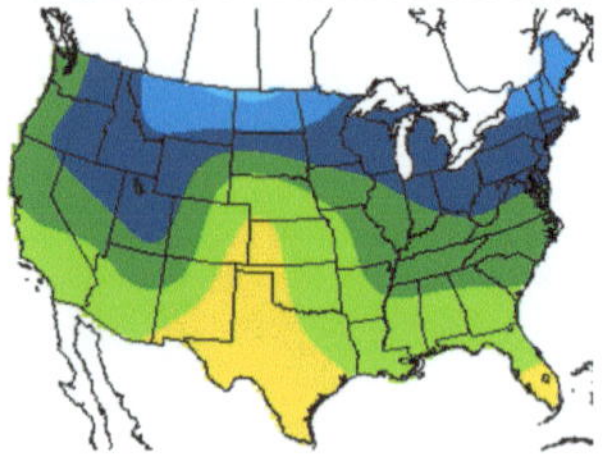

MINIMUM TEMPERATURE

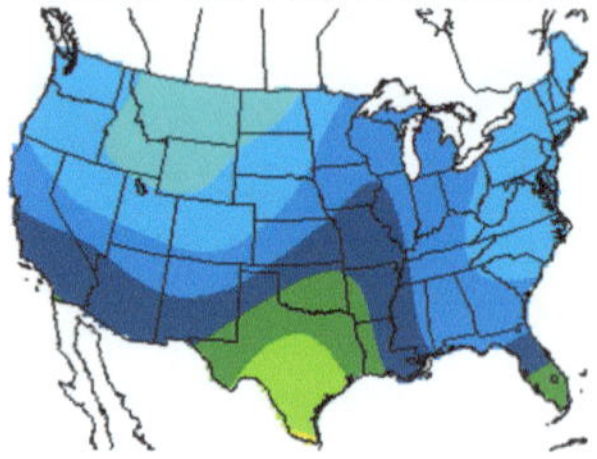

FEBRUARY 23 SUNDAY

ISOBARIC

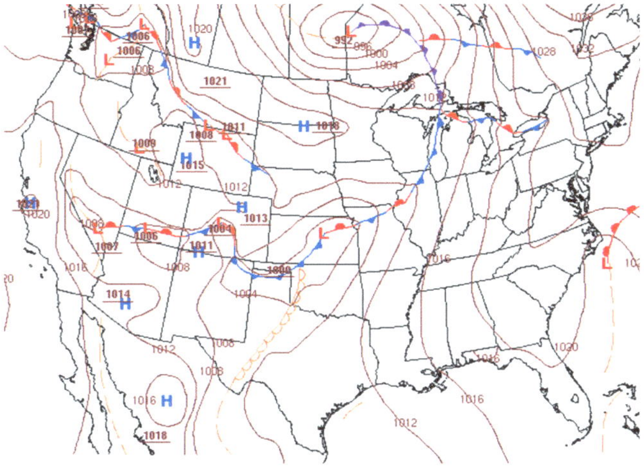

RAIN DISTRIBUTION (Inches)

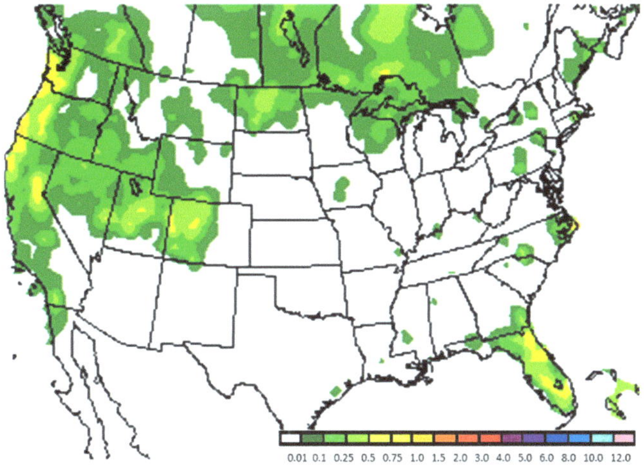

MAXIMUM TEMPERATURE

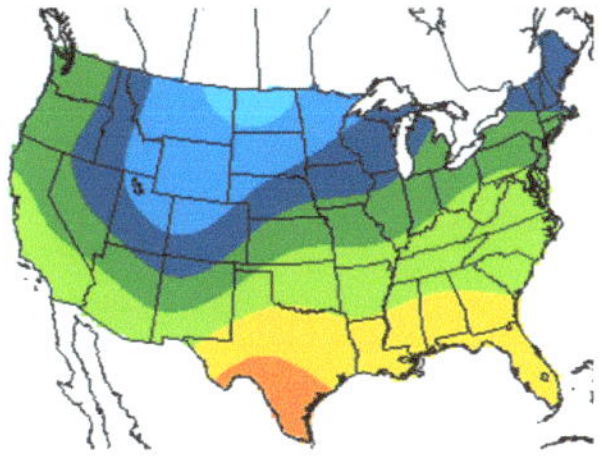

MINIMUM TEMPERATURE

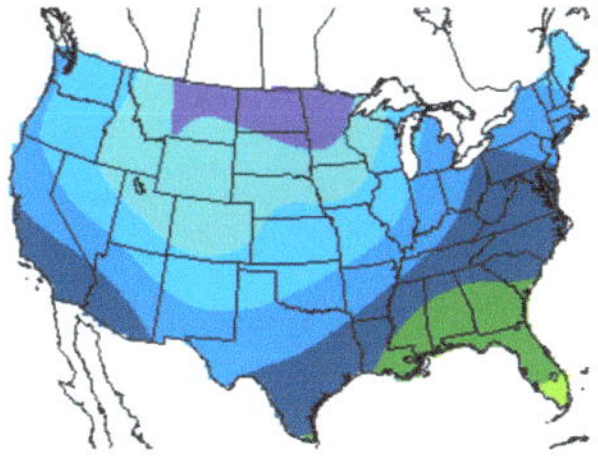

02/23/2025

An intense depression threatens north of the North Great Plains.

Widespread precipitation continues throughout the northwest of the country from the Pacific Northwest and California, east through the Great Basin and down through the Rocky Mountains.

Wintry rain also follows the US/Canadian border states across the North Great Plains to the Great Lakes.

Rain in Florida and spotty showers in the Metropolitan to the Northeast.

Mostly dry elsewhere.

It may become stormy over the central Great Plains moving eastward over the next few days, with blizzard conditions likely to develop in parts of the Midwest.

Scattered thunderstorms elsewhere, possible tornado activity may continue to be a risk in the Deep South.

02/24/2025

A wet brings wintry precipitation over the Pacific Northwest through the northern Rockies and onto the North Great Plains, as well as extending heavy rain into northern California.

Another front extends from the Four Corners to Texas and up to the Great Lakes.

Precipitation prevails along this front from the Deep South to the Great Lakes, with heavy falls around the coastal Gulf of Mexico and heavy snow over the Inland Midwest to the Great lakes.

Wintry showers, some of sleet or snow, over the Northeast and around the Chesapeake area, with isolated showers in Florida. Mostly dry elsewhere.

High winds in the Northeast region over the next couple of days may bring disruptions to transit corridors and infrastructure.

FEBRUARY 24 MONDAY

ISOBARIC

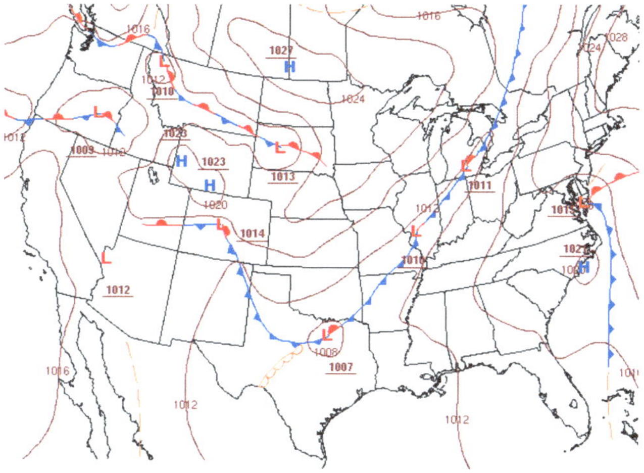

RAIN DISTRIBUTION (Inches)

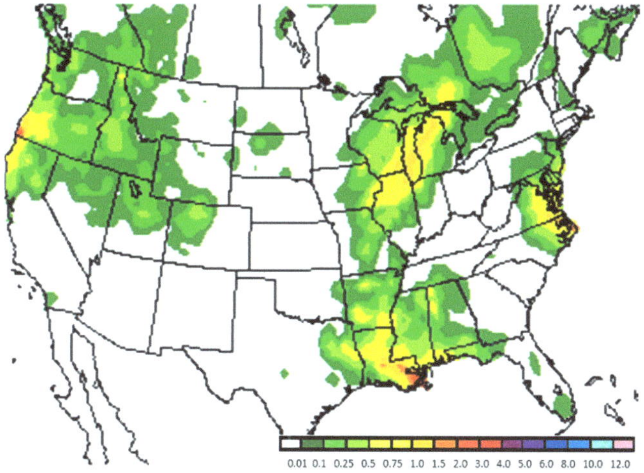

MAXIMUM TEMPERATURE

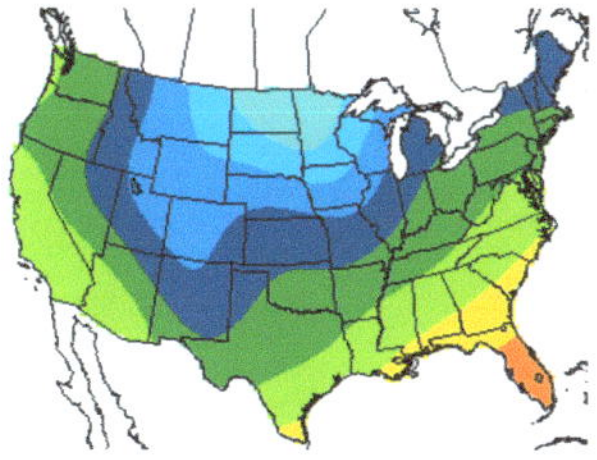

MINIMUM TEMPERATURE

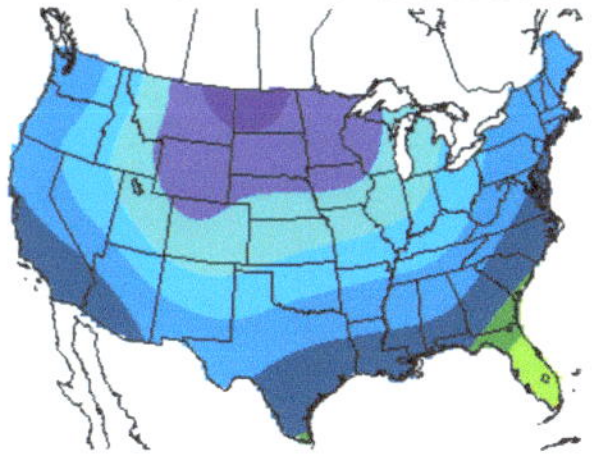

FEBRUARY 25 TUESDAY

ISOBARIC

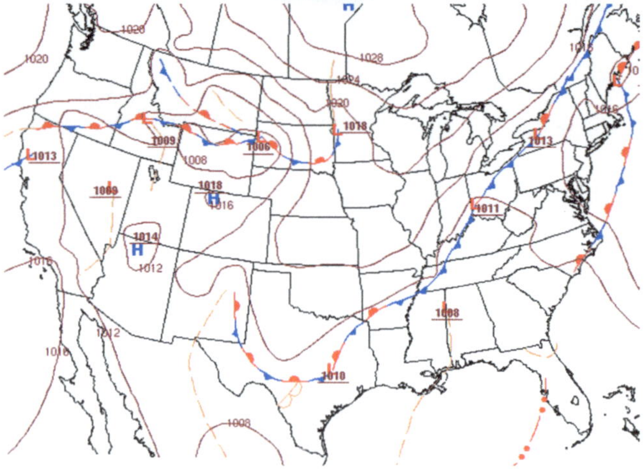

RAIN DISTRIBUTION (Inches)

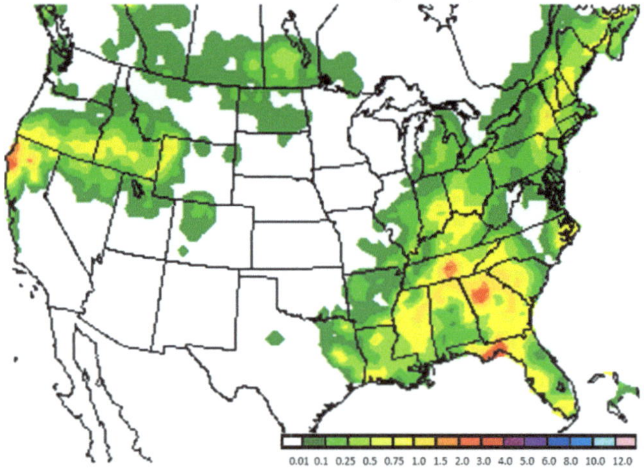

MAXIMUM TEMPERATURE

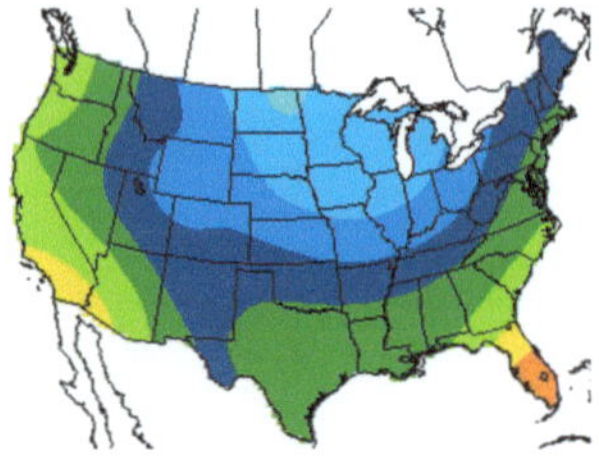

MINIMUM TEMPERATURE

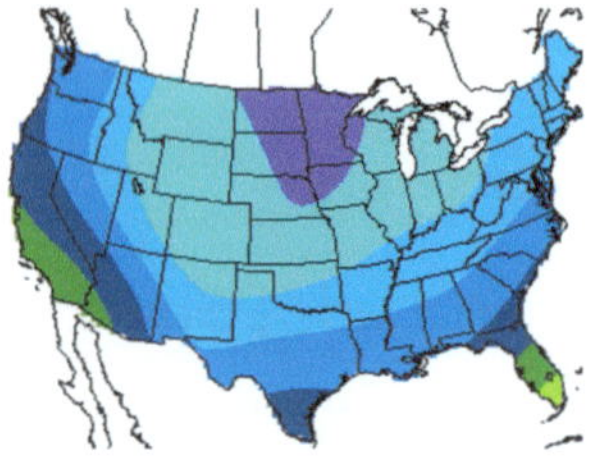

02/25/2025

A wet front crosses Texas, up through the Ozarks, along the Appalachians and into the Northeast.

This front, combined with a blast of Arctic air, brings very cold wintry conditions and an intense storm into the Northeast to the Inland Midwest. Heavy snow likely, especially Vermont to New York. Wintry rain and sleet to the south and east.

Widespread rain in the Deep South to Florida and around the Carolinas and in the far northeast of Texas.

Rain in Oregon, northern California and adjacent states, with snow likely in the North Great Plains.

Mostly dry elsewhere.

02/26/2025

A cold front crosses the Great Basin and up into a depression immediately to the west of the Great Lakes.

Another cold front flows runs from the Gulf of Mexico, across central Florida and northwards to the Northeast, parallel to the eastern coasts.

This front brings wintry precipitation, including heavy snow, from the Northeast to the Coastal Southeast and along the Appalachians and into the Carolinas, with showers into Georgia and the southern half of Florida.

Further shower activity also prevails in the Mississippi Basin into northern Texas.

Patches of precipitation down through the Pacific Northwest, California, the Great Basin and along the Rocky Mountains, as well as outbreaks of rain and occasional snow in the Great Plains.

Mostly dry around the Great Lakes, the far north Pacific Northwest. Isolated showers elsewhere.

Chance of very strong winds in Wyoming.

Showers in Hawaii over the next few days.

FEBRUARY 26 WEDNESDAY

ISOBARIC

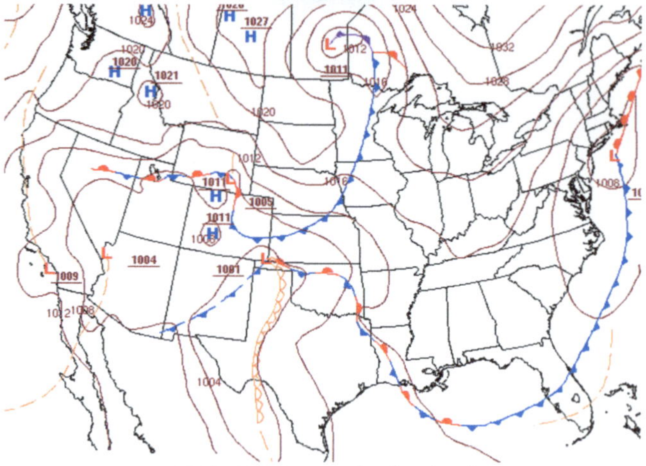

RAIN DISTRIBUTION (Inches)

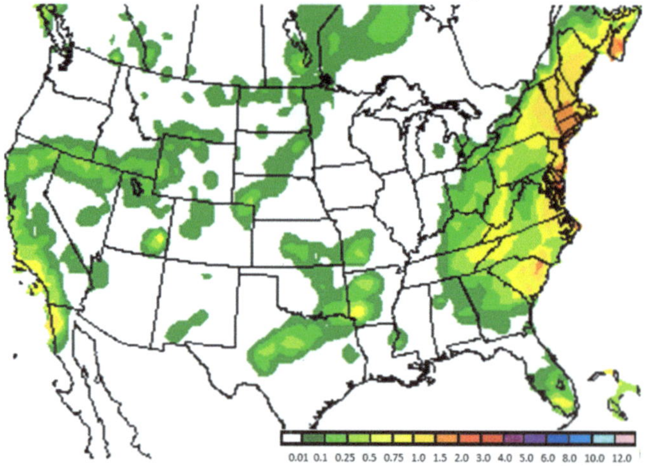

MAXIMUM TEMPERATURE

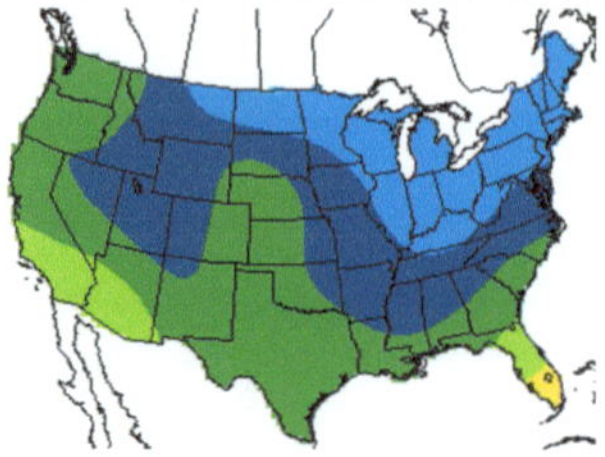

MINIMUM TEMPERATURE

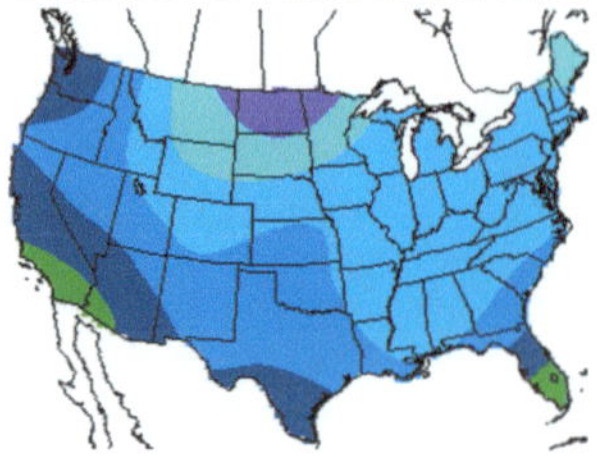

FEBRUARY 27 THURSDAY

ISOBARIC

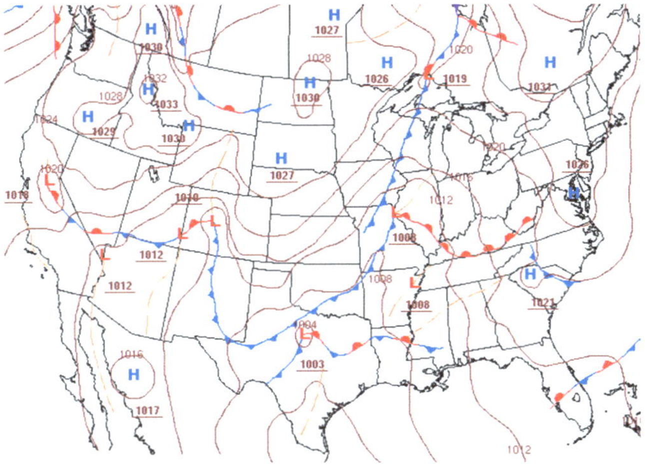

RAIN DISTRIBUTION (Inches)

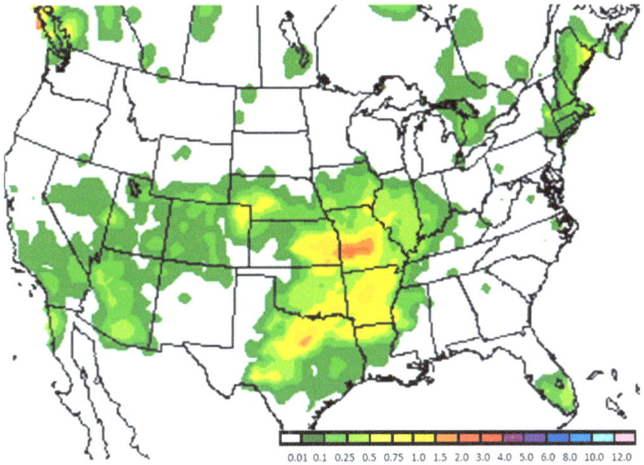

MAXIMUM TEMPERATURE

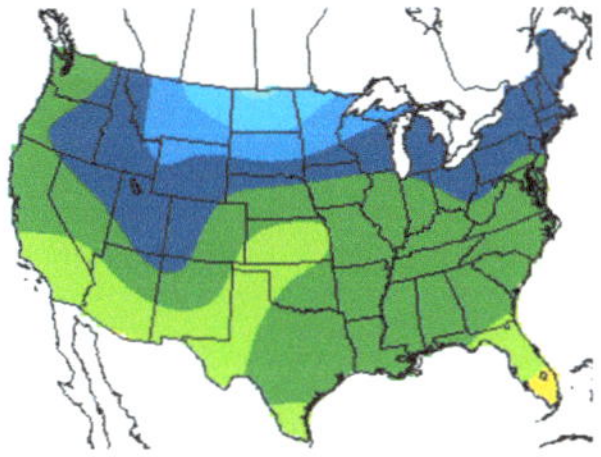

MINIMUM TEMPERATURE

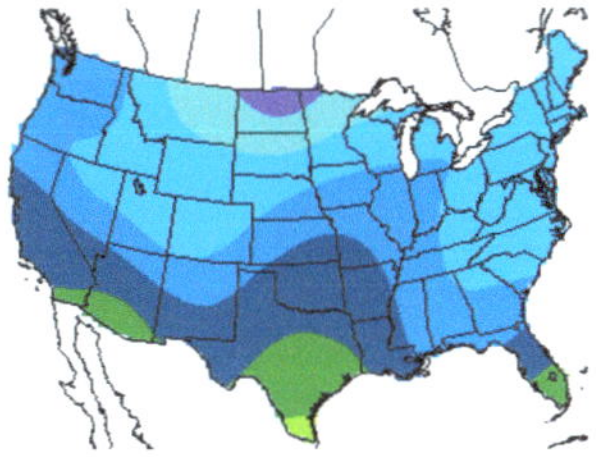

02/27/2025

Rain over southern California, the Great Basin and down into the Southwest and the Four Corners states.

Rain across the southern parts of the country and up to the Great Lakes along the east of the Great Plains.

Heavy falls through Texas to eastern Deep South and the Ozarks to the southwest of the Inland Midwest. Heavy falls also likely up through the Mississippi Basin.

Frost patches in sheltered parts of central Florida.

Scattered thunderstorms across the south, possibly generating tornado activity. Higher risk from Louisiana to South Carolina.

A warmer change may begin in the Midwest. Showers in southern Florida. Squalls likely in Hawaii.

02/28/2025

An intense depression forms over the Great Lakes and Northeast. Cold front through the east, Inland Midwest and eastern half of the Deep South. High pressure elsewhere.

Widespread falls throughout the east, and southwest to the Rocky Mountains and South Great Plains.

Precipitation may be particularly heavy in the Deep South with thunderstorms, flash flooding, and possible tornadoes, particularly Louisiana and South Carolina.

Heavy wintry falls including snow throughout the Great Lakes region to the Inland Midwest. Snow from Indiana to Nebraska as well as further west across the South Great Plains into Colorado.

Mostly dry elsewhere, including in the western half of Texas and over the west of the country from the Pacific coasts to the Rocky Mountain foothills and over the North Great Plains.

A cold northerly flow may see some frost patches appear in Florida over the next couple of days.

FEBRUARY 28 FRIDAY

ISOBARIC

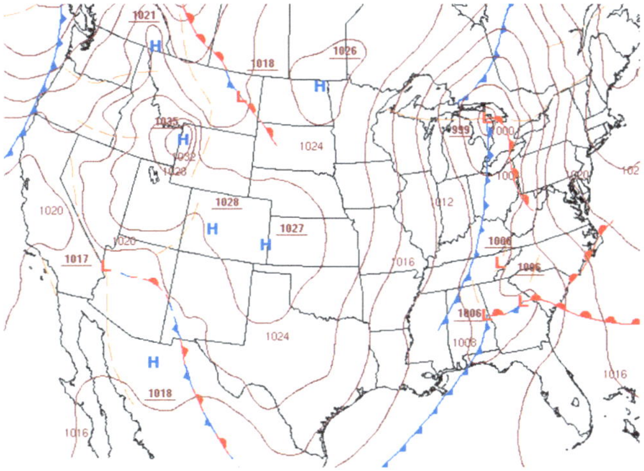

RAIN DISTRIBUTION (Inches)

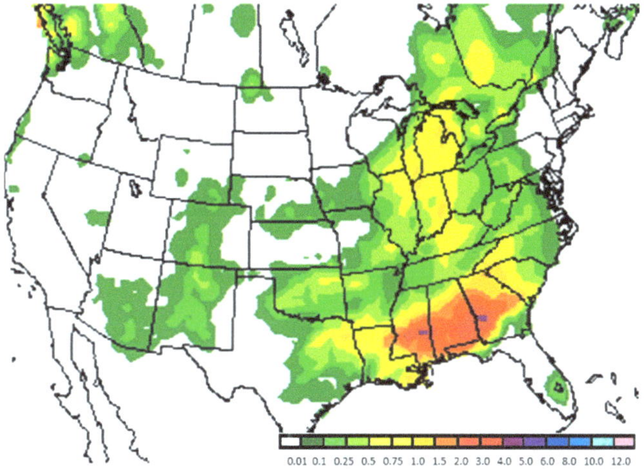

MAXIMUM TEMPERATURE

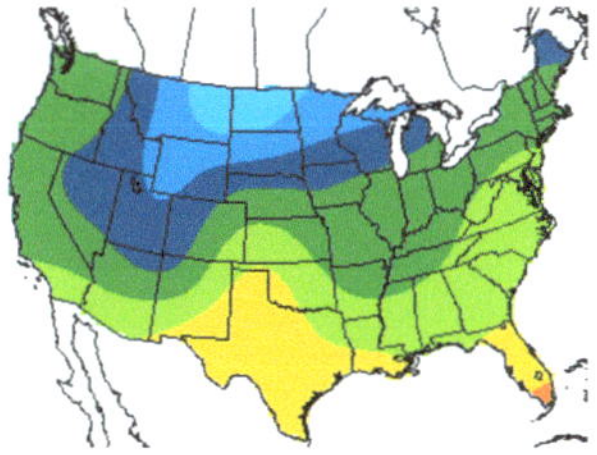

MINIMUM TEMPERATURE

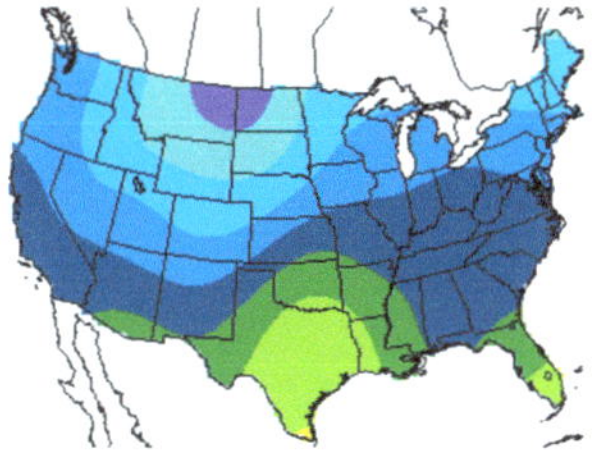

MARCH

PHASES OF THE MOON

1st	Perigee #7
2nd	Crossing Equator
6th	First Quarter
7th	Northern Declination
14th	Full Moon
15th	Crossing Equator
17th	Apogee
22nd	Third Quarter
22nd	Southern Declination
29th	New Moon
29th	Crossing Equator
30th	Perigee #4

1 Sat	2 Sun	3 Mon	4 Tue	5 Wed	6 Thur	7 Fri	8 Sat	9 Sun	10 Mon	11 Tue	12 Wed	13 Thur	14 Fri	15 Sat	16 Sun
P7	XhN				1Q	^							F	XhS	

17 Mon	18 Tue	19 Wed	20 Thur	21 Fri	22 Sat	23 Sun	24 Mon	25 Tue	26 Wed	27 Thur	28 Fri	29 Sat	30 Sun	31 Mon
A					3Q V							N XhN	P4	

MARCH 1 SATURDAY

ISOBARIC

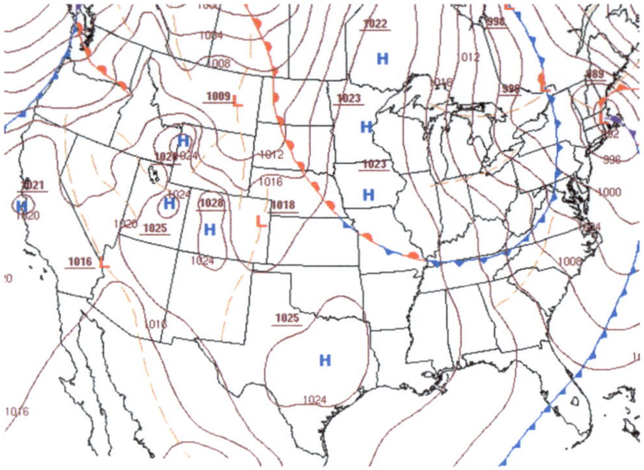

RAIN DISTRIBUTION (Inches)

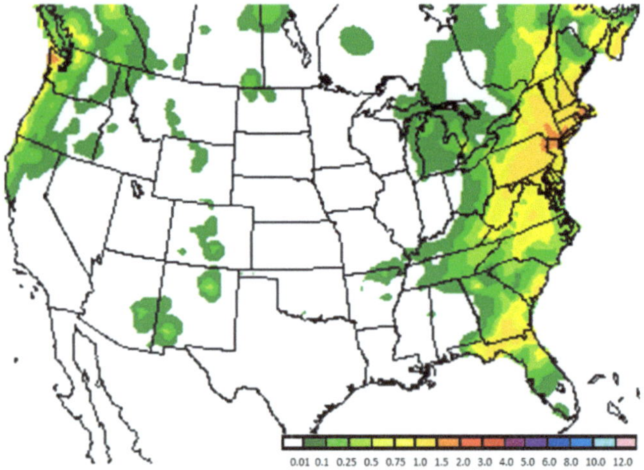

MAXIMUM TEMPERATURE / MINIMUM TEMPERATURE

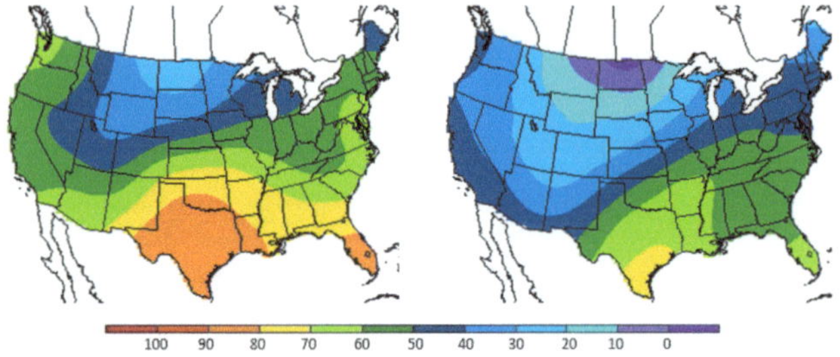

03/01/2025

Showers and pockets of rain in the Pacific Northwest and down into the northern parts of California with the odd shower likely to also appear in the Southwest to southern Rocky Mountains.

In the east, widespread and sometimes wintry precipitation continues east of the Appalachians with rain activity reaching into Florida and snow persisting at times around the Great Lakes to the Northeast.

Mostly dry elsewhere.

It may be particularly cold and windy across the Midwest to the Northeast, while a warmer spell returns to Florida as a milder southeasterly flow moves up the eastern coasts.

03/02/2025

Outbreaks of rain in the Pacific and Inland Northwest region as well as in the north of California.

Wintry and blustery precipitation, including snow and sleet, contracts in the east to linger mostly around the Great Lakes to the Northeast and only the odd showery outbreak in the southeast of the country.

Mostly dry elsewhere.

Heavy snow expected in Oregon in the next few days.

MARCH 2 SUNDAY

ISOBARIC

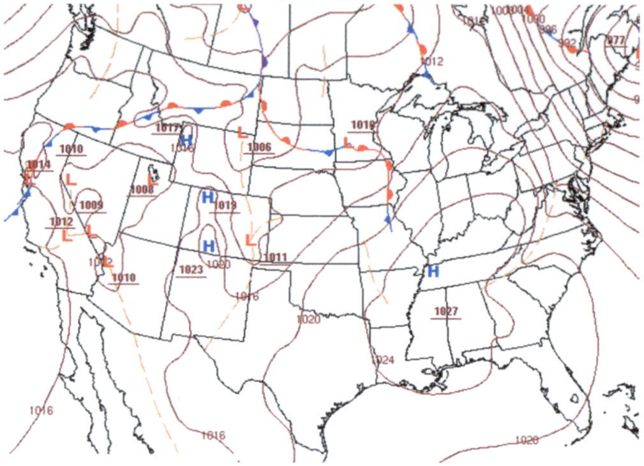

RAIN DISTRIBUTION (Inches)

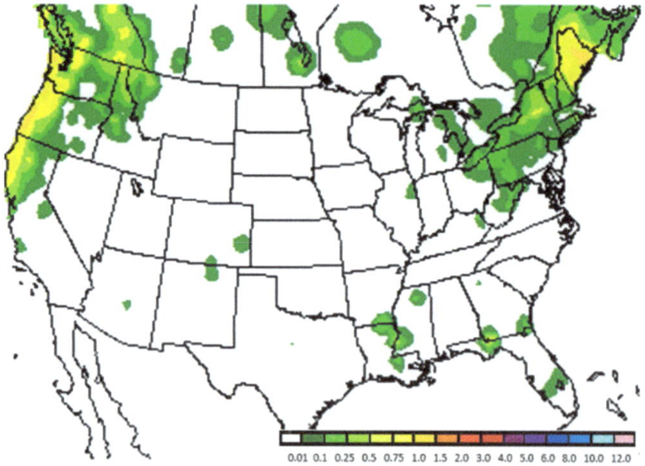

MAXIMUM TEMPERATURE / MINIMUM TEMPERATURE

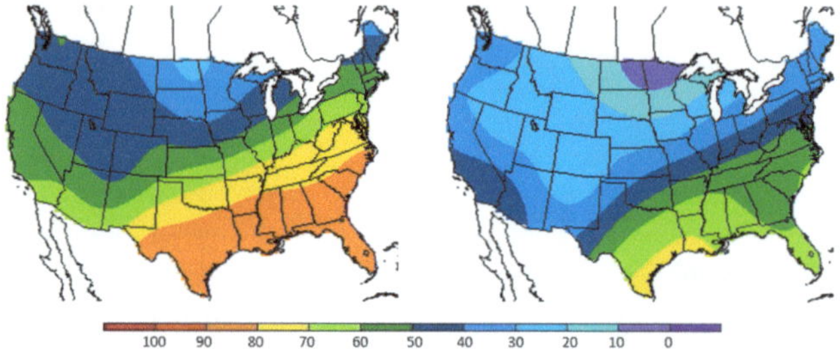

MARCH 3 MONDAY

ISOBARIC

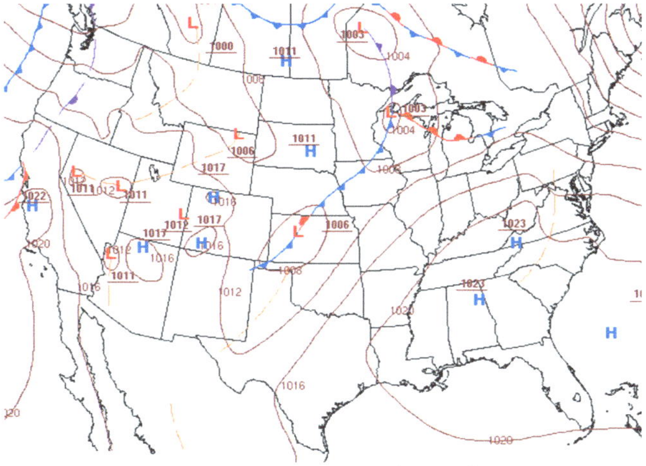

RAIN DISTRIBUTION (Inches)

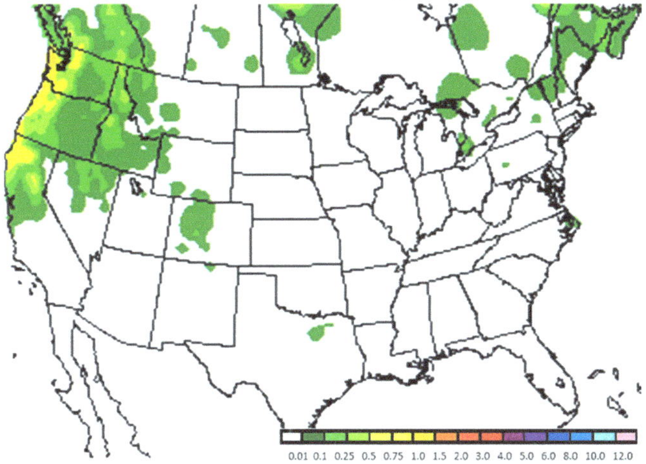

MAXIMUM TEMPERATURE

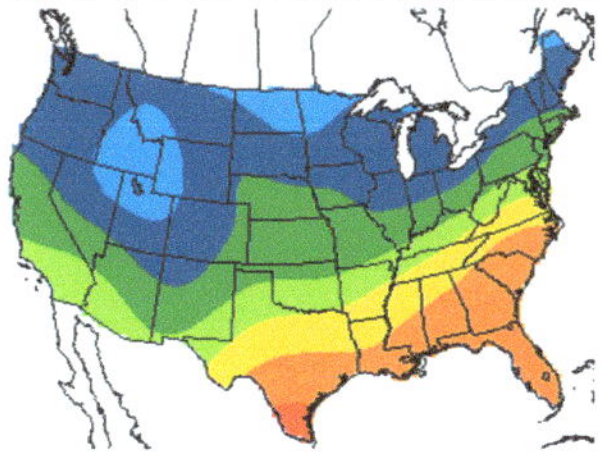

MINIMUM TEMPERATURE

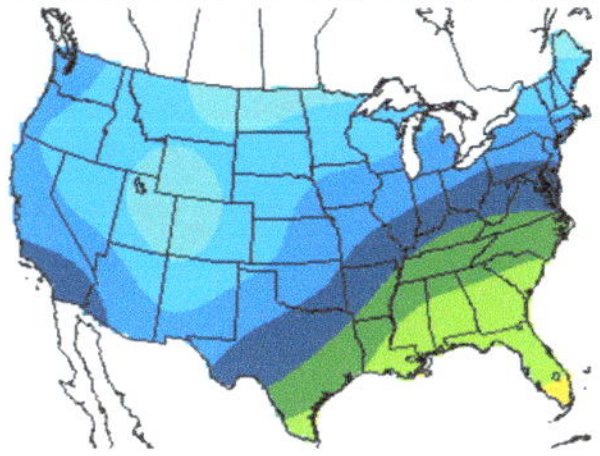

03/03/2025

Precipitation spreads from the Pacific Northwest and northern California and up into the Rocky Mountains.

Only scarce precipitation expected in the Northeast to the Great Lakes.

Most of the remainder of the country may be dry at this time.

It may be slightly warmer soon in the Northeast, with only a small chance of snow in the New York area.

There is the chance of a late winter storm forming over the Midwest to the Northeast over the next few days, with snowfall possibly spreading later into the Great North Plains.

03/04/2025

Unsettled conditions prevail in the west spreading widespread precipitation from the Pacific and Inland Northwest to northern California and across the Great Basin and up into the Rocky Mountains.

There may also be an outbreak of rain for a time around the Ozarks.

Mostly dry elsewhere.

It may be mainly mild and breezy across the Great Plains and the southern states.

While mild winds may prevail in the Northeast, they are likely to freshen to very strong levels especially across Virginia and Wisconsin in the next few days.

MARCH 4 TUESDAY

ISOBARIC

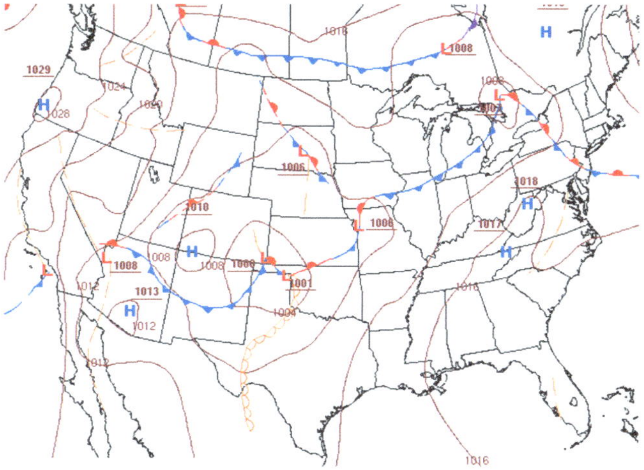

RAIN DISTRIBUTION (Inches)

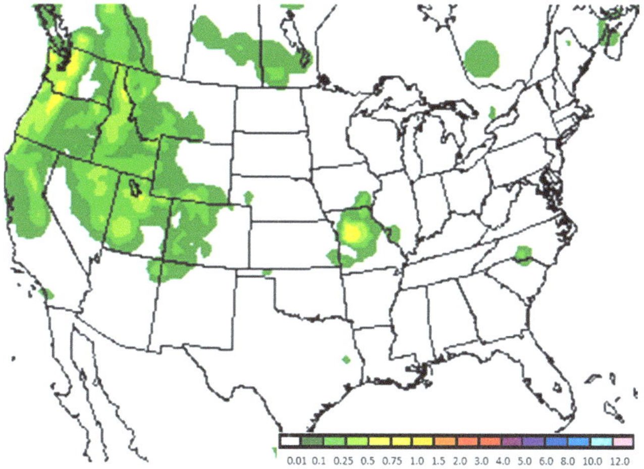

MAXIMUM TEMPERATURE

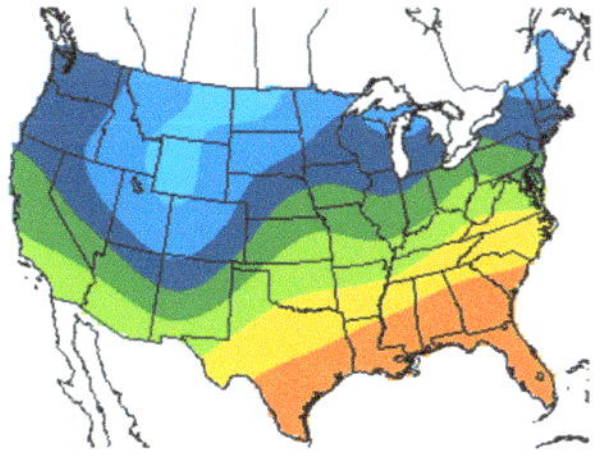

MINIMUM TEMPERATURE

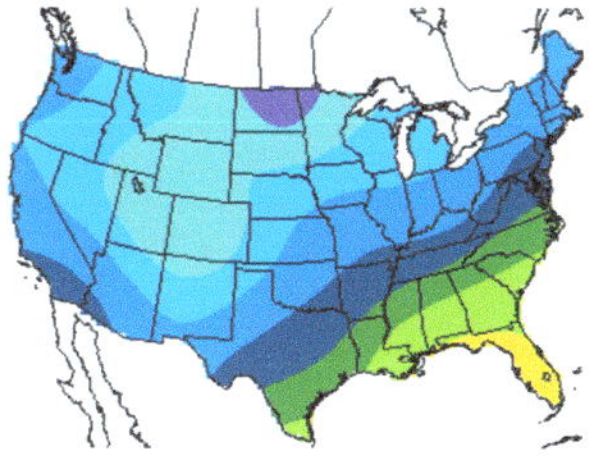

MARCH 5 WEDNESDAY

ISOBARIC

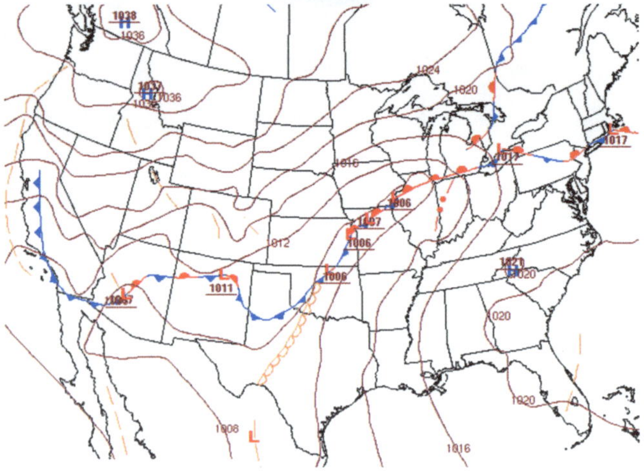

RAIN DISTRIBUTION (Inches)

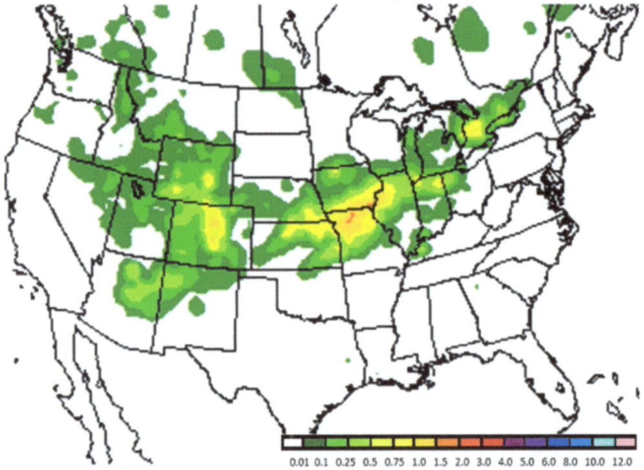

MAXIMUM TEMPERATURE MINIMUM TEMPERATURE

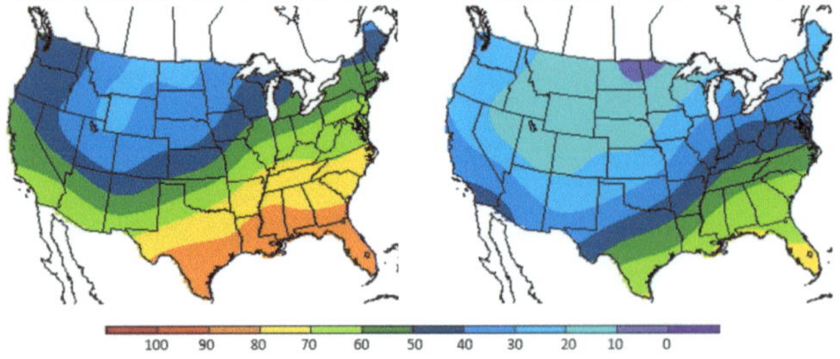

03/05/2025

Precipitation clears along the Pacific coastal states and spread over the Rocky Mountains, across the Great Plains and into the Inland Midwest and up into the Great Lakes region.

It should be mostly dry elsewhere.

It may be warmer than of recent in the Nebraska region.

There is a chance of tornadoes developing on the High Plains from Kansas to Texas, and dust storms may appear in the Southwest.

It is likely to be very cold but dry in Alaska over the next couple of days.

Showery with occasional heavy outbursts in Hawaii.

03/06/2025

Very stormy conditions develop in the southeast states, especially affecting Arkansas, Louisiana, and Mississippi, and rain is likely to be very heavy in the Deep South.

Heavy snow may fall in the Great Lakes to Wisconsin and cross the Inland Midwest with freezing rain and snow possibly damaging infrastructure.

Widespread precipitation in the Four Corners states and over the southern half of the Rocky Mountains.

Mostly dry elsewhere.

Chance of tornadoes across the Mississippi Basin, including into Louisiana and Arkansas.

MARCH 6 THURSDAY

ISOBARIC

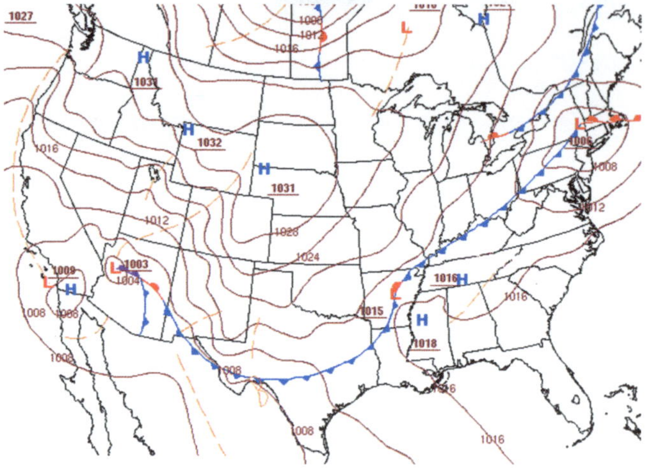

RAIN DISTRIBUTION (Inches)

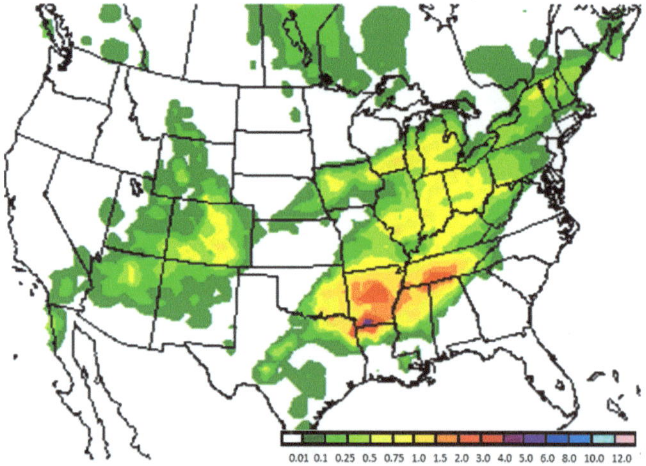

MAXIMUM TEMPERATURE MINIMUM TEMPERATURE

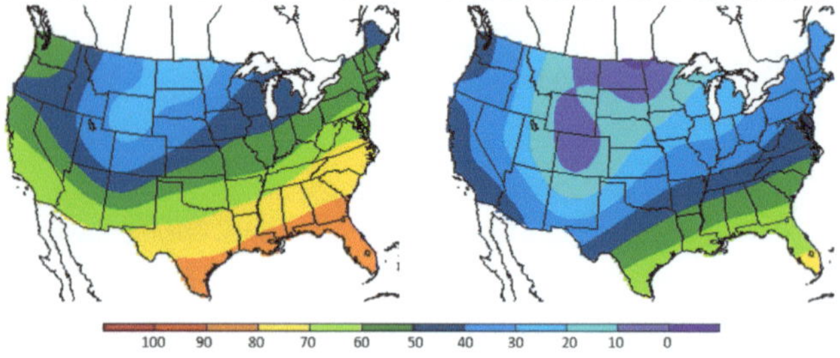

MARCH 7 FRIDAY

ISOBARIC

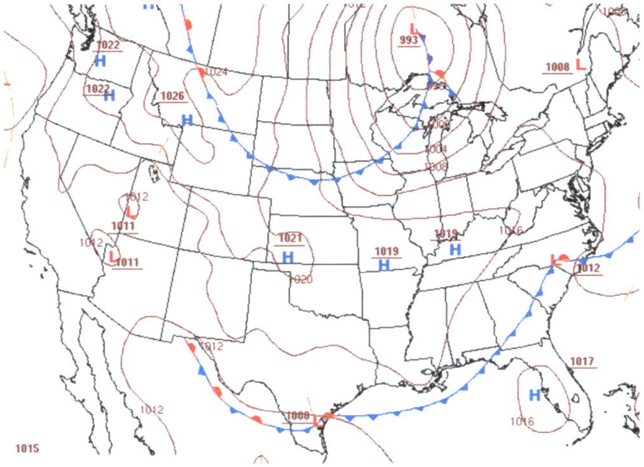

RAIN DISTRIBUTION (Inches)

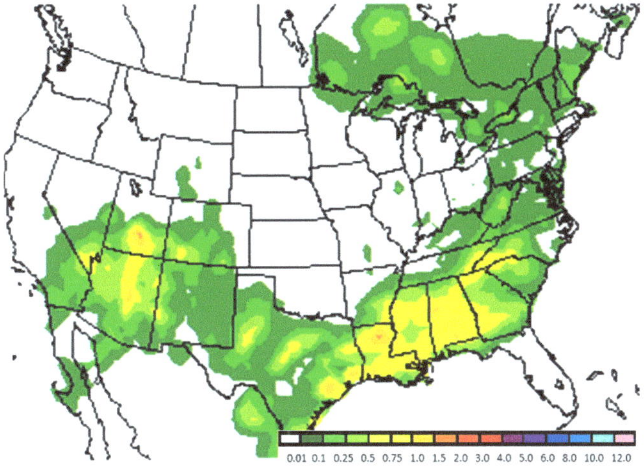

MAXIMUM TEMPERATURE

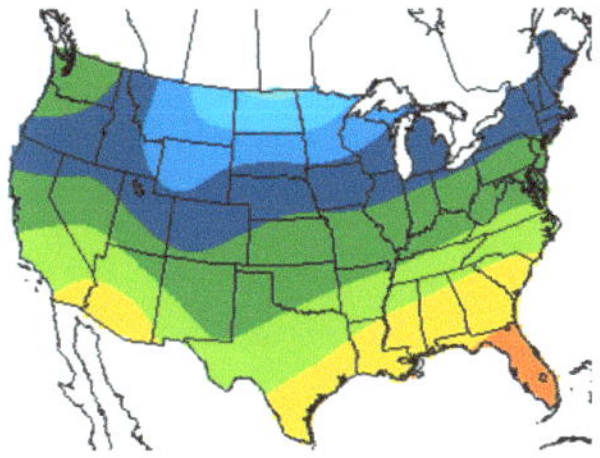

MINIMUM TEMPERATURE

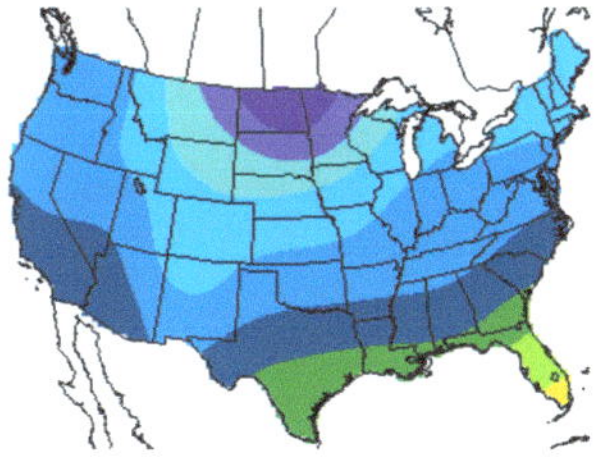

03/07/2025

An intensifying depression builds in strength immediately north of the Great Lakes bringing a return to wintry precipitation, including heavy snow, in the Northeast to the Great Lakes and in parts of the Midwest

Rain activity also prevails along the Appalachians to the Atlantic Coast.

Widespread rain activity prevails throughout the south of the country from the Pacific Southwest in southern California to the Coastal Southeast, including throughout Texas and with heavy rain at times in the Deep South.

Mostly dry through the Inland Midwest, across the Great Plains and in the northwest quarter of the country as well as in Florida.

03/08/2025

The depression to the Northeast of the country continues to bring widespread storms through the Northeast regions to the Great Lake with possible blizzard conditions developing for a time.

Heavy snow may also fall in northern parts of the Inland Midwest.

Widespread rain also prevails throughout the south parts of the country from southern California across the Southwest, throughout Texas and into the Deep South.

The stormy conditions may also produce pockets of flooding and tornadoes in Texas around this time.

The storms on the Four Corner states may also produce heavy wintry precipitation in Nevada, possibly falling as snow.

Mostly dry elsewhere, including in Georgia and Florida.

MARCH 8 SATURDAY

ISOBARIC

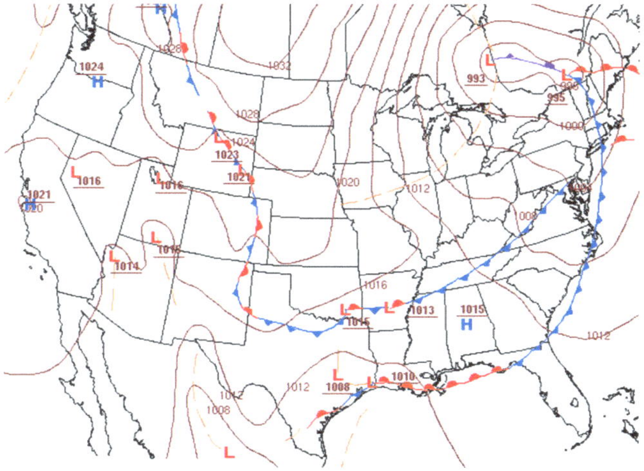

RAIN DISTRIBUTION (Inches)

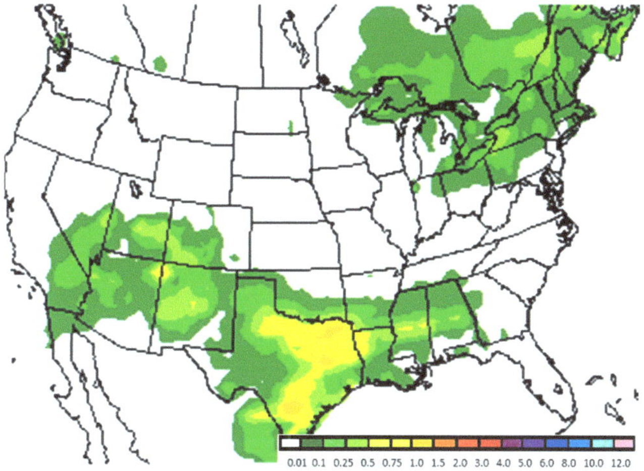

MAXIMUM TEMPERATURE

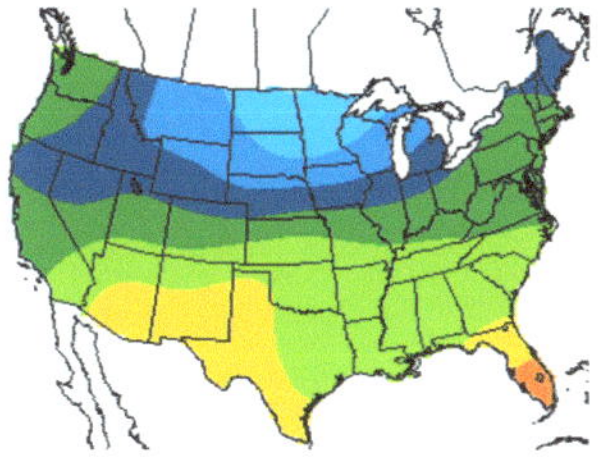

MINIMUM TEMPERATURE

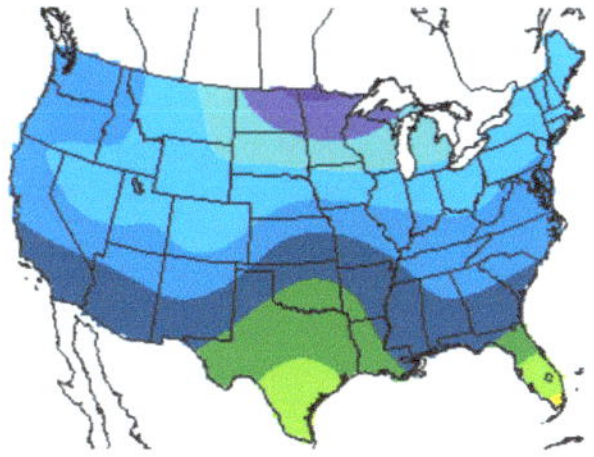

ISOBARIC

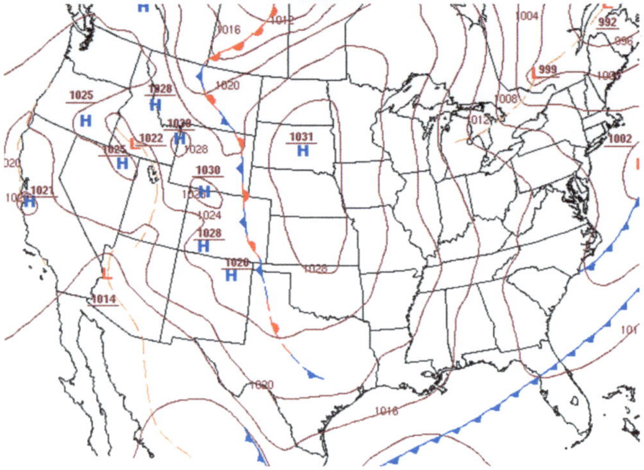

RAIN DISTRIBUTION (Inches)

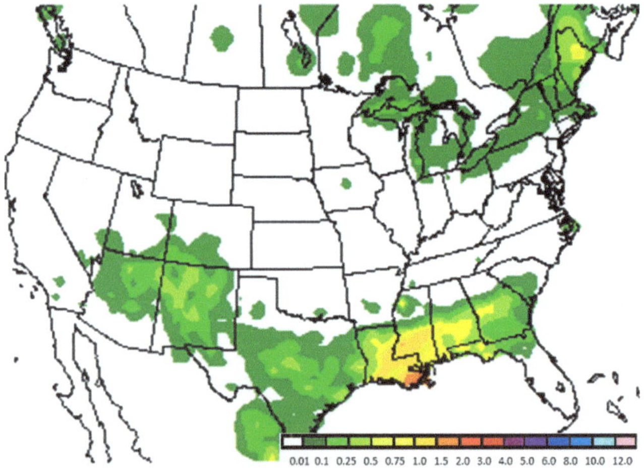

MAXIMUM TEMPERATURE

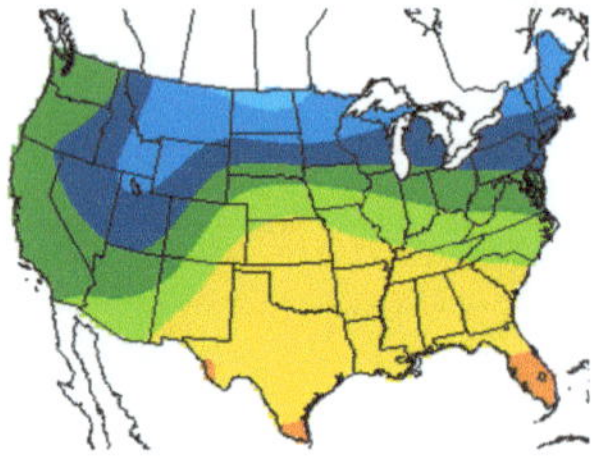

MINIMUM TEMPERATURE

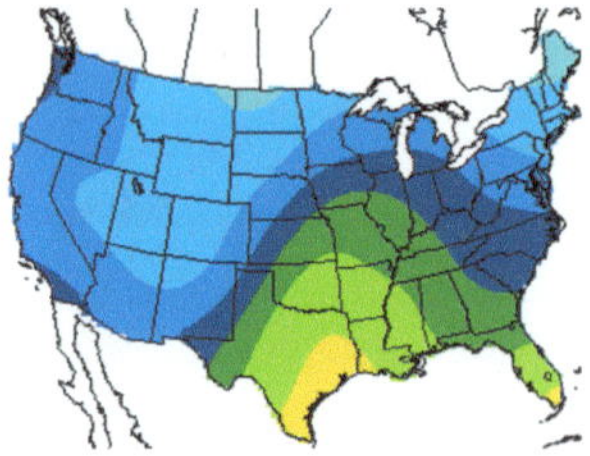

03/09/2025

Wintry precipitation eases around the Great Lakes to the Northeast.

The widespread rain across the south of the country from the Southwest to the Deep South continues, with heavy falls at times long the Gulf of Mexico coastal fringes

Mostly dry elsewhere.

Unusually warmer than normal temperatures may spread across the south from Florida to Texas and through the Gulf of Mexico.

In Alaska, despite it being mostly dry, it is also likely to be very cold with below normal temperatures.

03/10/2025

Stormy conditions may continue in the northern parts of the Great Lakes to the Northeast, and into the Inland Midwest.

These regions may continue to see snow and possible blizzards for several more days and include the chance of disruptions to travel and damage to infrastructure, especially around the Upper Midwest.

It may also be very cold and windy across the North Great Plains, where heavy snow may also prevail.

Wet conditions expected to linger around the Southwest and the Rio Grande Valley in Texas, as well as in Florida.

Mostly dry elsewhere.

ISOBARIC

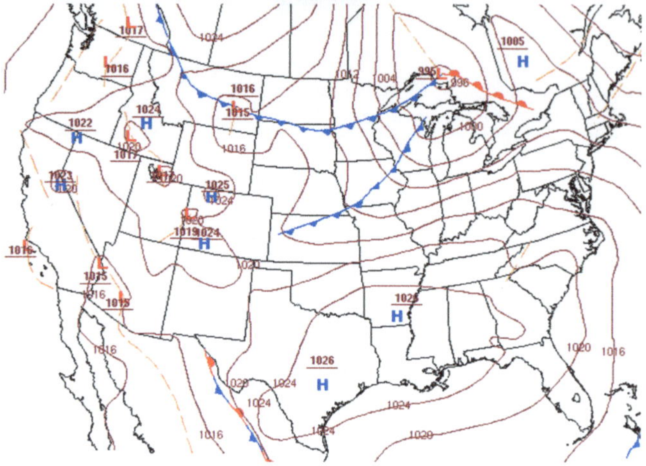

RAIN DISTRIBUTION (Inches)

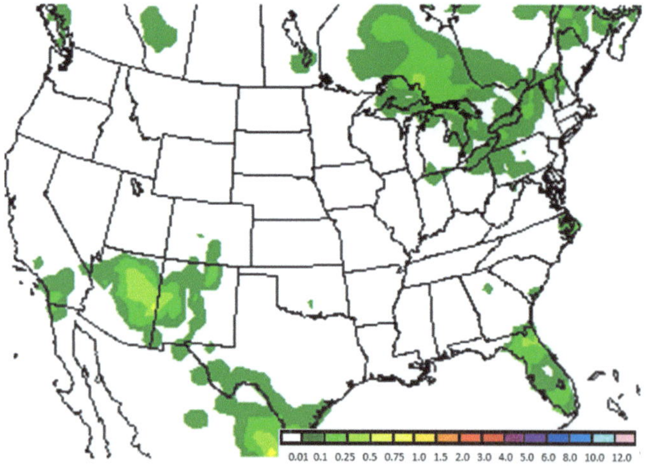

MAXIMUM TEMPERATURE

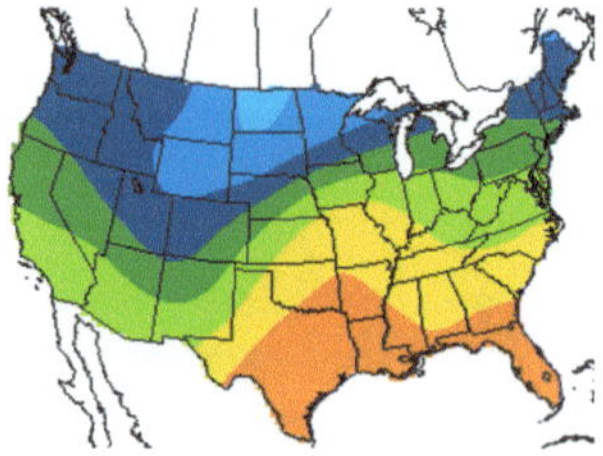

MINIMUM TEMPERATURE

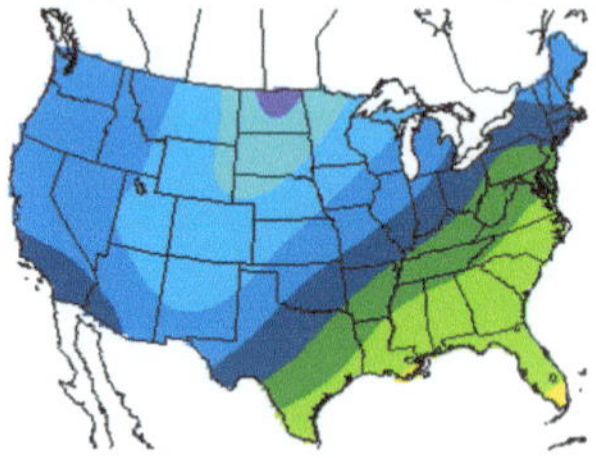

MARCH 11 TUESDAY

ISOBARIC

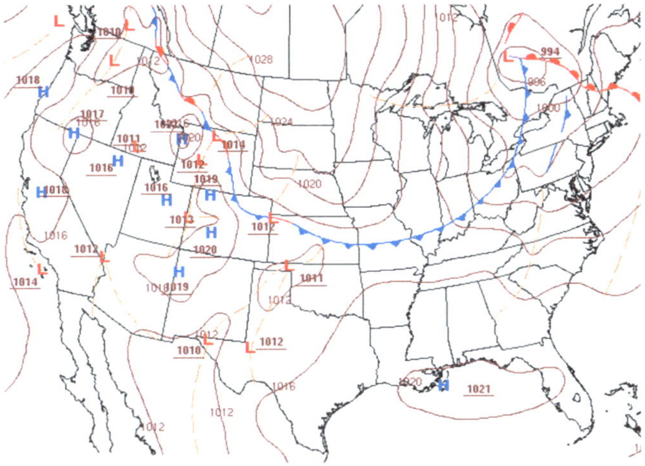

RAIN DISTRIBUTION (Inches)

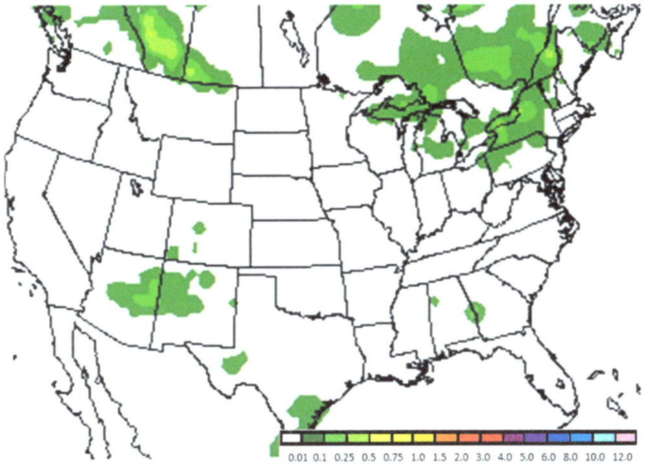

MAXIMUM TEMPERATURE MINIMUM TEMPERATURE

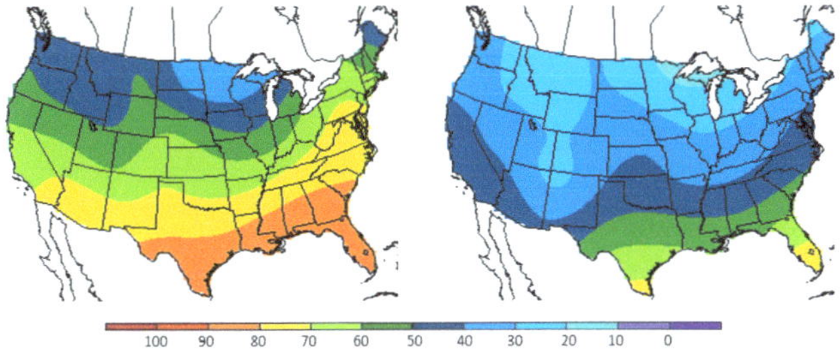

03/11/2025

It is likely to be mostly dry everywhere as rain activity clears throughout the south of the country today with just the odd shower lingering in the Southwest, Rio Grande Valley.

The only other exception may be the continued stormy conditions over the Northeast of the country.

Wind gusts may bring blizzard conditions for the next couple of days in parts of the Midwest including the Dakotas and Iowa to Michigan, Minnesota and Nebraska, with sizeable snowdrifts possible.

There is a chance of tornadoes and wind damage around this time in in the Deep South.

03/12/2025

Outbreaks of showers expected in the Pacific Northwest to the North Great Plains.

Rain activity may also return in the Southwest to Texas.

Further wintry rainfall is also likely to continue in the Northeast to the Great Lakes, and with the cold Artic airflow once more prevailing, with the chance of further snow into the region, especially around Maine.

Mostly dry elsewhere.

Drier than normal in Hawaii.

Snow expected in the southeastern parts of Alaska.

MARCH 12 WEDNESDAY

ISOBARIC

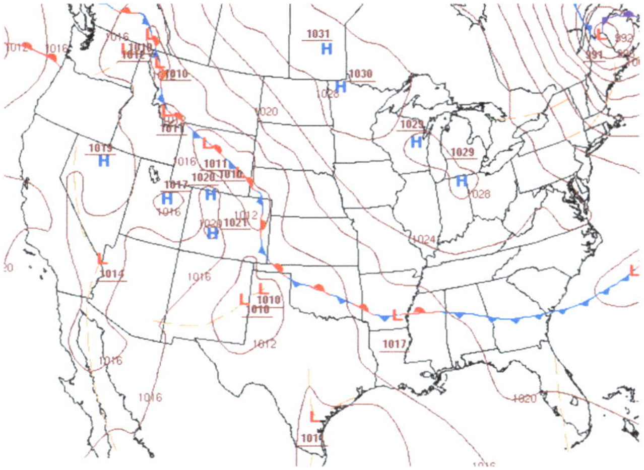

RAIN DISTRIBUTION (Inches)

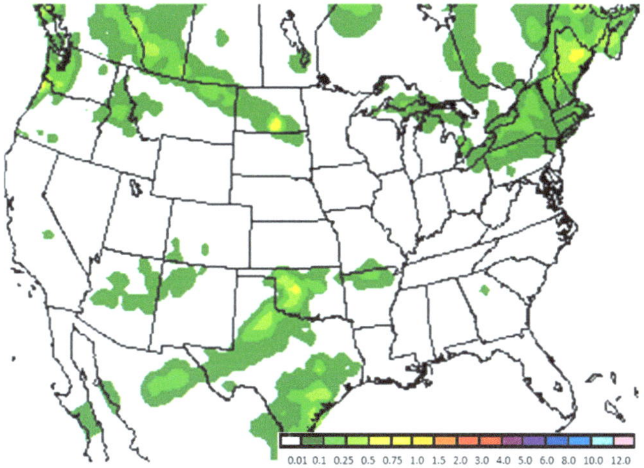

MAXIMUM TEMPERATURE MINIMUM TEMPERATURE

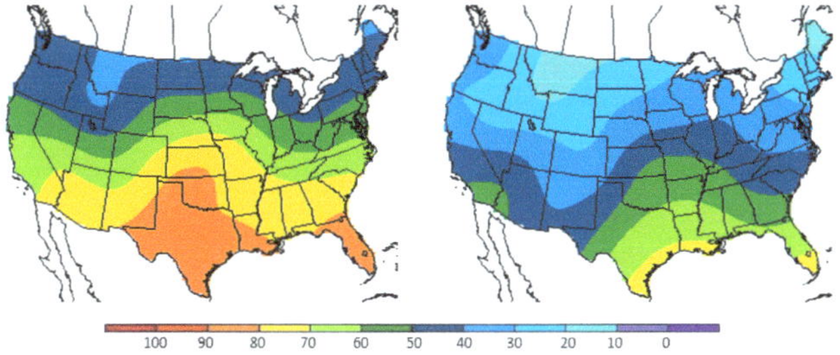

MARCH 13 THURSDAY

ISOBARIC

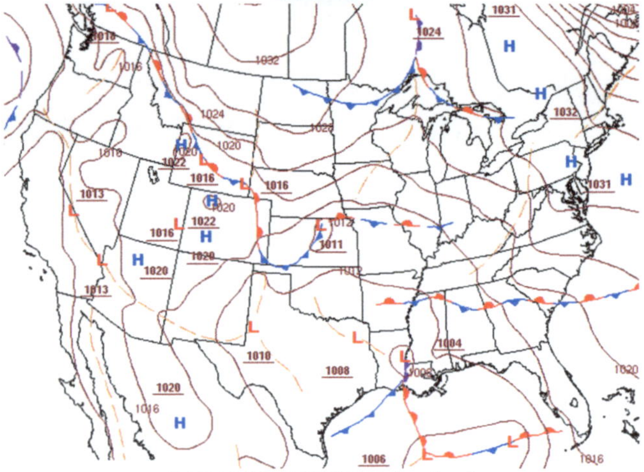

RAIN DISTRIBUTION (Inches)

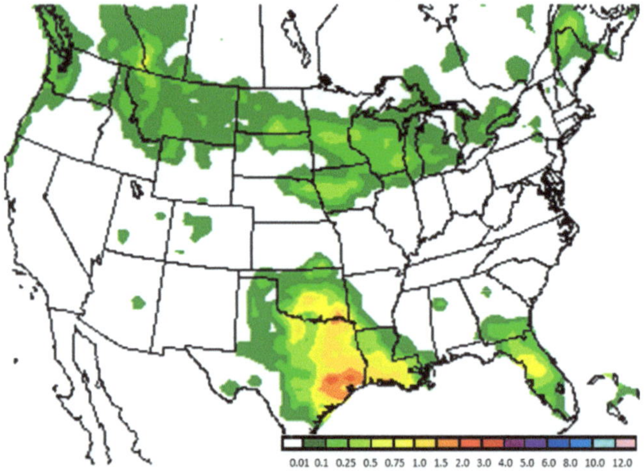

MAXIMUM TEMPERATURE MINIMUM TEMPERATURE

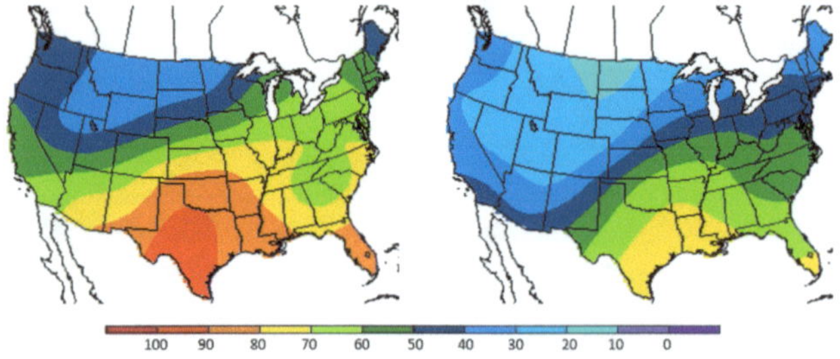

03/13/2025

Precipitation spreads across the north of the country from the Pacific Northwest over the northern Rocky Mountains and the North Great Plains to the Great Lakes, while easing in the Northeast.

Much of this precipitation is likely to arrive as snowstorms especially in the Inland Midwest to the Great Lakes regions as well as in the Dakotas to Colorado.

Widespread rain is also expected to move eastward, clearing in the Southwest while spreading across all of Texas and into the west of the Deep South and through to Florida.

Mostly dry elsewhere.

MARCH 14 FRIDAY

03/14/2025

Widespread precipitation builds across most of the eastern half of the country from the Ozarks, the Deep South, and Florida and up the Appalachians across most of the Inland Midwest and returning over the Great Lakes to the Northeast.

Rain also spreads down through the Pacific Northwest into northern California and eastward up into the northern half of the Rocky Mountains and onto the North Great Plains.

Much of this precipitation in the north of the country may fall as snow.

Mostly dry elsewhere.

It is also likely to be very cold in Texas over the next couple of days.

California may be unusually warmer than normal for this time of year.

ISOBARIC

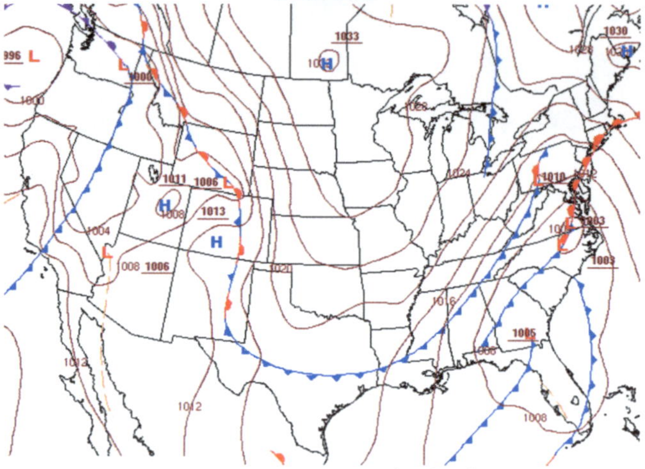

RAIN DISTRIBUTION (Inches)

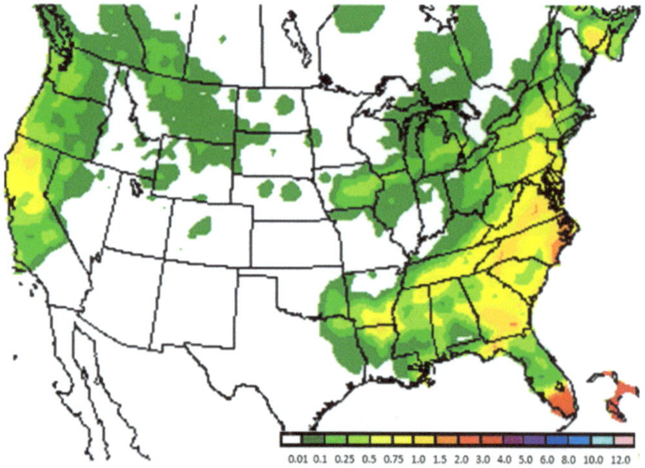

MAXIMUM TEMPERATURE MINIMUM TEMPERATURE

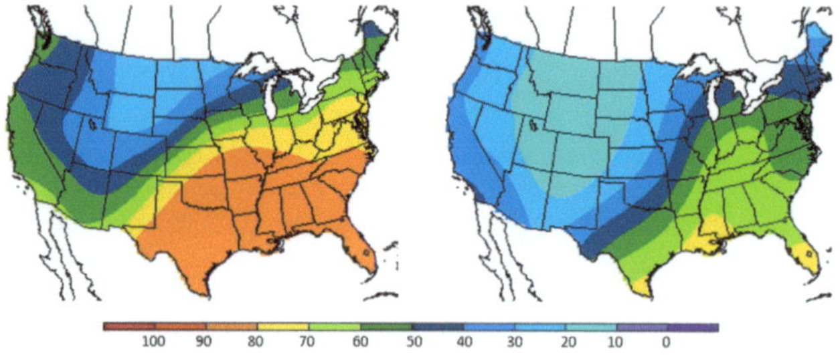

MARCH **15** SATURDAY

ISOBARIC

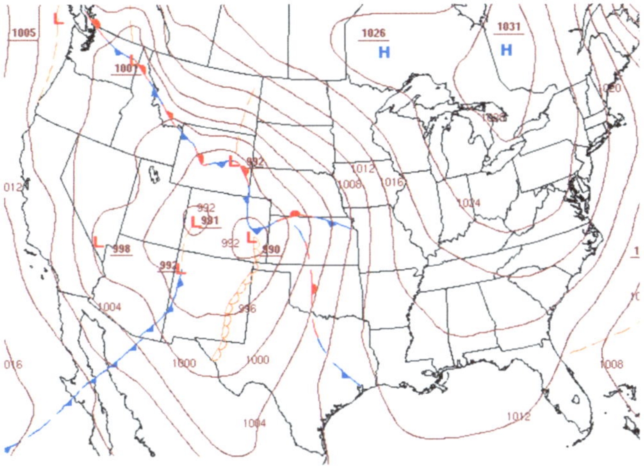

RAIN DISTRIBUTION (Inches)

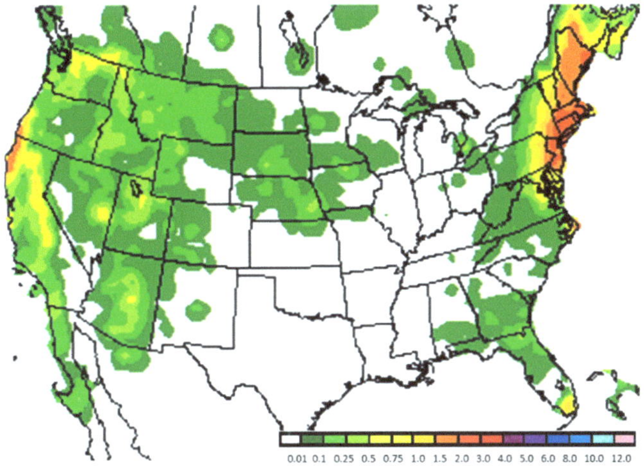

MAXIMUM TEMPERATURE

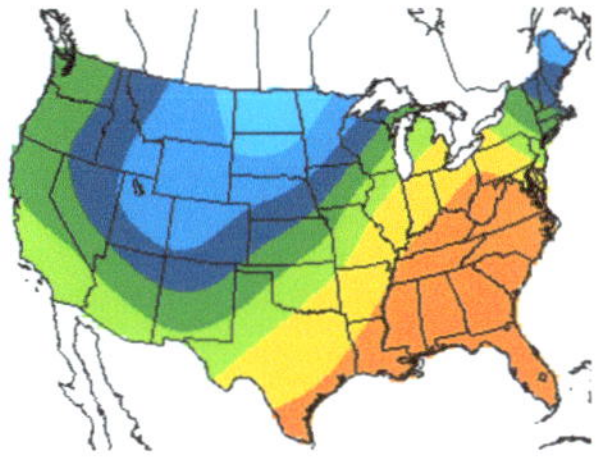

MINIMUM TEMPERATURE

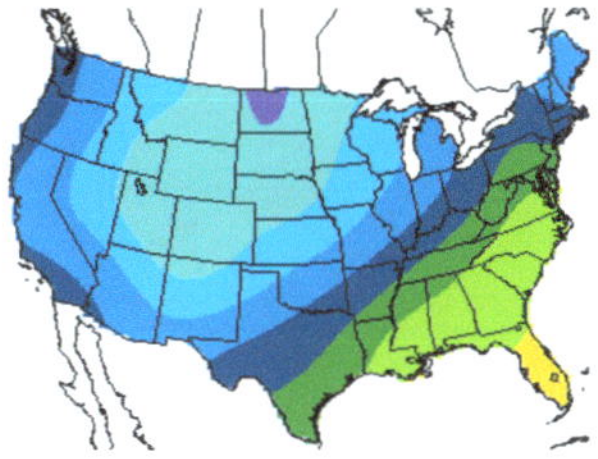

03/15/2025

Widespread heavy precipitation, some of it as snow in the north expected along the eastern seaboard, covering most regions east of the Appalachians, from the Northeast to Florida, with possible problems to infrastructure developing.

Similarly, widespread rain spreads across the west of the country from the Pacific coastal regions, through the Great Basin and western Southwest, up along the Rocky Mountains, and into the northern half of the Great Plains.

Snow still likely North Great Plains to the Rockies.

Most of the western half of the Southwest, the Great South Plains, Texas and the western half of the Deep South to the central Inland Midwest may be mostly dry.

Further snow is likely in the southeast of Alaska.

03/16/2025

A large depression forms over the mid-section of the country, bringing strong and unsettled winds throughout central regions from the Rocky Mountains in the west to the Appalachians in the east.

Chance of an occluded front forming around the South Great Plains as this depression intensifies.

Most precipitation clears in the east from the Atlantic coasts to the eastern Inland Midwest, through the Appalachians and down to the Deep South and Florida.

Widespread precipitation elsewhere, with heavy falls, including snow, likely over the Great Plains to Texas and up into the western Great Lakes and western parts of the Inland Midwest to the Ozarks and Texas.

Temperatures remain very cold in the Great Lakes and Northeast states over the next few days.

Heavy snow in the southeast of Alaska continues.

MARCH **16** SUNDAY

ISOBARIC

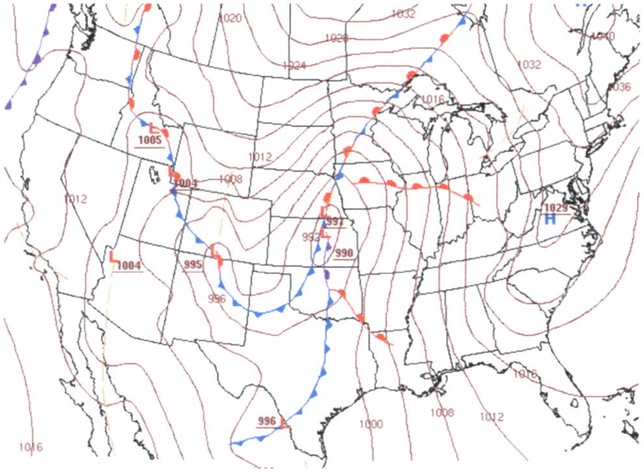

RAIN DISTRIBUTION (Inches)

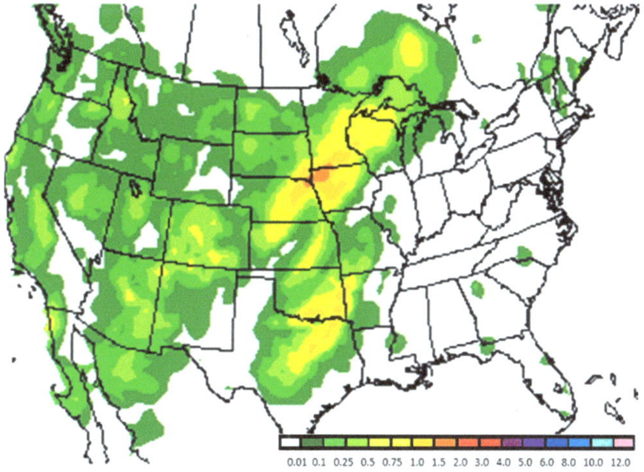

MAXIMUM TEMPERATURE

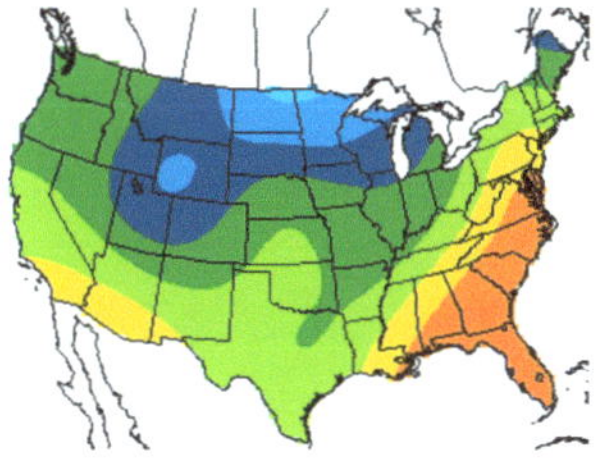

MINIMUM TEMPERATURE

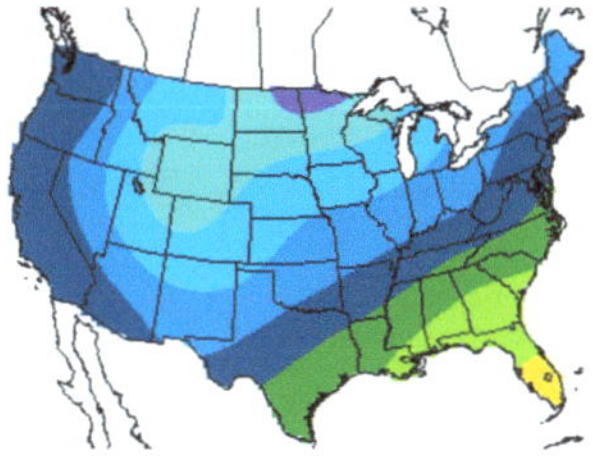

MARCH 17 MONDAY

ISOBARIC

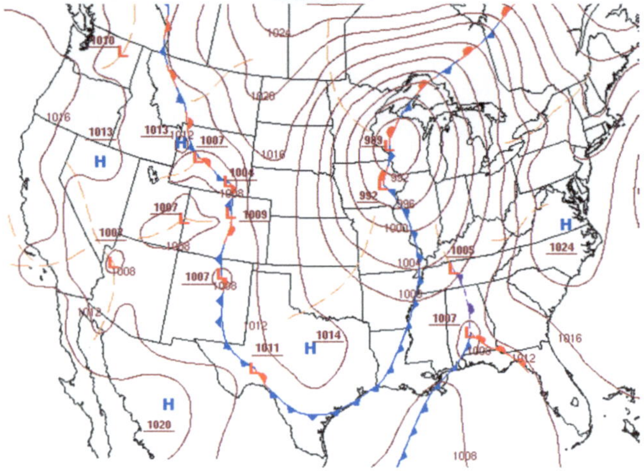

RAIN DISTRIBUTION (Inches)

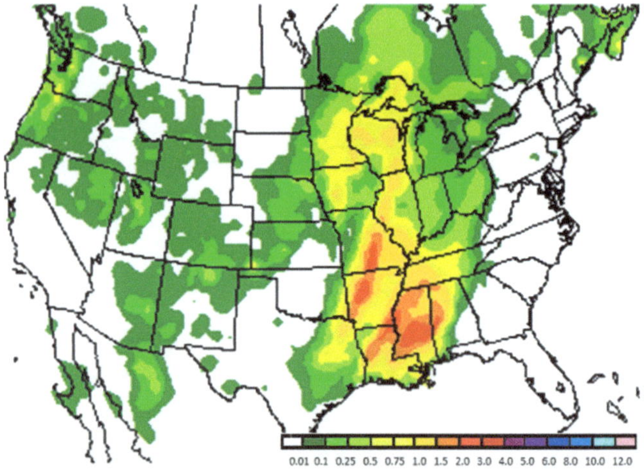

MAXIMUM TEMPERATURE

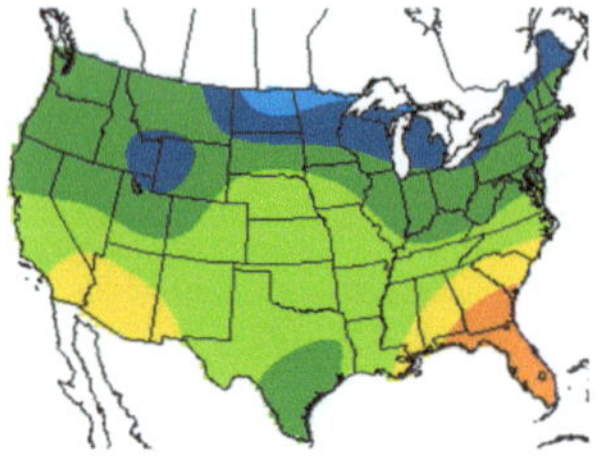

MINIMUM TEMPERATURE

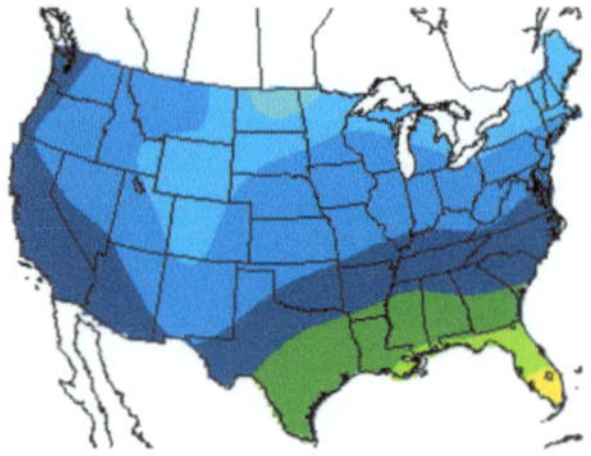

03/17/2025

An intense depression drifts across the country in an easterly trajectory, with an occluded front leaving the Great Plains and moving into the Deep South.

Most of the Atlantic coastal states from the Northeast to Florida, and California, may be mostly dry, while widespread precipitation prevails elsewhere.

Storms from the Appalachians to the Great Plains, with bursts from the Rockies to the Pacific Northwest, through the Great Basin and the Southwest.

Further snow is likely in the Midwest and Great Lakes, and heavy rain may cause flooding in the Deep South to the Ozarks.

Warmer temperatures around the Mississippi Basin may encourage snowmelts also adding to a flooding risk along many rivers systems in Ohio to Pennsylvania regions over the coming days.

03/18/2025

The large depression that has brought unsettled conditions throughout the eastern half of the country the past few days drifts further north.

Widespread precipitation continues to move eastward, clearing the east of the Great Plains, while spreading into the Atlantic coastal regions.

Heavy falls likely in the Deep South, and snow still possible around the Great Lakes to the Inland Midwest and Northeast regions.

Precipitation may continue in the Pacific Northwest, Great Basin and the western half of the Southwest, and up into the Rocky Mountains, but with rain activity becoming sparse in the west of the Great Plains.

Mainly dry elsewhere.

Chance of heavy snow in the southeast of Alaska over the next couple of days.

MARCH 18 TUESDAY

ISOBARIC

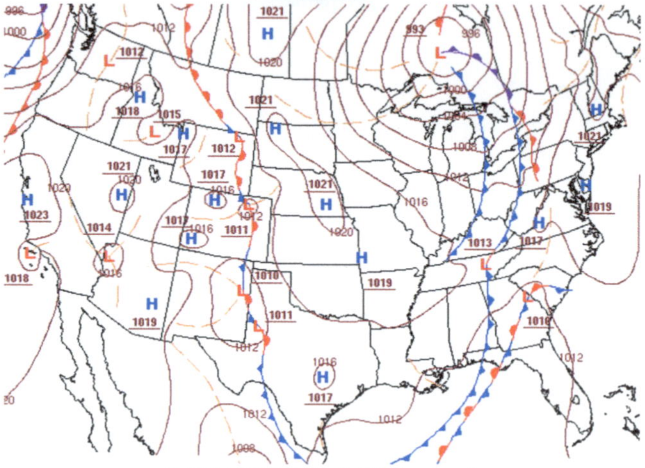

RAIN DISTRIBUTION (Inches)

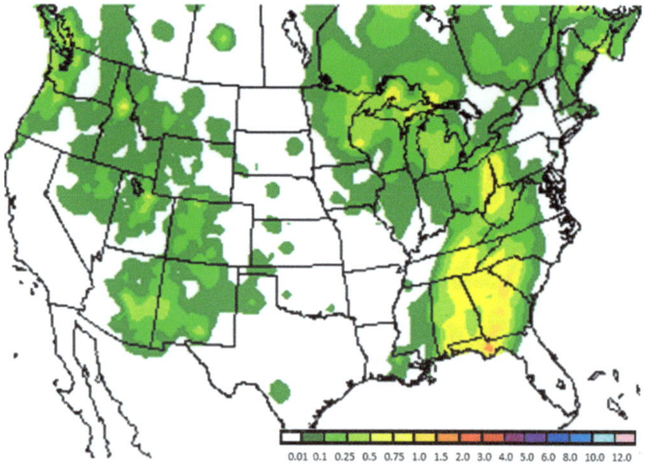

MAXIMUM TEMPERATURE

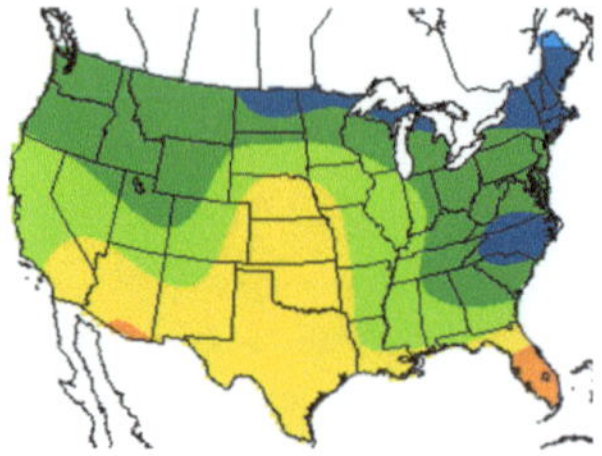

MINIMUM TEMPERATURE

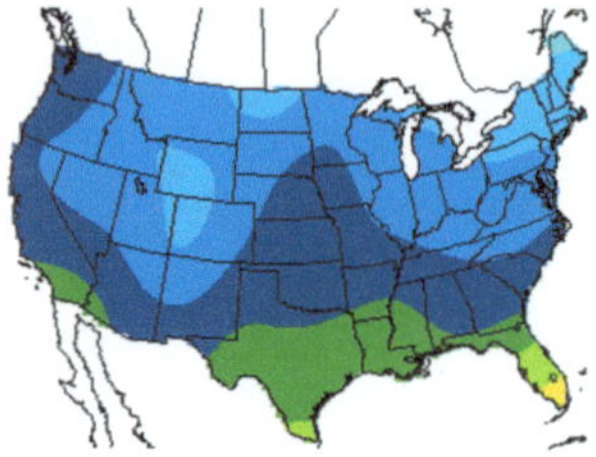

MARCH 19 WEDNESDAY

ISOBARIC

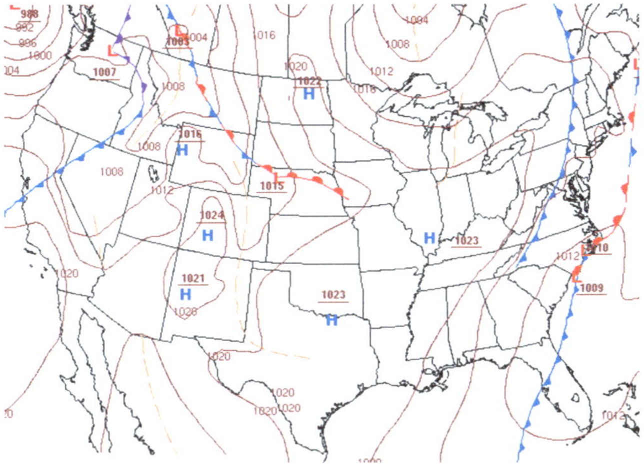

RAIN DISTRIBUTION (Inches)

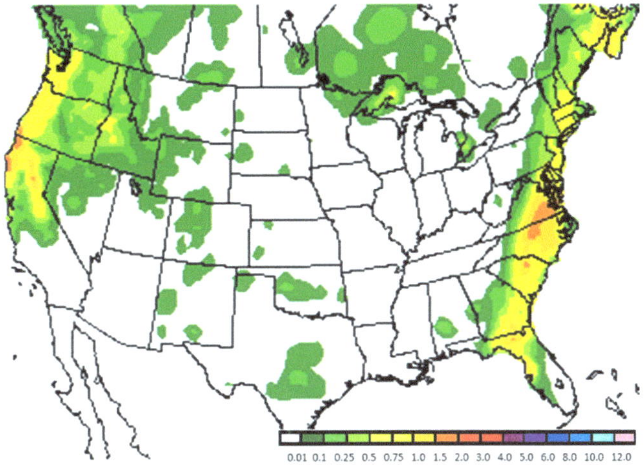

MAXIMUM TEMPERATURE

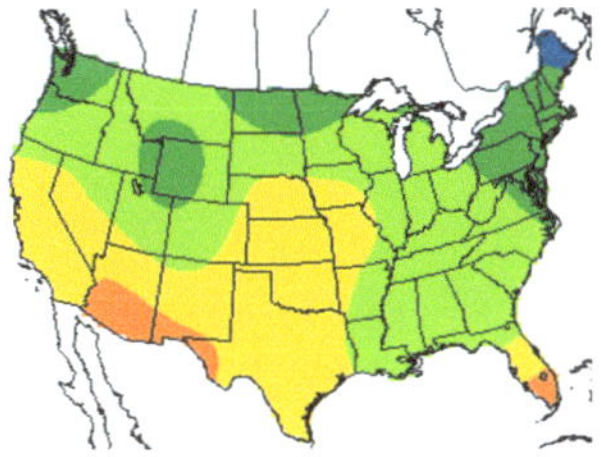

MINIMUM TEMPERATURE

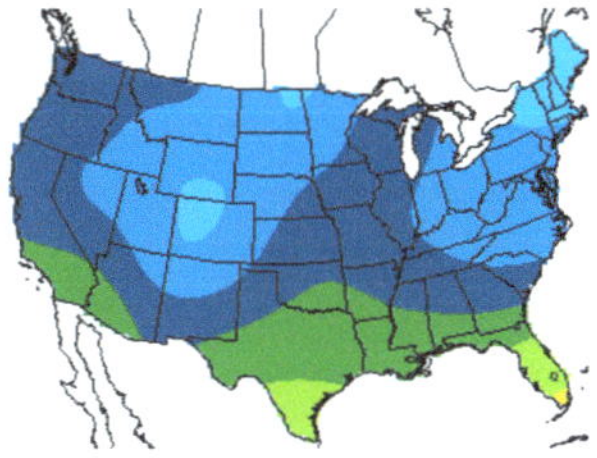

03/19/2025

Heavy precipitation prevails through the Atlantic seaboard states from the Northeast to Florida, with further snow in the Northeast, to heavy rain from the Coastal Southeast region to Florida.

Patches of wintry precipitation may also continue around the Great Lakes.

Rain activity, occasionally heavy, prevails across the Pacific Northwest to northern California including into the Sierra-Nevada and Cascade Ranges, and across the Great Basin, with showers extending into the Rocky Mountains.

Mostly dry elsewhere with just the odd lingering showers.

Widespread showers may also prevail across the western and central Hawaiian islands over the next couple of days.

03/20/2025

Outbreaks of heavy precipitation is still likely down through the Atlantic coastal fringes from the Northeast to Florida.

Similarly, widespread rain is likely to prevail across the northwest quarter of the country from the Pacific Northwest to the Rocky Mountains and south into northern California.

Mostly dry elsewhere.

The snowmelts around this time may be causing some flooding along river systems, with the Illinois river systems particularly at risk.

MARCH 20 THURSDAY

ISOBARIC

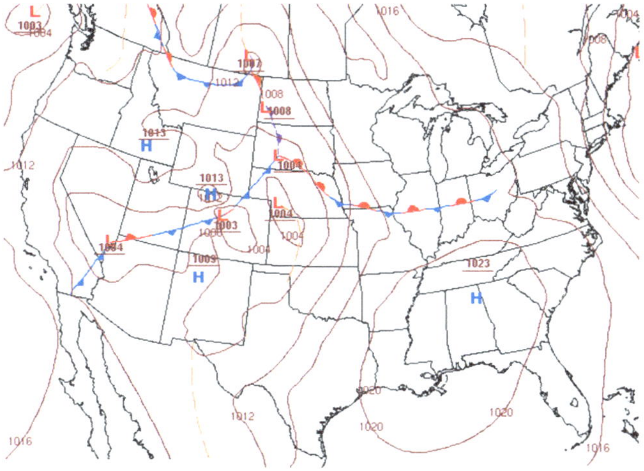

RAIN DISTRIBUTION (Inches)

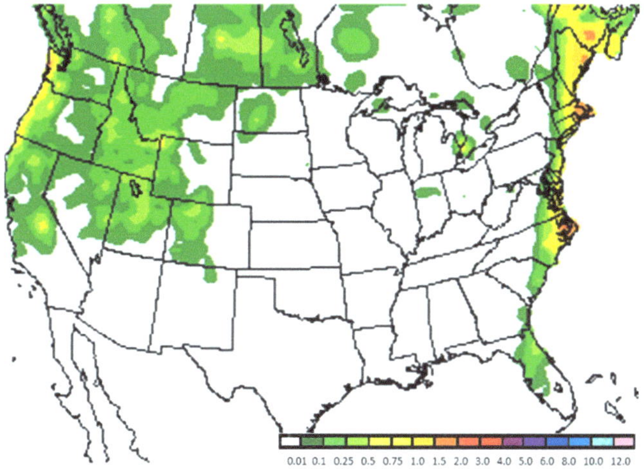

MAXIMUM TEMPERATURE

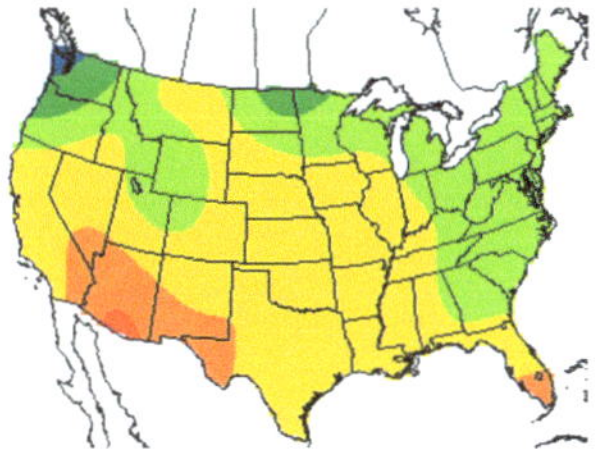

MINIMUM TEMPERATURE

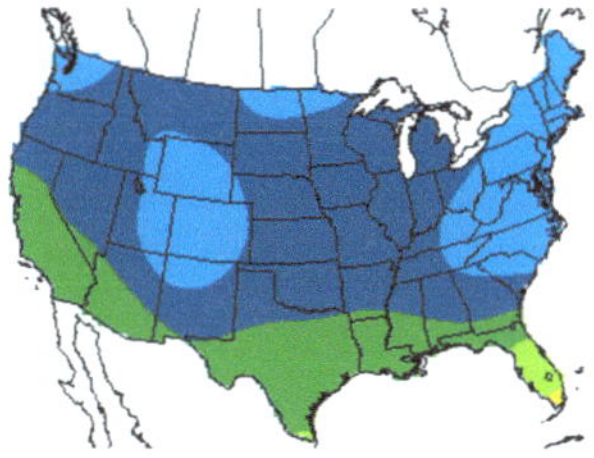

MARCH 21 FRIDAY

ISOBARIC

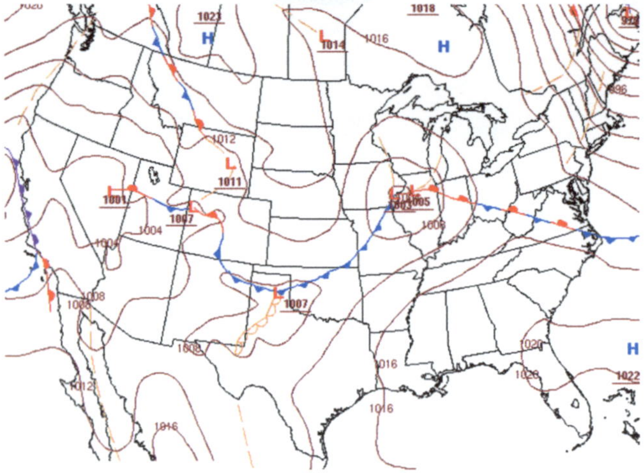

RAIN DISTRIBUTION (Inches)

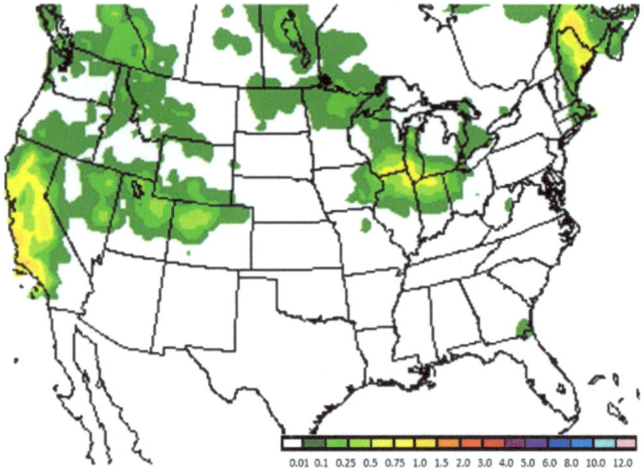

MAXIMUM TEMPERATURE

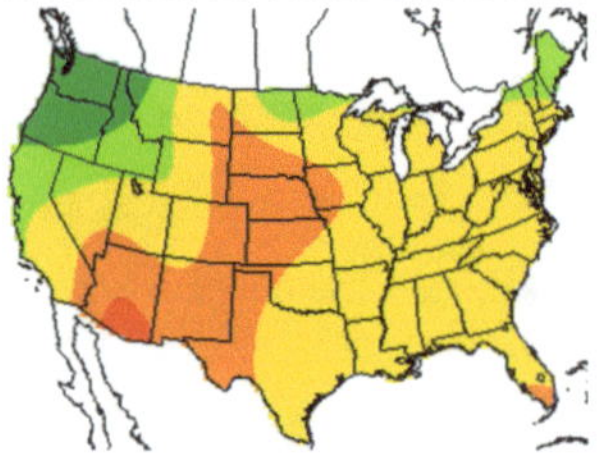

MINIMUM TEMPERATURE

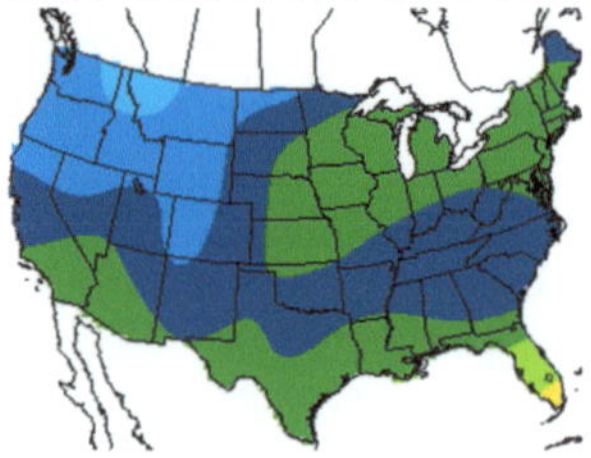

03/21/2025

Another intense depression builds over the Northeast.

Precipitation persist around the Great Lakes to the North Great Plains still, as well as in the far north of the Northeast, around the Pacific Northwest to the Great Basin and northern half of the Rocky Mountains.

Outbreaks of heavy rain may also move down through the Pacific Southwest.

Mostly dry elsewhere.

It may become very dry in southern California and the Southwest, with a heightened risk of fire prevailing.

There may be scattered thunderstorms at times in the Gulf of Mexico and into Texas over the next few days.

03/22/2025

Widespread precipitation spreads through the southwest of the country from the Pacific Southwest, up through the Great Basin over the Rocky Mountains and onto the North Great Plains.

Further precipitation may also prevail once more through the Great Lakes and into the Inland Midwest.

There may also a storm developing in the Gulf of Mexico and threatening Texas in the next few days.

Mostly dry elsewhere.

It may be unusually warm in the South Dakota to Nevada region before a cooler change arrives in a few days.

MARCH 22 SATURDAY

ISOBARIC

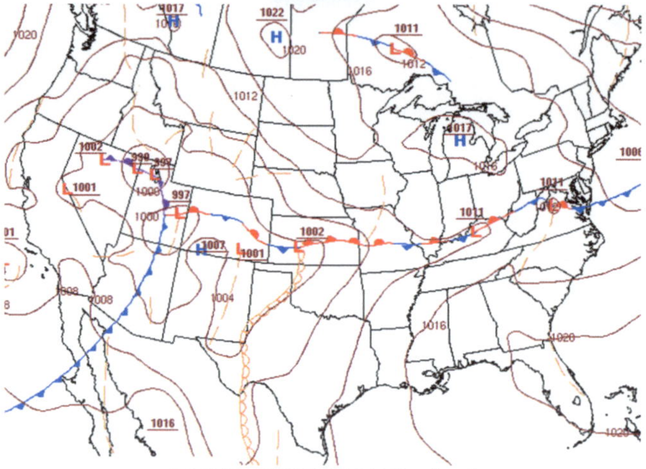

RAIN DISTRIBUTION (Inches)

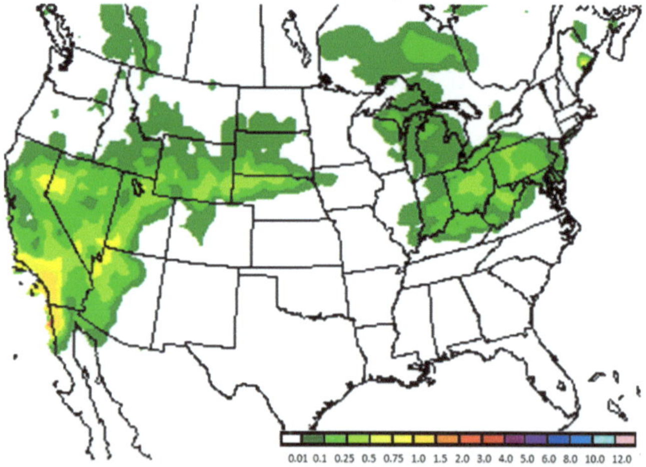

MAXIMUM TEMPERATURE

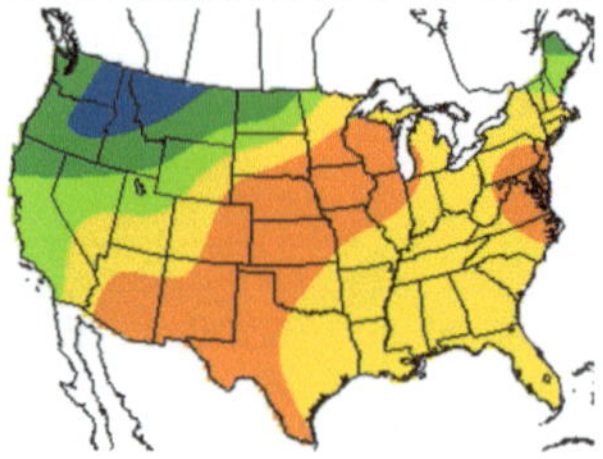

MINIMUM TEMPERATURE

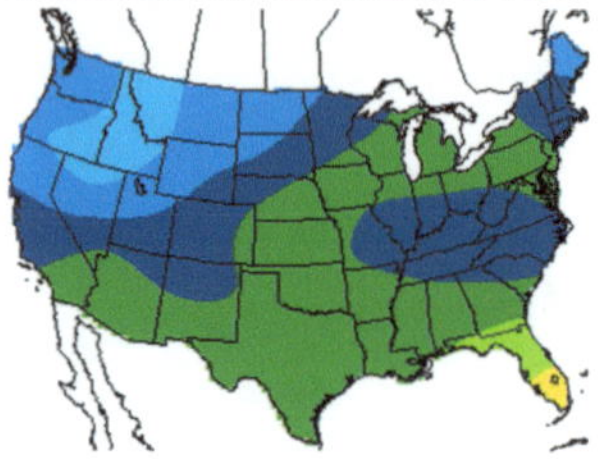

MARCH 23 SUNDAY

ISOBARIC

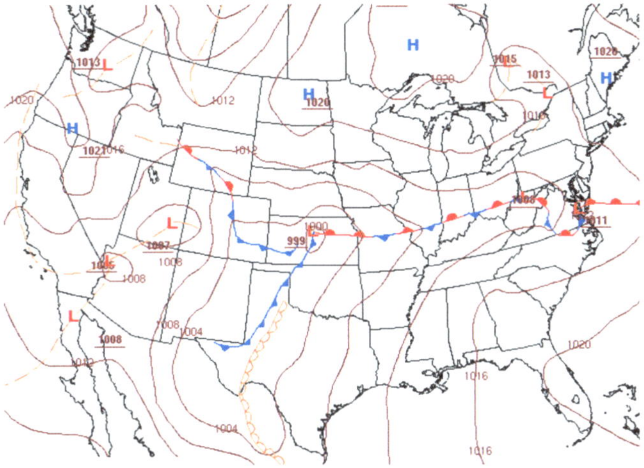

RAIN DISTRIBUTION (Inches)

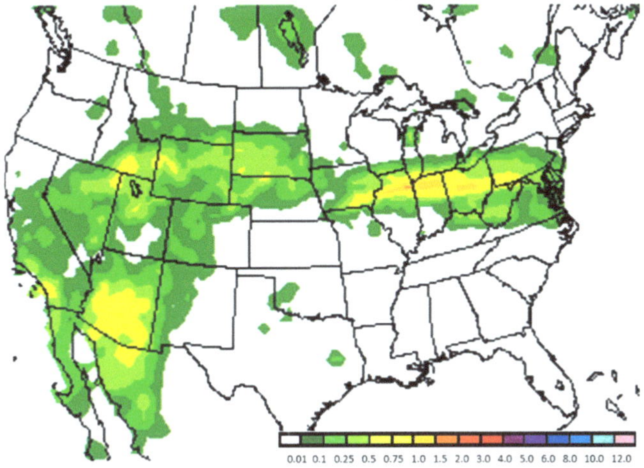

MAXIMUM TEMPERATURE   MINIMUM TEMPERATURE

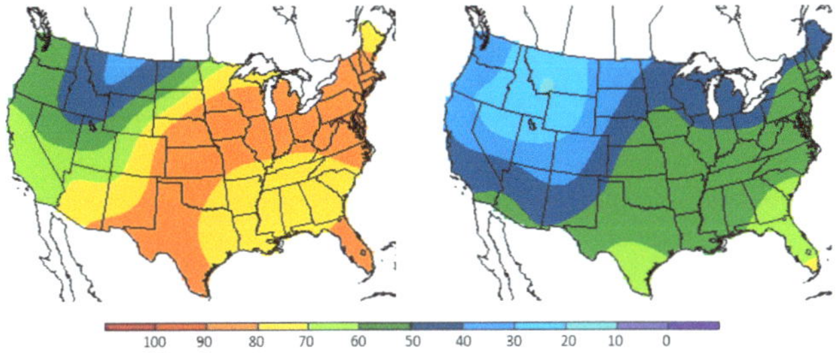

03/23/2025

Widespread precipitation prevails through the western half of the country from the Pacific Southwest and Great Basin, into the western half of the Southwest, and up through the Rocky Mountains.

Rain also spreads across the North Great Plains and in a line through the Inland Midwest to the adjacent Atlantic Coasts.

Mostly dry elsewhere, including in the Northeast, and Pacific Northwest.

A band of thunderstorms may move up from the Gulf of Mexico into the Southwest, with the chance of flash flooding developing.

Hawaii may experience unusually high temperatures in the state.

03/24/2025

A depression may continue to build over the eastern half of the country with cold fronts passing through the south of the country running from Texas and into the Deep South.

Most of the northwest quarter of the country, and the Canadian border states to the Great Lakes may be mostly dry, as well as in the southeast of the country from the Deep South and Coastal Southeast to Florida.

Widespread rain activity may prevail elsewhere.

A storm system over the Midwest may draw colder air into the south and east of the country, producing heavy rains, especially across Texas, through the Ozarks and Inland Midwest.

Heavy showers expected in Hawaii.

MARCH 24 MONDAY

ISOBARIC

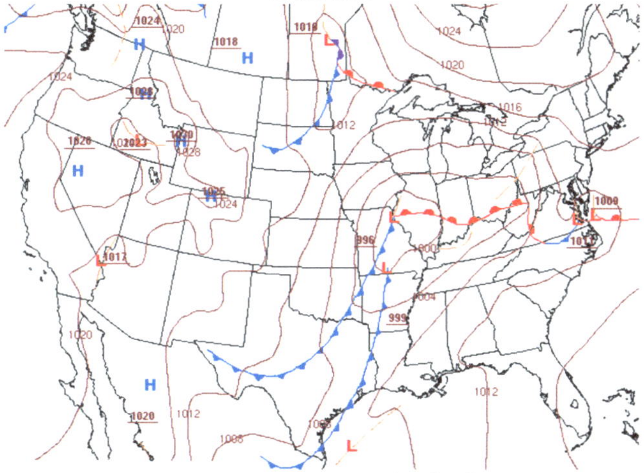

RAIN DISTRIBUTION (Inches)

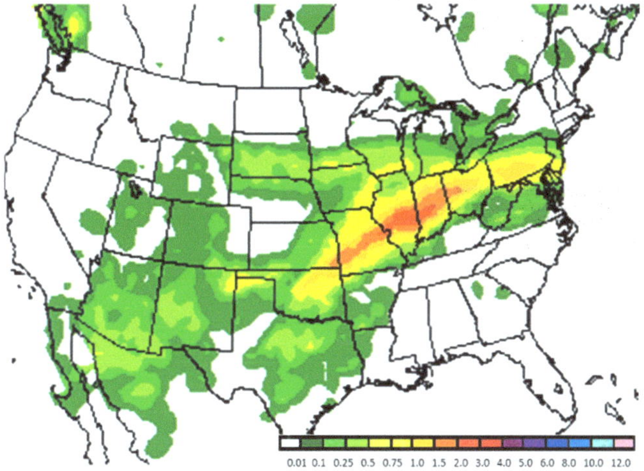

MAXIMUM TEMPERATURE   MINIMUM TEMPERATURE

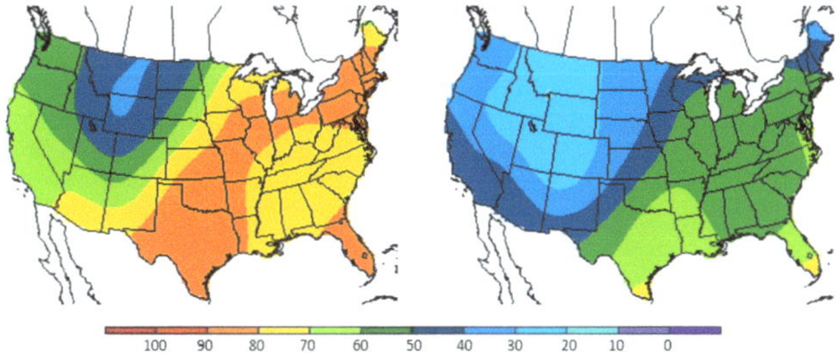

ISOBARIC

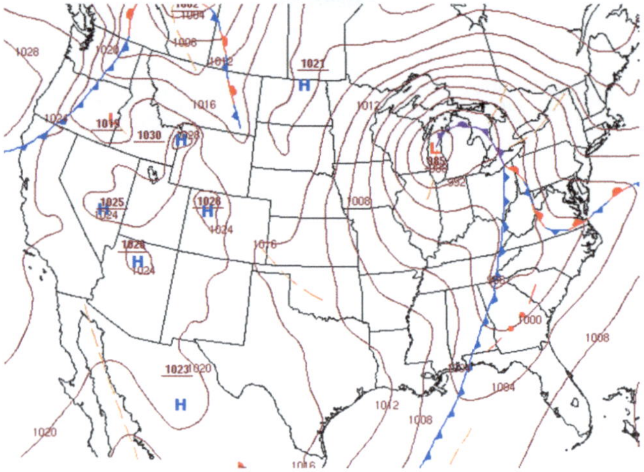

RAIN DISTRIBUTION (Inches)

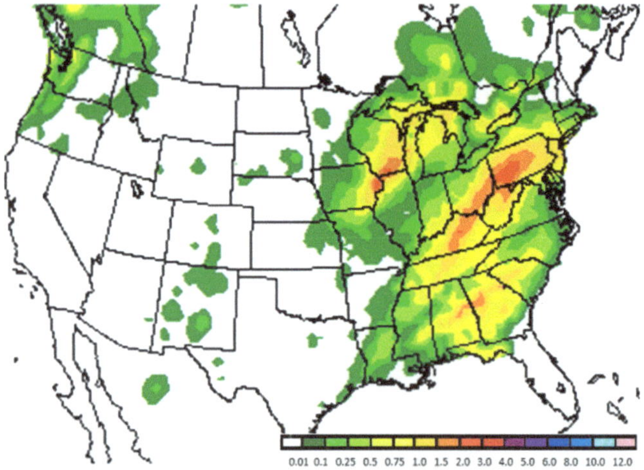

MAXIMUM TEMPERATURE

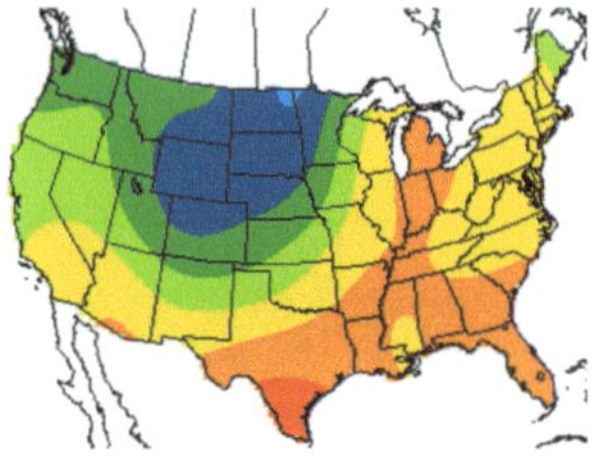

MINIMUM TEMPERATURE

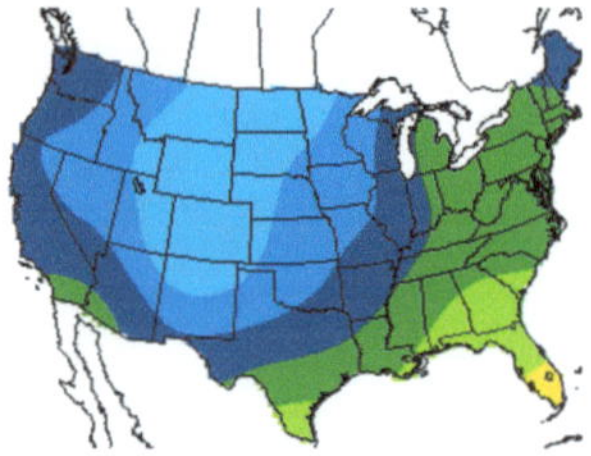

03/25/2025

A large depression continues to intensify in the northeast of the country, with an occluded front over the Great Lakes and a wet front moving down from the Great Lakes, through the Appalachians, across the Deep South and into the Gulf of Mexico.

This brings a cold change throughout the northeast quarter of the country, with further snow and wintry rain into the Northeast to the Great Lakes once more.

Widespread precipitation, often heavy, also prevails throughout the eastern half of the country from the Great Lakes and Midwest to the Deep South and Atlantic coastal states.

Similarly, some showers are likely to move across the Pacific Northwest and some spotty outbreaks of rain likely in the Southwest.

Mostly dry elsewhere, including in Florida.

03/26/2025

An intense depression may remain almost stalled over the northeast quarter of the country, with very cold temperatures over the north-central US, including in the Dakotas.

It may remain wet from the Great Lakes and Northeast, through the Midwest and down into Florida, with heavy precipitation likely in the Northeast, particularly around the Megalopolis districts, including the chance of further snow.

Heavy rain likely for a time in Florida.

Precipitation also prevails in the Pacific Northwest to the northern Rocky Mountains, and briefly brush into the northern parts of California.

Mostly dry elsewhere.

It may be unusually warm in the Southwest, including Arizona over the next couple of days.

ISOBARIC

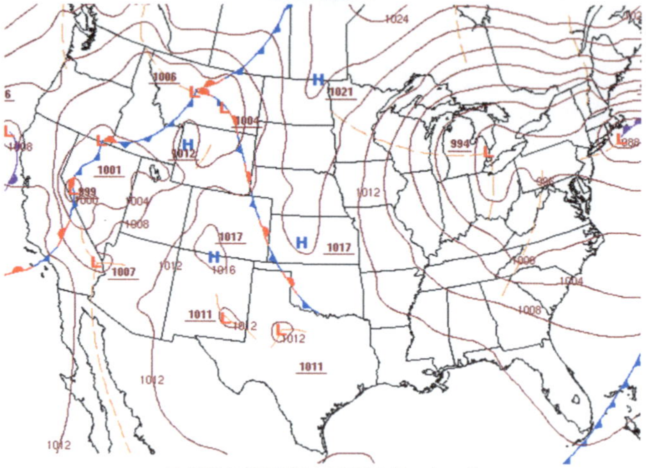

RAIN DISTRIBUTION (Inches)

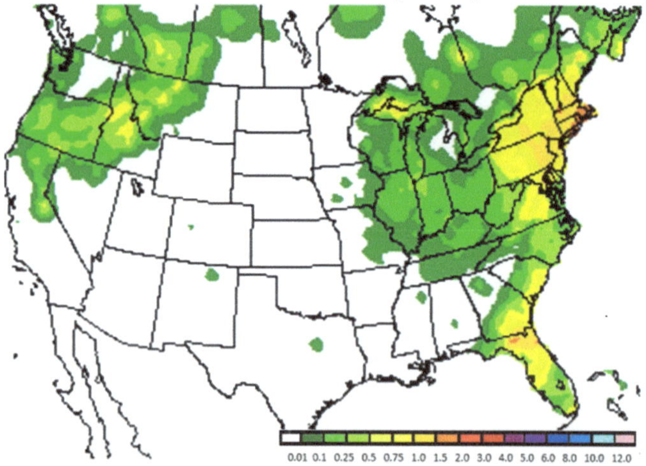

MAXIMUM TEMPERATURE

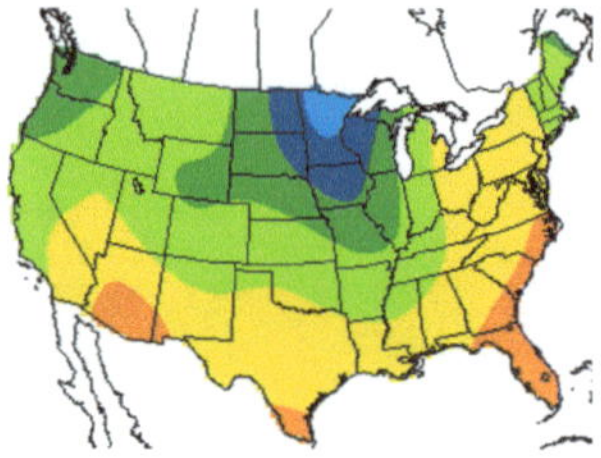

MINIMUM TEMPERATURE

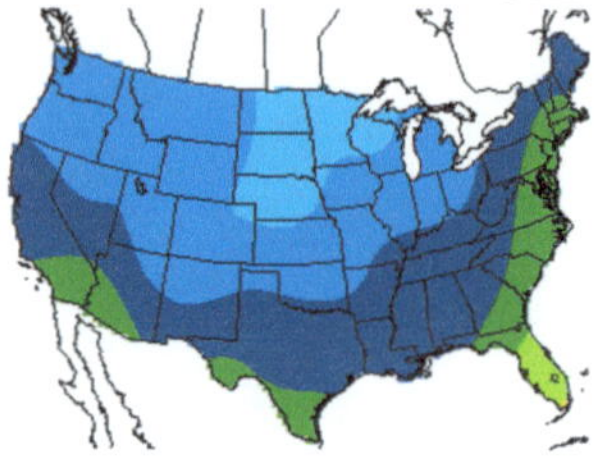

MARCH 27 THURSDAY

ISOBARIC

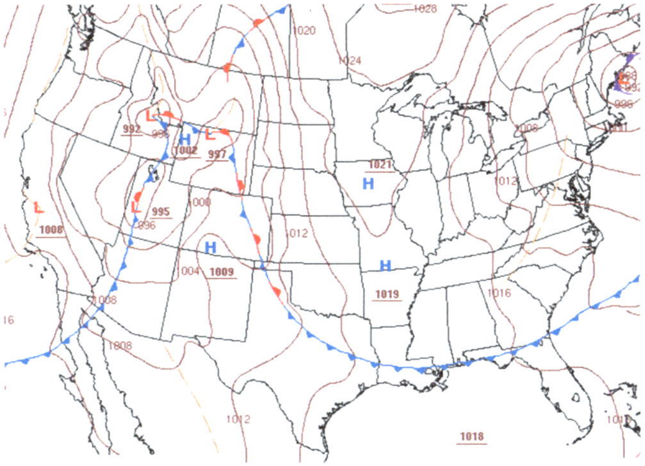

RAIN DISTRIBUTION (Inches)

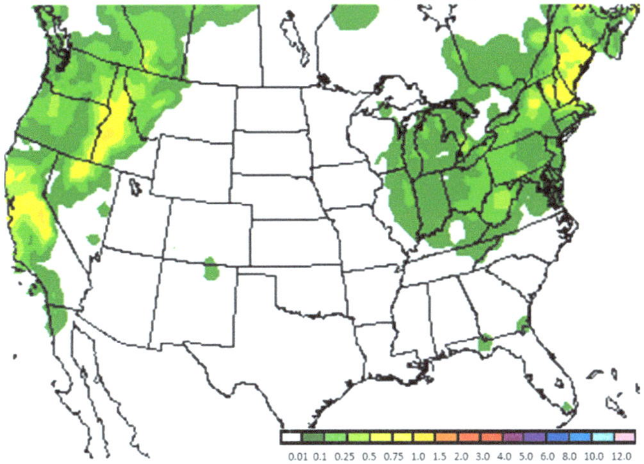

MAXIMUM TEMPERATURE MINIMUM TEMPERATURE

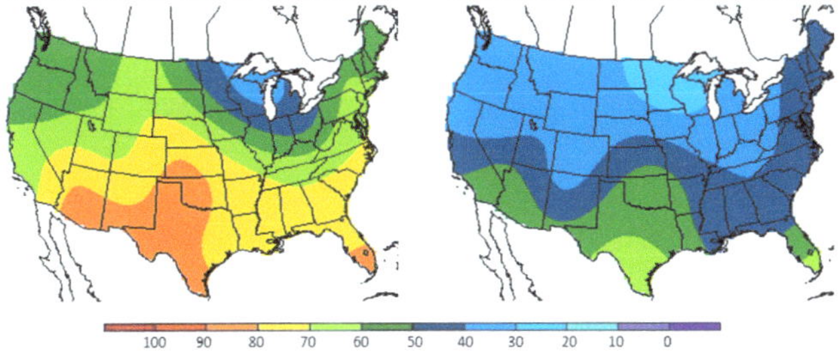

03/27/2025

The intense depression that had developed over the Northeast of the country moves further east into the Atlantic Ocean, with storm activity continuing in the Northeast to the Great Lakes and Inland Midwest to the north of the Appalachians, with a mixture of rain, sleet, and snow likely, particularly in the Northeast.

Rain activity builds over the northwest of the country from the Pacific and Inland Northwest regions and spread down through much of California.

Mostly dry elsewhere.

Temperatures likely to persist above normal in Nevada to the Arizona.

03/28/2025

Rain activity spreads across the west from the Pacific coasts to the foothills of the Rocky Mountains, including through the Southwest and Great Basin.

Precipitation eases in the Northeast, with only some wet spells remaining through the east of the Midwest and northern Appalachians and up into the Northeast.

Mostly dry elsewhere with strong and unsettled winds through the Great Plains to the Rocky Mountains.

Very warm temperatures may prevail in southern California, despite the rain.

MARCH 28 FRIDAY

ISOBARIC

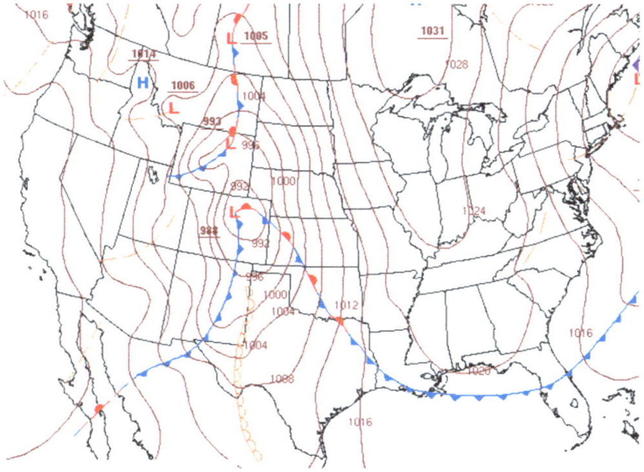

RAIN DISTRIBUTION (Inches)

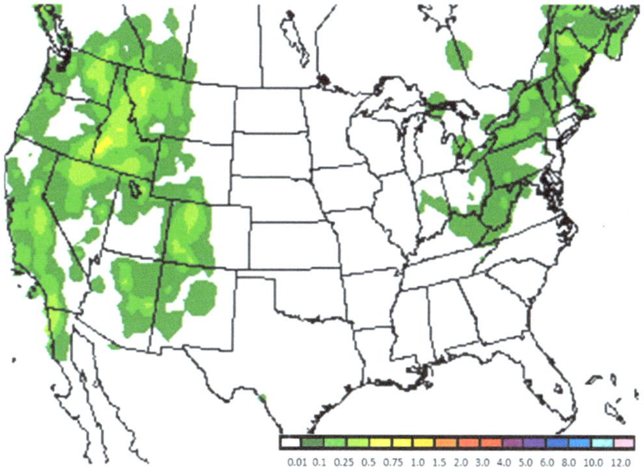

MAXIMUM TEMPERATURE MINIMUM TEMPERATURE

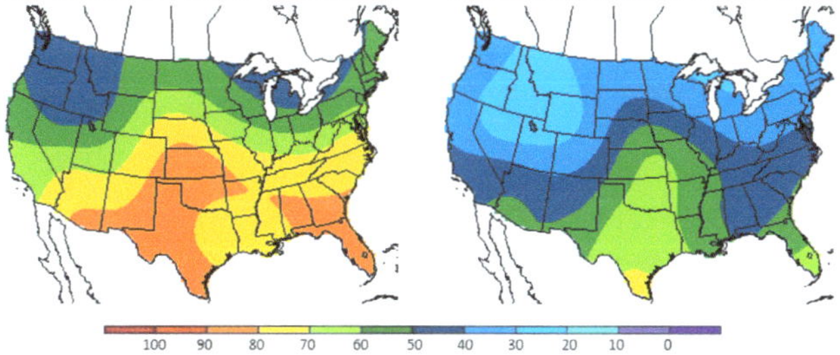

MARCH 29 SATURDAY

ISOBARIC

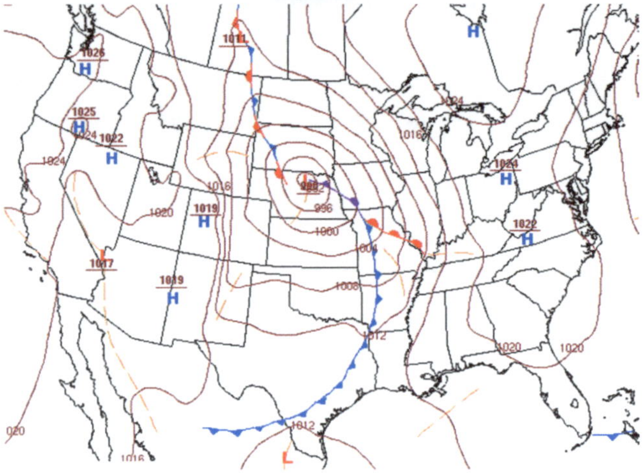

RAIN DISTRIBUTION (Inches)

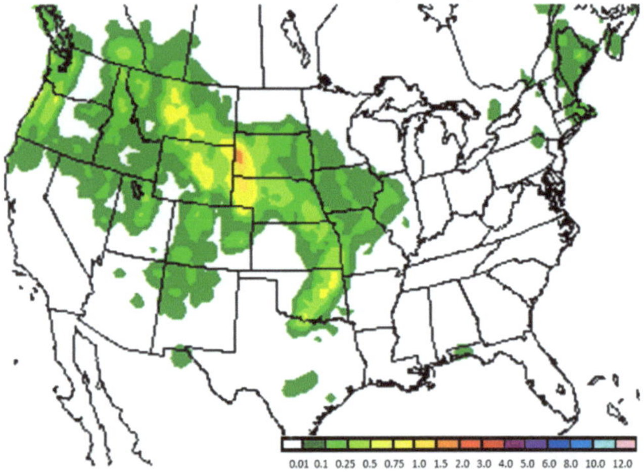

MAXIMUM TEMPERATURE / MINIMUM TEMPERATURE

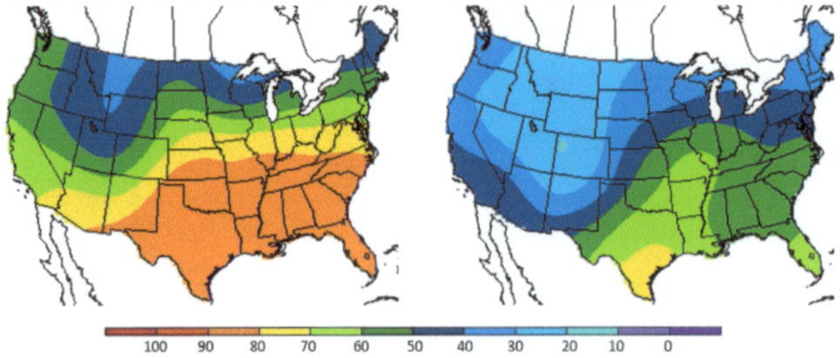

03/29/2025

The depression that formed over the Rocky Mountains develops an occluded front as it moves eastward across the Great Plains.

Rain activity clears out of the Pacific Southwest, while outbreaks of rain persist in the Pacific Northwest, and the odd shower still may linger in the Southwest.

Rain spreads across the Rocky Mountains and throughout the Great Plains extending later into the west of the Inland Midwest.

Pockets of rain linger in the far north of the Northeast.

Mostly dry elsewhere.

03/30/2025

Widespread shower activity continues across the Pacific Northwest with heavier rain prevailing through the Rocky Mountains and the North Great Plains regions, and spreading into the Inland Midwest and around the Great Lakes.

A band of rain also extends along the western edge of the Appalachians to the west of the Deep South.

Thunderstorms and outbreaks of heavy rain are possible over the Central Plains and into the Northeast, including in Kansas, possibly.

Mostly dry elsewhere.

MARCH 30 SUNDAY

ISOBARIC

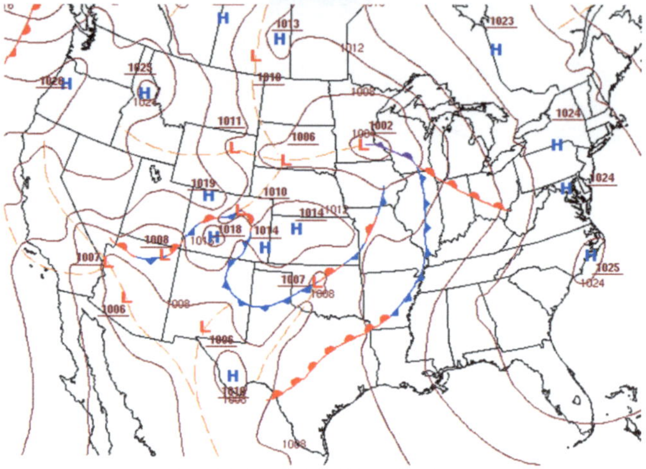

RAIN DISTRIBUTION (Inches)

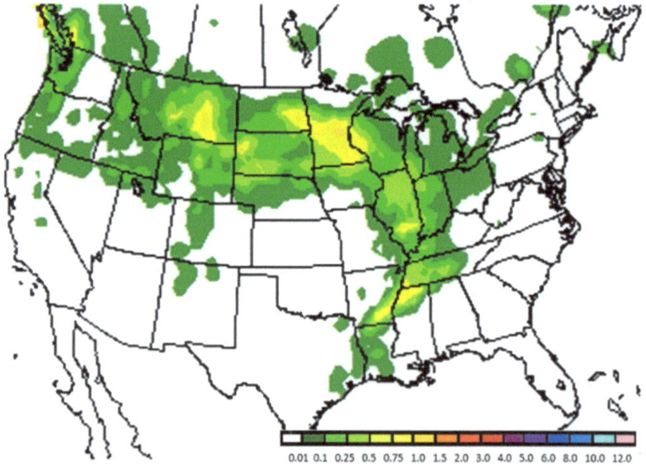

MAXIMUM TEMPERATURE / MINIMUM TEMPERATURE

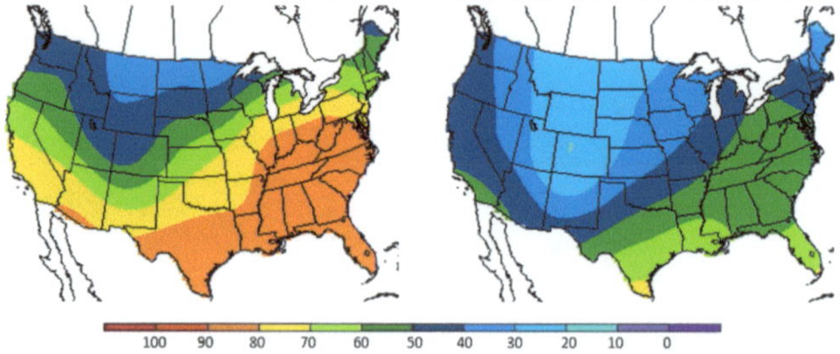

MARCH 31 MONDAY

ISOBARIC

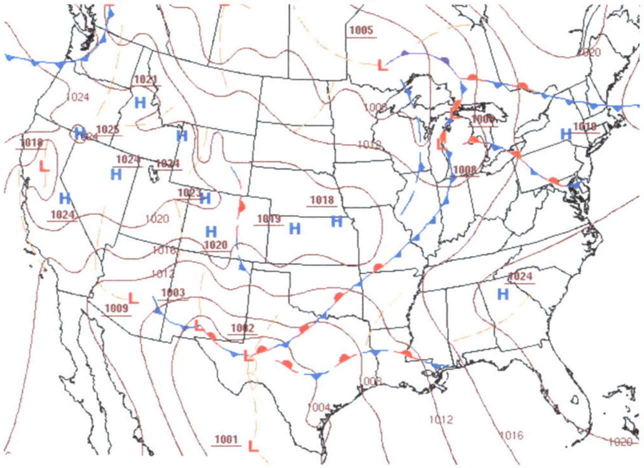

RAIN DISTRIBUTION (Inches)

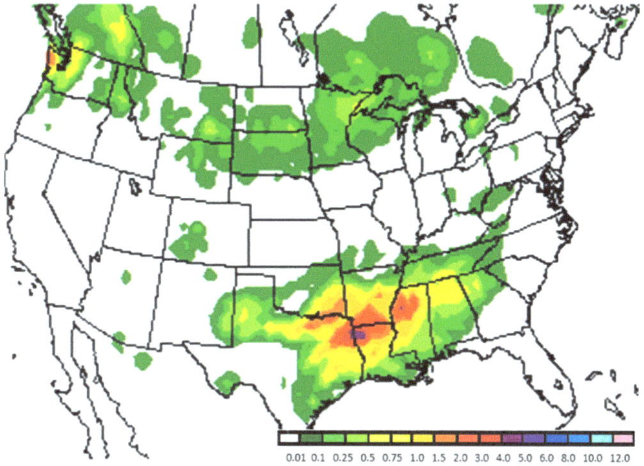

MAXIMUM TEMPERATURE

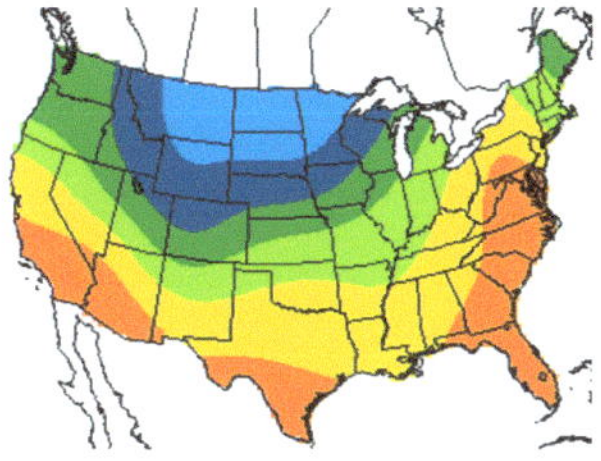

MINIMUM TEMPERATURE

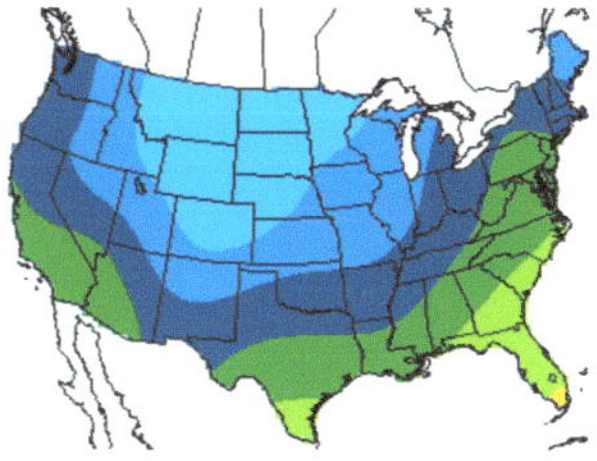

03/31/2025

Showers continue across the north of the country from the Pacific Northwest, over the North Great Plains and east to the Great Lakes, with the chance of heavy precipitation still likely in the Illinois to Michigan states over the next couple of days.

Heavy rain moves into the Deep South to northern parts of Texas, while lighter rain prevails in the Mississippi Valley.

Mostly dry elsewhere.

APRIL

PHASES OF THE MOON

3rd	Northern Declination
5th	First Quarter
11th	Crossing Equator
13th	Full Moon
13th	Apogee
18th	Southern Declination
21st	Third Quarter
26th	Crossing Equator
27th	New Moon
27th	Perigee #3

1	2	3	4	5	6	7	8	9	10	11	12	13	14	15	16	17	18	19	20	21	22	23	24	25	26	27	28	29	30
Tue	Wed	Thur	Fri	Sat	Sun	Mon	Tue	Wed	Thur	Fri	Sat	Sun	Mon	Tue	Wed	Thur	Fri	Sat	Sun	Mon	Tue	Wed	Thur	Fri	Sat	Sun	Mon	Tue	Wed
		^		1Q						XhS		F A					V			3Q					XhN	N P3			

ISOBARIC

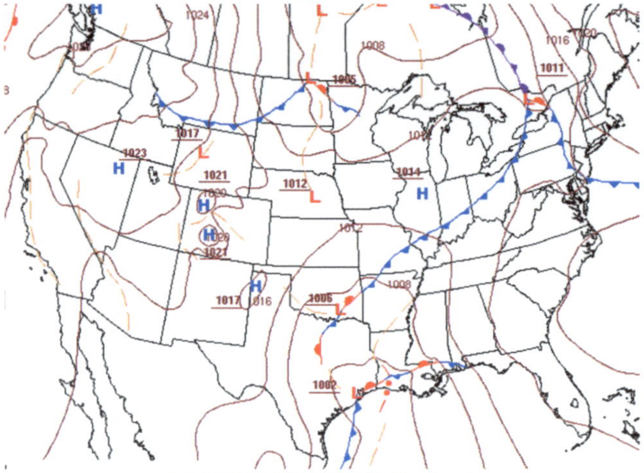

RAIN DISTRIBUTION (Inches)

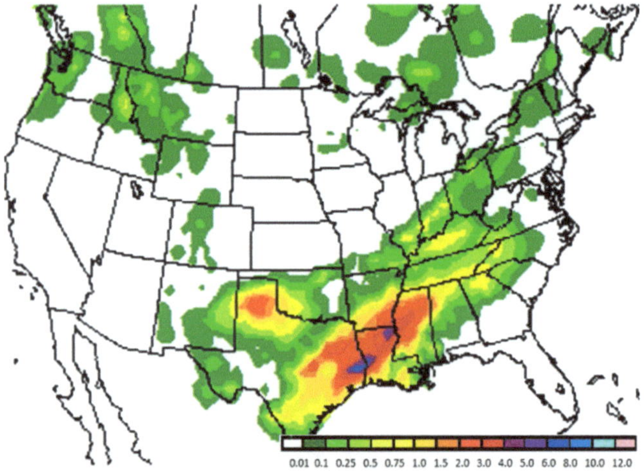

MAXIMUM TEMPERATURE / MINIMUM TEMPERATURE

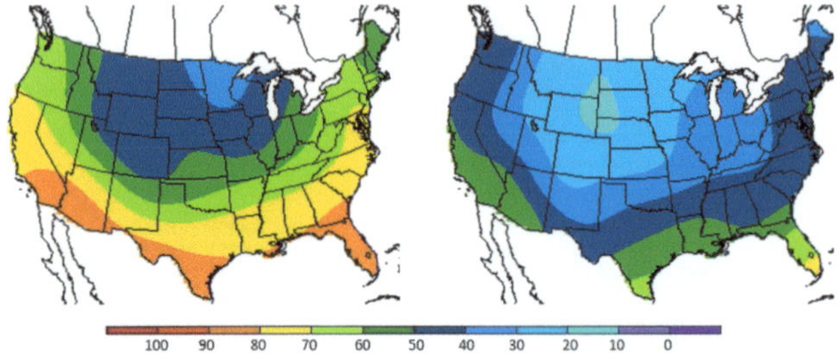

04/01/2025

Widespread rain activity expected through the south and east of the country over the next few days.

Significant shower activity likely across the eastern half of the Southwest, through the Gulf of Mexico regions across Texas and into the Deep South at first, spreading up through the Appalachians and into the Northeast later, with rain particularly heavy from Texas to the western half of the Deep South.

Some pockets of rain also likely to prevail in the Pacific Northwest to the Rocky Mountains, and isolated showers around the Great Lakes.

Mostly dry elsewhere.

04/02/2025

Widespread rain and storm activity almost everywhere over the east of the country, with heavy falls in particular around the Deep South.

Aggressive thunderstorm activity accompanied by strong winds may result in tornado activity developing over the Southwest, across Texas and into the Midwest and Northeast.

Mostly dry elsewhere except for some lingering showers across the North Great Plains.

ISOBARIC

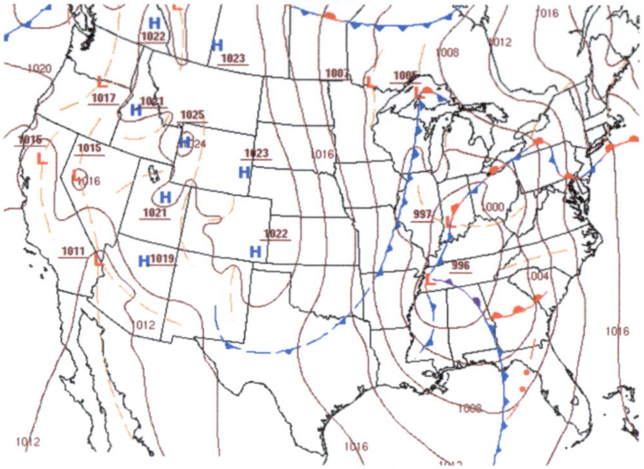

RAIN DISTRIBUTION (Inches)

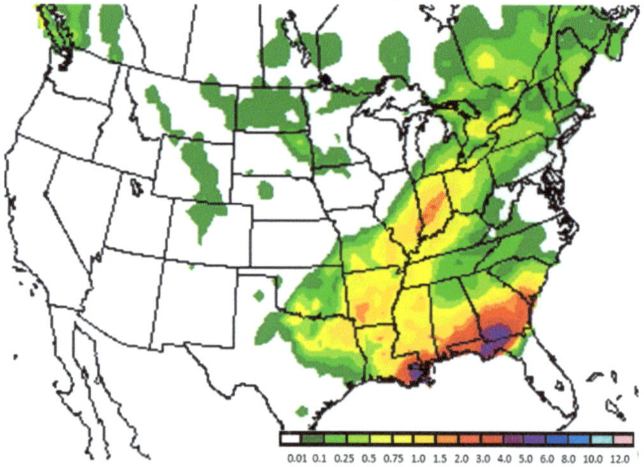

MAXIMUM TEMPERATURE / MINIMUM TEMPERATURE

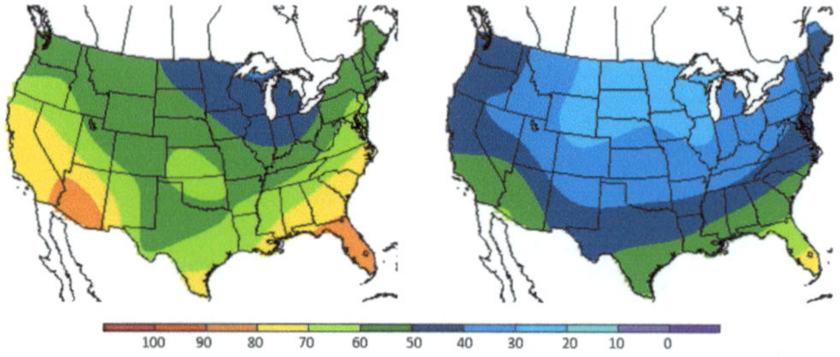

ISOBARIC

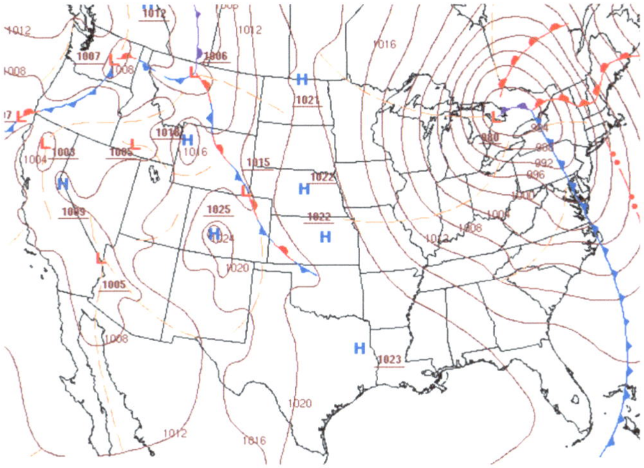

RAIN DISTRIBUTION (Inches)

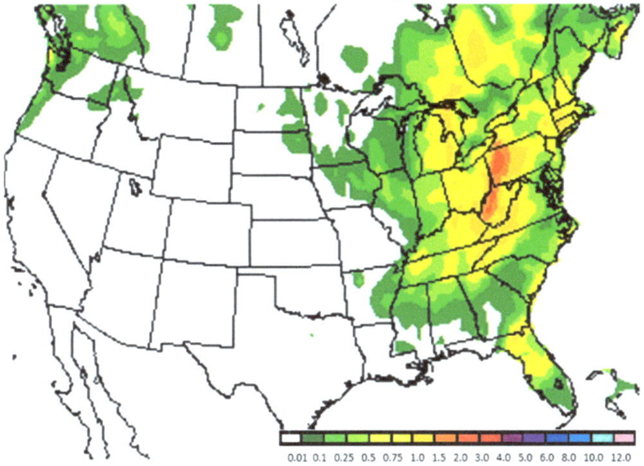

MAXIMUM TEMPERATURE

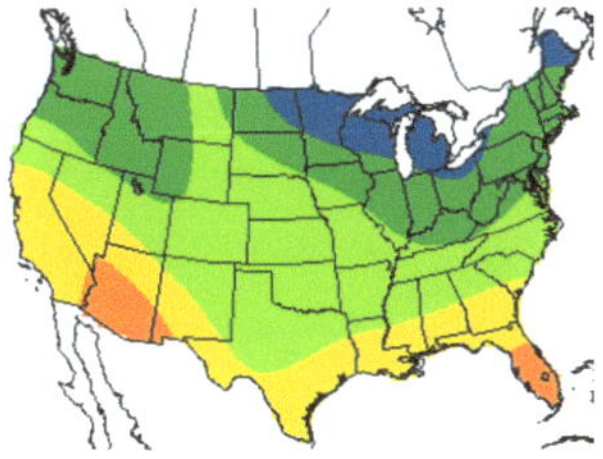

MINIMUM TEMPERATURE

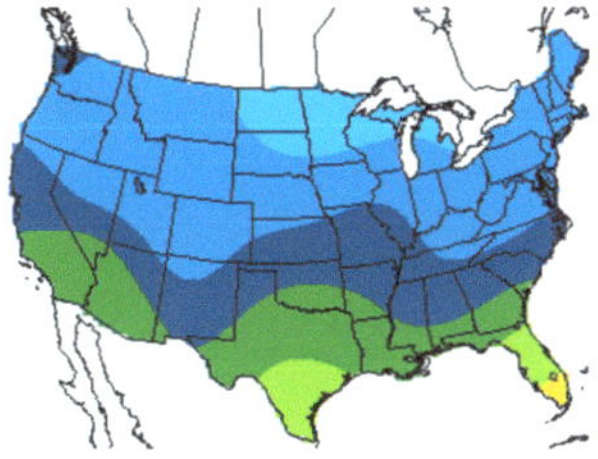

04/03/2025

An intense depression develops in the northeast of the country centered near the Great Lakes and stretching its influence into the Deep South and Florida to the south, and east from the Midwest to the Atlantic coast.

Widespread rain and storm activity continues and spread throughout the east of the country, becoming more severe over the Appalachians to the Northeast.

Rain activity eases to mostly dry in the west from the Great Plains to the Pacific coasts, except for some showers in the coastal fringes of the Pacific Northwest.

There is also still the chance of tornadoes forming around the Midwest to Colorado.

It may be unusually warm in California.

04/04/2025

Rain and storm activity eases over the east of the county, as the depression moves away to the north, with rain clearing to mostly dry in the Deep South, but still persisting over the Inland Midwest to Northeast and around the Great Lakes to the northern half of the Appalachians.

Showers prevail in the Pacific Northwest and down along the northern California coast over the next couple of days.

Mostly dry elsewhere.

There is likely to be a heightened tornado risk over the Great Plains to the Inland Midwest over the next several days as well as into Texas and west Nebraska regions.

It may be unusually warm Tennessee to Alabama over the next couple of days.

ISOBARIC

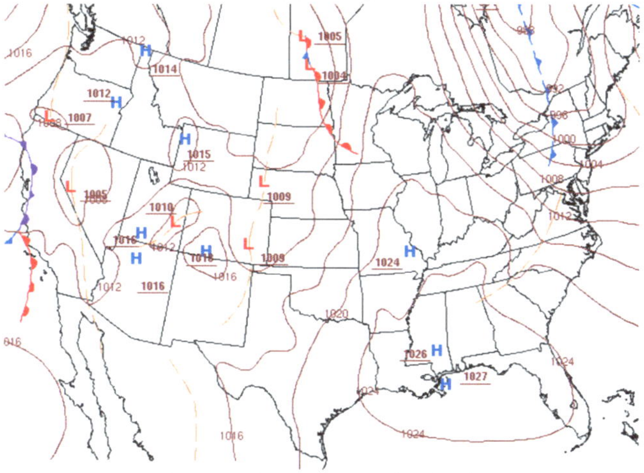

RAIN DISTRIBUTION (Inches)

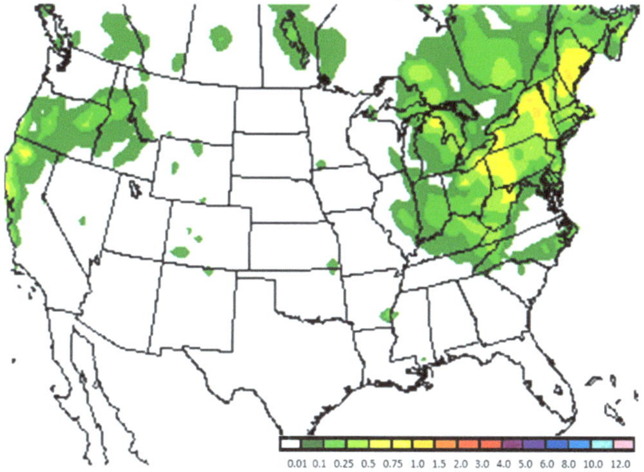

MAXIMUM TEMPERATURE

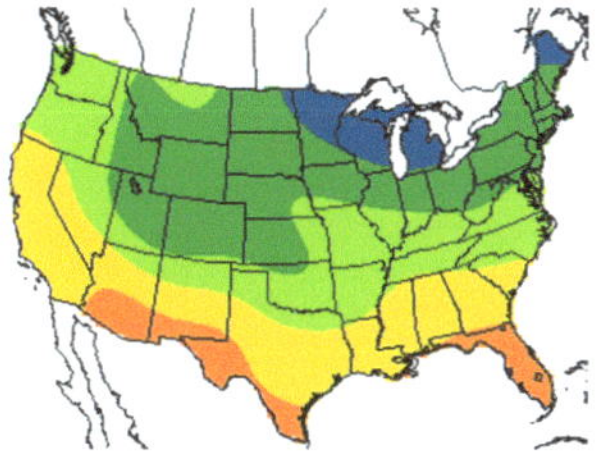

MINIMUM TEMPERATURE

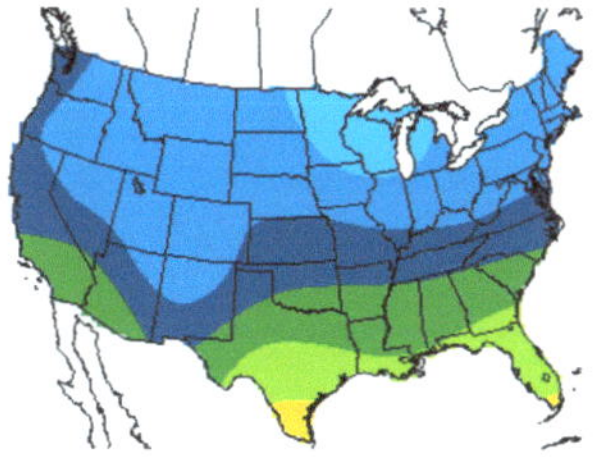

ISOBARIC

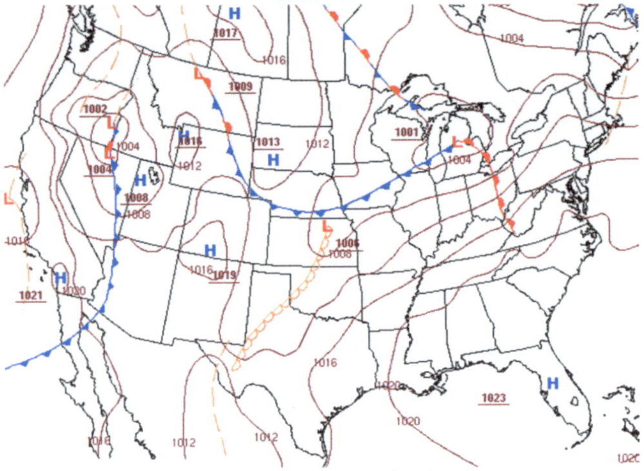

RAIN DISTRIBUTION (Inches)

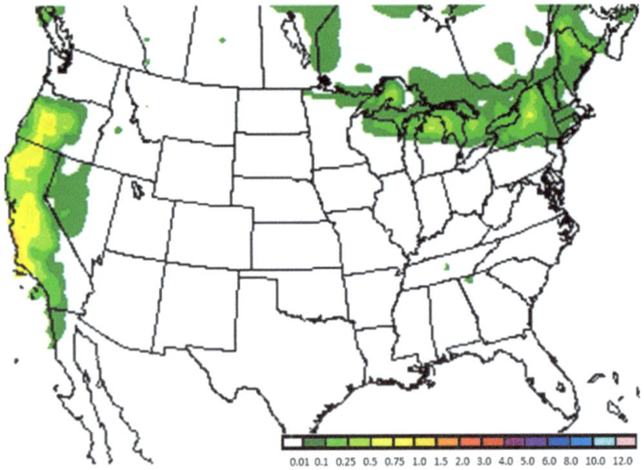

MAXIMUM TEMPERATURE

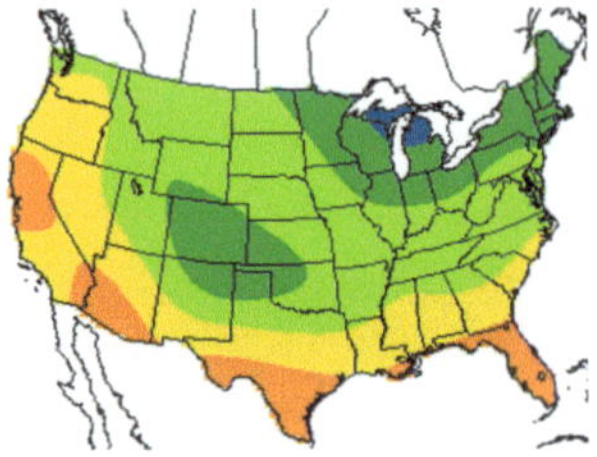

MINIMUM TEMPERATURE

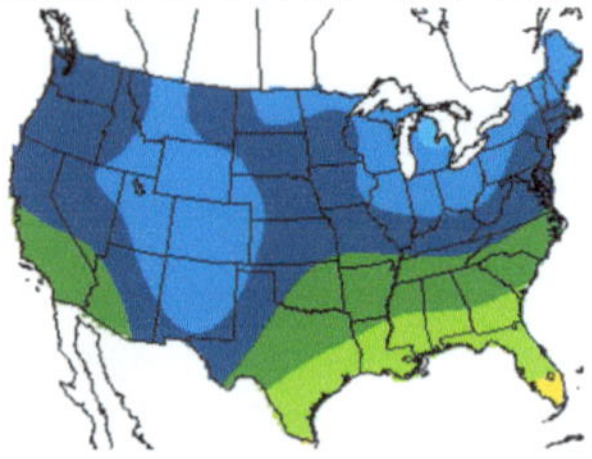

04/05/2025

Rain activity prevails in the west of the country along the coastal regions of the Pacific Northwest and California, including in the Sierra Nevada and Cascade Ranges.

Similarly, in the east of the country the rain continues to contract further north, persisting mostly around the Great Lakes to the Northeast.

Mostly dry elsewhere.

Hawaii may receive some light rain over the next couple of days.

It is likely to be very cold in Alaska.

04/06/2025

Scattered outbreaks of rain persist over the Pacific Northwest and California and spread for a time into the Great Basin and Rocky Mountains.

Wet conditions also continues to contract out of the east of the country, with only some rain activity lingering around the northern Appalachians to the Northeast region.

Mostly dry elsewhere.

There is a chance that a storm may develop over the Great North Plains during the next three to four days, bringing the chance of late heavy snowfalls to the regions.

ISOBARIC

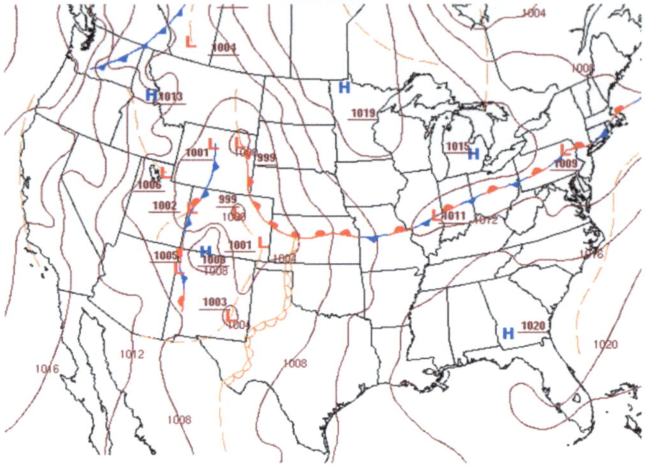

RAIN DISTRIBUTION (Inches)

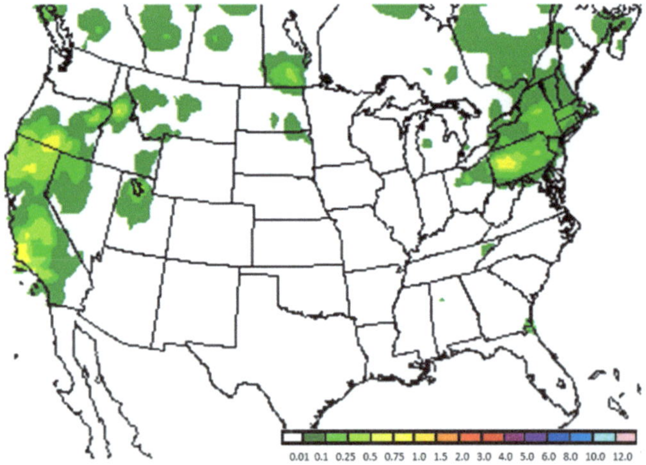

MAXIMUM TEMPERATURE

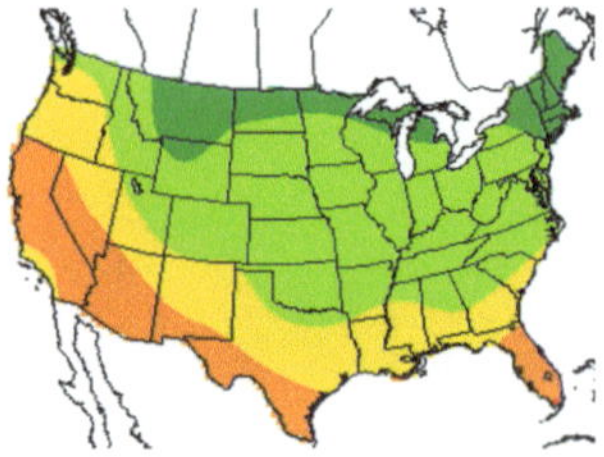

MINIMUM TEMPERATURE

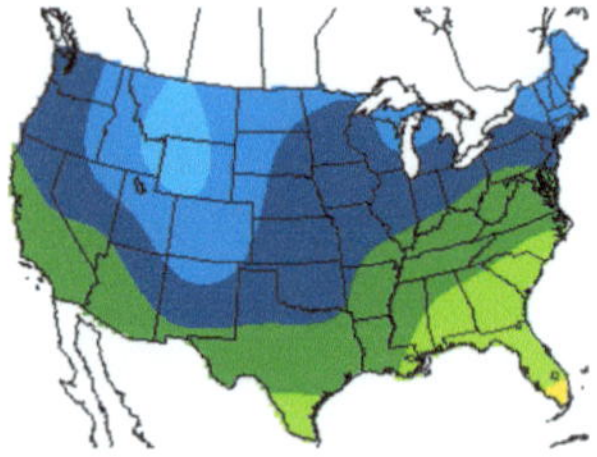

APRIL 7 MONDAY

ISOBARIC

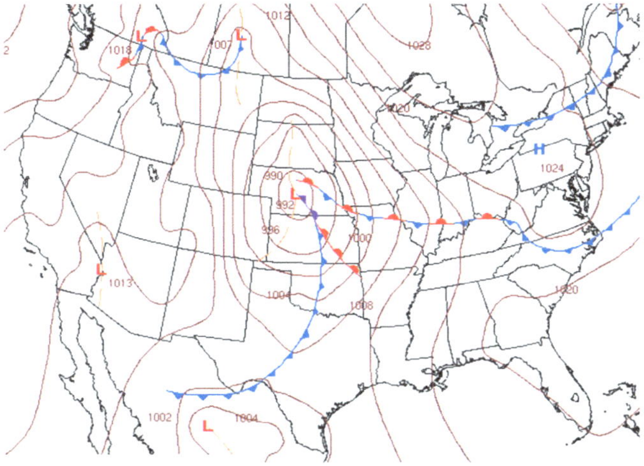

RAIN DISTRIBUTION (Inches)

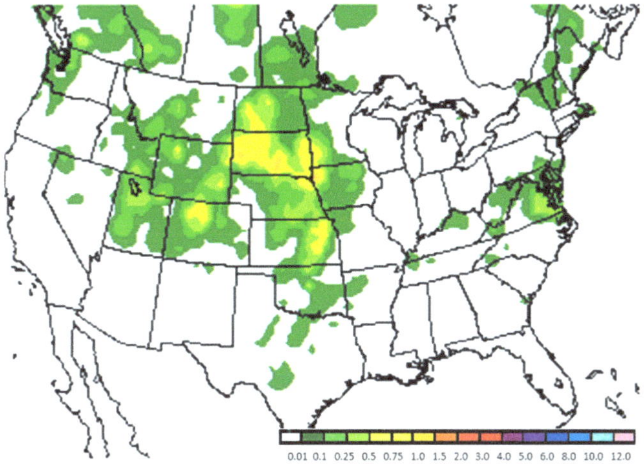

MAXIMUM TEMPERATURE

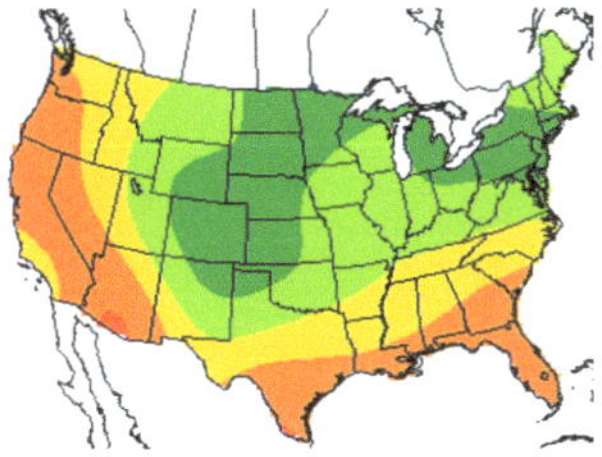

MINIMUM TEMPERATURE

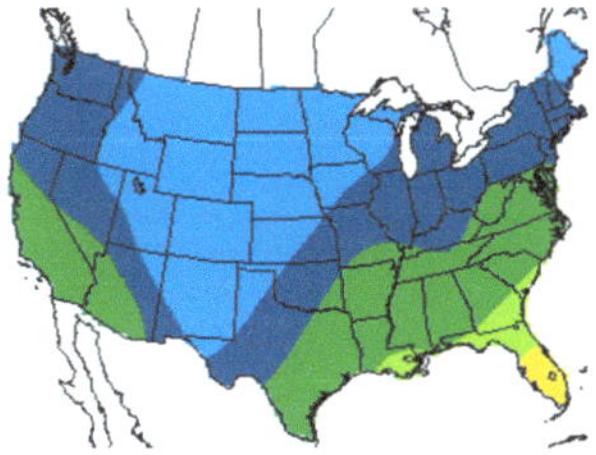

04/07/2025

A large, depression takes form over the Great Plains with an occluded front briefly developing over the central Great Plains.

Rain clears to isolated showers in the Pacific and Inland Northwest, while storm conditions develop across the Rocky Mountains and into the Great Plains.

These storms may bring a heightened risk of tornadoes over the Great Plains to north Texas, including in western parts of Nebraska and Colorado over the next couple of days.

Snow may return over the North Great Plains states, including in Wyoming and Montana.

Some showers linger in the Northeast to the Megalopolis regions.

Mostly dry elsewhere.

04/08/2025

The depression that built over the Great Plains in the past couple of days moves onto the Great Lakes bearing an occluded front into just south of the Great Lakes, and extending a cold front through the Ozarks into the north of Texas.

Widespread storm activity prevails from the Rocky Mountains in the west, to the Appalachians in the east, with heavy rain and thunderstorms likely especially around the Inland Midwest regions.

A heightened risk for tornado activity may still prevail, particularly around the Great Plains to the Inland Midwest.

Chance of further snow on the North Great Plains.

Mostly dry elsewhere.

APRIL 8 TUESDAY

ISOBARIC

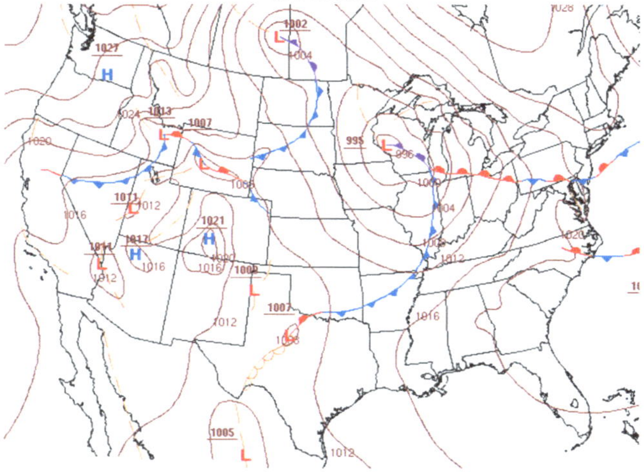

RAIN DISTRIBUTION (Inches)

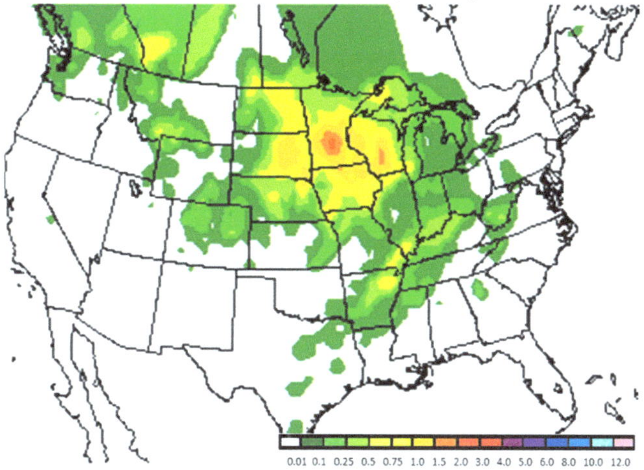

MAXIMUM TEMPERATURE

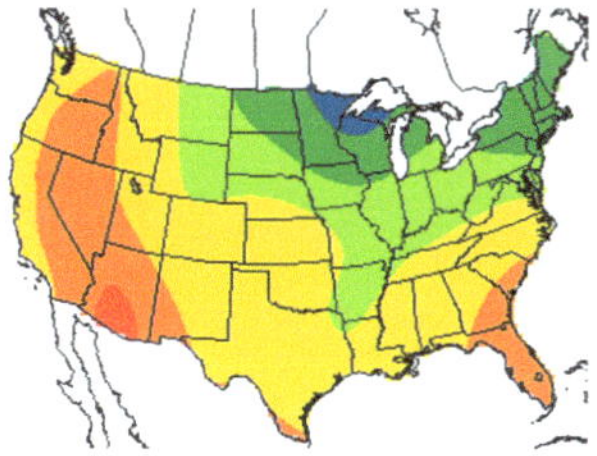

MINIMUM TEMPERATURE

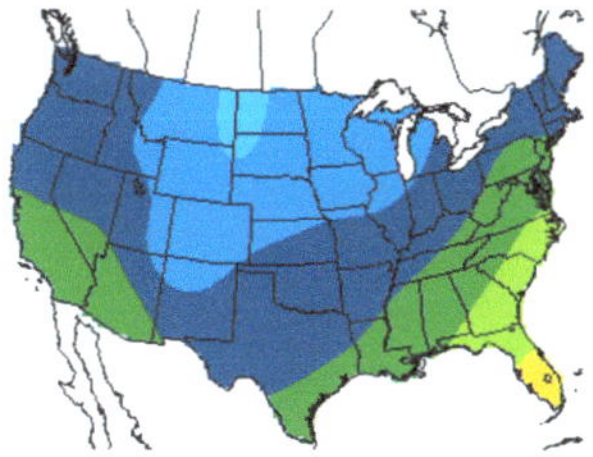

APRIL 9 WEDNESDAY

ISOBARIC

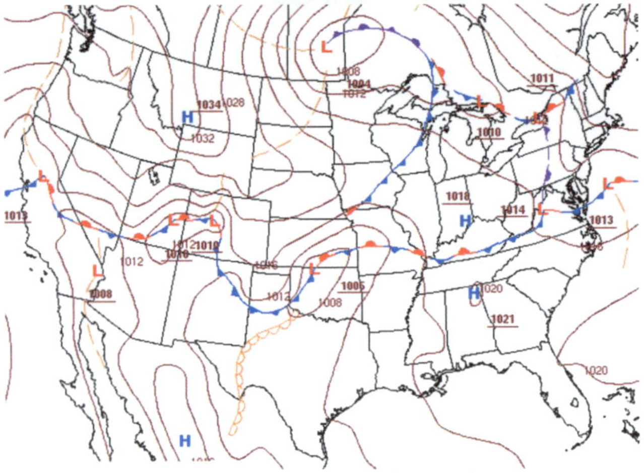

RAIN DISTRIBUTION (Inches)

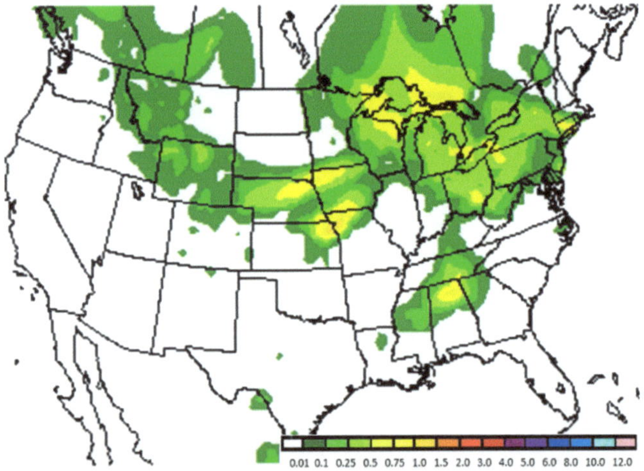

MAXIMUM TEMPERATURE MINIMUM TEMPERATURE

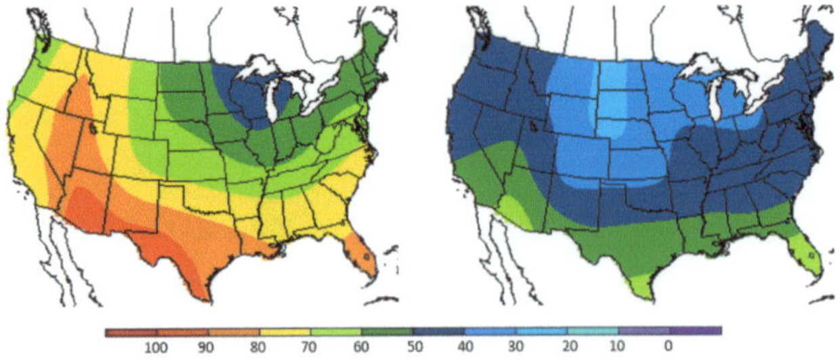

04/09/2025

Precipitation prevails from the Rocky Mountains in the west, down through the center of the Great Plains, across the Inland Midwest, over the Great Lakes and into the southern half of the Northeast regions, and rain may still prevail through the Appalachian.

Very heavy rain and possible thunderstorms may occur in Oklahoma over the next couple of days.

Mostly dry elsewhere.

Temperatures may be unusually warmer than normal in Alaska.

04/10/2025

Widespread precipitation continues across the east of the country with heavy rain around the Inland Midwest and possibly as far west as South Dakota.

Outbreaks of rain also likely in the southern half of the Northeast and as far south into the Deep South States, while remaining mostly dry throughout the Atlantic coastal states from the Megalopolis to Florida, and along the coastal fringes of the Deep South.

Some showers may linger in the Inland Northwest and along the Rocky Mountains into central Texas.

Mostly dry elsewhere.

APRIL 10 THURSDAY

ISOBARIC

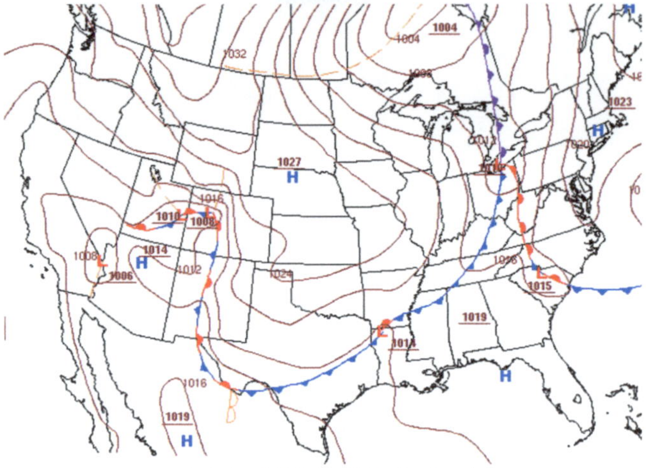

RAIN DISTRIBUTION (Inches)

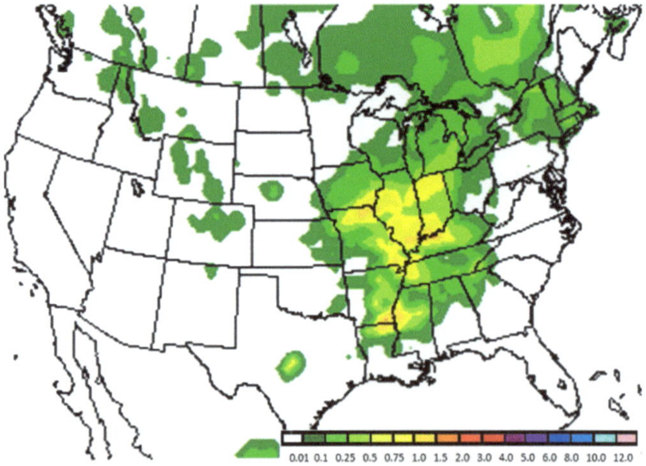

MAXIMUM TEMPERATURE MINIMUM TEMPERATURE

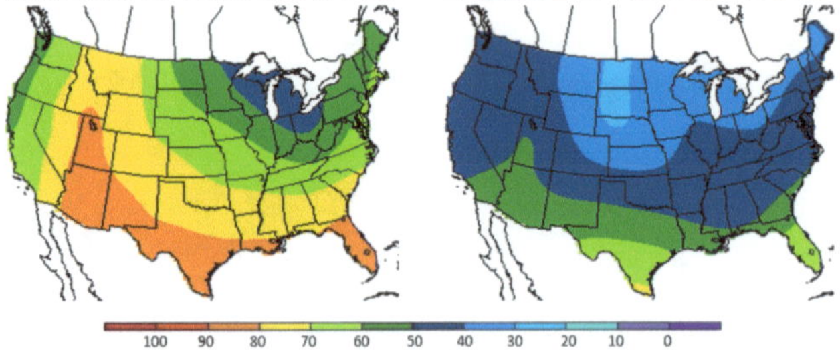

ISOBARIC

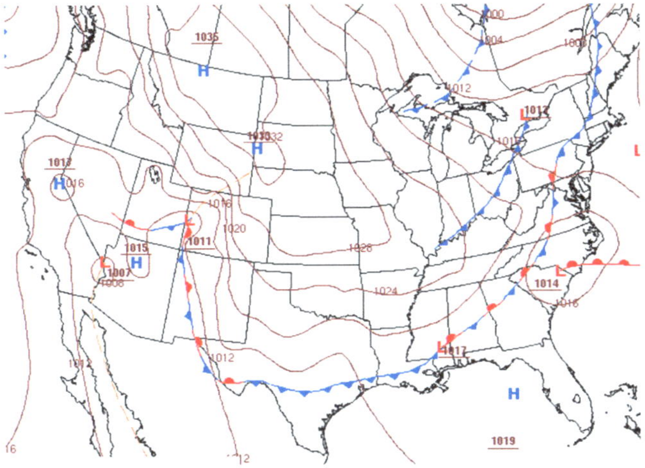

RAIN DISTRIBUTION (Inches)

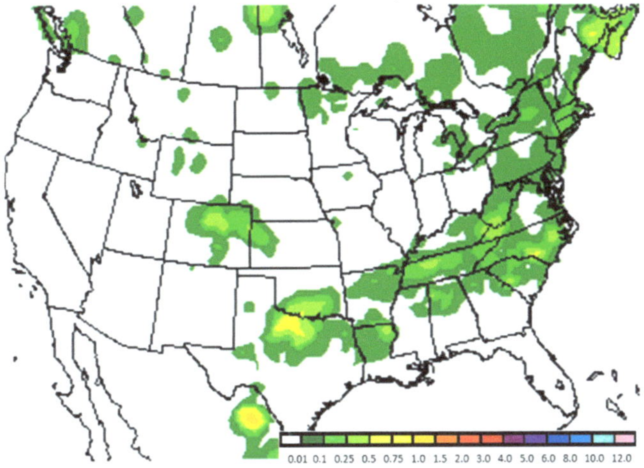

MAXIMUM TEMPERATURE / MINIMUM TEMPERATURE

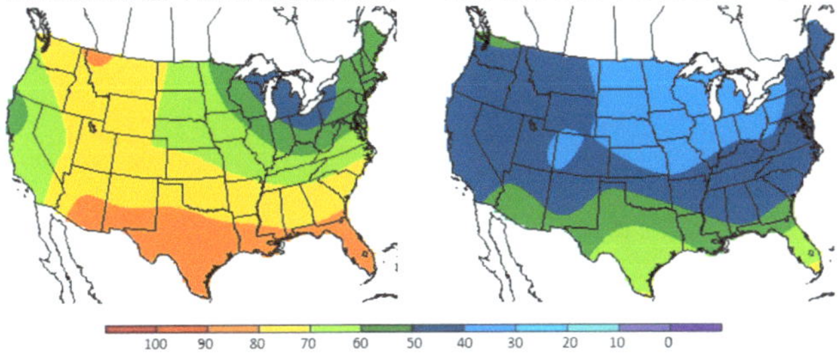

04/11/2025

Rain activity clears throughout the Inland Midwest to the Great Lakes, but persist in the Northeast and down through the Appalachians into the northern half of the Deep South, as well as into the Carolinas.

Rain is also likely in the Mississippi Delta around this time.

Some showers may also linger in the Rocky Mountains to the South Great Plains and into the northern half of Texas.

Mostly dry elsewhere.

Showers in Hawaii over the next couple of days.

04/12/2025

Some showers return around the coasts of Washington and Oregon, and linger around the south Rocky Mountains, while showers prevail over the Great Lakes to the Northeast.

Widespread stormy conditions with heavy rain and thunderstorms prevail over the south, from Texas and the Deep South to the Coastal Southeast.

Mostly dry elsewhere, including in Florida.

There is a chance of a return of further snow across Montana to South Dakota, before moving into Minnesota over the next couple of days.

ISOBARIC

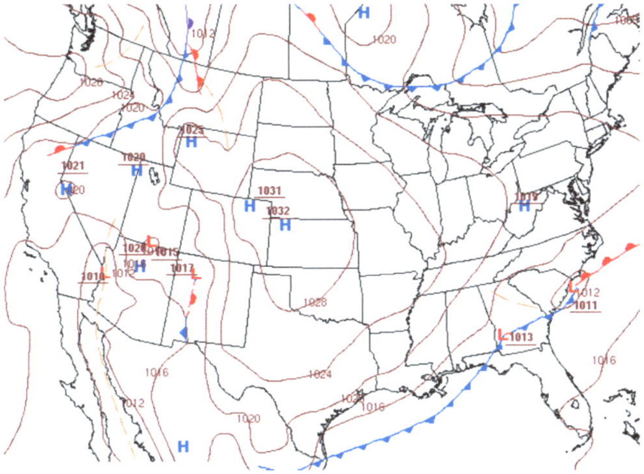

RAIN DISTRIBUTION (Inches)

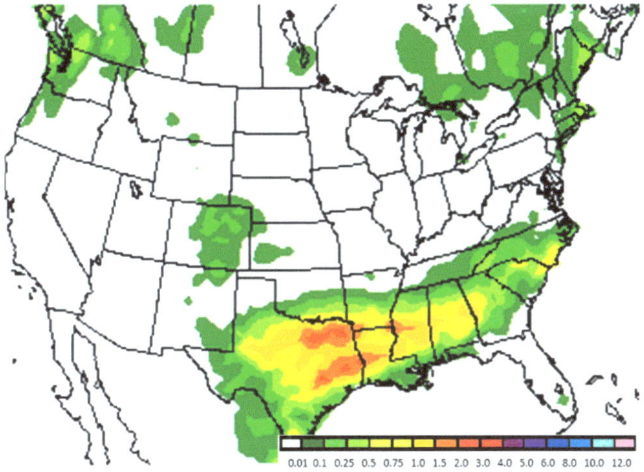

MAXIMUM TEMPERATURE / MINIMUM TEMPERATURE

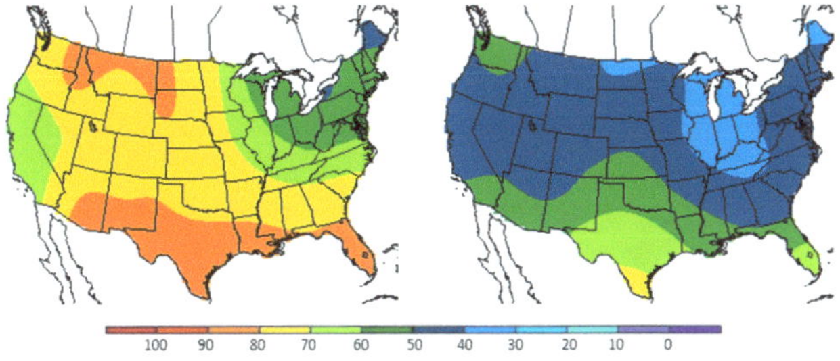

APRIL 13 SUNDAY

ISOBARIC

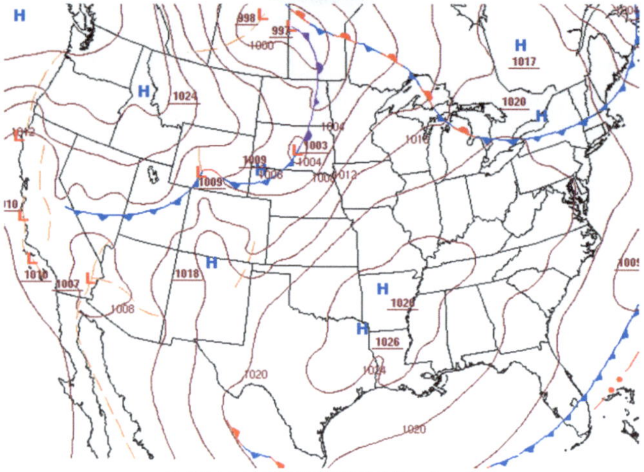

RAIN DISTRIBUTION (Inches)

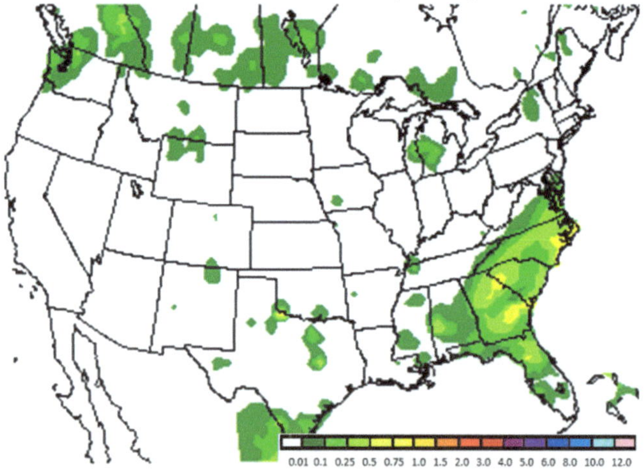

MAXIMUM TEMPERATURE / MINIMUM TEMPERATURE

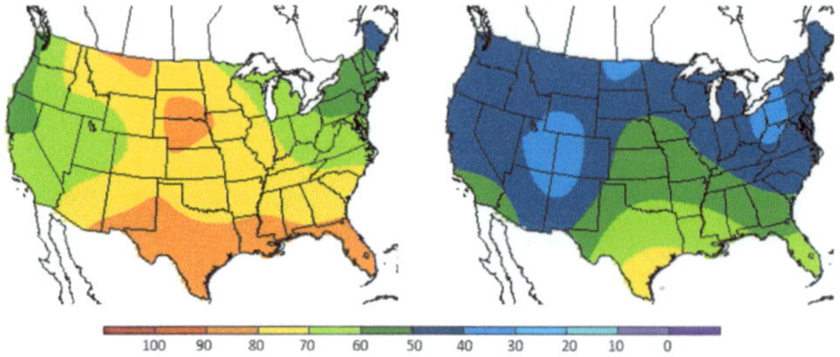

04/13/2025

The recent precipitation in the east of the country clears over most of Texas and the western half of the Deep South, but continues east of the Appalachians throughout the Coastal Southeast and into Florida, as well as the southern tip of Texas.

Mostly dry elsewhere, except for the chance of isolated showers in the Rocky Mountains and South Great Plains, and around the Great Lakes to the Northeast.

It may be very cold in the Pacific Northwest, while unusually warm in Ohio to Tennessee and Missouri.

04/14/2025

A depression that has formed to the north of the Great Lakes brings heavy rain back into the region and may spread further south and west over the next few days.

Rain may also continue in the southern tip of Texas around the Rio Grande Valley.

Late wintry precipitation expected in Michigan, with some of it possibly falling as snow once more.

Mostly dry elsewhere.

Unusually high maximum temperatures may develop in the Deep South over the next couple of days, particularly in Georgia.

APRIL 14 MONDAY

ISOBARIC

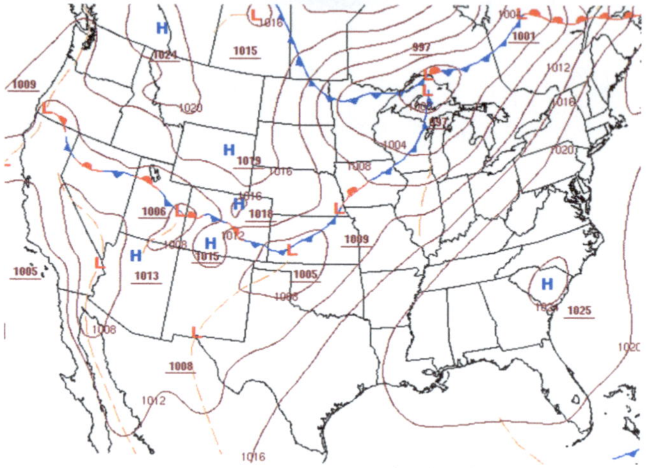

RAIN DISTRIBUTION (Inches)

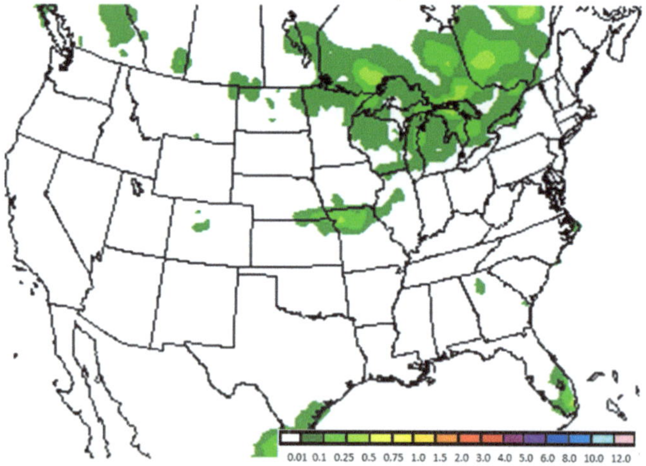

MAXIMUM TEMPERATURE / MINIMUM TEMPERATURE

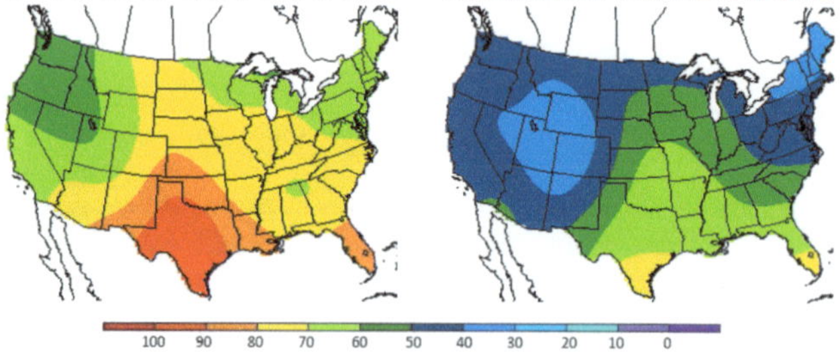

APRIL 15 TUESDAY

ISOBARIC

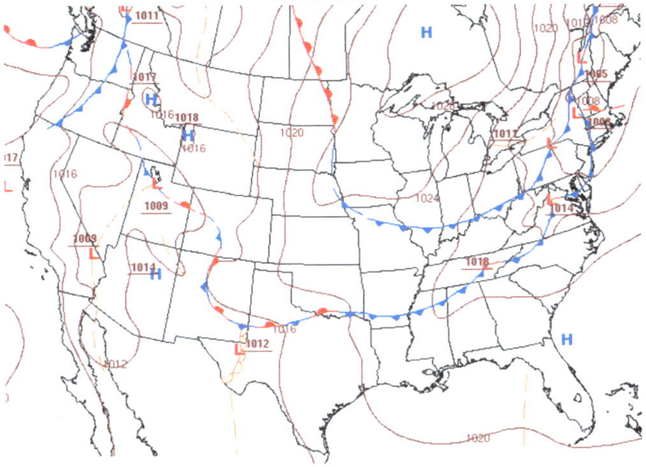

RAIN DISTRIBUTION (Inches)

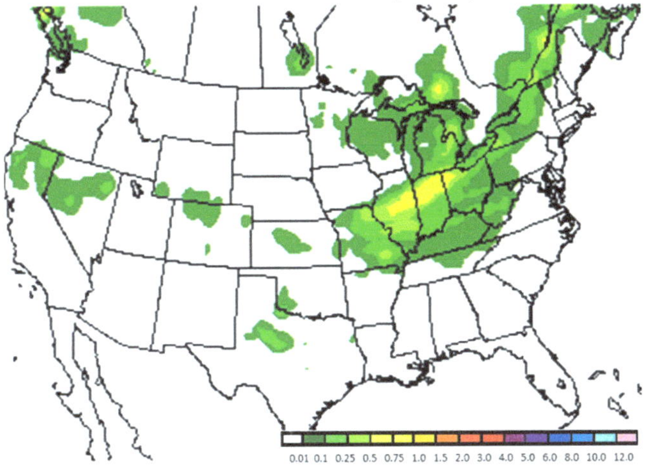

MAXIMUM TEMPERATURE

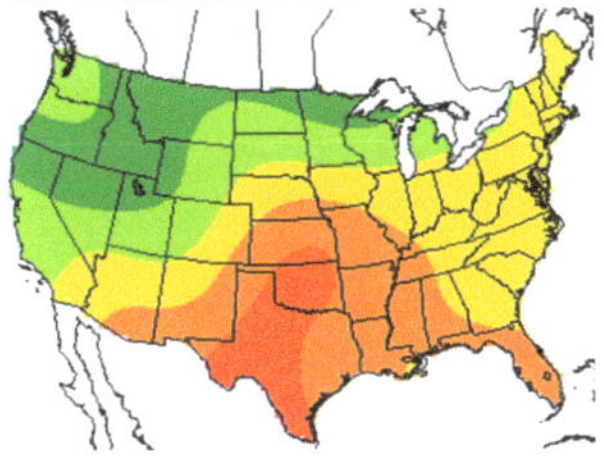

MINIMUM TEMPERATURE

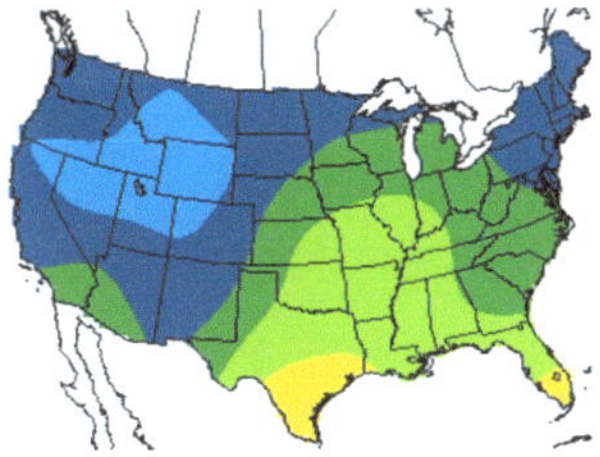

04/15/2025

Rain activity in the east of the country expands from the Great Lakes and the Northeast and down into the Inland Midwest to just north of the Appalachians.

Another line of showers cross north California, the Great Basin and Rockies into the South Great Plains and north Texas.

Mostly dry elsewhere.

A temperature change may move over the country becoming warmer in the west of the Rocky Mountains, while colder over the east from the Great Plains to the Atlantic coast.

04/16/2025

Scattered showers prevail in the Pacific Northwest, as well as increase through the Great Basin, over the Rocky Mountains and into the South Great Plains, and possibly also into the Ozarks.

Precipitation also prevails over the Northeast to the Coastal Southeast, with the chance of some further snow likely in the far Northeast down to Kentucky over the next couple of days.

Mostly dry elsewhere.

Heavy precipitation in the southeast of Alaska likely.

APRIL 16 WEDNESDAY

ISOBARIC

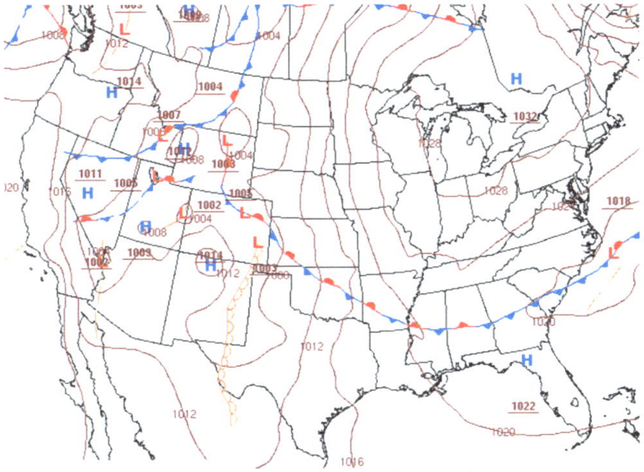

RAIN DISTRIBUTION (Inches)

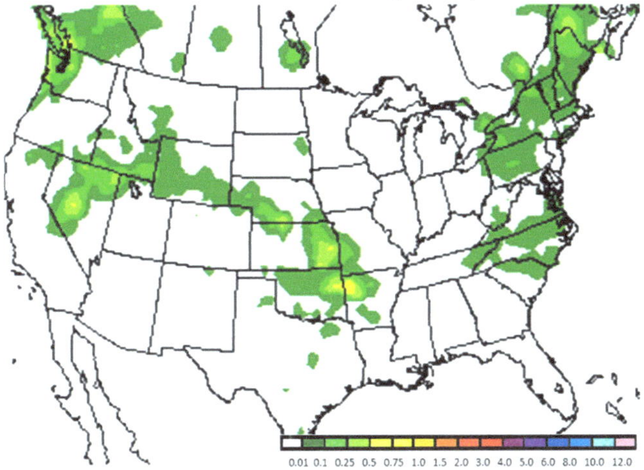

MAXIMUM TEMPERATURE

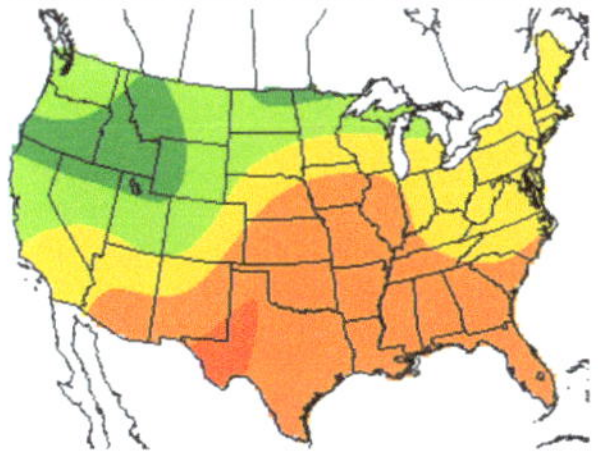

MINIMUM TEMPERATURE

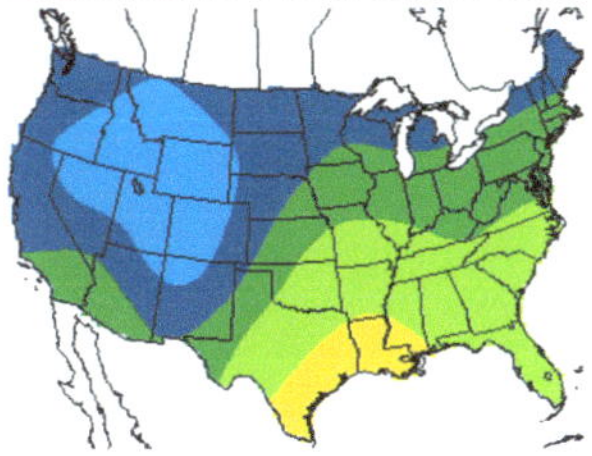

APRIL **17** THURSDAY

ISOBARIC

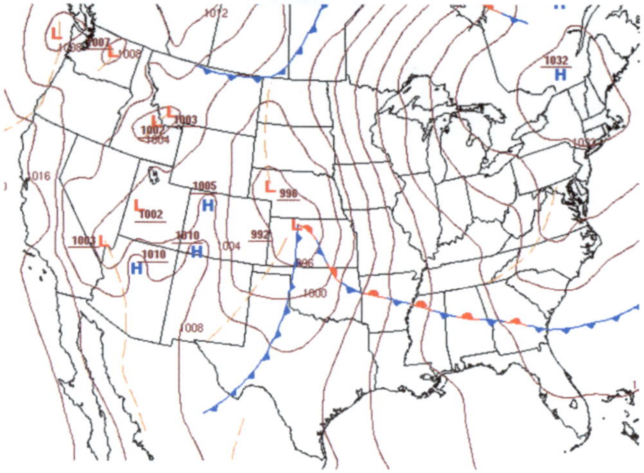

RAIN DISTRIBUTION (Inches)

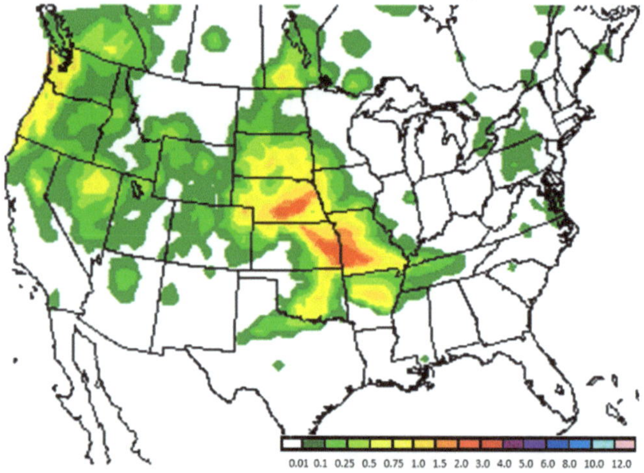

MAXIMUM TEMPERATURE

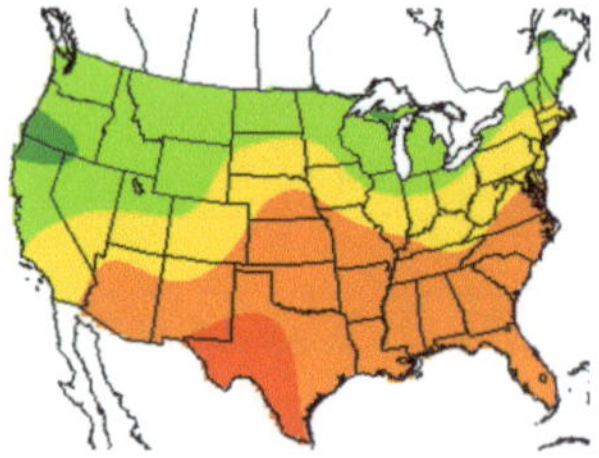

MINIMUM TEMPERATURE

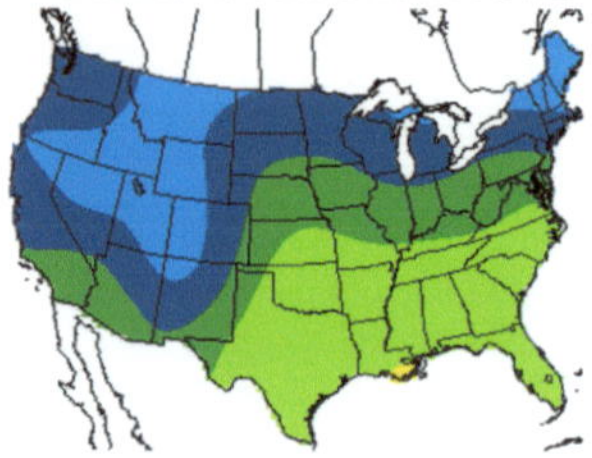

04/17/2025

A cold change sweeps across the Great Plains to the east coast over the next few days bringing further widespread precipitation west of the Appalachians.

Heavy snow may also return into northern and central parts of Texas.

Heavy rain activity prevails over much of the Great Plains, in the Ozarks and down into the northern parts of the Deep South.

Widespread rain also spreads down through the west of the country from the Pacific Northwest and over the Rocky Mountains and into northern California and across the Great Basin.

A large high centered over the Northeast brings mostly dry and clearing conditions throughout the northeast quarter of the country.

Mostly dry through the south of the country from southern California to the Deep South and Florida.

04/18/2025

Mostly dry along the eastern coastal regions from the Northeast to Florida and around the coastal fringes of the Gulf of Mexico.

Some precipitation prevails elsewhere.

A cold change continues in many parts of the southeast quarter with it being unusually cold in the central and southern Great Plains, as well as into the lower Midwest and Southeast.

Heavy precipitation in the Ozarks to the northwest of the Deep South may see flooding problems along local river systems over the next few days.

APRIL **18** FRIDAY

ISOBARIC

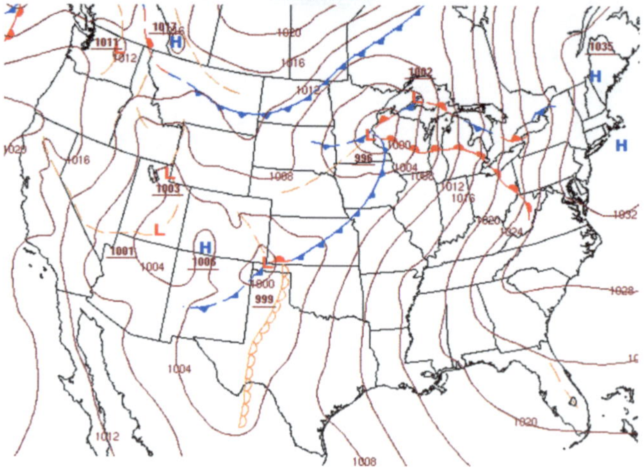

RAIN DISTRIBUTION (Inches)

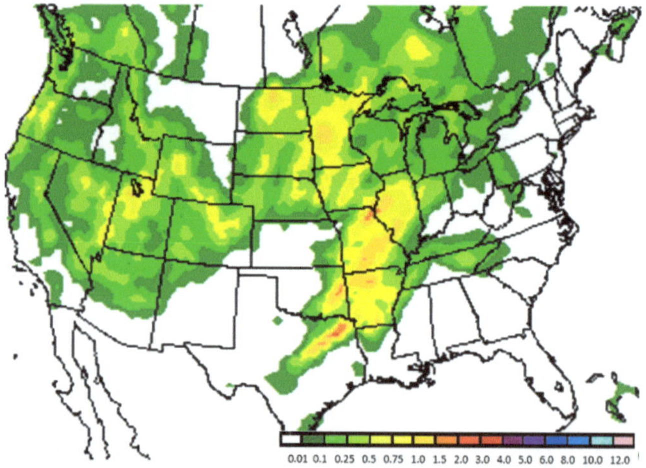

MAXIMUM TEMPERATURE

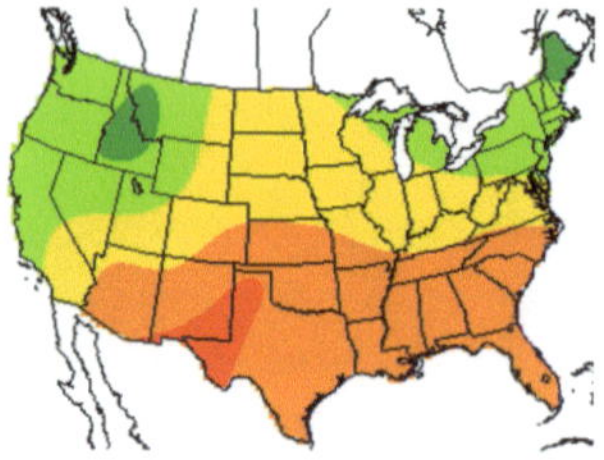

MINIMUM TEMPERATURE

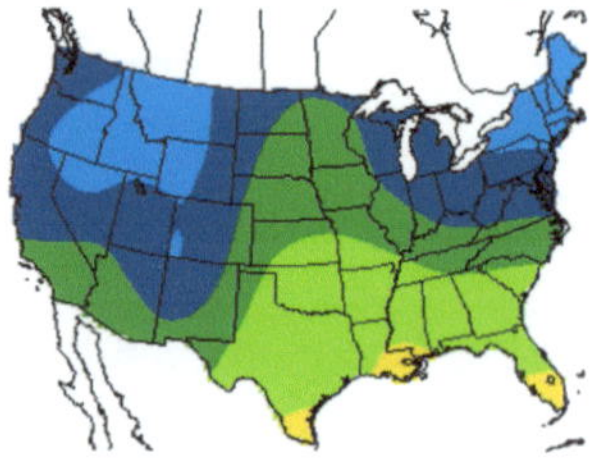

APRIL 19 SATURDAY

ISOBARIC

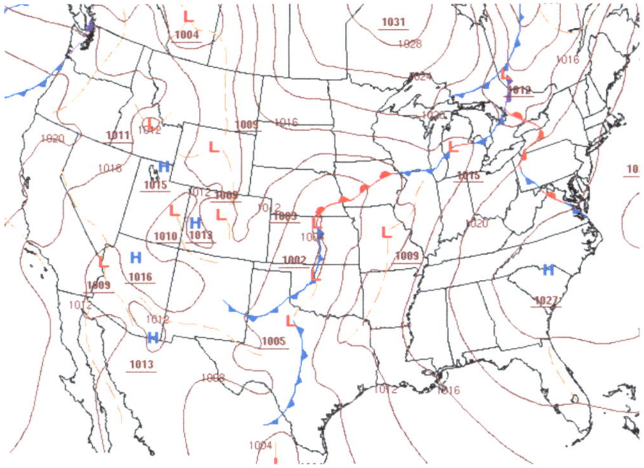

RAIN DISTRIBUTION (Inches)

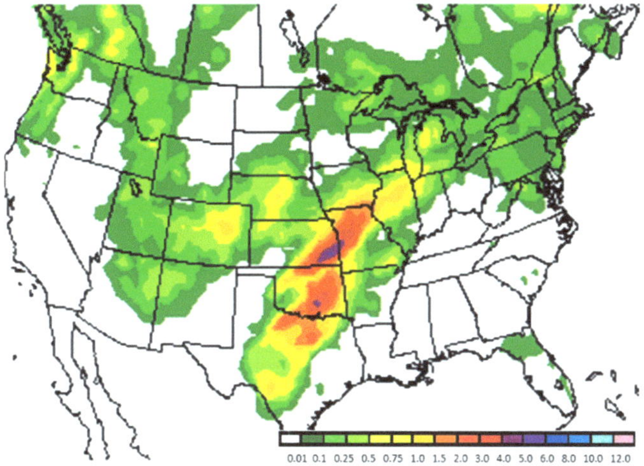

MAXIMUM TEMPERATURE MINIMUM TEMPERATURE

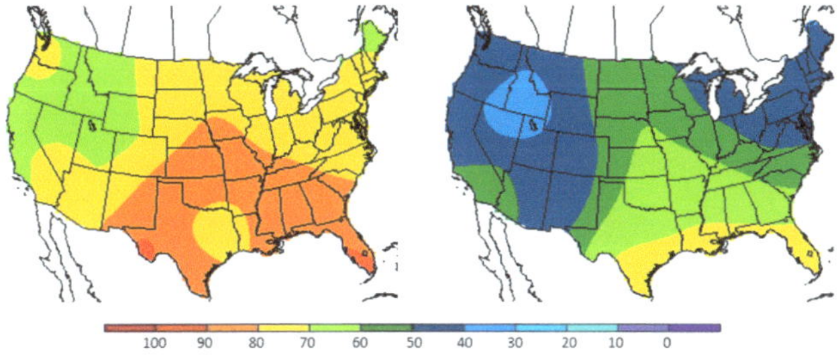

04/19/2025

Precipitation clears in California and the west of the Great Basin, but persists elsewhere over the western half of the country to the Great Plains.

Similarly, widespread precipitation also prevails from Texas and up into the Midwest and throughout the Great Lakes to the Northeast.

Precipitation may be very heavy around the Ozarks to the Inland Midwest, possibly causing localized flooding especially along the river systems.

Mostly dry elsewhere.

It may be unusually warm in Alaska.

04/20/2025

A wintry storm system is likely to cover the North Great Plains and northern parts of the Inland Midwest to the Great Lakes and Northeast over the next couple of days, once more bringing heavy snow to some parts of the central and southern Great Plains.

Heavy rain is also likely across the southern and eastern Midwest Plains with widespread showers reaching further south down through the eastern half of Texas into the west of the Deep South.

Further precipitation continues in the Pacific Northwest to Inland Northwest and down through the Rocky Mountains.

Mostly dry elsewhere.

APRIL 20 SUNDAY

ISOBARIC

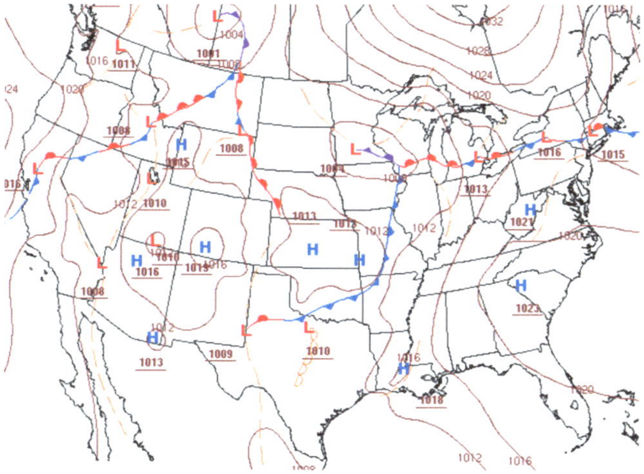

RAIN DISTRIBUTION (Inches)

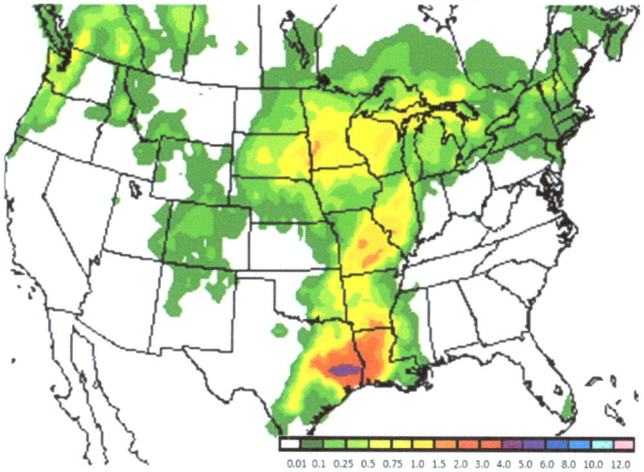

MAXIMUM TEMPERATURE MINIMUM TEMPERATURE

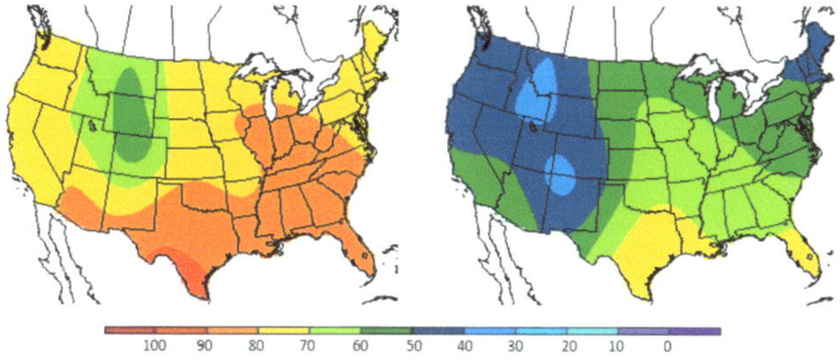

APRIL 21 MONDAY

ISOBARIC

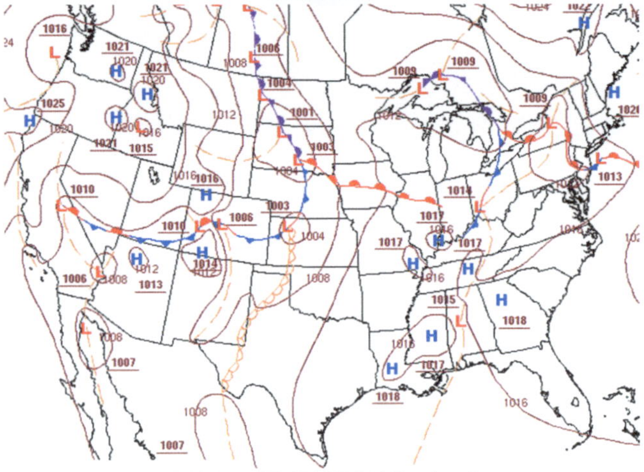

RAIN DISTRIBUTION (Inches)

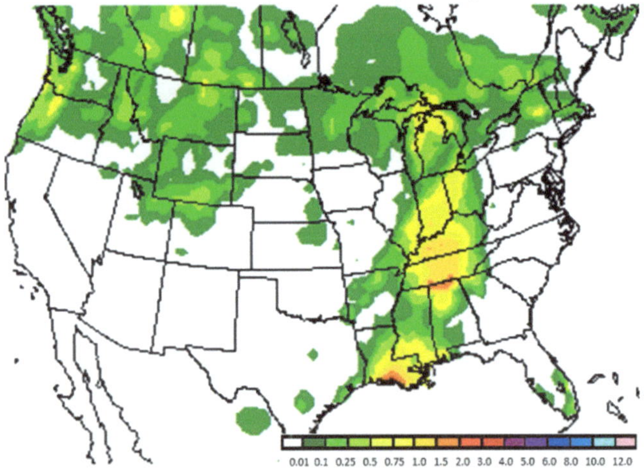

MAXIMUM TEMPERATURE / MINIMUM TEMPERATURE

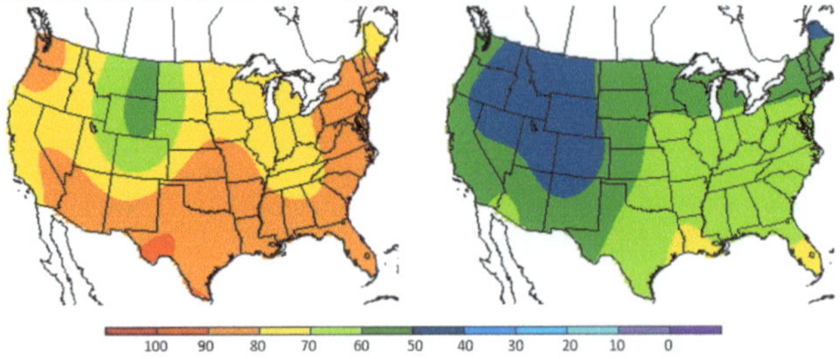

04/21/2025

Widespread precipitation continues across the north of the country from the Pacific Northwest and Rocky Mountains, across the North Great Plains to the Great Lakes.

Widespread precipitation also continues down through the Inland Midwest to the Gulf of Mexico, mostly west of the Appalachians.

There is a chance of heavy snow in the Inland Midwest states, including in Iowa, Minnesota, Michigan and Illinois once more.

Mostly dry from California to Texas and in the South Great Plains and down through the eastern coastal states.

04/22/2025

The passage of an occluded front brings widespread precipitation over the Great Plains to the Great Lakes and up into the Northeast, with heavy falls likely in the Ozarks.

Widespread precipitation also prevails east of the Appalachian foothills to the Atlantic coast, from the Megalopolis to Florida, with showers into the Deep South.

Some heavy showers prevail in the Pacific and Inland Northwest to the Rocky Mountains.

Mostly dry elsewhere.

Snow may continue to fall in the Great Lakes to the Inland Midwest as well as in the Northeast.

APRIL 22 TUESDAY

ISOBARIC

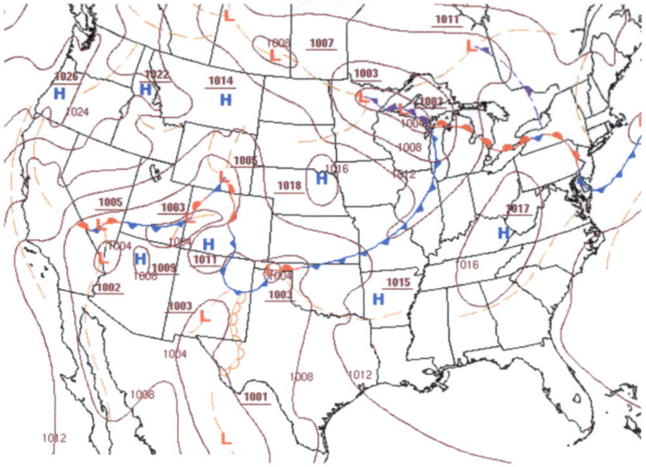

RAIN DISTRIBUTION (Inches)

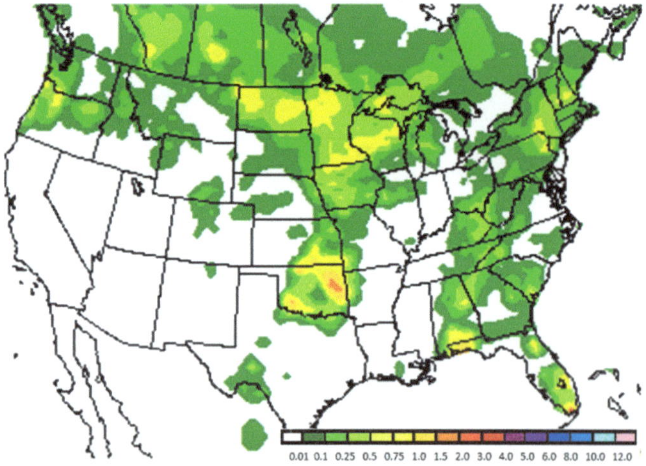

MAXIMUM TEMPERATURE / MINIMUM TEMPERATURE

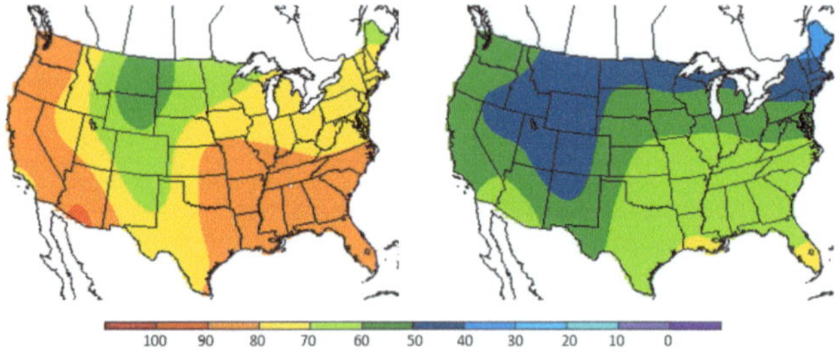

APRIL 23 WEDNESDAY

ISOBARIC

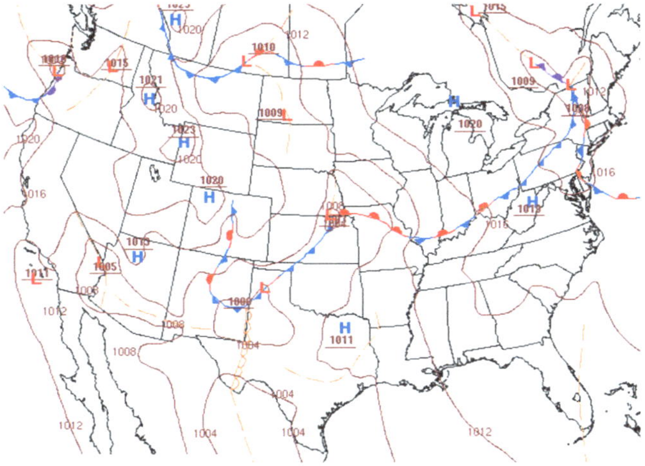

RAIN DISTRIBUTION (Inches)

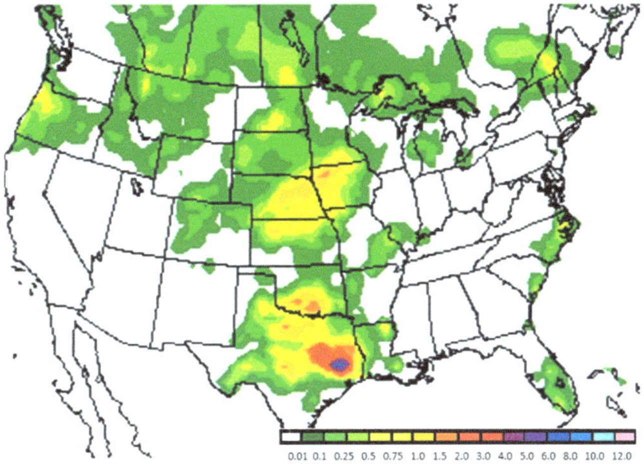

MAXIMUM TEMPERATURE

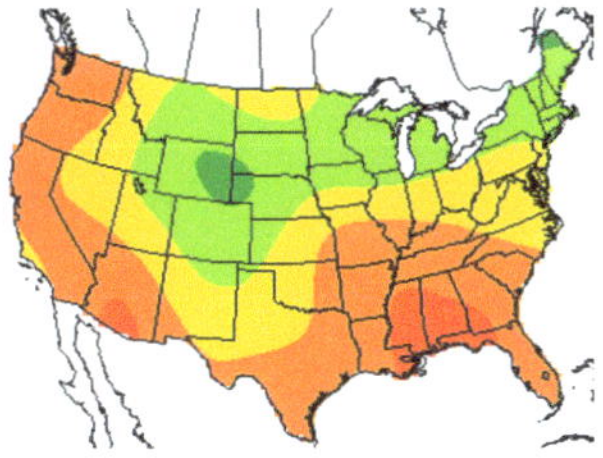

MINIMUM TEMPERATURE

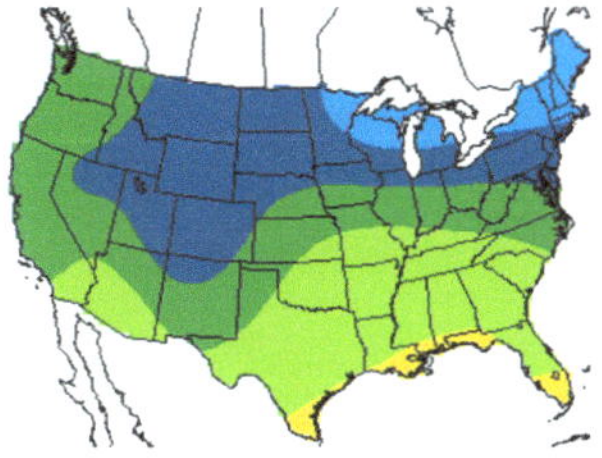

04/23/2025

A storm over the center of the country develops, followed by another blast of colder air.

Precipitation prevails through the Pacific and Inland Northwest to the Rocky Mountains, and down through the midsection of the country.

Widespread heavy precipitation prevails throughout the Great Plains and into Texas.

There is a chance of further heavy snow in the Great Plains over the next couple of days, as well as a series of thunderstorms, with hail, and gusty winds, through the center of the country and in Texas to the Deep South.

Some showers in the Great Lakes to the Northeast, and in coastal parts from the Carolinas to Florida.

Mostly dry elsewhere.

04/24/2025

A depression forms to the north of the Great Lakes, with an associated occluded front near its center, with widespread storm activity prevailing throughout the Great Plains to the Appalachians and south into Texas and the Deep South.

Storm activity may intensify as it crosses the Appalachians nearing the Atlantic coast, with high winds and torrential rain prevailing, and snow in the higher reaches of the Northeast.

A mixture of snow and rain may prevail from the Great Lakes and into the Inland Midwest.

Mostly dry east of the Appalachians to the Megalopolis and in the southwest quarter of the country.

APRIL 24 THURSDAY

ISOBARIC

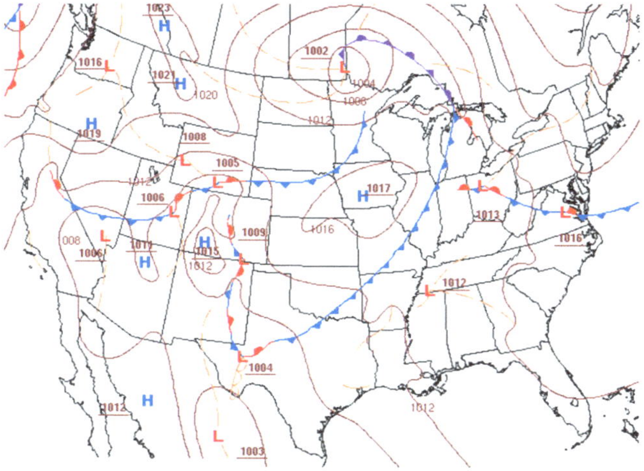

RAIN DISTRIBUTION (Inches)

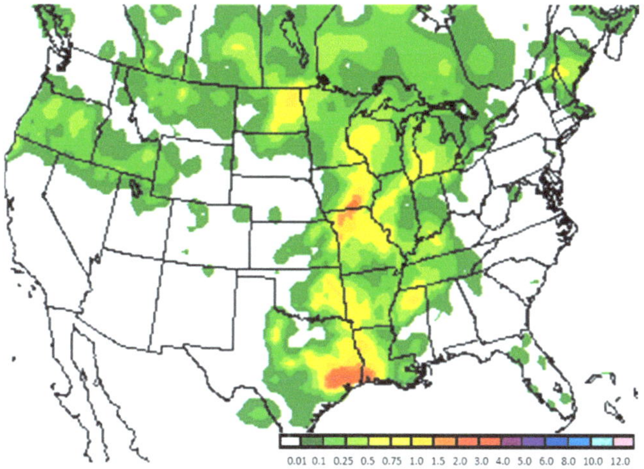

MAXIMUM TEMPERATURE

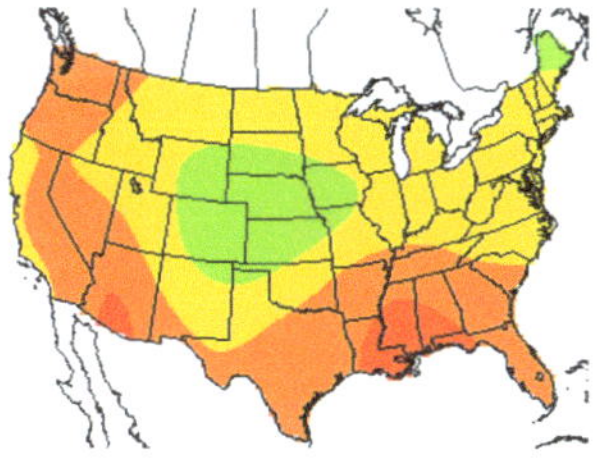

MINIMUM TEMPERATURE

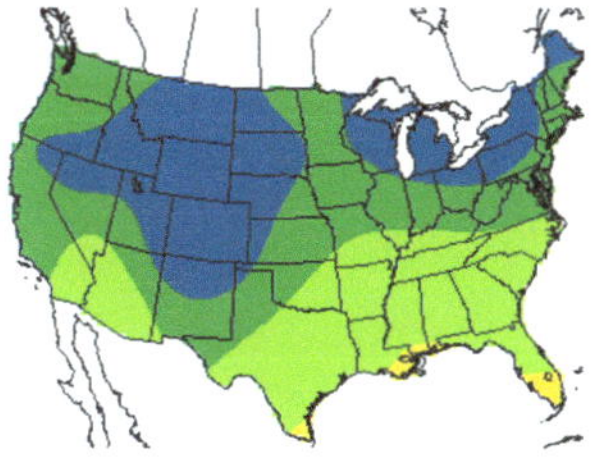

APRIL 25 FRIDAY

ISOBARIC

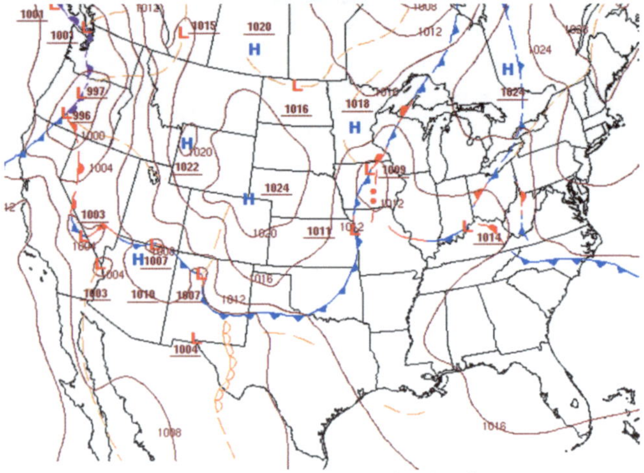

RAIN DISTRIBUTION (Inches)

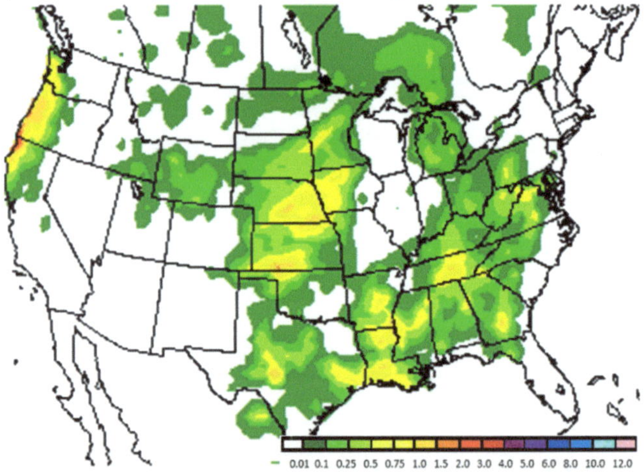

MAXIMUM TEMPERATURE / MINIMUM TEMPERATURE

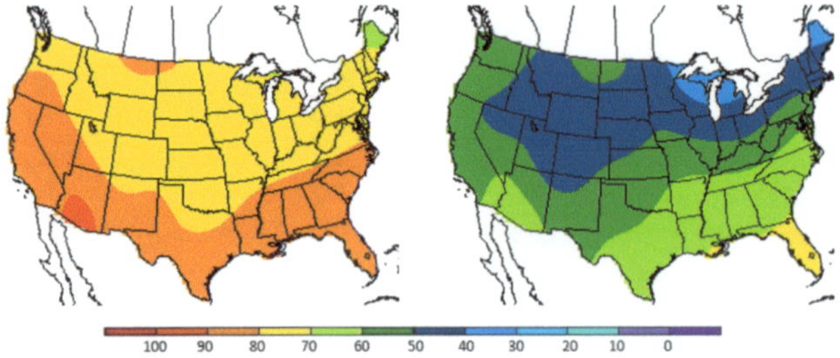

04/25/2025

Widespread precipitation prevails over most of the eastern half of the country from the Great Plains and Texas to the Atlantic Coast, although it may be mostly dry from Florida to the Carolina coastal fringes, and in the Northeast.

There is a chance of more heavy snowfall and high winds in the New York region.

Heavy rain also prevails in the Pacific Northwest and across the Rocky Mountains with very heavy falls at times along the Oregon and northern California coasts.

Mostly dry elsewhere.

04/26/2025

A cold front maintains widespread storm activity, gusty winds and precipitation throughout the east of the country from Texas, over the Inland Midwest to the Great Lakes and Northeast, with very heavy falls in northern Texas to the Ozarks and in the Carolinas.

Heavy precipitation also prevails in the west from the Pacific and Inland Northwest to the Great Basin and into the northern half of California, including throughout the Sierra Nevada and Cascade ranges.

Mostly dry elsewhere.

It may be unusually cold at this time in parts of the Northeast, Deep South, and around the Great Lakes.

APRIL 26 SATURDAY

ISOBARIC

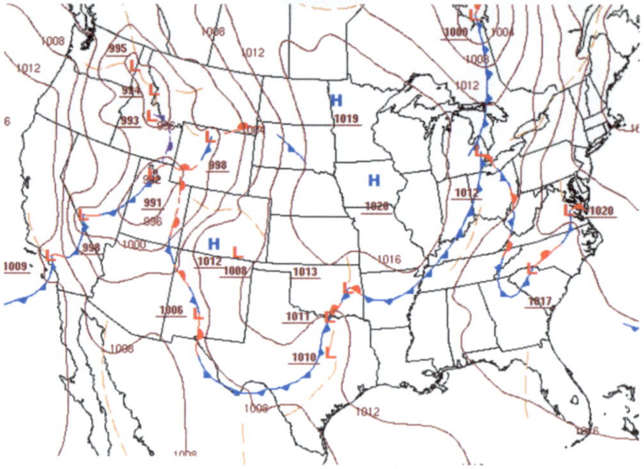

RAIN DISTRIBUTION (Inches)

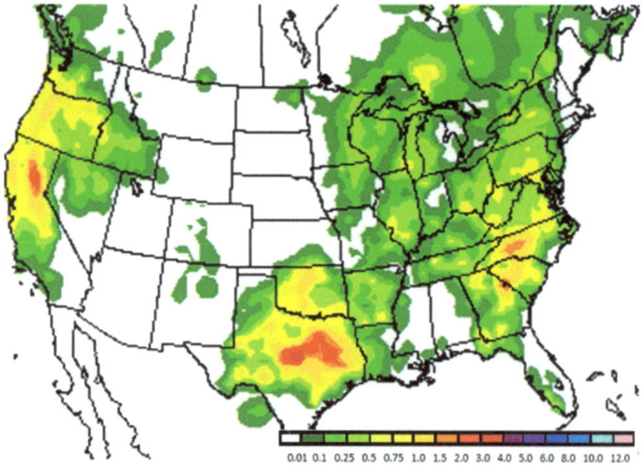

MAXIMUM TEMPERATURE / MINIMUM TEMPERATURE

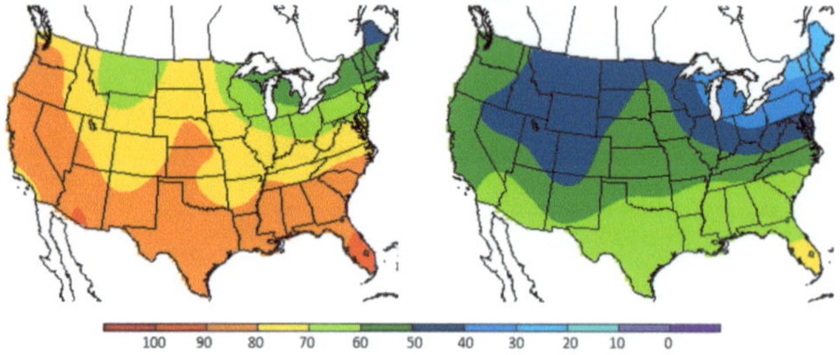

APRIL 27 SUNDAY

ISOBARIC

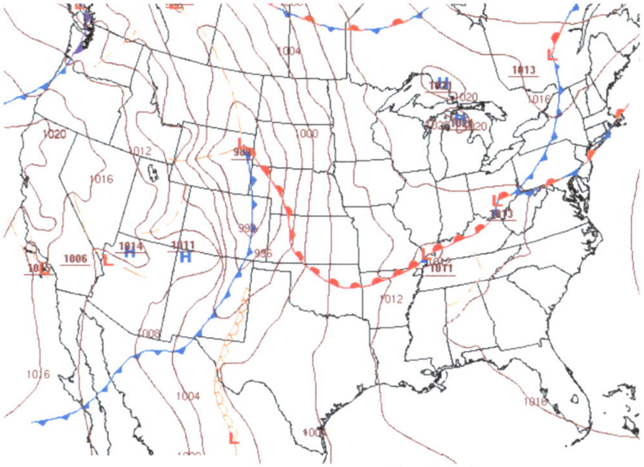

RAIN DISTRIBUTION (Inches)

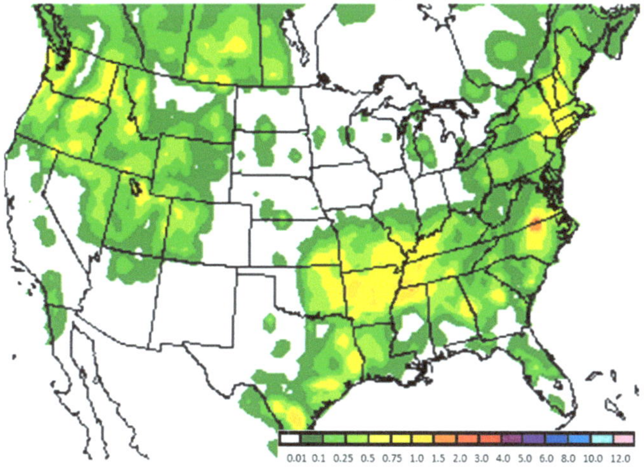

MAXIMUM TEMPERATURE / MINIMUM TEMPERATURE

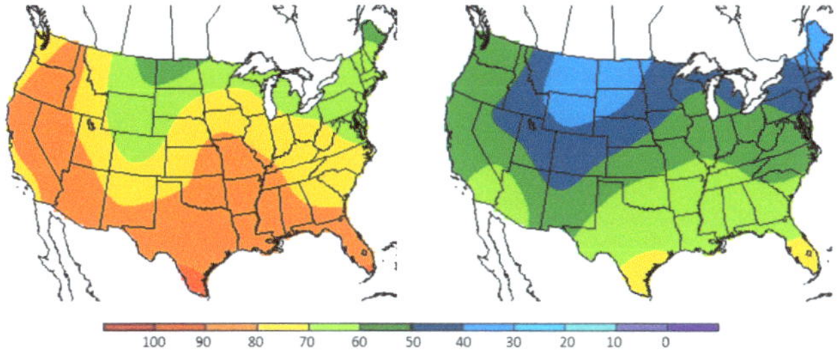

04/27/2025

Widespread precipitation prevails from the Pacific Northwest to the Great Basin and Rocky Mountains in the west, and from the Northeast to Texas, including into Florida in the east.

Only the Southwest and western parts of the Great Plains to the Great Lakes and surrounding Midwest states can expect to escape some precipitation.

There is a chance of flooding developing along the Northeast river systems over the next couple of days.

A series of storms are expected to cross the south-central regions of the country over the next few days, and once more, bring the chance of further snow, this time around the Great North Plains.

04/28/2025

An intense depression with a couple of occluded front strengthens over the North Great Plains brings a passage of storms and widespread precipitation over the northern half of the country.

Precipitation may be particularly heavy in the Great Plains to the Inland Midwest to the Ozarks.

Precipitation also spreads east across the Appalachians to the adjacent Atlantic coasts.

This storm system may bring further snow and stormy winds across the North Great Plains over the next couple of days.

Some showers prevail around the Great Lakes to the Northeast.

Mostly dry from California, through the Southwest and Texas through the Gulf states.

Heavy rain on the Big Island in Hawaii.

APRIL 28 MONDAY

ISOBARIC

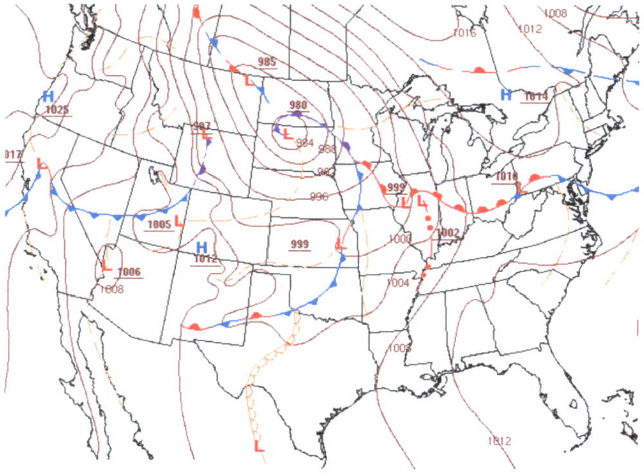

RAIN DISTRIBUTION (Inches)

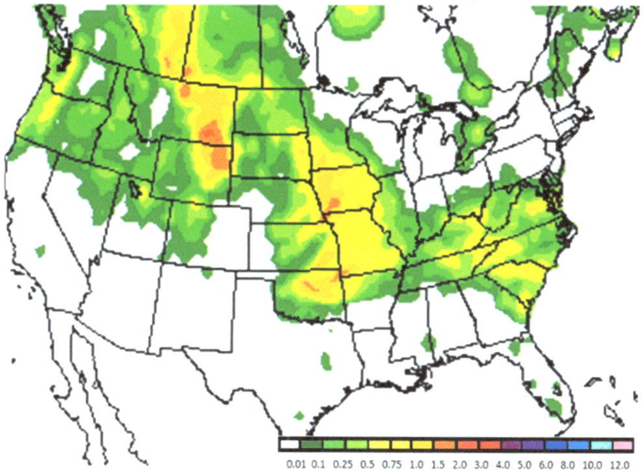

MAXIMUM TEMPERATURE / MINIMUM TEMPERATURE

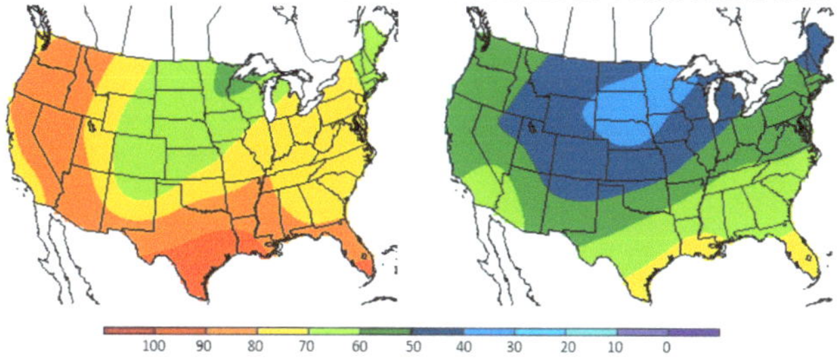

APRIL 29 TUESDAY

ISOBARIC

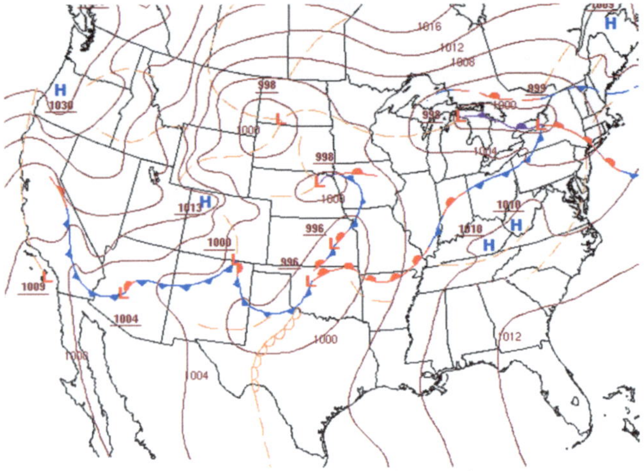

RAIN DISTRIBUTION (Inches)

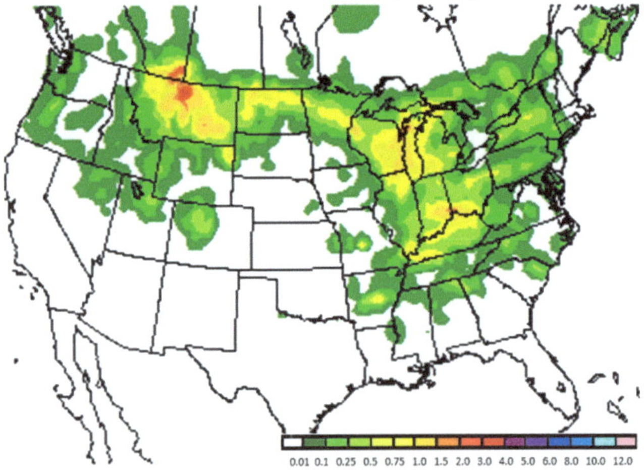

MAXIMUM TEMPERATURE / MINIMUM TEMPERATURE

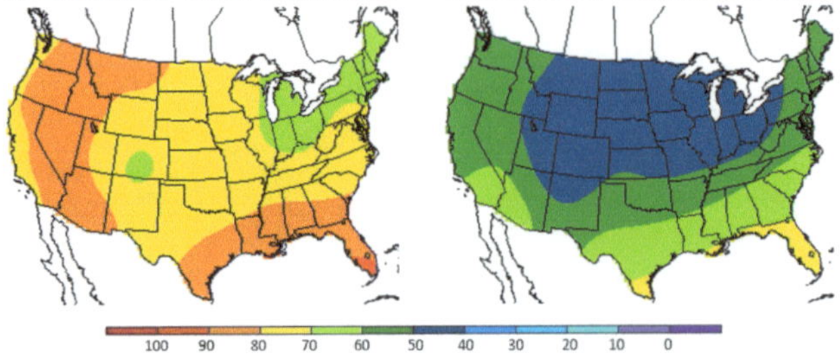

04/29/2025

A weakening depression with an occluded front moves into the northeast of the country extending widespread storm activity over the Great Lakes and Inland Midwest to parts of the Northeast, while persisting over the North Great Plains to the northern Rocky Mountains.

Showers in the Pacific Northwest.

Mostly dry elsewhere.

04/30/2025

Outbreaks of precipitation continue across the north and east of the country.

Precipitation persists from the northern Rocky Mountains, over the North Great Plains and throughout the Inland Midwest to the Coastal Southeast regions, with heavy rain in the Appalachians.

The Northeast can expect widespread rain, while only isolated showers may prevail around the Great Lakes to the Megalopolis.

Mostly dry elsewhere.

APRIL 30 WEDNESDAY

ISOBARIC

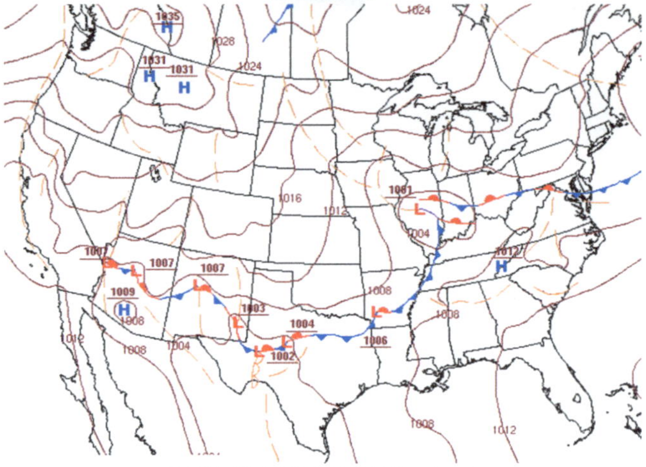

RAIN DISTRIBUTION (Inches)

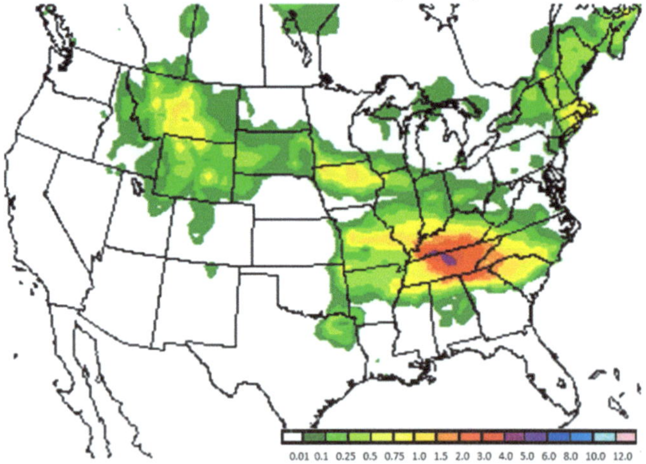

MAXIMUM TEMPERATURE / MINIMUM TEMPERATURE

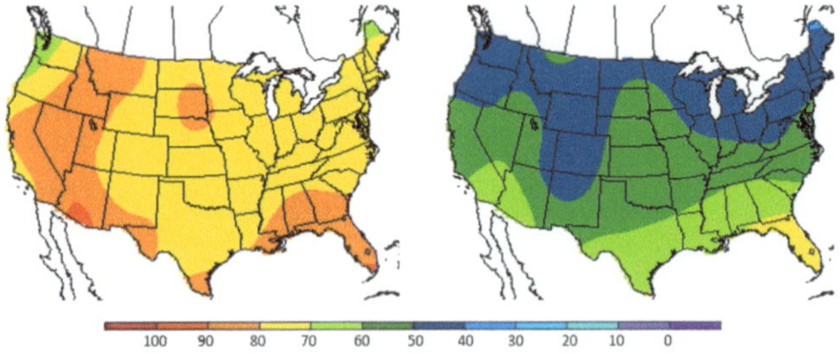

MAY

PHASES OF THE MOON

1st	Northern Declination
4th	First Quarter
9th	Crossing Equator
11th	Apogee
12th	Full Moon
15th	Southern Declination
20th	Third Quarter
23rd	Crossing Equator
26th	Perigee #5
27th	New Moon
28th	Northern Declination

1 Thur	2 Fri	3 Sat	4 Sun	5 Mon	6 Tue	7 Wed	8 Thur	9 Fri	10 Sat	11 Sun	12 Mon	13 Tue	14 Wed	15 Thur	16 Fri	17 Sat	18 Sun	19 Mon	20 Tue	21 Wed	22 Thur	23 Fri	24 Sat	25 Sun	26 Mon	27 Tue	28 Wed	29 Thur	30 Fri	31 Sat
^			1Q					XhS		A	F			V					3Q			XhN			P5	N	^			

MAY 1 THURSDAY

ISOBARIC

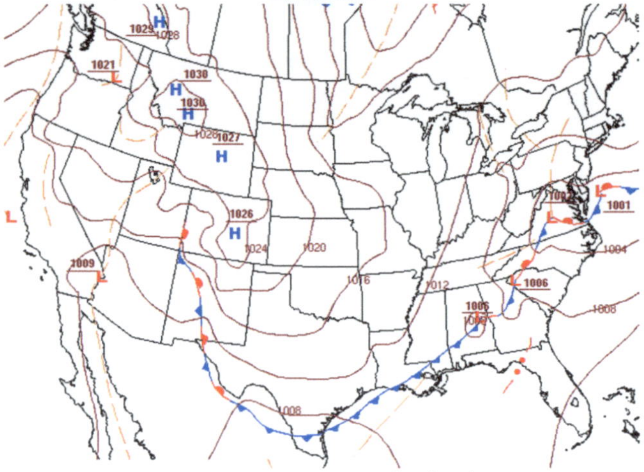

RAIN DISTRIBUTION (Inches)

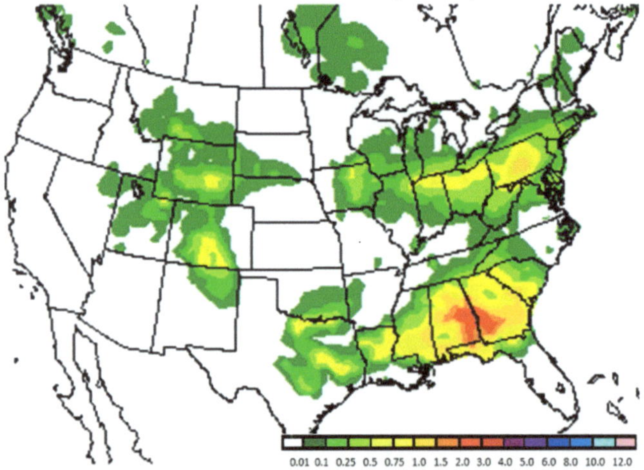

MAXIMUM TEMPERATURE

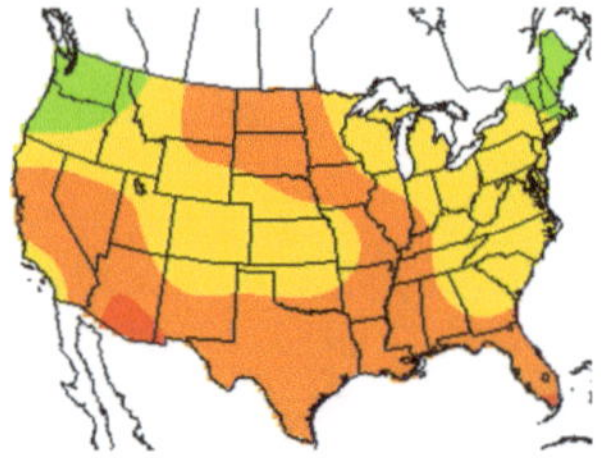

MINIMUM TEMPERATURE

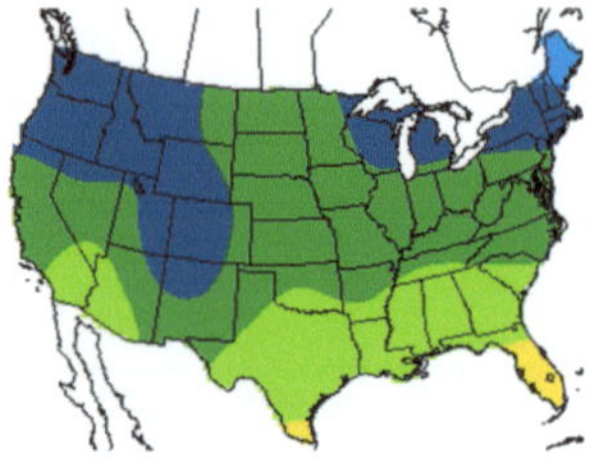

05/01/2025

Heavy rain and accompanying thunderstorm activity in the Inland Midwest and Great Lakes to the Megalopolis, as well as from the east of Texas, across Deep South, and up into the Coastal Southeast regions.

Outbreaks of rain also persist over the Rocky Mountains to the Four Corner states.

Mostly dry elsewhere.

Storms may see the development of tornadoes in Texas, while some storms are also likely to bring some heavy falls for a time in Colorado.

It may be unusually mild in Alaska.

05/02/2025

A depression forms over the Great Lakes bringing widespread rain activity into the region and into the Upper Midwest, as well as down through the east of the country from the Megalopolis to Florida, east of the Appalachians.

Some showers also prevail around the Four Corner states into the southern Rocky Mountains, with isolated showers into central Texas.

Mostly dry elsewhere.

MAY 2 FRIDAY

ISOBARIC

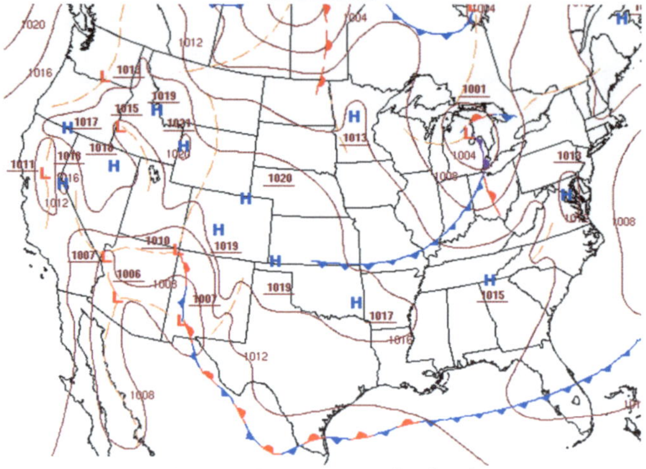

RAIN DISTRIBUTION (Inches)

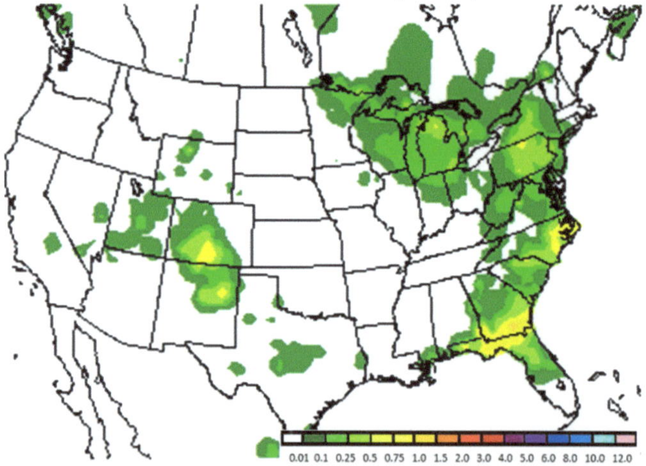

MAXIMUM TEMPERATURE

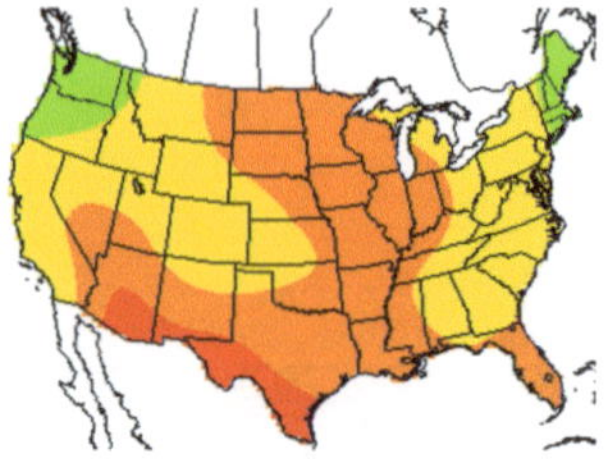

MINIMUM TEMPERATURE

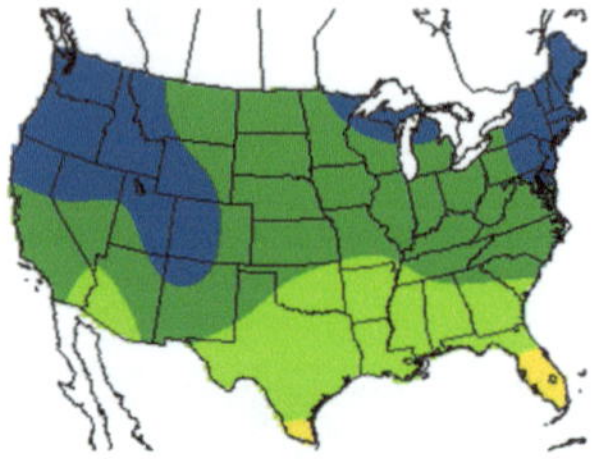

MAY 3 SATURDAY

ISOBARIC

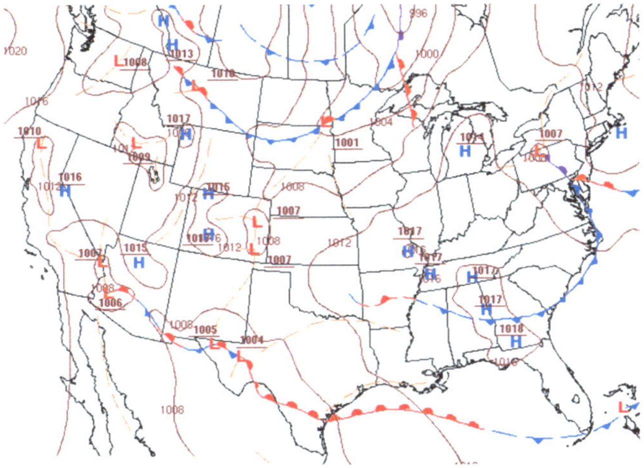

RAIN DISTRIBUTION (Inches)

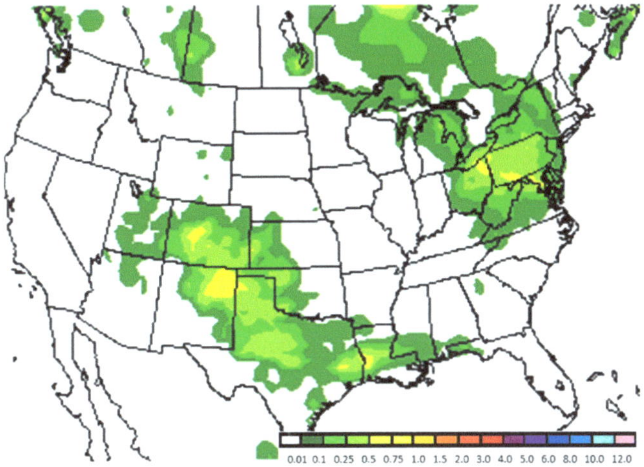

MAXIMUM TEMPERATURE

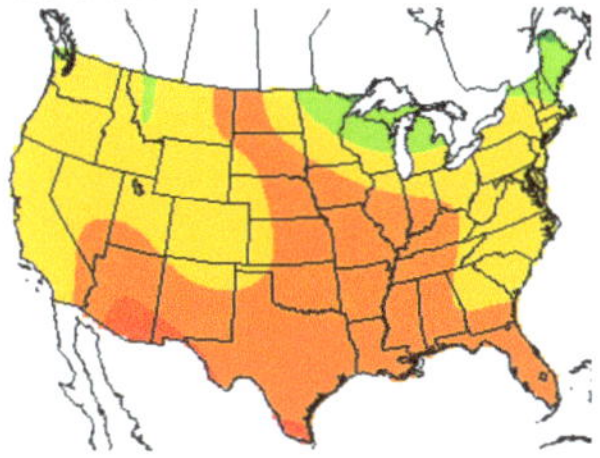

MINIMUM TEMPERATURE

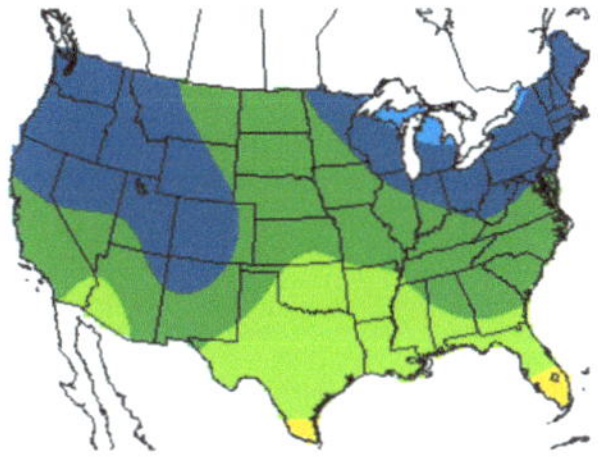

05/03/2025

Rain persists in the northeast quarter of the country from the Great Lakes to the Northeast regions and down into the northern Appalachians and Coastal Southeast, while clearing in east of the Deep South to Florida.

Outbreaks of rain and possible severe thunderstorms in the southern half of the Rocky Mountains, across the central and southern Great Plains and into northern parts of Texas to the west and coastal Deep South.

There is also the chance of accompanying heavy rain with the thunderstorms, particularly in Nebraska.

Mostly dry elsewhere.

MAY 4 SUNDAY

05/04/2025

Some outbreaks of rain linger over the Northeast to the Great Lakes, while heavy precipitation spreads northwards up through the Great Plains as well as throughout most of the Deep South.

Chance of wintry precipitation, including snow, in the Rockies to Colorado.

There is a chance of supercell thunderstorms and possible tornadoes moving up from northern Mexico and into Texas.

Mostly dry elsewhere.

ISOBARIC

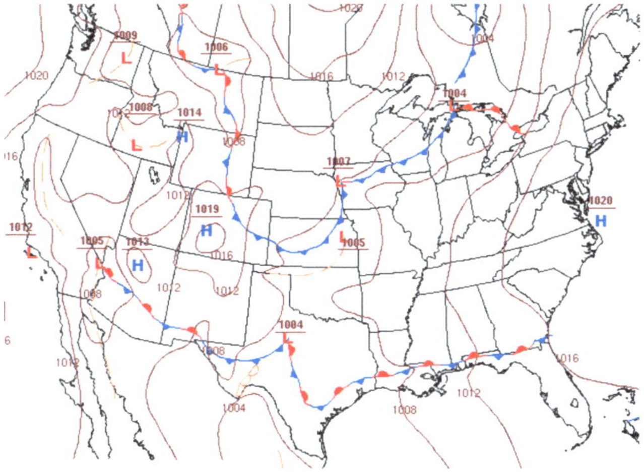

RAIN DISTRIBUTION (Inches)

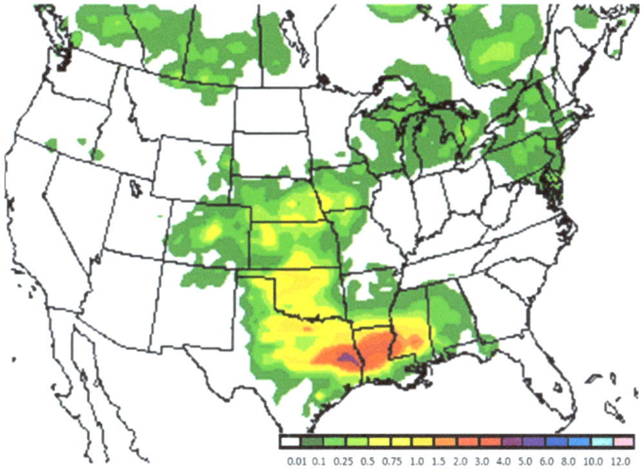

MAXIMUM TEMPERATURE

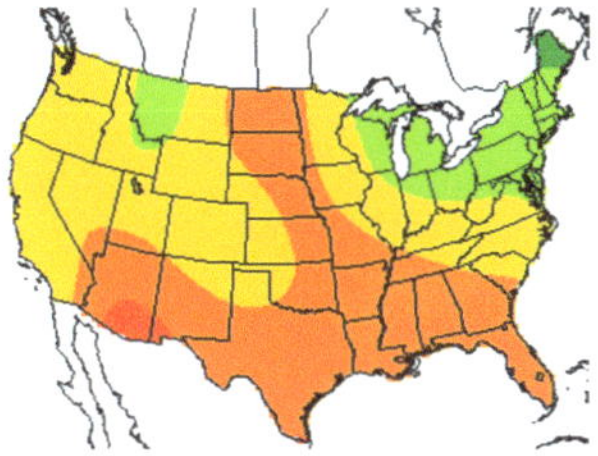

MINIMUM TEMPERATURE

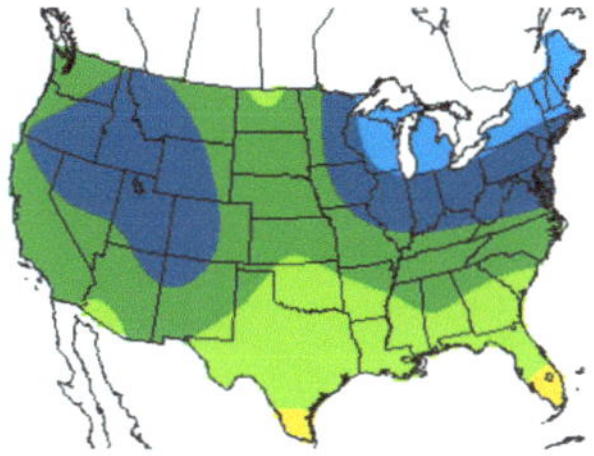

MAY 5 MONDAY

ISOBARIC

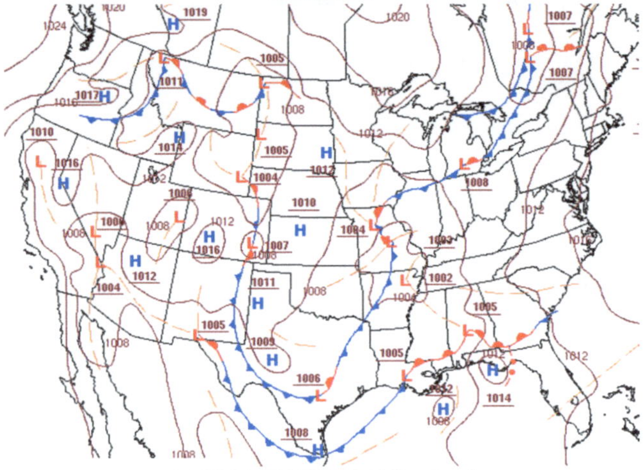

RAIN DISTRIBUTION (Inches)

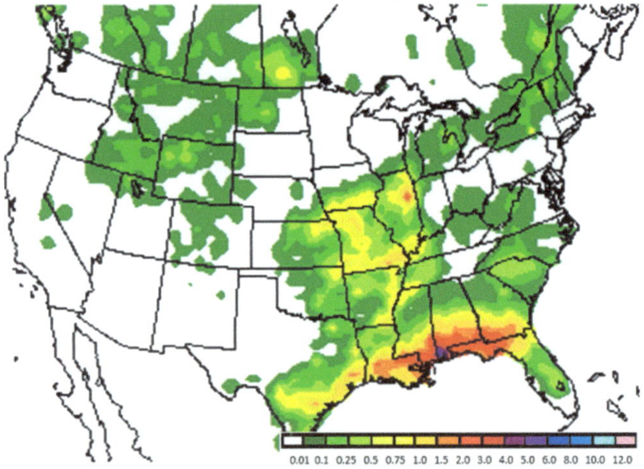

MAXIMUM TEMPERATURE

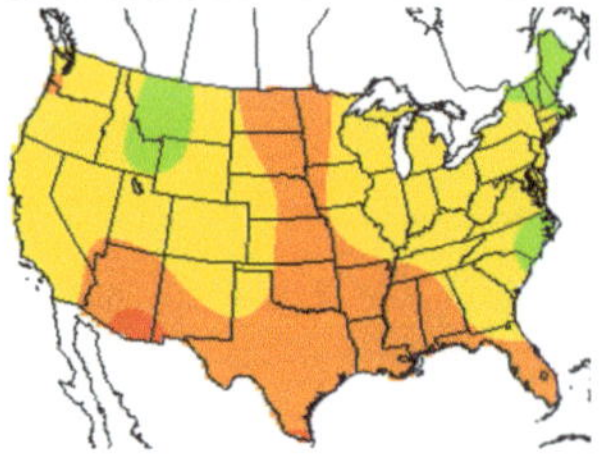

MINIMUM TEMPERATURE

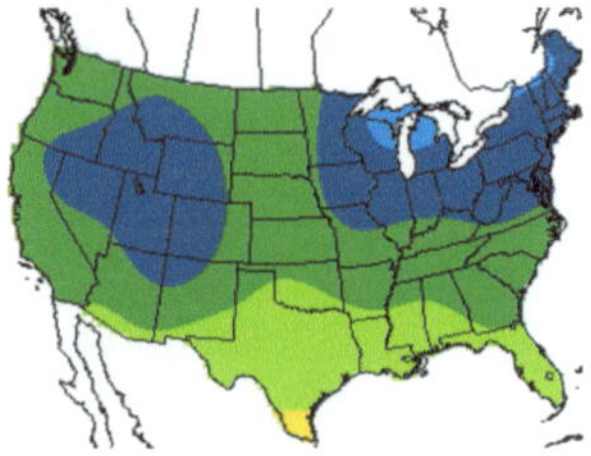

05/05/2025

Widespread outbreaks of rain Nebraska to Colorado and across the Central Plains and around the Great Lakes to the Northeast.

Heavy precipitation elsewhere throughout the eastern half of the country, from the eastern half of Texas to the Inland Midwest, as well as throughout the Deep South and southern Appalachians.

Chance of flooding in the Midwest river systems, particularly in Iowa, where flooding may continue to be an issue for several more days.

Precipitation may be particularly heavy along the Gulf of Mexico coastal regions.

In the west, outbreaks of showers and rain prevail over the Inland Northwest to the Rocky Mountains.

Mostly dry elsewhere.

Unusually cool in Hawaii.

05/06/2025

Widespread rain and scattered thunderstorms prevail throughout the eastern half of the country, from the Great Plains to the Atlantic Coast and from the Northeast to Florida.

Rain may be heavy at times, particularly in the Coastal Southeast, especially in the Carolinas.

Similarly, rain activity spreads in the west from the Inland Northwest and through the Rocky Mountains and onto the Great Plains to the east.

Mostly dry elsewhere although some brief pockets of showers may briefly sprinkle in the Great Basin.

MAY 6 TUESDAY

ISOBARIC

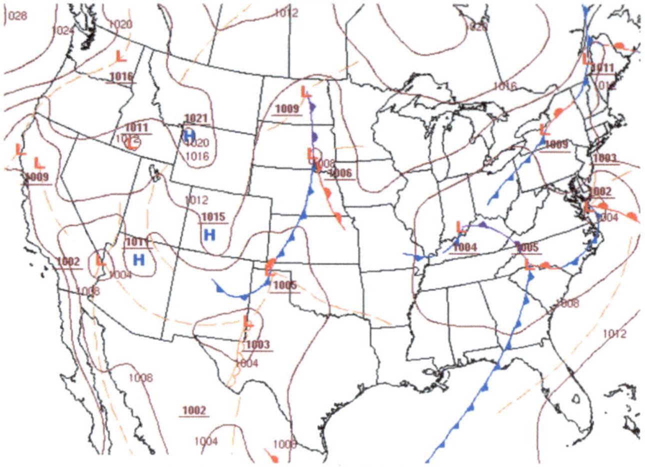

RAIN DISTRIBUTION (Inches)

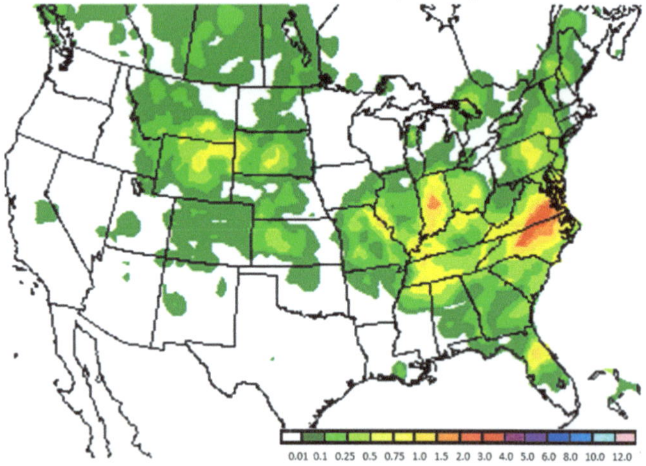

MAXIMUM TEMPERATURE

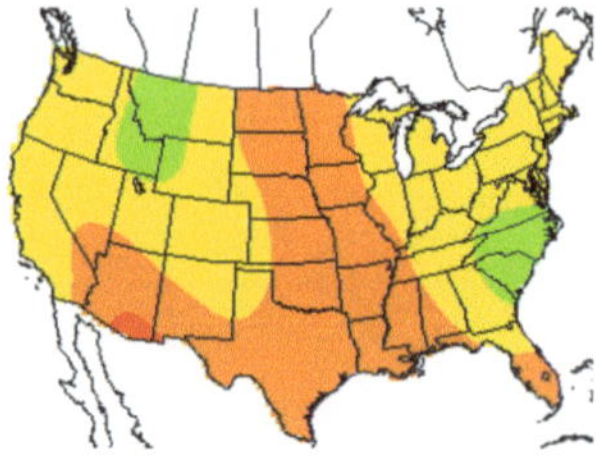

MINIMUM TEMPERATURE

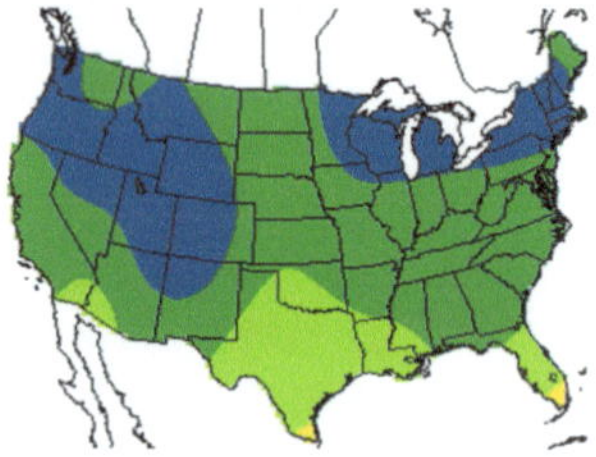

MAY 7 WEDNESDAY

ISOBARIC

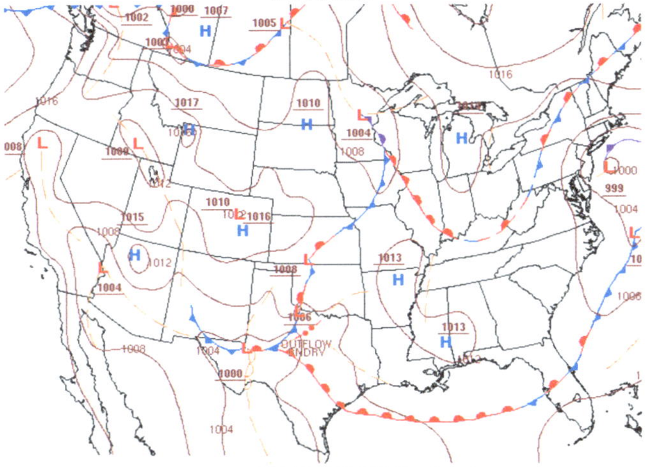

RAIN DISTRIBUTION (Inches)

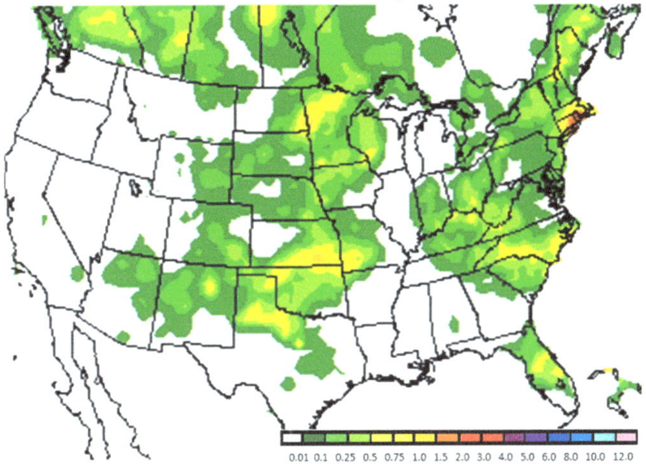

MAXIMUM TEMPERATURE MINIMUM TEMPERATURE

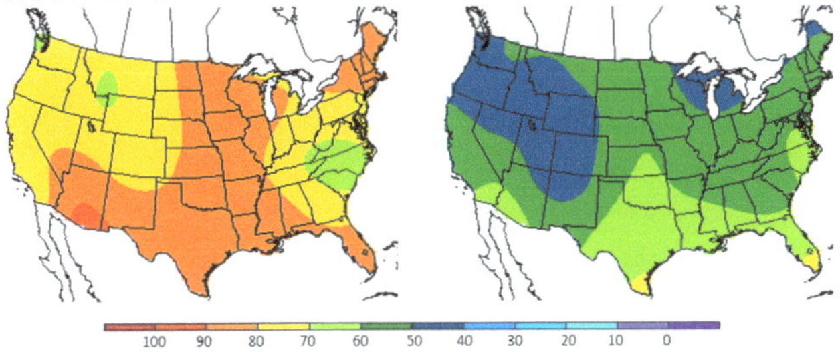

05/07/2025

Heavy rain returns to the Northeast, with some very heavy downpours possible in the Megalopolis.

Widespread precipitation continues to spread across the central parts of the country from the Rocky Mountains in the west and across the Great Plains as well as into the northern parts of Texas and the Four Corner states.

Showers in Florida.

Mostly dry elsewhere, including clearing to mostly dry in the Deep South.

MAY 8 THURSDAY

ISOBARIC

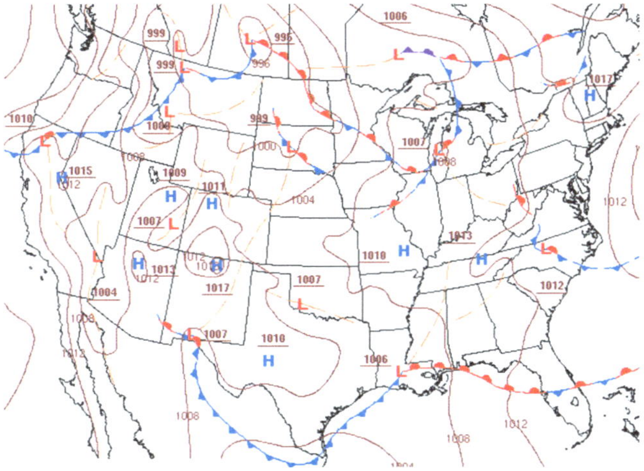

RAIN DISTRIBUTION (Inches)

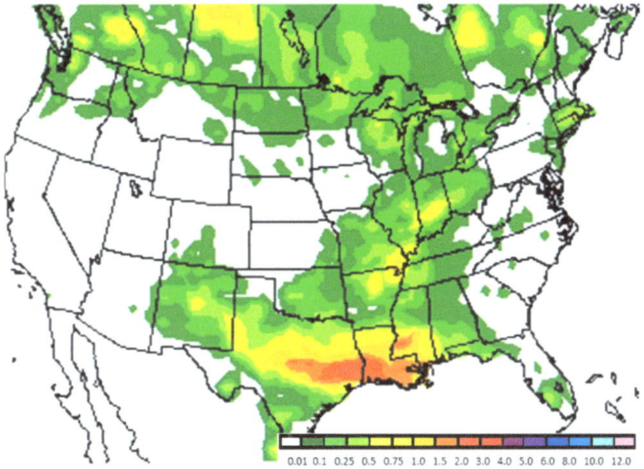

MAXIMUM TEMPERATURE MINIMUM TEMPERATURE

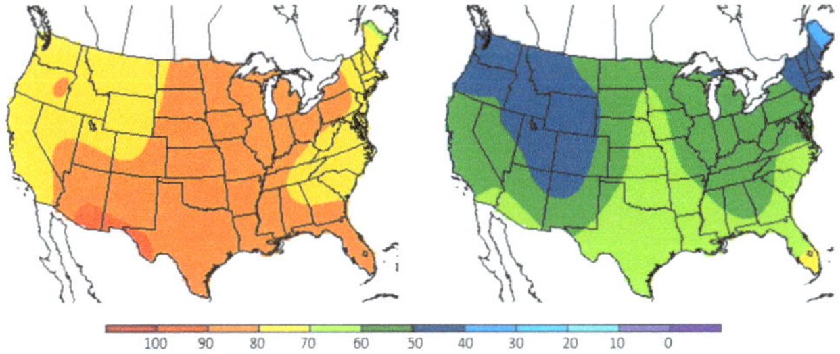

05/08/2025

Widespread precipitation continues to move in an easterly tract out of the central Great Plains and into the Inland Midwest and Great Lakes, with showers into the Northeast, and across the North Great Plains to the Pacific Northwest.

Further south, widespread rain prevails from the Southwest, across Texas and into the Deep South, with very heavy rain likely through central parts of Texas to the southern half of the Deep South.

Some isolated showers may linger around Florida.

Mostly dry everywhere else, including east of the Appalachians.

MAY **9** FRIDAY

ISOBARIC

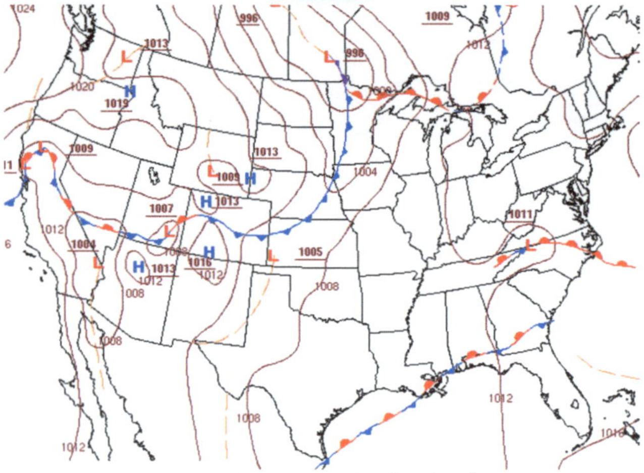

RAIN DISTRIBUTION (Inches)

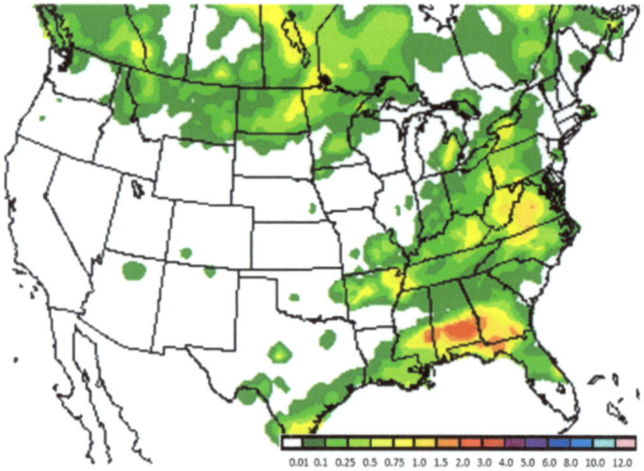

MAXIMUM TEMPERATURE

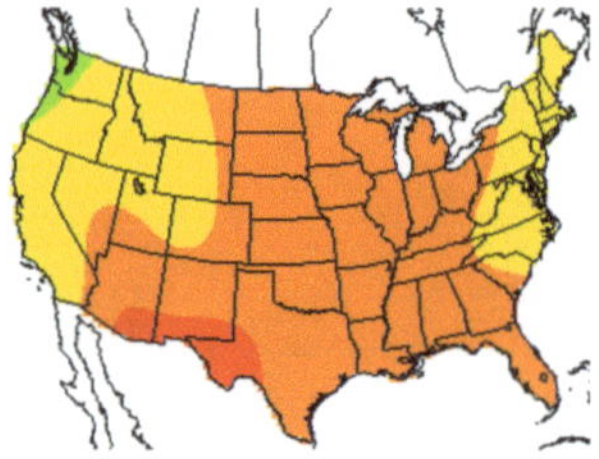

MINIMUM TEMPERATURE

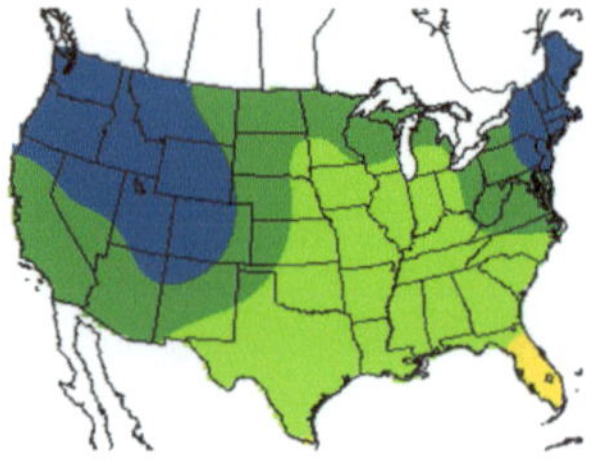

05/09/2025

Widespread rain activity continues to drift eastward, clearing out of the central and southern Great Plains and central parts of the Inland Midwest, while intensifying from the eastern Midwest down through the Appalachians to the Atlantic coasts.

Precipitation is likely to be particularly heavy in the Deep South and around the Carolinas.

Showers also into the Gulf of Mexico coastal areas from Texas to Florida.

Some rain will also continue in the far north from the Inland Northwest to the border states of the North Great Plains.

Mostly dry elsewhere.

05/10/2025

Widespread rain activity in the east of the country continues to contract closer to the Atlantic coast, clearing throughout the Inland Midwest, while persisting in the east from the Appalachians to the Coastal Southeast and southward into Florida and the eastern half of the Deep South.

Occasional showers linger in the Pacific Northwest, along the norther border regions to the Great Lakes, and up through the Northeast.

Mostly dry elsewhere, although there may be a pocket of rain lingering around the South Great Plains.

MAY **10** SATURDAY

ISOBARIC

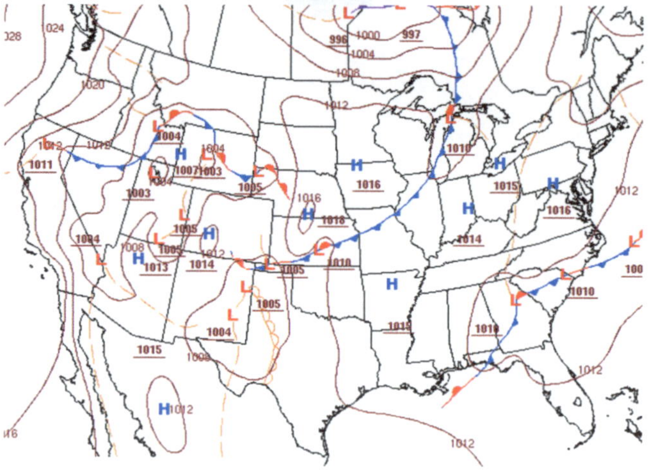

RAIN DISTRIBUTION (Inches)

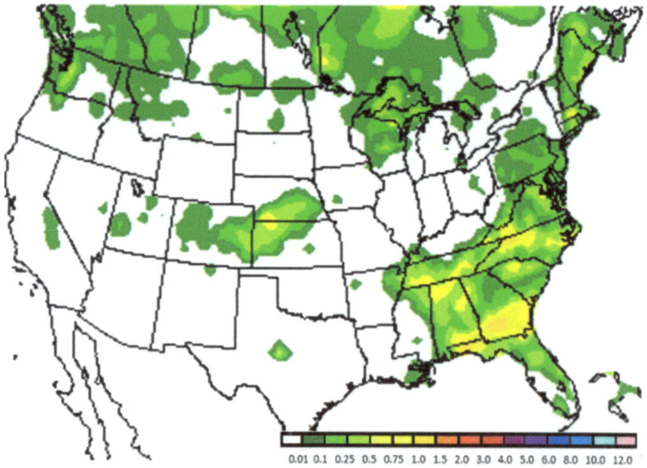

MAXIMUM TEMPERATURE

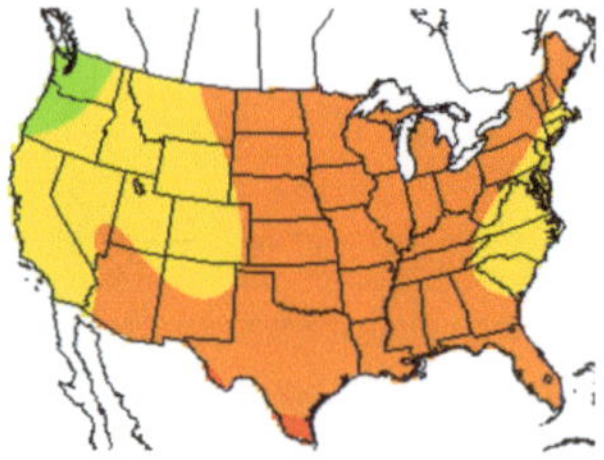

MINIMUM TEMPERATURE

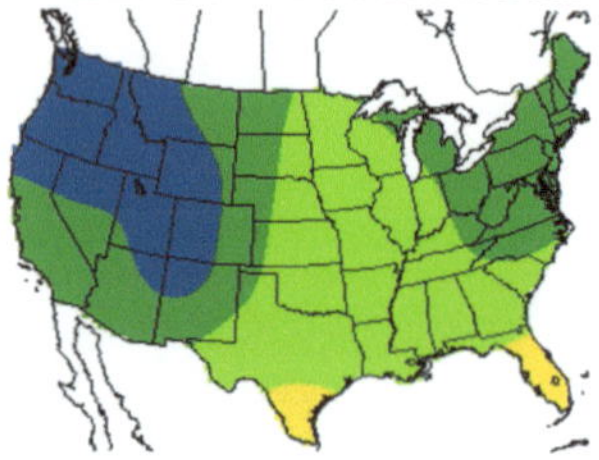

ISOBARIC

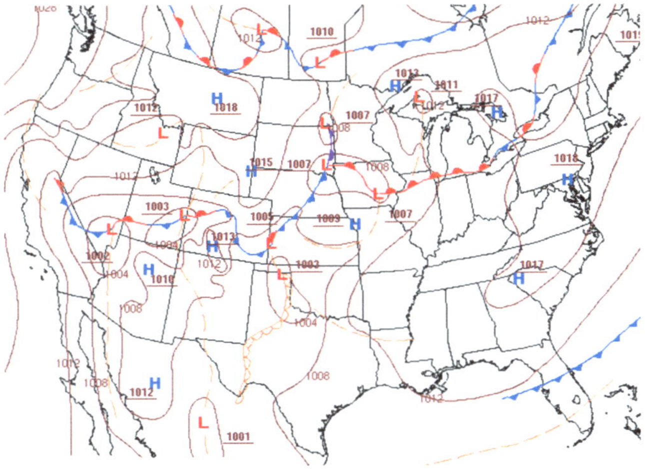

RAIN DISTRIBUTION (Inches)

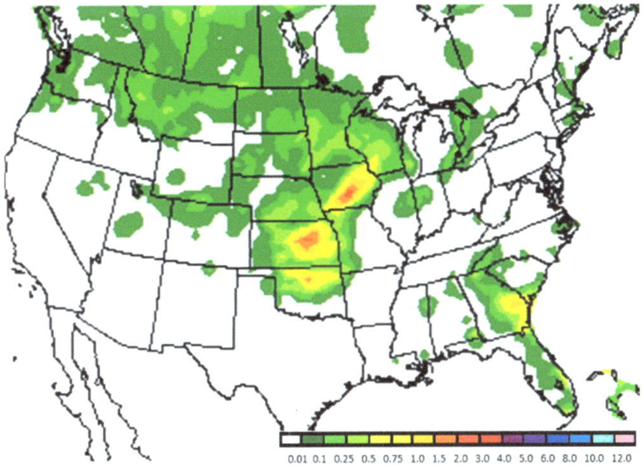

MAXIMUM TEMPERATURE

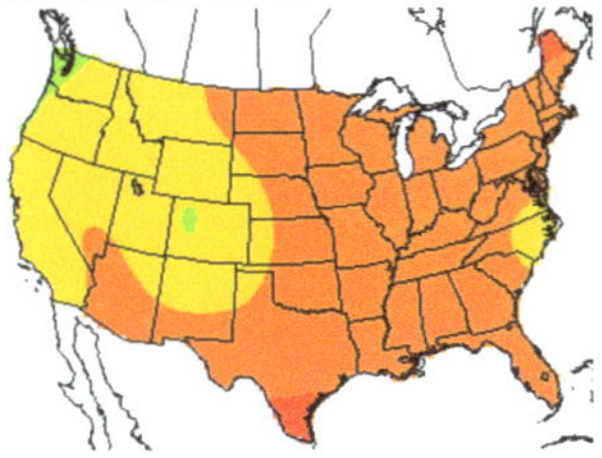

MINIMUM TEMPERATURE

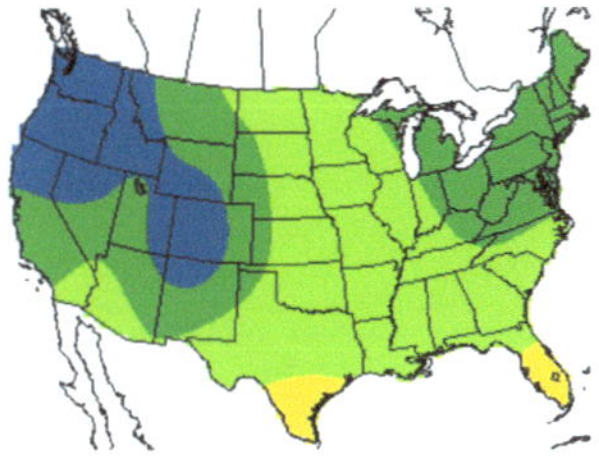

05/11/2025

Stormy conditions across the Great Plains and into the west and north of the Inland Midwest, as well as in the Pacific Northwest and in patches down through the Rocky Mountains.

These stormy conditions may affect the Four Corner States, including into New Mexico and the southwest of Colorado at first, later spreading into Texas and Montana in the next couple of days.

Some heavy showers around Georgia to Florida, and isolated showers in the Northeast and Great Lakes.

Mostly dry elsewhere.

05/12/2025

Widespread rain and scattered storms prevail from the Rocky Mountains over the central Great Plains, and into the Great Lakes and Inland Midwest, with some showery pockets up into the Northeast.

These stormy conditions are likely to spread into the South Great Plains as well as over to the Mid-Atlantic states over the next couple of days, including in West Virginia and Montana.

Outbreaks of rain are also likely in the Inland Northwest, and eastern parts of the Great Basin.

Mostly dry elsewhere.

ISOBARIC

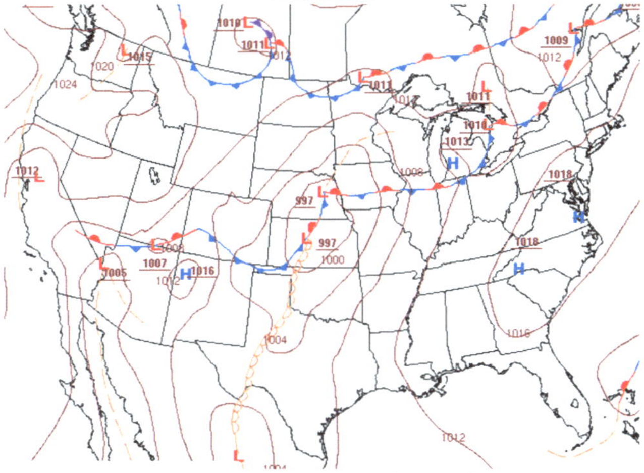

RAIN DISTRIBUTION (Inches)

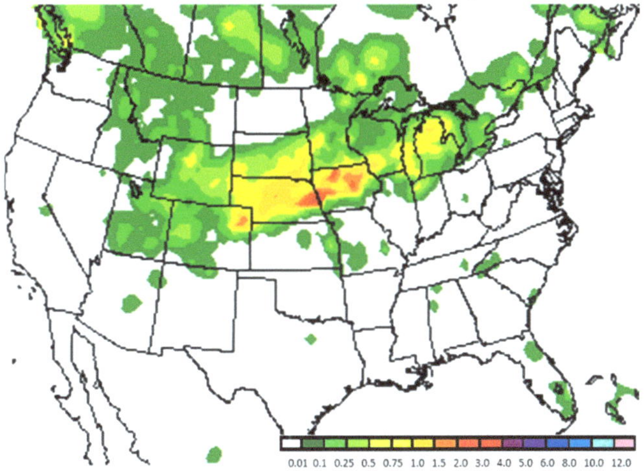

MAXIMUM TEMPERATURE

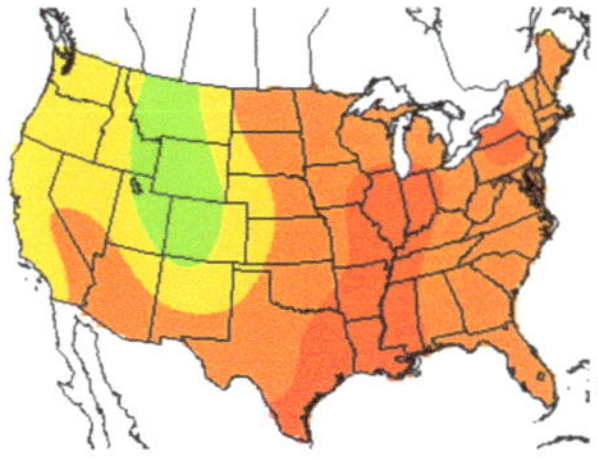

MINIMUM TEMPERATURE

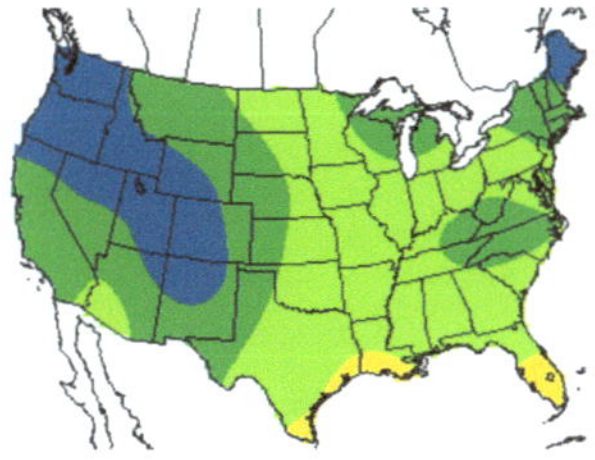

MAY 13 TUESDAY

ISOBARIC

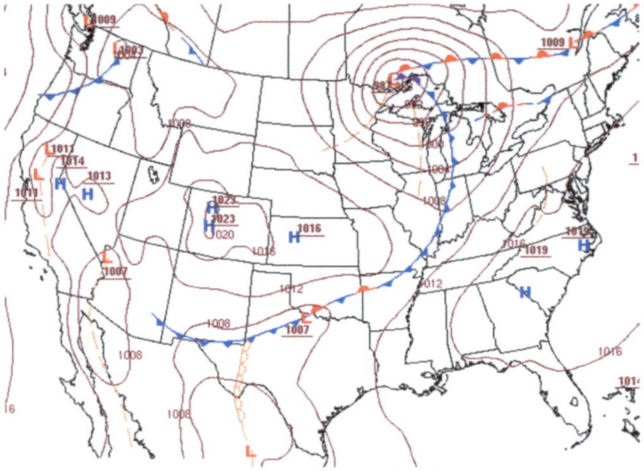

RAIN DISTRIBUTION (Inches)

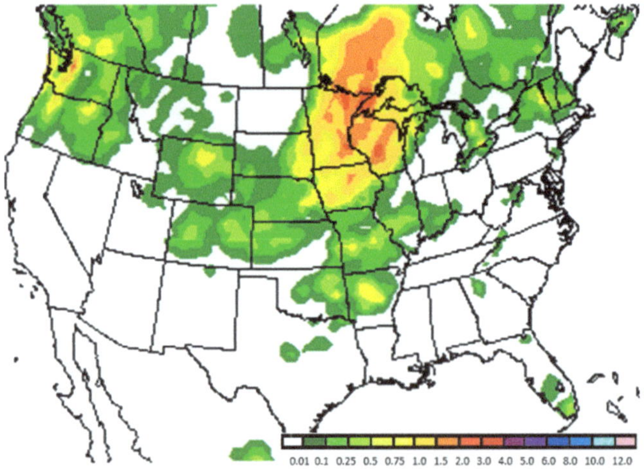

MAXIMUM TEMPERATURE

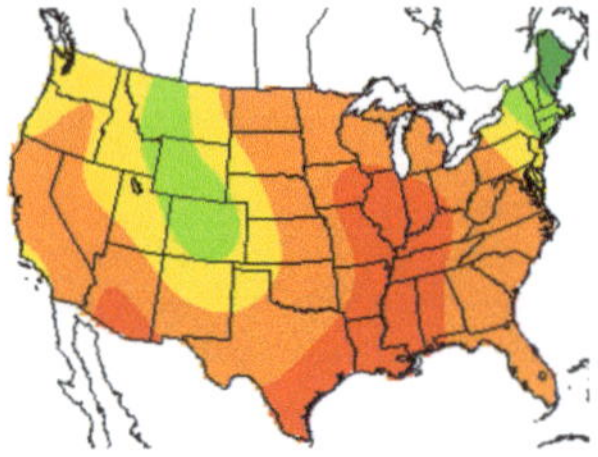

MINIMUM TEMPERATURE

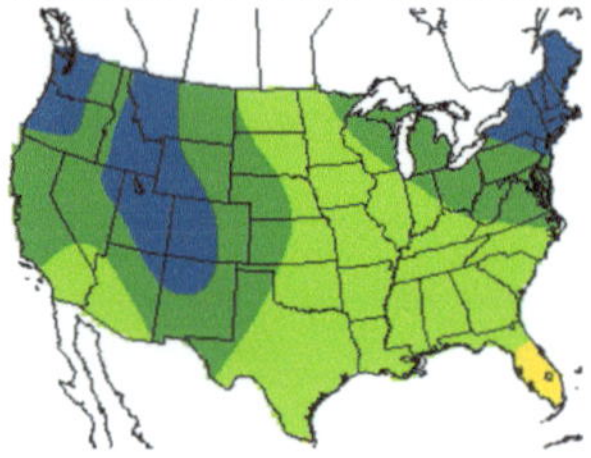

05/13/2025

A deep depression forms over the Great Lakes and Inland Midwest bringing a stormy air over the northern half of the country.

Widespread rain from the Pacific Northwest, through the Rocky Mountains and the Great Plains to the Inland Midwest and Great Lakes, with particularly heavy rain spreading across the south and west of the Corn Belt.

Rainfall may also be particularly heavy in the Dakotas, especially along the Iowa-Nebraska border.

Outbreaks of rain may also reach the Ozarks and the west of the Deep South.

Mostly dry elsewhere. Some showers linger around the Florida Peninsula, as well as in the Northeast.

It may be unusually warm in Hawaii.

05/14/2025

Precipitation spreads out of the Inland Midwest and central Great Plains as the depression associated with the storm yesterday tracks further north, but with widespread rain still persisting from the Pacific Northwest, along the northern borders of the North Great Plains to the Great Lakes and into the Northeast.

Widespread storms and rain through Texas and the Deep South and up along the Appalachians to the Metropolitan and showers likely in southern Florida.

Mostly dry elsewhere.

There is a chance of tornado activity in Oklahoma to Kansas region over the next couple of days as well as in many areas from Texas to South Dakota.

There is also the chance of a snowstorm developing in the northwest mountain regions particularly affecting Wyoming over the next couple of days.

Possibility of some flooding east of the Great Plains to the western parts of the Corn Belt.

MAY 14 WEDNESDAY

ISOBARIC

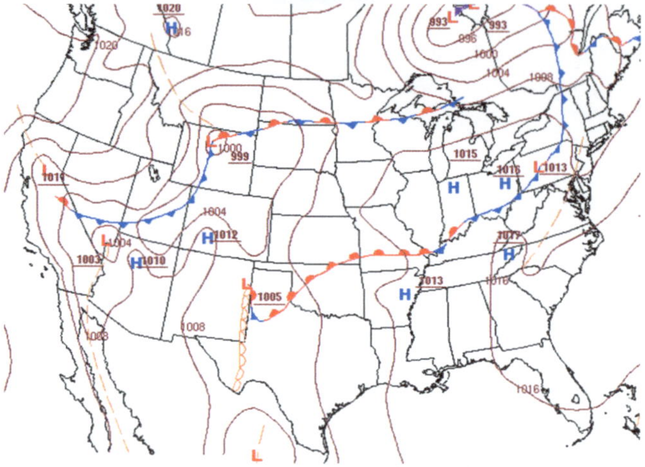

RAIN DISTRIBUTION (Inches)

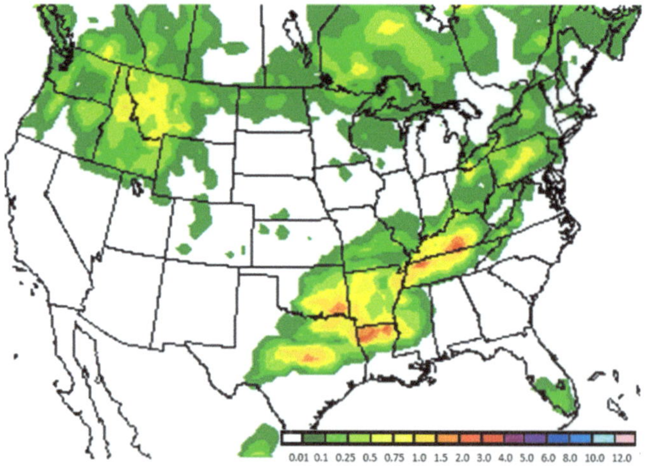

MAXIMUM TEMPERATURE

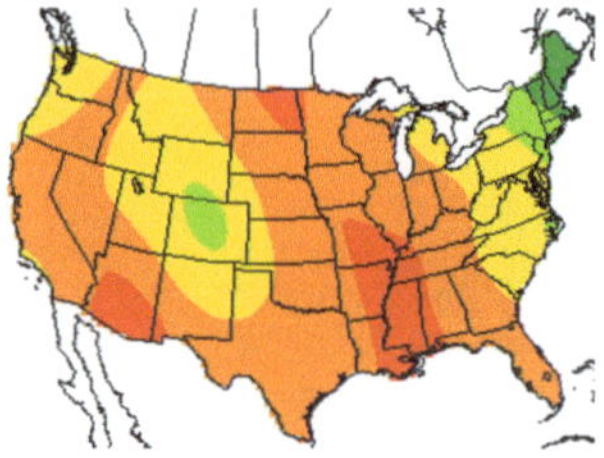

MINIMUM TEMPERATURE

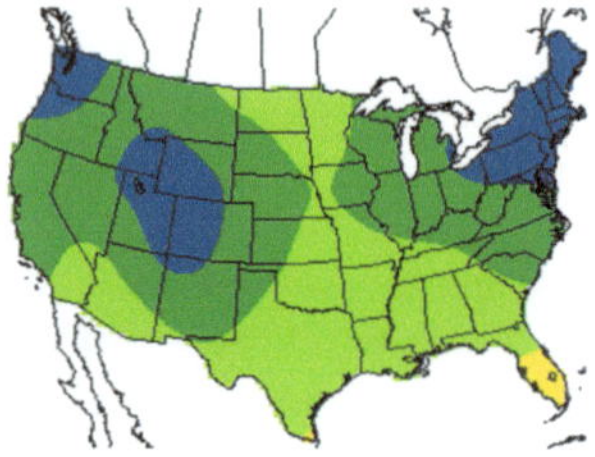

MAY 15 THURSDAY

ISOBARIC

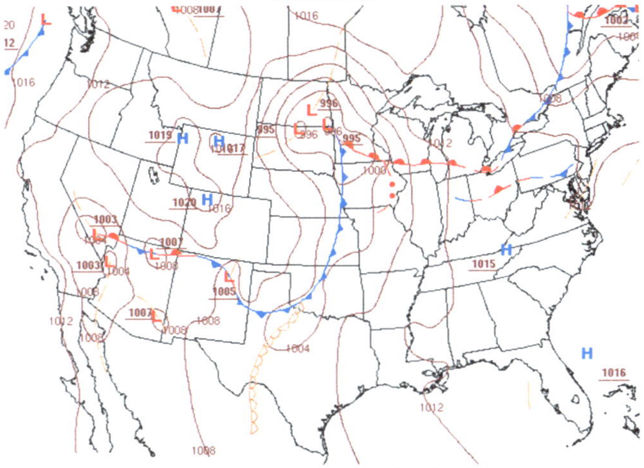

RAIN DISTRIBUTION (Inches)

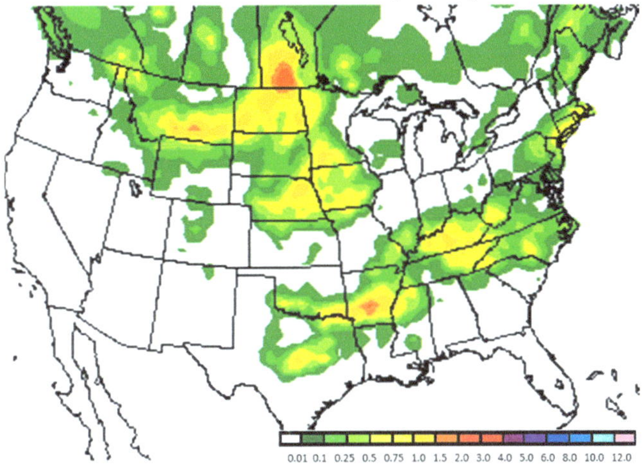

MAXIMUM TEMPERATURE

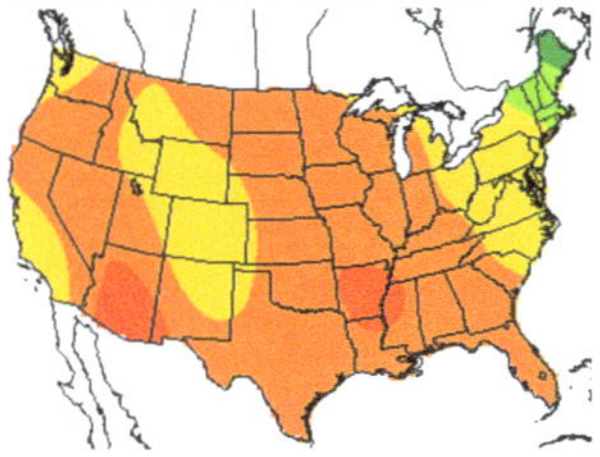

MINIMUM TEMPERATURE

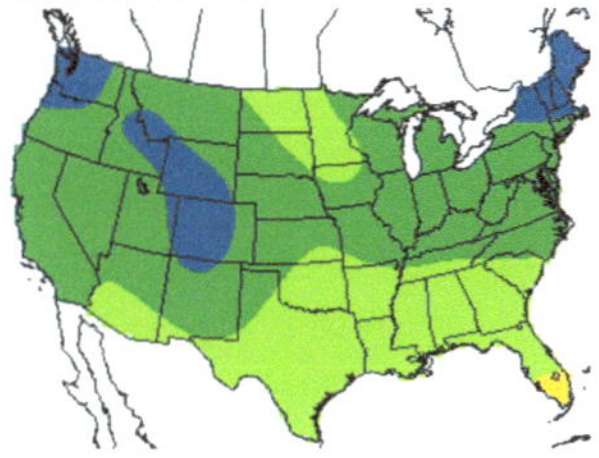

05/15/2025

Widespread precipitation continues across the Great Plains to the Inland Northwest and northern Rocky Mountains, as well as from Texas and up along the Appalachians to the Coastal Southeast regions.

Rain returns to the Northeast, but clears around the Great Lakes and through much of the Inland Midwest.

There is still a chance of snow in Wyoming around this time, as well as heavy rain in South Dakota in particular over the next couple of days.

Mostly dry elsewhere.

There may be outbreaks of torrential rain across much of the mid-section regions of the country over the next few days, especially in South Dakota at first, then Kansas by the following day, with the chance of widespread areas of flooding occurring.

05/16/2025

An occluded front and joins a depression that has drifted northwards out of the Great Plains and into the border regions north of the Great Lakes.

Heavy rain in the eastern Great Plains to the southwest of the Inland Midwest, including in Kansas and Missouri, as well as in the Dakotas to the north.

Heavy rain also through the southern half of the Appalachians and into the northeast of Texas and across the northern half of the Deep South to the Carolinas.

Showers from the Pacific Northwest to the Rockies.

Mostly dry elsewhere.

Chance of late light snow in the Colorado Rockies in the next couple of days.

A sub-tropical storm may form in the Atlantic Ocean, which is likely to bring further unsettled conditions over the east of the country in the coming days.

MAY 16 FRIDAY

ISOBARIC

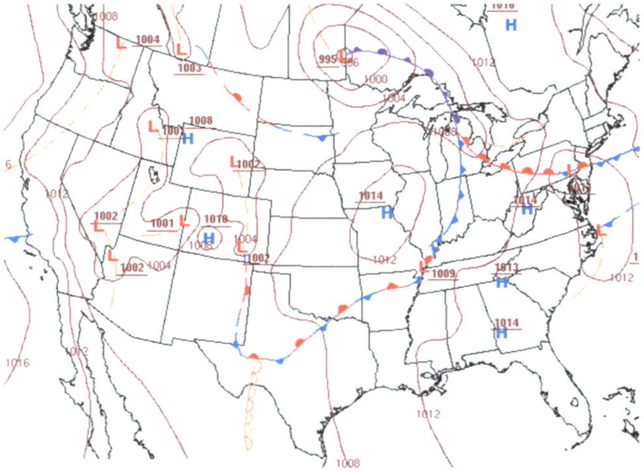

RAIN DISTRIBUTION (Inches)

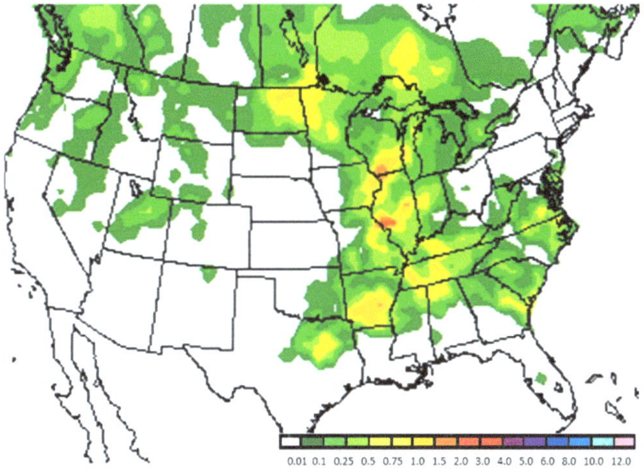

MAXIMUM TEMPERATURE

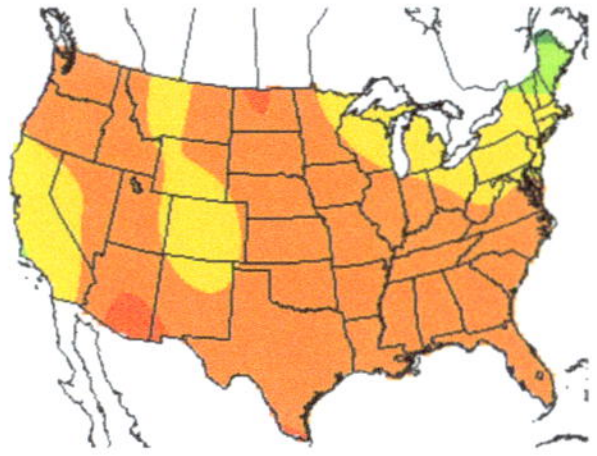

MINIMUM TEMPERATURE

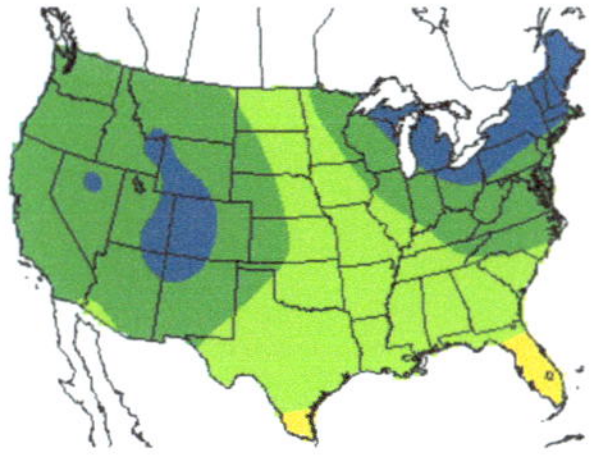

ISOBARIC

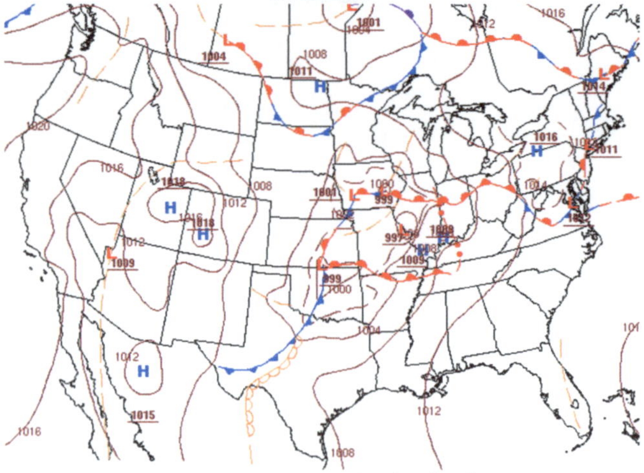

RAIN DISTRIBUTION (Inches)

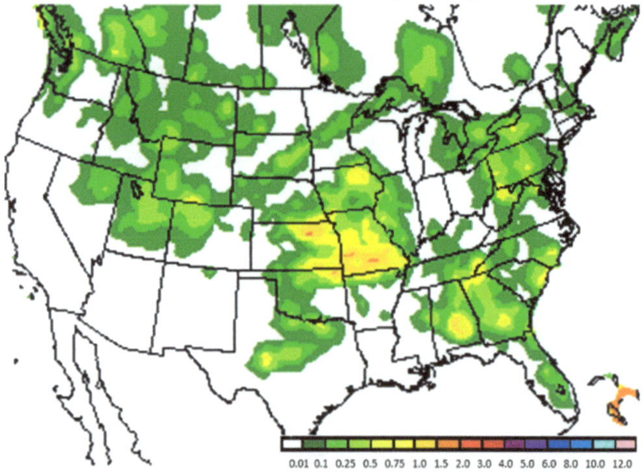

MAXIMUM TEMPERATURE

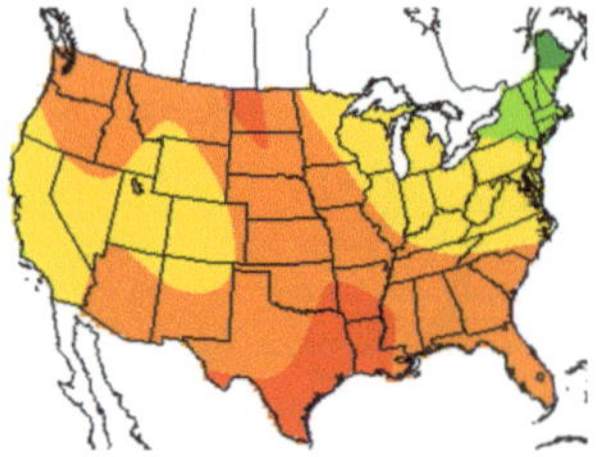

MINIMUM TEMPERATURE

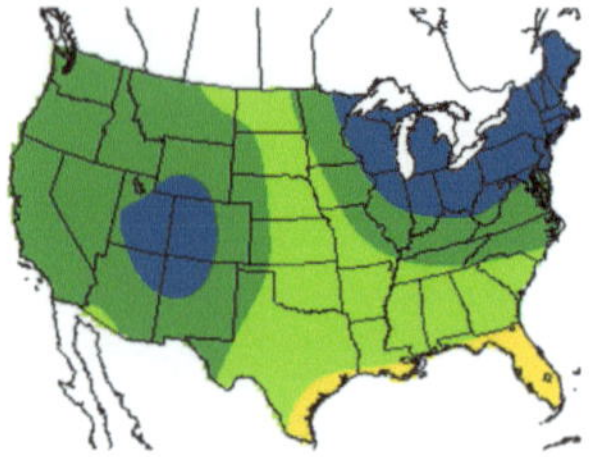

05/17/2025

Outbreaks of rain persist in the Northwest and down into the Great Basin and over the Rocky Mountains, as well as across the Great Plains, and in patches around the Great Lakes to the Megalopolis regions.

Outbreaks of rain, heavy at times, also prevails over the South Great Plains and into the north of Texas as well as into the Deep South and Florida peninsula, with showers up into the Coastal Southeast and through the Appalachians.

Rain may be very heavy around the Ozarks to Texas as well as moving into Oklahoma over the next few days.

Showers may also persist in parts of the Northeast.

Mostly dry elsewhere.

05/18/2025

A depression forms over the Great Lakes bringing further widespread and frequently heavy rain to the northeast quarter of the country, from the Appalachians to the Northeast, as well as spreading into the northern half of the Deep South and over the central and North Great Plains.

This heavy rain may cause flooding along the Missouri and Arkansas River systems over the next few days.

Isolated showers extend into the northern Rocky Mountains.

Mostly dry elsewhere.

ISOBARIC

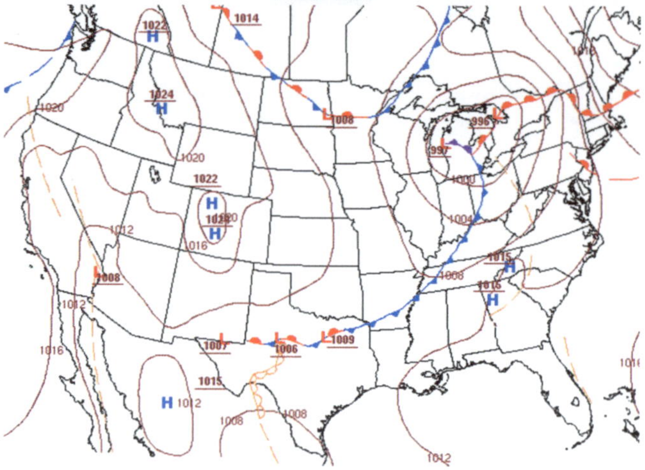

RAIN DISTRIBUTION (Inches)

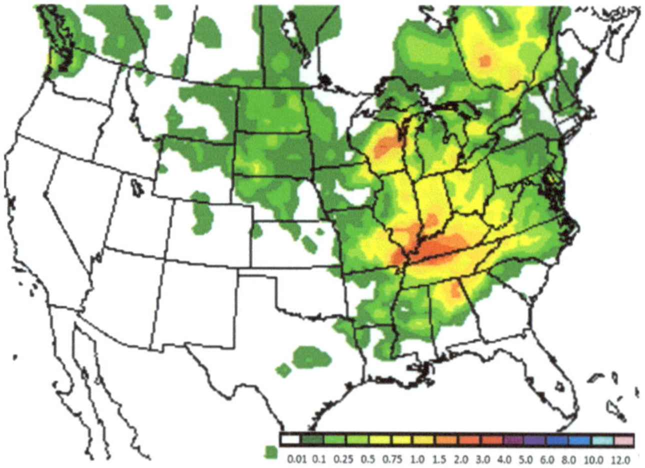

MAXIMUM TEMPERATURE

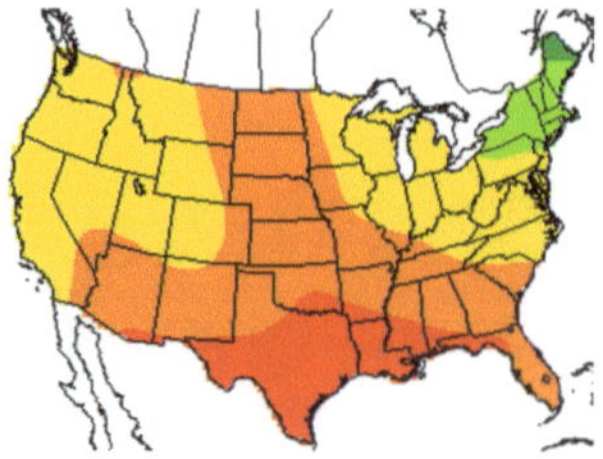

MINIMUM TEMPERATURE

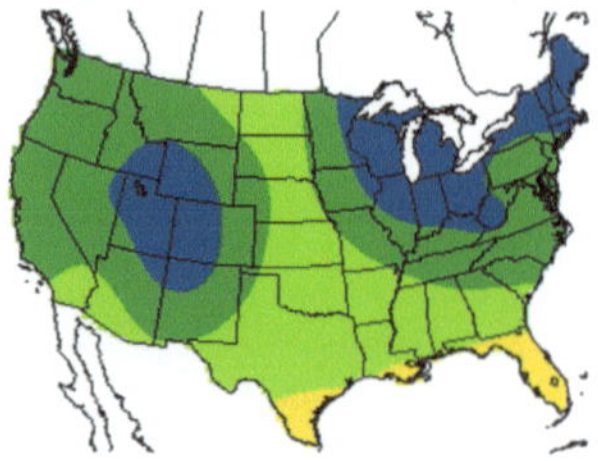

ISOBARIC

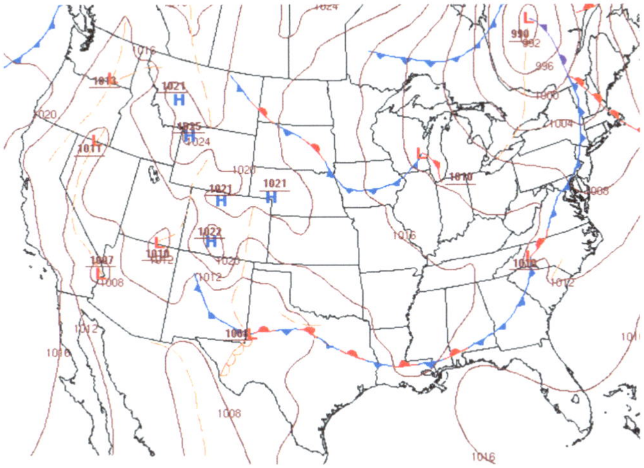

RAIN DISTRIBUTION (Inches)

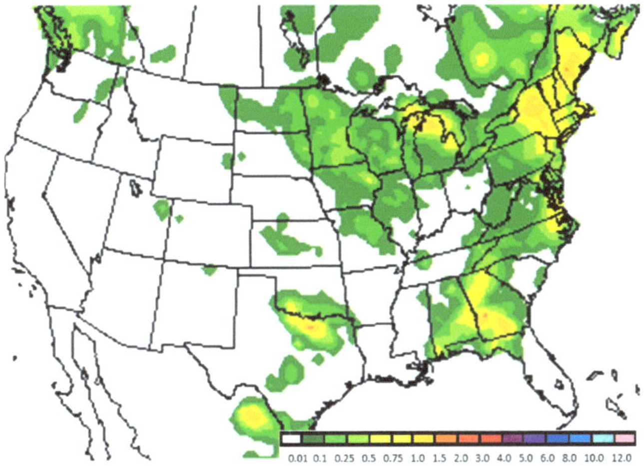

MAXIMUM TEMPERATURE MINIMUM TEMPERATURE

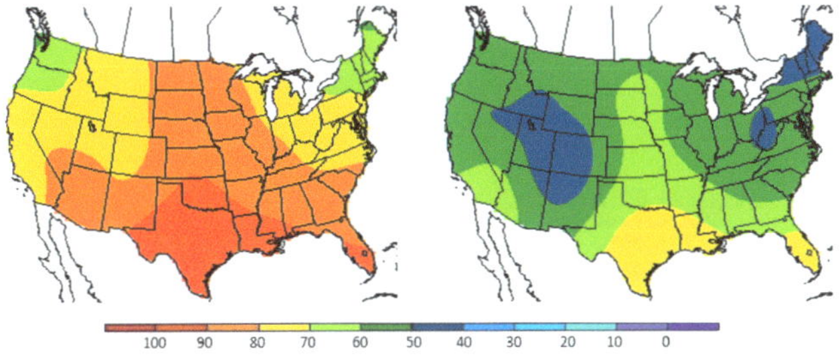

05/19/2025

The depression in the northeast of the country drifts further north away from the Great Lakes region.

Widespread rain and storms continue in the east from the North Great Plains, across the Great Lakes and into the Northeast, as well as down through the states east of the Appalachians to the east half of the Deep South.

Rain eases to showers in the southern parts of the Inland Midwest to the Ozarks, and become isolated in the central and southern Great Plains.

Chance of some brief rain bursts in Texas.

Mostly dry elsewhere.

Chance of a tropical storm watch off the Florida coasts, but it may swiftly decay in the next few days with only a few showers making landfall.

05/20/2025

Widespread outbreaks of rain return over much of the country, with only the Southwest region to the western half of Texas, and the northern Appalachians to the Coastal Southeast regions being mostly dry.

All other parts of the country can expect some precipitation, with storms and some heavy falls likely in a line from the northern Rocky Mountains, across the North Great Plains, down through the western half of the Inland Midwest into the eastern parts of Texas and the Deep South to north Florida, and rain elsewhere.

ISOBARIC

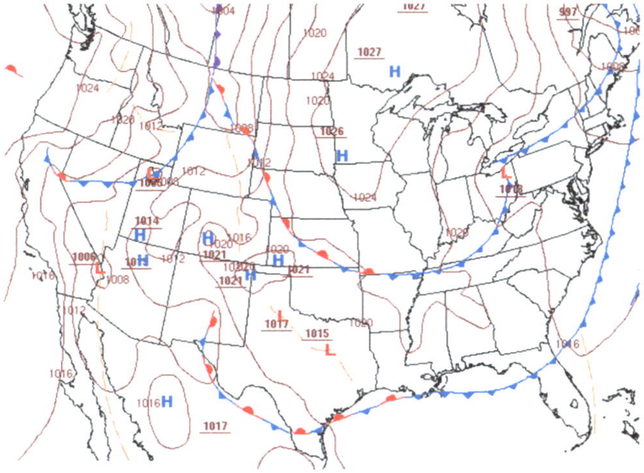

RAIN DISTRIBUTION (Inches)

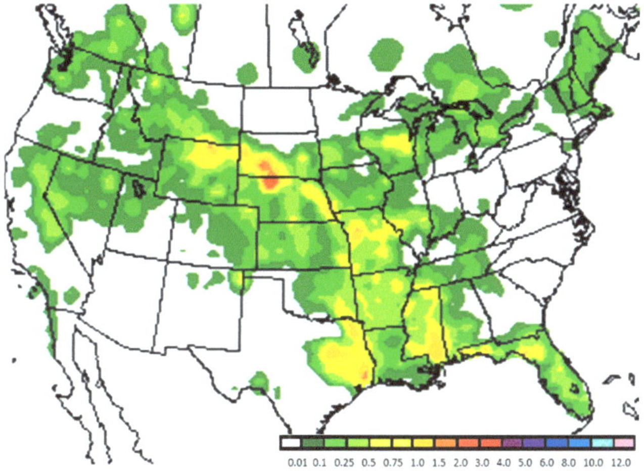

MAXIMUM TEMPERATURE MINIMUM TEMPERATURE

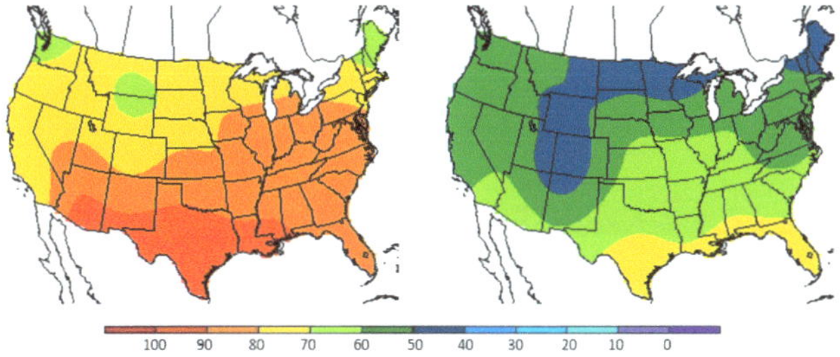

ISOBARIC

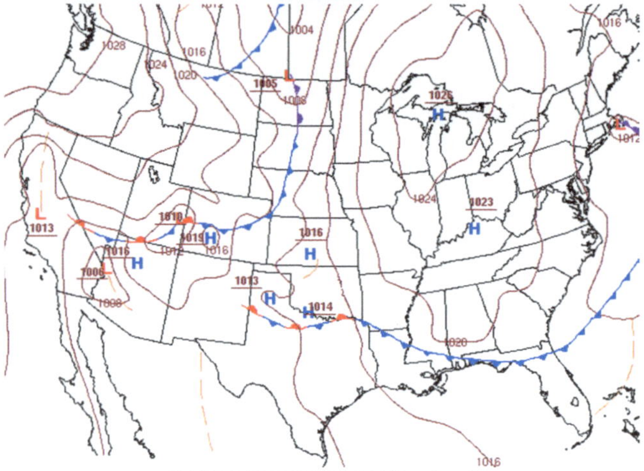

RAIN DISTRIBUTION (Inches)

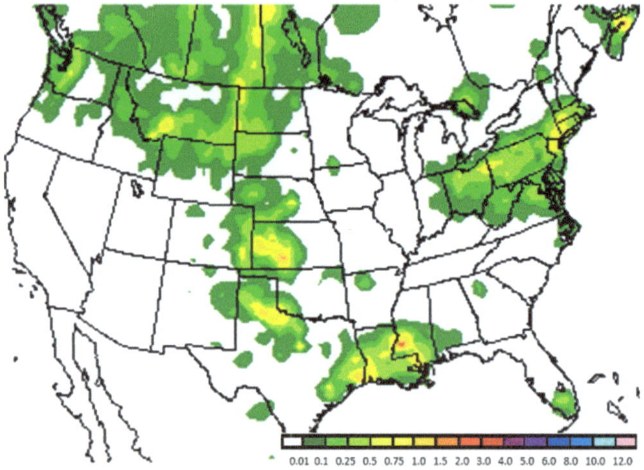

MAXIMUM TEMPERATURE **MINIMUM TEMPERATURE**

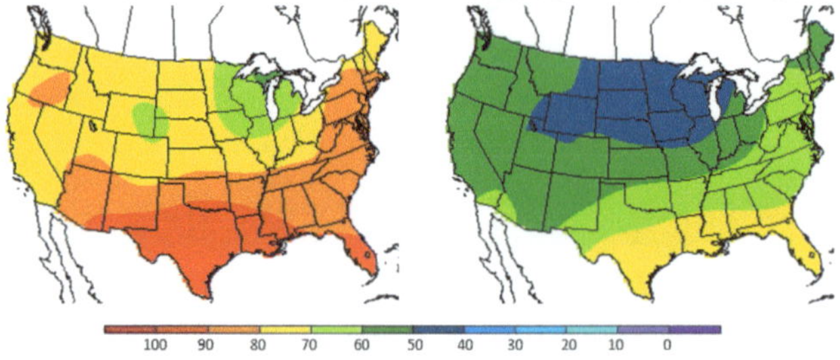

05/21/2025

Rain clears in many regions but persists in the Pacific Northwest and down through the Rocky Mountains, as well as around the northwest of the Great Plains to northern Texas.

Rain also persists in the south from around the Gulf of Mexico regions, from Texas to the Deep South, and in the north from the east of the Inland Midwest to the Northeast Megalopolis regions.

Mostly dry elsewhere.

05/22/2025

Large band of storms and rain, often heavy, prevails down through the midsection of the country, with mostly dry conditions prevailing to the west and east of the storm band.

All regions west of the Rocky Mountains can expect to be mostly warm and dry, and similarly, all regions east of the Great Plains may also be warm and dry.

The only exceptions in the east may be some coastal showers along the Coastal Southeast region to the Northeast, and around the southern half of Florida.

Heavy rain on the Great Plains around this time may see flooding along the Grand River system in Missouri.

Showers likely in Alaska.

ISOBARIC

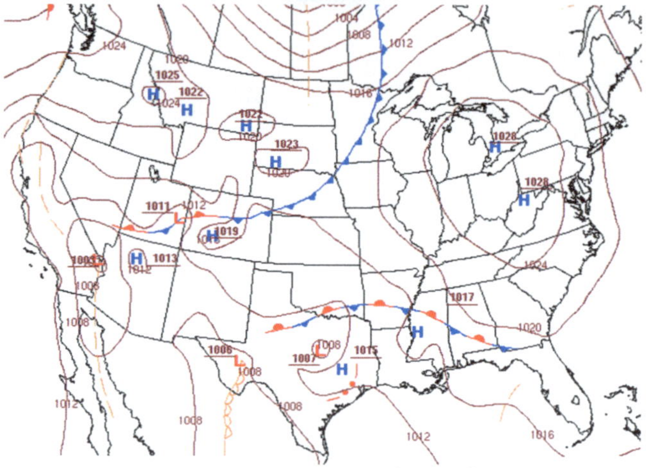

RAIN DISTRIBUTION (Inches)

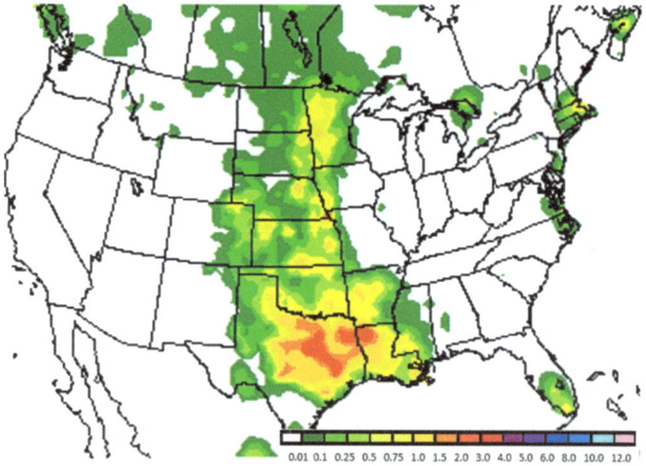

MAXIMUM TEMPERATURE **MINIMUM TEMPERATURE**

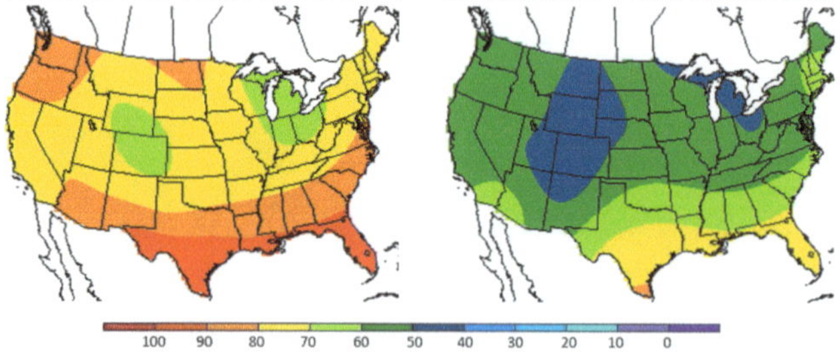

MAY 23 FRIDAY

ISOBARIC

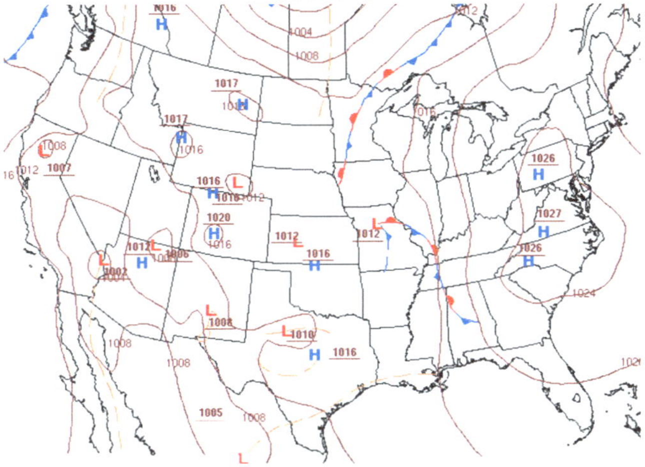

RAIN DISTRIBUTION (Inches)

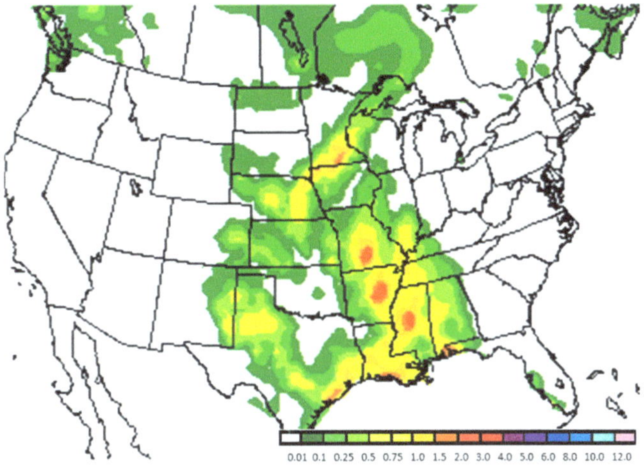

MAXIMUM TEMPERATURE / MINIMUM TEMPERATURE

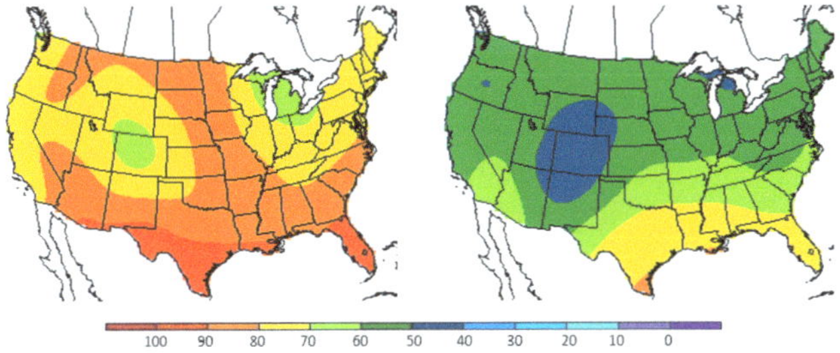

05/23/2025

Bands of rain continue to drift back and forth across the midsection of the country, with heavy rain particularly in the Deep South to the Ozarks, around the coastal Gulf regions, and in the western half of the Inland Midwest.

Some showers in Maine to New Hampshire, in the far northwest of Washington and along the northern borders of North Dakota to Lake Superior.

Mostly dry elsewhere.

Outbreaks of rain return in southern Alaska over most of this coming week.

05/24/2025

Scattered rain activity prevails through the Great Plains and down into Texas.

Widespread and occasionally heavy rain also prevails from the Great Lakes southwards through the Inland Midwest, across most of the Appalachians and into the Deep South.

Showers in the Pacific and Inland Northwest to the northern Rocky Mountains and in the far Northeast.

Mostly dry elsewhere, including along the Atlantic coast from the southern Northeast to Florida, although some showers may brush the southern tip of the Florida Keys.

There is a chance of snow in southern parts of Alaska over the next few days.

MAY 24 SATURDAY

ISOBARIC

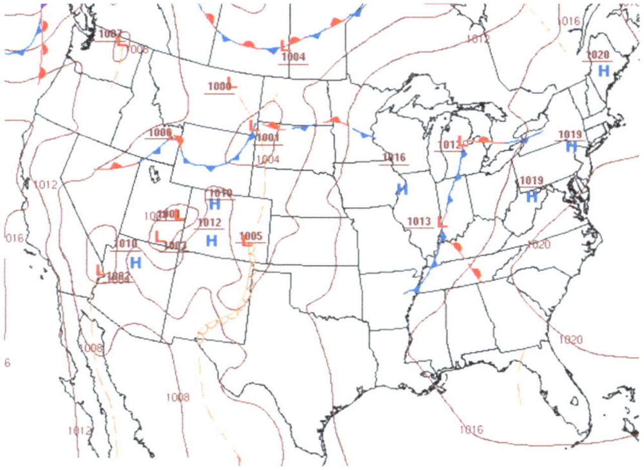

RAIN DISTRIBUTION (Inches)

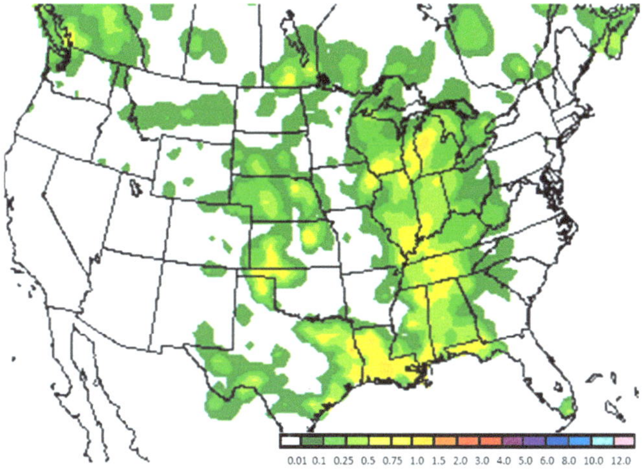

MAXIMUM TEMPERATURE / MINIMUM TEMPERATURE

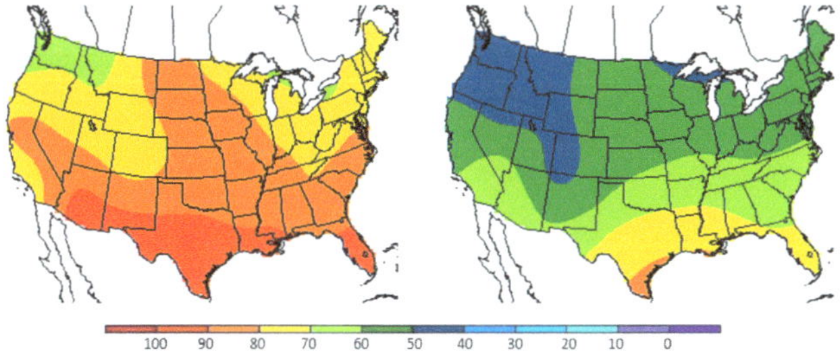

MAY 25 SUNDAY

ISOBARIC

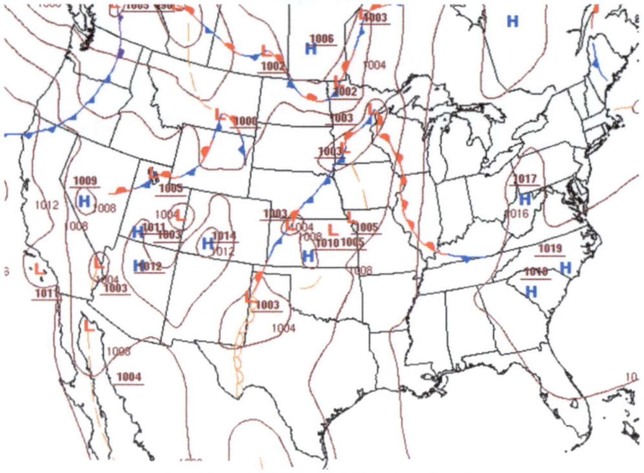

RAIN DISTRIBUTION (Inches)

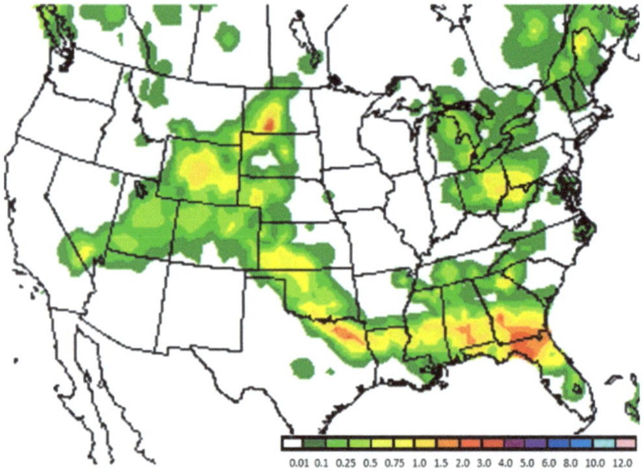

MAXIMUM TEMPERATURE

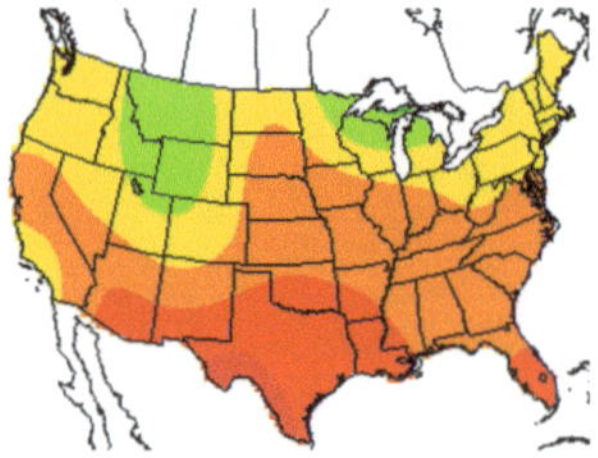

MINIMUM TEMPERATURE

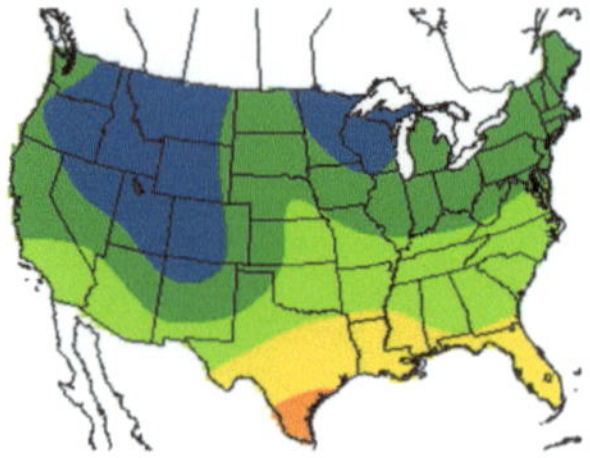

05/25/2025

Outbreaks of rain prevail from the Great Lakes and eastern Inland Midwest to the Northeast, while widespread rain, often heavy, prevails from the Four Corner states and Rocky Mountains, across the Great Plains and into the Deep South and Florida.

Rain may be heavy at times in the Deep South to the northern half of Florida, as well as some heavy showers likely in Arizona.

Mostly dry elsewhere.

05/26/2025

Widespread rain moves into the northwest quarter of the country bringing frequent and heavy falls at times across the Pacific Northwest and Inland Northwest, to the North Great Plains, with some showers spreading into the northern Rocky Mountains.

Heavy rain is also likely to return through the Great Lakes region into the Dakotas.

Heavy precipitation also prevails in the southeast of the country from the Ozarks and the northeast of Texas to the Deep South and Carolinas, with showers around the Gulf coastal fringes.

Mostly dry elsewhere.

MAY 26 MONDAY

ISOBARIC

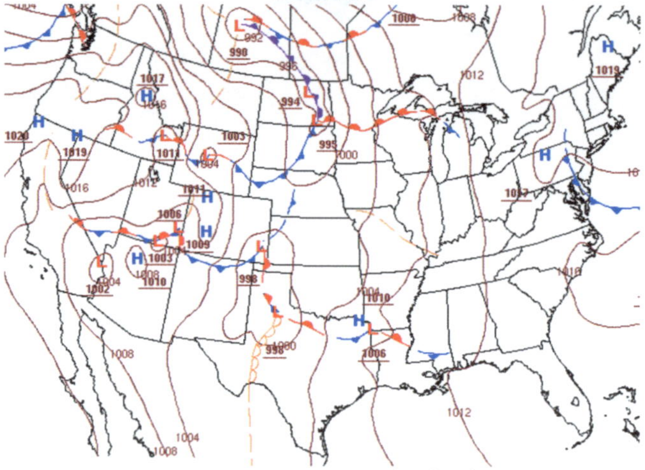

RAIN DISTRIBUTION (Inches)

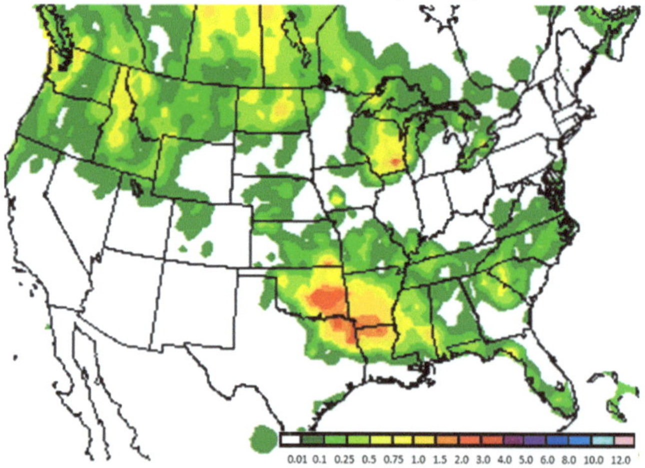

MAXIMUM TEMPERATURE

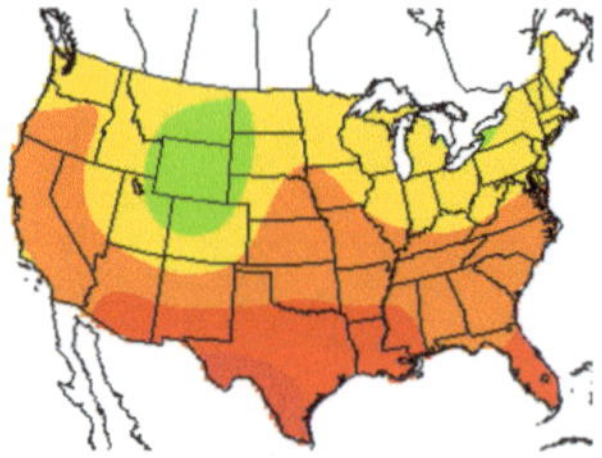

MINIMUM TEMPERATURE

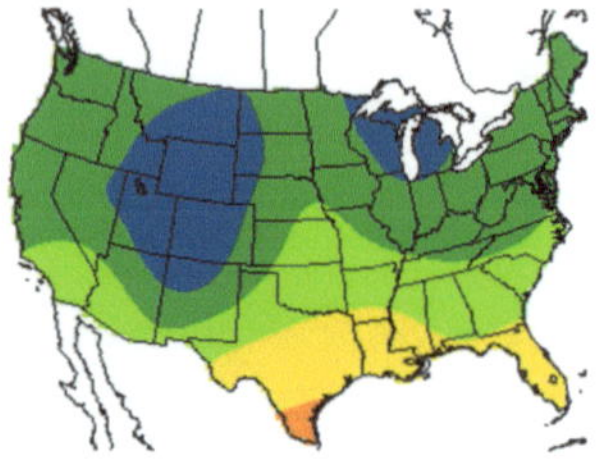

MAY 27 TUESDAY

ISOBARIC

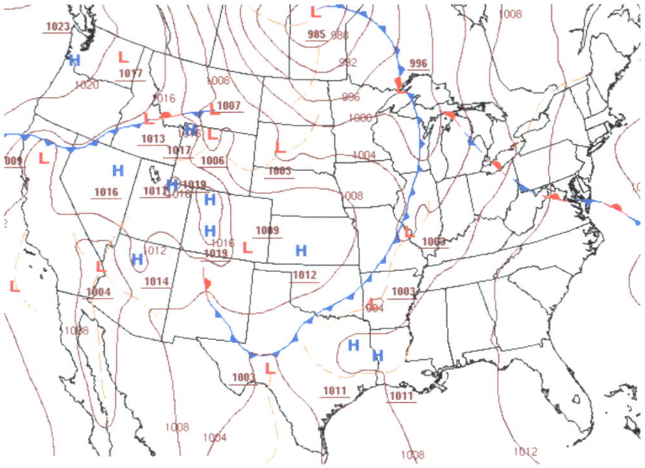

RAIN DISTRIBUTION (Inches)

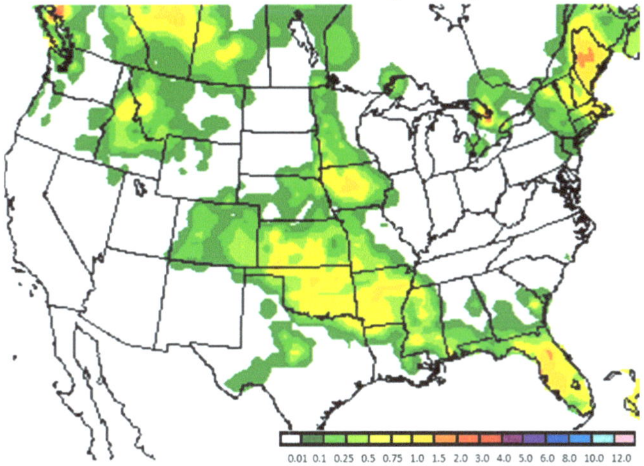

MAXIMUM TEMPERATURE MINIMUM TEMPERATURE

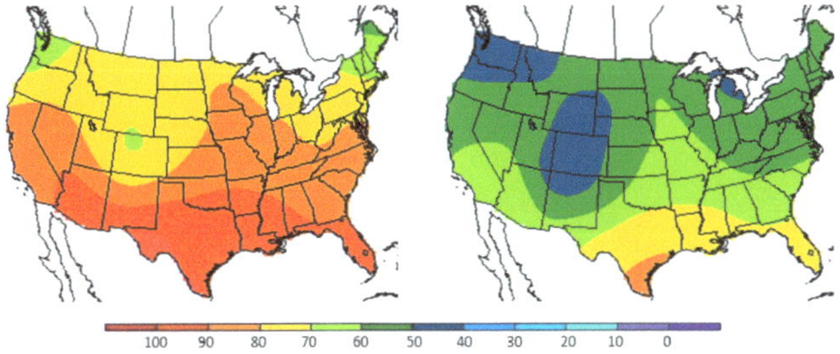

05/27/2025

Rain in the Pacific Northwest eases but persist in the Inland Northwest to the northern Rocky Mountains.

Precipitation persists from the southern Rocky Mountains and through the South Great Plains into the Deep South and Florida, as well as extending up through the Dakotas.

Heavy rain in the Northeast with showers to Lake Erie.

Mostly dry elsewhere.

05/28/2025

Widespread rain prevails across the eastern half of the country from the Inland Midwest and Texas to the Atlantic coastal regions with particularly heavy rain at times over the northern half of the Appalachians as well as in pockets around the Great Lakes to the Northeast, and in the eastern half of Texas.

Mostly dry elsewhere from the Great Plains to the Pacific coasts, and only isolated showers in Florida.

A return to wintry precipitation may also occur in southern parts of Alaska and persist over the next couple of days.

MAY 28 WEDNESDAY

ISOBARIC

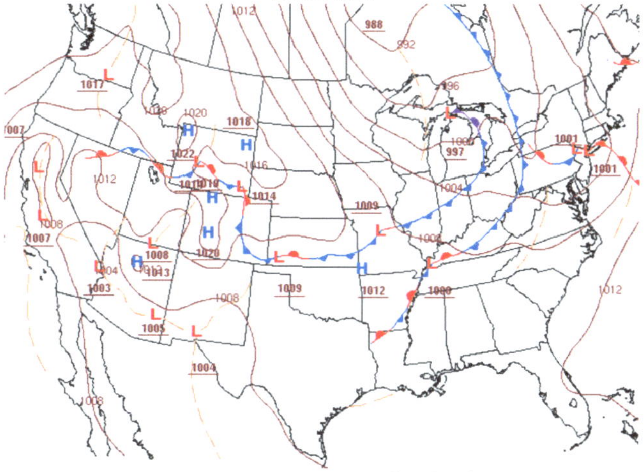

RAIN DISTRIBUTION (Inches)

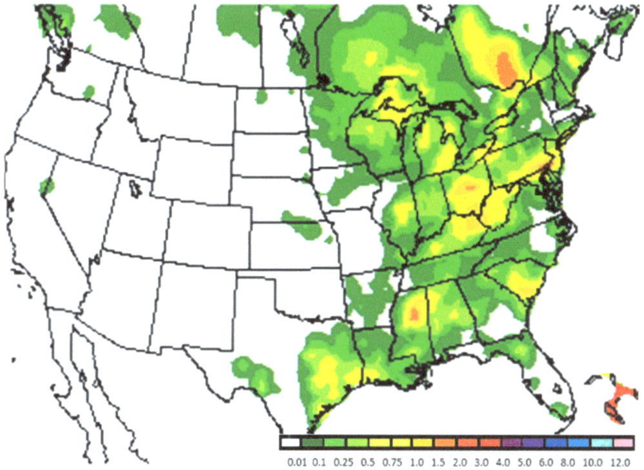

MAXIMUM TEMPERATURE MINIMUM TEMPERATURE

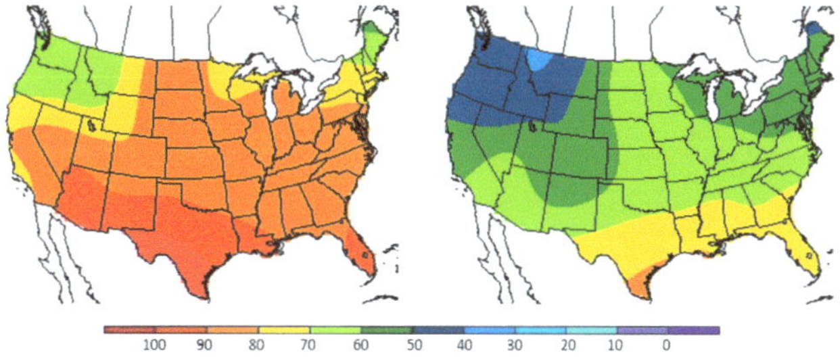

MAY **29** THURSDAY

ISOBARIC

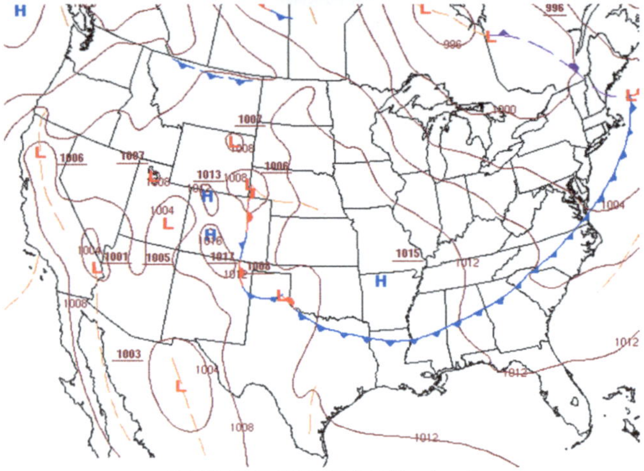

RAIN DISTRIBUTION (Inches)

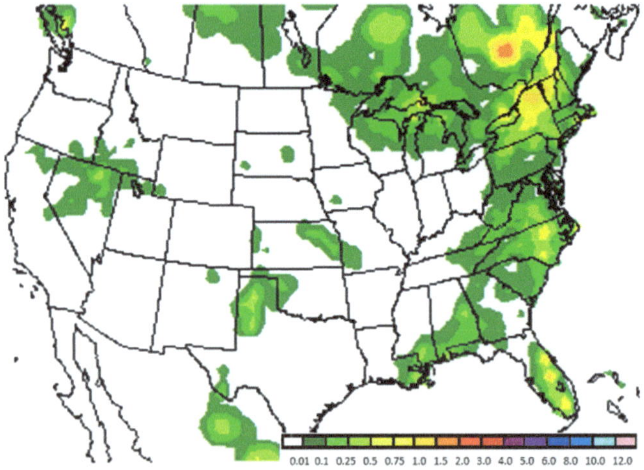

MAXIMUM TEMPERATURE / MINIMUM TEMPERATURE

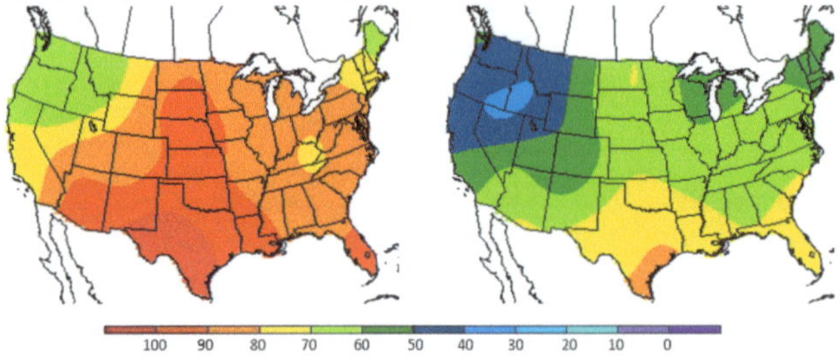

05/29/2025

The cold fronts that brought widespread rain over the eastern half of the country in the past couple of days continues to tract eastward, with rain clearing over through the Inland Midwest, as well as in the Texas to the western half of the Deep South.

Rain activity continues to be frequent and heavy at times in the Great Lakes to the Northeast and down through the Atlantic coastal regions east of the Appalachians, through the Coastal Southeast and return once more into Florida.

A pocket of rain may return in the Great Basin region.

Mostly dry elsewhere, although some showers may also linger on the Great Plains.

05/30/2025

Showers in the Great Basin region, across the central Rocky Mountains heavier rain throughout the Great Plains, as well as showers extending into the northern parts of the Southwest, and outbreaks of heavy rain into the Ozarks.

Rain is also likely to continue throughout the Northeast and Great Lakes region, as well as some outbreaks of showers in the southern half of Florida.

Some precipitation may be wintry over the North Great Plains to the Inland Midwest and fall as snow in places.

Mostly dry in Texas and in most regions from the Appalachians to the Atlantic Coast, as well as clearing to mostly dry in the Deep South.

It may be mostly dry in the Pacific and Inland Northwest, and continue to be dry down through California to Arizona.

MAY **30** FRIDAY

ISOBARIC

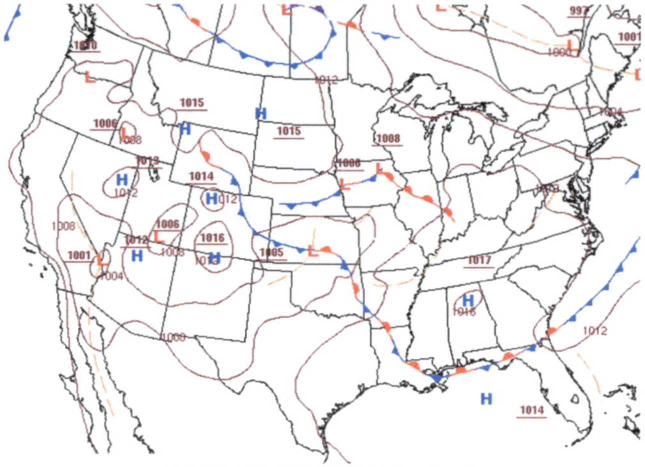

RAIN DISTRIBUTION (Inches)

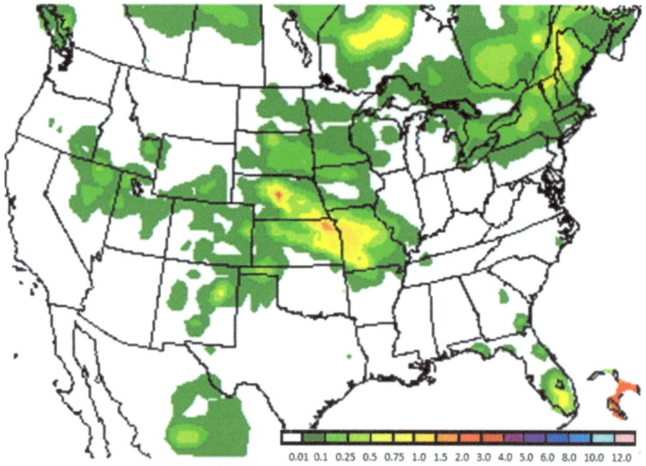

MAXIMUM TEMPERATURE / MINIMUM TEMPERATURE

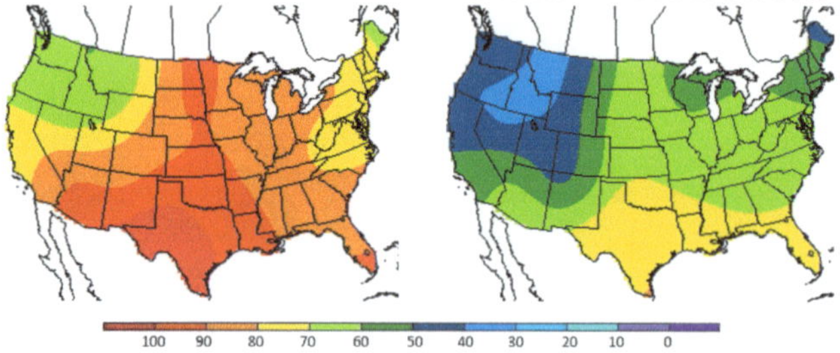

MAY 31 SATURDAY

ISOBARIC

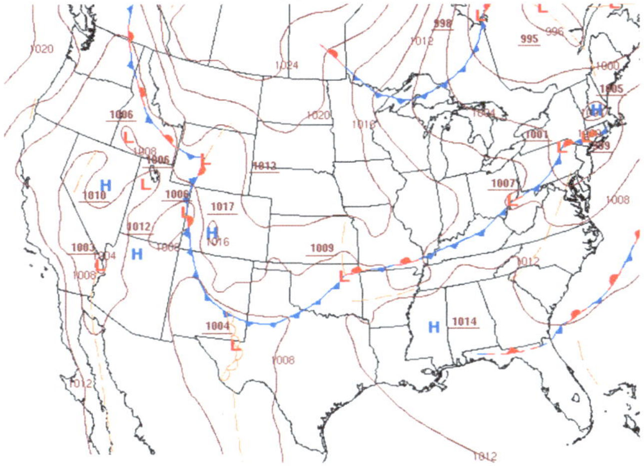

RAIN DISTRIBUTION (Inches)

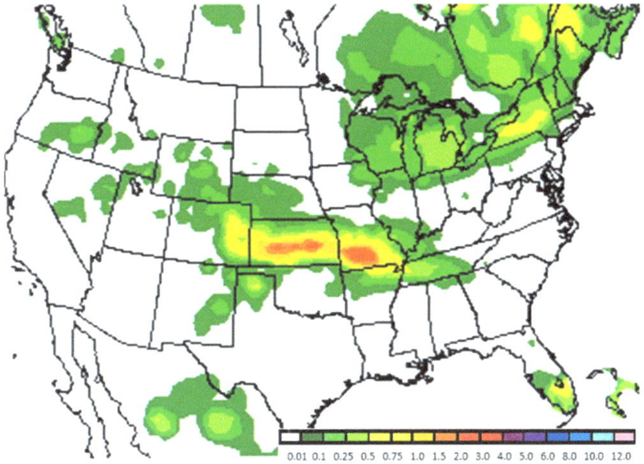

MAXIMUM TEMPERATURE

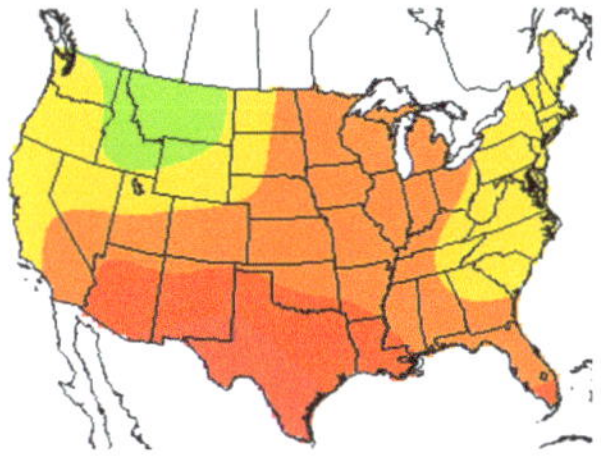

MINIMUM TEMPERATURE

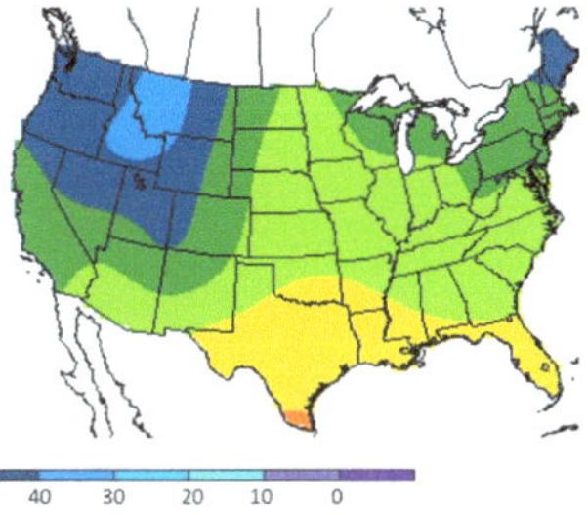

05/31/2025

Widespread rain still prevails from the Great Lakes and northern parts of the Inland Midwest to the Northeast.

Showers over the southern Rocky Mountains.

Outbreaks of heavy rain through the South Great Plains, across the Ozarks and into the southern parts of the Appalachians.

Some rain may also linger around the southern parts of Florida, and showers into the Great Basin.

Mostly dry elsewhere.

A cooler spell in the northeast of the country may see some patches of frost around the Great Lakes, Northeast, North Great Plains, and Upper Midwest.

JUNE

PHASES OF THE MOON

1st	Northern Declination
4th	First Quarter
9th	Crossing Equator
11th	Apogee
12th	Full Moon
15th	Southern Declination
20th	Third Quarter
23rd	Crossing Equator
26th	Perigee #5
27th	New Moon
28th	Northern Declination

1 Sun	2 Mon	3 Tue	4 Wed	5 Thur	6 Fri	7 Sat	8 Sun	9 Mon	10 Tue	11 Wed	12 Thur	13 Fri	14 Sat	15 Sun	16 Mon	17 Tue	18 Wed	19 Thur	20 Fri	21 Sat	22 Sun	23 Mon	24 Tue	25 Wed	26 Thur	27 Fri	28 Sat	29 Sun	30 Mon
		1Q		XhS		A				F V							3Q	XhN				P8		N ^					

JUNE 1 SUNDAY

ISOBARIC

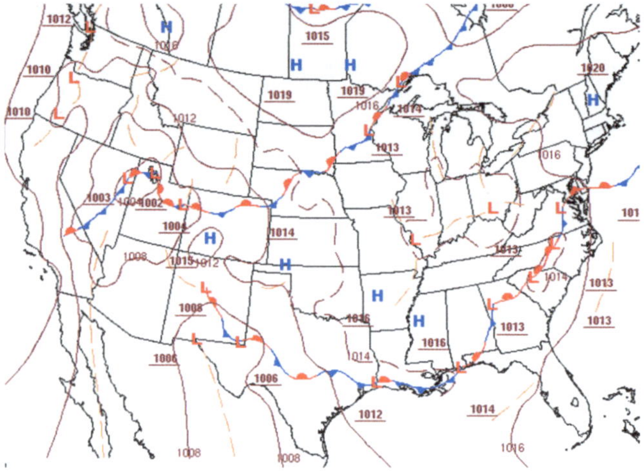

RAIN DISTRIBUTION (Inches)

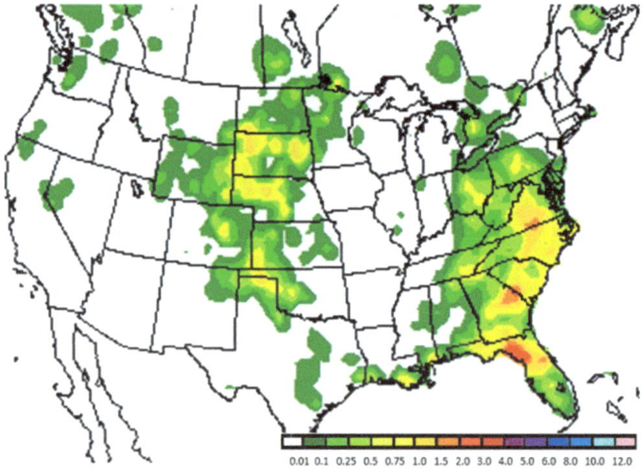

MAXIMUM TEMPERATURE

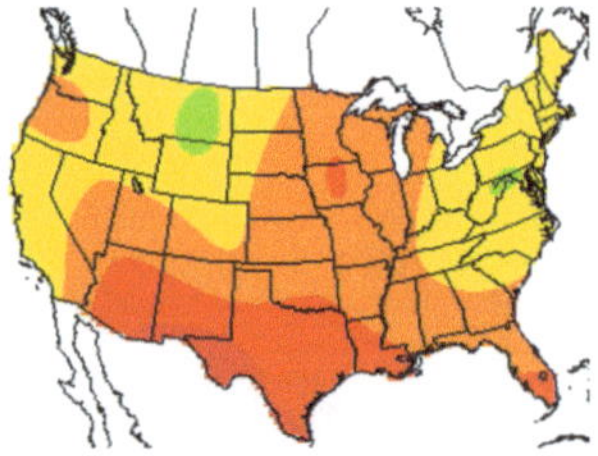

MINIMUM TEMPERATURE

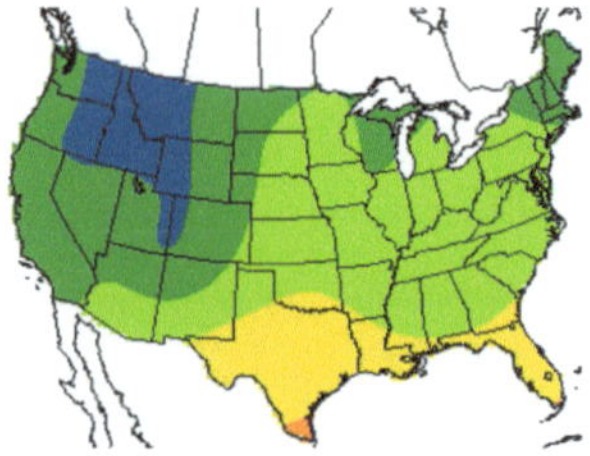

06/01/2025

Widespread rain prevails over much of the eastern parts of the country from the Atlantic Coast to the western side of the Appalachians and extending from the eastern Great Lakes to Florida, with some heavy falls likely along the Atlantic coastal fringes of the Coastal Southeast to the northern half of Florida.

Storm activity moves onto the Great Plains, with the chance of heavy wintry precipitation in South Dakota to Montana, possibly bringing more snow into the region, particularly in Montana.

A pool of cold air in the northwest may see snow up into the northern Rocky Mountains, and freezing temperatures may prevail around Idaho and Nevada over the next couple of days.

Mostly dry elsewhere.

Heavy rain in southern parts of Alaska.

06/02/2025

Widespread rain may prevail throughout the east of the country from the Northeast to Florida and all regions east of the Appalachians.

Similarly, widespread rain prevails over the north of the country from the Inland Northwest to the North Great Plains.

Rain and accompanying thunderstorm activity may also increase around this time in the South Great Plains to northern Texas regions.

Chance of isolated showers in the Pacific Northwest to northern California.

Mostly dry elsewhere.

Temperatures may be very cold over the next couple of days in the North Great Plains, and into the Northwest, including in Montana, Yellowstone, and North Dakota.

JUNE 2 MONDAY

ISOBARIC

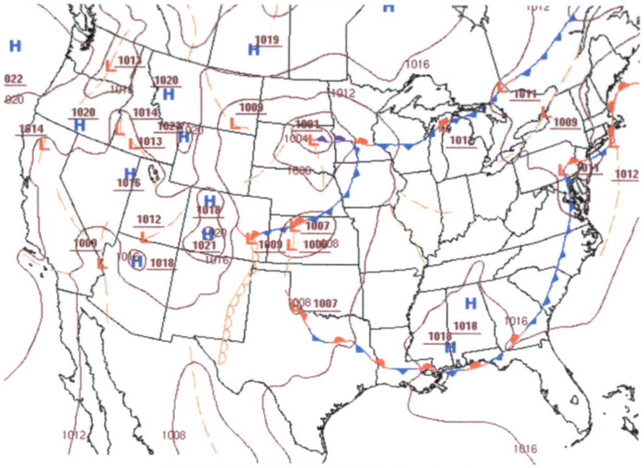

RAIN DISTRIBUTION (Inches)

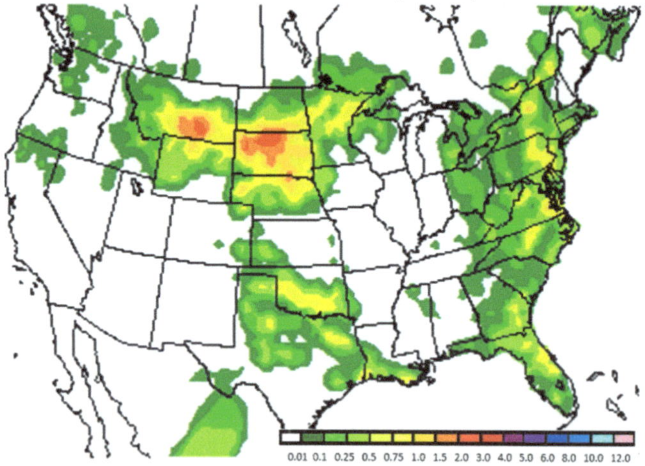

MAXIMUM TEMPERATURE

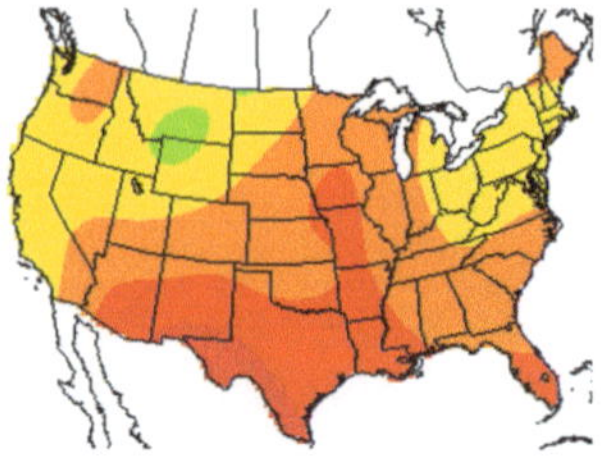

MINIMUM TEMPERATURE

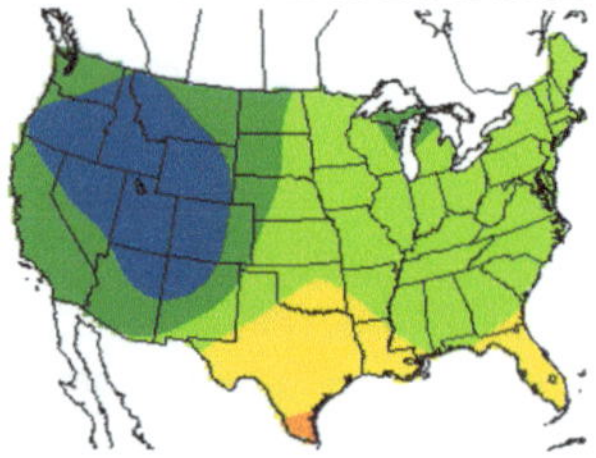

JUNE 3 TUESDAY

ISOBARIC

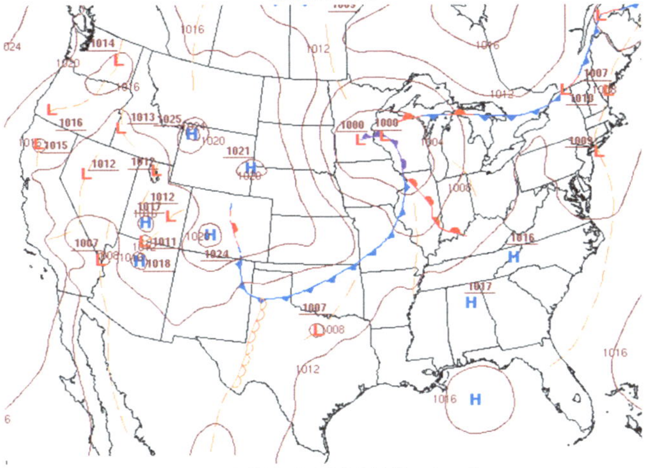

RAIN DISTRIBUTION (Inches)

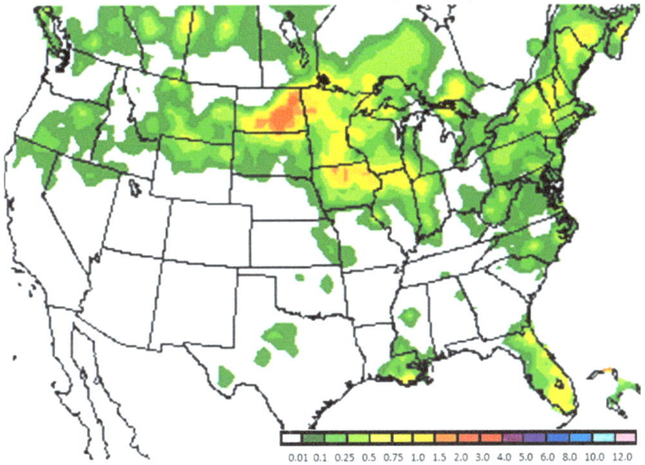

MAXIMUM TEMPERATURE

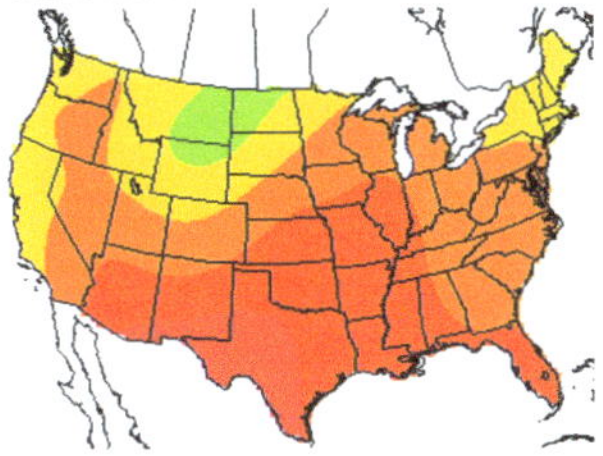

MINIMUM TEMPERATURE

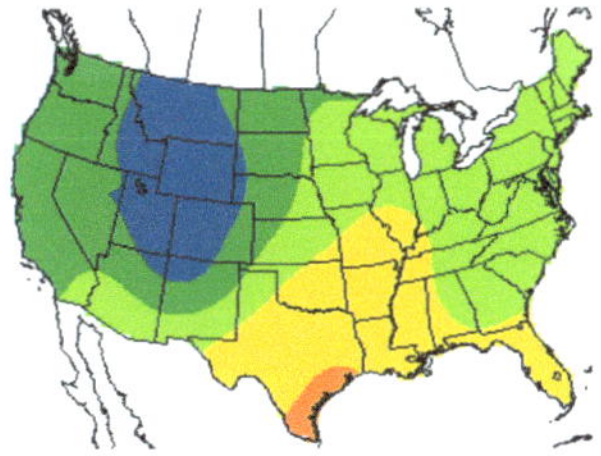

06/03/2025

Showers prevail in the Pacific Northwest to the Great Basin, with heavier precipitation and possible thunderstorms over the northern Rocky Mountains and throughout the North Great Plains to the Great Lakes and Inland Midwest.

Widespread precipitation also prevails through the Northeast to the Coastal Southeast, as well as in Florida, and with isolated showers in Texas to the coastal regions of the Deep South.

Mostly dry elsewhere.

Poor weather likely to continue over the next few days in the east of the country, possibly producing further heavy rain, large hail, pockets of flooding and possible tornadic activity in the Upper Mississippi regions.

There is a chance of patchy frosts still occurring in the North Great Plains to the Inland Midwest regions, and these are likely persist for a few more days.

Chance of some rain in Hawaii.

06/04/2025

A depression over the Great Lakes strengthens bringing widespread rain and storm activity throughout the east of the country, from the Northeast and Great Lakes southward through the Inland Midwest, over the Appalachians and into the Deep South and Florida.

Showers may also prevail in the west of the country from the Pacific Northwest to the central parts of the Rocky Mountains.

Mostly dry elsewhere.

The recent rains in the Great Plains, including Kansas, may see the development of flooding along some Kansas river systems.

JUNE 4 WEDNESDAY

ISOBARIC

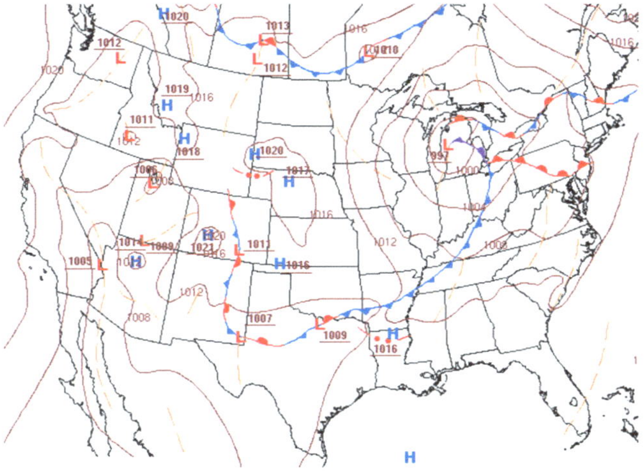

RAIN DISTRIBUTION (Inches)

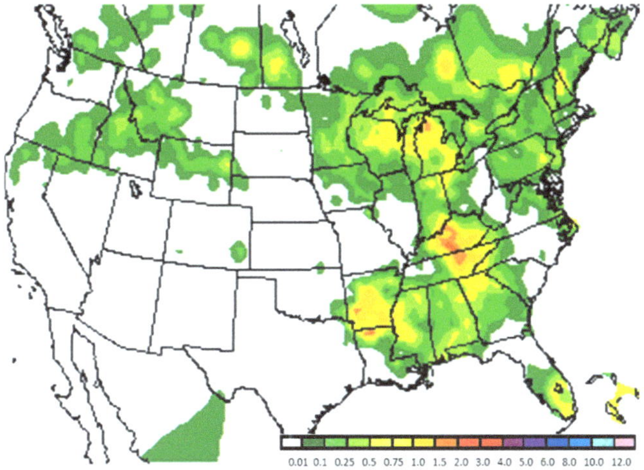

MAXIMUM TEMPERATURE

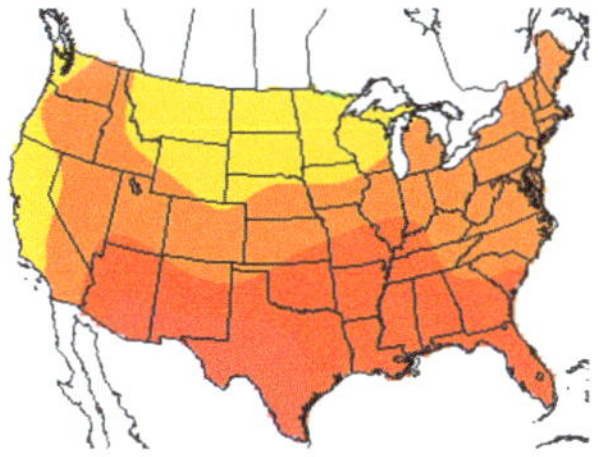

MINIMUM TEMPERATURE

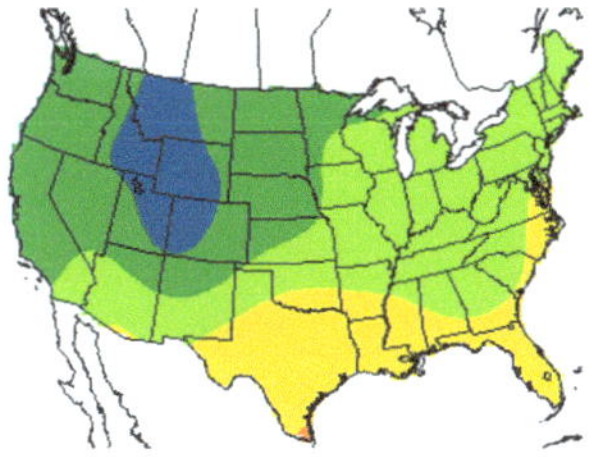

ISOBARIC

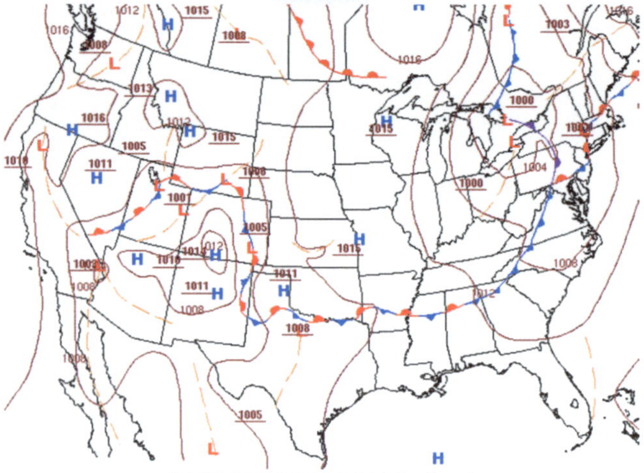

RAIN DISTRIBUTION (Inches)

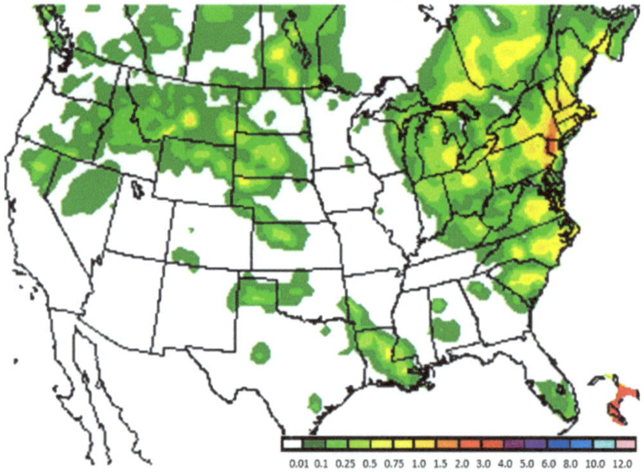

MAXIMUM TEMPERATURE

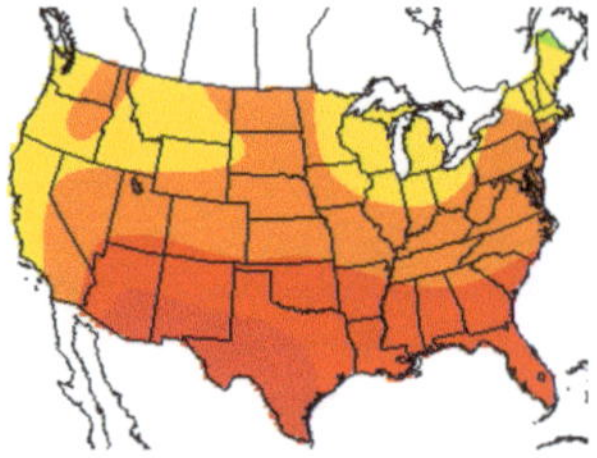

MINIMUM TEMPERATURE

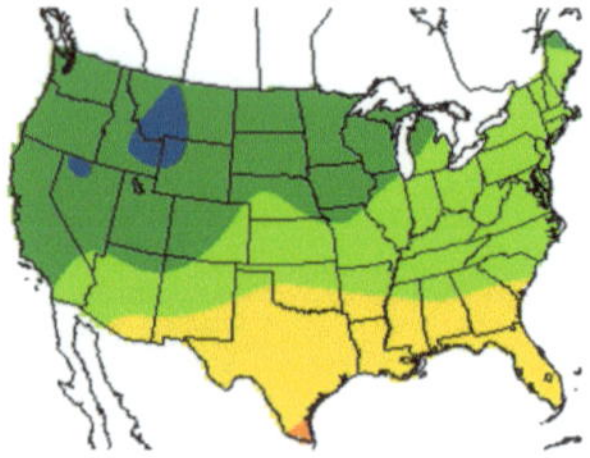

06/05/2025

Widespread rain continues over the Great Lakes to the Northeast regions and down through the Inland Midwest, across the northern Appalachians and into the Coastal Southeast, with some showers still lingering around the south of Florida and across the Deep South region.

Outbreaks of rain across the Pacific Northwest to the Great Basin and further east across the northern Rocky Mountains and onto the Great Plains.

Mostly dry elsewhere.

It may be unusually warmer than average in central parts of Alaska.

06/06/2025

The occluded front of the past couple of days may still linger around the Northeast, maintaining widespread rain activity throughout the region and south to the Megalopolis but with rain clearing from the Coastal Southeast to Florida and through the Inland Midwest to the Great Lakes.

Scattered outbreaks of rain from the Inland Northwest, over the Rocky Mountains and down throughout the northern and central parts of the Great Plains, with heavy showers reaching into the southwest of the Deep South.

Mostly dry elsewhere, although some showers may appear in northern California, and in the Southwest.

ISOBARIC

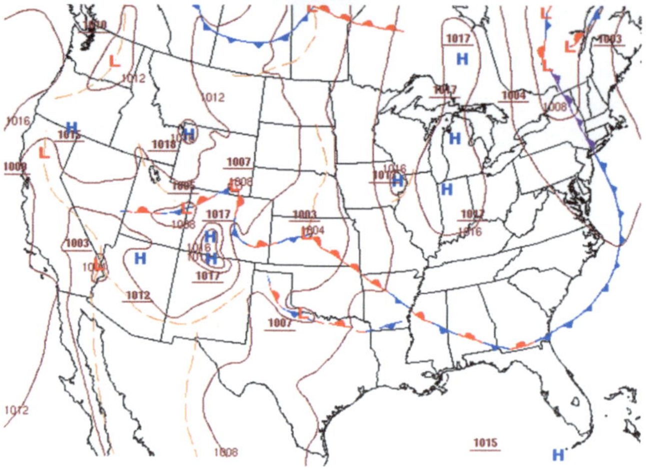

RAIN DISTRIBUTION (Inches)

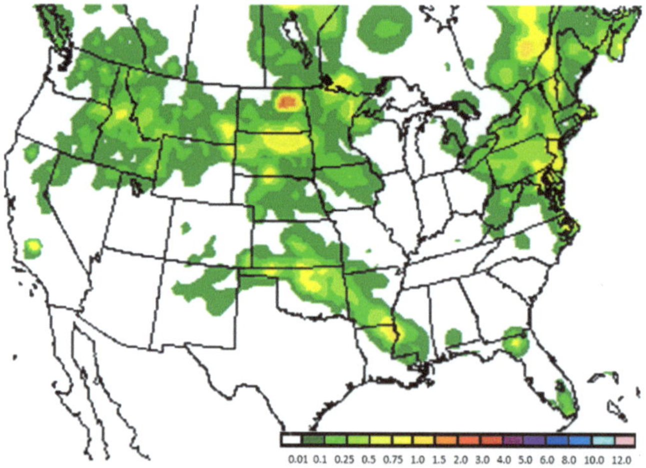

MAXIMUM TEMPERATURE

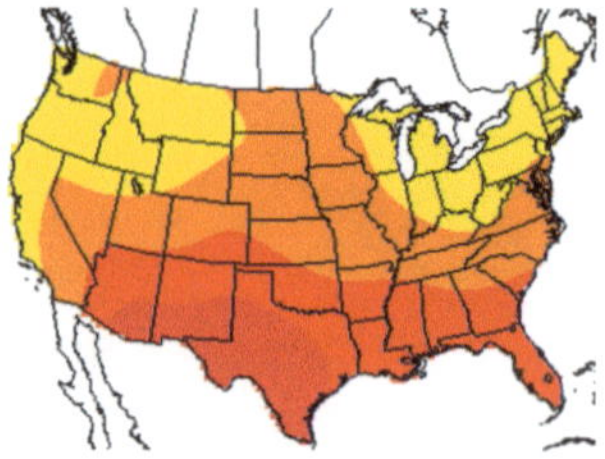

MINIMUM TEMPERATURE

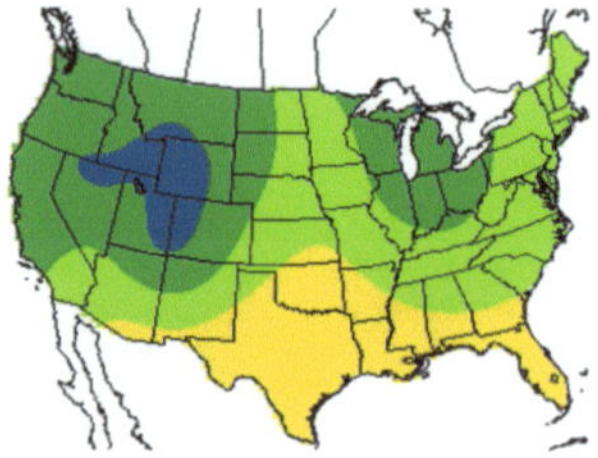

JUNE 7 SATURDAY

ISOBARIC

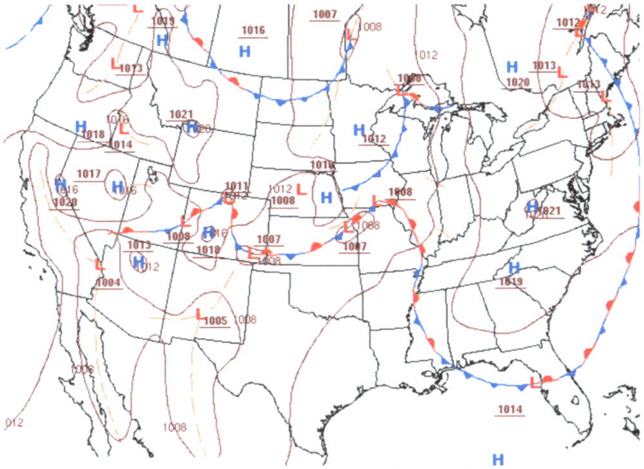

RAIN DISTRIBUTION (Inches)

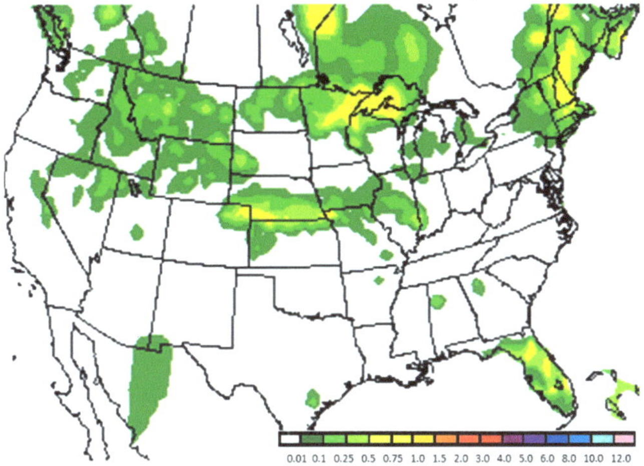

MAXIMUM TEMPERATURE / MINIMUM TEMPERATURE

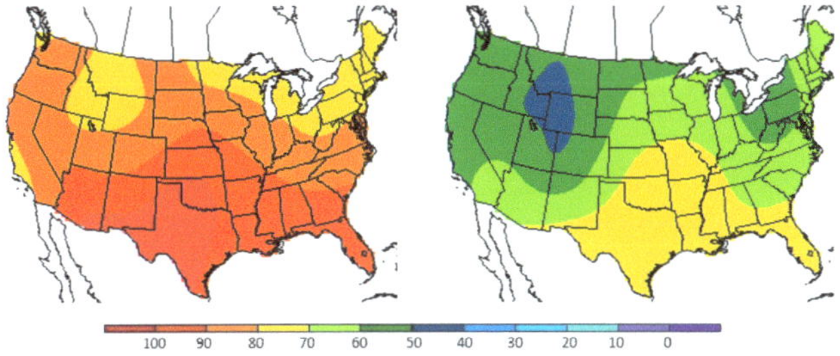

06/07/2025

Rain activity should continue to clear through the east of the country around this time, except for further outbreaks of heavy rain likely in the Northeast region.

Showers around the eastern parts of the Great Lakes, while more significant rain is likely in the western Great Lakes areas to northern parts of the North Great Plains, across the northern Rockies, into the Inland Northwest and northern parts of the Great Basin.

Some rain may also linger over the central parts of the Great Plains, while rain may return throughout Florida.

Mostly dry elsewhere.

JUNE 8 SUNDAY

ISOBARIC

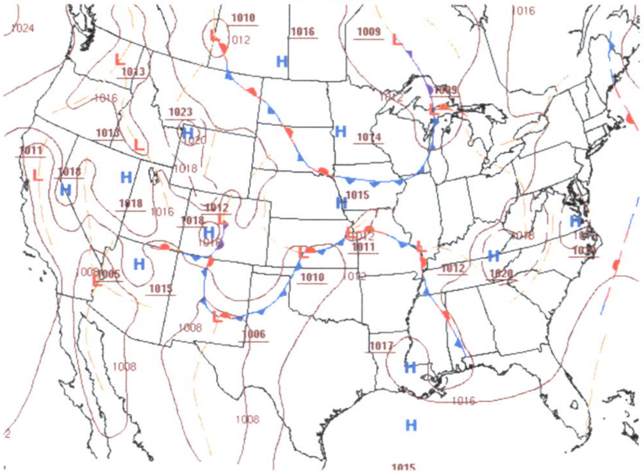

RAIN DISTRIBUTION (Inches)

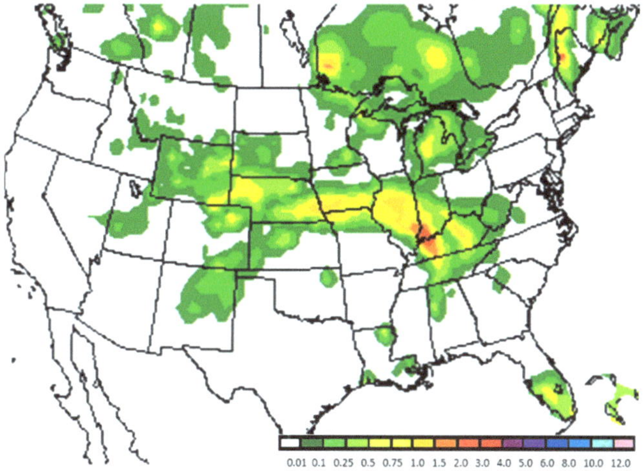

MAXIMUM TEMPERATURE / MINIMUM TEMPERATURE

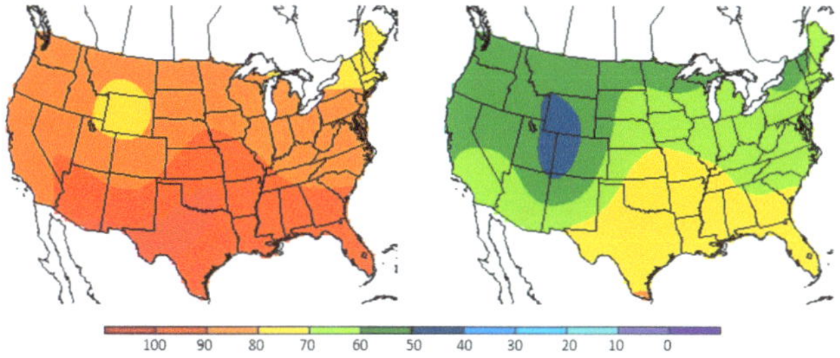

06/08/2025

Showers, some heavy, still prevail over the Northeast while rain spreads once more over the Great Lakes and into the Inland Midwest to the Appalachians, and west over the central and southern Great Plains and up into the southern half of the Rocky Mountains.

Rains may be particularly heavy in Nebraska around this time, while the cooler air may see the return of heavy snowfall in some places in Montana.

Some showers may extend for a time into the northern parts of the Southwest, while spotty pockets of showers may affect the Deep South.

Rain is likely to continue over southern Florida.

Mostly dry elsewhere.

JUNE 9 MONDAY

ISOBARIC

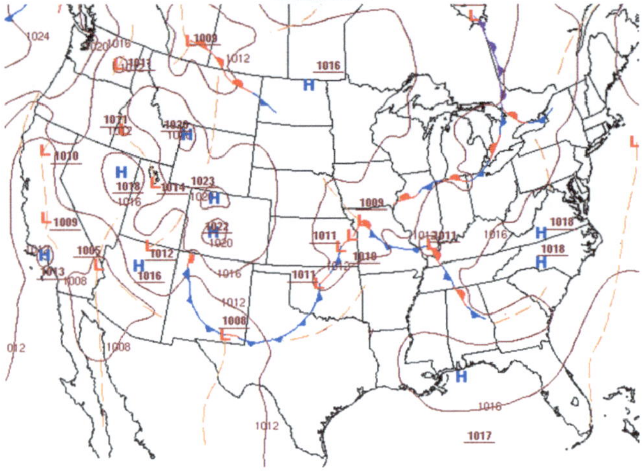

RAIN DISTRIBUTION (Inches)

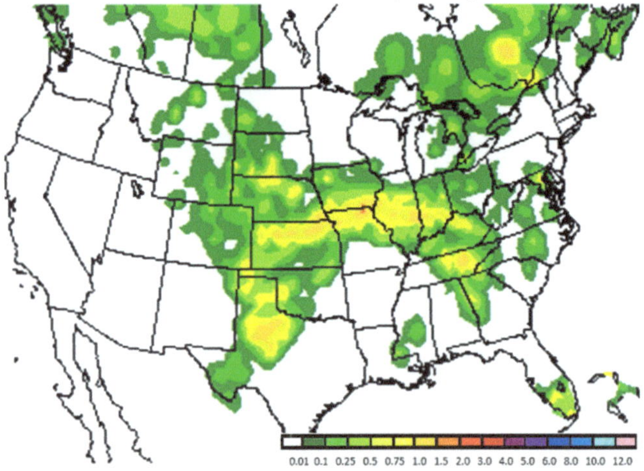

MAXIMUM TEMPERATURE

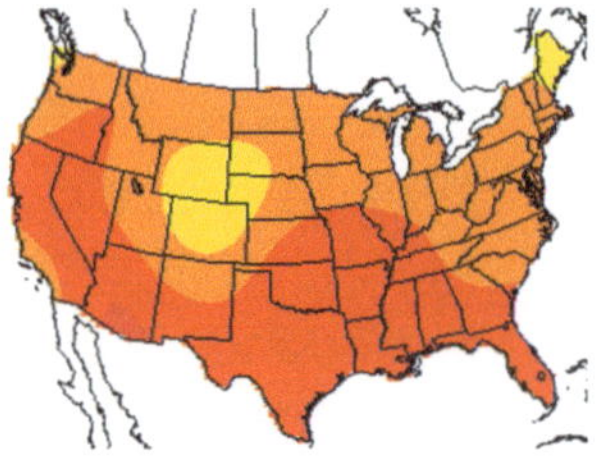

MINIMUM TEMPERATURE

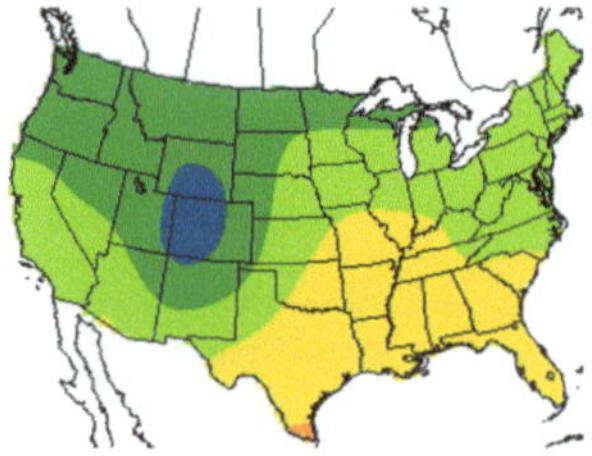

06/09/2025

Widespread rain, often heavy, prevails throughout the Great Plains to the upper Midwest, with particularly heavy rain likely for a time in South Dakotas.

Some rain also persists over the Great Lakes to the Northeast regions, as well as down through the Appalachians, with showers spreading into the Coastal Southeast region.

Showers likely in the northern Rocky Mountains with heavier rain in the south of the ranges.

Heavy rain is also likely in the north of Texas, as well as through the southern parts of the Florida peninsula.

Mostly dry elsewhere.

Heavy rain likely in southern Alaska.

06/10/2025

Widespread rain spreads once more throughout the eastern half of the country from the central Great Plains to the Atlantic coast, and from the Northeast to the Deep South.

Rain can also be expected in the western half of Texas, and showers persist in southern Florida.

Some showers may linger in the Inland Northwest region to the west of the North Great Plains.

Mostly dry elsewhere.

JUNE 10 TUESDAY

ISOBARIC

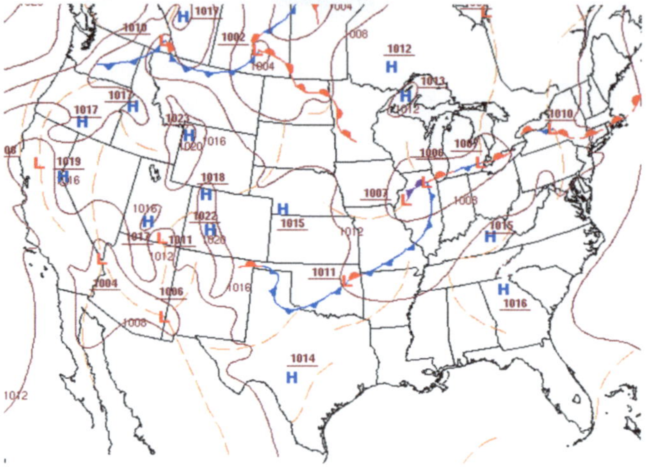

RAIN DISTRIBUTION (Inches)

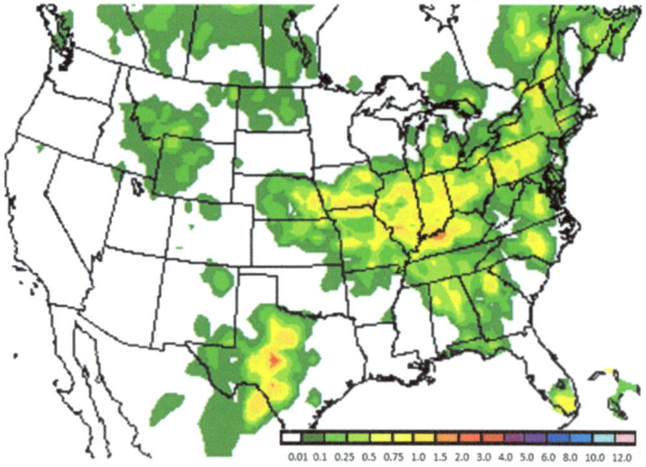

MAXIMUM TEMPERATURE

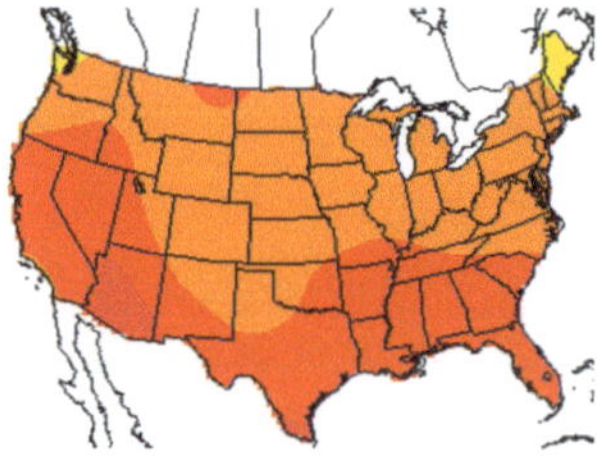

MINIMUM TEMPERATURE

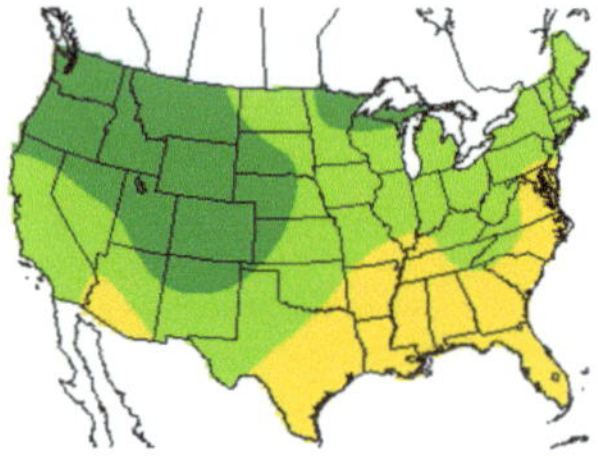

JUNE 11 WEDNESDAY

ISOBARIC

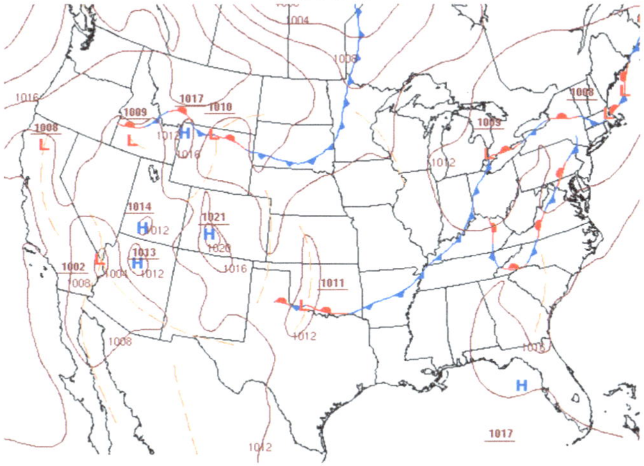

RAIN DISTRIBUTION (Inches)

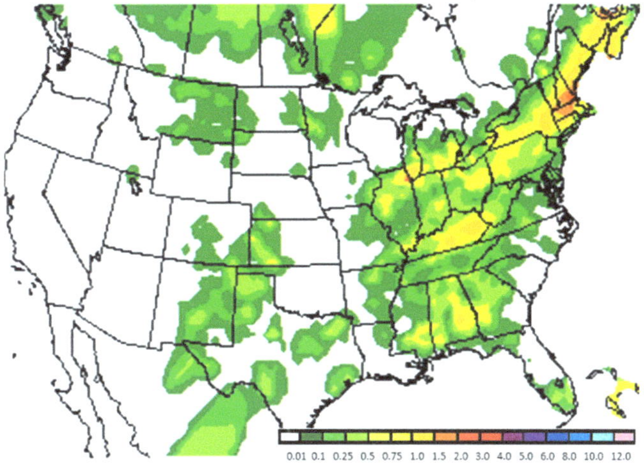

MAXIMUM TEMPERATURE

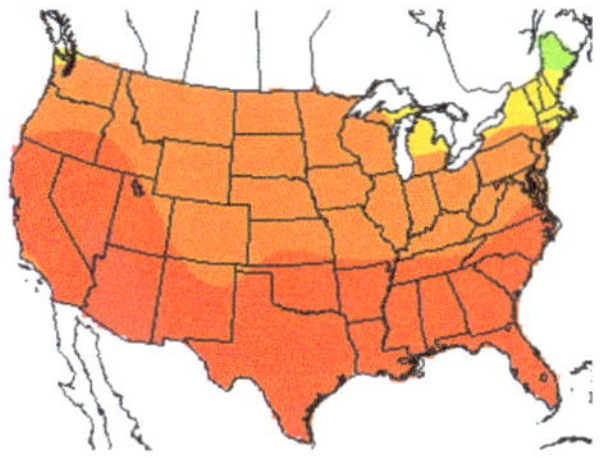

MINIMUM TEMPERATURE

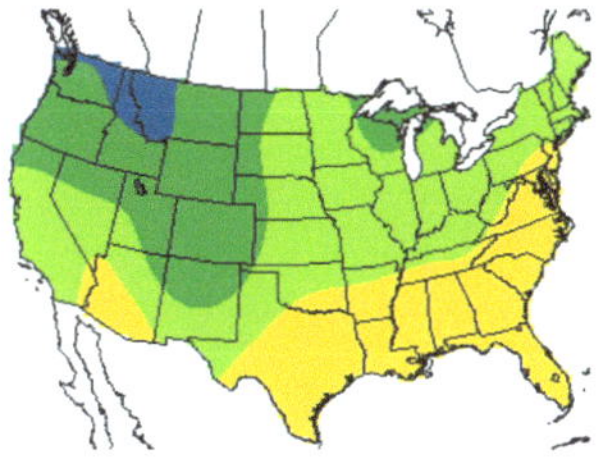

06/11/2025

Widespread heavy rain continues throughout the east of the country from the Northeast down through the Great Lakes and into the Inland Midwest and further south into the Deep South.

Rain may be particularly heavy at times in the Minnesota start around this time.

Similarly, showers may also continue to linger around the Florida Peninsula and outbreaks of showers prevail over Texas to the Southwest as well as in the North Great Plains.

Mostly dry elsewhere.

There is a chance of a tropical storm forming in the Gulf of Mexico threatening Florida over the next few days.

06/12/2025

Rain activity intensifies over the Inland Northwest, over the northern Rocky Mountains, and across the Great North Plains, while easing around the eastern half of the Great Lakes.

Further rain prevails from the Northeast to the Coastal Southeast, and down through the Appalachians to the Deep South, and eastern Texas.

Some showers likely in the west of Texas to the New Mexico.

Mostly dry elsewhere.

A tropical storm may cross Florida bringing heavy rain and strong winds in its path, as well as bringing the heavy rain into the Deep South, including in Georgia, over the next couple of days, and outbreaks of rain through the Gulf of Mexico and over Texas.

Mostly dry elsewhere.

JUNE 12 THURSDAY

ISOBARIC

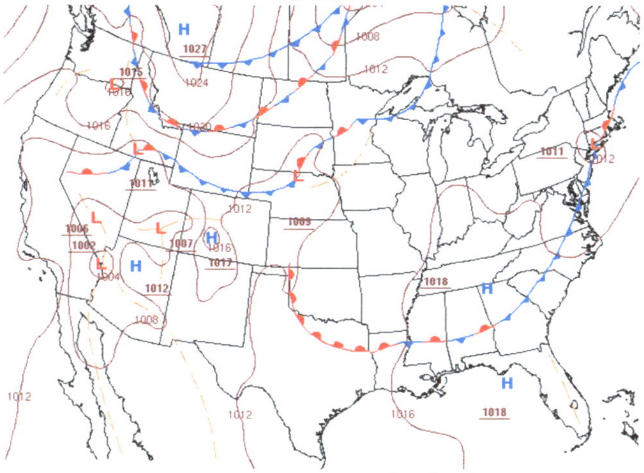

RAIN DISTRIBUTION (Inches)

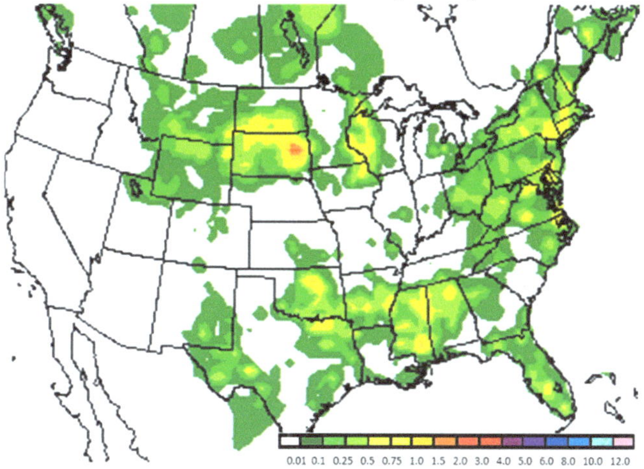

MAXIMUM TEMPERATURE

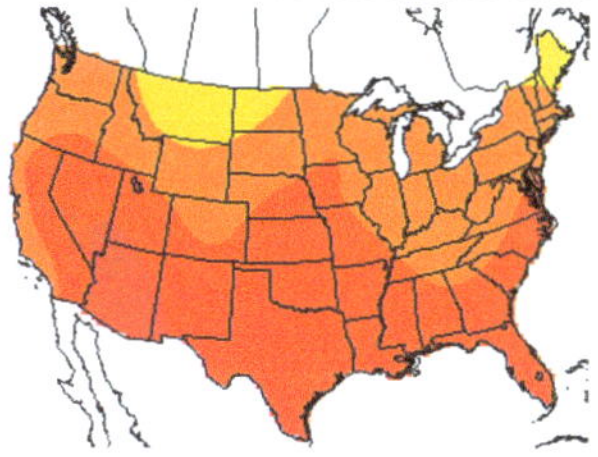

MINIMUM TEMPERATURE

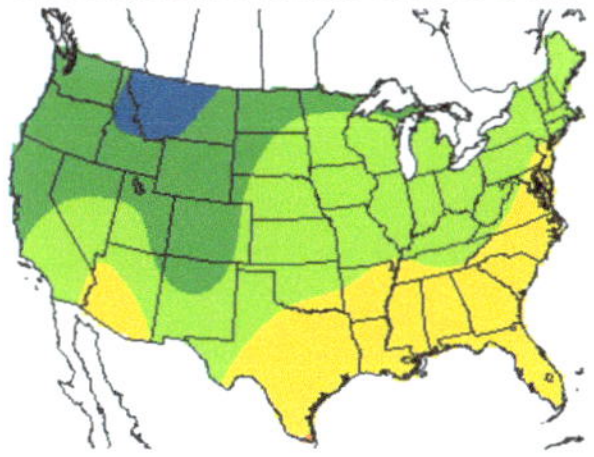

JUNE **13** FRIDAY

ISOBARIC

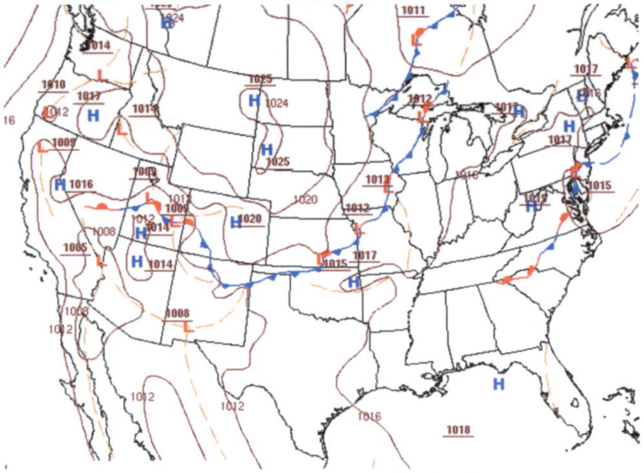

RAIN DISTRIBUTION (Inches)

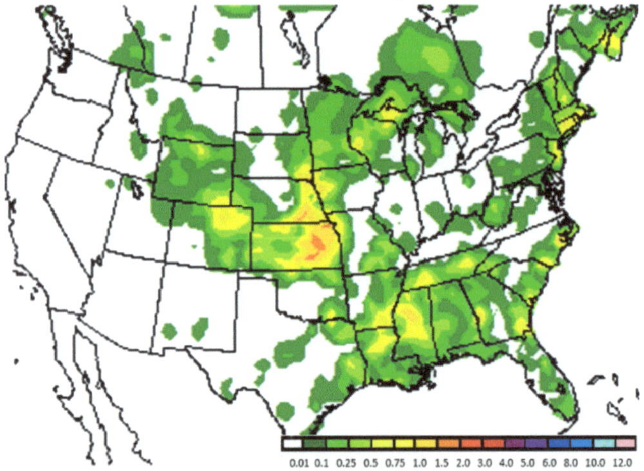

MAXIMUM TEMPERATURE

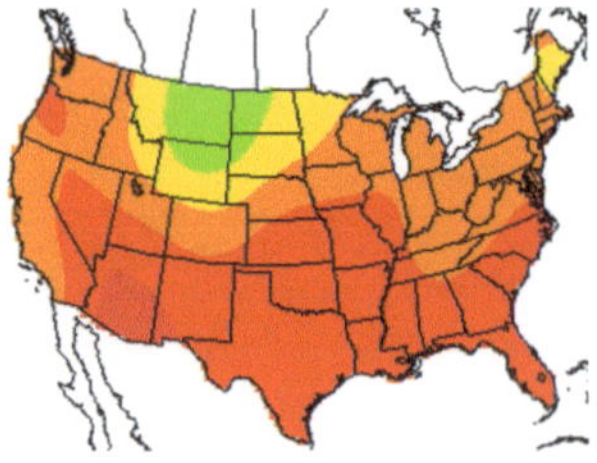

MINIMUM TEMPERATURE

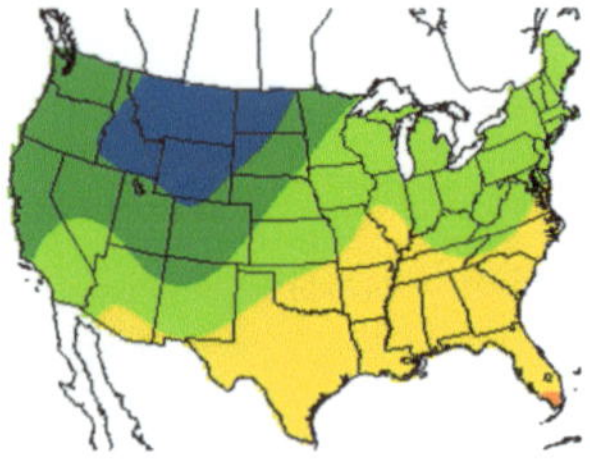

06/13/2025

A tropical storm near Florida brings heavy rain through the Gulf of Mexico states from Florida to eastern Texas and up throughout the Deep South.

Scattered rain activity also continues in the Northeast region to the Appalachians.

Some heavy showers likely in the northern parts of the Inland Midwest. Rain over the Great Lakes and Plains, and up into the Rocky Mountains.

Mostly dry elsewhere.

06/14/2025

Rain may clear to spotty showers over the Northeast, but become more persistent and heavy in the south from the South Great Plains across to the Coastal Southeast and down through Florida and the Deep South, and around the Gulf of Mexico coastal areas from Texas to Florida, possibly as a result of a tropical storm influence in the Atlantic Ocean near Florida.

Rain can also be expected up through the southern half of the Rocky Mountains and isolated showers at times into the Great Basin.

Mostly dry elsewhere.

A cooler change in the Northwest over the next couple of days, with storms returning and including heavy rain and the chance of snow in the higher reaches.

JUNE **14** SATURDAY

ISOBARIC

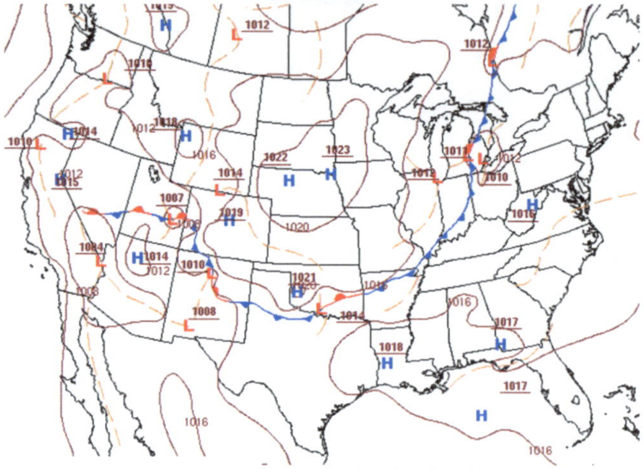

RAIN DISTRIBUTION (Inches)

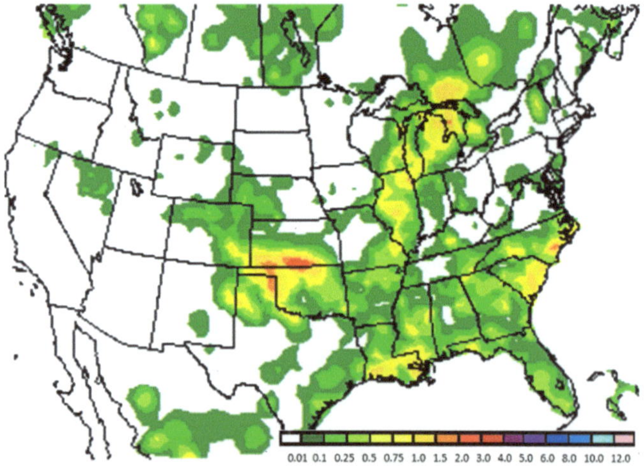

MAXIMUM TEMPERATURE

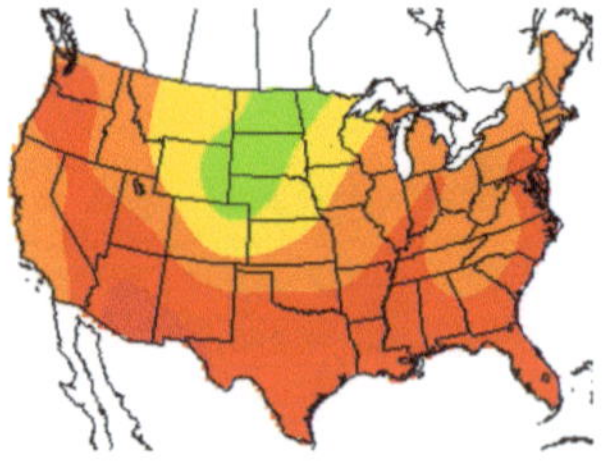

MINIMUM TEMPERATURE

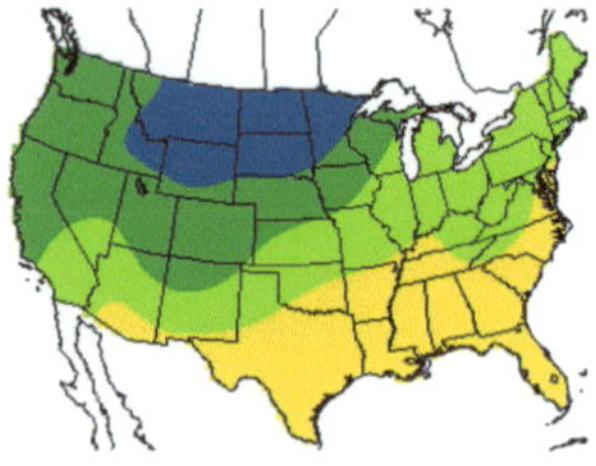

JUNE 15 SUNDAY

ISOBARIC

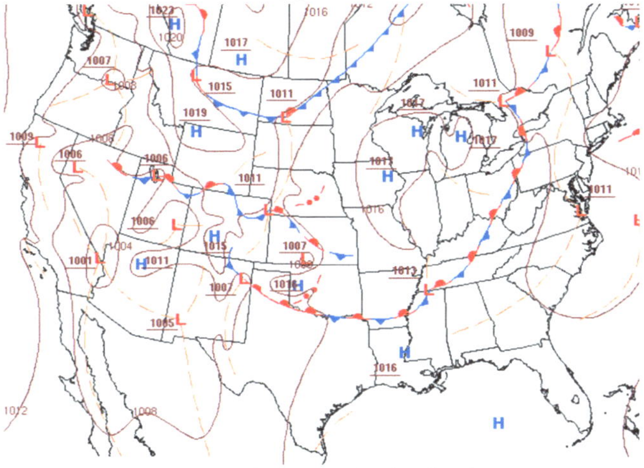

RAIN DISTRIBUTION (Inches)

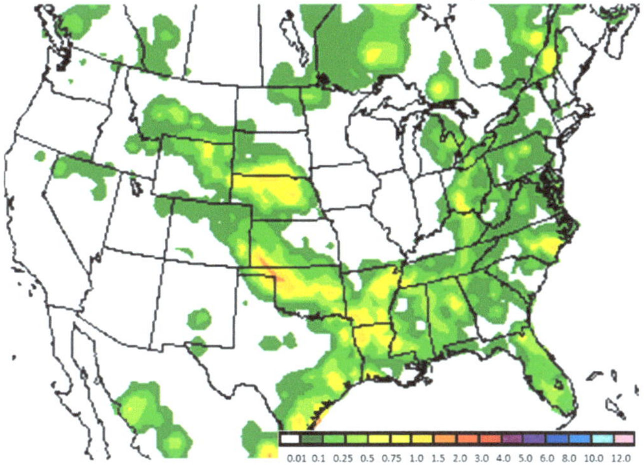

MAXIMUM TEMPERATURE

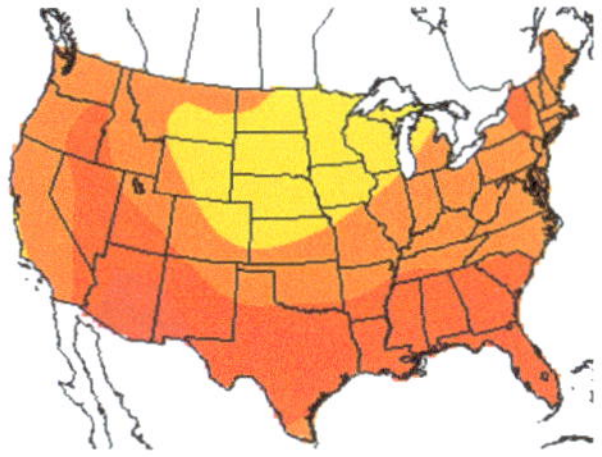

MINIMUM TEMPERATURE

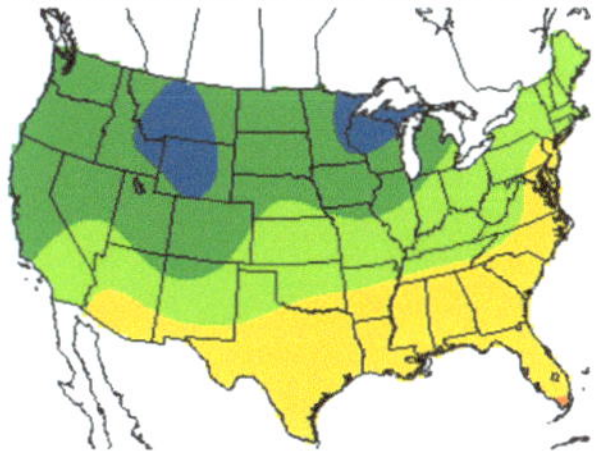

06/15/2025

Precipitation prevails over much of the east and south of the country, although the Great Lakes to the west of the Inland Midwest will be mostly dry.

Rain may be heavy at times in the Great Plains to the Ozarks, as well as around the Gulf of Mexico coastal fringes of Texas to Florida.

Rain is also likely up through the Coastal Southeast and Appalachians and around the eastern Great Lakes to the south and west of the Northeast regions.

Chance of snow returning in some parts of Utah over the next couple of days.

Mostly dry in the west from the Pacific Northwest to the Southwest, with just the isolated shower in the Great Basin.

06/16/2025

Most regions in the southwest of the country can expect to be mostly dry as well as in the north of Texas and over the central Appalachians, while some rain activity prevails elsewhere.

Much of the rain in the North Great Plains to the Northeast will accompany thunderstorms and supercells crossing the regions, with these storms possibly bringing the development of tornadoes in some places, followed later by cooler temperatures.

Rain activity may be very heavy for a time in Idaho and Montana in particular.

Very strong winds through Wyoming to Colorado and down into Texas over the next couple of days.

JUNE 16 MONDAY

ISOBARIC

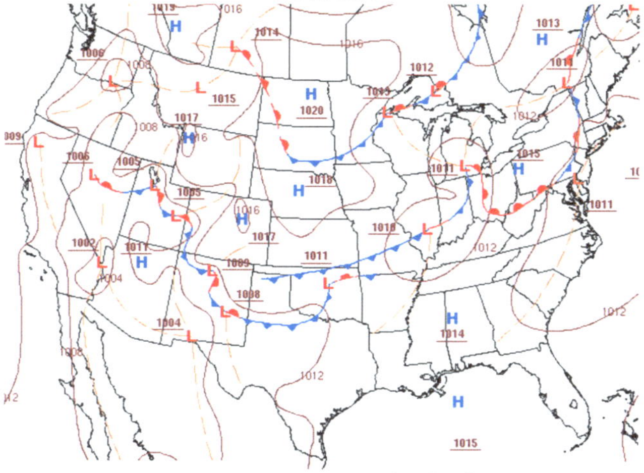

RAIN DISTRIBUTION (Inches)

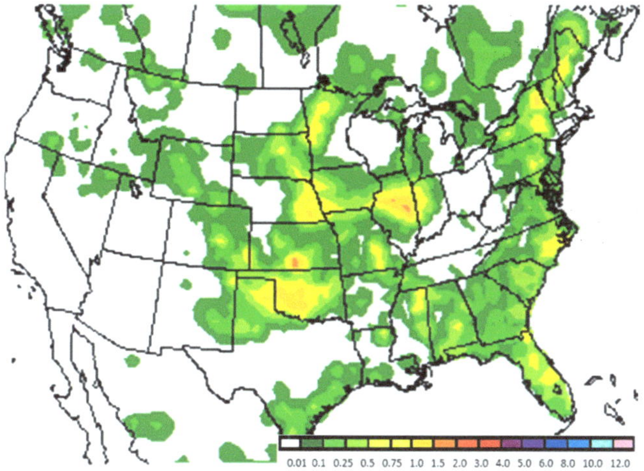

MAXIMUM TEMPERATURE

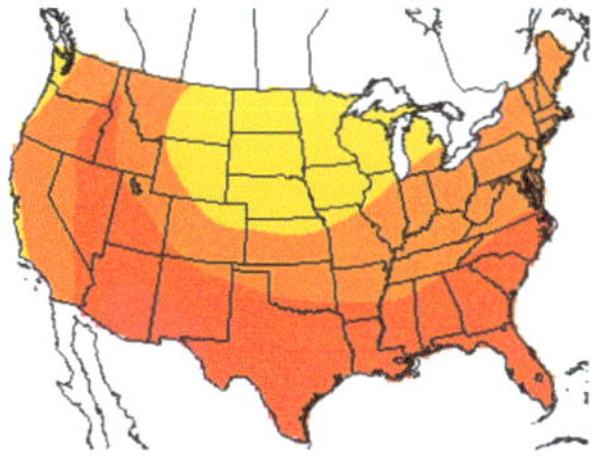

MINIMUM TEMPERATURE

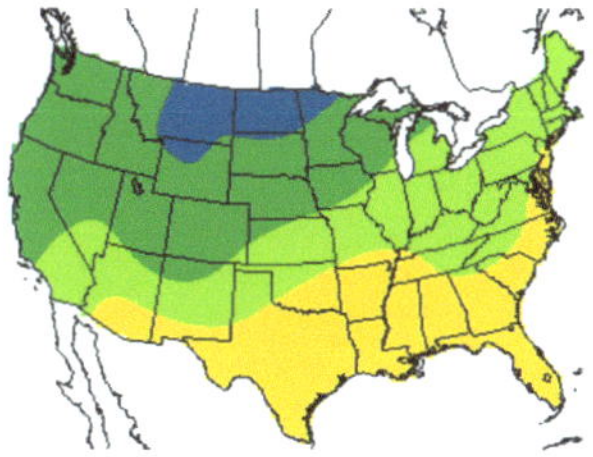

JUNE 17 TUESDAY

ISOBARIC

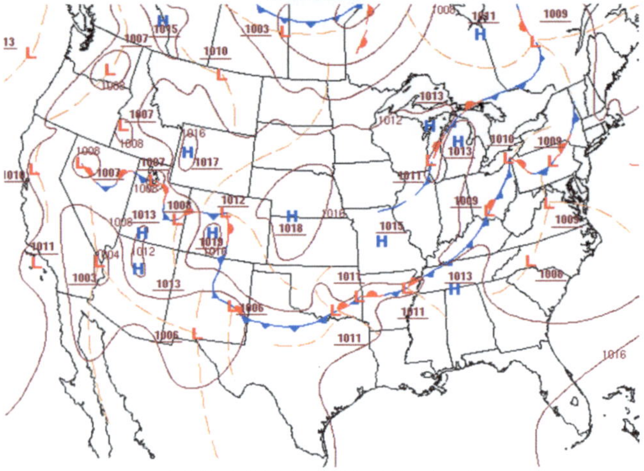

RAIN DISTRIBUTION (Inches)

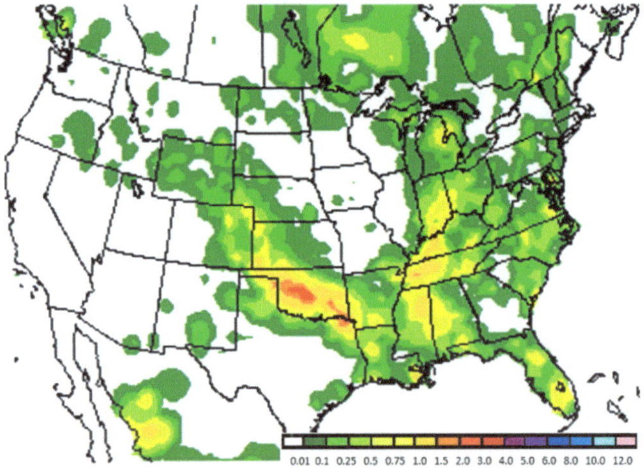

MAXIMUM TEMPERATURE / MINIMUM TEMPERATURE

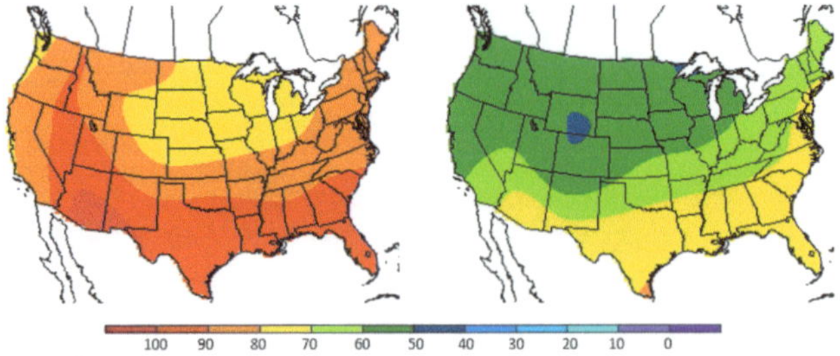

06/17/2025

Mostly dry west of the Rocky Mountains and in the Great Basin and Southwest, as well as in the western half of the Inland Midwest and western Texas, except for some isolated showers, mostly in the Inland Midwest to the northern Rocky Mountains.

Widespread precipitation elsewhere.

Supercell thunderstorms may occur at times, with the state of Wisconsin particularly at risk, and including the chance of tornadoes forming.

Rain is likely to be particularly heavy in the South Great Plains to the Ozarks, with the chance of pockets of localized flooding developing.

06/18/2025

Intense rain activity can be expected throughout the east of the country, from the Northeast to Florida and down through the Appalachians and throughout the Deep South into the northeast of Texas and southern parts of the South Great Plains.

Some rain may also prevail across the northern Great Lakes to Minnesota, while some showers prevail over the Pacific and Inland Northwest regions.

Mostly dry elsewhere.

A cooler change in the Great Plains to the Rocky Mountains may see the brief appearance of frosts on the central Great Plains.

JUNE 18 WEDNESDAY

ISOBARIC

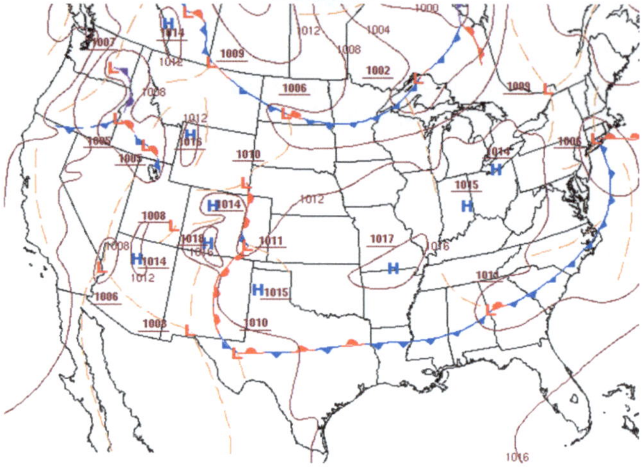

RAIN DISTRIBUTION (Inches)

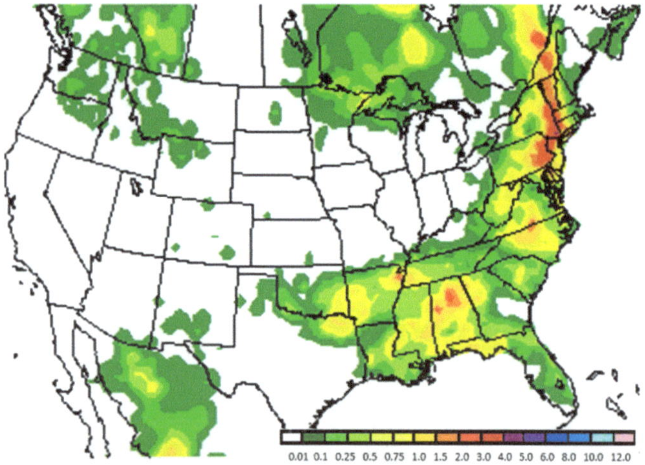

MAXIMUM TEMPERATURE / MINIMUM TEMPERATURE

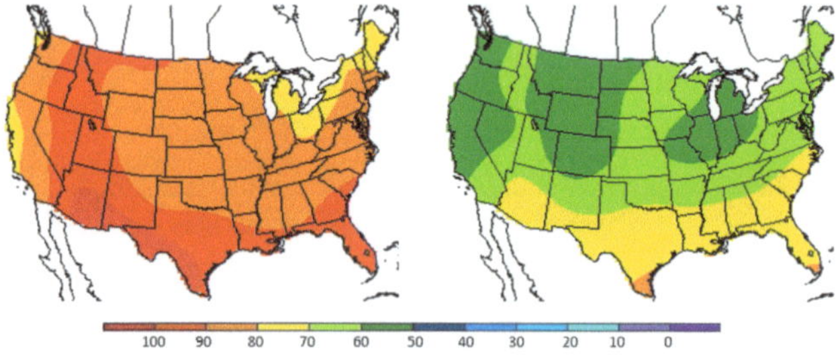

JUNE **19** THURSDAY

ISOBARIC

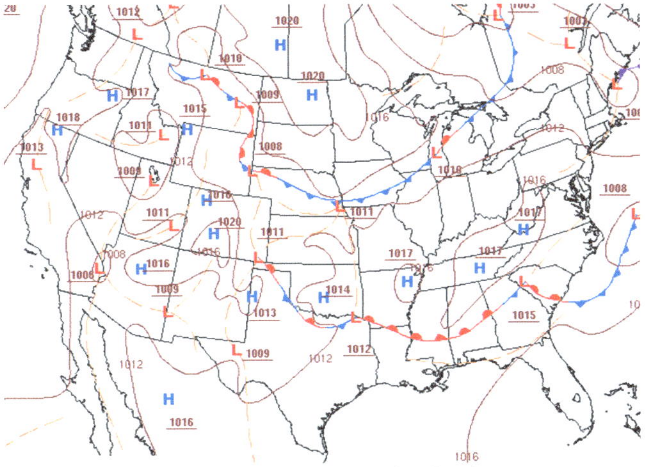

RAIN DISTRIBUTION (Inches)

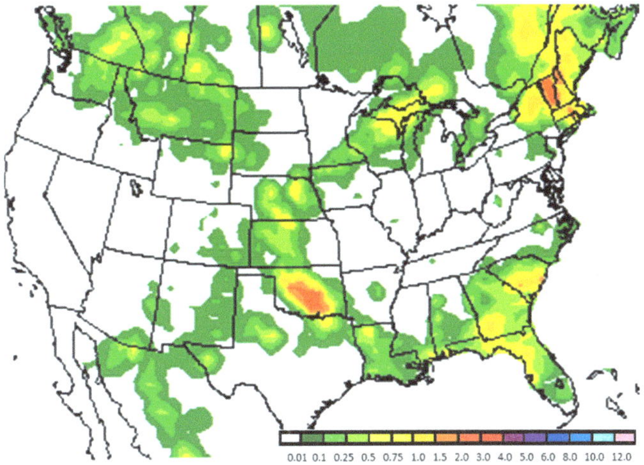

MAXIMUM TEMPERATURE / MINIMUM TEMPERATURE

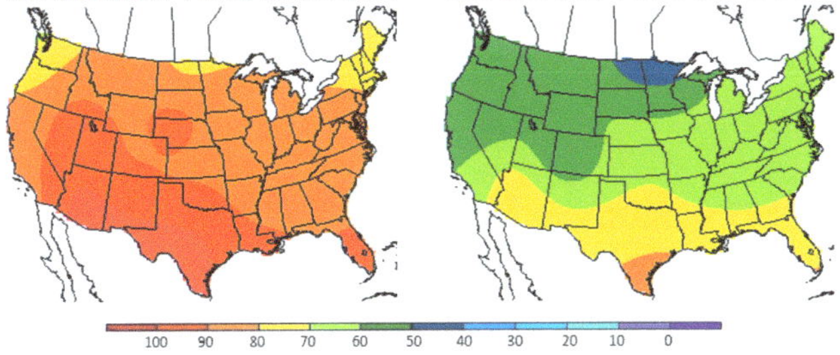

06/19/2025

Heavy rain expected in the Northeast, and outbreaks of rain from the Great Lakes to the central Great Plains, south into Texas and across the Deep South and Florida to South Carolina.

Rain from Washington to the North Great Plains, and showers in the Southwest.

Mostly dry elsewhere, including the Ozarks, southern Texas, Oregon, and through the Appalachians.

It may be unusually warm in Hawaii.

06/20/2025

Heavy rain across the North Great Plains including in Missouri, Nebraska, South Dakota, Oklahoma, as well as in the Southwest, particularly in Arizona.

Outbreaks of rain are also likely in the Great Lakes to the Northeast including the chance of heavy rain at times in Pennsylvania to Virginia.

Rain activity may ease but still linger in the Deep South and Florida around this time.

Chance of flooding along some of the river systems within the Great Plains.

Mostly dry elsewhere.

JUNE **20** FRIDAY

ISOBARIC

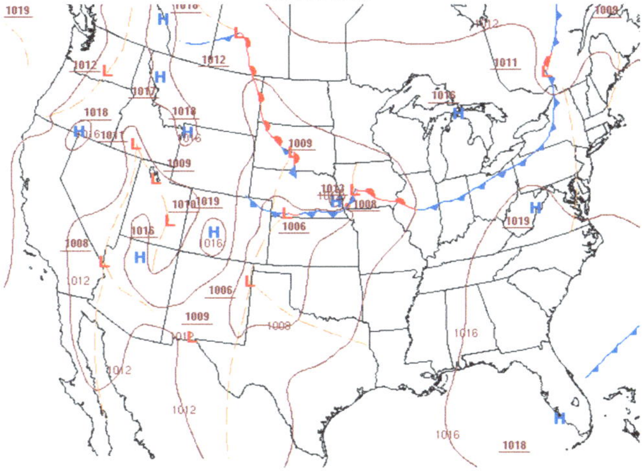

RAIN DISTRIBUTION (Inches)

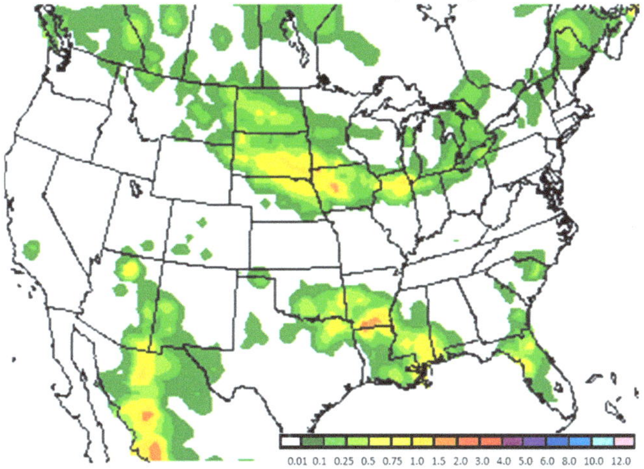

MAXIMUM TEMPERATURE / MINIMUM TEMPERATURE

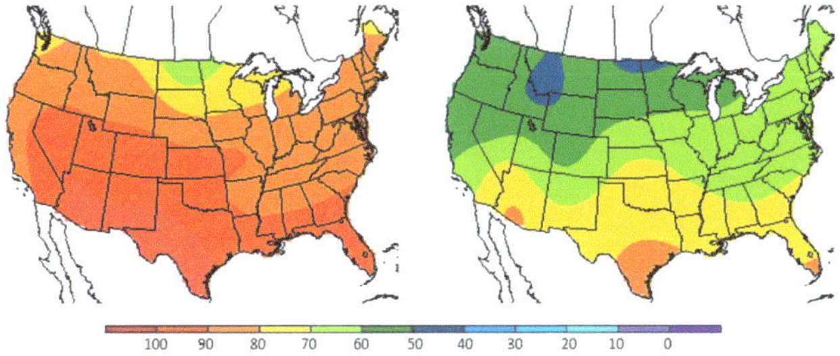

JUNE 21 SATURDAY

ISOBARIC

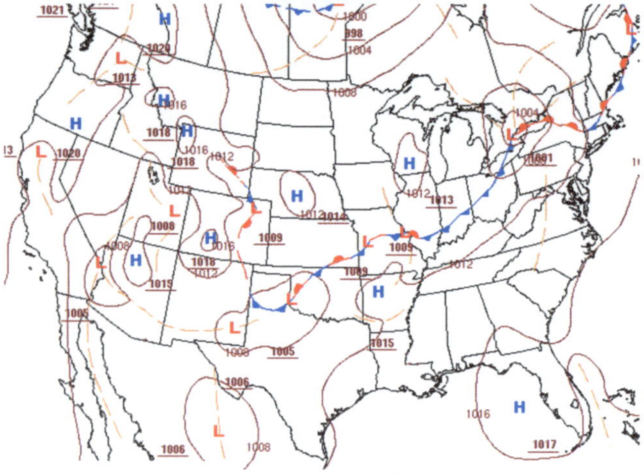

RAIN DISTRIBUTION (Inches)

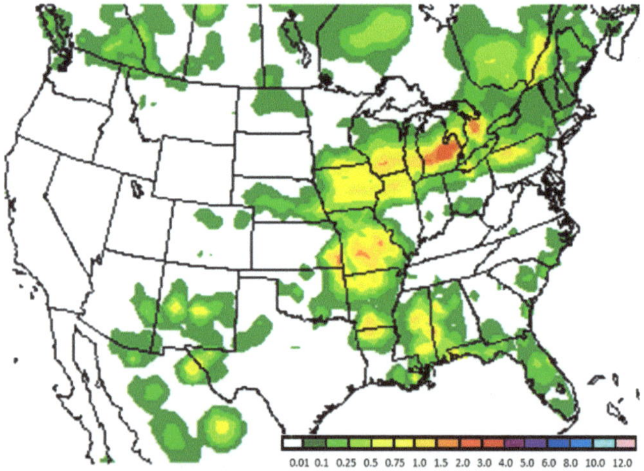

MAXIMUM TEMPERATURE

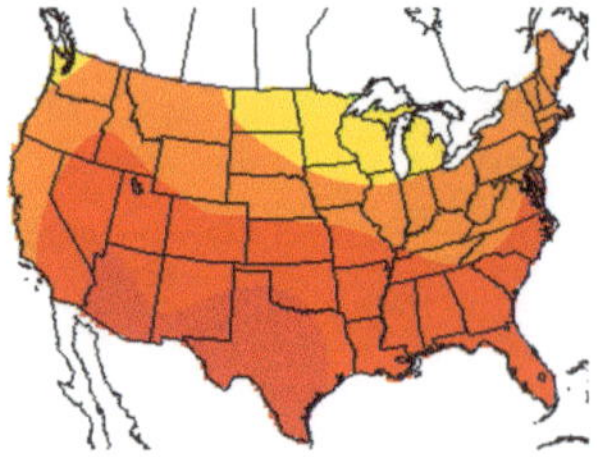

MINIMUM TEMPERATURE

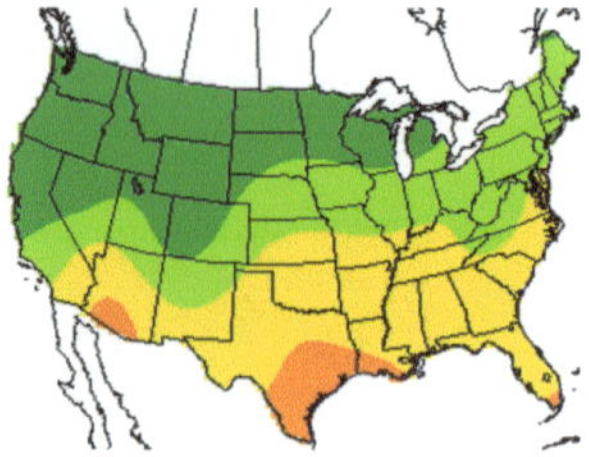

06/21/2025

Widespread rain may return over the Northeast and down through the eastern Great Lakes and the Inland Midwest, as well as south through the Ozarks and throughout most of the Deep South into Florida.

Some showers may also prevail in the Southwest region and into the west of Texas, as well as along the coasts in the Coastal Southeast region.

There may also be a chance of heavy rain in Arizona.

Mostly dry elsewhere, including the Coastal Southeast.

06/22/2025

Mostly dry in the west from the Pacific coasts to the Great Basin and into the southern parts of the Rocky Mountains, as well as through most of Texas and in the southern parts of the Inland Midwest.

Some outbreaks of rain elsewhere.

Rain may be heavy at times in the Coastal Southeast region, as well as in parts of the Deep South and in the Ozarks.

JUNE 22 SUNDAY

ISOBARIC

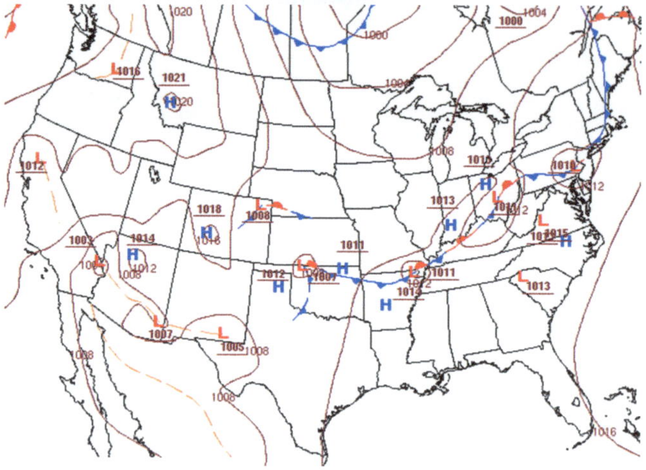

RAIN DISTRIBUTION (Inches)

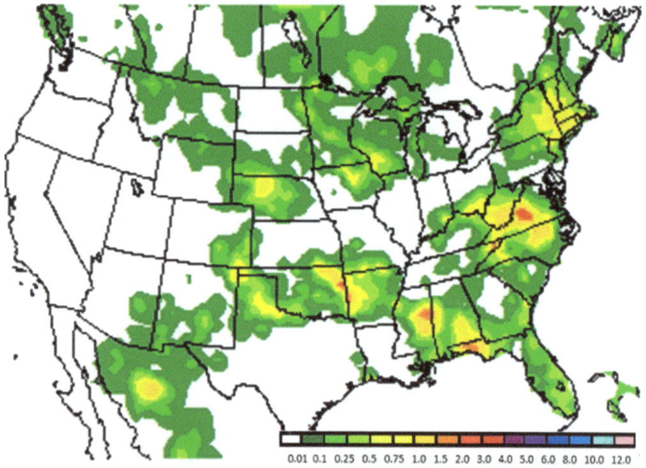

MAXIMUM TEMPERATURE

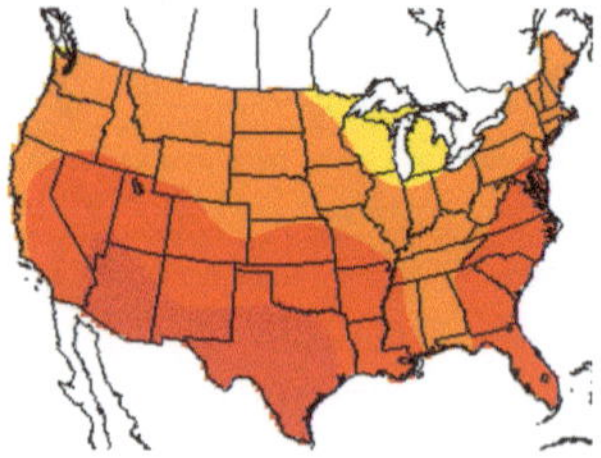

MINIMUM TEMPERATURE

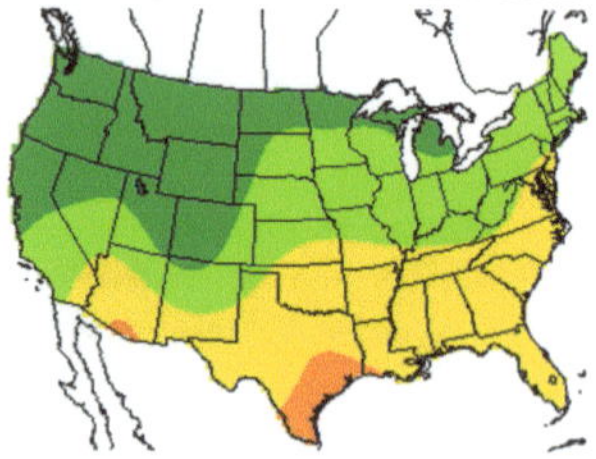

JUNE 23 MONDAY

ISOBARIC

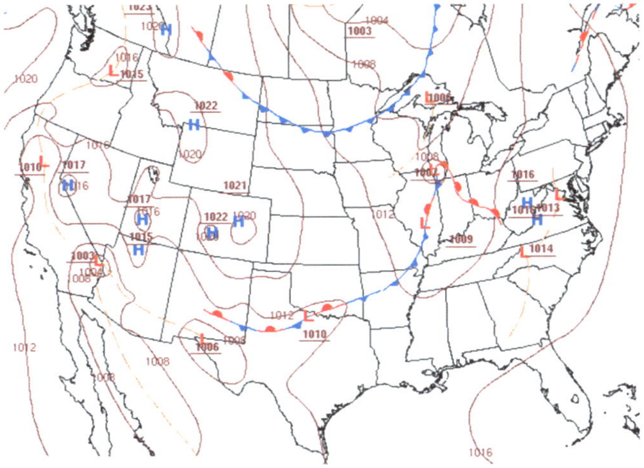

RAIN DISTRIBUTION (Inches)

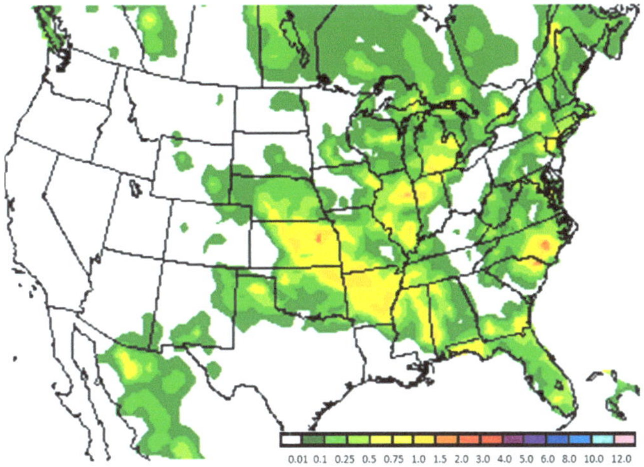

MAXIMUM TEMPERATURE MINIMUM TEMPERATURE

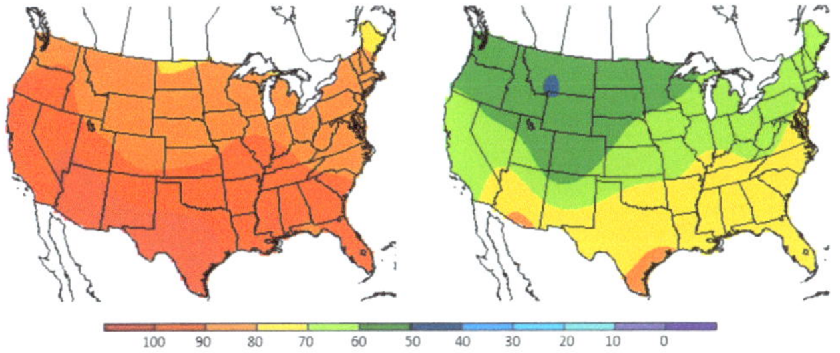

06/23/2025

Mostly dry over the western half of the country west of the Rocky Mountains to the Pacific Ocean, as well as in the southern half of Texas and in a line passing down the Appalachians.

Some showers in the Southwest and the North Great Plains, while widespread and occasionally heavy rain may prevail elsewhere.

Rain may be particularly heavy in some parts of the South Great Plains to the Inland Midwest, as well as in some parts of the Coastal Southeast region.

Occasional showers are likely in Hawaii over the next three to four days, mostly on the Big Island.

06/24/2025

Widespread rain may prevail throughout the east of the country, from the Northeast to the Deep South including in Florida and throughout the Great Lakes and Inland Midwest regions.

Some showers and outbreaks of rain may also prevail in the Southwest and northern parts of Texas as well as down through the eastern fringes of the southern Rocky Mountains spreading briefly into central parts of the Great Plains.

Mostly dry elsewhere.

JUNE 24 TUESDAY

ISOBARIC

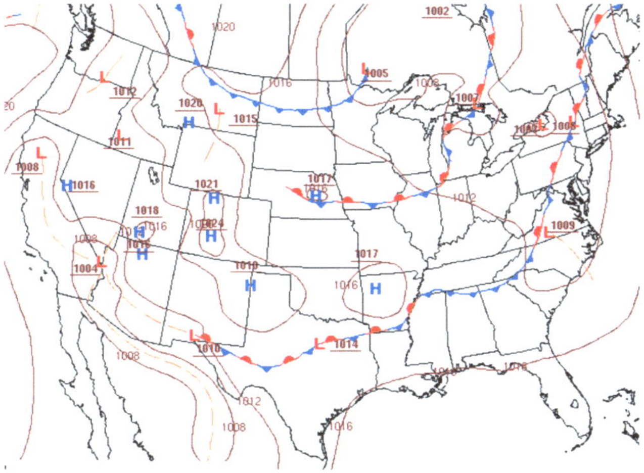

RAIN DISTRIBUTION (Inches)

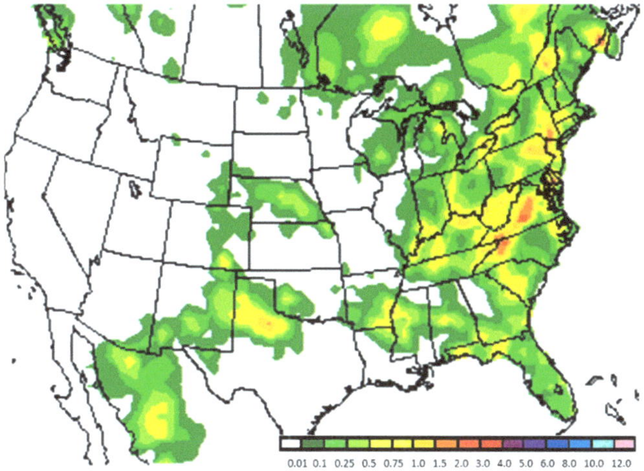

MAXIMUM TEMPERATURE MINIMUM TEMPERATURE

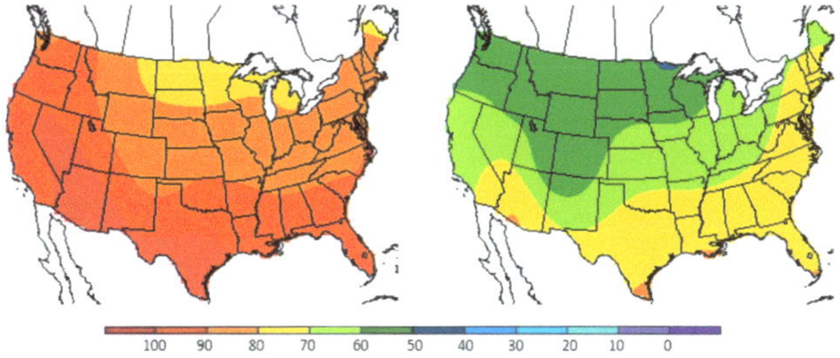

ISOBARIC

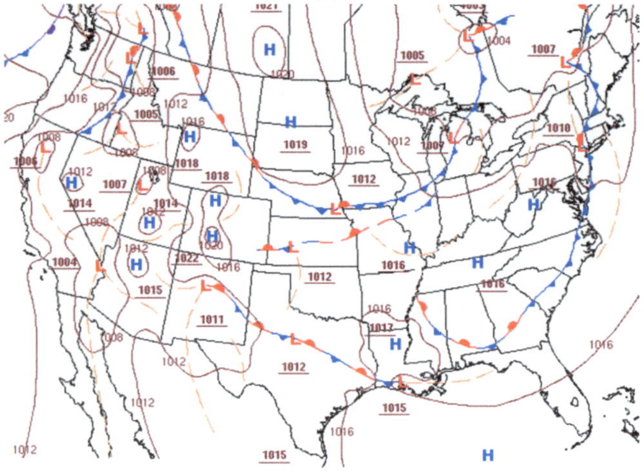

RAIN DISTRIBUTION (Inches)

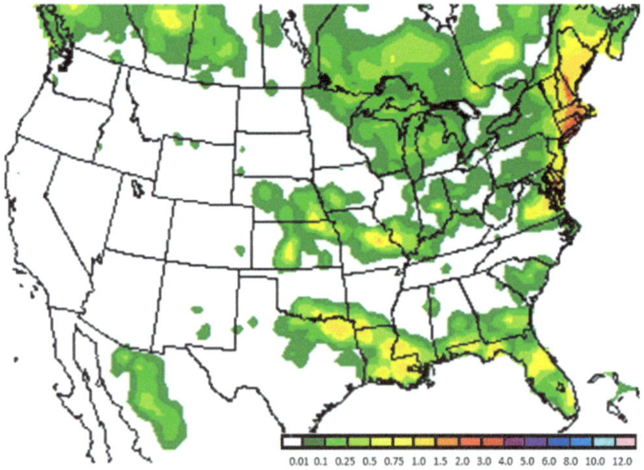

MAXIMUM TEMPERATURE MINIMUM TEMPERATURE

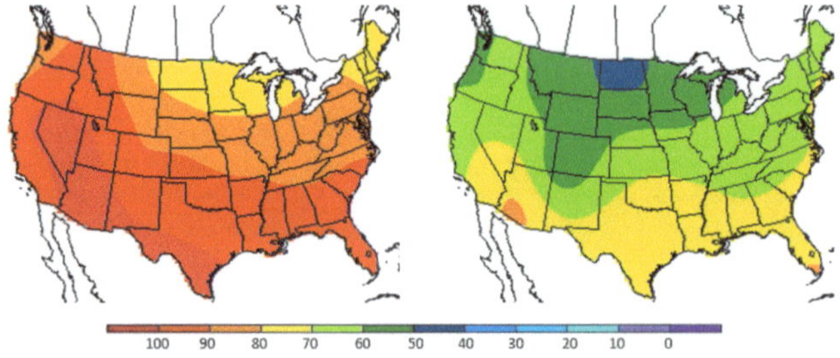

06/25/2025

Rain prevails in the northeast of the country from the Great Lakes and the central Great Plains to the Atlantic coasts, with heavy falls in the Northeast and down into the Megalopolis.

Rain also persists along the coastal Deep South regions to throughout Florida and into the northeast of Texas.

Mostly dry elsewhere.

06/26/2025

Precipitation begins to clear in the east of the country, but with outbreaks of rain still persisting through the central Great Plains, across the Ozarks and over southern parts of the Inland Midwest, as well as around the Great Lakes and the western side of the Northeast and down into the Appalachians.

Some showers and pockets of rain persist around Florida and there is the chance of an occasional shower in the Southwest and into the Great Basin, with isolated showers possibly reaching into the northern Rocky Mountains for a time.

There may be severe thunderstorms with hail and accompanying strong winds across the central and eastern parts of Montana around this time.

Mostly dry elsewhere.

ISOBARIC

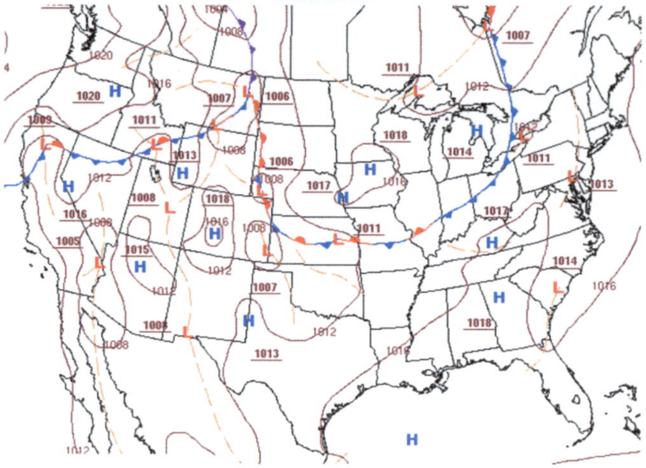

RAIN DISTRIBUTION (Inches)

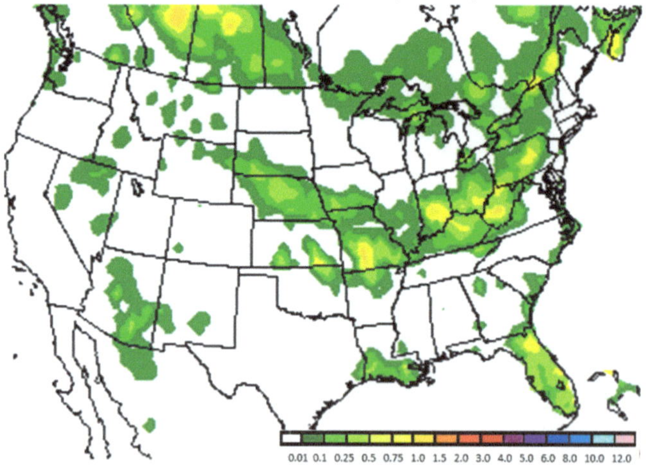

MAXIMUM TEMPERATURE MINIMUM TEMPERATURE

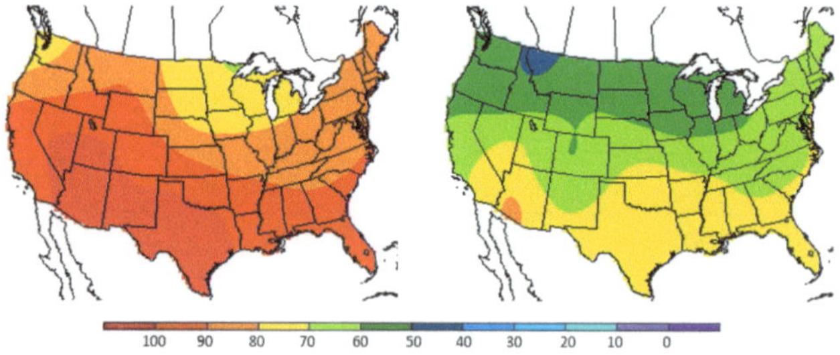

JUNE 27 FRIDAY

ISOBARIC

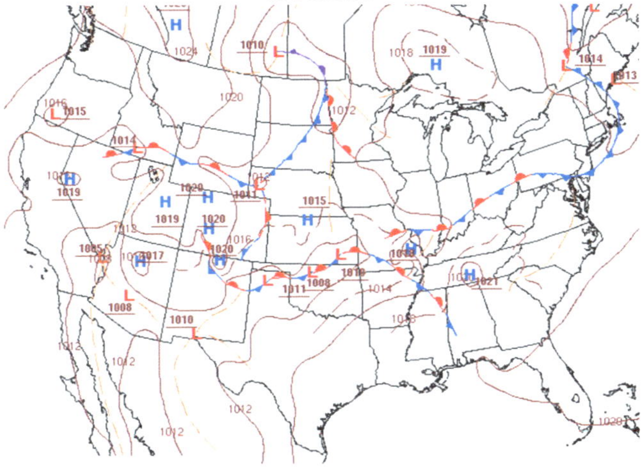

RAIN DISTRIBUTION (Inches)

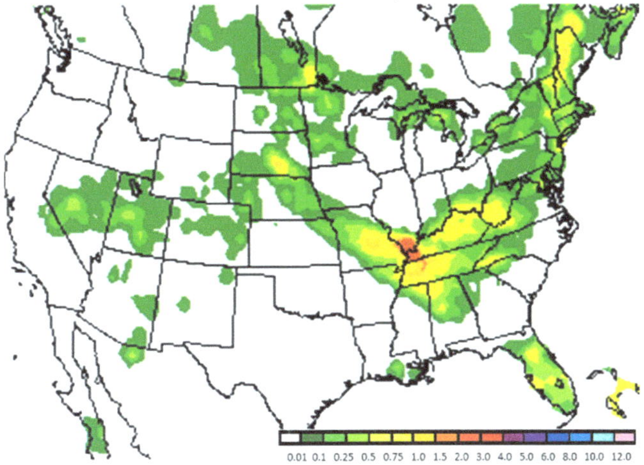

MAXIMUM TEMPERATURE

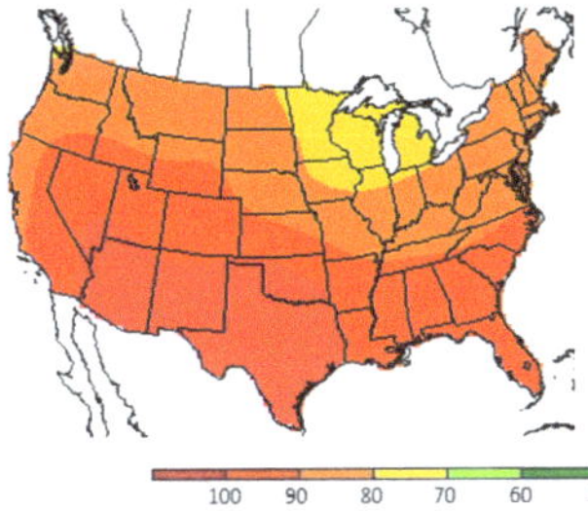

MINIMUM TEMPERATURE

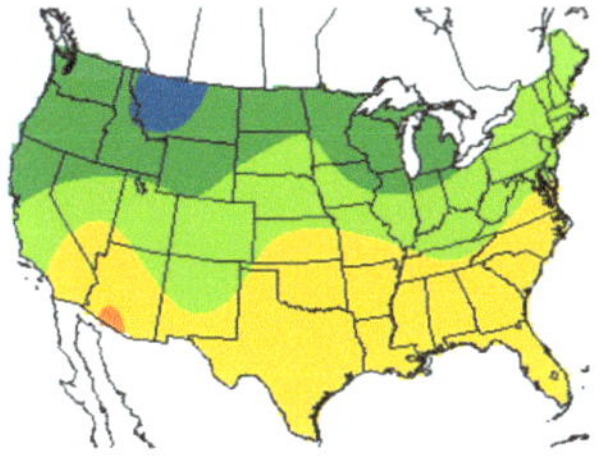

06/27/2025

Precipitation becomes more widespread once more throughout the Northeast with showers over the Great Lakes and into the North Great Plains.

Some outbreaks of heavy rain expected through the central Great Plains and down into the Ozarks and along the southern areas of the Inland Midwest and Corn Belt to over the Appalachians.

Showers in the Great Basin with isolated showers into the southern Rocky Mountains and into Arizona.

Showers persist in Florida.

Mostly dry elsewhere.

06/28/2025

Widespread rain returns across the Great Lakes and into the North Great Plains regions while clearing throughout the Northeast.

A band of rain also crosses the country from the Great Basin over the southern Rocky Mountains and onto the South Great Plains, through the Ozarks and southern Appalachians to the Coastal Southeast regions.

Some rain continues around Florida around this time as well, and some showers may brush into the Southwest for a time.

Mostly dry elsewhere.

JUNE 28 SATURDAY

ISOBARIC

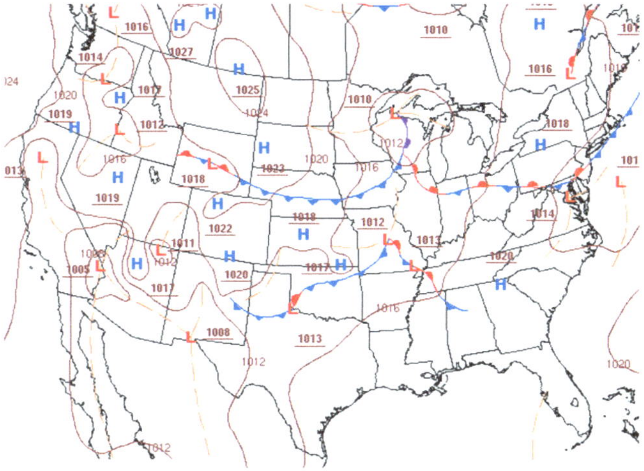

RAIN DISTRIBUTION (Inches)

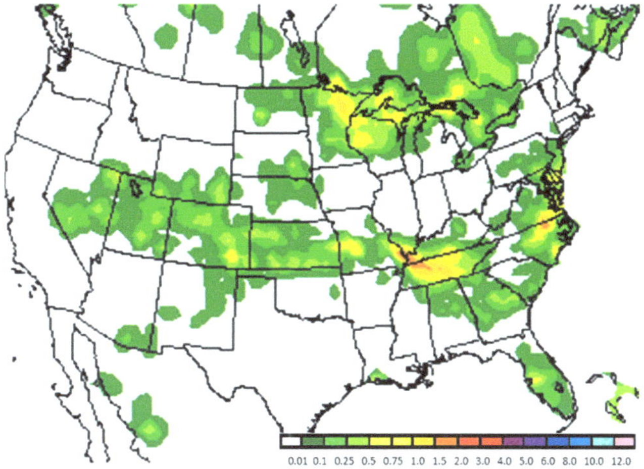

MAXIMUM TEMPERATURE

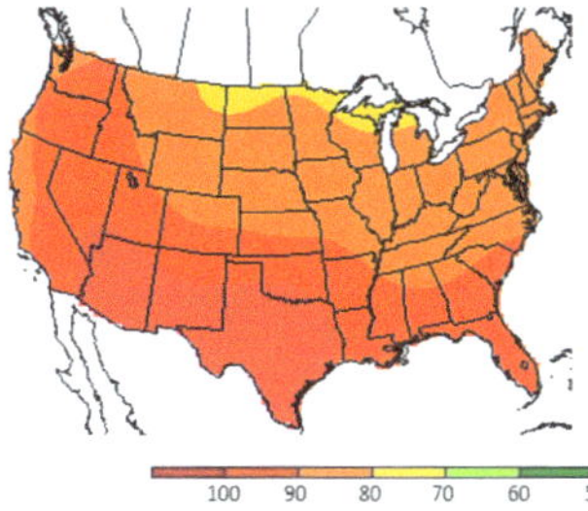

MINIMUM TEMPERATURE

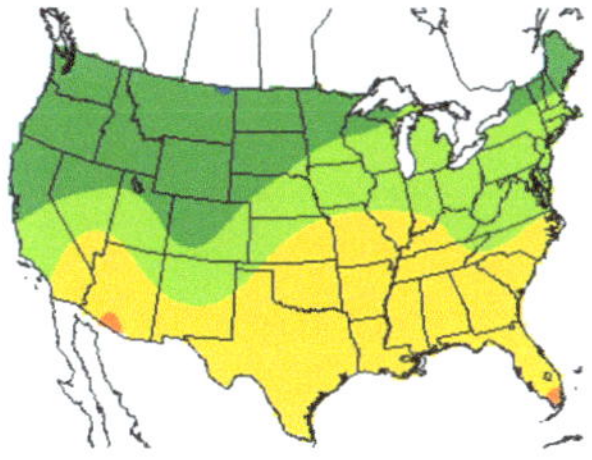

JUNE 29 SUNDAY

ISOBARIC

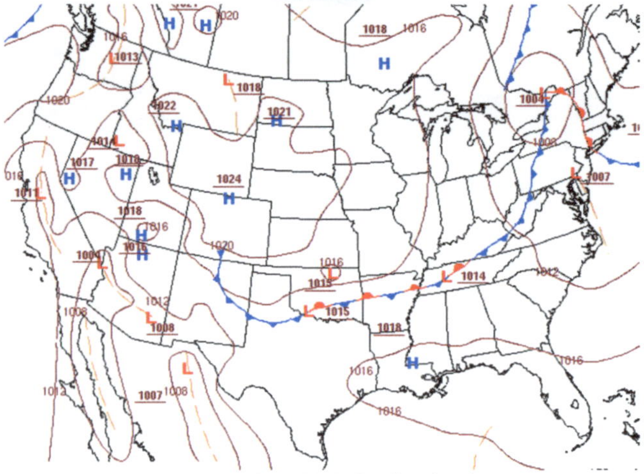

RAIN DISTRIBUTION (Inches)

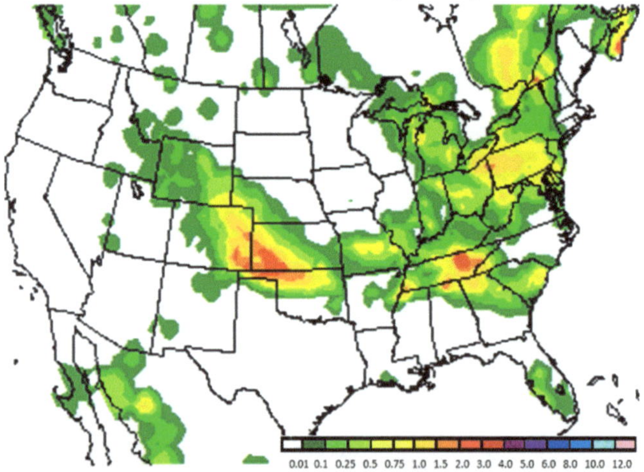

MAXIMUM TEMPERATURE

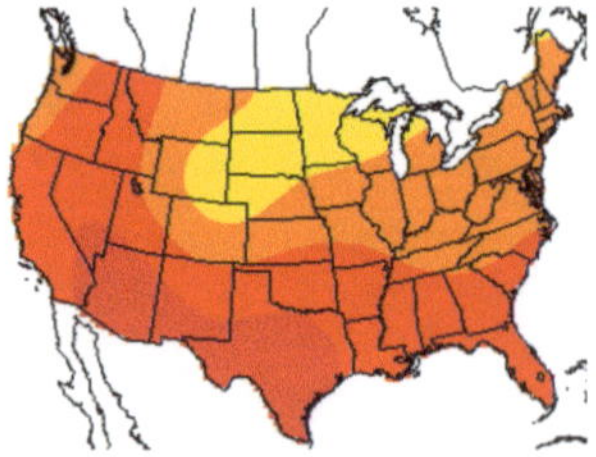

MINIMUM TEMPERATURE

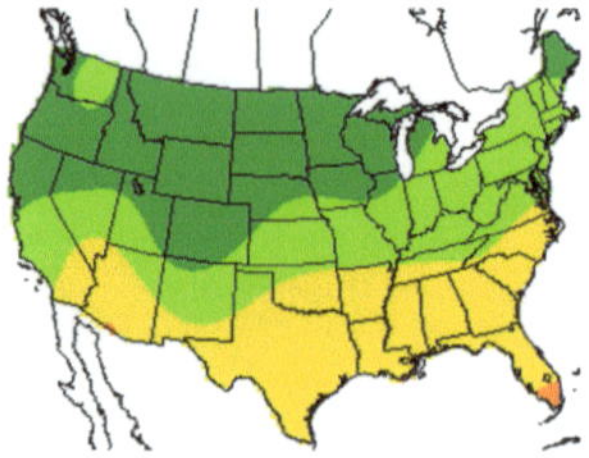

06/29/2025

Widespread rain throughout the east of the country from the Northeast to the Coastal Southeast regions including down through the Appalachians and westward through the South Great Plains and Ozarks, and some showers in Florida.

Heavy rain also over the Great South Plains and up into the Rocky Mountains with showers extending into the Inland Northwest.

An intense band of thunderstorms with hail and strong wind gusts may accompany heavy rain across the central and southern Great Plains over the next couple of days, with the severe storms possibly starting in Oklahoma, moving into New Mexico, later into Texas and finally into Arkansas.

There is also a chance of heavy rain in Mississippi around this time, as well as in the New York region.

Mostly dry elsewhere.

It may be unusually warm in Alaska.

06/30/2025

Widespread heavy rain prevails in the Northeast as well as through the Deep South, with showers remaining around the southern half of Florida as well as over the Great Lakes and the Great Plains.

Outbreaks of rain in the Southwest region and expanding up through the Rocky Mountains.

Mostly dry elsewhere, although there may be the chance of some isolated showers in the Great Basin.

JUNE 30 MONDAY

ISOBARIC

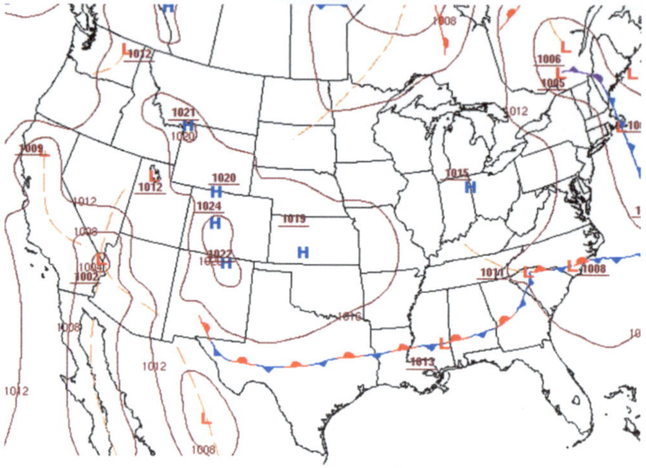

RAIN DISTRIBUTION (Inches)

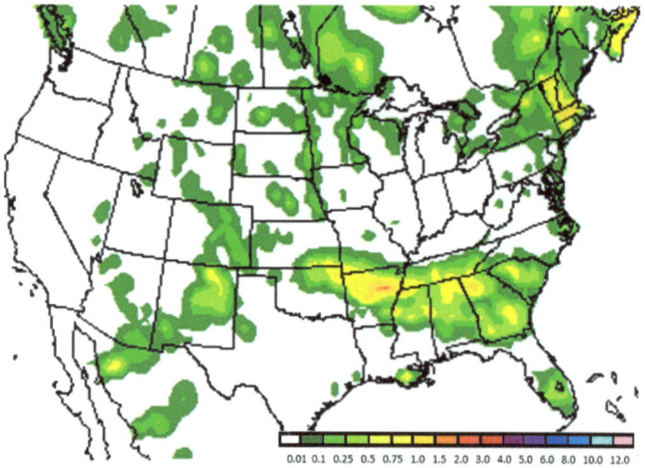

MAXIMUM TEMPERATURE

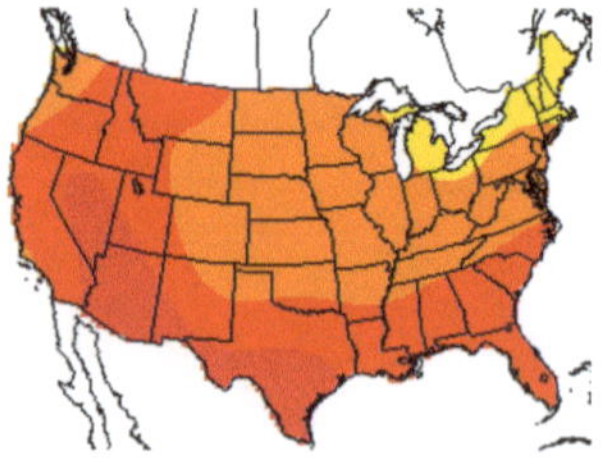

MINIMUM TEMPERATURE

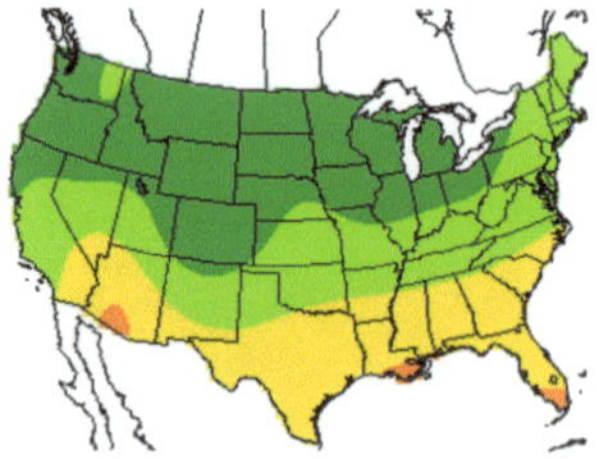

JULY

PHASES OF THE MOON

2nd	First Quarter
2nd	Crossing Equator
5th	Apogee
9th	Southern Declination
10th	Full Moon
17th	Crossing Equator
18th	Third Quarter
20th	Perigee #11
22nd	Northern Declination
24th	New Moon
30th	Crossing Equator

1 Tue	2 Wed	3 Thur	4 Fri	5 Sat	6 Sun	7 Mon	8 Tue	9 Wed	10 Thur	11 Fri	12 Sat	13 Sun	14 Mon	15 Tue	16 Wed	17 Thur	18 Fri	19 Sat	20 Sun	21 Mon	22 Tue	23 Wed	24 Thur	25 Fri	26 Sat	27 Sun	28 Mon	29 Tue	30 Wed	31 Thur
	1Q XhS			A				V	F							XhN	3Q		P11		^		N						XhS	

JULY 1 TUESDAY

ISOBARIC

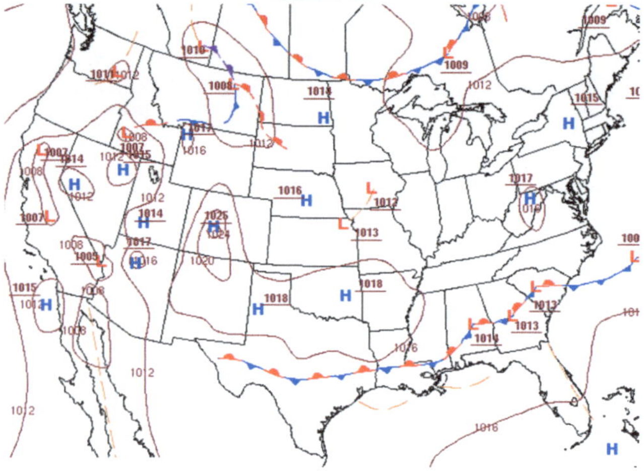

RAIN DISTRIBUTION (Inches)

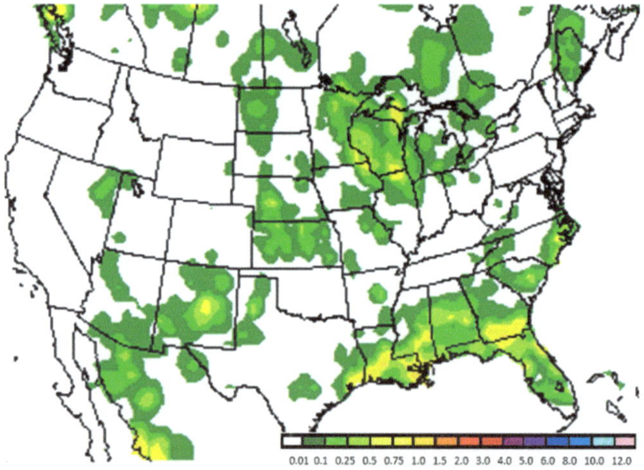

MAXIMUM TEMPERATURE

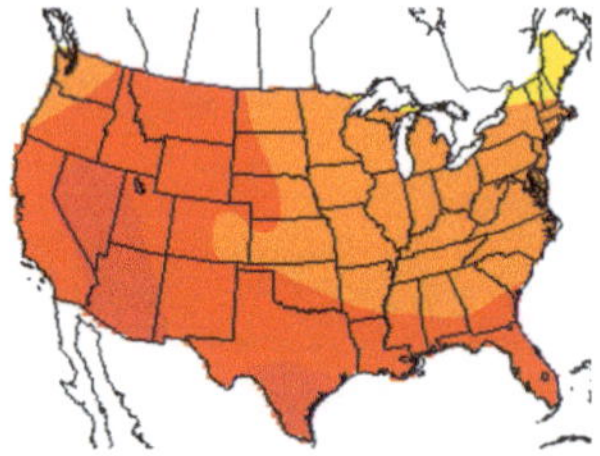

MINIMUM TEMPERATURE

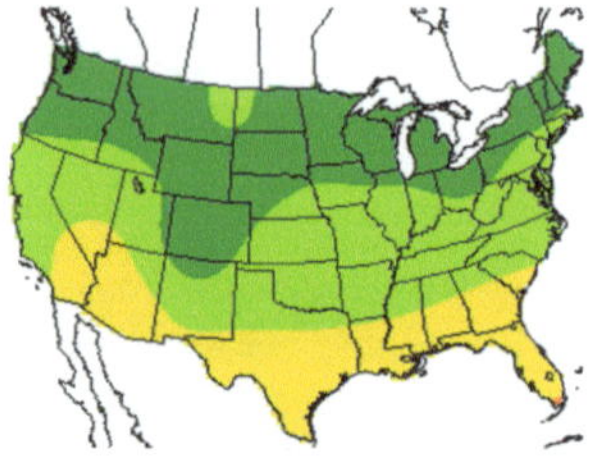

07/01/2025

High pressure throughout most of the country from the east of the Great Plains to the Inland Midwest.

Showers from the Southwest across the Great Plains and up to the Great Lakes regions.

A stationary front over the South of the country brings rain to the Deep South and Florida.

Some showers in the Northeast region.

Mostly dry elsewhere.

07/02/2025

High pressure over most of the country.

Outbreaks of rain Great Lakes to the Inland Midwest.

Showers up into the Northeast.

Showers over the Rocky Mountains and Great Plains.

Heavy rain through the Coastal Southeast to Florida.

Chance of thunderstorms in southern Texas.

Mostly dry elsewhere.

JULY 2 WEDNESDAY

ISOBARIC

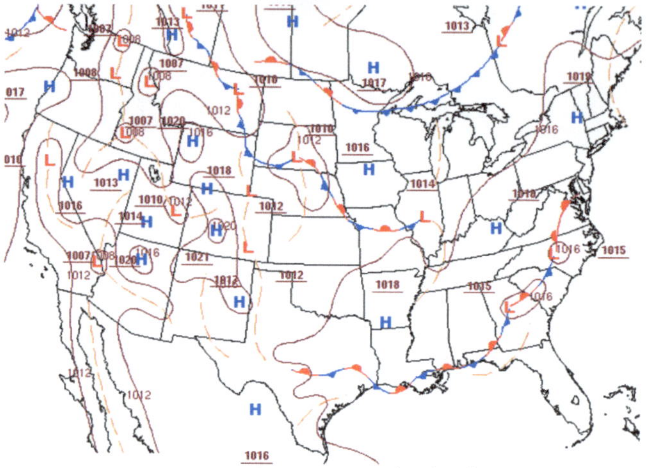

RAIN DISTRIBUTION (Inches)

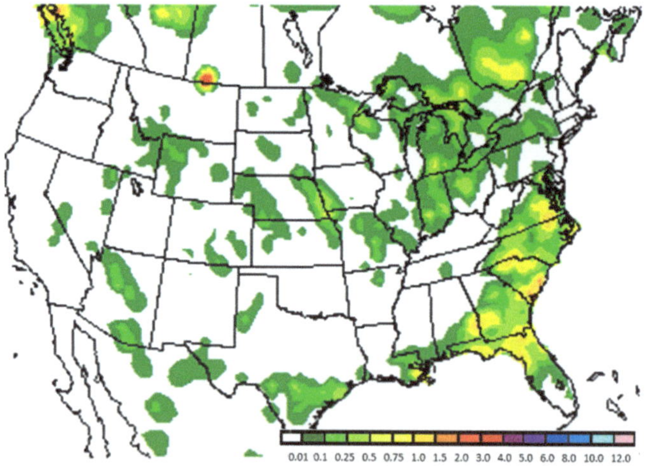

MAXIMUM TEMPERATURE

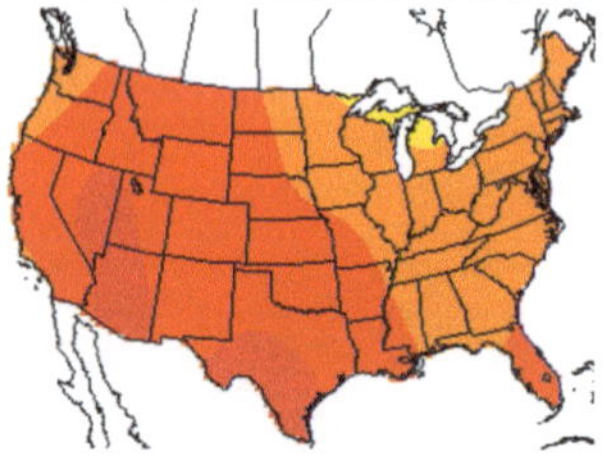

MINIMUM TEMPERATURE

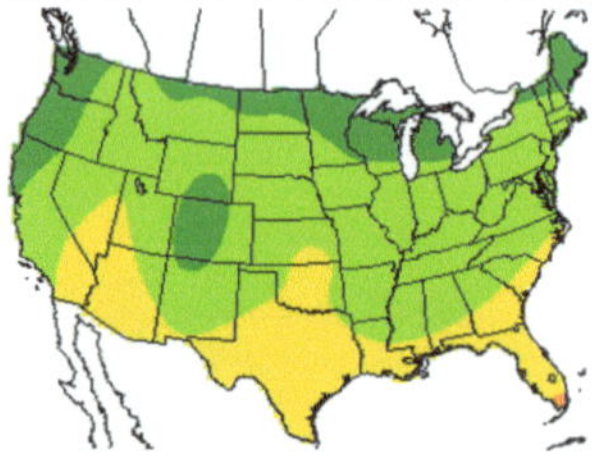

JULY 3 THURSDAY

ISOBARIC

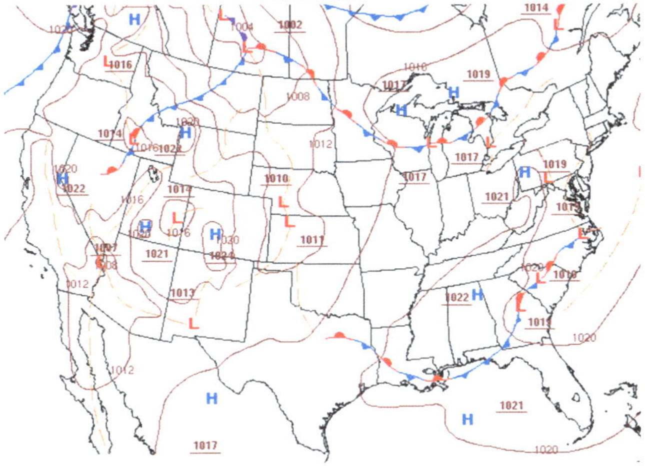

RAIN DISTRIBUTION (Inches)

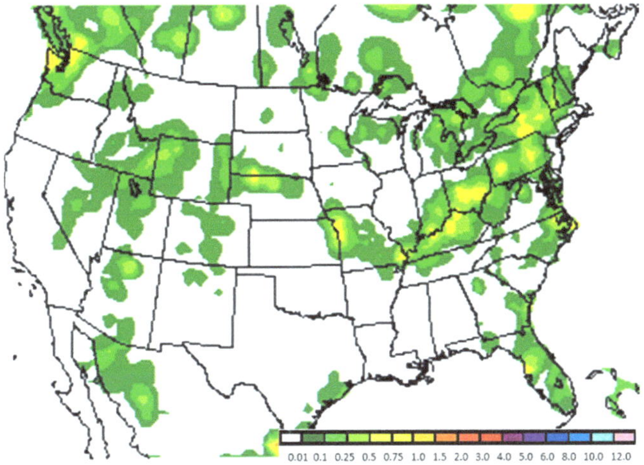

MAXIMUM TEMPERATURE

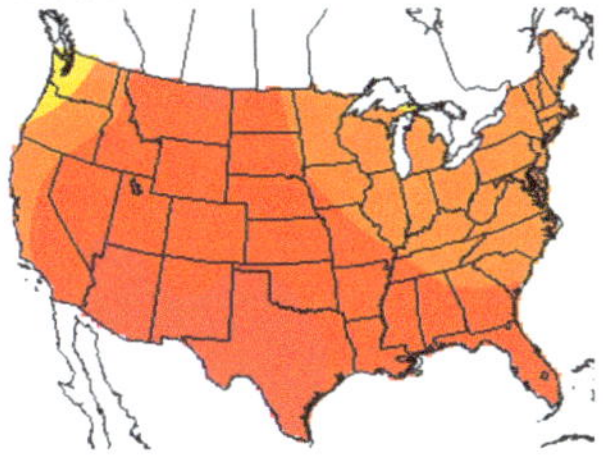

MINIMUM TEMPERATURE

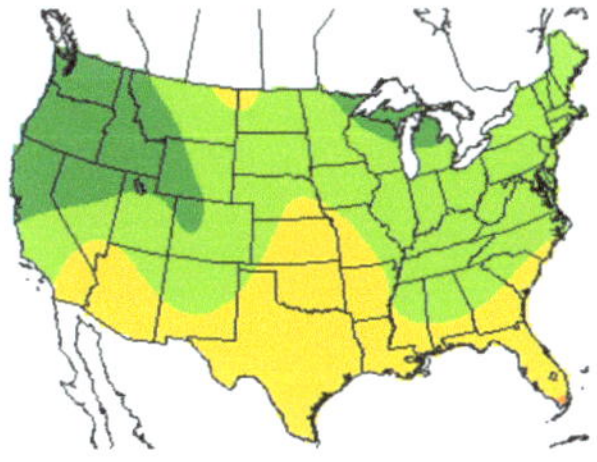

07/03/2025

Low pressure from the Southwest to the Rocky Mountains and into the Great Plains brings outbreaks of rain.

A cold front brings rain into the Pacific Northwest.

A stationary front over the Great Lakes brings rain around the Great Lakes to the Inland Midwest north of the Appalachians.

Showers into the Northeast region.

Further showers in Florida.

Chance of further thunderstorms in Texas.

Mostly dry elsewhere.

07/04/2025

Building high pressure clears most rain across the western half of the country.

Widespread rain around the Great Lakes to the Northeast.

Heavy rain likely in the Metropolitan region.

Some showers over Florida.

Mostly dry elsewhere.

Showery in Hawaii.

Showers in Alaska, including Anchorage.

JULY 4 FRIDAY

ISOBARIC

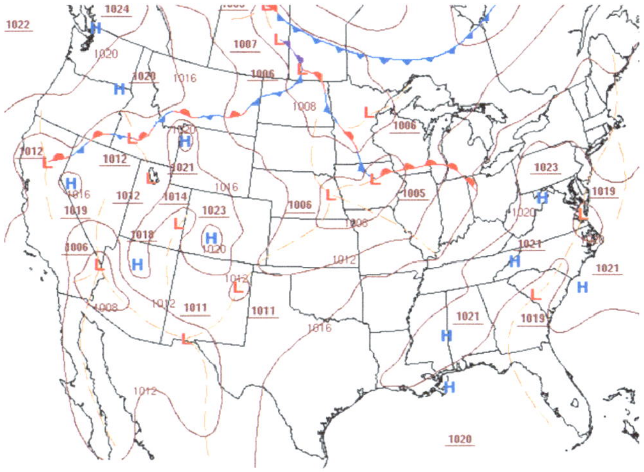

RAIN DISTRIBUTION (Inches)

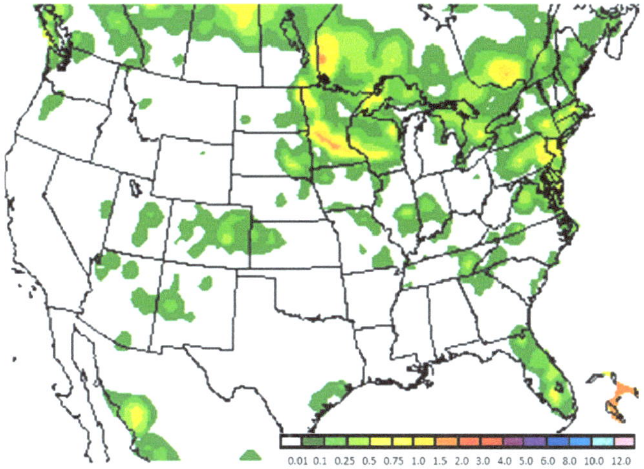

MAXIMUM TEMPERATURE

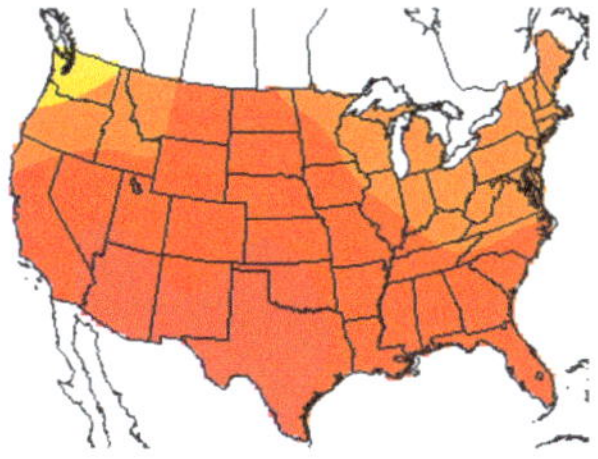

MINIMUM TEMPERATURE

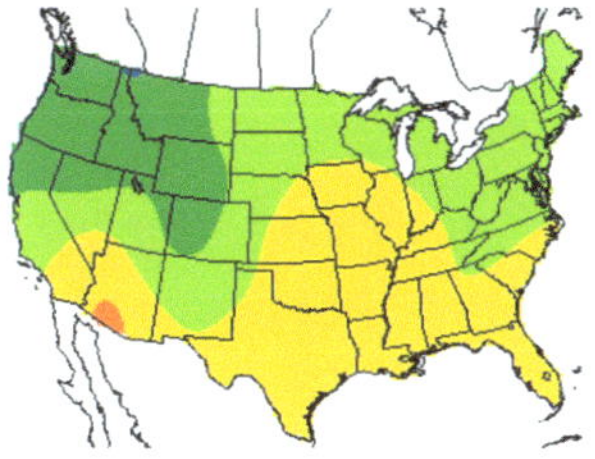

ISOBARIC

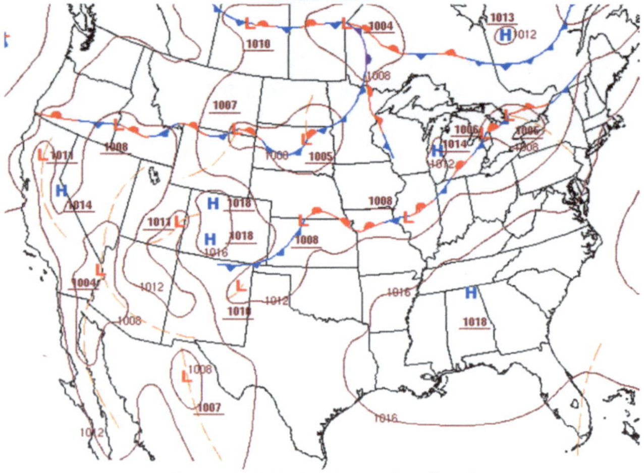

RAIN DISTRIBUTION (Inches)

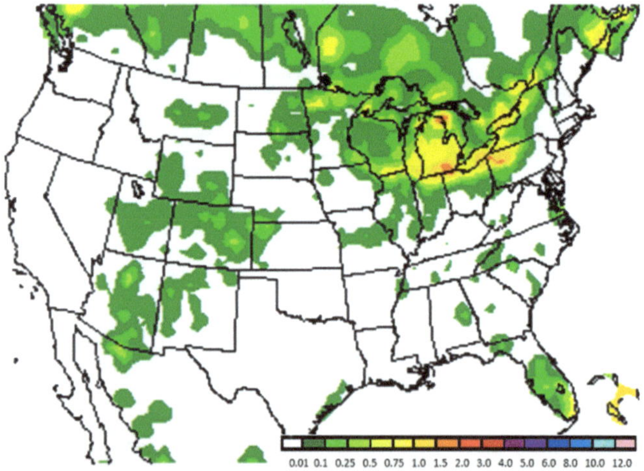

MAXIMUM TEMPERATURE

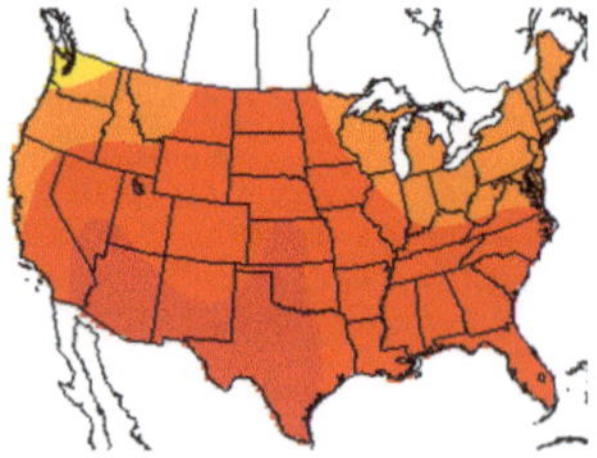

MINIMUM TEMPERATURE

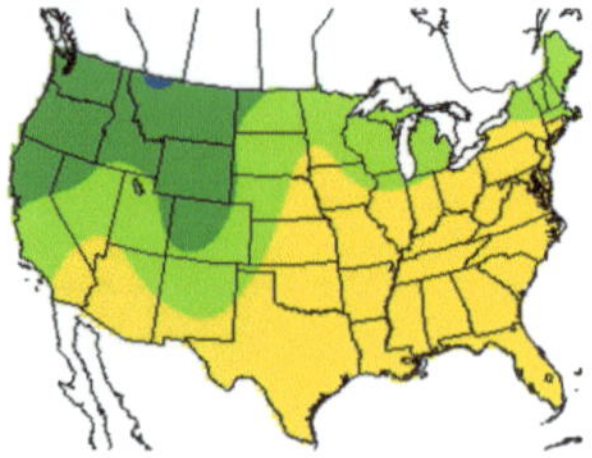

07/05/2025

Widespread rain North Great Plains to the east of the Northeast and into the Inland Midwest.

Rain from the Southwest into the Rocky Mountains.

Lingering showers in the Coastal Southeast to Florida.

Mostly dry elsewhere.

Chance of rain in Kansas developing flooding along the Verdigris River system in the next few days.

07/06/2025

Rain through the Rocky Mountains to the Southwest region.

Heavy rain around the Great Lakes, Inland Midwest and throughout the Northeast.

Showers in the Coastal Southeast to Florida.

Mostly dry elsewhere.

It may become cooler in Oregon and Montana over the next few days.

ISOBARIC

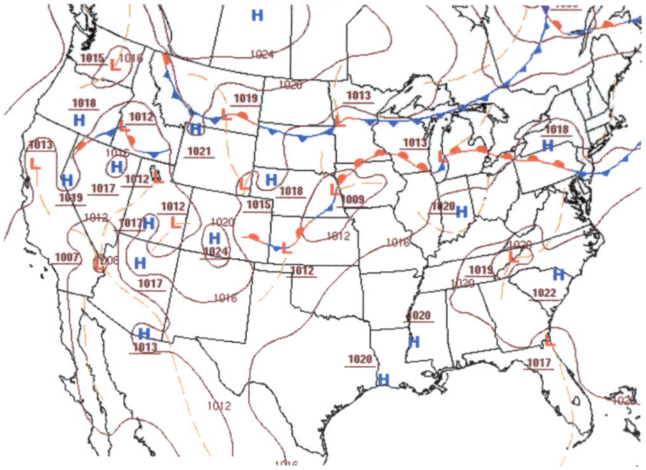

RAIN DISTRIBUTION (Inches)

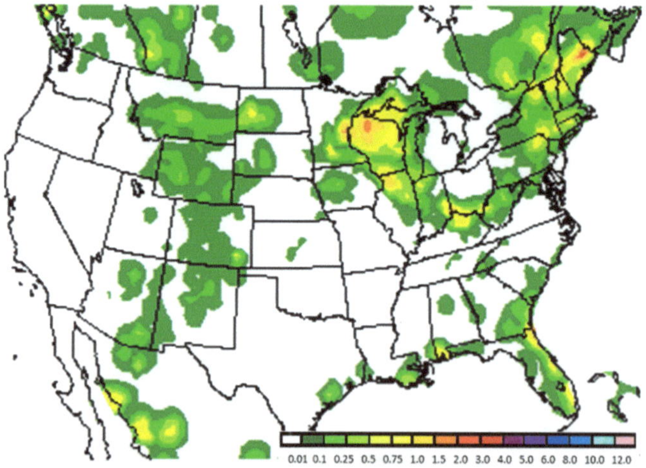

MAXIMUM TEMPERATURE

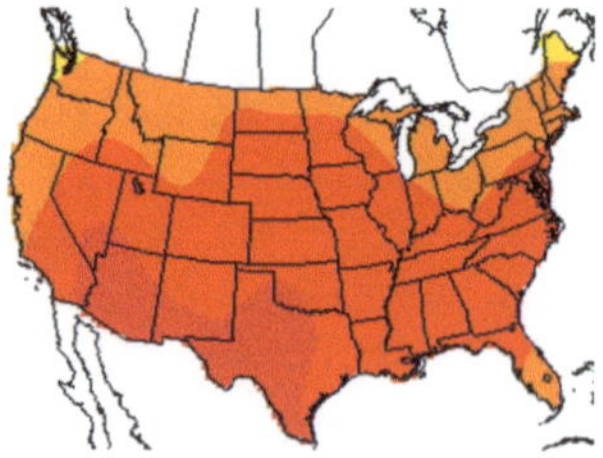

MINIMUM TEMPERATURE

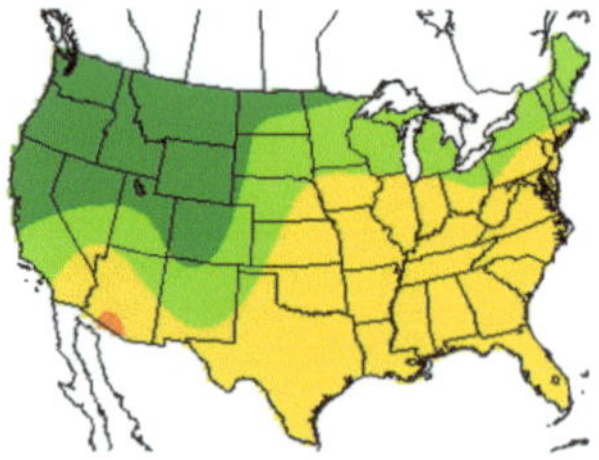

ISOBARIC

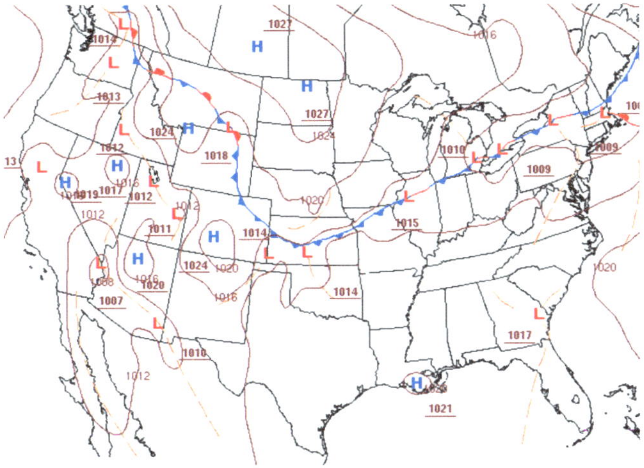

RAIN DISTRIBUTION (Inches)

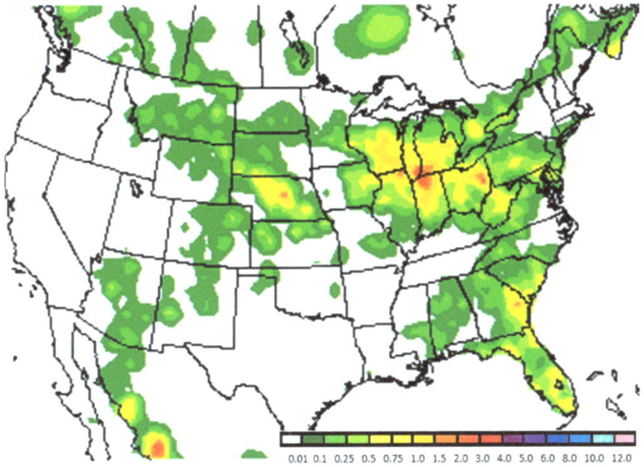

MAXIMUM TEMPERATURE

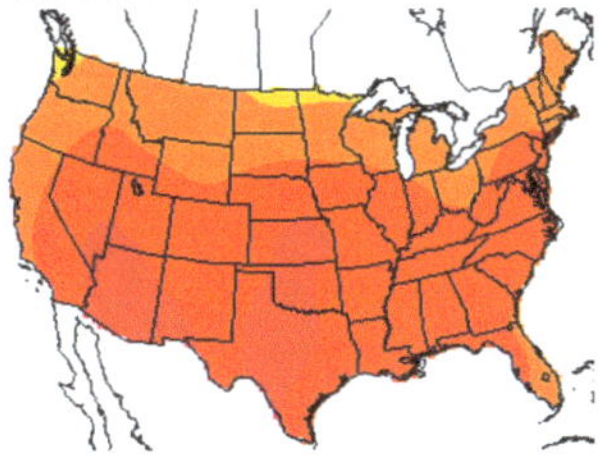

MINIMUM TEMPERATURE

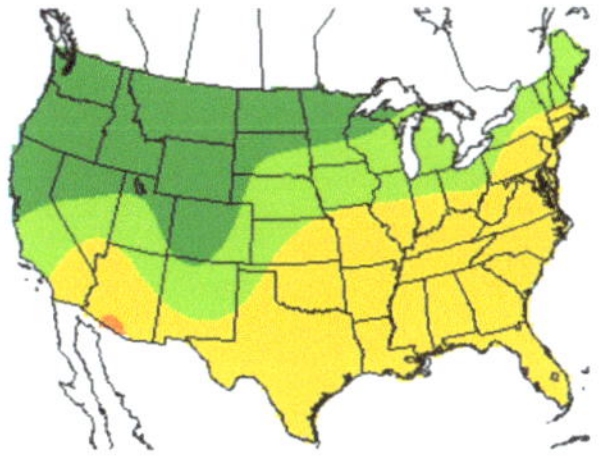

07/07/2025

Widespread rain from the Inland Northwest, across the Plains, and throughout the Inland Midwest to the Atlantic Coast.

Low pressure spreads rain through the Coastal Southeast and Florida.

Outbreaks of showers from the Southwest region up through the Rocky Mountains.

Mostly dry elsewhere.

Chance of flooding downstream through some Texan river systems over the next few days.

07/08/2025

Heavy rain in the south Inland Midwest, Appalachians, and Florida with showers into the Great Lakes.

Heavy rain in the Northeast.

Showers in the Southwest and through the Rocky Mountains to Inland Pacific Northwest, as well as in the Deep South.

Mostly dry elsewhere.

Warmer over Oregon to Montana, later spreading to the Great Plains and Northeast over the next few days.

ISOBARIC

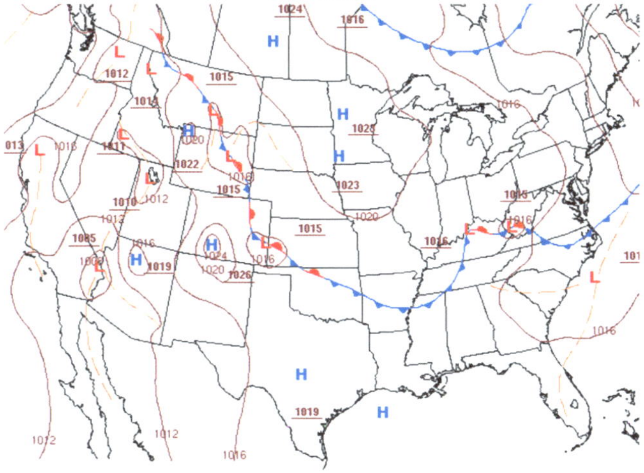

RAIN DISTRIBUTION (Inches)

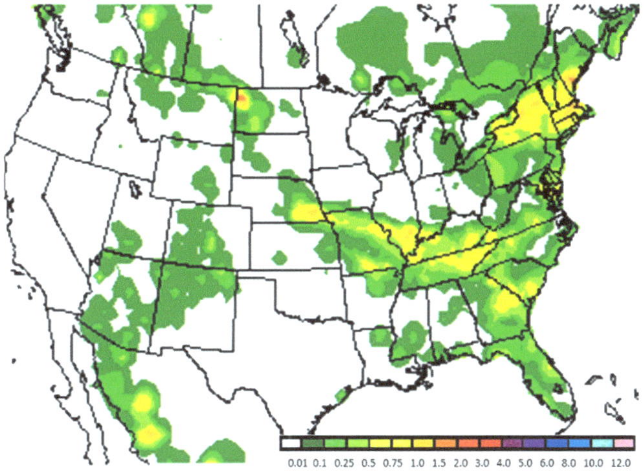

MAXIMUM TEMPERATURE

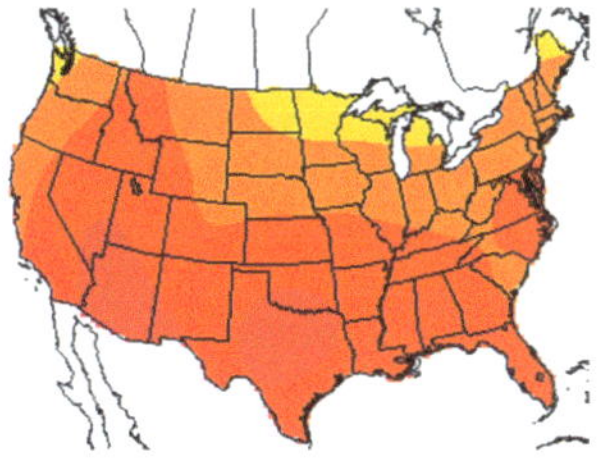

MINIMUM TEMPERATURE

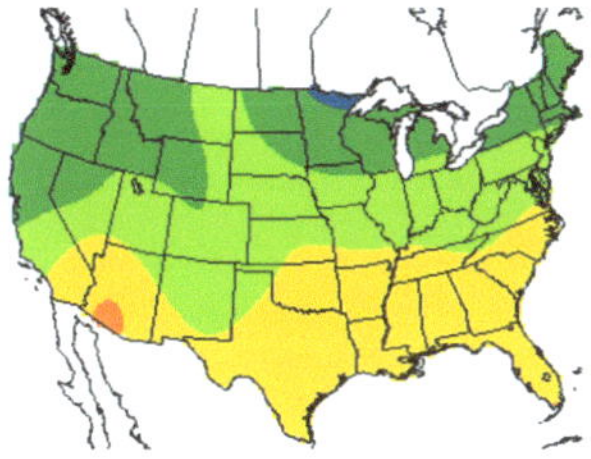

JULY 9 WEDNESDAY

ISOBARIC

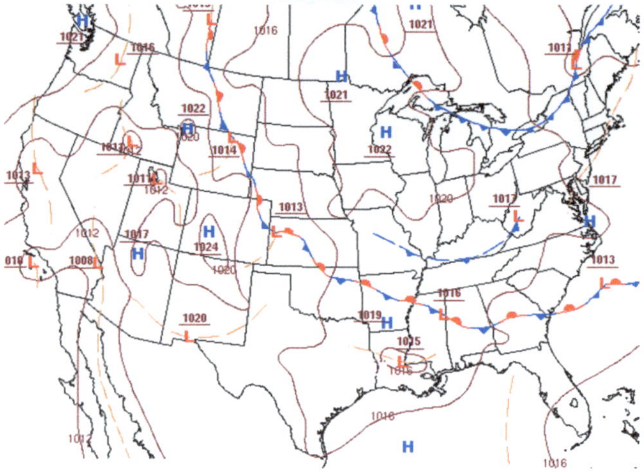

RAIN DISTRIBUTION (Inches)

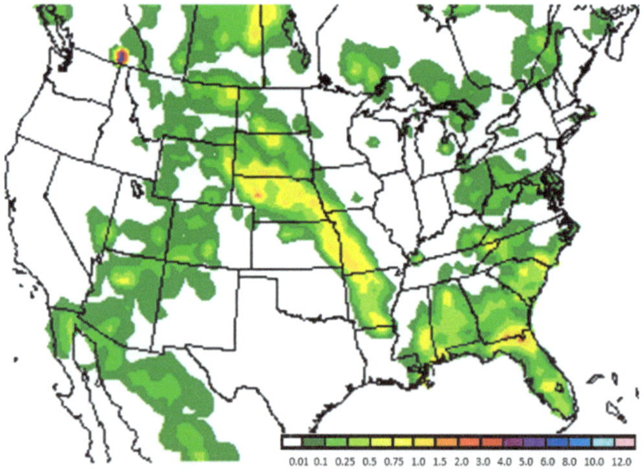

MAXIMUM TEMPERATURE / MINIMUM TEMPERATURE

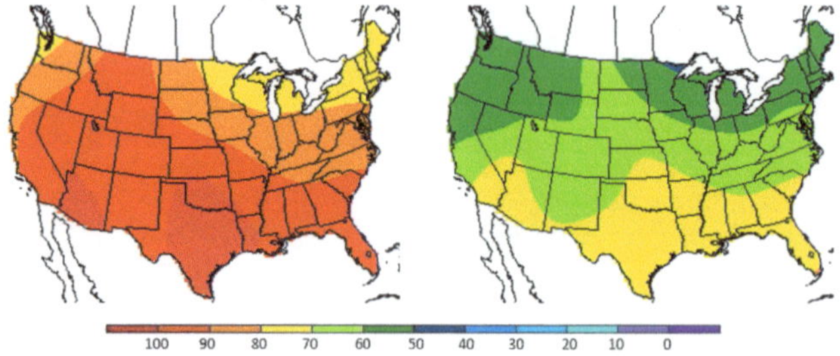

07/09/2025

Heavy rain through the Great Plains to the Coastal Southeast.

Rain into the Rocky Mountains and showers in the Southwest.

Occasional showers Great Lakes to the Northeast.

Heavy showers Coastal Southeast to Florida and into the Deep South.

Heavy downpours at times in the Ozarks.

Mostly dry elsewhere.

Unusually hot in the Great Plains.

07/10/2025

High pressure dominates the country.

Rain from the Rocky Mountains and Great Basin to the Southwest.

Rain eases over the Great Plains to the Deep South.

Showers in the Northeast.

Rain Coastal Southeast to Florida.

Mostly dry elsewhere.

Chance of flooding along the Wichita and Leon River systems.

JULY 10 THURSDAY

ISOBARIC

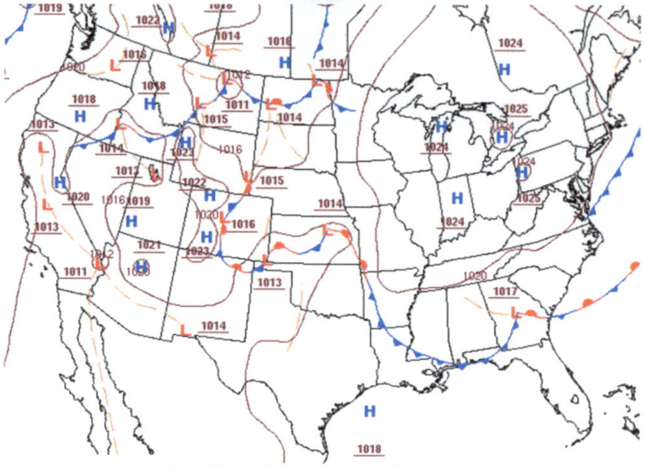

RAIN DISTRIBUTION (Inches)

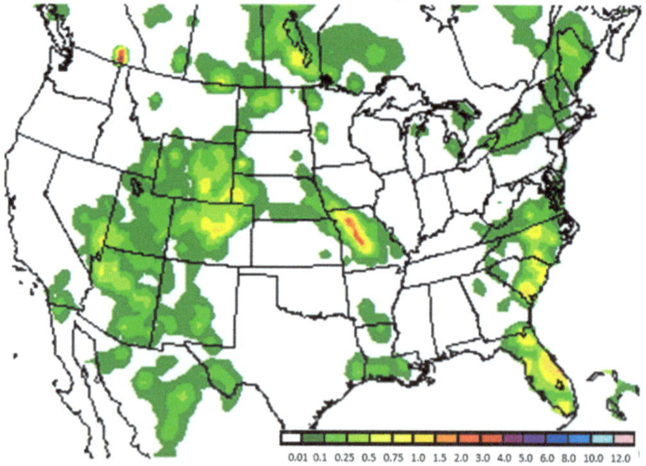

MAXIMUM TEMPERATURE / MINIMUM TEMPERATURE

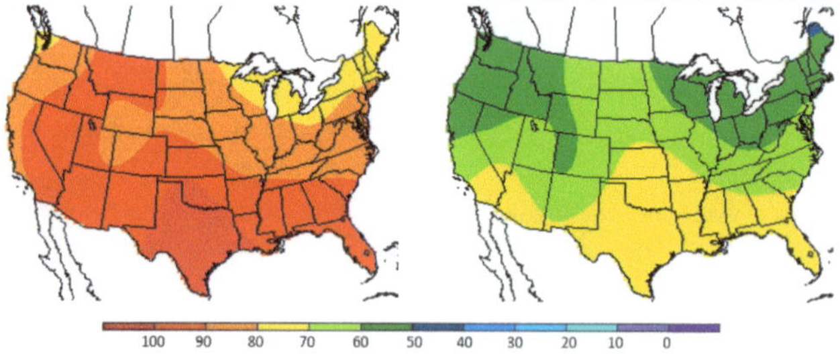

JULY **11** FRIDAY

ISOBARIC

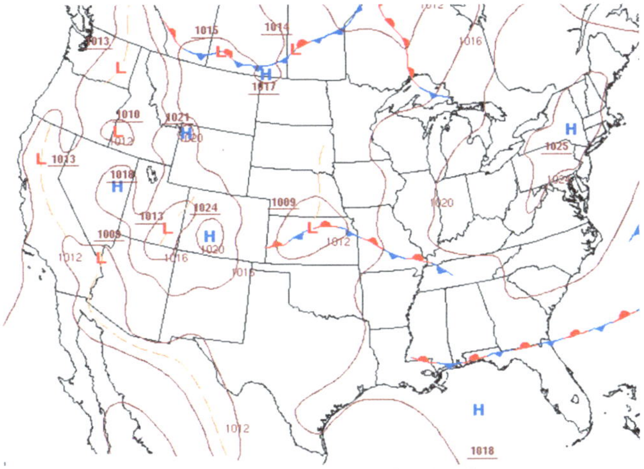

RAIN DISTRIBUTION (Inches)

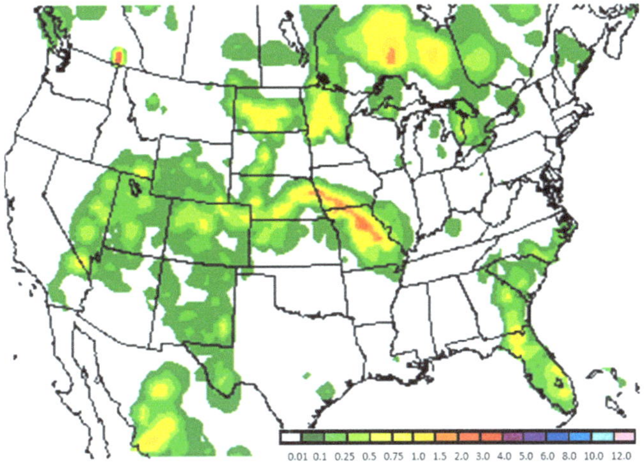

MAXIMUM TEMPERATURE / MINIMUM TEMPERATURE

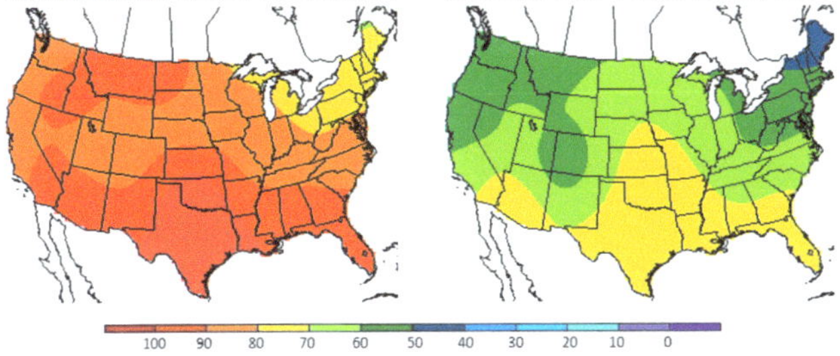

07/11/2025

High pressure weakens west of the Rocky Mountains.

Precipitation from the Great Basin to over the Rocky Mountains and into the Southwest.

Stationary front with heavy rain over the South Great Plains.

Some rain around the North Great Plains to the Great Lakes.

Isolated showers in the Northeast.

Heavy showers Coastal Southeast to Florida.

Chance of flooding in the Verdigris River system.

Mostly dry elsewhere.

Very hot in the Dakotas.

JULY **12** SATURDAY

ISOBARIC

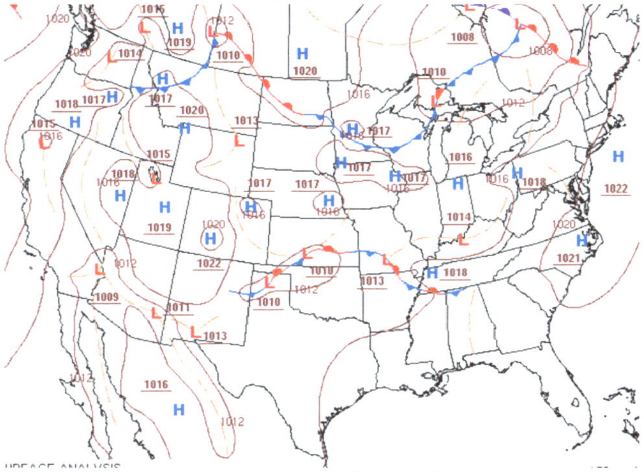

RAIN DISTRIBUTION (Inches)

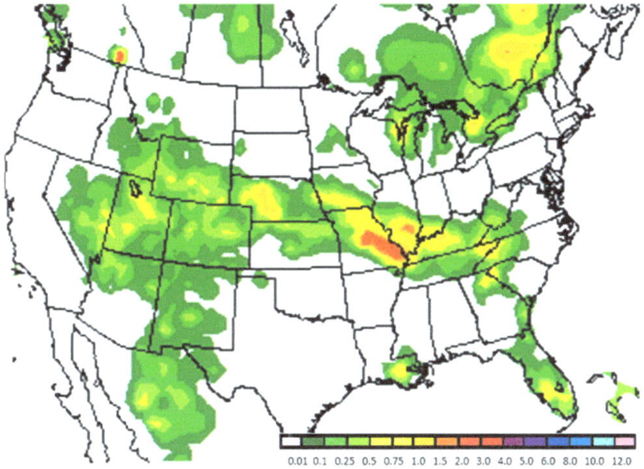

MAXIMUM TEMPERATURE / MINIMUM TEMPERATURE

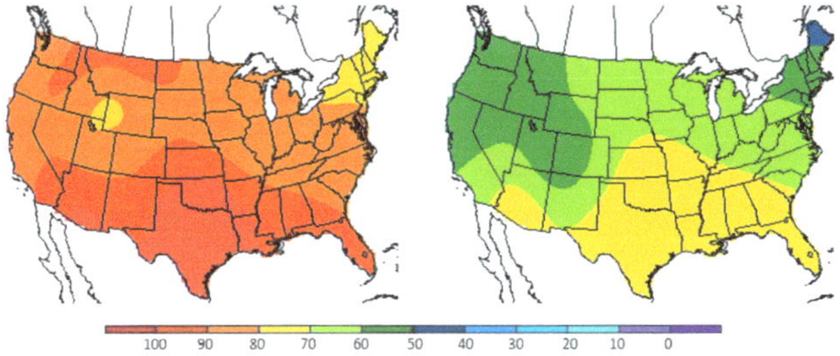

07/12/2025

High pressure builds over the country.

Widespread precipitation from the Great Basin, over the Rocky Mountains, across the central Great Plains to the Appalachians.

Showers into the Southwest.

Outbreaks of rain Great Lakes, with showers western parts of the Northeast.

Some rain in Florida.

Mostly dry elsewhere.

Rain becomes heavier and more widespread around the western Gulf Coast regions of Texas around this time.

JULY 13 SUNDAY

ISOBARIC

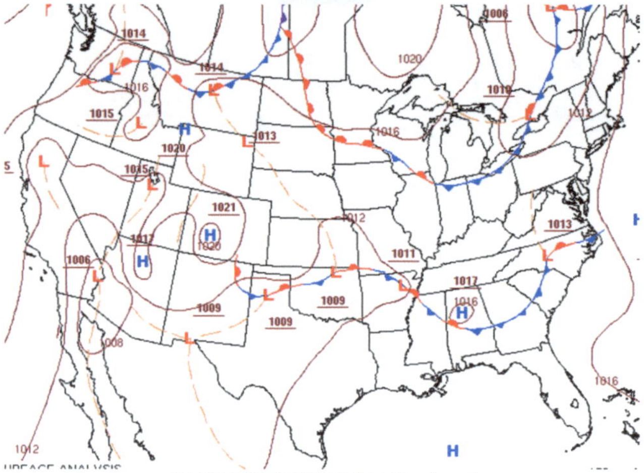

RAIN DISTRIBUTION (Inches)

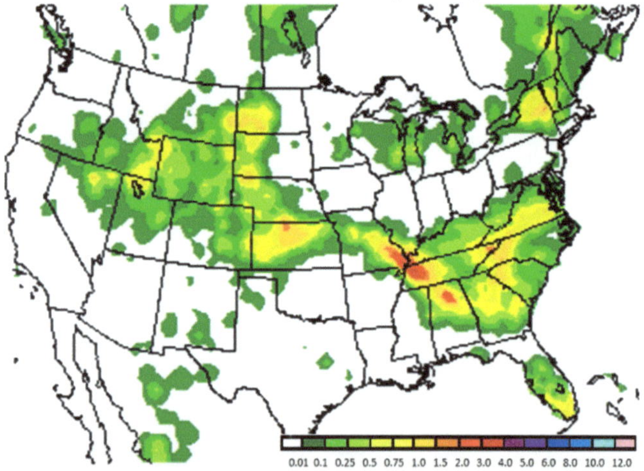

MAXIMUM TEMPERATURE MINIMUM TEMPERATURE

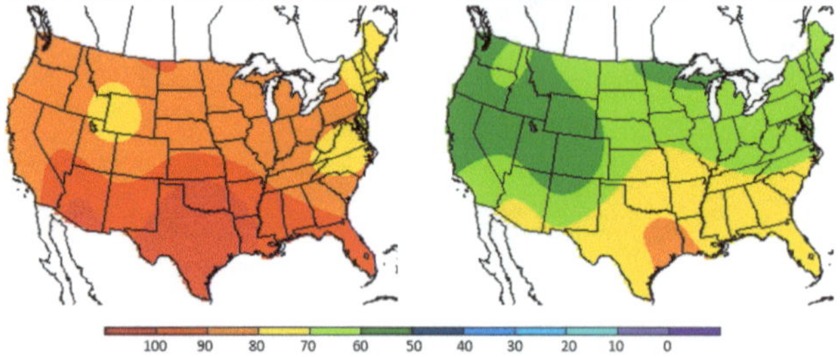

07/13/2025

High pressure weakens.

Isolated showers in the Southwest.

Fronts bring widespread precipitation from the Great Basin in the west, to the Coastal Southeast and Deep South.

Heavy downpours in the Ozarks.

Heavy rain western half of North Great Plains.

Showers Great Lakes.

Rain in the Northeast, showers in Florida.

Mostly dry elsewhere.

Chance of thunderstorms in Alaska.

07/14/2025

A complex of fronts cross the country.

Precipitation from the Pacific Northwest, through the Rocky Mountains and Great Plains, into the Deep South and Florida.

Rain along the Atlantic coasts from Florida to the Metropolitan and widespread in the Northeast.

Heavy rain in the Ozarks and North Great Plains.

Mostly dry elsewhere.

Heat wave temperatures in the west including California to Nevada.

JULY 14 MONDAY

ISOBARIC

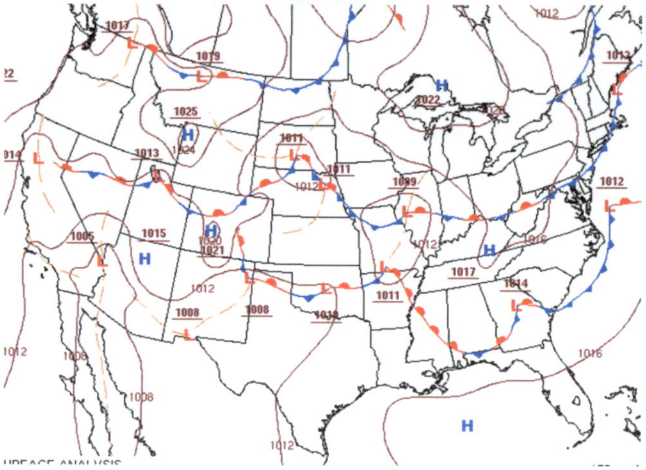

RAIN DISTRIBUTION (Inches)

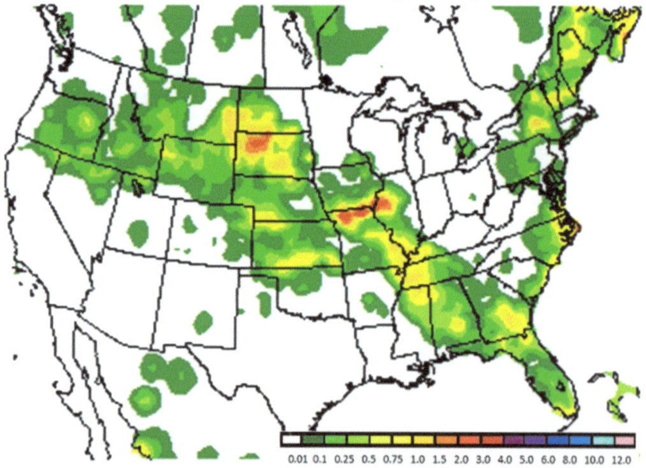

MAXIMUM TEMPERATURE MINIMUM TEMPERATURE

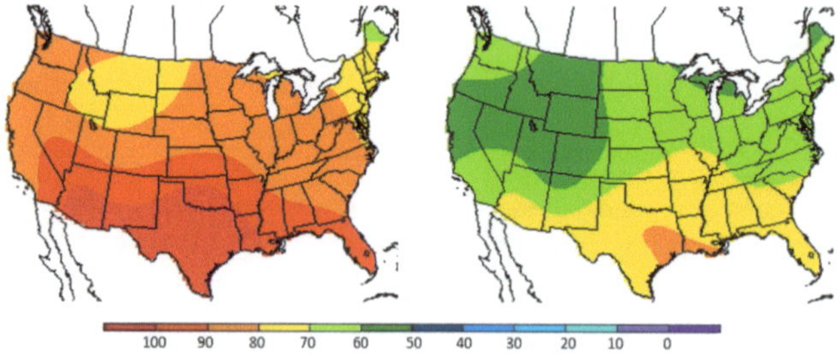

JULY 15 TUESDAY

ISOBARIC

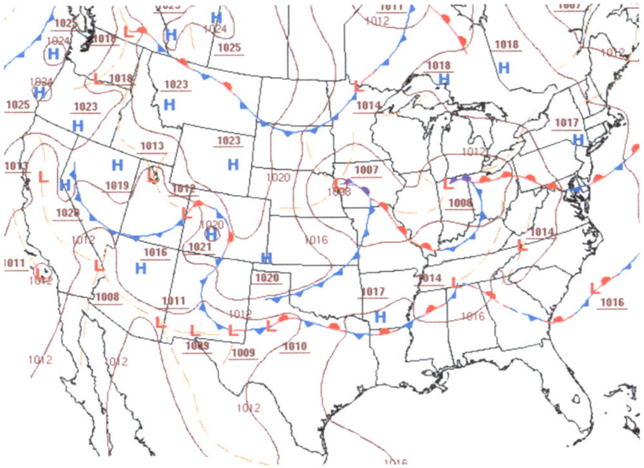

07/15/2025

A complexity of cold fronts prevail.

Widespread rain from the Pacific Northwest to the northern Rocky Mountains.

Precipitation throughout the Great Plains to the Inland Midwest, heavy at times.

Showers over the Deep South to Florida.

Rain in the Northeast Highlands.

Showers in Georgia and Mississippi.

Mostly dry elsewhere.

RAIN DISTRIBUTION (Inches)

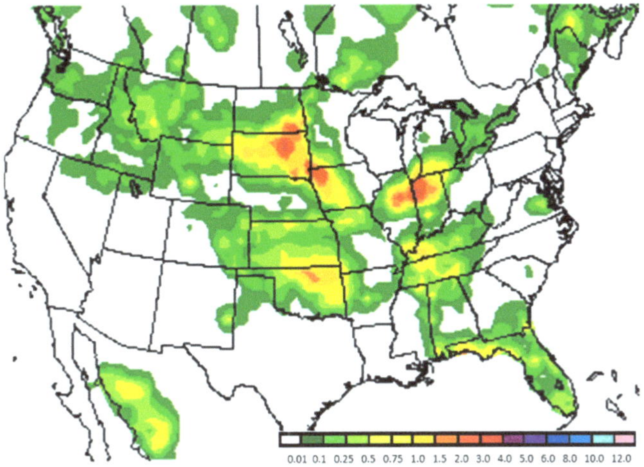

MAXIMUM TEMPERATURE

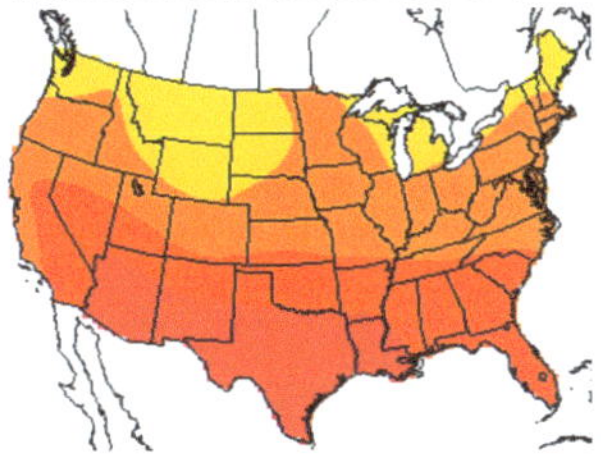

MINIMUM TEMPERATURE

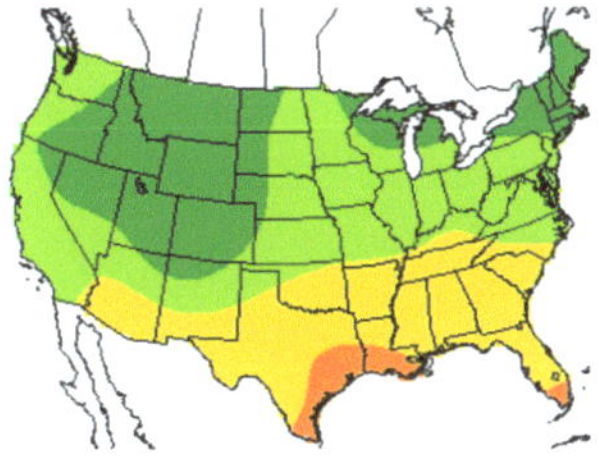

07/16/2025

A complex of fronts cross the country from the Great Basin to the Great Lakes.

Rain persists from the Pacific Northwest, through the Rocky Mountains and across the Great Plains.

Heavy rain Inland Midwest to the Northeast Megalopolis, down through the Appalachians to the Deep South and Florida.

Mostly dry elsewhere.

Heat wave temperatures may spread across the Northern Rockies to the North Great Plains, including across Montana.

Unusually warm temperatures in central Alaska.

JULY 16 WEDNESDAY

ISOBARIC

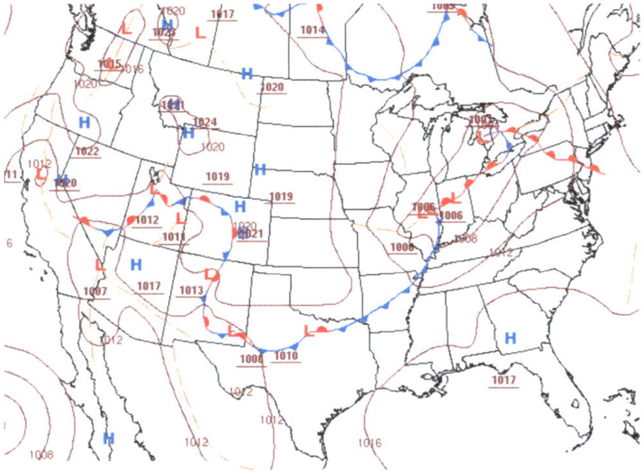

RAIN DISTRIBUTION (Inches)

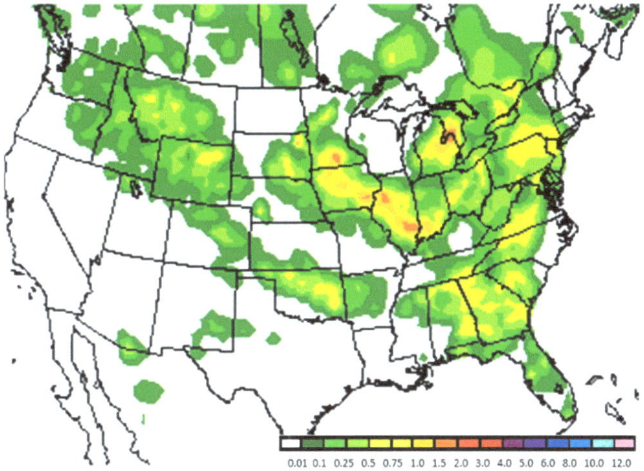

MAXIMUM TEMPERATURE

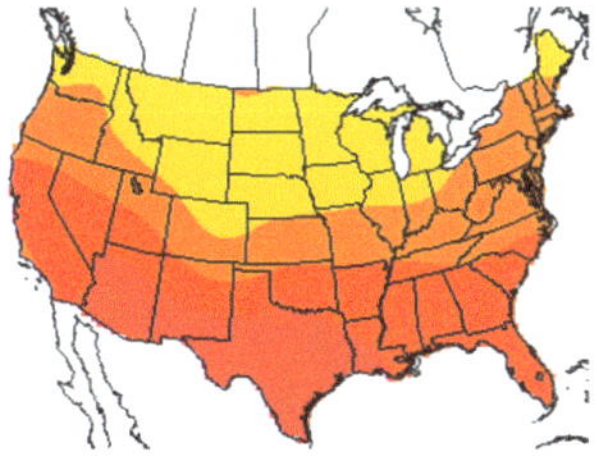

MINIMUM TEMPERATURE

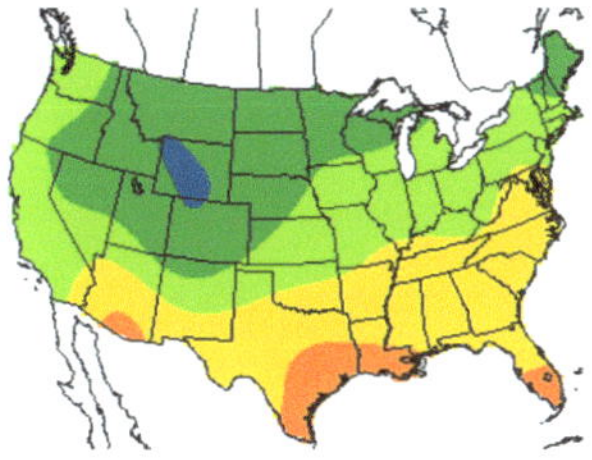

JULY 17 THURSDAY

ISOBARIC

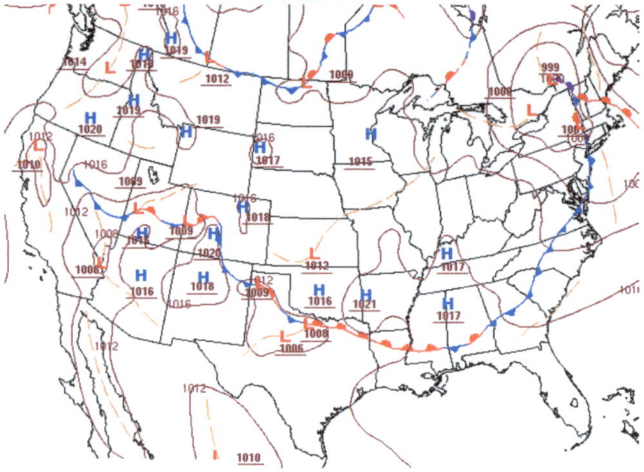

RAIN DISTRIBUTION (Inches)

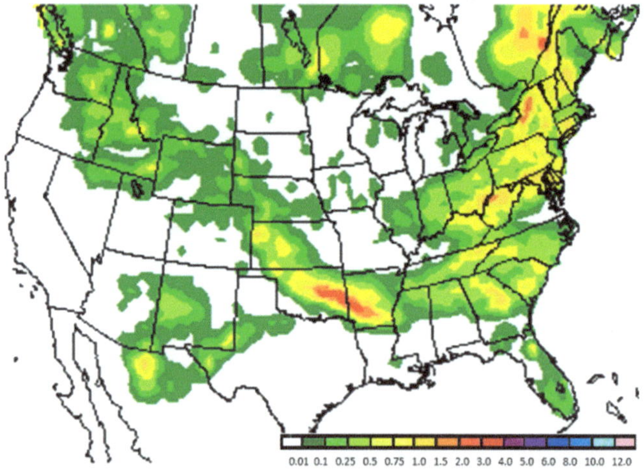

MAXIMUM TEMPERATURE / MINIMUM TEMPERATURE

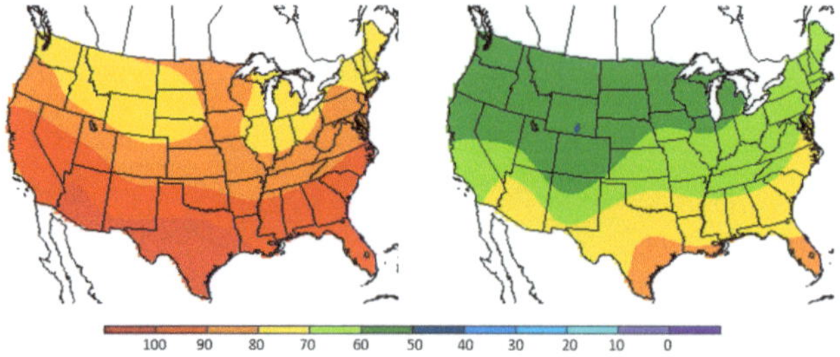

07/17/2025

Wet fronts over south of country, with rain through the Deep South to the Coastal Southeast.

Showers into Florida.

Heavy rain Inland Midwest to the Northeast.

Showers Great Lakes.

Continued rain from the Pacific Northwest and Great Basin to the Rocky Mountains and Great Plains.

Showers in the Southwest.

Mostly dry elsewhere.

Very hot in South Dakota.

Unusually warm in Alaska.

07/18/2025

Wet front over south of country with precipitation from the Great Plains to the Deep South.

Showers in Florida and around the Great Lakes.

Heavy rain in the Northeast and Georgia.

Some showers Rocky Mountains

Rain outbreaks in the Southwest.

Mostly dry elsewhere.

Very hot in South Dakota, Minnesota and Wisconsin and Arizona.

Heavy rain in the Northeast region over the next few days.

JULY 18 FRIDAY

ISOBARIC

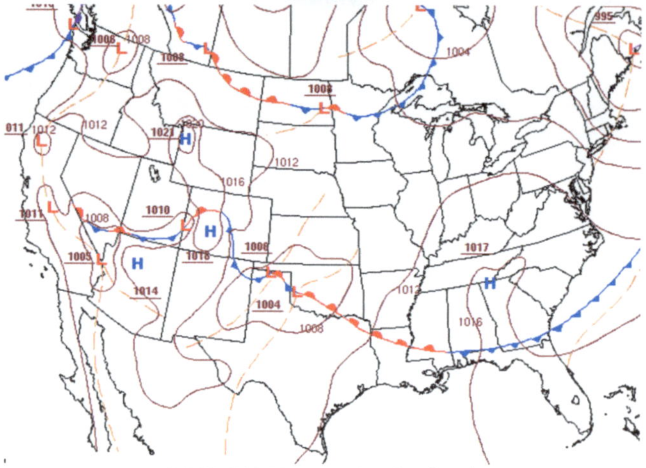

RAIN DISTRIBUTION (Inches)

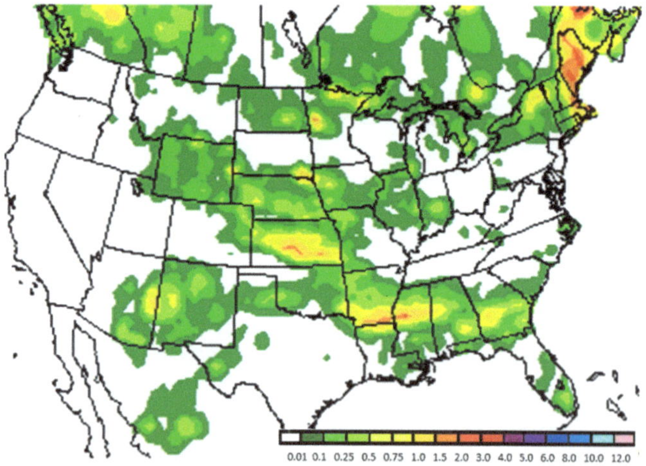

MAXIMUM TEMPERATURE / MINIMUM TEMPERATURE

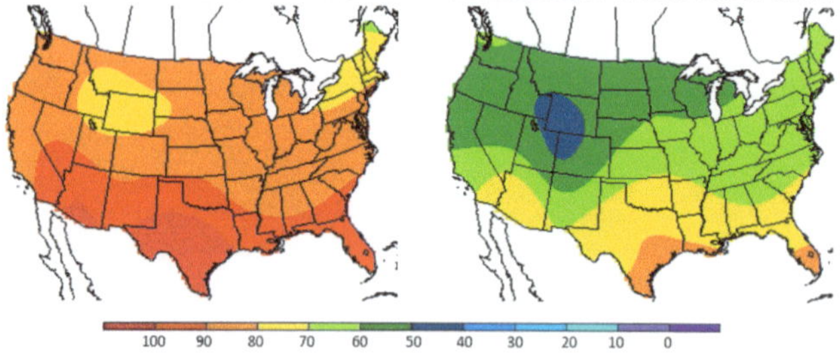

JULY 19 SATURDAY

ISOBARIC

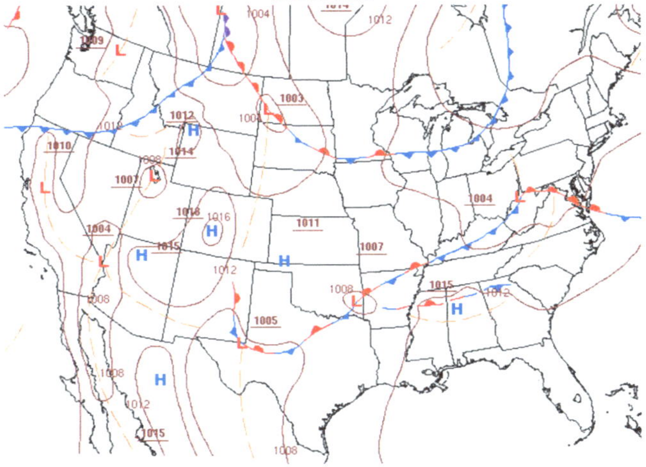

RAIN DISTRIBUTION (Inches)

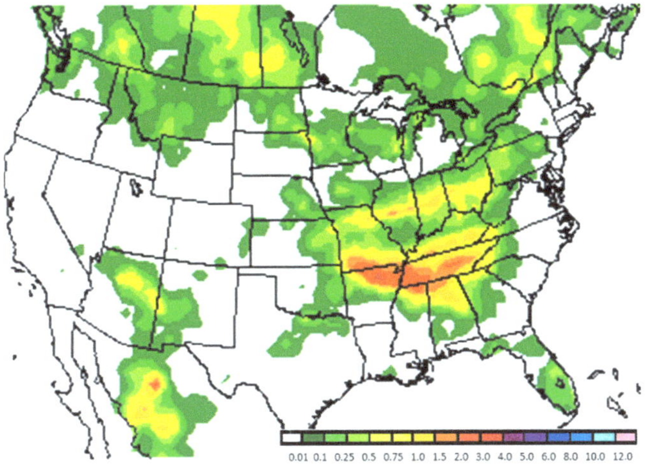

MAXIMUM TEMPERATURE   MINIMUM TEMPERATURE

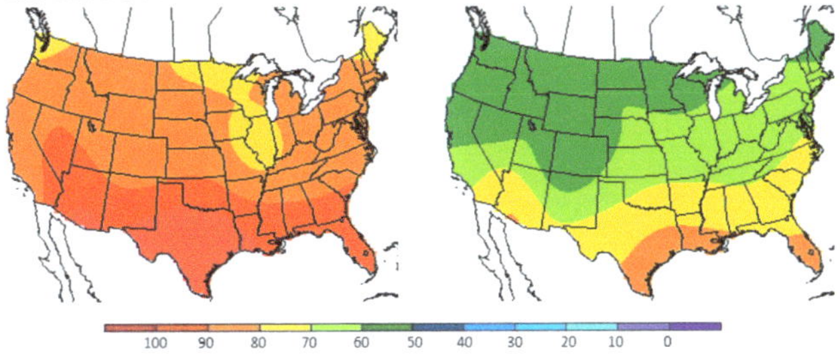

07/19/2025

High pressure builds Great Basin to the Great Plains.

Precipitation Pacific and Inland Northwest.

Heavy rain Inland Midwest to the Deep South.

Showers Great Lakes to the Northeast.

Bands of heavy showers likely to cross the Northeast.

Dry along the Atlantic coasts except some showers in Florida.

Spells of heavy rain in the Southwest.

Mostly dry elsewhere.

Chance of heavy rain in Arkansas over the next couple of days.

JULY 20 SUNDAY

ISOBARIC

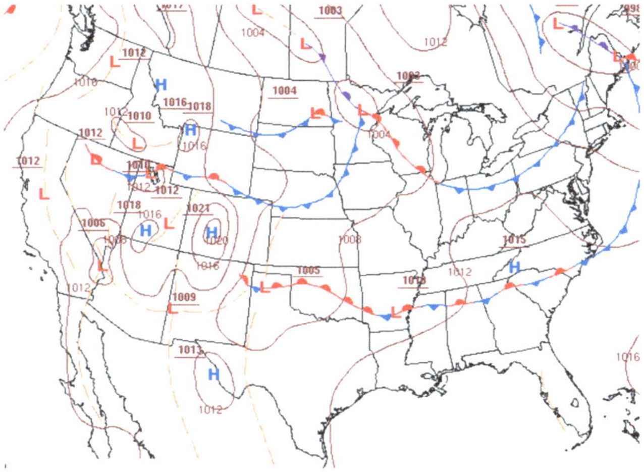

RAIN DISTRIBUTION (Inches)

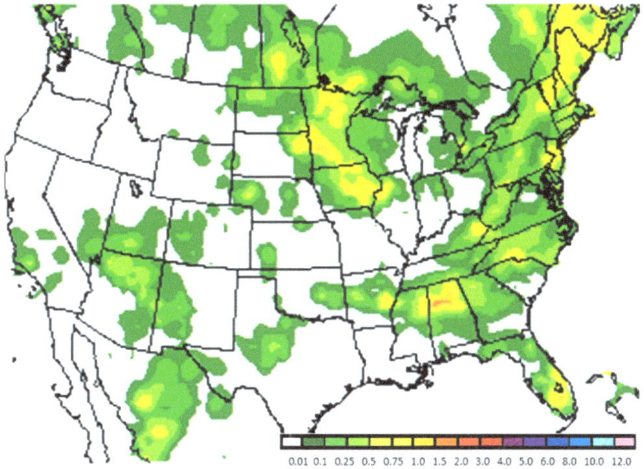

MAXIMUM TEMPERATURE   MINIMUM TEMPERATURE

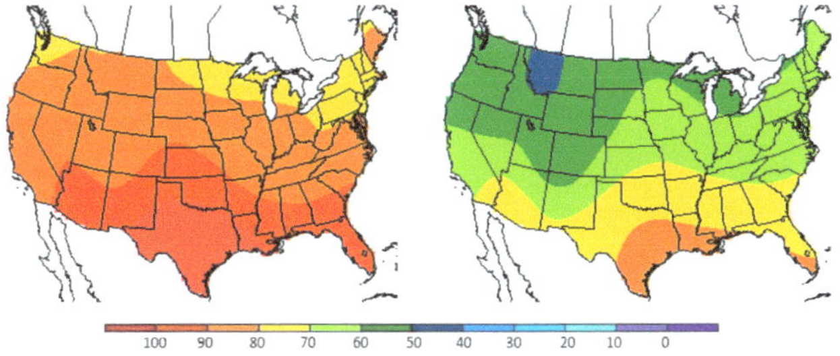

07/20/2025

A complex of fronts cross the country.

Heavy rain from the Great Lakes to the Northeast and down through the Atlantic States from the Megalopolis to Florida.

Chance of heavy rain in Mississippi.

Rain into the Deep South with showers into Texas.

Lingering showers in the Rocky Mountains, and showers from the Southwest to the Great Basin.

Mostly dry elsewhere.

Very hot once more in the north and west of the country, including in Washington and Oregon.

Gusty winds in North Dakota.

JULY 21 MONDAY

ISOBARIC

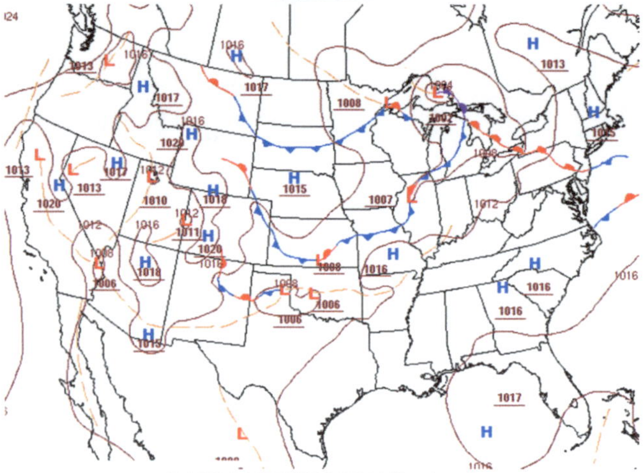

RAIN DISTRIBUTION (Inches)

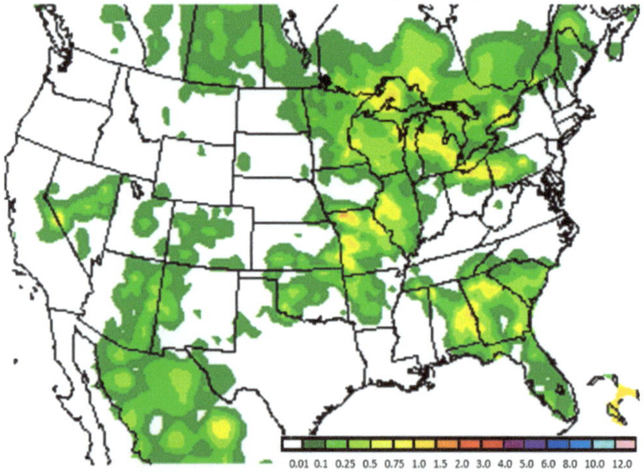

MAXIMUM TEMPERATURE

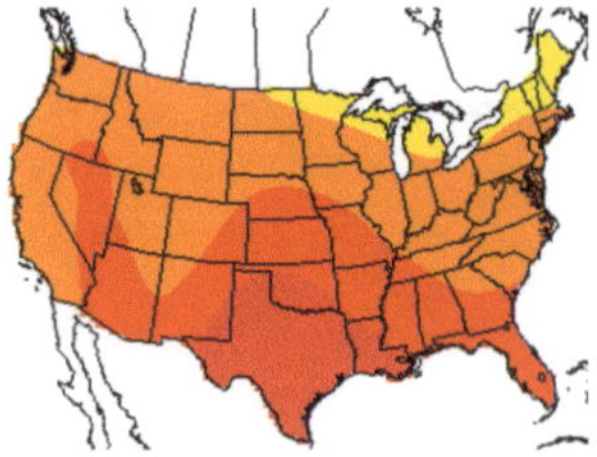

MINIMUM TEMPERATURE

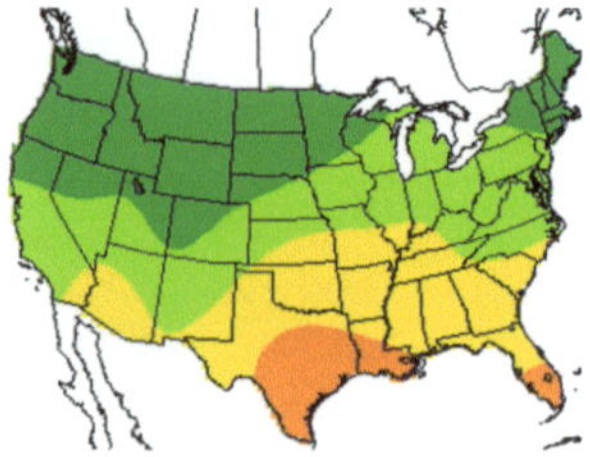

07/21/2025

High pressure prevails.

Widespread precipitation Great Lakes to the Inland Midwest.

Heavy rain Deep South to Florida.

Showers in the Great Plains.

Outbreaks of rain Great Basin to the Southwest.

Mostly dry elsewhere.

Heavy rain possible in Vermont as well as in California over the next few days.

07/22/2025

Wet fronts cross the country.

Precipitation in the Great Plains and Inland Midwest.

Heavy rain from the Northeast, through the Appalachians to the Deep South.

Mostly dry Coastal Southeast.

Showers in Florida and the Great Basin to the Southwest and southern Rocky Mountains.

Mostly dry elsewhere.

Very high maximum temperatures likely in Montana over the next few days.

JULY 22 TUESDAY

ISOBARIC

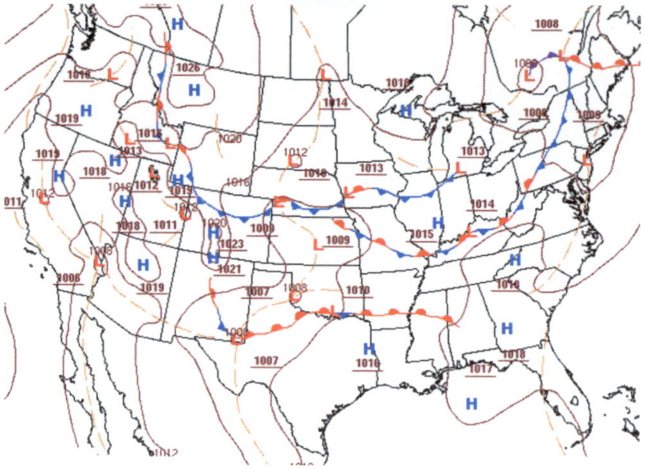

RAIN DISTRIBUTION (Inches)

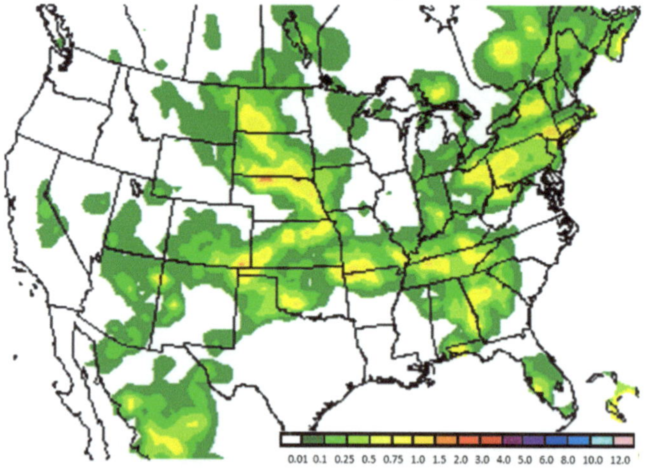

MAXIMUM TEMPERATURE

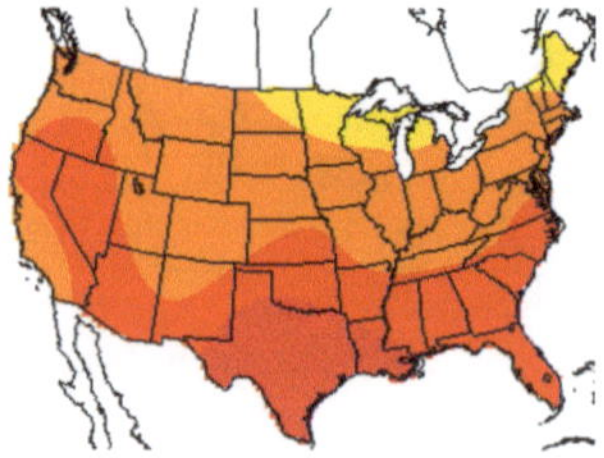

MINIMUM TEMPERATURE

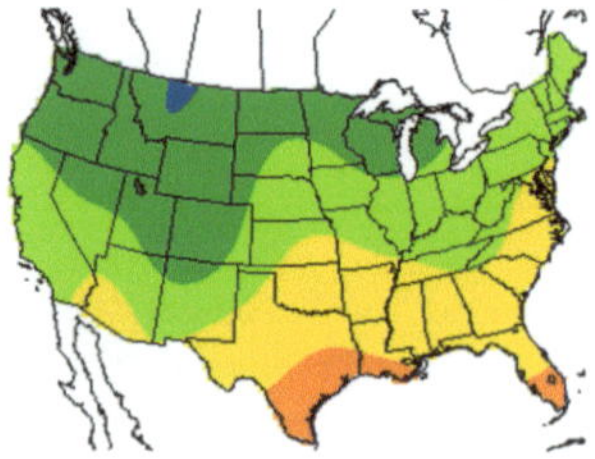

JULY 23 WEDNESDAY

07/23/2025

Wet fronts through the midsection and southern states.

Precipitation through the Great Plains to the Inland Midwest, heavy at times.

Heavy rain in the Ozarks.

Rain in the Northeast.

Showers Deep South to Florida, and the Pacific Southwest to the Southwest states.

Mostly dry elsewhere.

Unusually cool in Ohio and Iowa over the next couple of days.

ISOBARIC

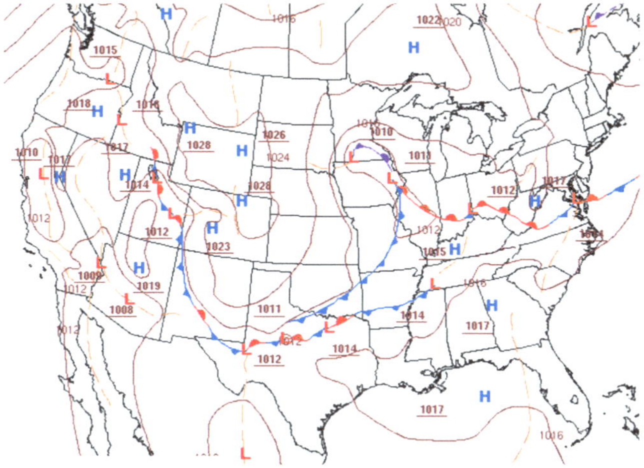

RAIN DISTRIBUTION (Inches)

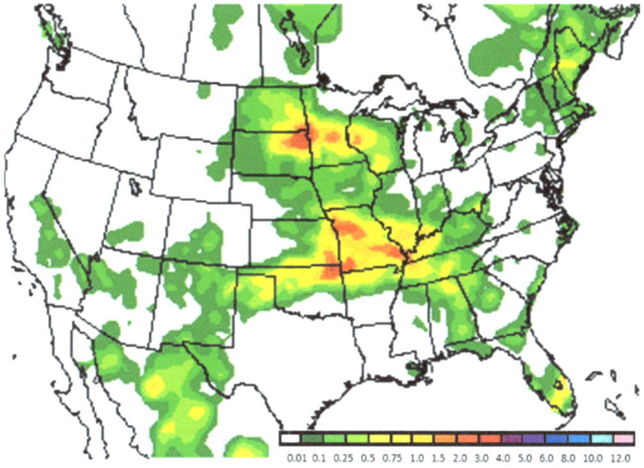

MAXIMUM TEMPERATURE

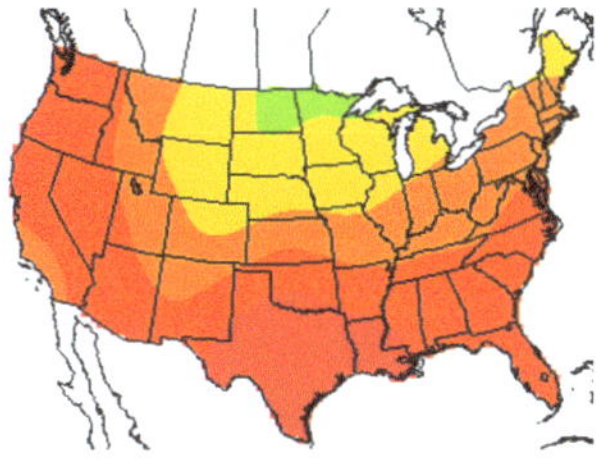

MINIMUM TEMPERATURE

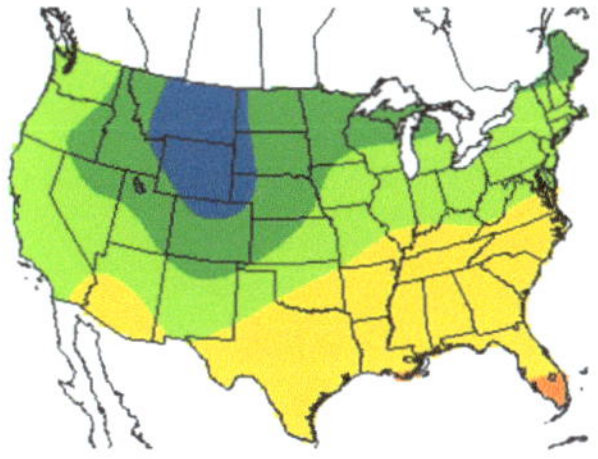

JULY 24 THURSDAY

07/24/2025

Widespread precipitation east of the Great Plains.

Showers in the Northeast to Florida.

Heavy rain south of Great Lakes.

Showers throughout the Pacific Southwest, with rain into the Southwest.

Mostly dry elsewhere.

Heavy rain in southwestern parts of Texas may see some pockets of localized flooding over the next couple of days.

ISOBARIC

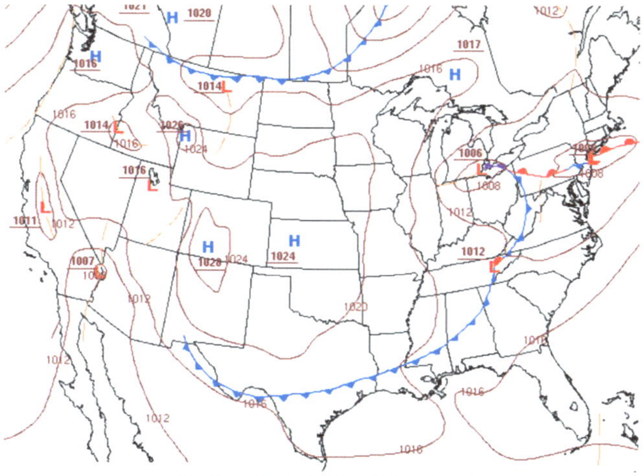

RAIN DISTRIBUTION (Inches)

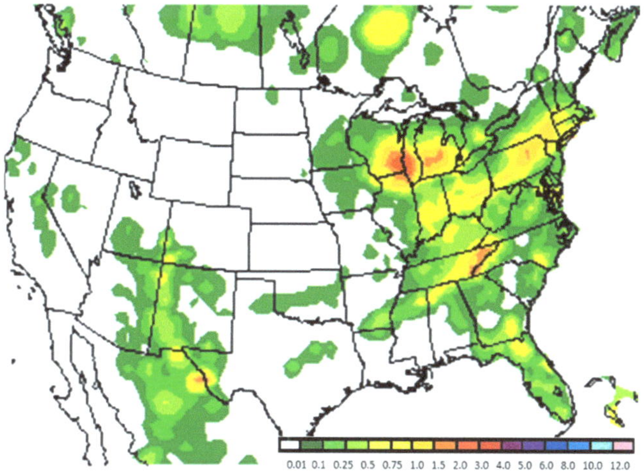

MAXIMUM TEMPERATURE

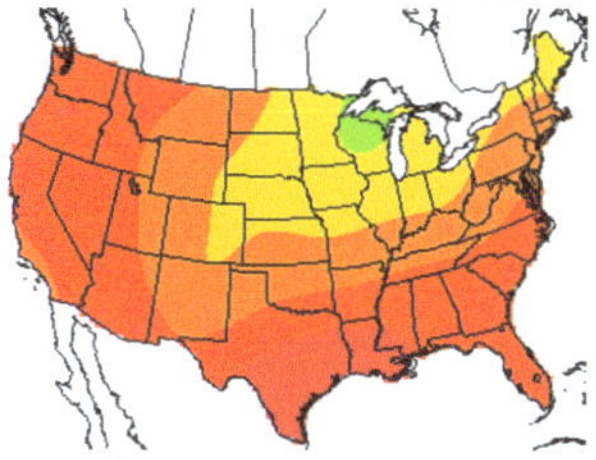

MINIMUM TEMPERATURE

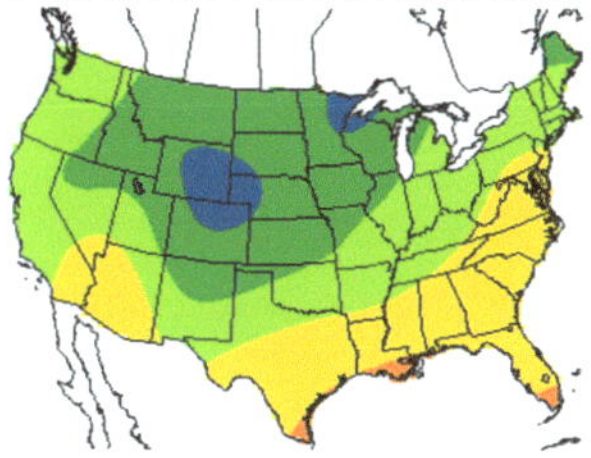

JULY **25** FRIDAY

ISOBARIC

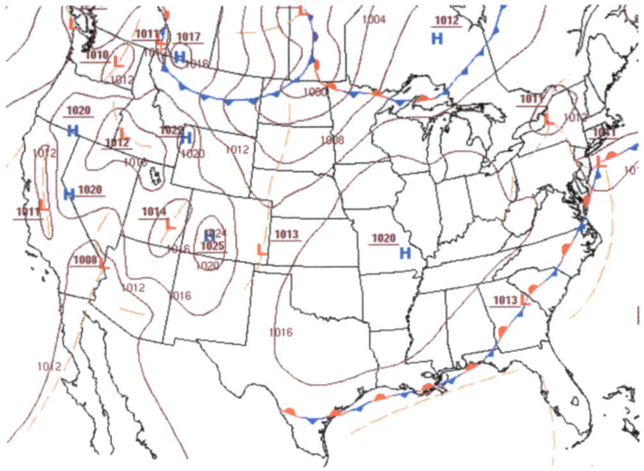

RAIN DISTRIBUTION (Inches)

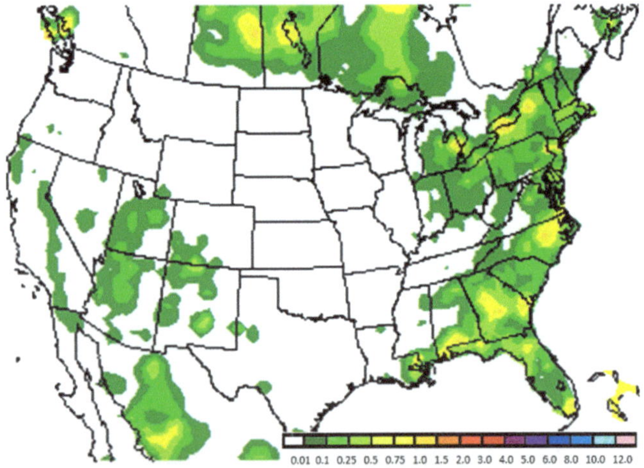

MAXIMUM TEMPERATURE

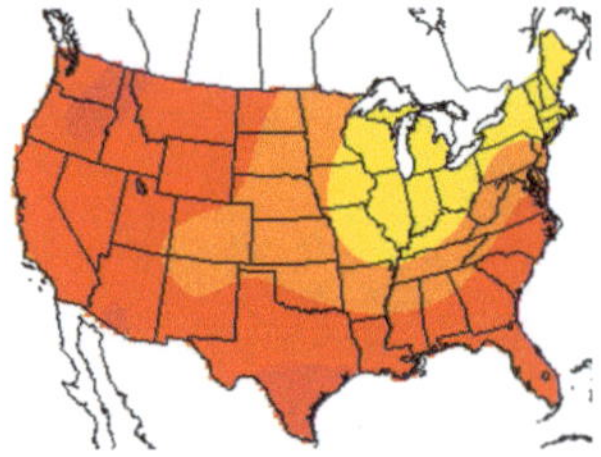

MINIMUM TEMPERATURE

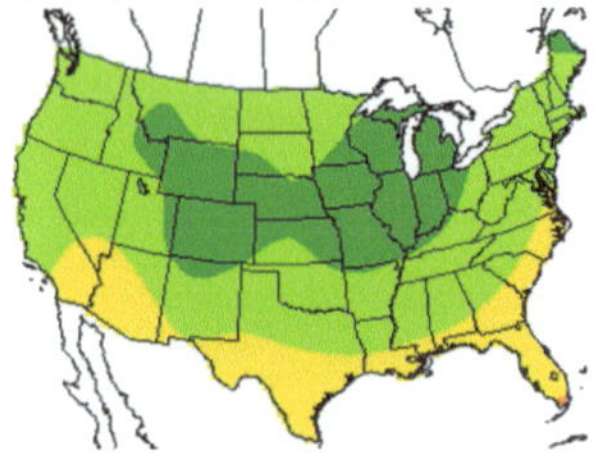

07/25/2025

High pressure over the eastern half of the country.

Wet front along the eastern seaboard brings precipitation from the Northeast Megalopolis to Florida and into the east of the Inland Midwest.

Some showers in the southwest of the country from the Pacific South West to the Southwest.

Mostly dry elsewhere.

Chance of wildfires in Oregon, Utah and Washington states.

Very high temperatures in many regions in the west and south of the country including in Utah, Montana, and South Dakota.

Chance of heavy showers in Alaska.

07/26/2025

Large depression center north of Great Lakes.

Widespread rain Great Lakes to southern half of the Northeast.

Outbreaks of heavy rain coastal fringes of the Coastal Southeast to Florida.

Some showers in the southeast of country from the Pacific coast into the Great Basin and the Southwest.

Mostly dry elsewhere.

JULY **26** SATURDAY

ISOBARIC

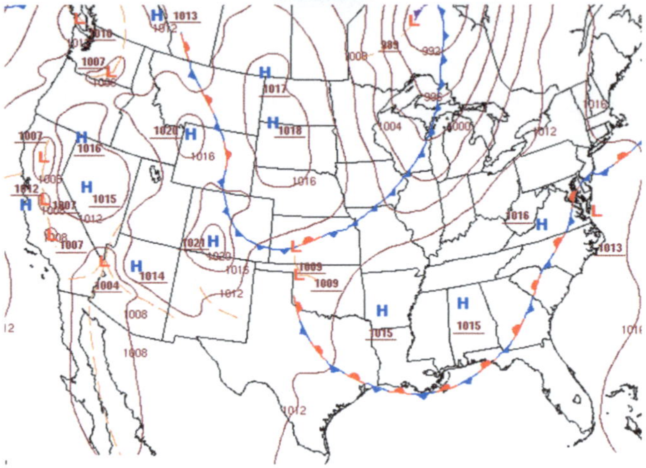

RAIN DISTRIBUTION (Inches)

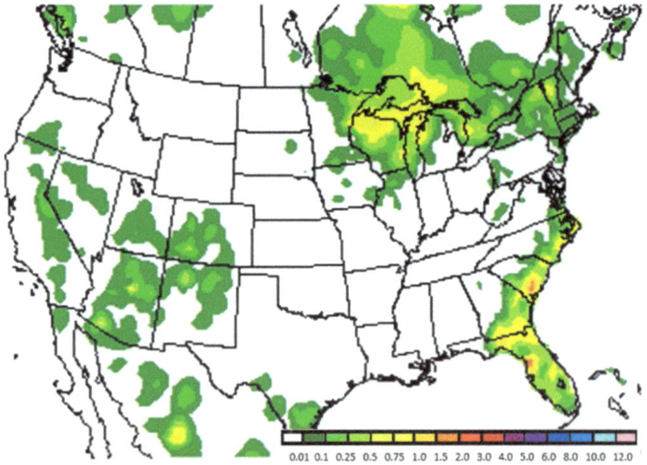

MAXIMUM TEMPERATURE

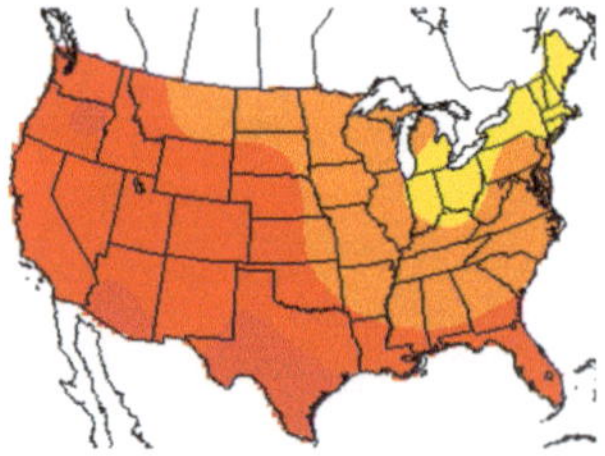

MINIMUM TEMPERATURE

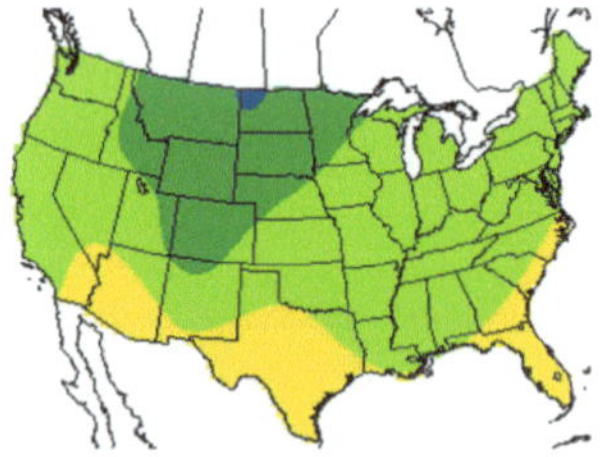

ISOBARIC

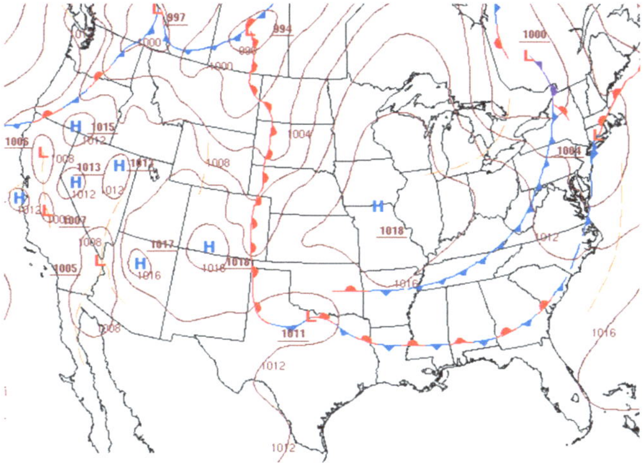

RAIN DISTRIBUTION (Inches)

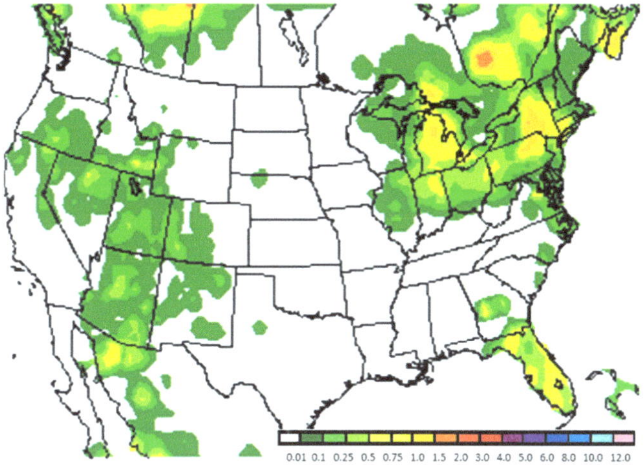

MAXIMUM TEMPERATURE MINIMUM TEMPERATURE

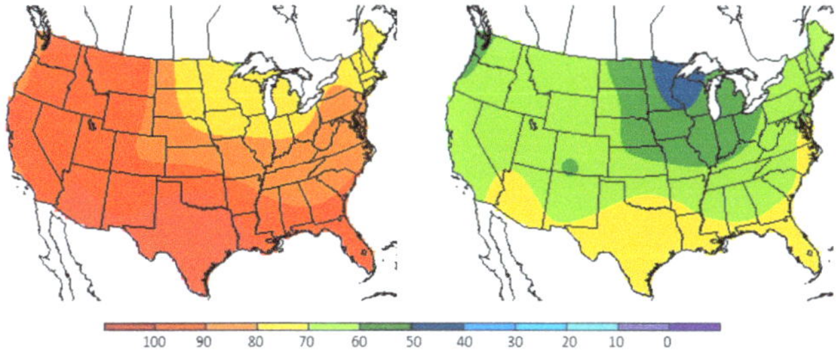

07/27/2025

High pressure over the midsection, east of the Rocky Mountains.

Widespread precipitation builds in west of country from the Pacific Northwest, through the Great Basin and into the Southwest.

Precipitation from the Inland Midwest and Great Lakes to the Northeast and northern Appalachians.

Heavy showers in Florida and chance of heavy rain in Missouri.

Mostly dry elsewhere.

Very heavy rainfall may occur in a few Midwestern pockets over the next couple of days, including in Iowa and Wisconsin.

07/28/2025

High pressure over the east of the country.

Rain from the Inland Northwest to the Southwest.

Heavy rain in the Northeast to the Megalopolis.

Some showers in Florida.

Heavy rain likely in North Dakota, New York, Iowa.

Mostly dry elsewhere.

Chance of heavy rain developing in northern California to the Pacific Northwest regions over the next few days.

Warmer than normal day in Alaska.

ISOBARIC

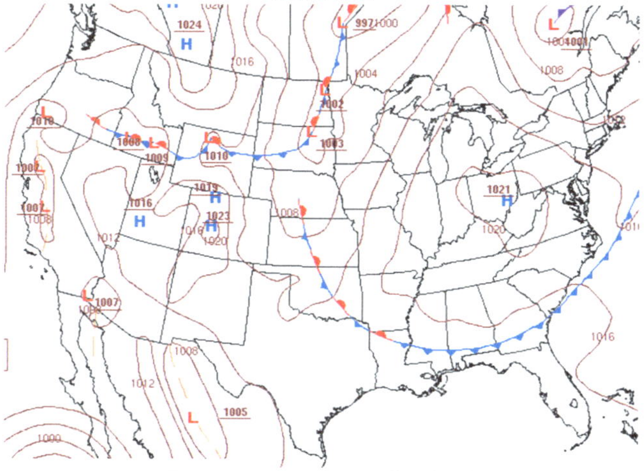

RAIN DISTRIBUTION (Inches)

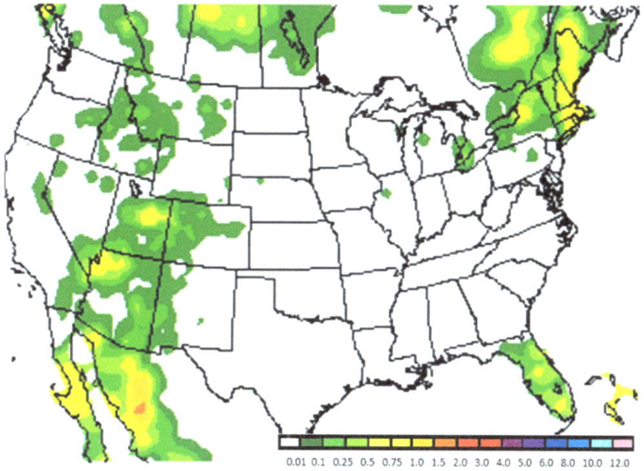

MAXIMUM TEMPERATURE MINIMUM TEMPERATURE

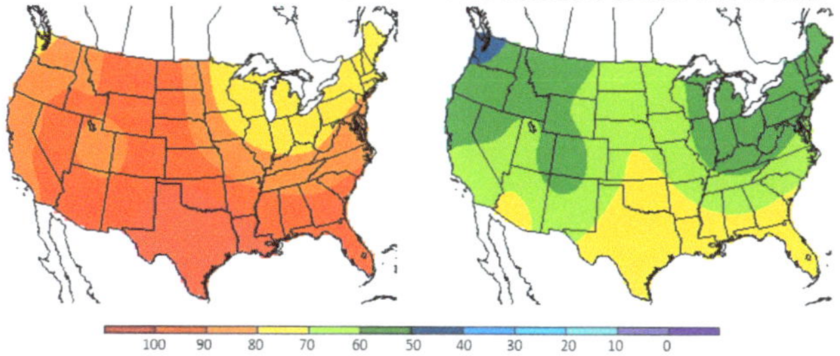

JULY 29 TUESDAY

ISOBARIC

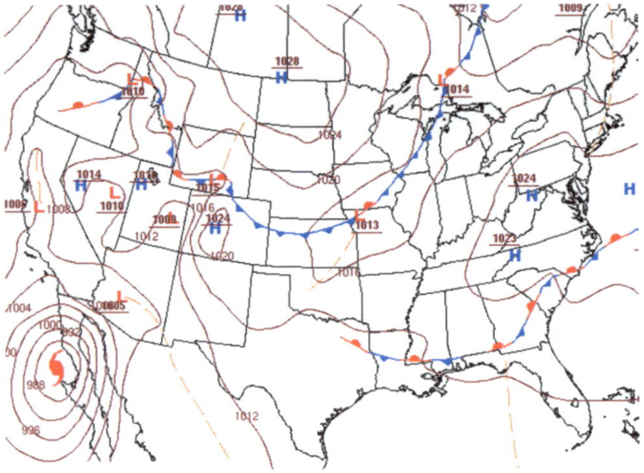

RAIN DISTRIBUTION (Inches)

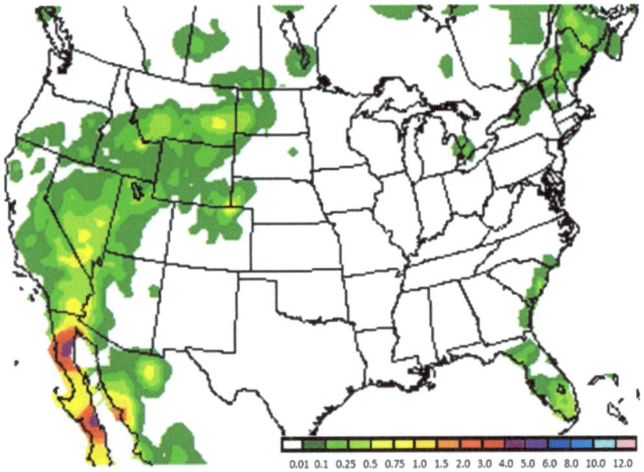

MAXIMUM TEMPERATURE / MINIMUM TEMPERATURE

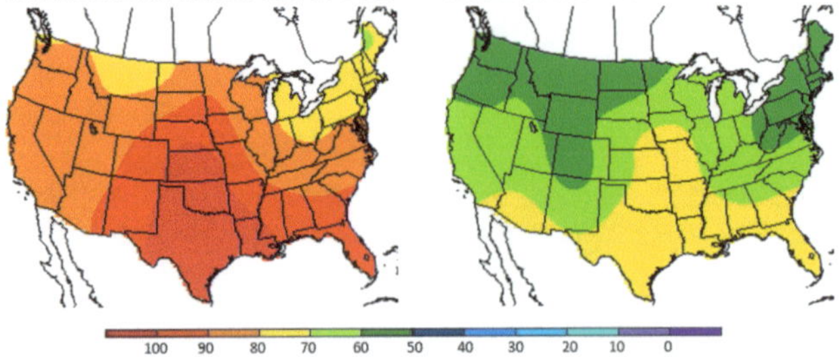

07/29/2025

High pressure and mostly dry over the eastern half of the country except for some showers in the Northeast and Florida to the coastal Carolinas.

A Pacific tropical storm near Mexico spreads rain up through the west, from the Pacific Southwest to the Great Basin and into the northern Rocky Mountains over the next few days.

Heavy rain clearing in New York.

Mostly dry elsewhere.

Thunderstorms with heavy rain likely in the Four Corner States, including in Colorado and into Arizona in a couple of days.

07/30/2025

Decaying tropical depression near Mexico increases low pressure over the Pacific coastal regions.

Widespread heavy precipitation in the Pacific Southwest to the Great Basin and into the north of the Rocky Mountains to the North Great Plains.

Rain into the western half of the Southwest.

Some showers in the Northeast Highlands.

Rain in Florida to coastal fringes of the Southeast.

Chance of heavy rain in Mississippi.

Mostly dry elsewhere.

A tropical depression to the south of Hawaii may bring outbursts of heavy rain and strong winds over the next couple of days, especially on the Big Island.

JULY 30 WEDNESDAY

ISOBARIC

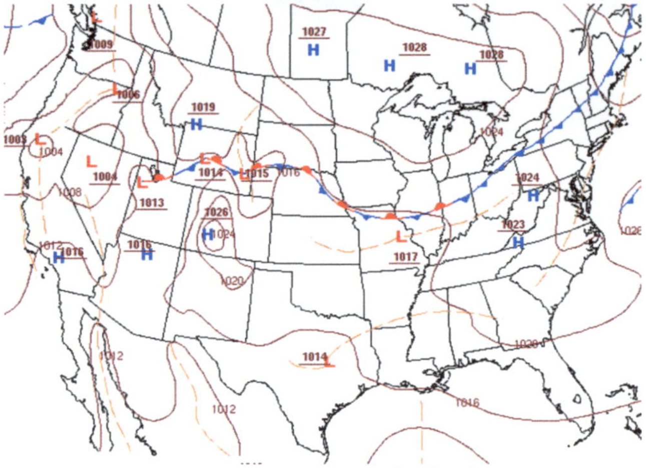

RAIN DISTRIBUTION (Inches)

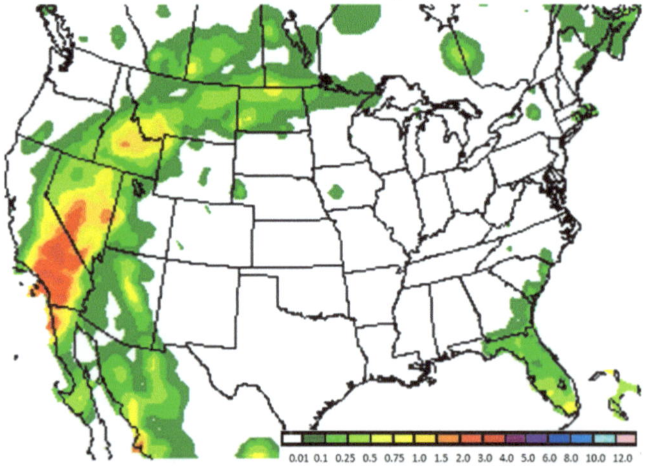

MAXIMUM TEMPERATURE / MINIMUM TEMPERATURE

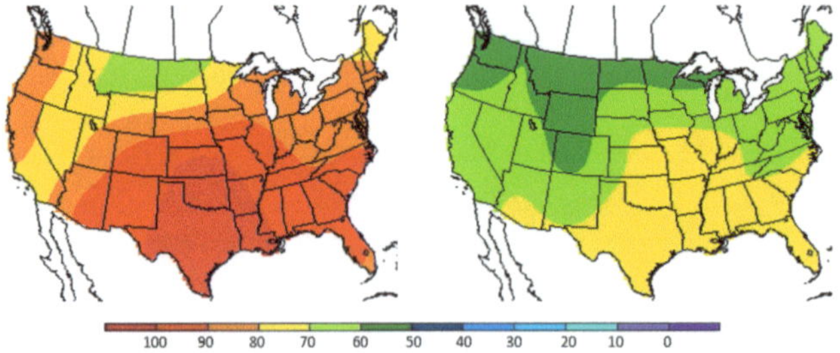

ISOBARIC

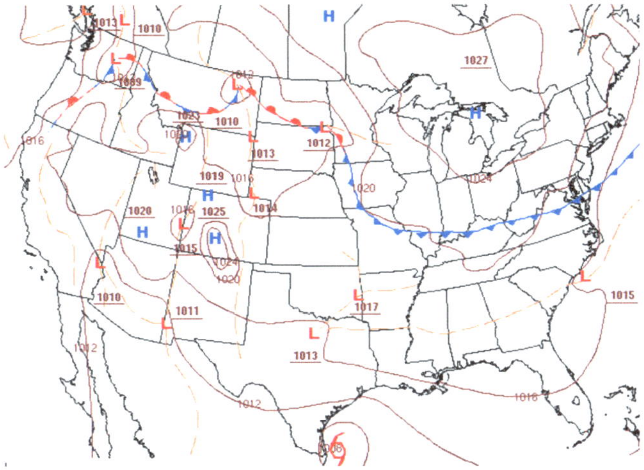

07/31/2025

Wet troughs west of the Rocky Mountains

Widespread precipitation from the north Rocky Mountains through the Great Plains and into the Southwest.

Showers in the Pacific Southwest.

Showers around the Gulf of Mexico coastal fringes from Florida to Texas.

Some showers around the Great Lakes.

Chance of heavy rain around this time in Florida and Arizona.

Mostly dry elsewhere.

RAIN DISTRIBUTION (Inches)

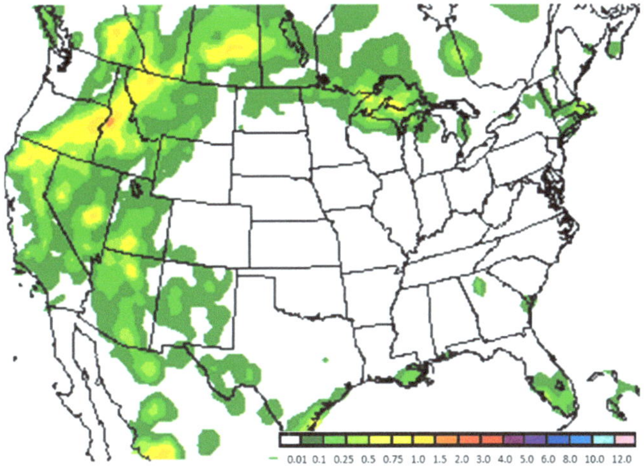

MAXIMUM TEMPERATURE

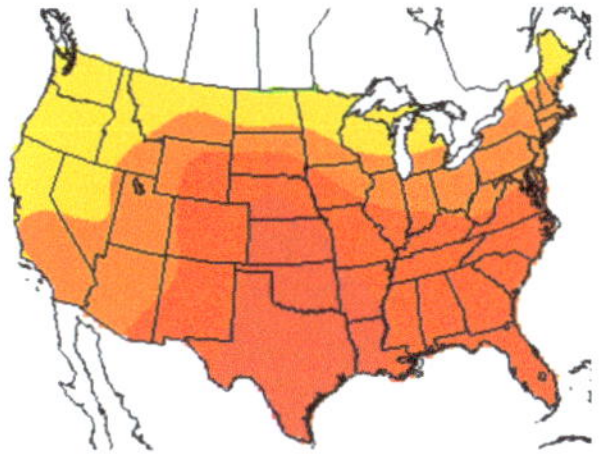

MINIMUM TEMPERATURE

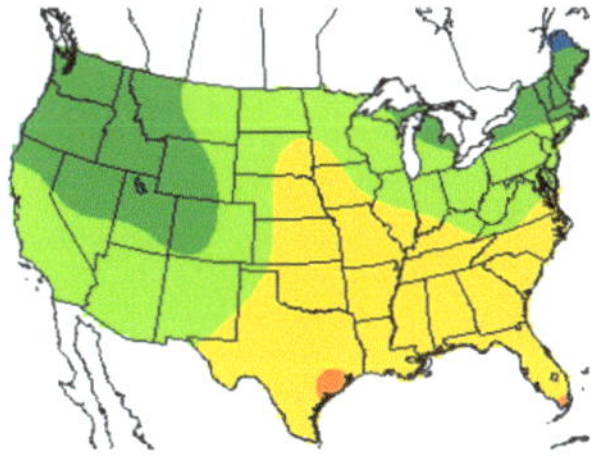

AUGUST

PHASES OF THE MOON

1st	First Quarter
1st	Apogee
5th	Southern Declination
9th	Full Moon
13th	Crossing Equator
14th	Perigee #12
16th	Third Quarter
18th	Northern Declination
23rd	New Moon
26th	Crossing Equator
29th	Apogee
31st	First Quarter

1 Fri	2 Sat	3 Sun	4 Mon	5 Tue	6 Wed	7 Thur	8 Fri	9 Sat	10 Sun	11 Mon	12 Tue	13 Wed	14 Thur	15 Fri	16 Sat	17 Sun	18 Mon	19 Tue	20 Wed	21 Thur	22 Fri	23 Sat	24 Sun	25 Mon	26 Tue	27 Wed	28 Thur	29 Fri	30 Sat	31 Sun
1Q A				V				F				XhN	P12		3Q		^					N			XhS			A		1Q

AUGUST 1 FRIDAY

ISOBARIC

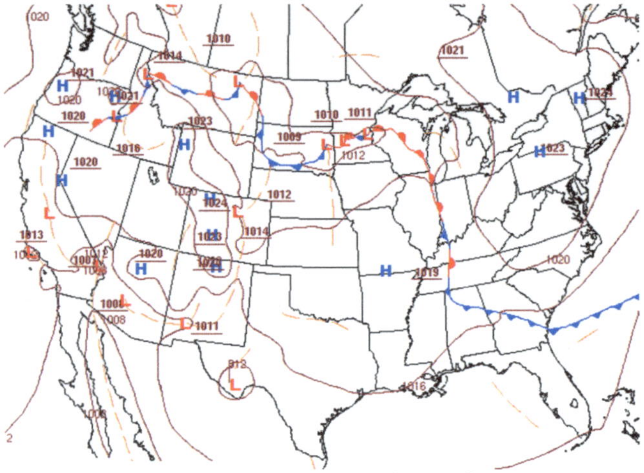

RAIN DISTRIBUTION (Inches)

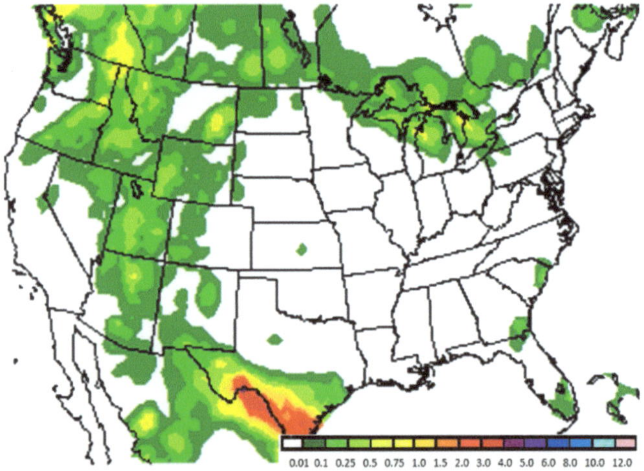

MAXIMUM TEMPERATURE

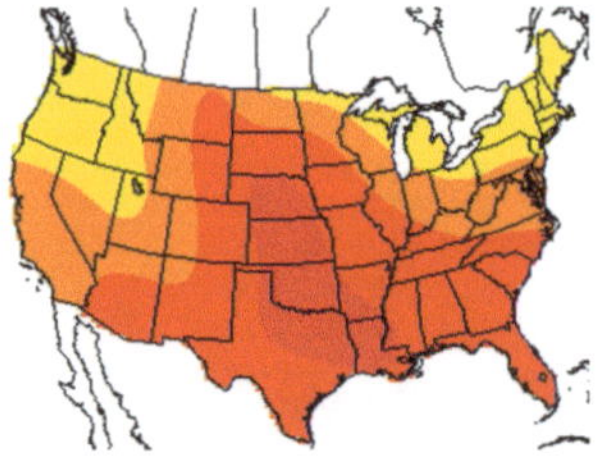

MINIMUM TEMPERATURE

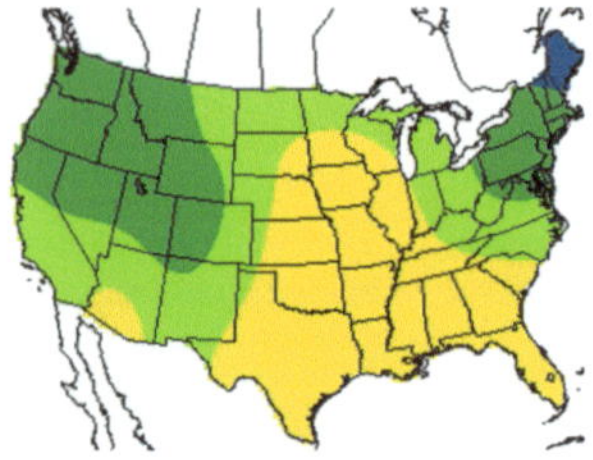

08/01/2025

Widespread rain Inland Northwest and adjacent Rocky Mountains, southward through the Great Basin and into the Southwest.

Heavy rain in southern half of Texas.

Showers around the Great Lakes.

Mostly dry elsewhere.

Chance of wildfires in Idaho and other parts of the northern Rocky Mountains over the next few days.

Chance of cooler temperatures across the south and east of the country.

Heavy showers likely to develop in the Midwest to the Southeast over the next couple of days.

08/02/2025

Rain embedded troughs spread widespread showers from the south of the Rocky Mountains, through the Southwest and into the western half of Texas, with heavy rain north of the Rio Grande Valley.

Isolated showers in Florida and the coastal fringes of the Deep South.

Mostly dry elsewhere.

Very hot in the Great North Plains including Montana over the next couple of days with high temperatures expanding into the Dakotas, Wyoming, and Nebraska.

Cooler spell in West Virginia to South Carolina.

AUGUST 2 SATURDAY

ISOBARIC

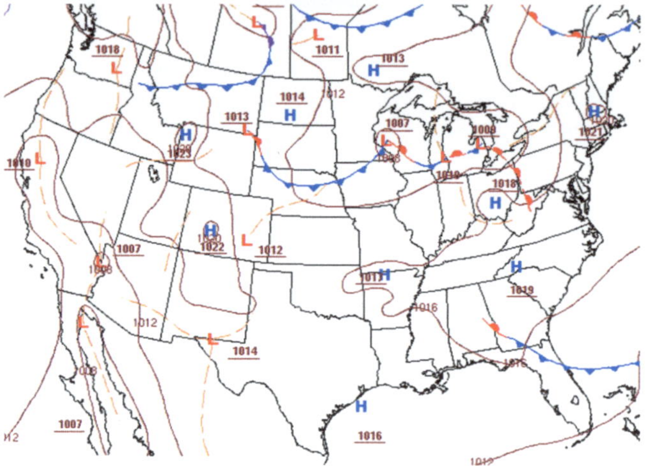

RAIN DISTRIBUTION (Inches)

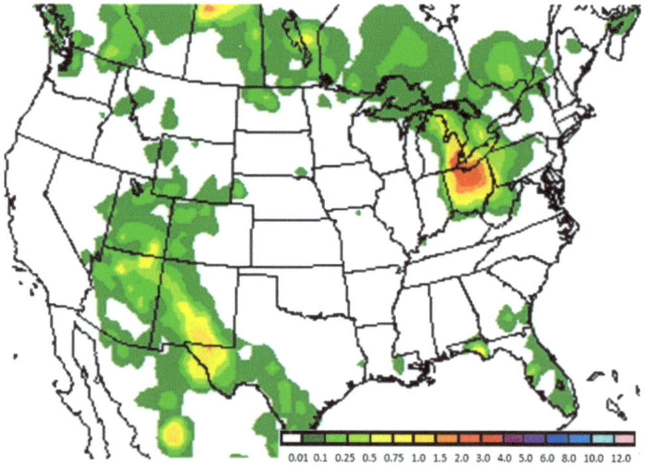

MAXIMUM TEMPERATURE

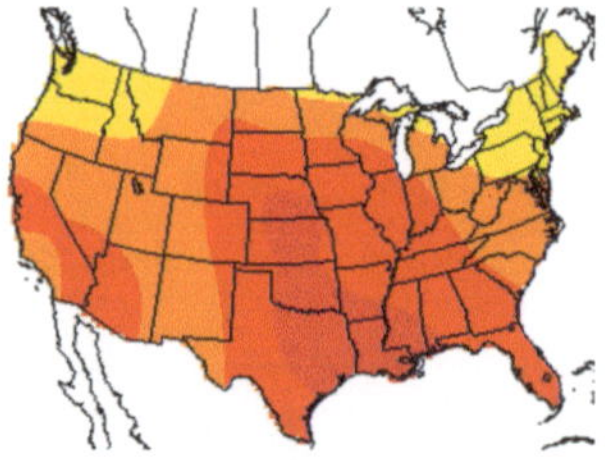

MINIMUM TEMPERATURE

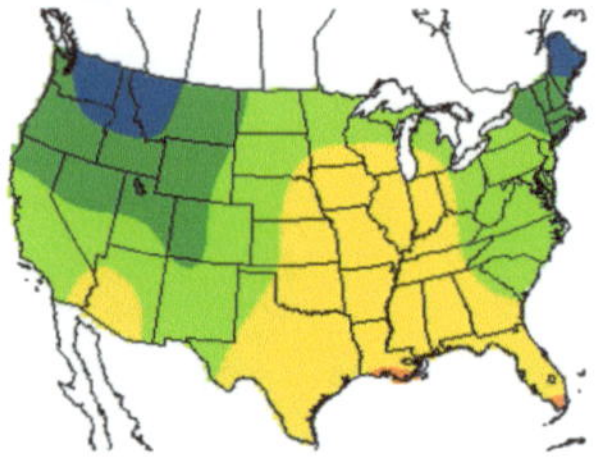

AUGUST **3** **SUNDAY**

AUGUST **4** **MONDAY**

ISOBARIC

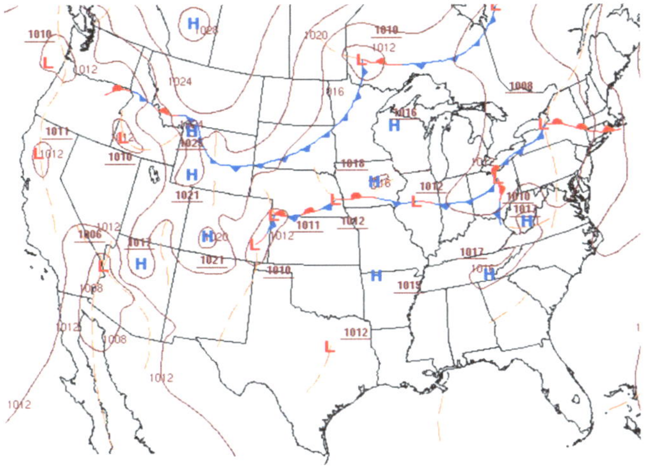

RAIN DISTRIBUTION (Inches)

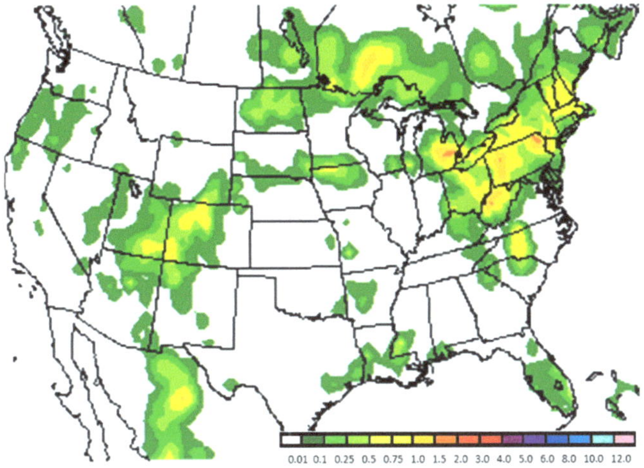

MAXIMUM TEMPERATURE / MINIMUM TEMPERATURE

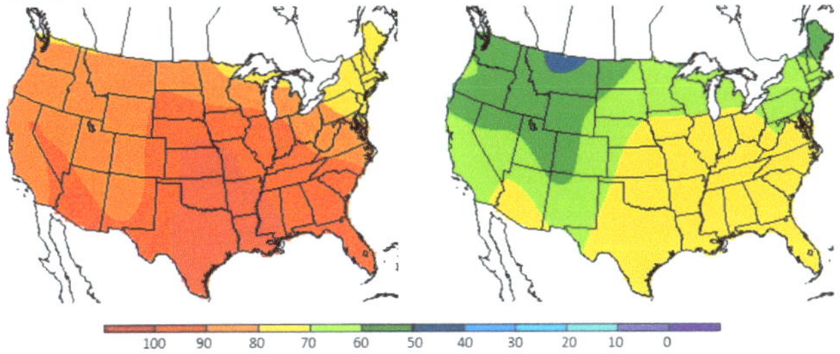

08/03/2025

Widespread outbreaks of rain North Dakota, the Great Lakes, eastern Inland Midwest and Northeast regions, and through the Appalachians with isolated pockets in the Coastal Southwest.

Showers linger in the southern Rocky Mountains to the Great Basin and Southwest, while clearing in Texas.

Some showers spread across the central Great Plains.

Rain in southern Florida and showers around the coastal fringes of the Deep South.

Mostly dry elsewhere.

Outbreaks of heavy showers in Oregon in the next couple of days.

Heavy rain in the south-central parts of Alaska over the next couple of days.

08/04/2025

A large wet front brings rain from the Rocky Mountains into the Southwest, across the southern and central Great Plains, and over the Inland Midwest to the Northeast.

Heavy falls likely central Great Plains to the Ozarks and central Appalachians.

Rain in the Great Lakes, heavy falls in Wisconsin.

Heavy rain in the east of the Northeast from the Megalopolis to New England.

Some showers in the Deep South.

Outbreaks of rain in the Pacific to Inland Northwest regions.

Mostly dry elsewhere.

Unusually hot in the Dakotas.

Heavy rain likely in Wyoming over the next few days.

ISOBARIC

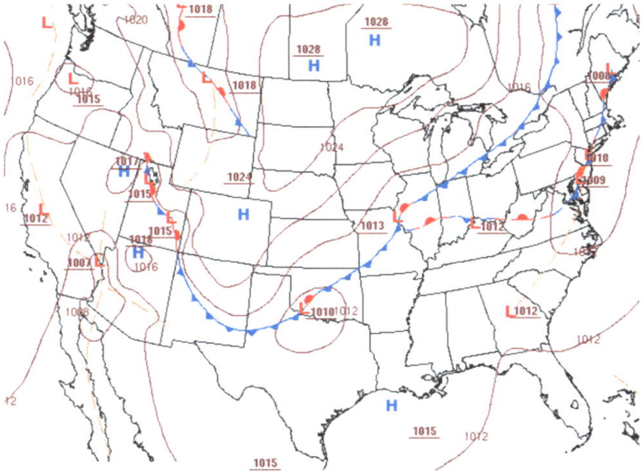

RAIN DISTRIBUTION (Inches)

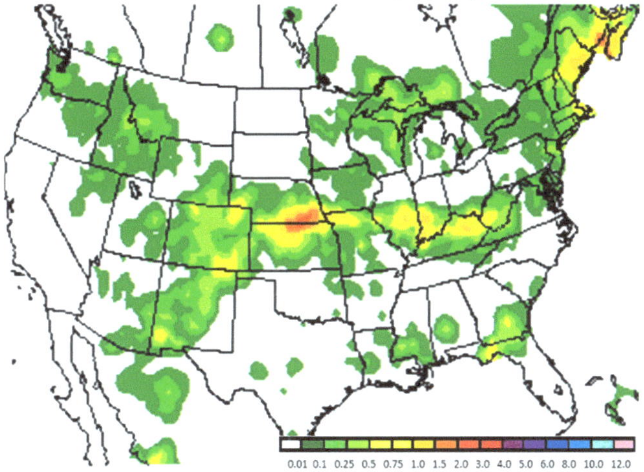

MAXIMUM TEMPERATURE / MINIMUM TEMPERATURE

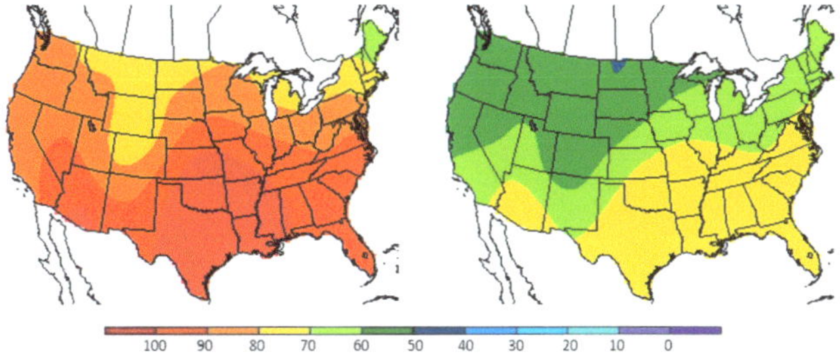

AUGUST 5 TUESDAY

ISOBARIC

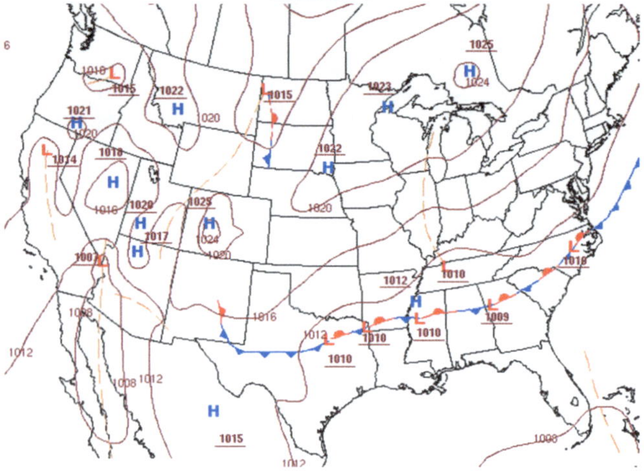

RAIN DISTRIBUTION (Inches)

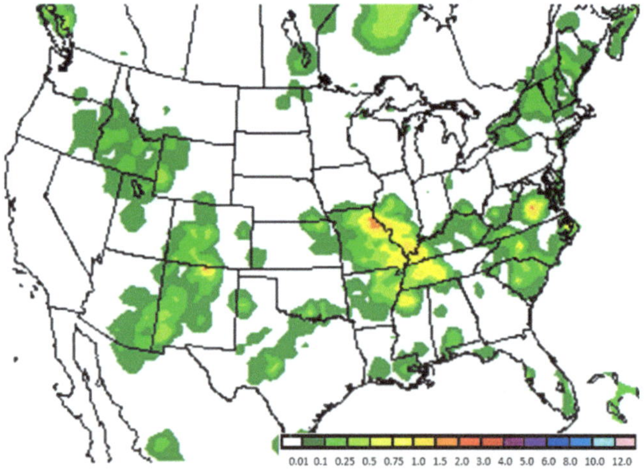

MAXIMUM TEMPERATURE **MINIMUM TEMPERATURE**

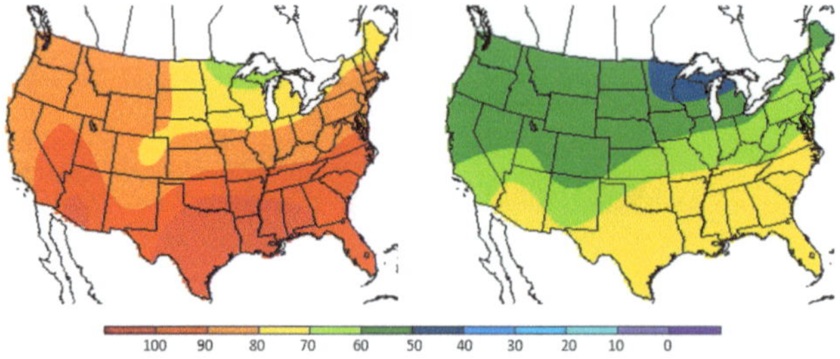

08/05/2025

High pressure dominates most of the country.

Rain clears to showers in the Northeast.

A wet front maintains showers in the Southwest, the north of Texas and the Deep South to the Coastal Southeast.

Heavy rain in the Ozarks and up into the southern half of the Inland Midwest, including possible heavy showers in Indiana and Iowa.

Outbreaks of heavy rain likely in Idaho to Wisconsin.

Mostly dry elsewhere.

08/06/2025

High pressure prevails over the northeast of the country with only a few isolated showers mostly in the Northeast region.

A wet front in the south from the Southwest to the Coastal Southeast brings further outbreaks of rain, with heavy falls in the Deep South including in Kentucky, and over the Appalachians to the Coastal Southeast.

Some showers prevail over the Rocky Mountains and along the western fringes of the Great Plains.

Chance of heavy rain in Wisconsin.

Mostly dry elsewhere.

AUGUST 6 WEDNESDAY

ISOBARIC

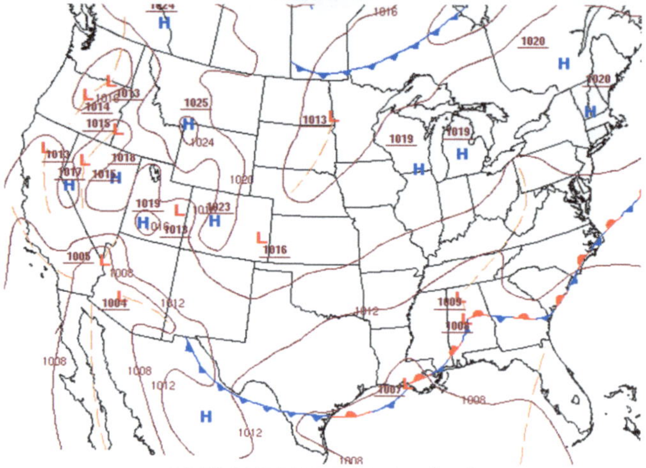

RAIN DISTRIBUTION (Inches)

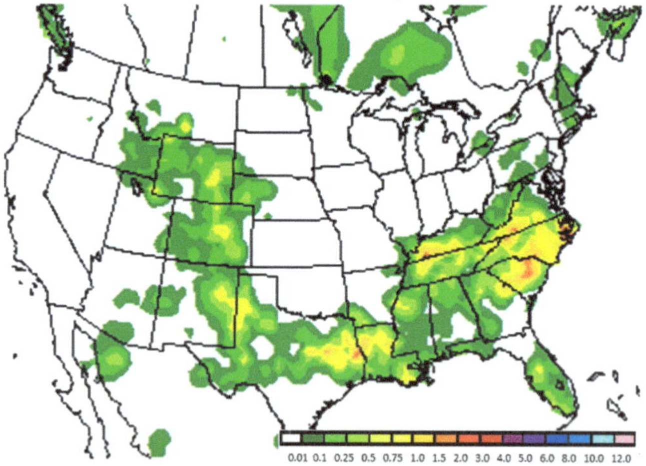

MAXIMUM TEMPERATURE **MINIMUM TEMPERATURE**

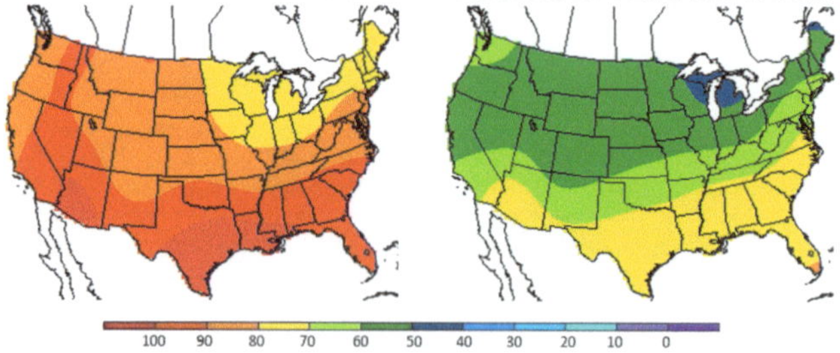

AUGUST **7** THURSDAY

AUGUST **8** FRIDAY

ISOBARIC

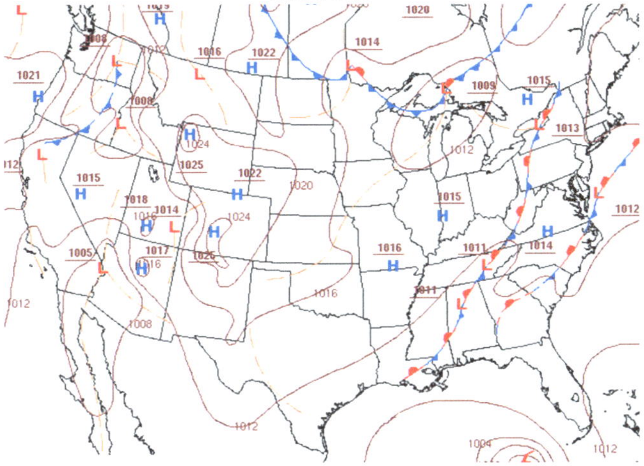

RAIN DISTRIBUTION (Inches)

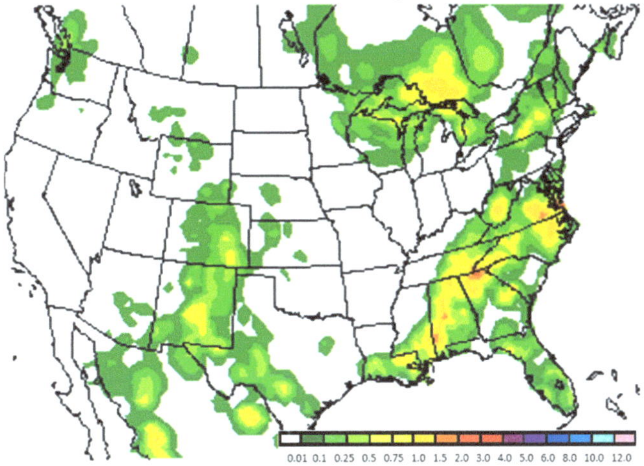

MAXIMUM TEMPERATURE

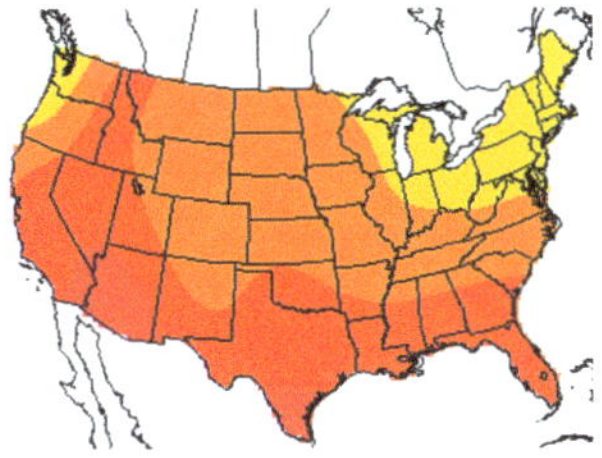

MINIMUM TEMPERATURE

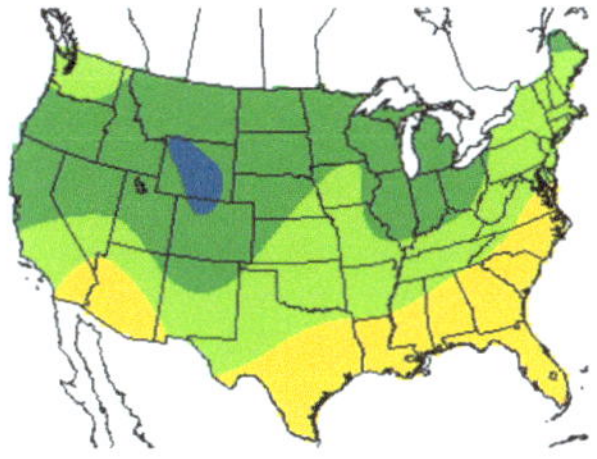

08/07/2025

Wet fronts over the southeast of the country brings widespread rain through the Deep South and Florida, to the Coastal Southeast and along the Appalachians.

Showers may be particularly heavy at times in Tennessee and Florida.

Another front brings the return of rain over the Great Lakes to the Northeast.

Showers prevail down through the Rocky Mountains into the Southwest and the western half of Texas.

Some showers around the Pacific Northwest.

Mostly dry elsewhere.

Chance of a tropical storm in the Gulf of Mexico.

Very warm in Hawaii.

08/08/2025

Tropical storm, possibly to hurricane force, brings widespread heavy rain and storm winds over Florida, and up into the Coastal Southeast regions and the eastern half of the Deep South.

Heavy rain also through all eastern states along the Atlantic coast over the next couple of days, including in Louisiana.

Widespread rain over the Great Lakes, the Northeast and upper Inland Midwest.

Rain in the Pacific Northwest to the Inland Northwest.

Rain clears to isolated showers in the Southwest.

Mostly dry elsewhere.

Thunderstorms and outbreaks of heavy rain likely over the next couple of days in the central and southern Atlantic States and heavy rain developing in Arizona.

Very hot in Montana.

ISOBARIC

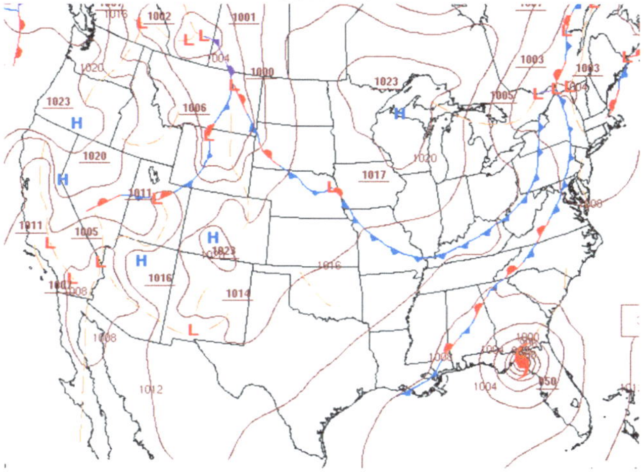

RAIN DISTRIBUTION (Inches)

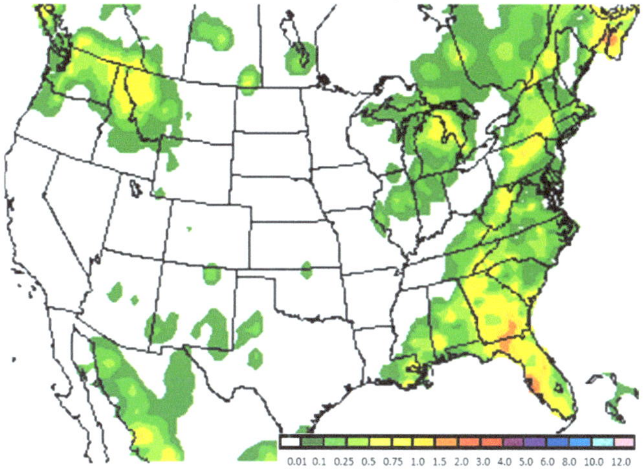

MAXIMUM TEMPERATURE

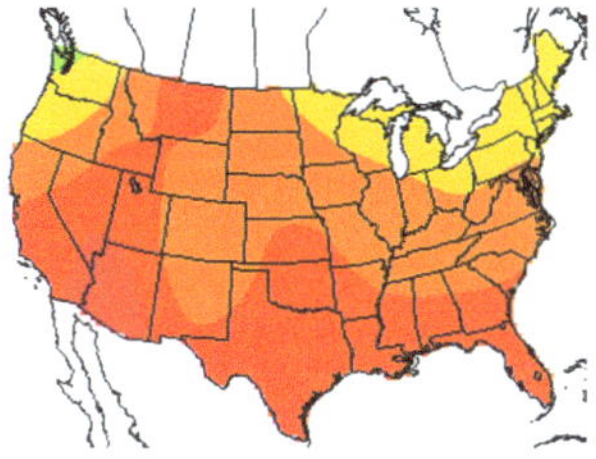

MINIMUM TEMPERATURE

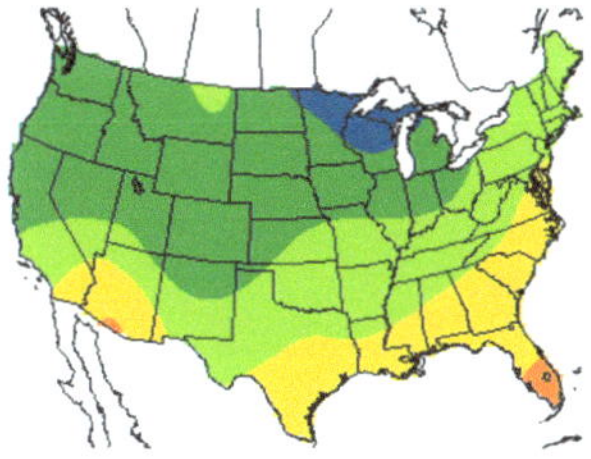

AUGUST 9 SATURDAY

ISOBARIC

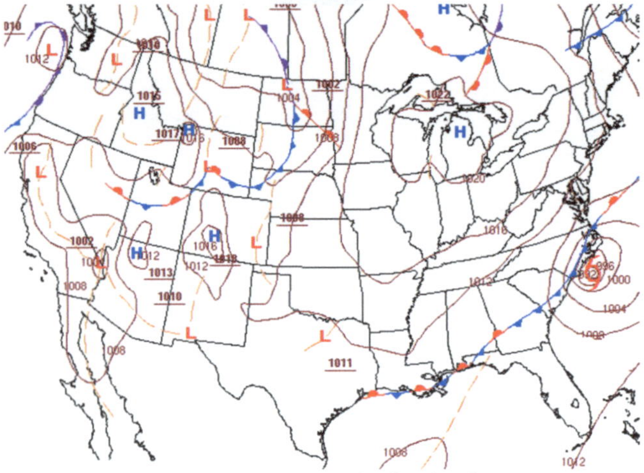

RAIN DISTRIBUTION (Inches)

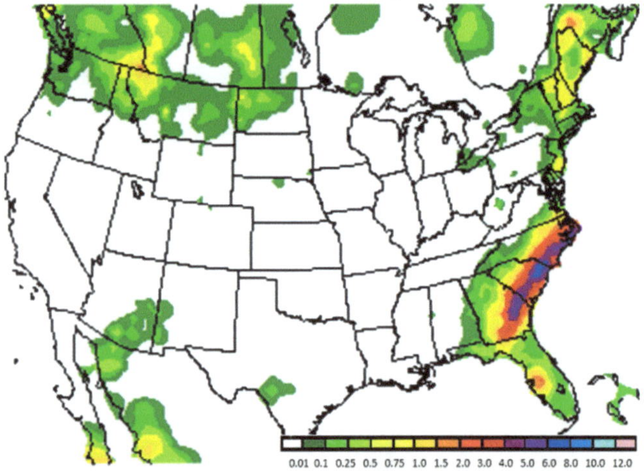

MAXIMUM TEMPERATURE

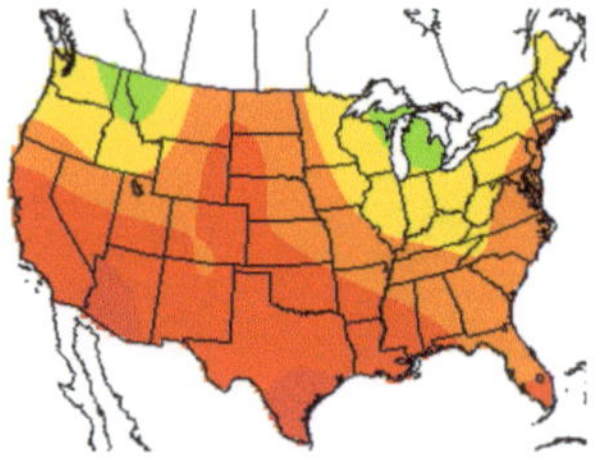

MINIMUM TEMPERATURE

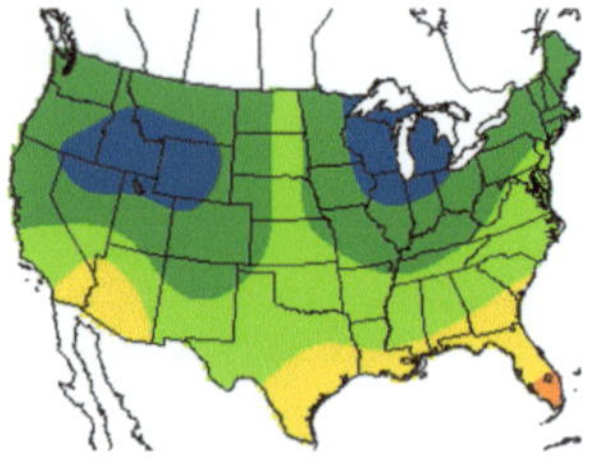

08/09/2025

A possible hurricane continues to bring widespread storm activity through the Coastal Southeast to Florida.

Widespread rain persists in the Northeast.

Heavy rain at times in the Pacific Northwest and along the northern border regions to the far northwest of the Great Plains.

Some showers in Arizona, heavy at times.

Mostly dry elsewhere.

08/10/2025

High pressure rebuilds over the east with rain clearing throughout the Great Lakes and the Northeast.

The Atlantic tropical storm of the last few days may pass near the Northeast coasts near Cape Cod, with very strong winds prevailing for a time.

A stationary front prevails over the south of the country from Texas to Georgia, bringing showers to the coastal fringes of the Deep South and widespread rain over Florida.

Pockets of rain may linger around the Carolinas coasts.

Rain in the west in the Pacific Northwest region as well as in the Great Basin and into Arizona where it may be heavy for a time.

Mostly dry elsewhere.

AUGUST 10 SUNDAY

ISOBARIC

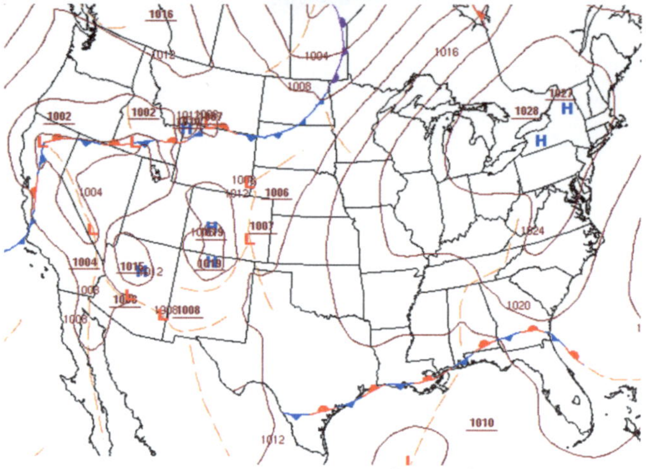

RAIN DISTRIBUTION (Inches)

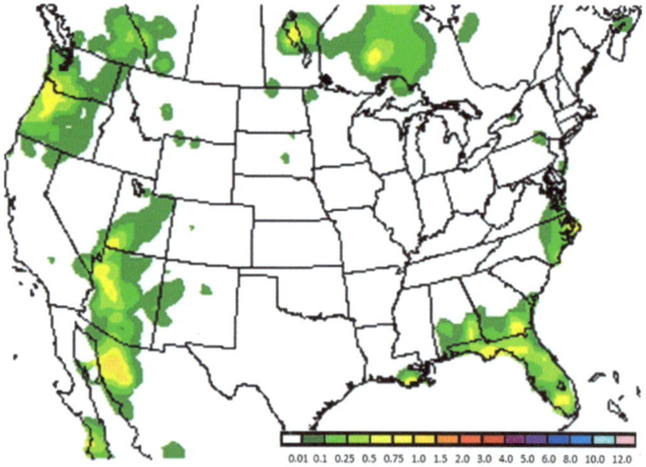

MAXIMUM TEMPERATURE

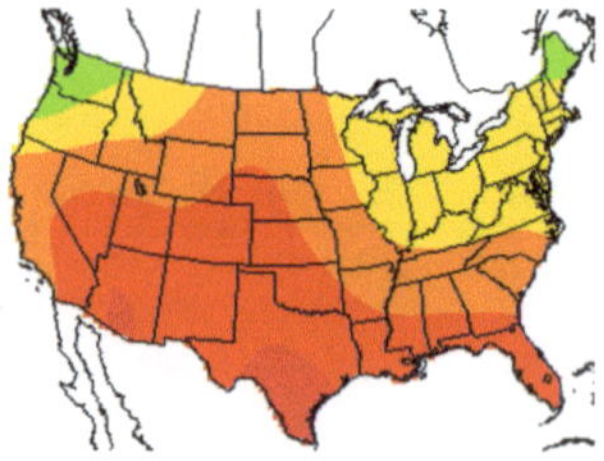

MINIMUM TEMPERATURE

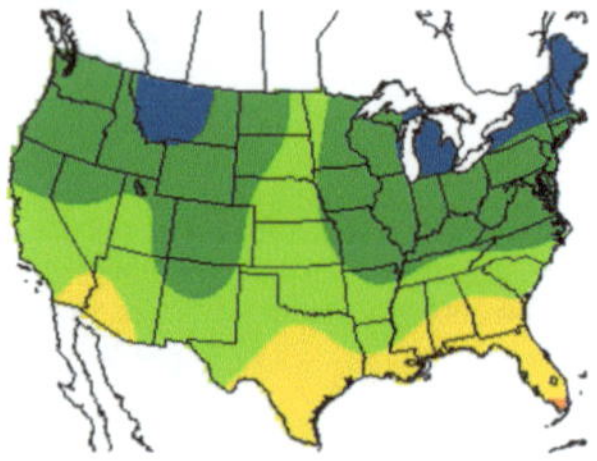

AUGUST **11** MONDAY

AUGUST **12** TUESDAY

ISOBARIC

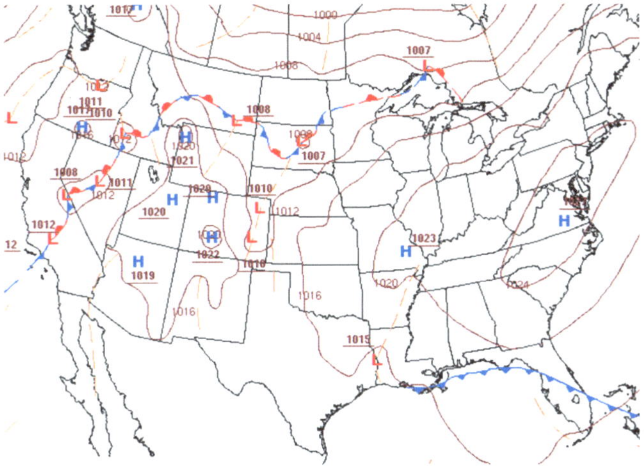

RAIN DISTRIBUTION (Inches)

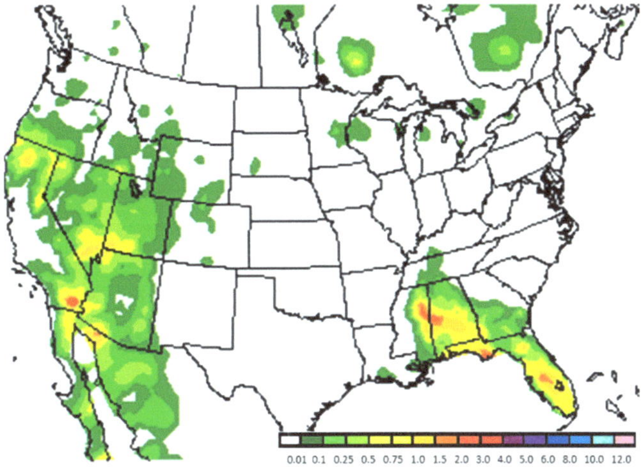

MAXIMUM TEMPERATURE

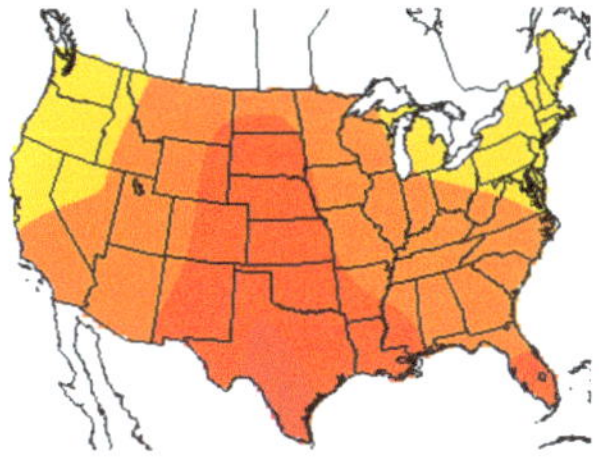

MINIMUM TEMPERATURE

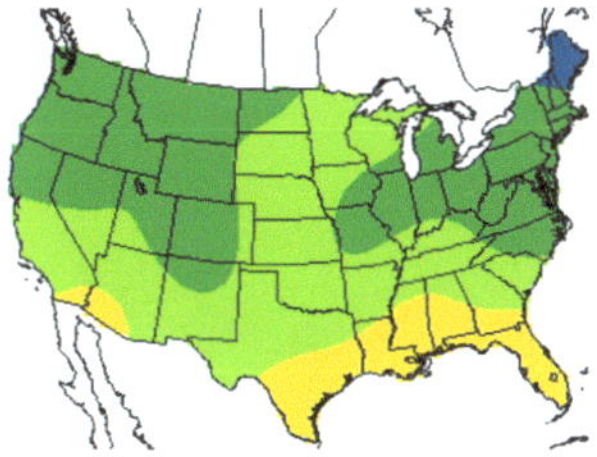

08/11/2025

High pressure over the northeast of the country maintains a mostly dry and pleasant air over the Great Lakes, the Northeast, the Inland Midwest and Coastal Southeast regions, as well as through the mid-section regions from the Great Plains to Texas.

A wet front brings rain, heavy at times, into Florida and the Deep South.

A cluster of troughs and fronts in the west of the country brings widespread rain from the Pacific coastal fringes to the foothills of the Rocky Mountains, including through Arizona and California.

Temperatures may be very high in Montana to the Dakotas over the next few days.

Heavy showers likely in Hawaii.

08/12/2025

Widespread rain prevails in the west from the Pacific Northwest to California and east into the Great Basin and Southwest, including Arizona.

Chance of thunderstorms in the Southwest.

Some showers over the Great Lakes, with isolated showers in the Northeast.

A wet front lingers in the south bringing further rain into the Deep South.

Some showers in the southern tip of Florida.

Mostly dry elsewhere.

Very hot in Pennsylvania and Alabama.

ISOBARIC

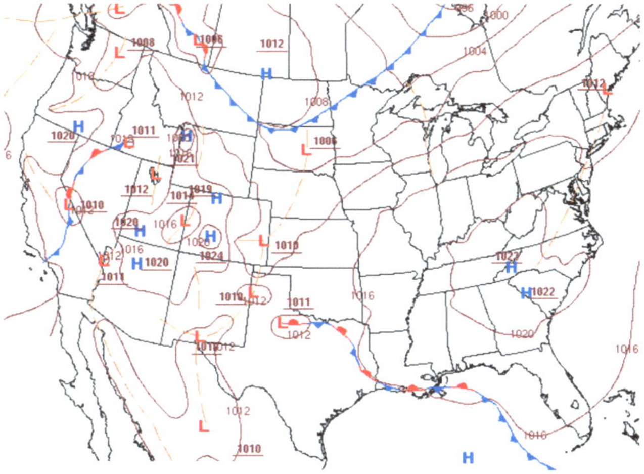

RAIN DISTRIBUTION (Inches)

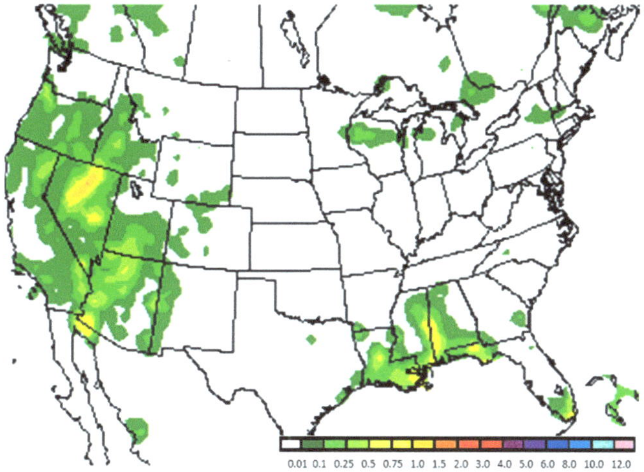

MAXIMUM TEMPERATURE

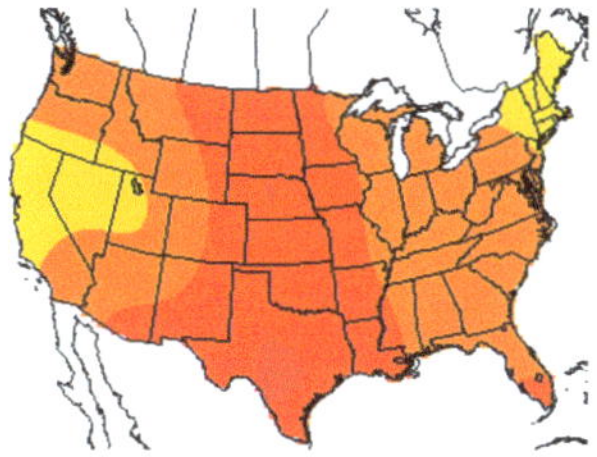

MINIMUM TEMPERATURE

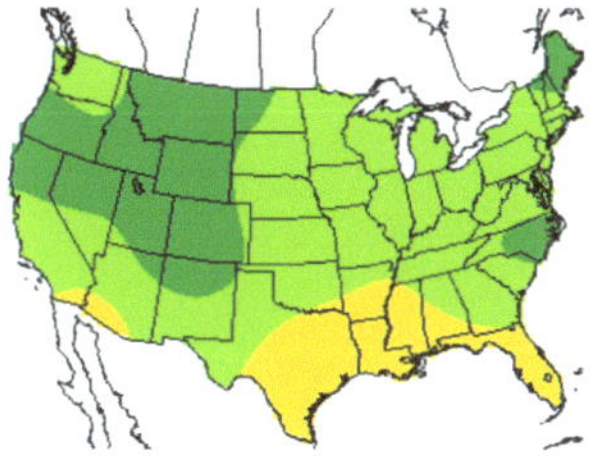

AUGUST 13 WEDNESDAY

ISOBARIC

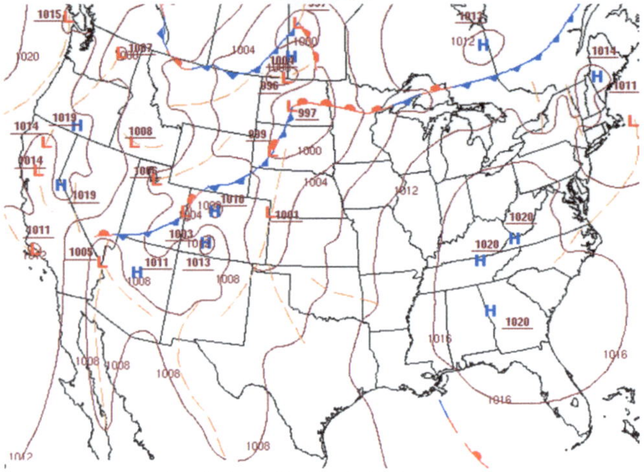

RAIN DISTRIBUTION (Inches)

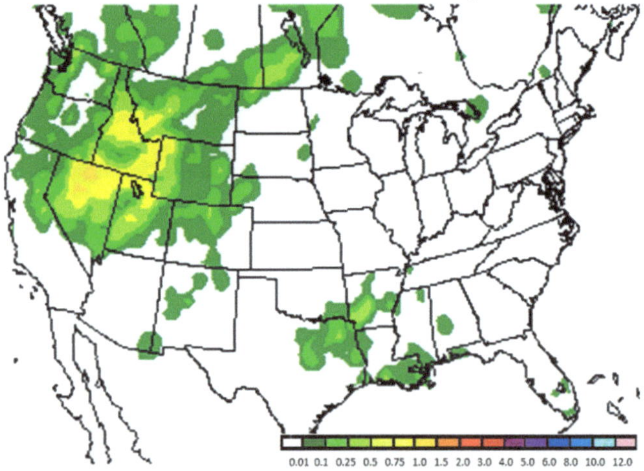

MAXIMUM TEMPERATURE

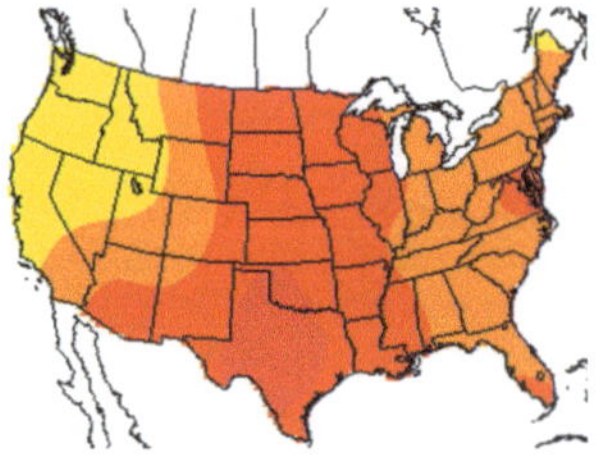

MINIMUM TEMPERATURE

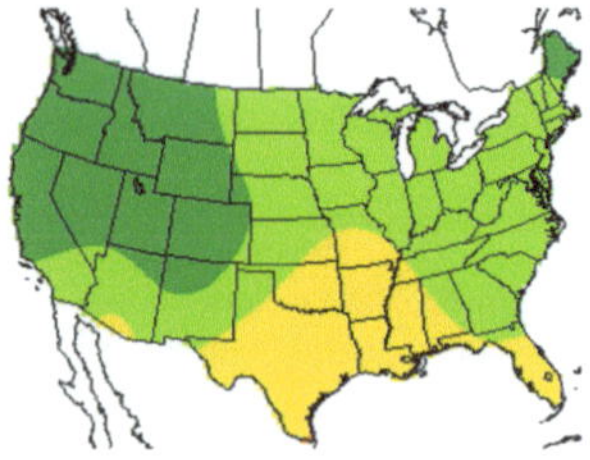

08/13/2025

Widespread rain throughout the northwest quarter of the country from the Great Basin and northern California to the northern border states, and eastward to the adjacent Rocky Mountains and northwest of the Great Plains.

Heavy rain likely at times in the Great Basin to the Inland Northwest.

Thunderstorms in the Dakotas.

Showers linger in the Southwest.

Showers in the northeast of Texas to the Deep South.

Mostly dry elsewhere.

Chance of thunderstorms and heavy rain developing in Georgia over the next couple of days.

08/14/2025

Wet fronts drift east into the middle of the country.

Widespread rain, often heavy or with thunderstorm activity in the Inland Northwest, the northern half of the Rocky Mountains, northern fringes of the Great Basin and North Great Plains.

Heavy rain likely in Minnesota.

Rain clears to mostly dry along the Pacific Coast.

Outbreaks of heavy rain and possible thunderstorms in the northeast of Texas to the western half of the Deep South including Georgia and into the Ozarks.

Mostly dry elsewhere.

Very hot in Alabama.

AUGUST 14 THURSDAY

ISOBARIC

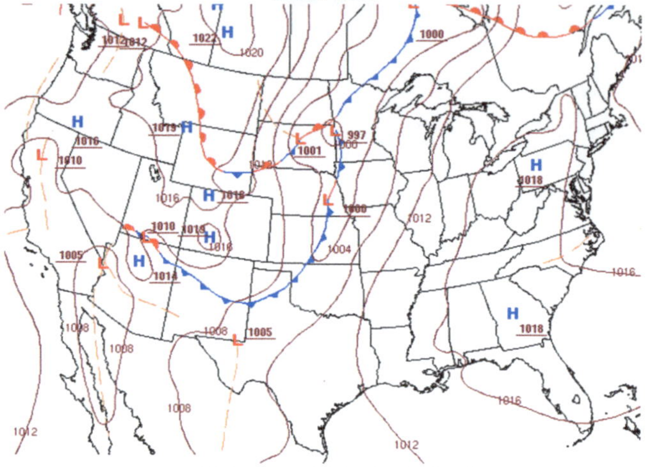

RAIN DISTRIBUTION (Inches)

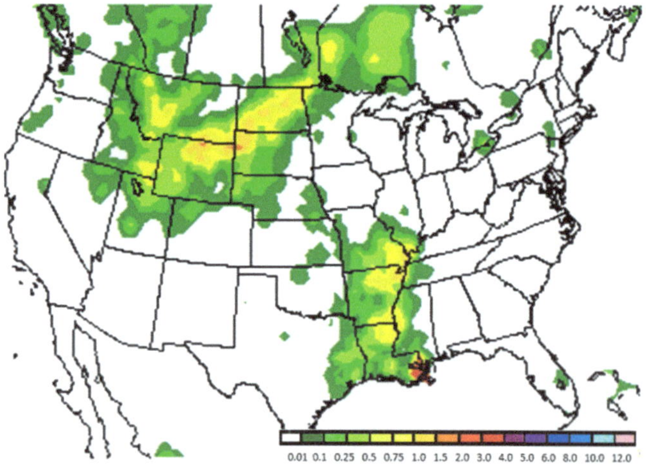

MAXIMUM TEMPERATURE

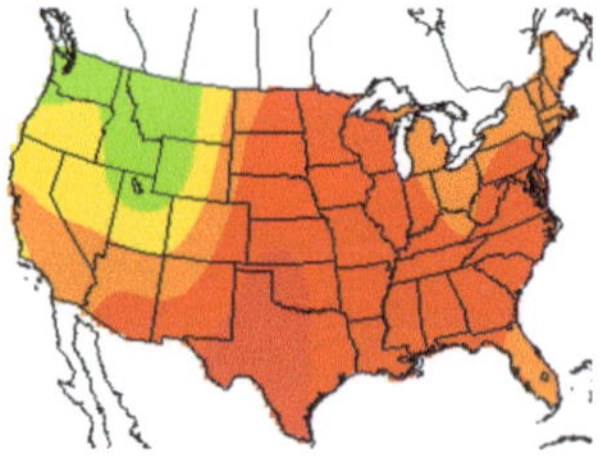

MINIMUM TEMPERATURE

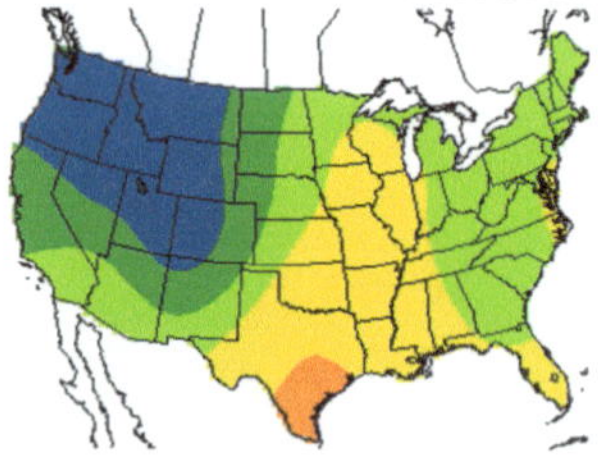

AUGUST 15 FRIDAY

ISOBARIC

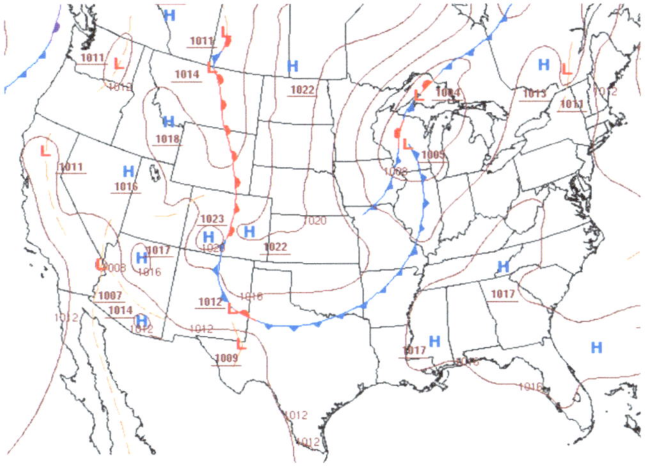

RAIN DISTRIBUTION (Inches)

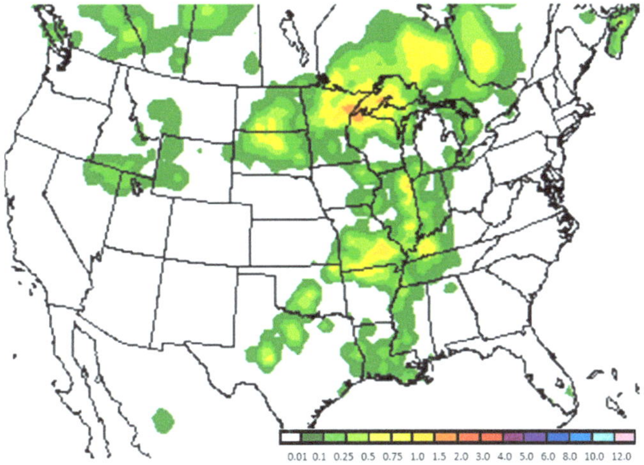

MAXIMUM TEMPERATURE

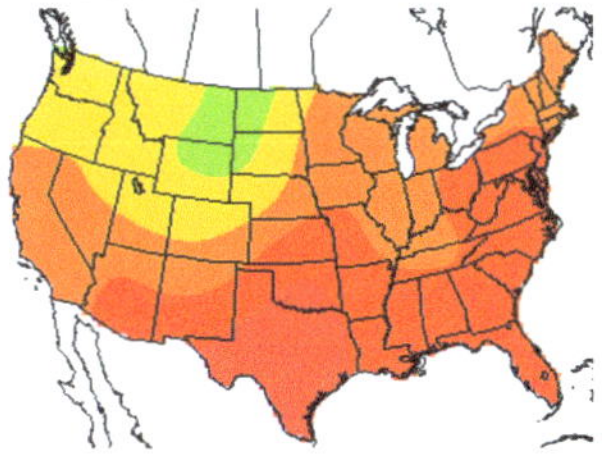

MINIMUM TEMPERATURE

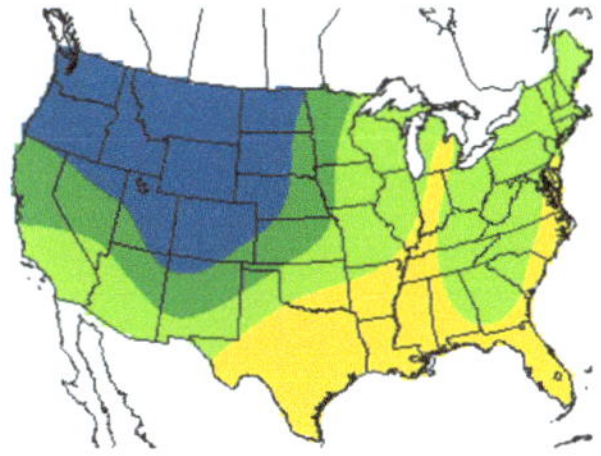

08/15/2025

A weak depression forms over the Great Lakes with wet fronts trailing through the Midwest into the north of Texas bringing widespread and often heavy rain along the front, including in the Ozarks.

Heavy rain also in the North Great Plains, with heavy falls likely in Minnesota.

Some showers linger in the northern Rocky Mountains.

Mostly dry elsewhere, including along the Atlantic coastal states from Florida to the Northeast.

Very hot over the next few days in the southeast, including in many parts of the Deep South, Coastal Southeast, and southern parts of the Inland Midwest.

AUGUST 16 SATURDAY

08/16/2025

A wet front lingers up through the east of the country from the Deep South to the Northeast bringing widespread rain along its front and possible thunderstorms at times.

Rain through the Deep South, up along the Appalachians, into the Inland Midwest and the Great Lakes, with showers into the Northeast.

Some showers linger in the Inland Northwest to the northern fringes of the Great Basin.

Mostly dry elsewhere.

ISOBARIC

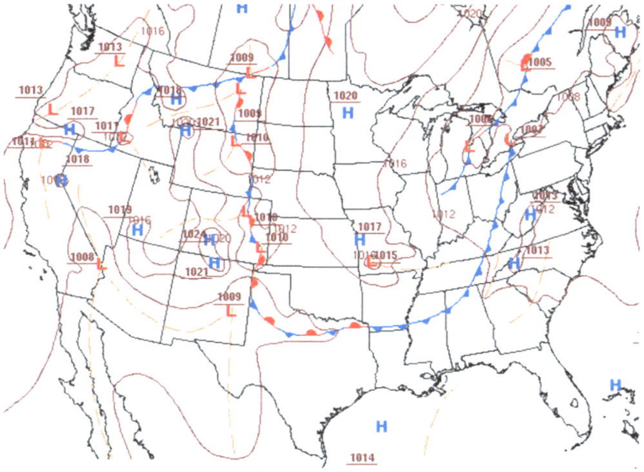

RAIN DISTRIBUTION (Inches)

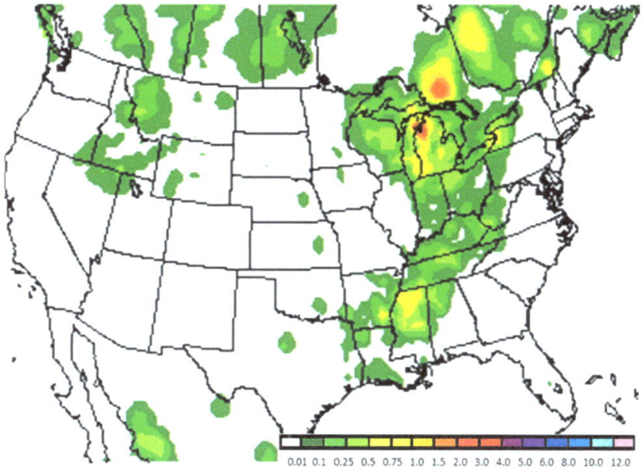

MAXIMUM TEMPERATURE

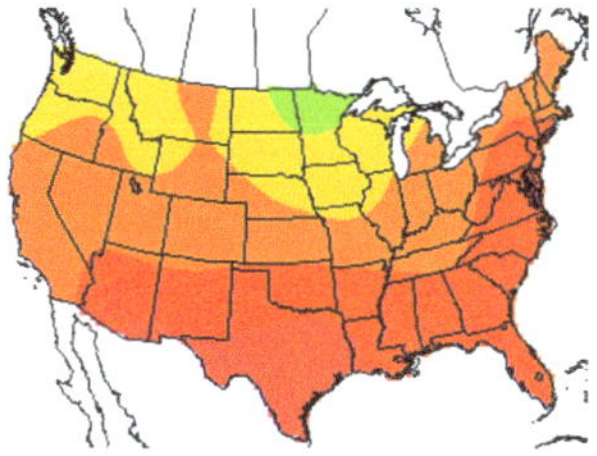

MINIMUM TEMPERATURE

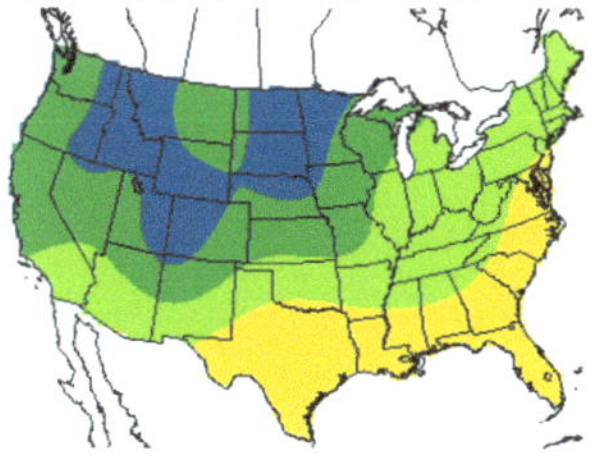

AUGUST 17 SUNDAY

ISOBARIC

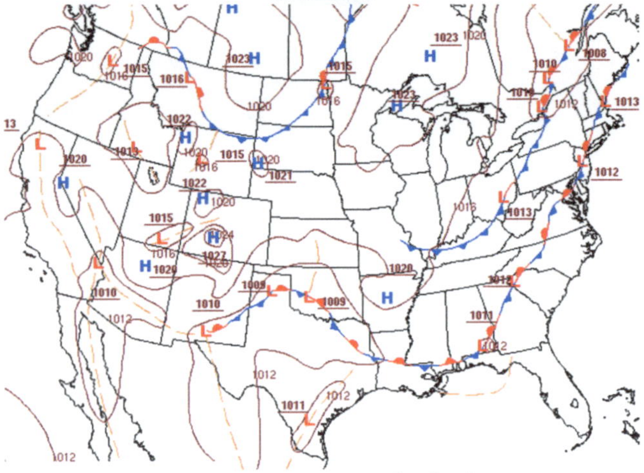

RAIN DISTRIBUTION (Inches)

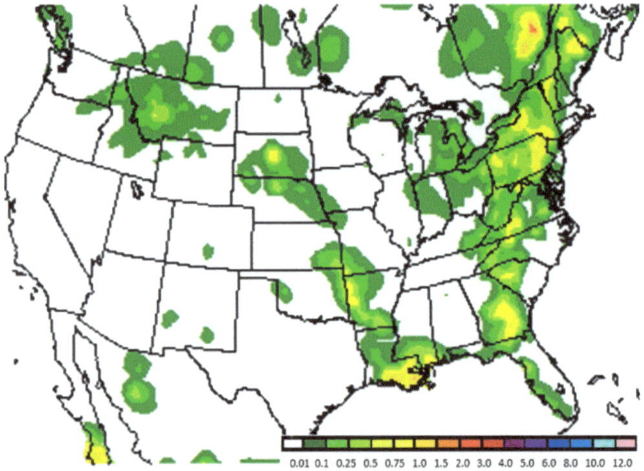

MAXIMUM TEMPERATURE

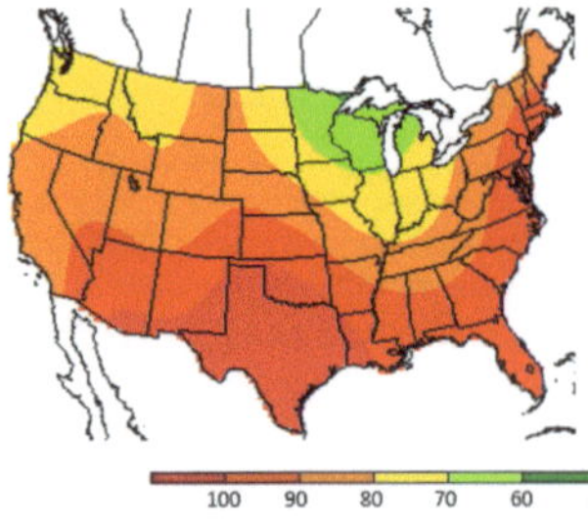

MINIMUM TEMPERATURE

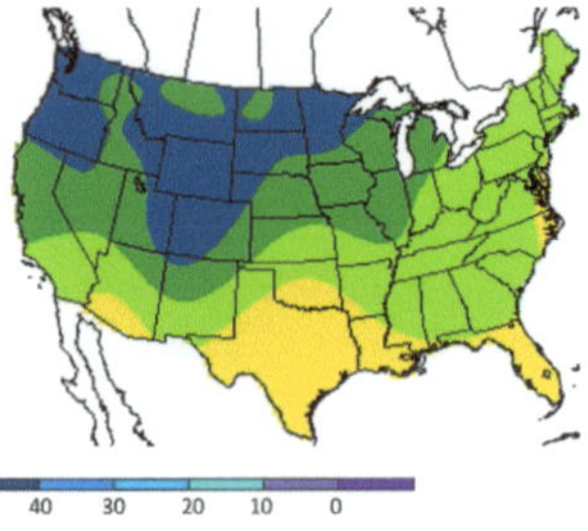

08/17/2025

Wet fronts continue to drift east towards the Atlantic Ocean shifting the rain band over the coastal states from the Northeast to the Coastal Southeast regions.

Thunderstorms and outbreaks of heavy rain at times in the Northeast.

Chance of heavy rain in Maine.

Rain eases over the western Great Lakes but persists in the east to the eastern half of the Inland Midwest.

Chance of heavy rain in Illinois and Ohio.

A trough in the Gulf of Mexico brings rain into the coastal regions from Florida to the Deep South and up into the Ozarks.

Showers in the central Great Plains to the Inland Northwest.

Mostly dry elsewhere.

Very hot in South Carolina and Indiana.

08/18/2025

A stationary front maintains rain over the east of the country from the Northeast to the Coastal Southeast regions.

Rain may be very heavy over the next few days in the New Jersey to New York region.

Rain also returns across Florida, as well as in the west of the Deep South and into the east of Texas.

Some showers in the Upper Midwest.

Spells of rain in the Inland Northwest across the northern Rockies and into the North Great Plains.

Mostly dry elsewhere.

Very hot over the next three days in the Deep South to the Inland Midwest.

AUGUST 18 MONDAY

ISOBARIC

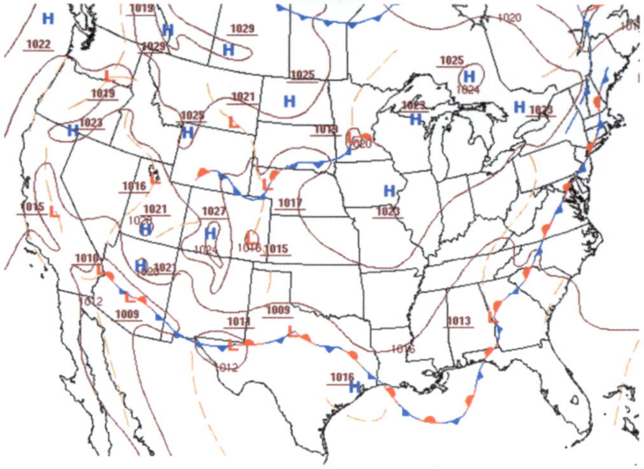

RAIN DISTRIBUTION (Inches)

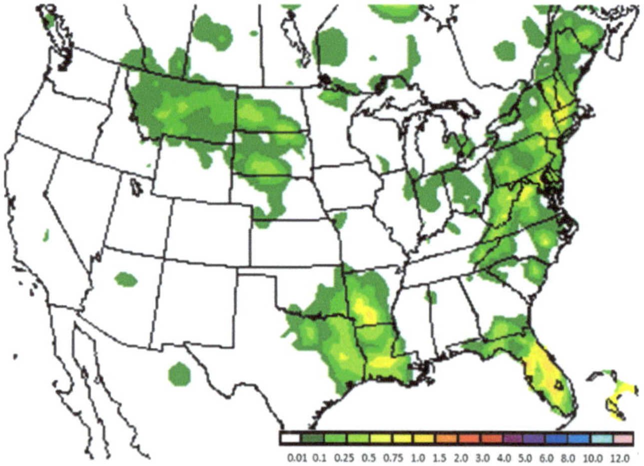

MAXIMUM TEMPERATURE

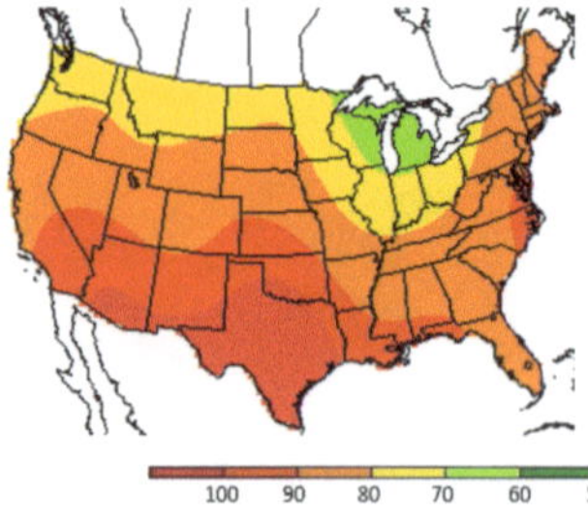

MINIMUM TEMPERATURE

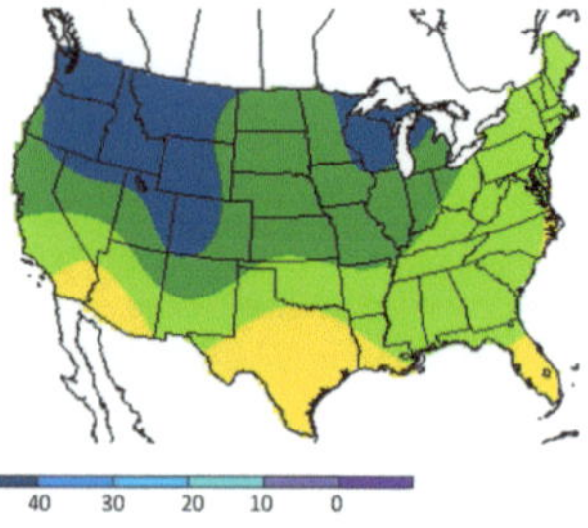

AUGUST 19 TUESDAY

ISOBARIC

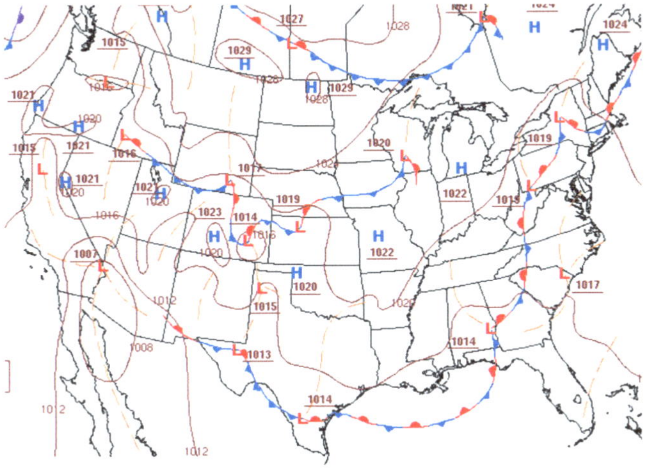

RAIN DISTRIBUTION (Inches)

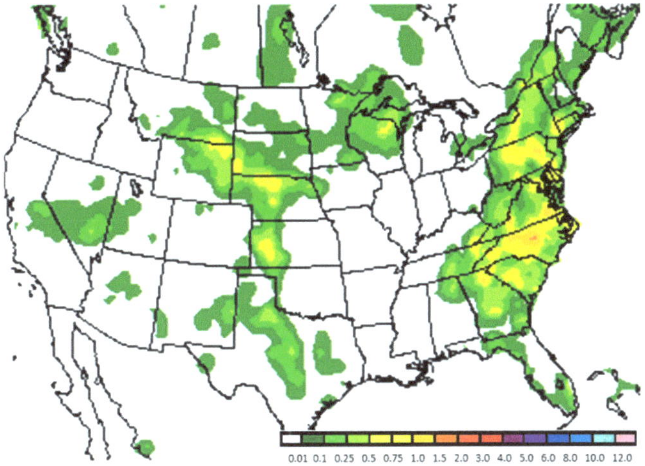

MAXIMUM TEMPERATURE

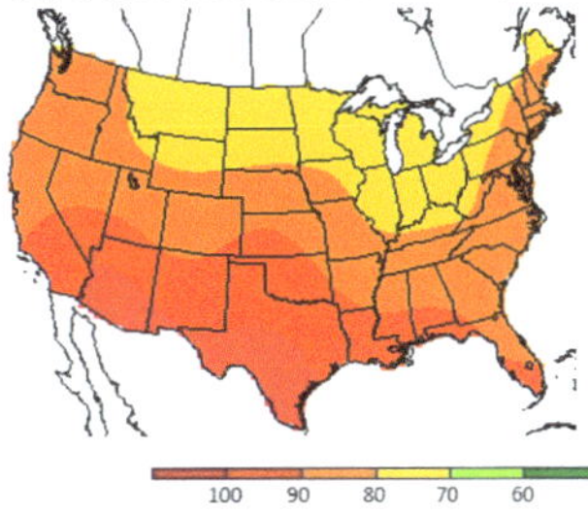

MINIMUM TEMPERATURE

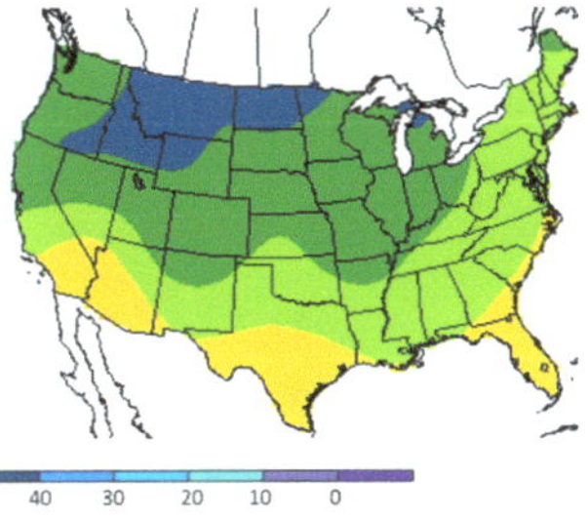

08/19/2025

Widespread rain along the Atlantic coasts, from the eastern half of the Deep South to the Northeast and up into the Appalachians.

To the west, outbreaks of rain from the North Great Plains to the Great Lakes and down through the eastern foothills of the Rocky Mountains and into Texas.

Some showers linger around the Great Basin to central California, with isolated showers into the Southwest.

Mostly dry elsewhere except for some showers around the Florida coasts.

It may be very hot in the Deep South to the Carolinas around this time.

08/20/2025

A stationary front over the east coastal regions continues to bring rain from the Northeast region to the Coastal Southeast and up into the east of the Appalachians.

Showers spread through Florida.

Another front crosses the mid-section of the country bringing showers into the Great Basin and widespread outbreaks of rain over the Rocky Mountains, across the Great Plains to the and upper Midwest to the Great Lakes, as well as south into Texas.

Rain may be particularly heavy in New Jersey.

Chance of showers in central California.

Mostly dry elsewhere.

AUGUST 20 WEDNESDAY

ISOBARIC

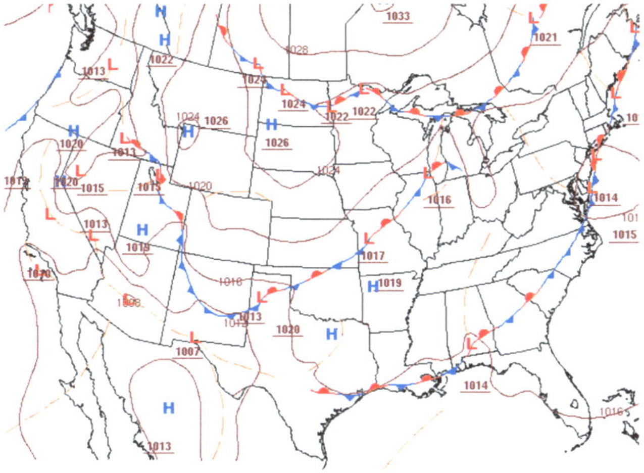

RAIN DISTRIBUTION (Inches)

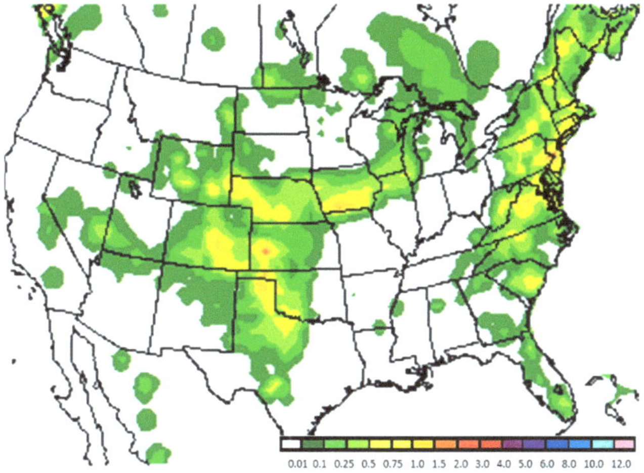

MAXIMUM TEMPERATURE

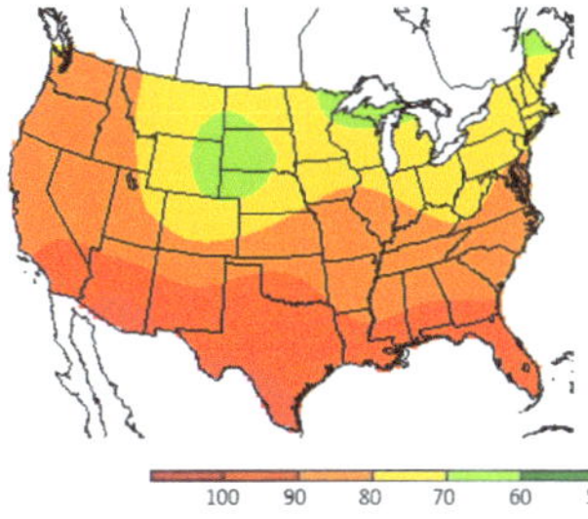

MINIMUM TEMPERATURE

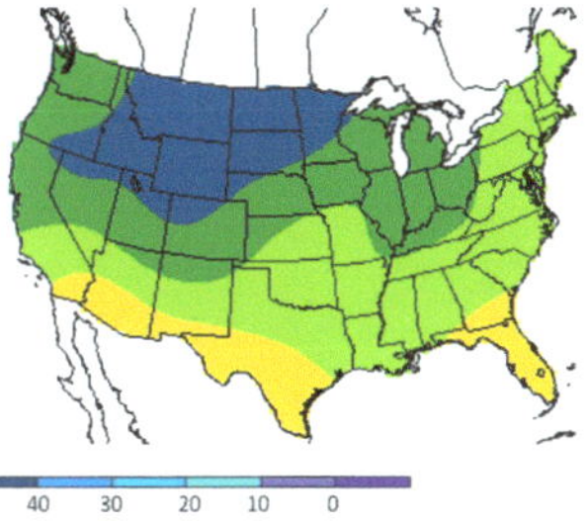

AUGUST **21** THURSDAY

ISOBARIC

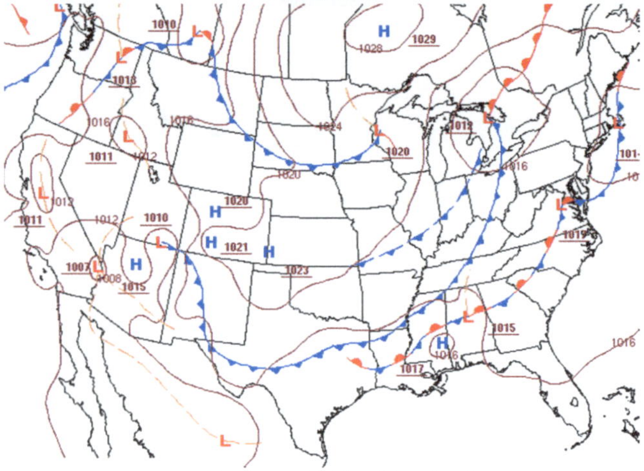

RAIN DISTRIBUTION (Inches)

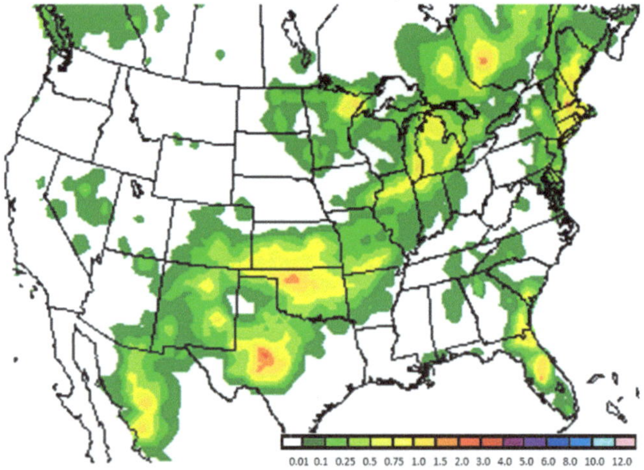

MAXIMUM TEMPERATURE

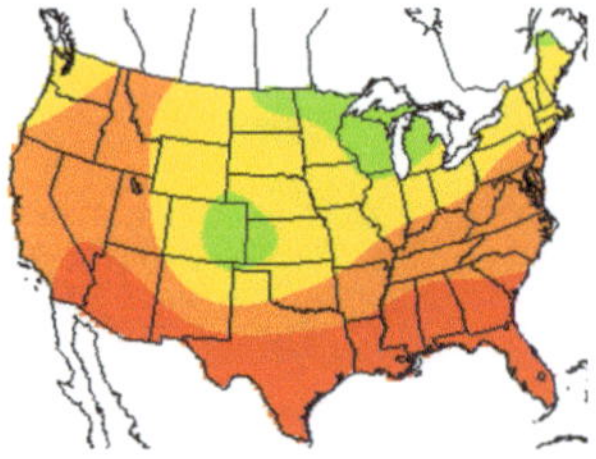

MINIMUM TEMPERATURE

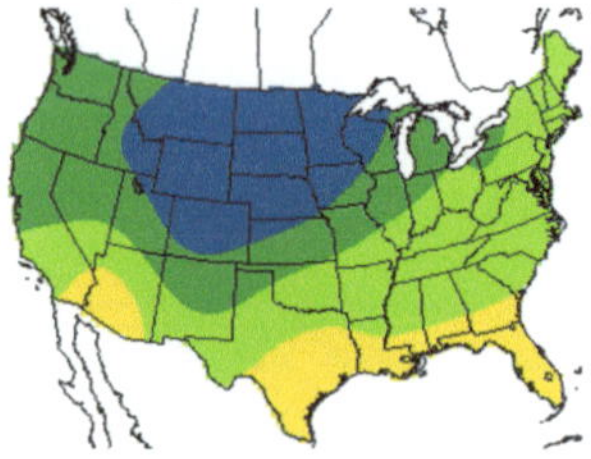

08/21/2025

Widespread rain from the Southwest and Texas, up through the South Great Plains, throughout the Inland Midwest and Great Lakes and into the Northeast.

Some isolated showers in the Great Basin.

Rain in Florida and heavy rain in the Megalopolis.

Heavy rain may develop in the Carolinas to Georgie over the next couple of days.

Mostly dry elsewhere.

Very hot in Texas, while cooler in the Northwest, particularly in Montana and Washington.

Chance of a tropical storm forming to the south of Hawaii bringing high winds and heavy rain over the State over the next few days.

It may be unusually warm in Alaska.

AUGUST **22** FRIDAY

ISOBARIC

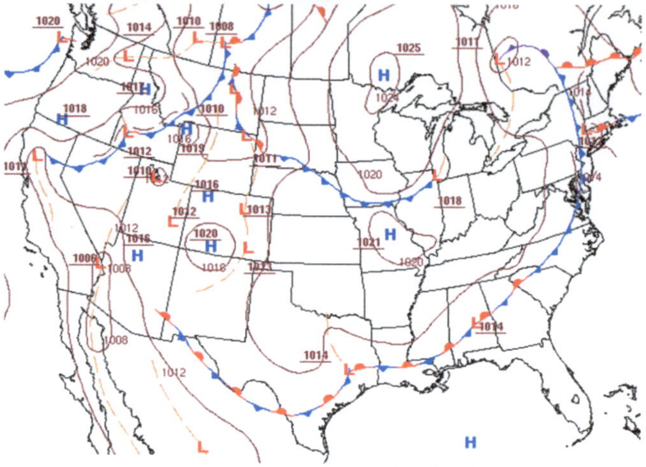

RAIN DISTRIBUTION (Inches)

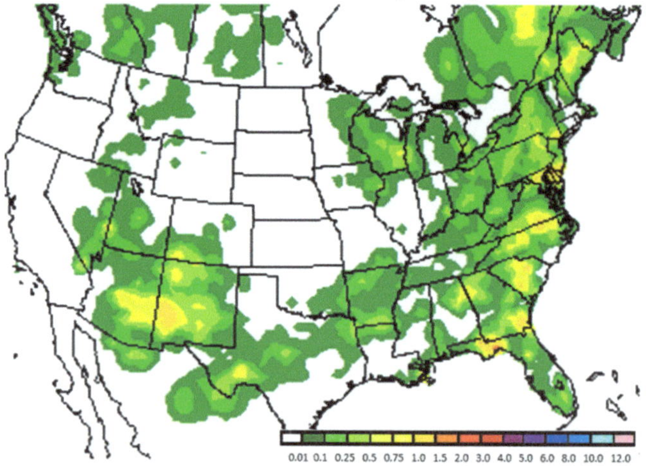

MAXIMUM TEMPERATURE

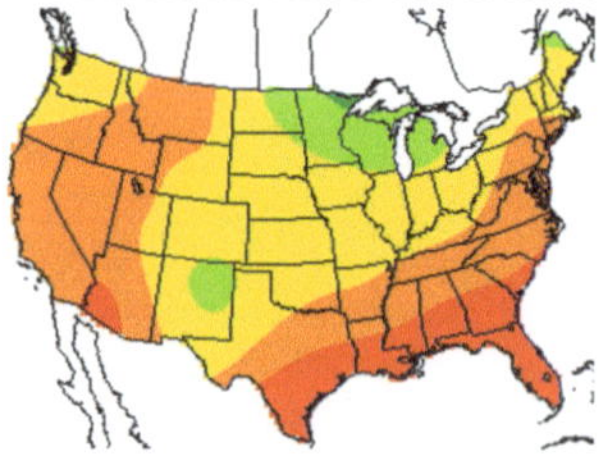

MINIMUM TEMPERATURE

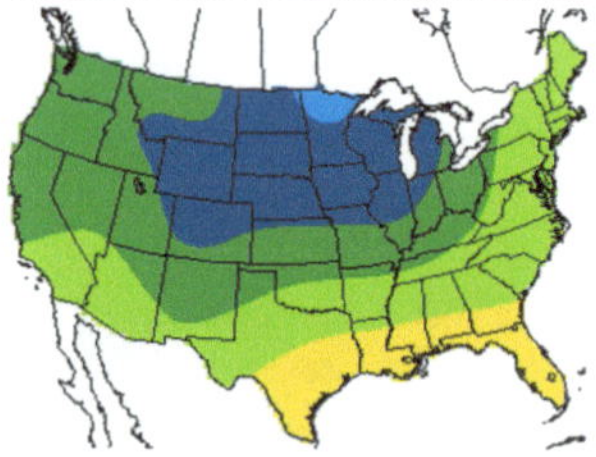

08/22/2025

Widespread rain over much of the east of the country from the Atlantic coast to the Inland Midwest and Great Lakes, throughout the Northeast, over the Appalachians, the Coastal Southeast and Florida.

Showers through the Deep South, across central Texas and widespread heavy rain into the Southwest, with showers up through the Great Basin.

Isolated showers in the Inland Northwest.

Mostly dry elsewhere.

Very hot in Texas to Arkansas, while unusually cooler in the Northwest from Washington to California.

High winds and heavy rain associated with a tropical storm may buffet Hawaii around this time.

Alaska may see sharp temperature differences across the state with some parts suddenly cooler than average, while other parts unusually warm.

AUGUST 23 SATURDAY

ISOBARIC

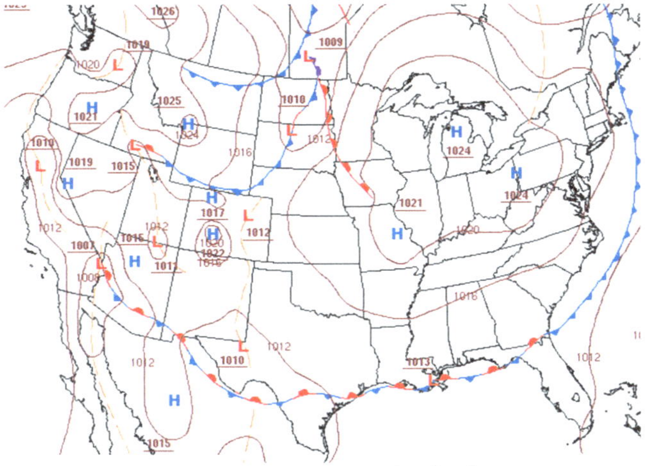

RAIN DISTRIBUTION (Inches)

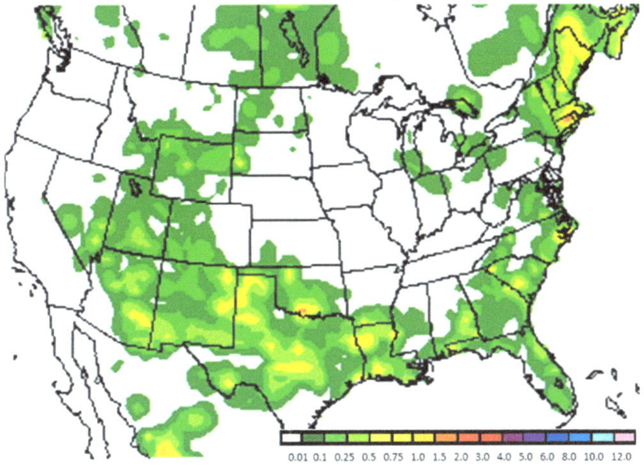

MAXIMUM TEMPERATURE

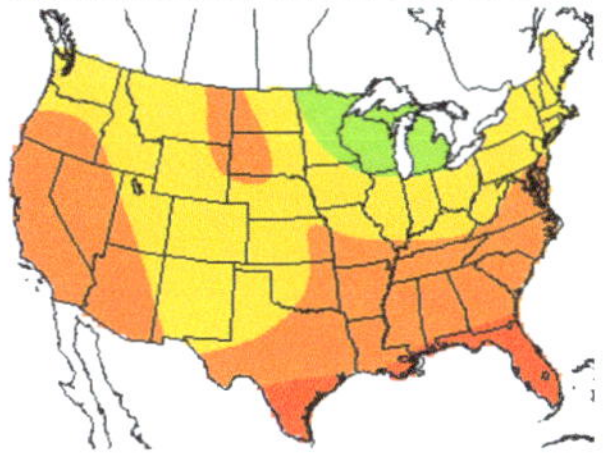

MINIMUM TEMPERATURE

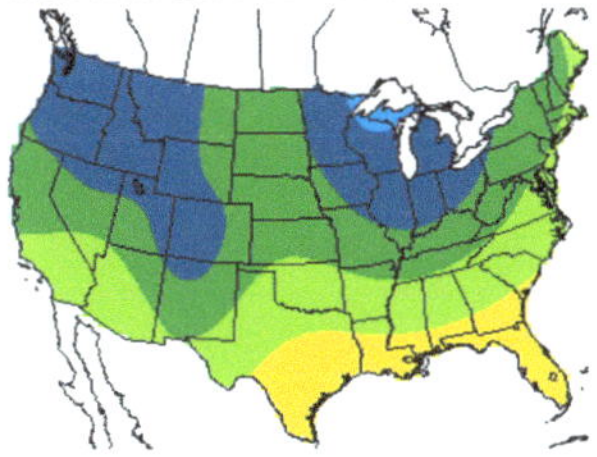

08/23/2025

A wet front brings widespread rain through the south of the country from the Great Basin, throughout the Southwest, Texas, southern fringes of the South Great Plains, over most of the Deep South, Florida and into the Coastal Southeast.

Rain also up through the Rocky Mountains to the North Great Plains.

Rain throughout the Northeast, while rain clears to isolated showers around the Great Lakes to the Inland Midwest.

Mostly dry elsewhere.

The tropical storm near Hawaii may begin to weaken.

08/24/2025

Wet fronts continue to bring widespread and often heavy rain from the Southwest to Texas and up into the south of the Great Plains.

Rain spreads from the Great Basin, through the Rocky Mountains, over the North Great Plains to the Dakotas and around the northern Great Lakes.

Pockets of rain in the Deep South with rain more widespread through Florida to the south of the Coastal Southeast.

Rain eases to mostly dry in the Northeast.

Mostly dry elsewhere.

Very high temperatures may prevail in the Deep South.

AUGUST 24 SUNDAY

ISOBARIC

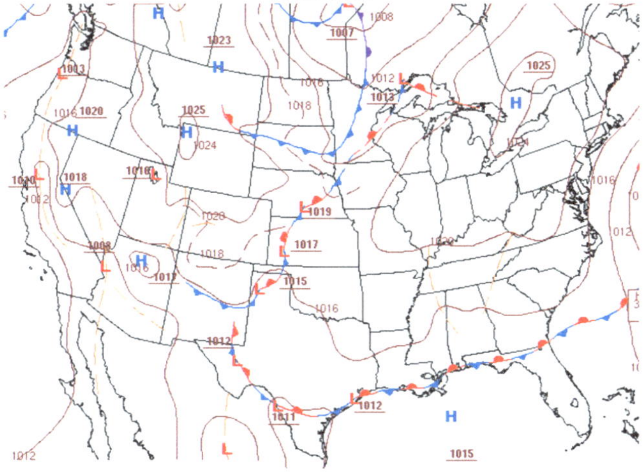

RAIN DISTRIBUTION (Inches)

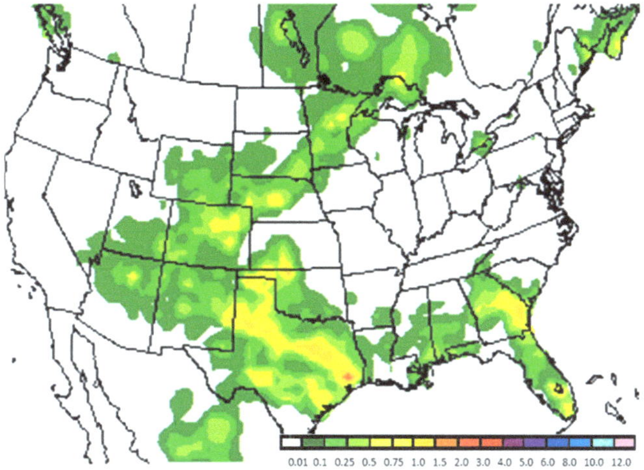

MAXIMUM TEMPERATURE

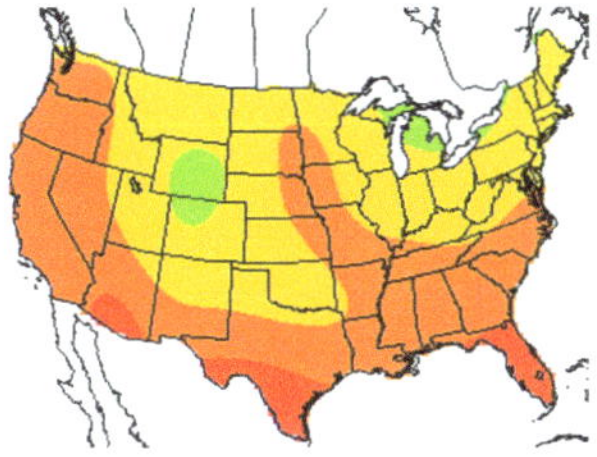

MINIMUM TEMPERATURE

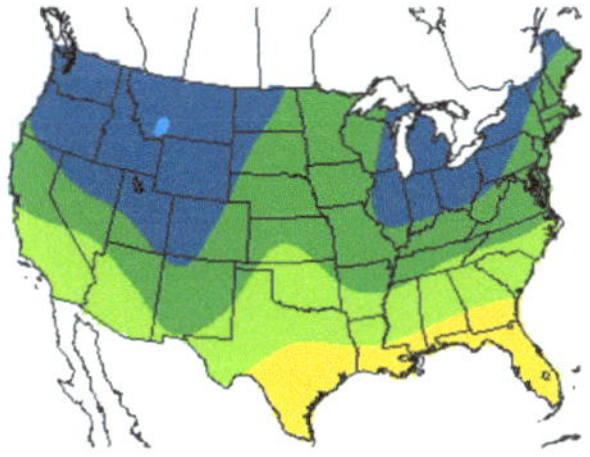

AUGUST 25 MONDAY

ISOBARIC

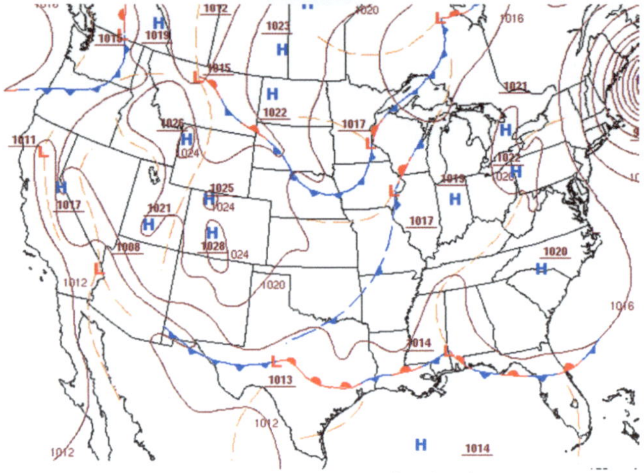

RAIN DISTRIBUTION (Inches)

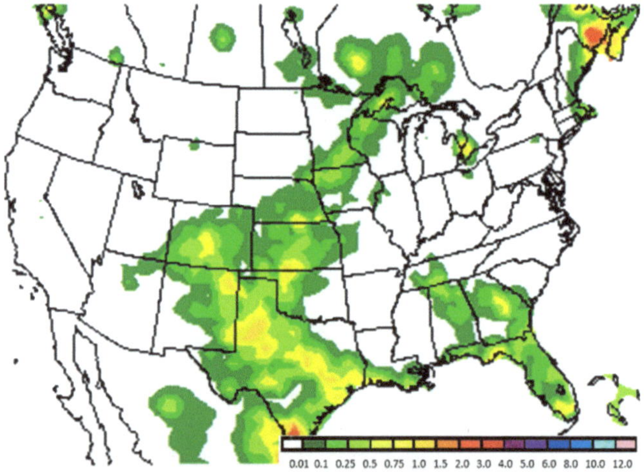

MAXIMUM TEMPERATURE

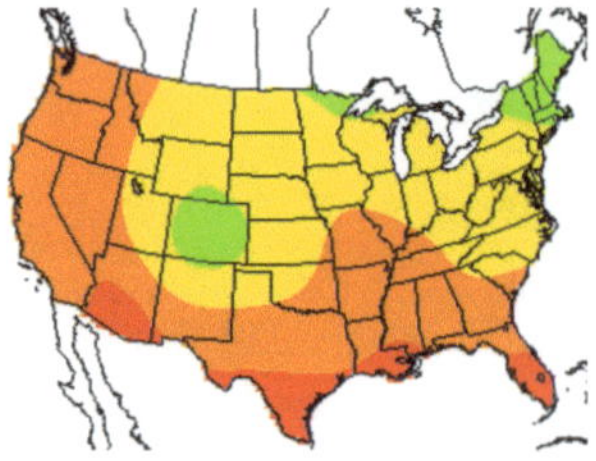

MINIMUM TEMPERATURE

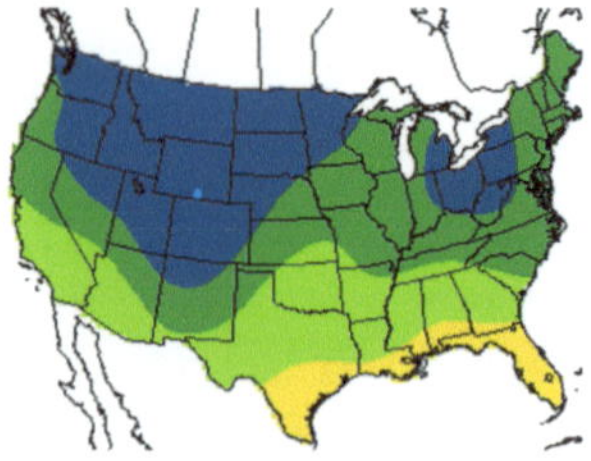

08/25/2025

Widespread rain, often heavy, prevails throughout the mid-section of the country from the Southwest and Texas up through the southern half of the Rocky Mountains, over the Great Plains and into the northwest of the Great Lakes.

An intense Atlantic depression to the east of the Northeast may reach hurricane strength and bring storm conditions into the north and east of the region.

A stationary front brings rain through the Florida Peninsula and Deep South.

Mostly dry elsewhere.

08/26/2025

A possible hurricane level storm near the northeast of the country brings very heavy rain and strong force winds over the northern half of the Northeast.

Cold fronts bring outbreaks of rain from the Great Lakes and through the Inland Midwest as well as down through the Appalachians to the Deep South.

Some rain in Florida and isolated showers lingering in the Southwest.

There is a chance of a tropical storm forming in the Gulf of Mexico that may make landfall in Texas for a brief time, spreading very heavy rain across the State and possibly into Oklahoma.

Mostly dry elsewhere.

AUGUST 26 TUESDAY

ISOBARIC

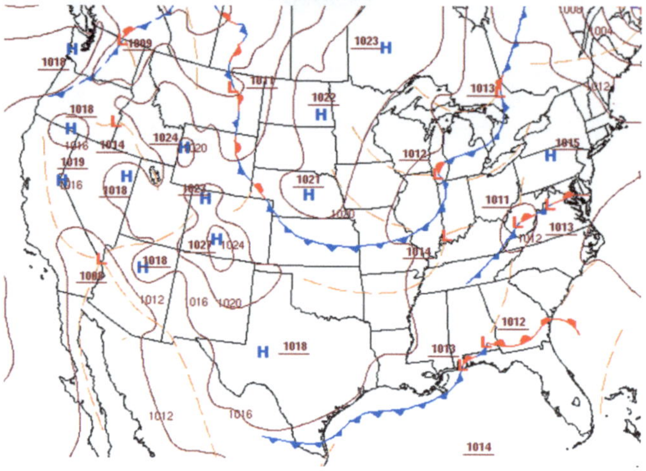

RAIN DISTRIBUTION (Inches)

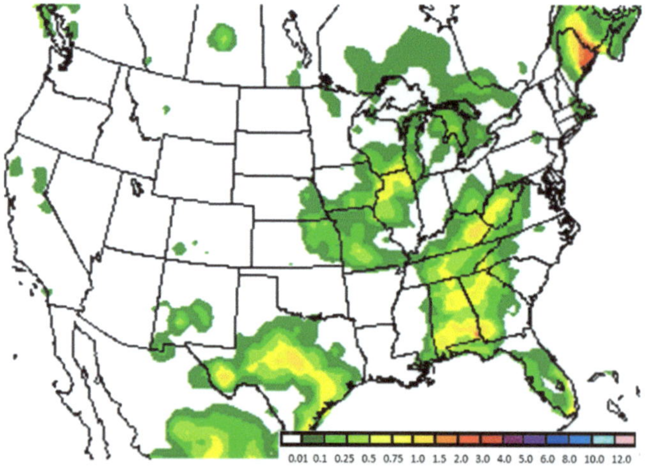

MAXIMUM TEMPERATURE

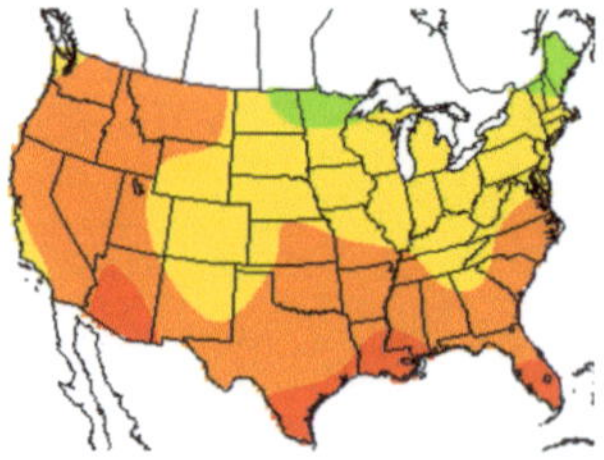

MINIMUM TEMPERATURE

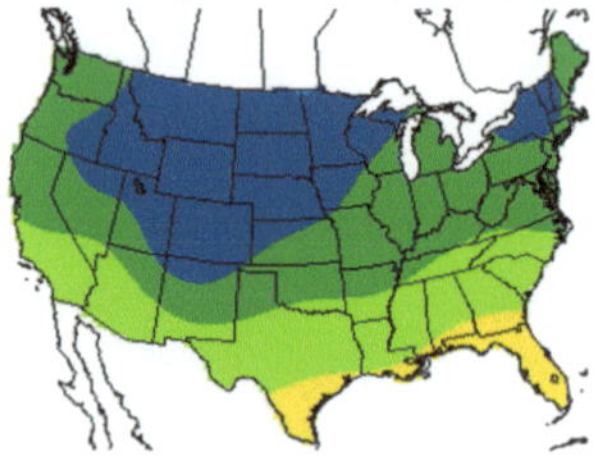

AUGUST 27 WEDNESDAY

ISOBARIC

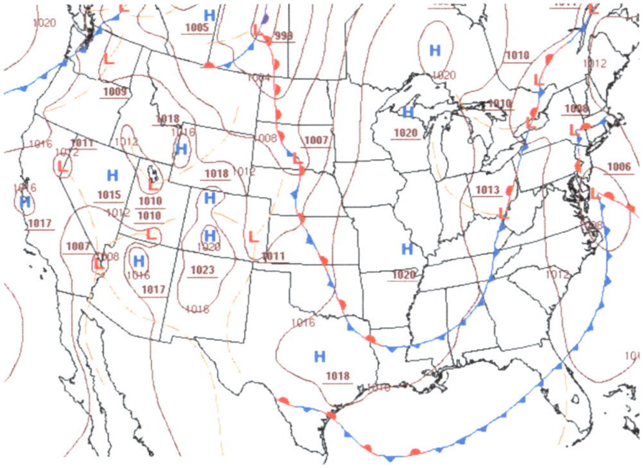

RAIN DISTRIBUTION (Inches)

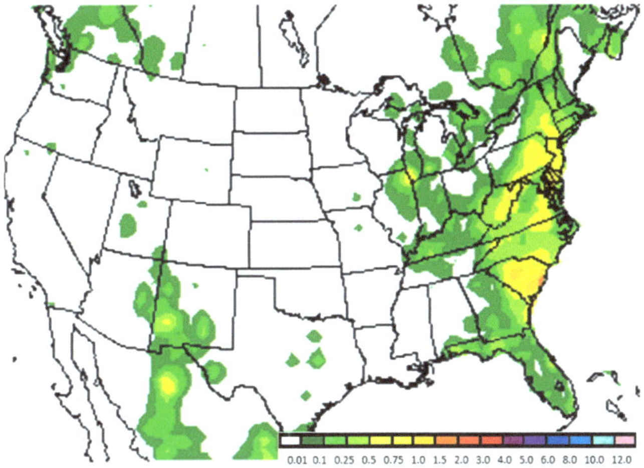

MAXIMUM TEMPERATURE / MINIMUM TEMPERATURE

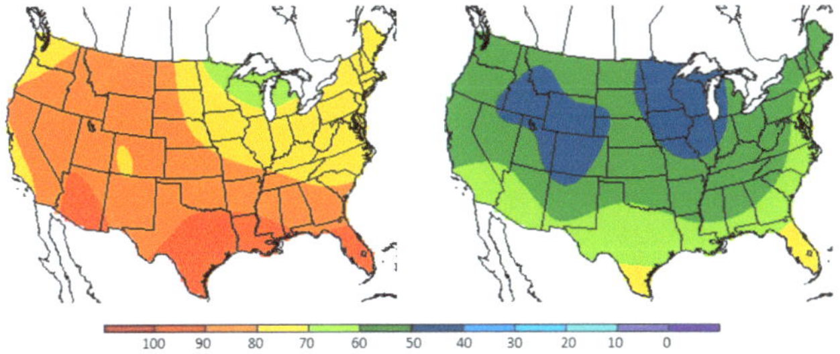

08/27/2025

Heavy rain from the tropical storm that may have made landfall in Texas in the past couple of days eases to mostly dry across the state, while a trough brings heavy rain in a band through the center of the Southwest.

Further widespread rain through the east of the country from the Northeast to Florida, and west into the Appalachians and Inland Midwest.

Some showers around the Great Lakes.

Chance of heavy rain in Minnesota over the next few days, possibly causing some flooding in the Root River catchment.

Mostly dry elsewhere.

Temperatures may be unusually high in Alaska over the next couple of days.

08/28/2025

The Atlantic storm off the northeast of the country may continue to linger near the Northeast bringing heavy rain and strong winds throughout, with possible damage to local infrastructure.

Showers also prevail in some parts of the Great Lakes to the Upper Midwest.

A decaying tropical depression to the south may still bring thunderstorms into the north of Texas, southern parts of the Great Plains, as well as in the Southwest.

Some spotty showers likely in the Great Basin.

Mostly dry elsewhere.

AUGUST 28 THURSDAY

ISOBARIC

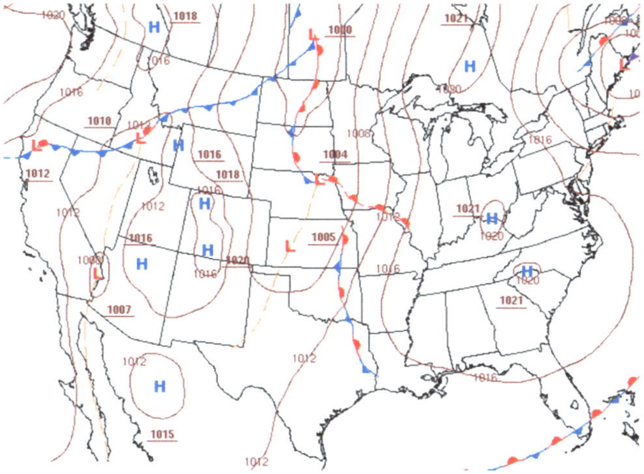

RAIN DISTRIBUTION (Inches)

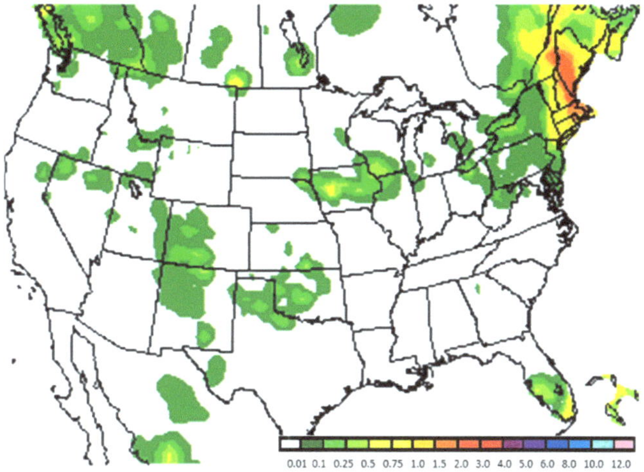

MAXIMUM TEMPERATURE / MINIMUM TEMPERATURE

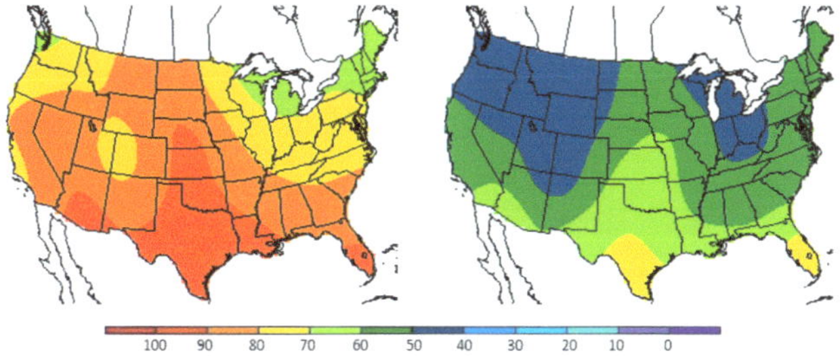

ISOBARIC

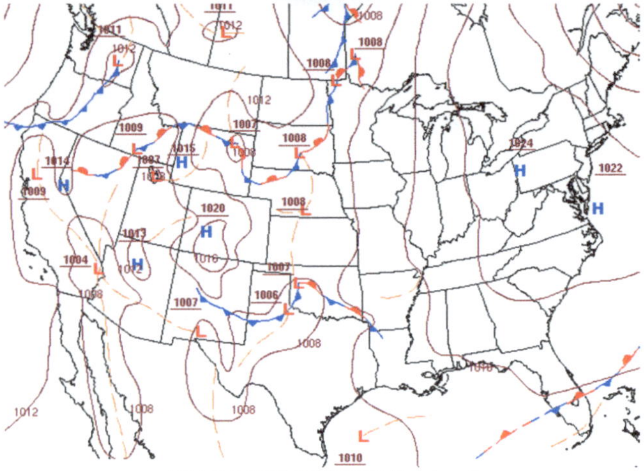

RAIN DISTRIBUTION (Inches)

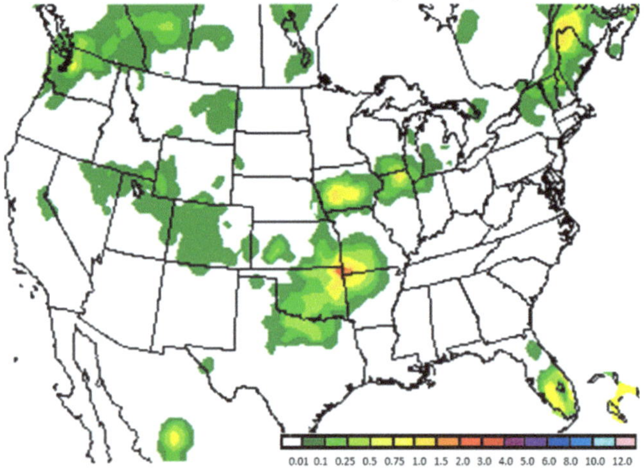

MAXIMUM TEMPERATURE MINIMUM TEMPERATURE

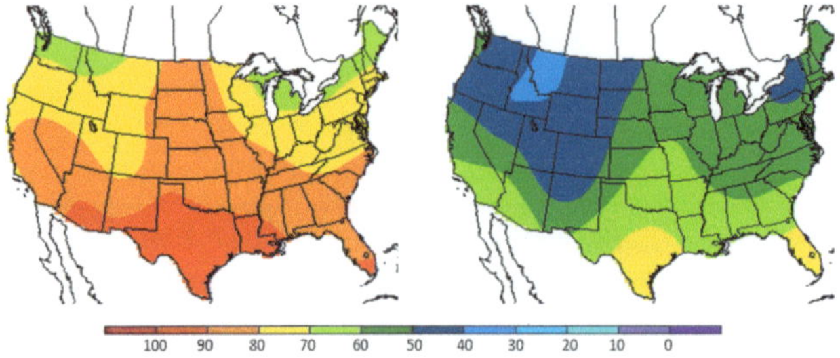

08/29/2025

High pressure builds over the east, easing rain in most eastern states, although heavy rain still likely in the Northeast, from the departing Atlantic storm.

Pockets of rain in the Upper Midwest to the Ozarks and into the north of Texas.

Showers across the southern Rockies into the Great Basin and some showers building in the Pacific Northwest to the North Great Plains.

Mostly dry elsewhere.

The recent passage of a tropical storm in the Oklahoma region may have brought damage to local infrastructure and there may be some flooding in the local river catchments in the Upper Midwest.

Very hot in Hawaii, and usually warm in Alaska over the next few days.

08/30/2025

Outbreaks of rain through the Pacific Northwest to the Inland Northwest, across the northern Rockies and into the North Great Plains.

Showers also spread into the Great Basin and down into the Southwest.

Some outbreaks of heavy rain into the Ozarks and up through the Inland Midwest, including in Missouri.

Rain also throughout Florida and along the coastal fringes of the east coast to the Carolinas.

Mostly dry elsewhere.

Heavy rain likely to spread into Wisconsin and Indiana over the next couple of days, possibly causing some flooding in local river catchments, including the Kickapoo and Root Rivers.

Very hot in New Mexico.

There is a chance of a funnel formation winds in Alaska.

ISOBARIC

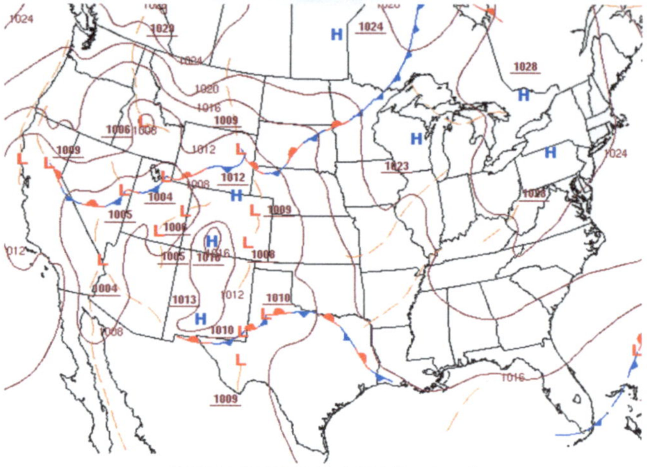

RAIN DISTRIBUTION (Inches)

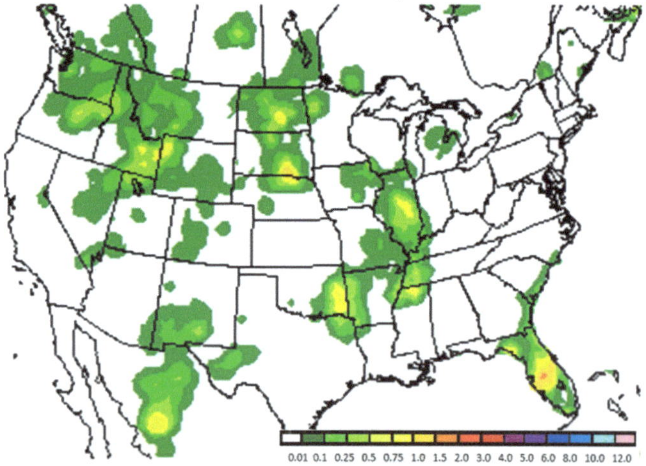

MAXIMUM TEMPERATURE MINIMUM TEMPERATURE

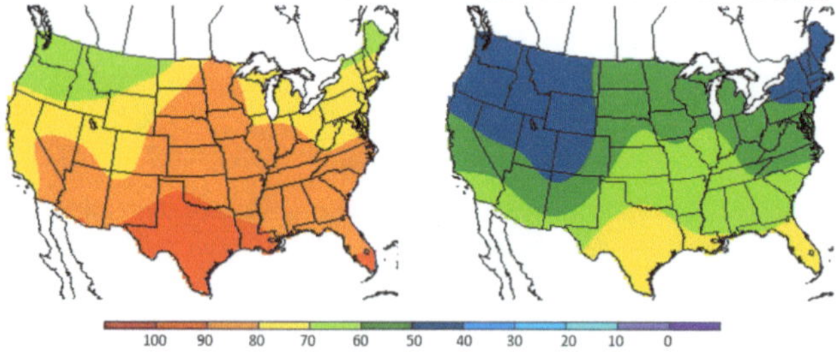

ISOBARIC

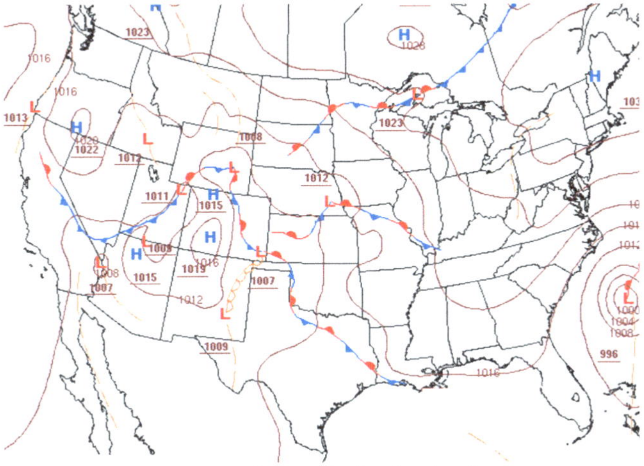

RAIN DISTRIBUTION (Inches)

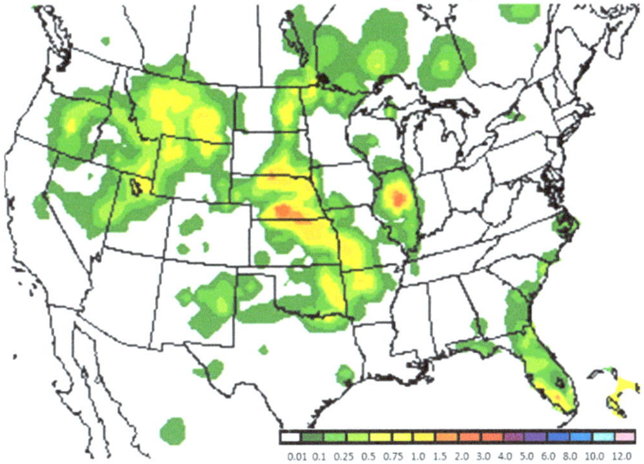

MAXIMUM TEMPERATURE

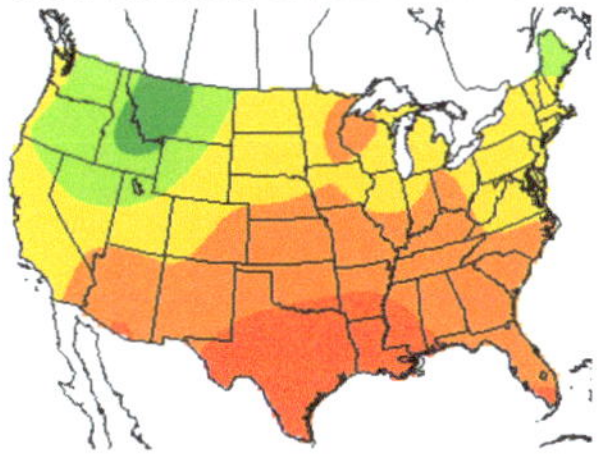

MINIMUM TEMPERATURE

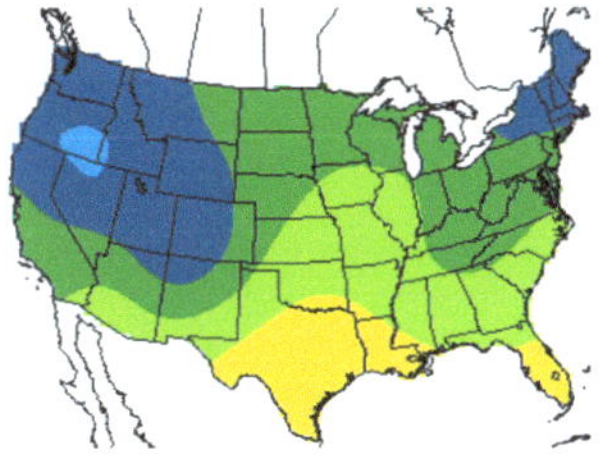

08/31/2025

Another Atlantic depression, possibly already to hurricane strength nears the southeast bringing heavy rain along the coasts from the Carolinas to widespread rain in Florida.

Wet fronts also bring widespread rain from the Great Basin and Pacific Northwest, and down throughout the Great Plains into the north of Texas into New Mexico, but with only showers down through the Rocky Mountains.

Heavy rain also likely in the Upper Midwest.

Mostly dry elsewhere.

Cooler temperatures in the Northwest, but very hot in North Carolina to Georgia over the next couple of days.

A hurricane may make landfall in Mexico bringing severe damage to infrastructure and crops.

SEPTEMBER

PHASES OF THE MOON

1st	Southern Declination
7th	Full Moon
9th	Crossing Equator
10th	Perigee #9
14th	Third Quarter
14th	Northern Declination
21st	New Moon
22nd	Crossing Equator
26th	Apogee
29th	First Quarter
29th	Southern Declination

1 Mon	2 Tue	3 Wed	4 Thur	5 Fri	6 Sat	7 Sun	8 Mon	9 Tue	10 Wed	11 Thur	12 Fri	13 Sat	14 Sun	15 Mon	16 Tue	17 Wed	18 Thur	19 Fri	20 Sat	21 Sun	22 Mon	23 Tue	24 Wed	25 Thur	26 Fri	27 Sat	28 Sun	29 Mon	30 Tue
V						F		XhN	P9				3Q ^							N	XhS				A			1Q V	

SEPTEMBER 1 MONDAY

ISOBARIC

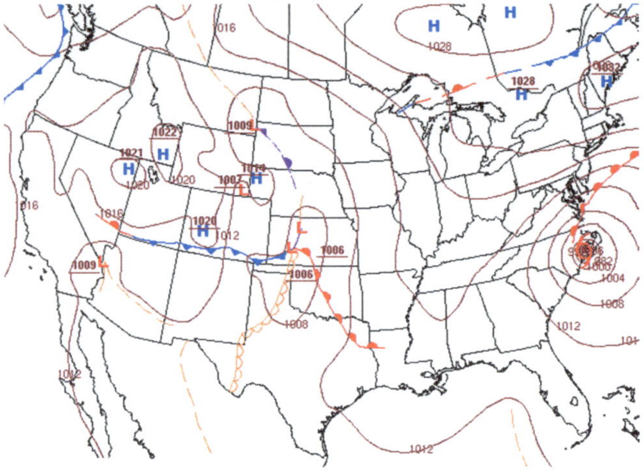

RAIN DISTRIBUTION (Inches)

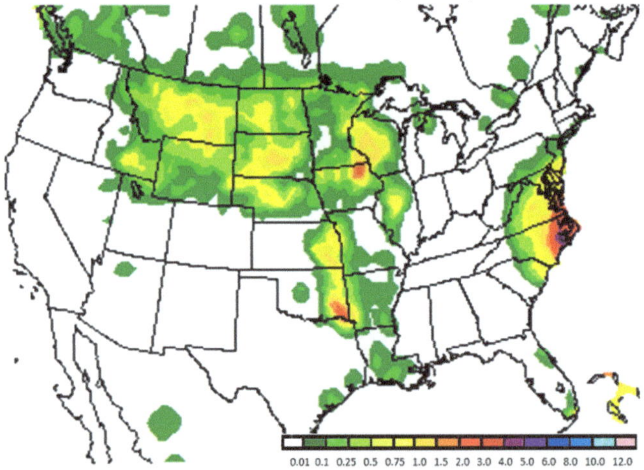

MAXIMUM TEMPERATURE MINIMUM TEMPERATURE

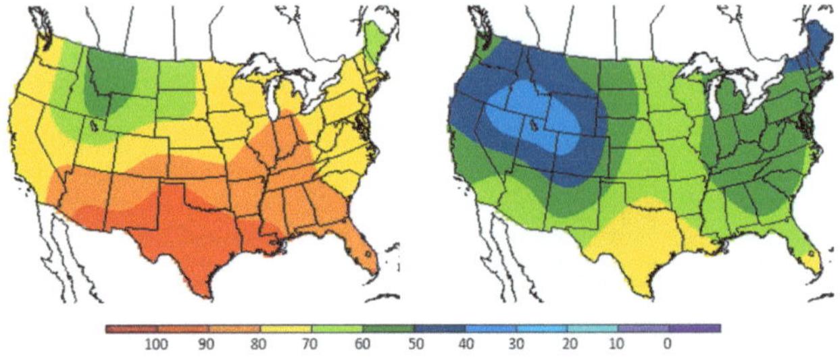

09/01/2025

An depression, which may be a named hurricane, make its way south along the Atlantic coast, bringing intense rain and winds from the Megalopolis to the Carolinas.

Widespread rain over northern parts from the Inland Northwest and northern Rocky Mountains, over the adjacent Great Plains and the western half of the Inland Midwest.

Very heavy rain in Wisconsin.

Rain also down the east of the South Great Plains and through the Ozarks to the west of the Deep South.

Some showers around the eastern Texas coastal fringes.

Mostly dry elsewhere.

Another hurricane may make landfall in Mexico, although it is likely to have considerably weakened.

Temperatures may be unusually high in Hawaii over the next few days.

SEPTEMBER 2 TUESDAY

ISOBARIC

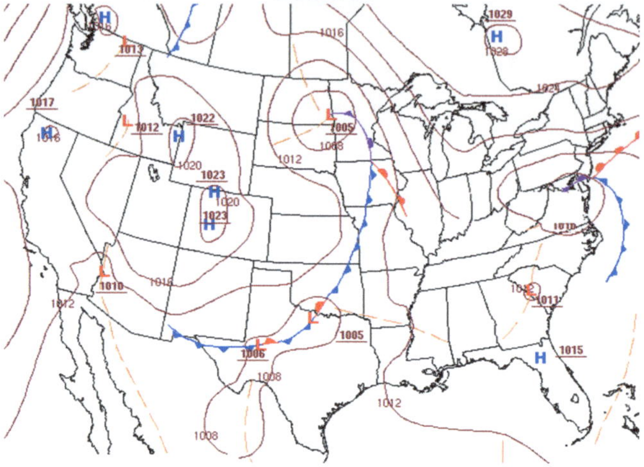

RAIN DISTRIBUTION (Inches)

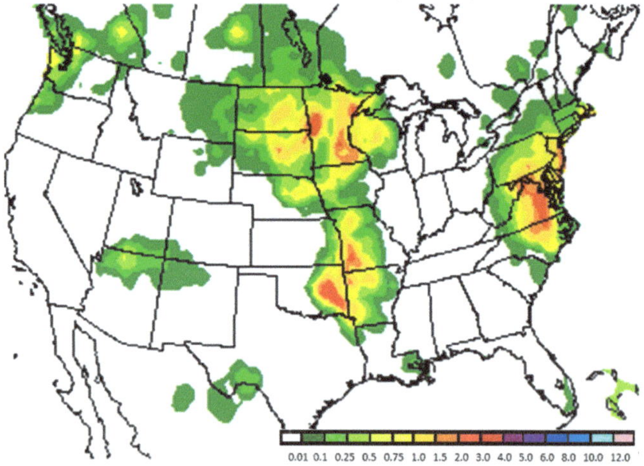

MAXIMUM TEMPERATURE MINIMUM TEMPERATURE

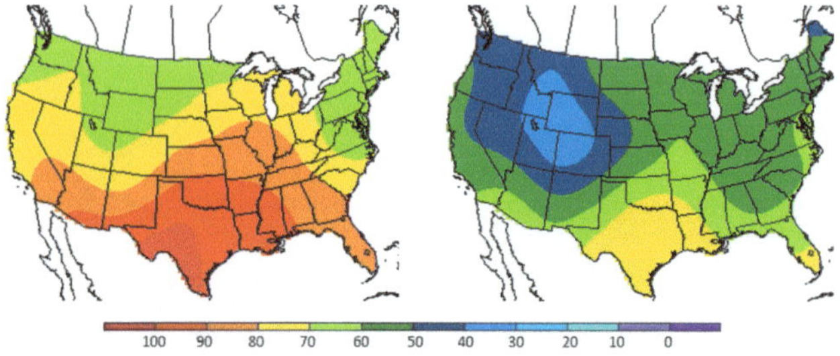

09/02/2025

A degrading Atlantic storm system continues to bring widespread and heavy rain in the eastern coasts regions from the Megalopolis to the Carolinas.

Another weak depression and associated fronts bring widespread heavy rain and storms to the North Great Plains and down through the west of the South Great Plains to the adjacent Inland Midwest states.

Very heavy rain in Iowa likely and heavy rain in Ohio may cause some flooding along the Sandusky River system.

Some outbreaks of rain also likely in the Pacific Northwest and in the Four Corners region.

Mostly dry elsewhere.

It may be very hot in Tennessee.

ISOBARIC

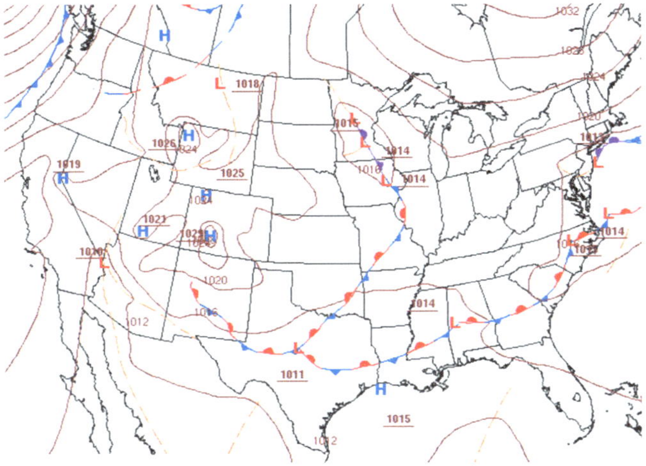

RAIN DISTRIBUTION (Inches)

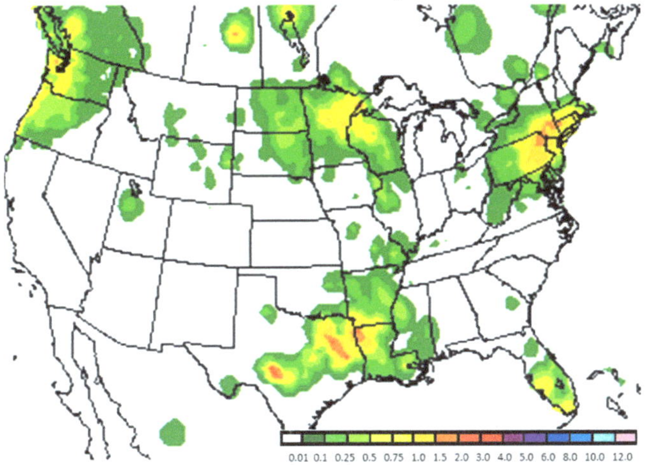

MAXIMUM TEMPERATURE

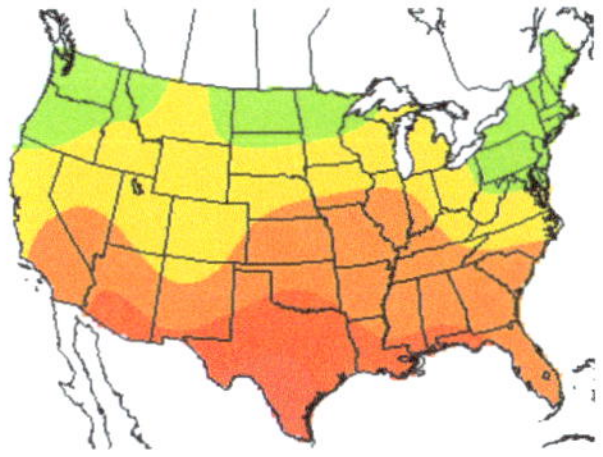

MINIMUM TEMPERATURE

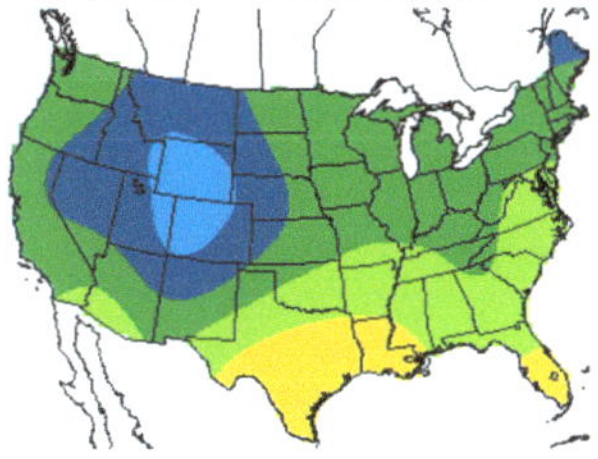

09/03/2025

An intense depression to the northwest of Washington may build to hurricane strength over the next couple of days and extends storms and heavy precipitation into the Pacific Northwest.

In the east, a stationary front brings rain into the North Great Plains to the Great Lakes and down through the west of the Inland Midwest to the northern half of Texas and into the west of the Deep South.

Heavy rain may persist in the lower Northeast through Megalopolis and into the northern Appalachians.

Showers in Florida.

Some heavy rain likely in Missouri to Arkansas, and there may be some flooding along the Fox River system in Wisconsin.

Mostly dry elsewhere.

A cooler airflow associated with a stationary front may move over the Deep South over the next few days.

09/04/2025

A depression lingering near the northeast of the country continues to expand heavy rain into the Pacific Northwest to northern California, with isolated showers possible into the Great Basin.

A wet front remains almost stationary over the south of the country bringing outbreaks of rain throughout most of Texas and through the Deep South.

A trough brings widespread showers over Florida.

Precipitation persists over the Megalopolis to the north of the Appalachians.

Rain eases in Minnesota, but continues further east over the Great Lakes and the Upper Midwest.

Mostly dry elsewhere.

Chance of heavy isolated showers in Hawaii.

ISOBARIC

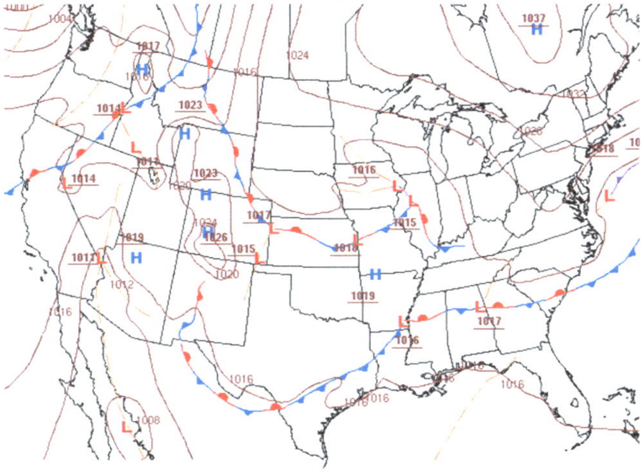

RAIN DISTRIBUTION (Inches)

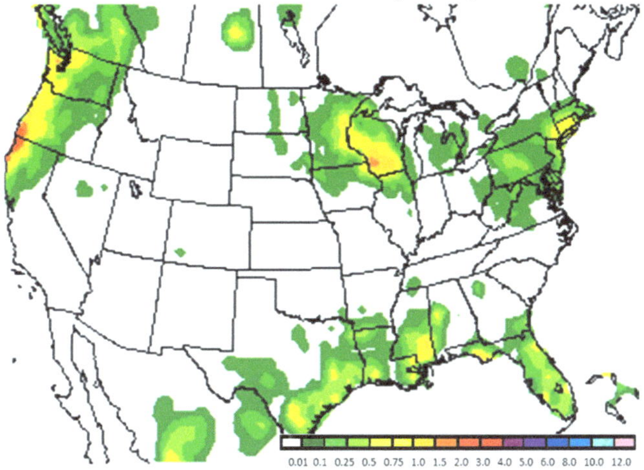

MAXIMUM TEMPERATURE

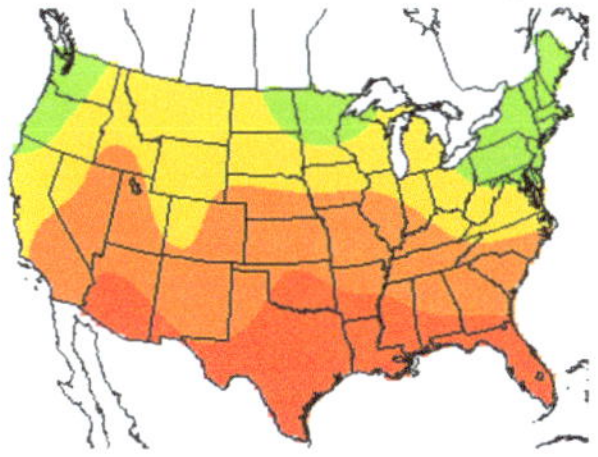

MINIMUM TEMPERATURE

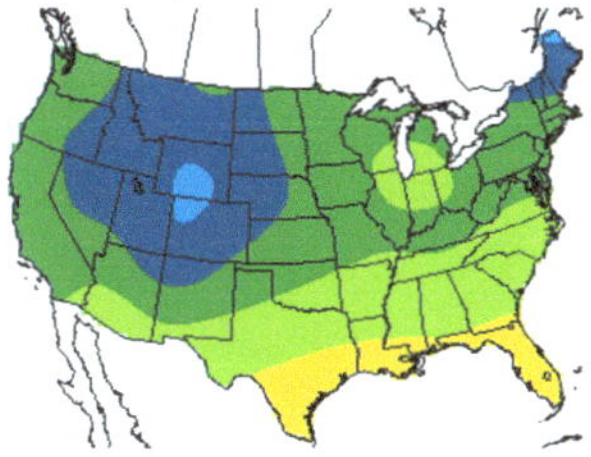

SEPTEMBER 5 FRIDAY

ISOBARIC

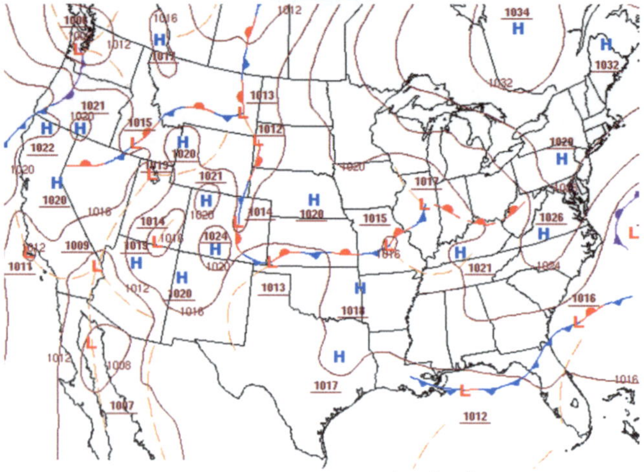

RAIN DISTRIBUTION (Inches)

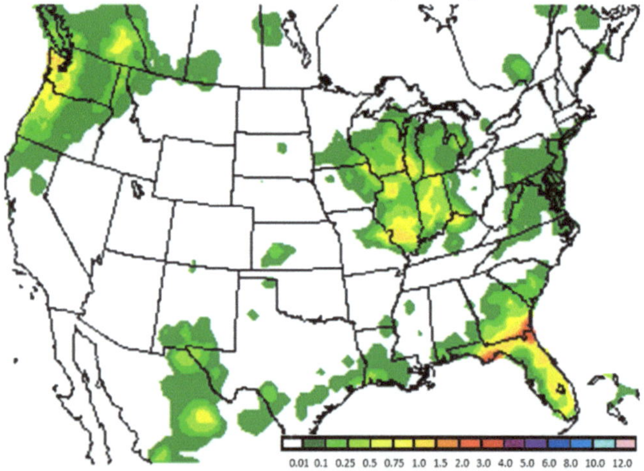

MAXIMUM TEMPERATURE MINIMUM TEMPERATURE

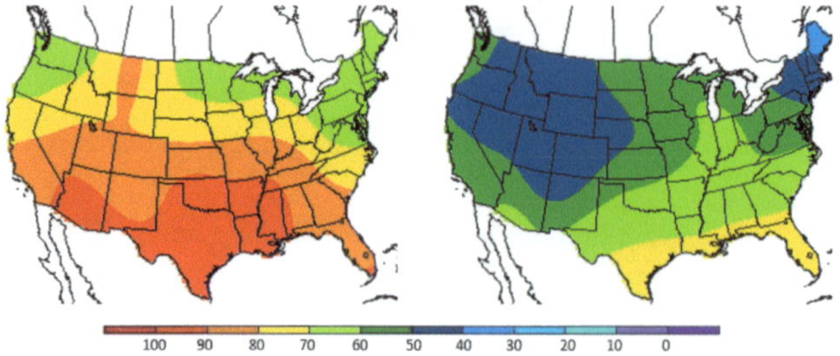

09/05/2025

A depression, possibly an ex-hurricane, makes landfall in Washington maintaining a stormy air over the Pacific Northwest to the northern fringes of California, with showers possibly spreading into Nevada for a time.

A band of rain persists along the south of the country bringing showers from New Mexico, through central parts of Texas, and along the coastal fringes of the Deep South.

A wet front brings heavy rain throughout Florida and up into South Carolina.

Further rain persists in the Great Lakes to the Inland Midwest, as well as showers in the Megalopolis to North Carolina.

Isolated shower in the Northeast.

Mostly dry elsewhere.

Shower activity throughout the coming week in Texas.

09/06/2025

Widespread rain persists in the Pacific Northwest to the Inland Northwest as well as showers possibly still lingering in the northern areas of California and Nevada.

Widespread rain across the Great Lakes and Inland Midwest regions and into the southern Appalachians.

Isolated showers in the Northeast and showers across Texas.

Rain in Florida to Georgia.

Mostly dry elsewhere.

A warm front crosses the Great Plains.

SEPTEMBER 6 SATURDAY

ISOBARIC

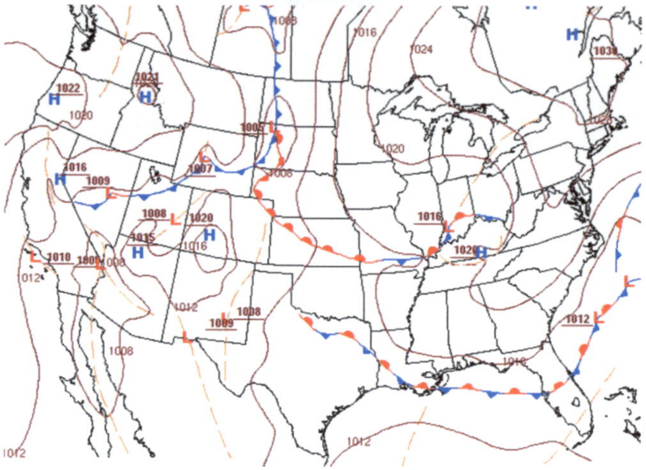

RAIN DISTRIBUTION (Inches)

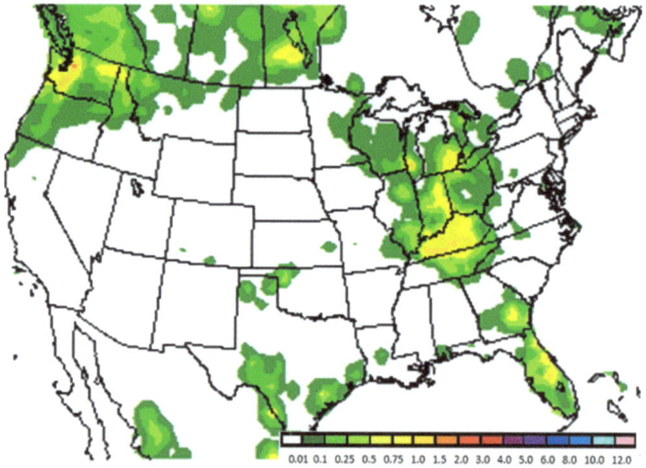

MAXIMUM TEMPERATURE MINIMUM TEMPERATURE

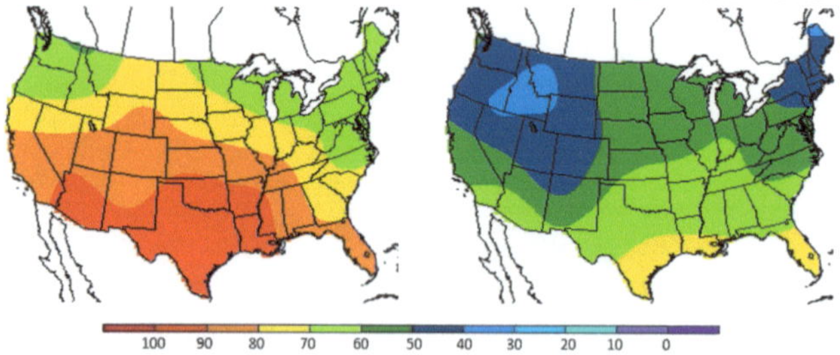

SEPTEMBER 7 SUNDAY

ISOBARIC

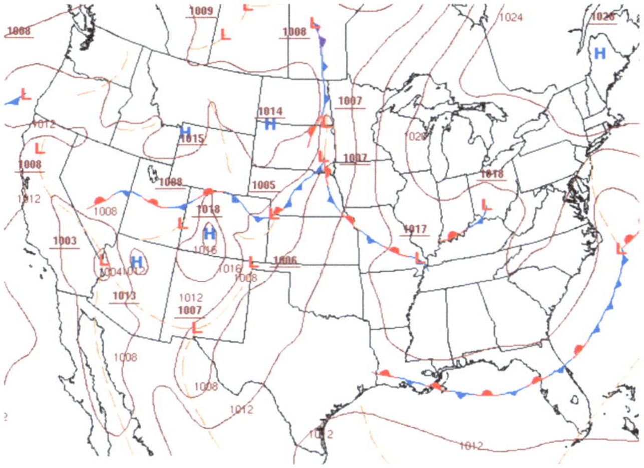

RAIN DISTRIBUTION (Inches)

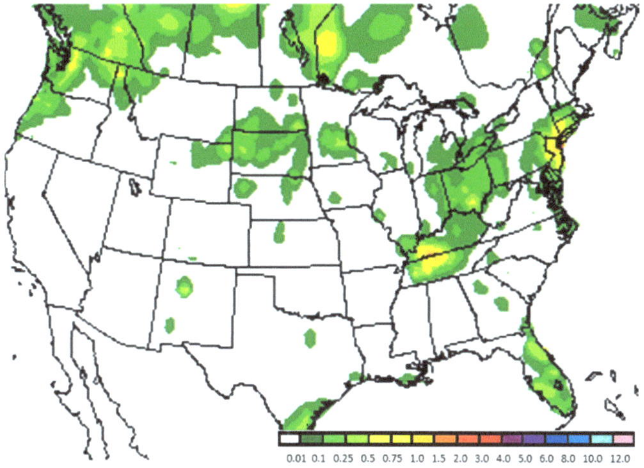

MAXIMUM TEMPERATURE

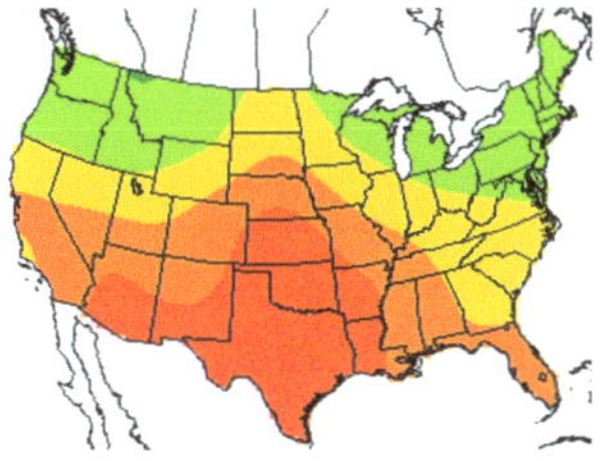

MINIMUM TEMPERATURE

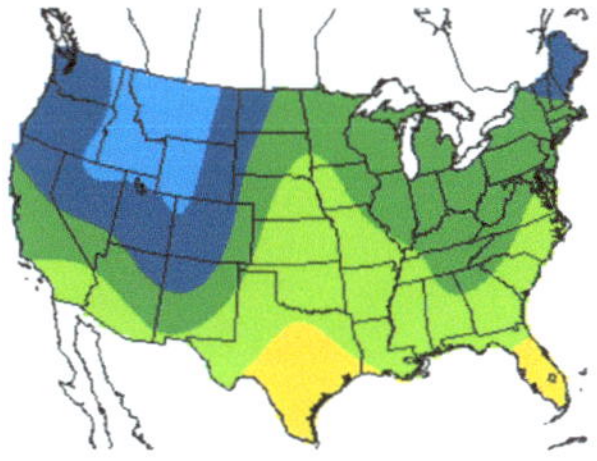

09/07/2025

Rain activity contracts to the north and west of the Pacific Northwest.

Wet fronts meet in the central Great Plains bringing outbreaks of showers mostly in the North Great Plains.

Low pressure maintains rain activity in the Inland Midwest to the Appalachians, while clearing to mostly dry over the Great Lakes.

Isolated showers persist in the Northeast, while outbreaks of heavy rain develop over the Megalopolis to the Atlantic coastal fringes of the Coastal Southeast.

Very hot and shower in Florida over the next few days.

Mostly dry elsewhere.

Temperatures settling back closer to normal in the Deep South.

Chance of precipitation in Alaska.

09/08/2025

Rain returns to the Pacific Northwest and Inland Northwest, as well as extending through the northern half of California and into the Great Basin.

Some showers linger around the southern parts of the Southwest along the New Mexico-Texas border.

Showers also along the eastern coastal fringes of Texas as well as across Florida.

Rain in the North Great Plains to the northern Great Lakes, while clearing throughout the Inland Midwest.

Heavy rain in the Megalopolis with showers into the northern Appalachians.

Mostly dry elsewhere.

Cooler in Indiana.

Very hot in the Southwest and southern California.

SEPTEMBER 8 MONDAY

ISOBARIC

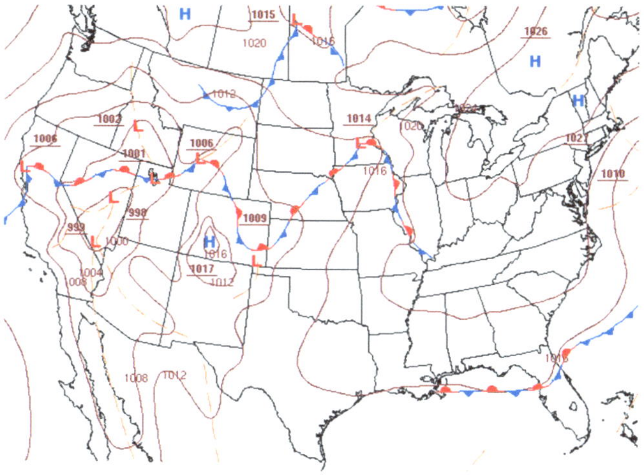

RAIN DISTRIBUTION (Inches)

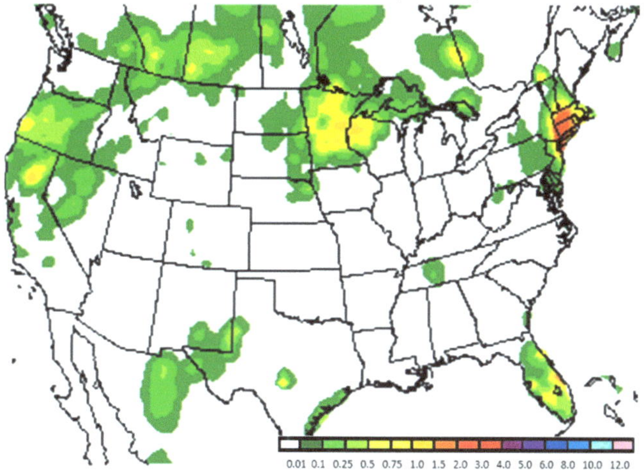

MAXIMUM TEMPERATURE

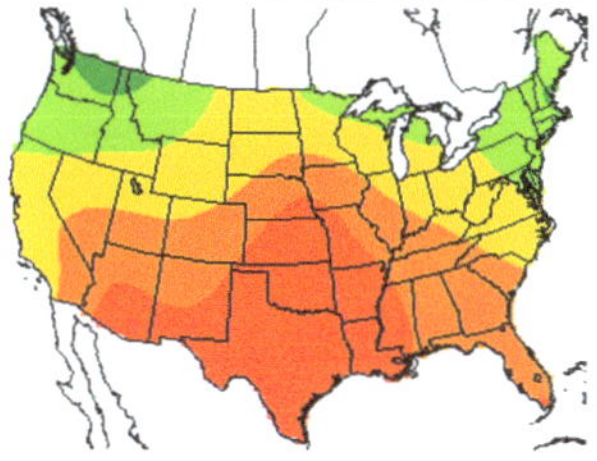

MINIMUM TEMPERATURE

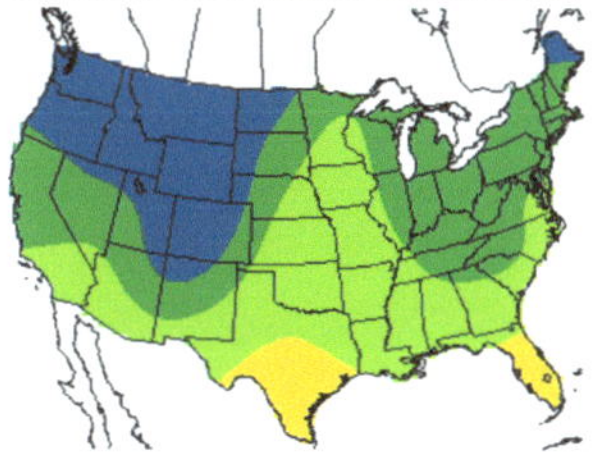

SEPTEMBER 9 TUESDAY

ISOBARIC

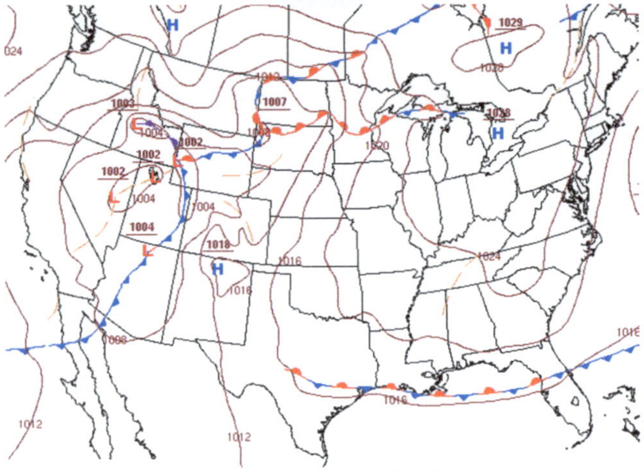

RAIN DISTRIBUTION (Inches)

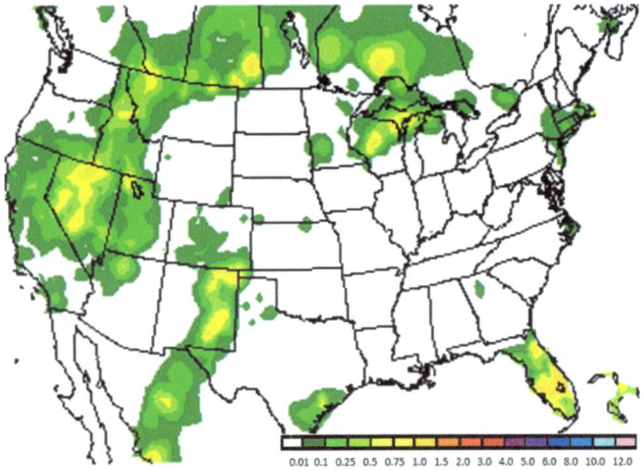

MAXIMUM TEMPERATURE

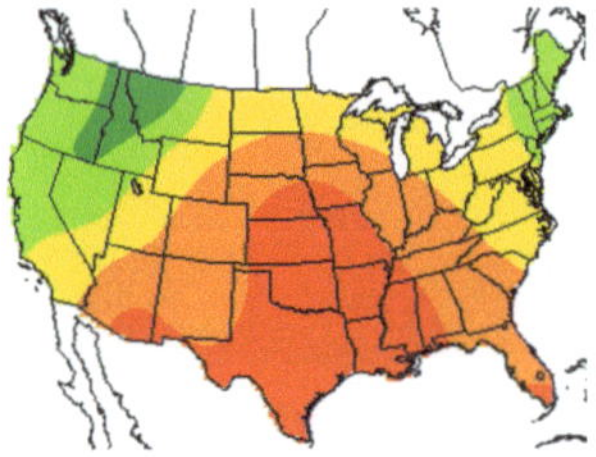

MINIMUM TEMPERATURE

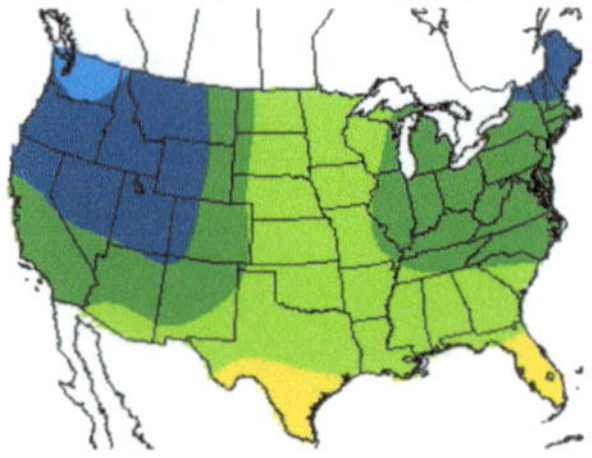

09/09/2025

Wet fronts associated with a low-pressure complex bring widespread rain from California, up through the Great Basin and into the Inland Northwest.

A band of rain prevails up through the center of the South West to the southern foothills of the Rocky Mountains.

Occasional showers around the Great Lakes.

A stationary front through the Gulf of Mexico brings some showers on the eastern coastal fringes of Texas and rain throughout Florida.

Some showers may linger around the Megalopolis.

Mostly dry elsewhere.

09/10/2025

A wet front flow up through the west from the Southwest, along the Rocky Mountains and onto the North Great Plains, bringing widespread rain to most regions west of the front, except in the Pacific Northwest.

Some showers also in the border region of New Mexico and Texas.

A slow moving front slips further south into the Gulf of Mexico, bringing further rain across Florida as well as through the southern half of Texas.

A large high prevails over the eastern half of the country with only some showers over the north of the Great Lakes.

Mostly dry elsewhere.

It may be very hot in California over the next couple of days.

SEPTEMBER 10 WEDNESDAY

ISOBARIC

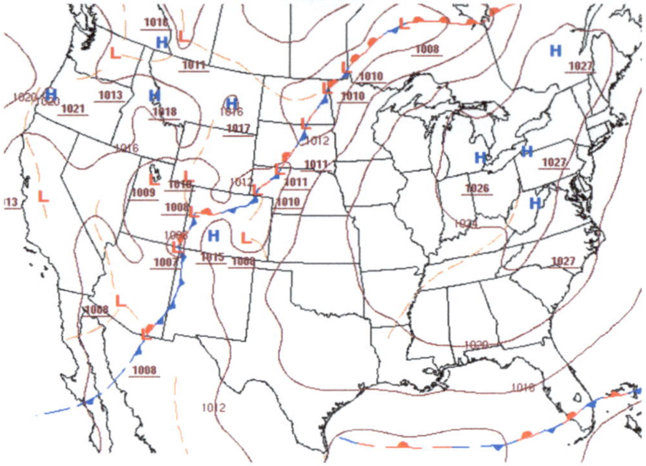

RAIN DISTRIBUTION (Inches)

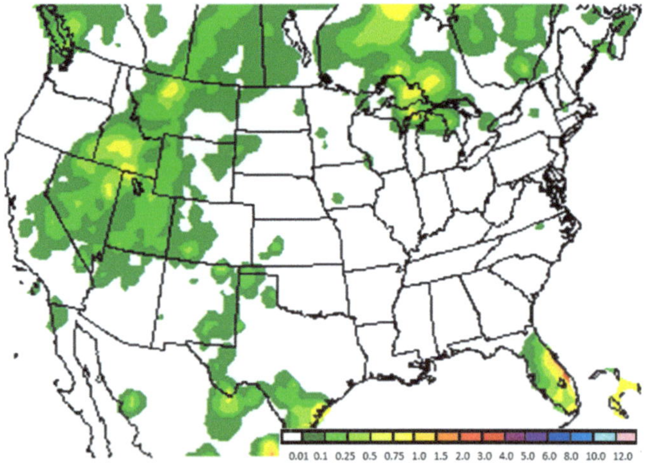

MAXIMUM TEMPERATURE

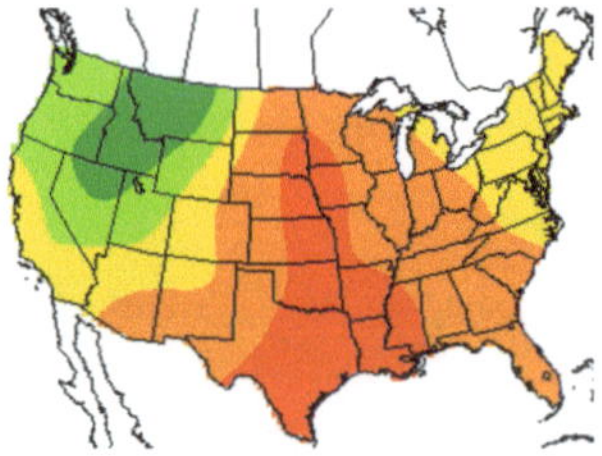

MINIMUM TEMPERATURE

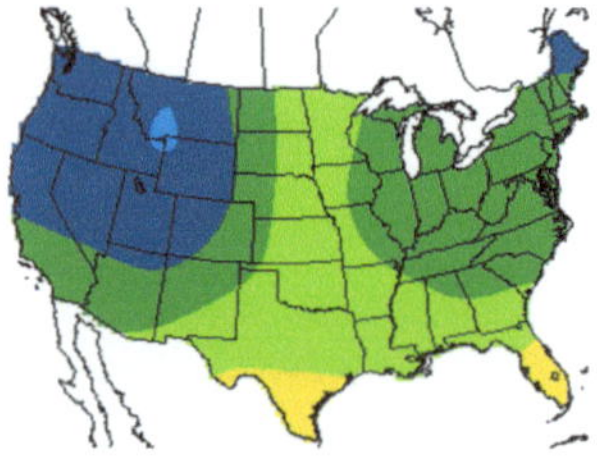

SEPTEMBER **11** THURSDAY

ISOBARIC

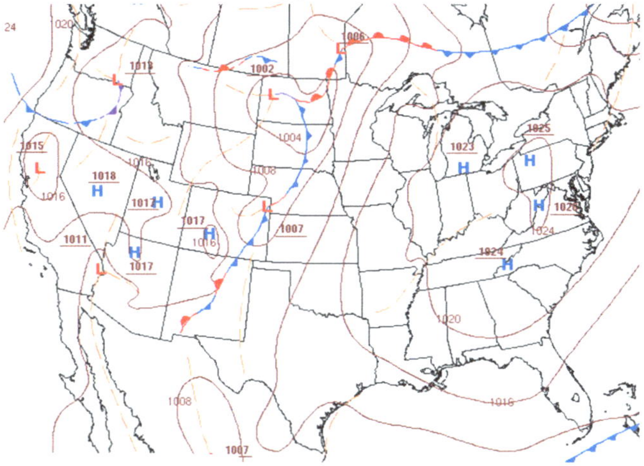

RAIN DISTRIBUTION (Inches)

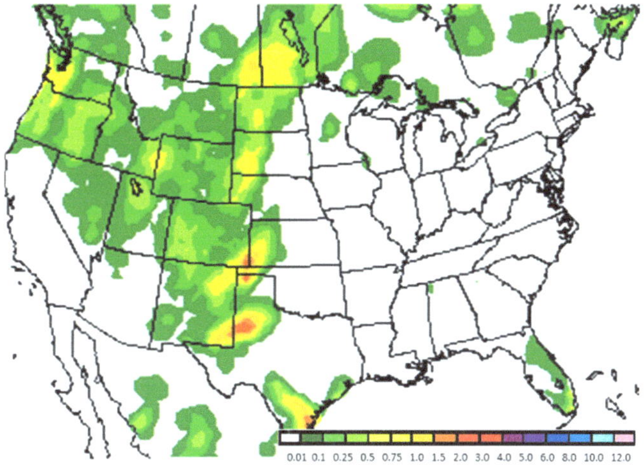

MAXIMUM TEMPERATURE

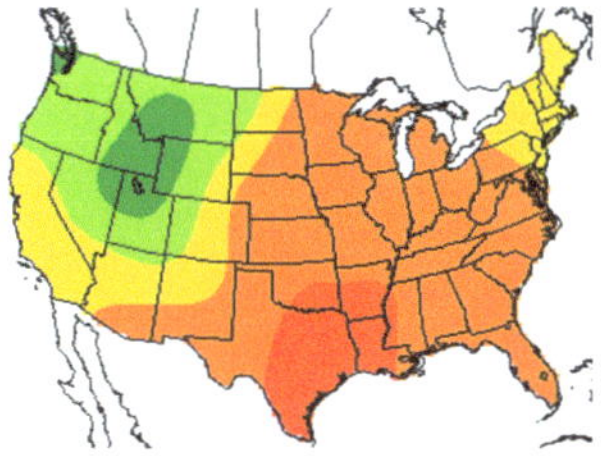

MINIMUM TEMPERATURE

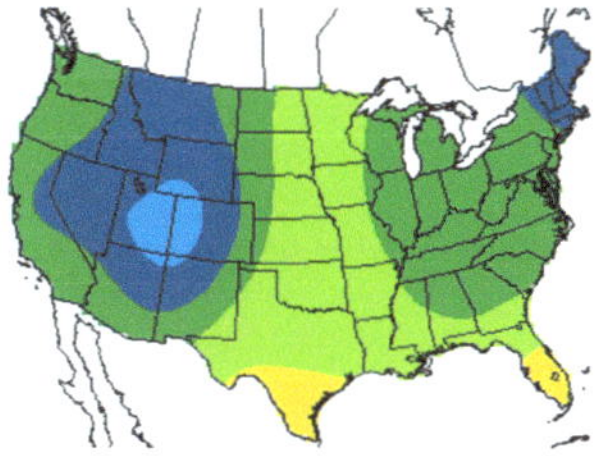

09/11/2025

A wet front and a series of low-pressure troughs cross the west of the country bringing widespread rain to almost all areas west of the front, except in California and Arizona, which may be mostly dry and very hot.

High pressure over the eastern half of the country brings mostly dry conditions through all areas east of the Great Plains.

Some rain in the southern half of Texas and brief showers in Florida.

Chance of isolated showers in the far north of the Northeast and over the northern border areas of the Great Lakes.

09/12/2025

Precipitation spreads eastward with some outbreaks of heavy rain over the Great Plains and into the west of the Inland Midwest regions and into the west of Texas.

Rain eases to isolated showers along the Pacific Northwest coasts but remains persistent in the Great Basin and through the Rocky Mountains.

Some isolated showers likely in the Northeast to the Great Lakes, as well as the eastern coasts of Florida.

Mostly dry elsewhere.

Chance of a hurricane forming in the Gulf of Mexico near Texas, with outbreaks of heavy rain in the coastal regions.

Unusually hot in Montana, as well as down through the intermountain regions of the West, including in California, over the next couple of days.

SEPTEMBER **12** FRIDAY

ISOBARIC

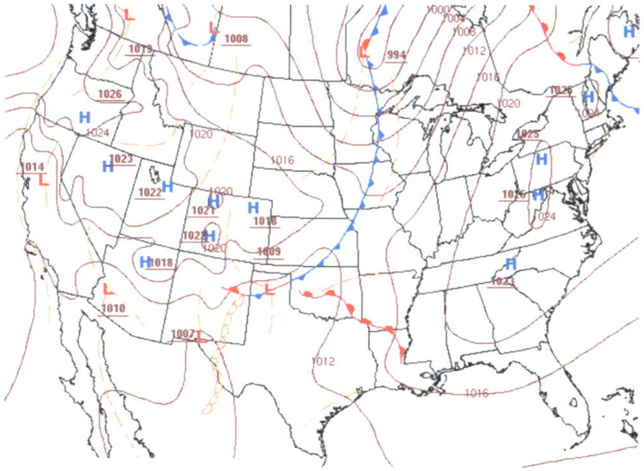

RAIN DISTRIBUTION (Inches)

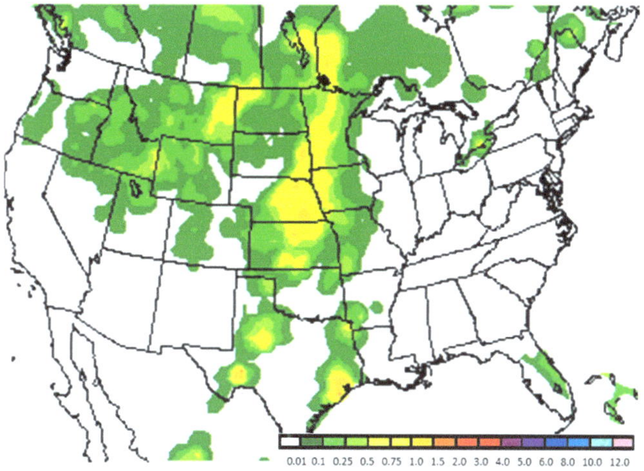

MAXIMUM TEMPERATURE

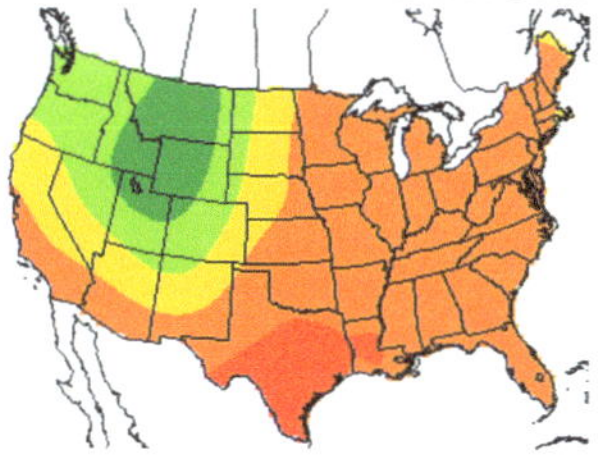

MINIMUM TEMPERATURE

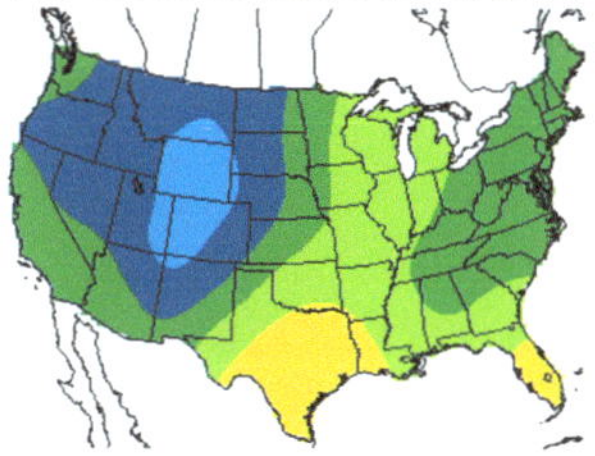

SEPTEMBER 13 SATURDAY

ISOBARIC

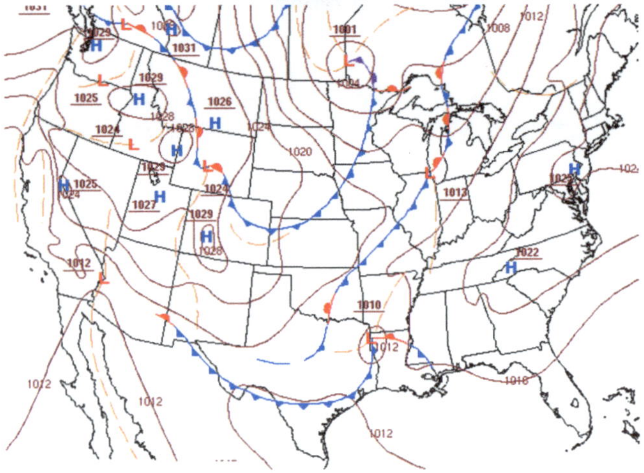

RAIN DISTRIBUTION (Inches)

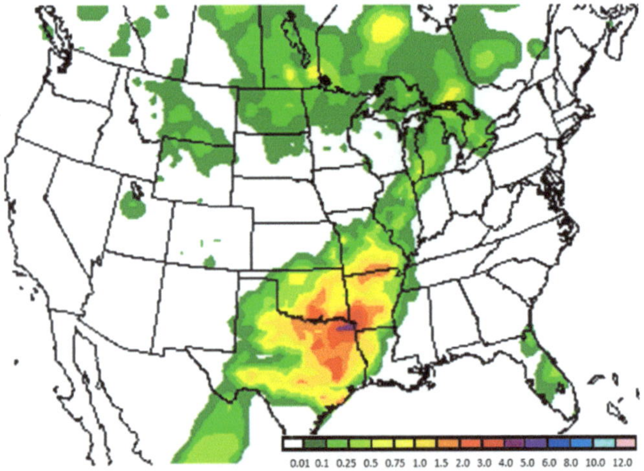

MAXIMUM TEMPERATURE / MINIMUM TEMPERATURE

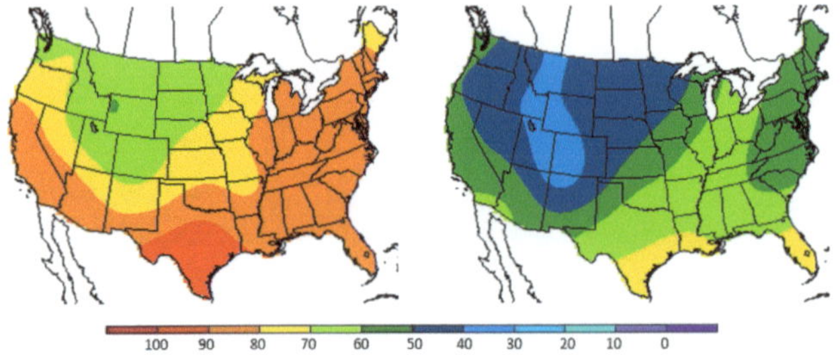

09/13/2025

A wet front continues to move east across the country, with rain clearing west of the Rocky Mountains except for some lingering around the Great North Plains.

Widespread rain crosses the country in a band from the Great Lakes, through the Inland Midwest and down through the South Great Plains into northern Texas.

Heavy rain in the Ozarks, Arkansas, Oklahoma and the northern half of Texas.

Some showers in Florida.

Mostly dry elsewhere.

Late season heat may persist in the Dakotas and it will be hot in Florida.

There is a chance of a hurricane rapidly strengthening near the south of Texas and affecting the state.

09/14/2025

The intense rain associated with the passage of a cold front continues to move eastwards and nears the Atlantic coast.

Widespread rain prevails from the Inland Midwest and Great Lakes, down through the west of the Appalachians and into the Deep South.

Some isolated showers lingering around the Rocky Mountains and Florida, and increasing threats of rain developing in the Atlantic coastal states.

Heavy downpours in Michigan to Illinois and Indiana.

A possible hurricane near Texas may produce heavy rain into the southern half of the State, resulting in wind damage and localized flooding in some places.

Mostly dry elsewhere.

SEPTEMBER 14 SUNDAY

ISOBARIC

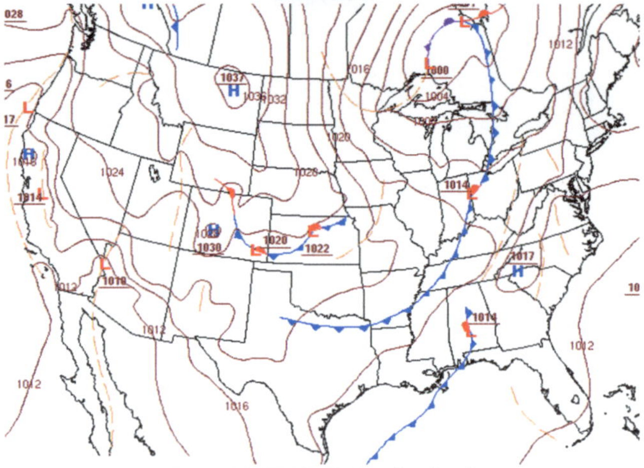

RAIN DISTRIBUTION (Inches)

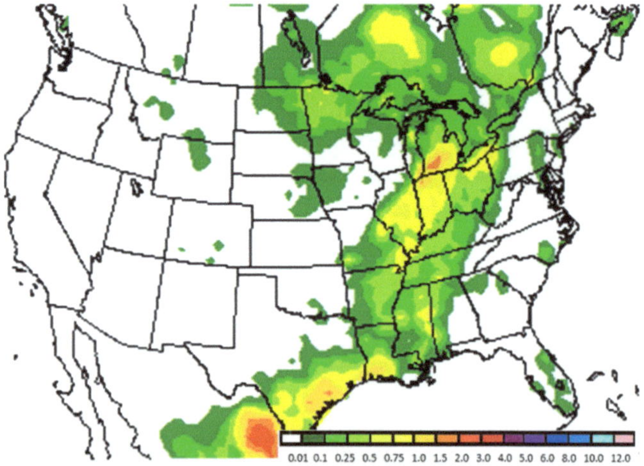

MAXIMUM TEMPERATURE / MINIMUM TEMPERATURE

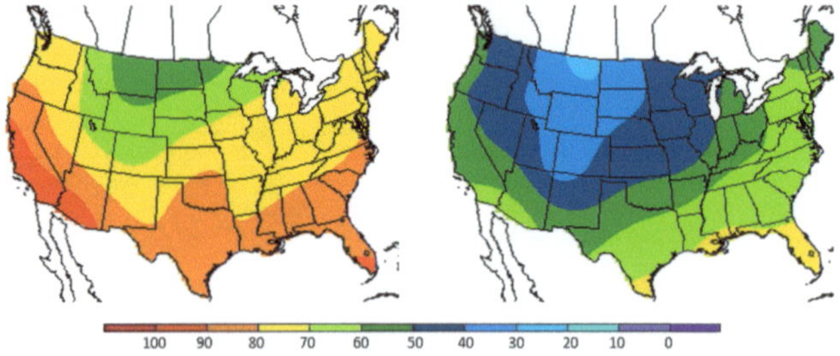

SEPTEMBER 15 MONDAY

ISOBARIC

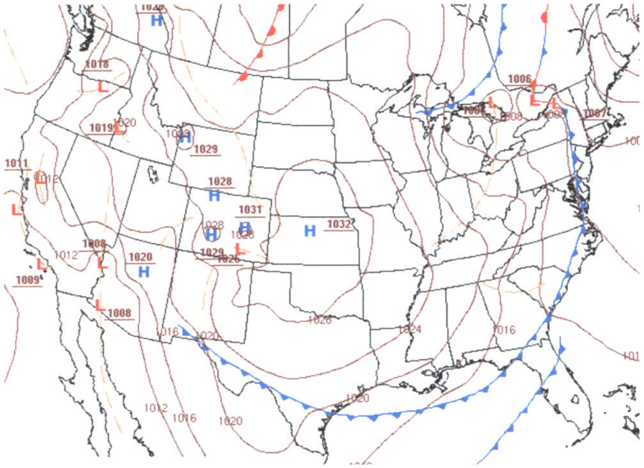

RAIN DISTRIBUTION (Inches)

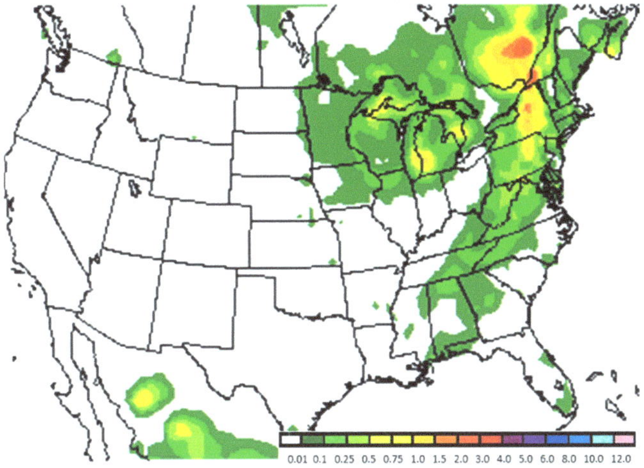

MAXIMUM TEMPERATURE

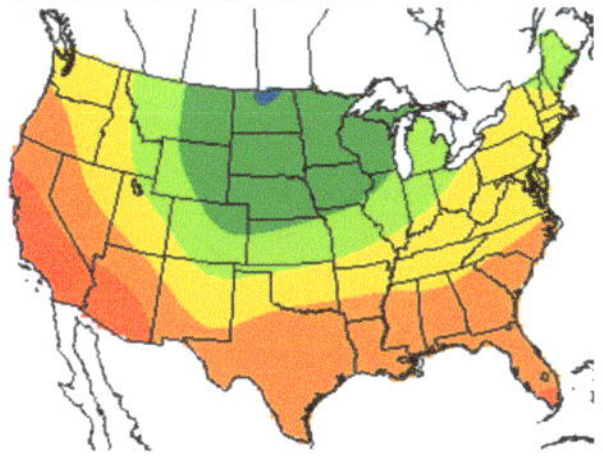

MINIMUM TEMPERATURE

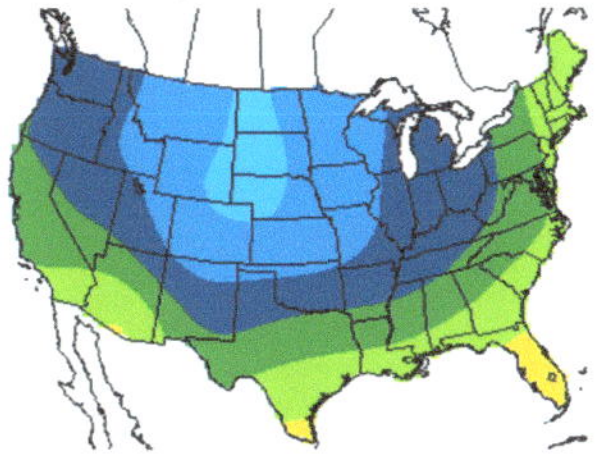

09/15/2025

High pressure builds over the midsection and the Rocky Mountains bringing mostly dry conditions throughout from the Pacific coast to the Great Plains.

Rain contracts northwards, clearing the Deep South to the Ozarks.

A front moves east across the Atlantic coast, with rain from the Appalachians up to the Megalopolis and into the Northeast, as well as westward throughout the Great Lakes and the Upper Midwest.

Mostly dry elsewhere, including along the Atlantic coasts of the Coastal Southeast.

It may be unusually hot in Kentucky and West Virginia over the next couple of days and unusually warm in Alaska.

The recent heavy rain in the Deep South, possibly associated with a hurricane passing through the Gulf of Mexico, may leave some localized flooding and damage to infrastructure, particularly in Louisiana.

SEPTEMBER 16 TUESDAY

09/16/2025

An intense Atlantic depression crosses into the Northeast and extends storms with heavy rain through the region and into the Great Lakes to the Megalopolis over the next couple of days, and showers down into the Coastal Southeast.

Some showers persist in Florida.

Mostly dry elsewhere.

Very warm on the Big Island in Hawaii.

ISOBARIC

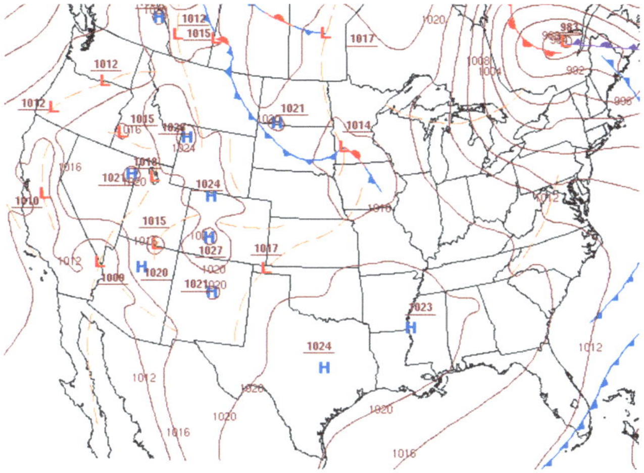

RAIN DISTRIBUTION (Inches)

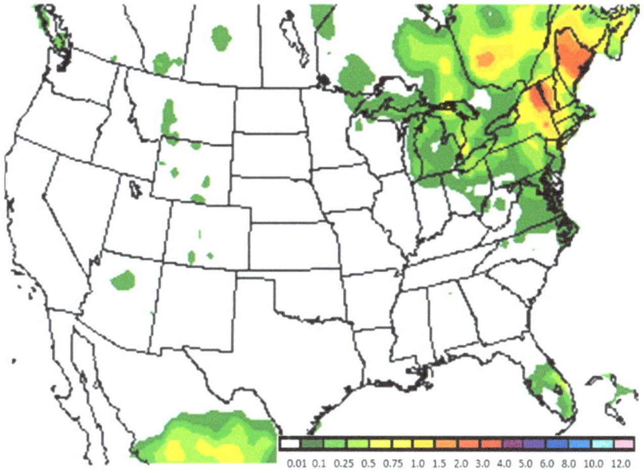

MAXIMUM TEMPERATURE

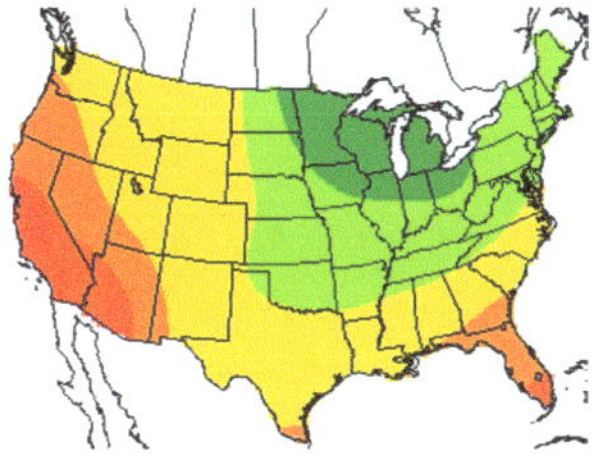

MINIMUM TEMPERATURE

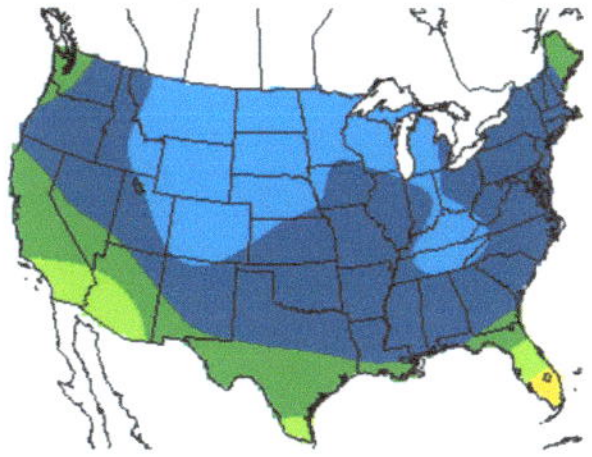

ISOBARIC

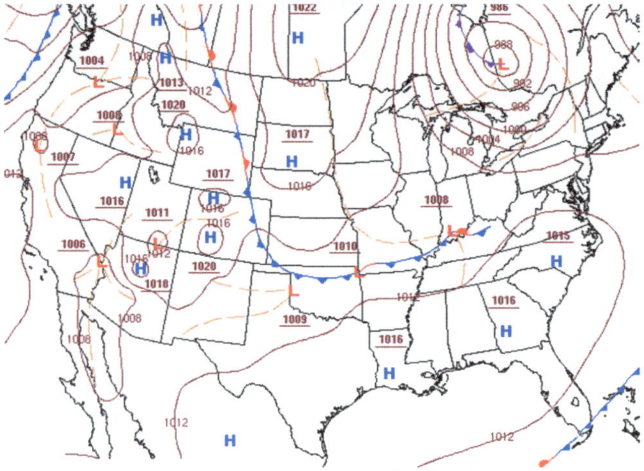

RAIN DISTRIBUTION (Inches)

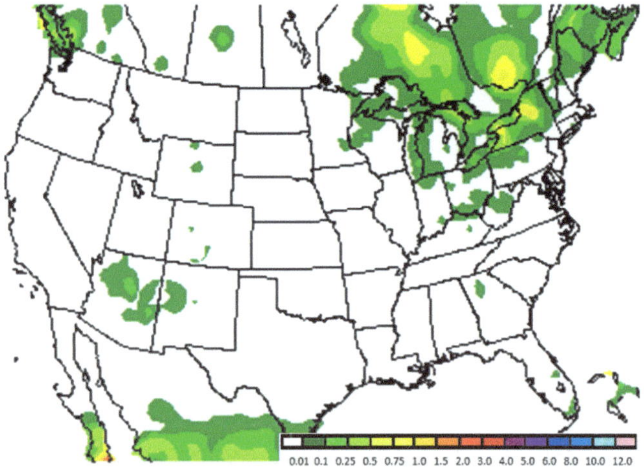

MAXIMUM TEMPERATURE MINIMUM TEMPERATURE

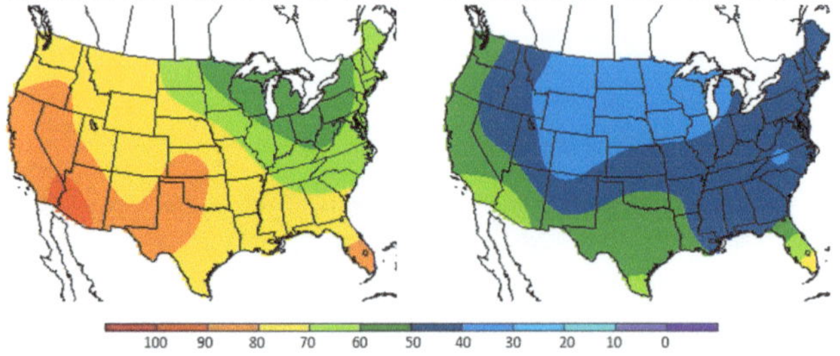

09/17/2025

A deep depression brings storms and heavy rain into the Northeast to the Great Lakes.

Some showers in the Inland Midwest, as well as around the Southwest.

Mostly dry elsewhere.

Despite the storms, it may be unusually hot in the Northeast over the next couple of days, as well as very hot in New York.

A cooler flow brings a break in the heatwave temperatures of recent in the Deep South regions, including in Alabama, Tennessee, and the Carolinas.

09/18/2025

The depression over the northeast of the country eases with shower activity decreasing in the Great Lakes to the Northeast and around the Coastal Southeast.

In the west, another low nears the Pacific Northwest returning scattered shower activity into the Pacific Northwest to the Inland Northwest regions.

Isolated showers in the Southwest and some rain around the southern tip of Texas.

Mostly dry elsewhere.

Extreme range of temperatures possible in the Northeast with very hot during maximums and unusually cool minimums.

Heavy showers likely in southern Alaska over the next couple of days.

ISOBARIC

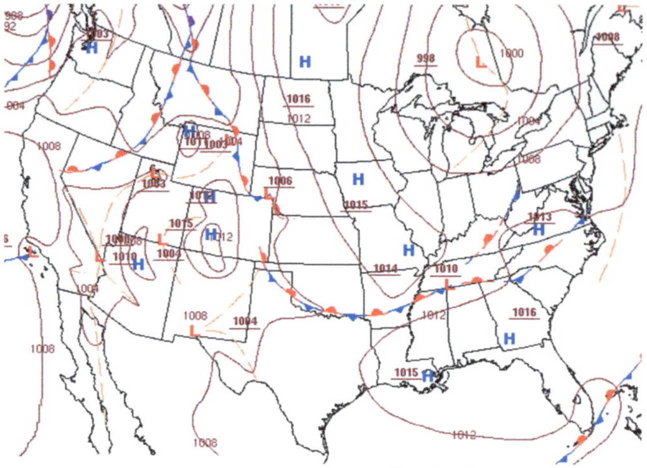

RAIN DISTRIBUTION (Inches)

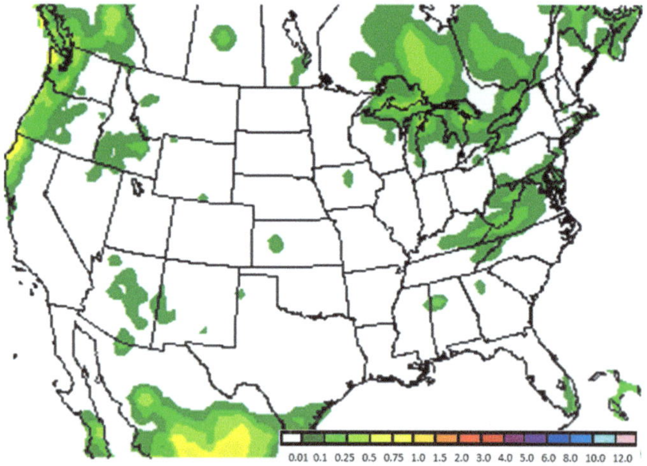

MAXIMUM TEMPERATURE MINIMUM TEMPERATURE

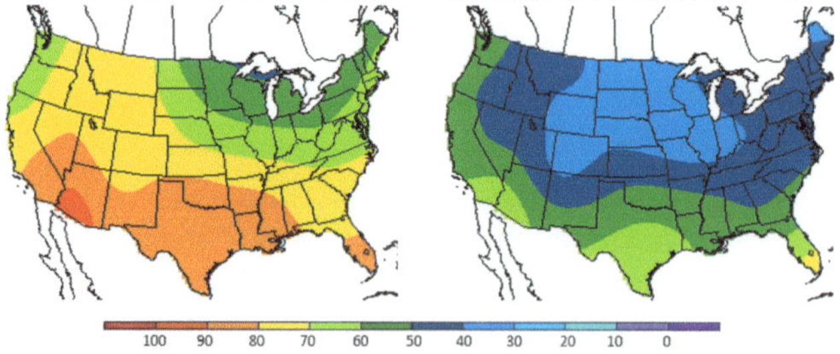

SEPTEMBER **19** FRIDAY

ISOBARIC

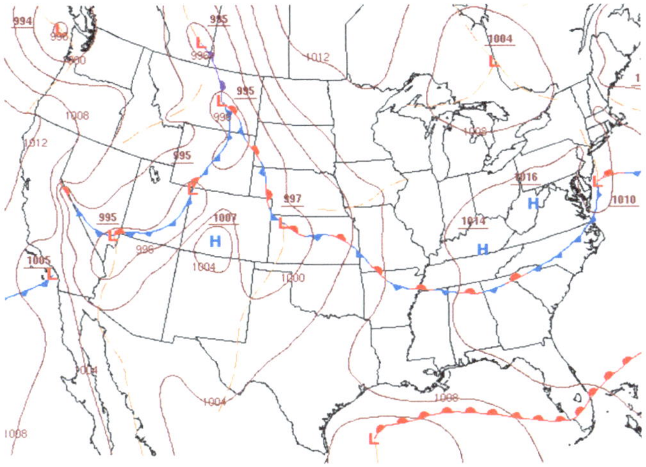

RAIN DISTRIBUTION (Inches)

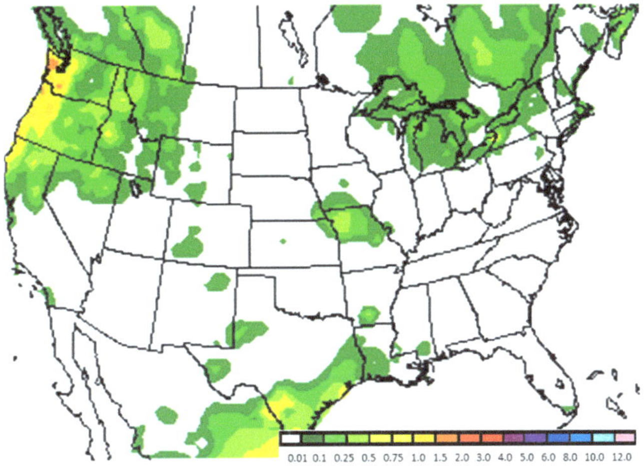

MAXIMUM TEMPERATURE | MINIMUM TEMPERATURE

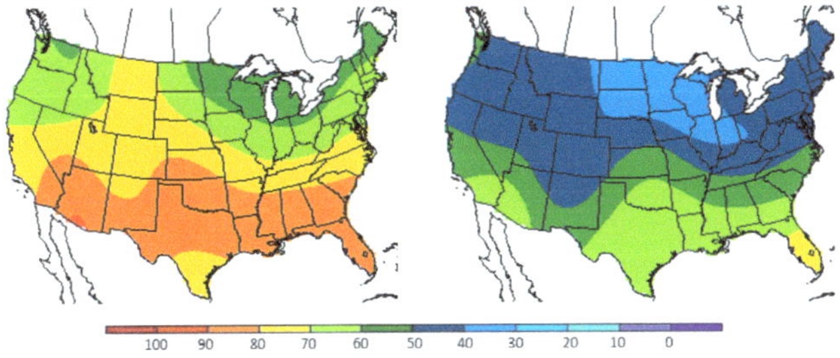

09/19/2025

A Pacific depression moves onto the northwest of the country, bringing storms and rain to the Pacific and Inland Northwest, northern California and the northwest the Rocky Mountains. Heavy rain along the coast from Washington to northern California.

Widespread shower activity persists over the Great Lakes, with isolated showers up into the Northeast.

There may be outbreaks of rain in the Ozarks.

Heavy rain likely in the south and east of Texas.

Mostly dry elsewhere.

Chance of a tropical storm making landfall near the Carolinas in the next couple of days, bringing stormy conditions into the State.

Unusually warm day in Alaska, with chance of heavy rain in the south.

09/20/2025

A large depression crosses the country from the Pacific coast to the mid-section bringing widespread and occasionally heavy rain from the Pacific Northwest and Great Basin across the Rocky Mountains and central Great Plains into the Midwest south of the Great Lakes.

Showers around the Great Lakes to the Northeast.

An Atlantic tropical storm makes landfall near the Carolinas producing heavy rain throughout the southeast quarter of the country from the Carolinas, through the Deep South and Florida and into the east of Texas, with chance of flash flooding in some areas.

Mostly dry elsewhere.

Unusually warm in the Carolinas to West Virginia, as well as in Washington in the northwest.

It may be unusually cooler in regions east of the Rockies, including Montana.

SEPTEMBER **20** SATURDAY

ISOBARIC

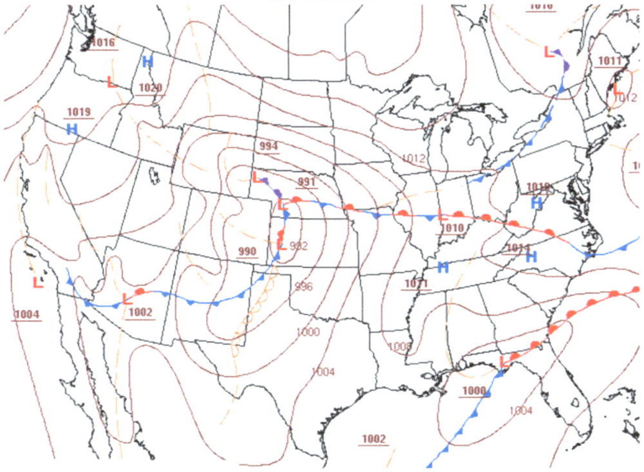

RAIN DISTRIBUTION (Inches)

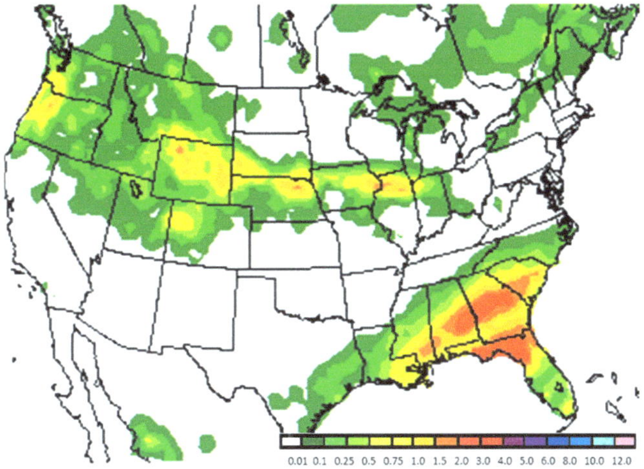

MAXIMUM TEMPERATURE | MINIMUM TEMPERATURE

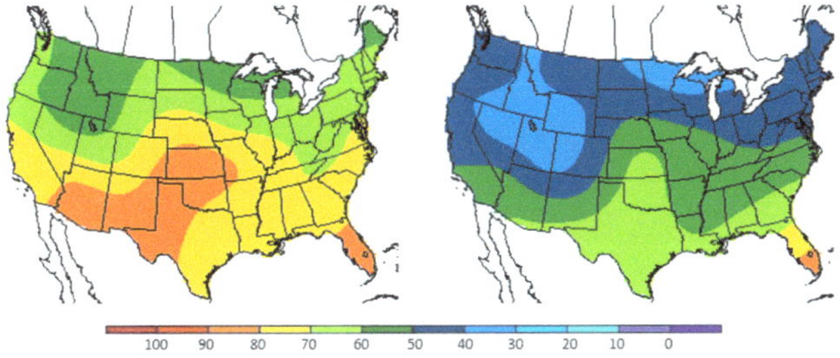

SEPTEMBER 21 SUNDAY

ISOBARIC

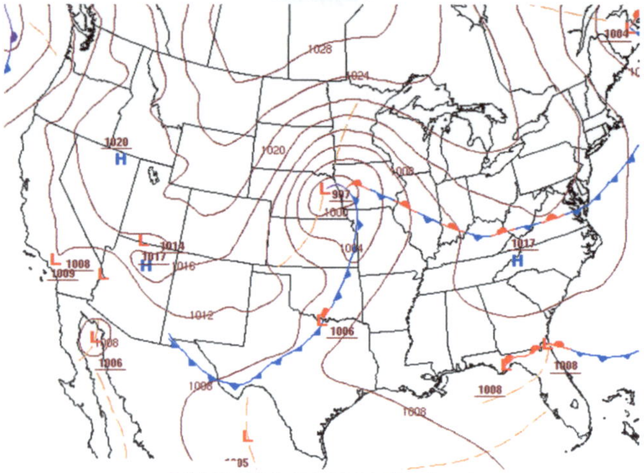

RAIN DISTRIBUTION (Inches)

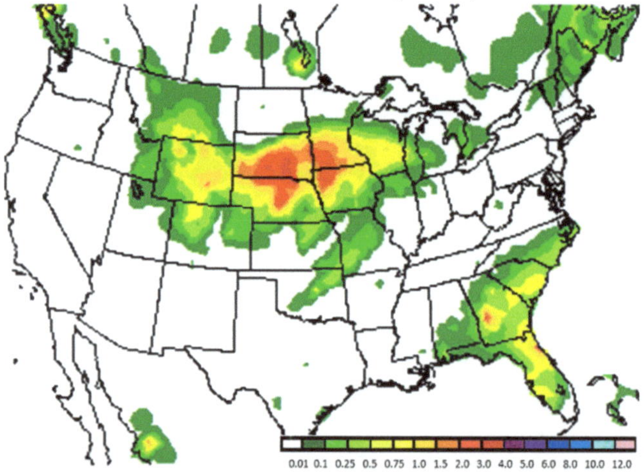

MAXIMUM TEMPERATURE

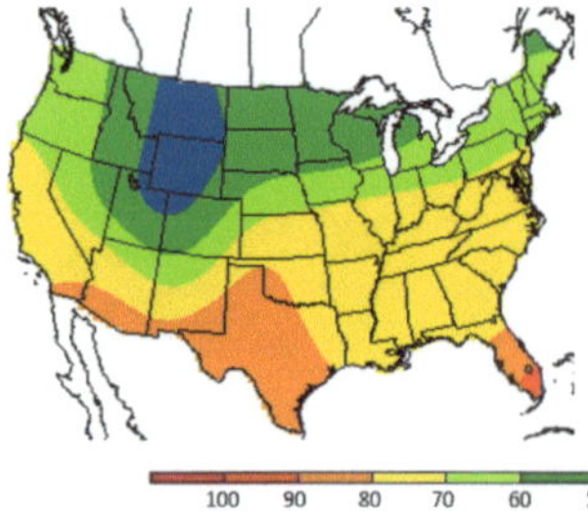

MINIMUM TEMPERATURE

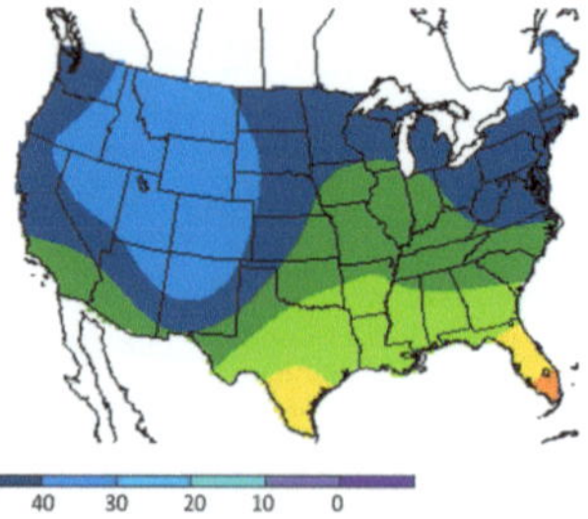

09/21/2025

A large depression moves slowly eastward bringing storms and widespread heavy rain from the Rocky Mountains, across the Great Plains and into the upper inland Midwest with showers down into the Ozarks.

Heavy rain in Kentucky and Tennessee over the next couple of days.

Another depression near Florida decays and produces widespread and heavy rain into the states from the Carolinas to Florida including in Georgia to Alabama.

Another Atlantic depression brings widespread rain into the Northeast region.

Mostly dry elsewhere.

Unusual low maximum temperatures in Texas and Kansas and Tennessee likely.

Unusually warmer in the Artic coastal fringes of Alaska.

09/22/2025

The eastward tracking depression slowly heading towards the Atlantic Ocean, centers over the Great Lakes and Inland Midwest spreading storms, heavy rain and strong winds from the North Great Plains and Great Lakes, across the Inland Midwest and over to the Coastal Southeast regions as far south as Georgia.

Another front brings rain into the Pacific Northwest.

Lingering showers in the Rocky Mountains, southern Florida and coastal fringes of Texas.

Mostly dry elsewhere.

A tropical depression in the Gulf of Mexico possibly upgraded to a Tropical storm.

SEPTEMBER 22 MONDAY

ISOBARIC

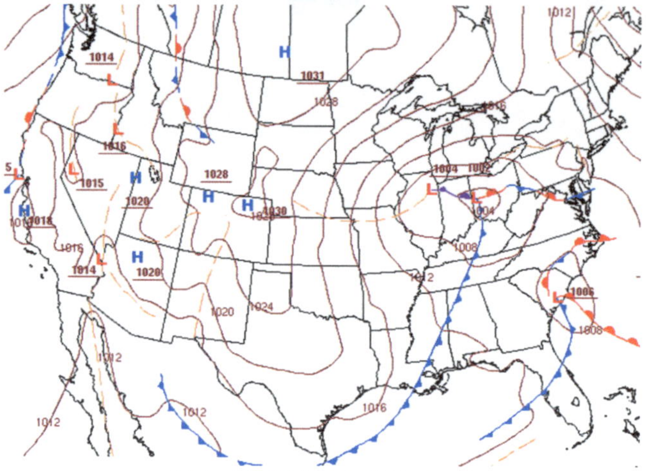

RAIN DISTRIBUTION (Inches)

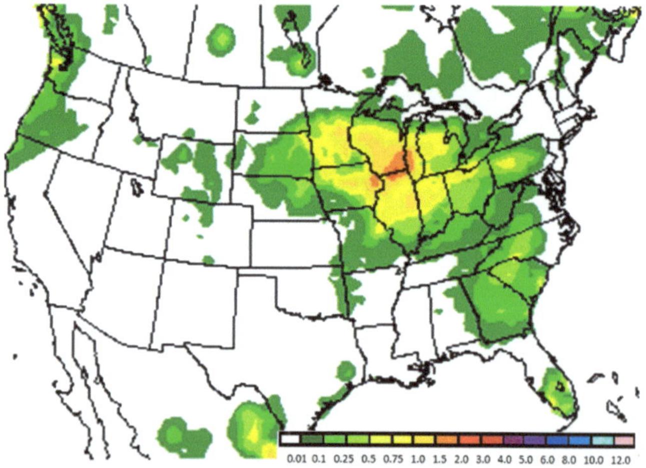

MAXIMUM TEMPERATURE

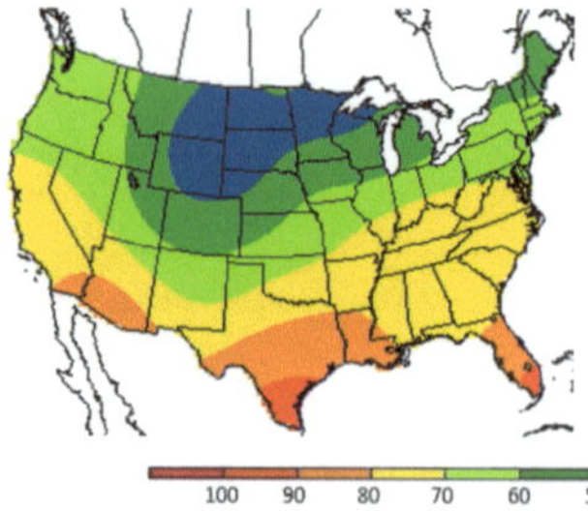

MINIMUM TEMPERATURE

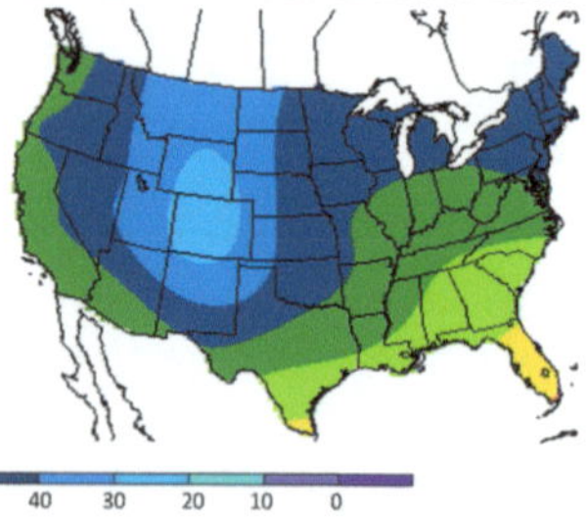

SEPTEMBER 23 TUESDAY

ISOBARIC

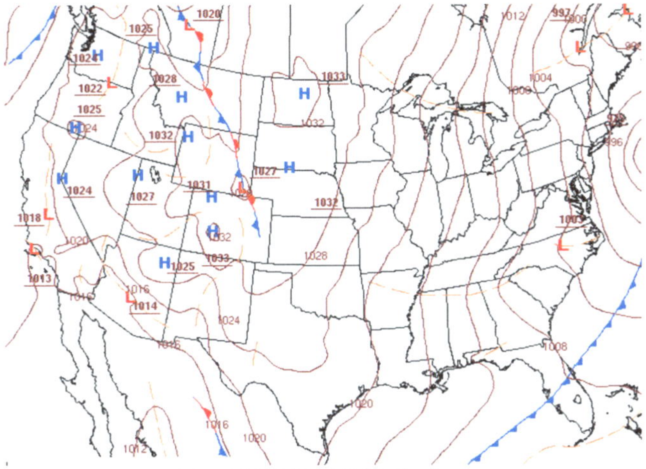

RAIN DISTRIBUTION (Inches)

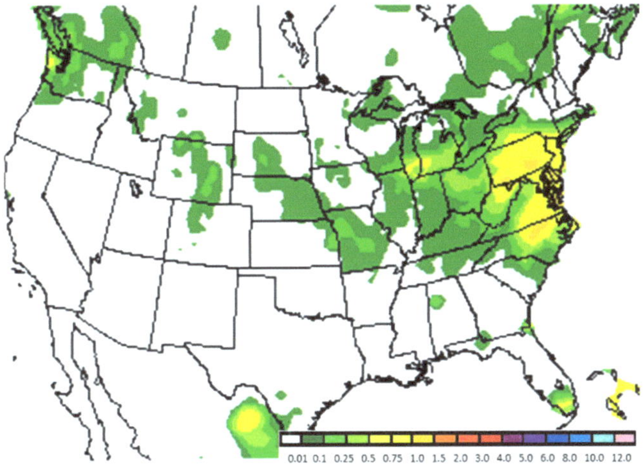

MAXIMUM TEMPERATURE

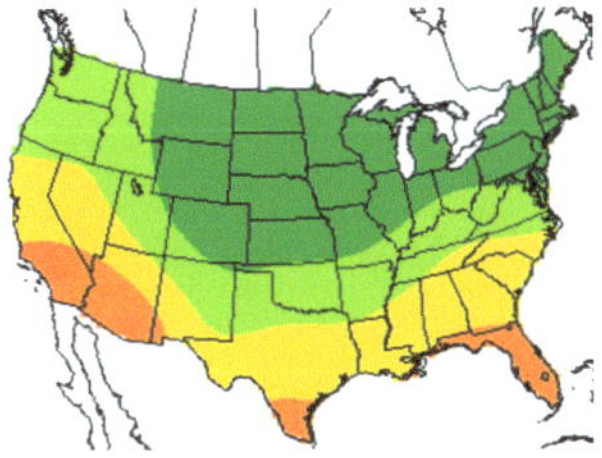

MINIMUM TEMPERATURE

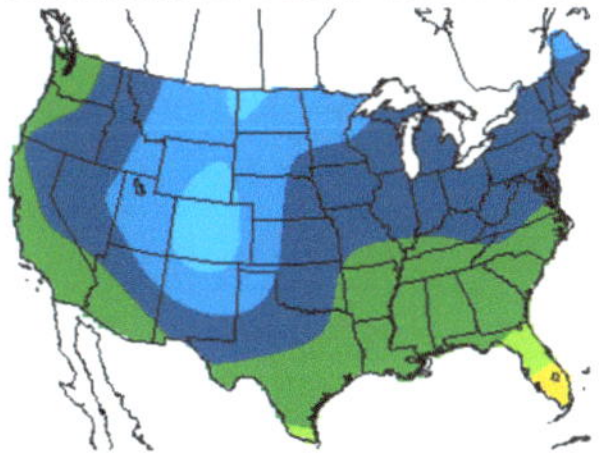

09/23/2025

The eastward tracking depression moves offshore into the Atlantic Ocean in the vicinity of the Megalopolis, shifting the widespread storm activity into the eastern quarter of the country, from the Great Lakes and Inland Midwest to the Atlantic coasts, and as far south as the Carolinas.

Showers linger in the Rocky Mountains to the central parts of the Great Plains and into Montana.

Showers in the far north of the Pacific Northwest with isolated showers sprinkling into the northern Rockies, and some showers in the south of Florida.

Mostly dry elsewhere.

A tropical storm that has formed near the Texan coast recently may be upgraded briefly to hurricane status as it threatens landfall near the State, with strong winds and heavy rain prevailing across the southeast areas.

However, this storm may be downgraded to a tropical storm as quickly as it was upgraded.

09/24/2025

Storm conditions continue to ease across the eastern half of the country, as the depression continues to drift deeper eastward into the Atlantic Ocean.

Widespread outbreaks of rain and strong winds prevail from the Northeast to the Deep South and west into the Inland Midwest and as far east as the Ozarks.

Chance that strong winds in the Carolinas may develop tornadoes as the storm lines move eastward offshore.

Some showers in the far north of the Pacific Northwest. Isolated showers in the northern Rocky Mountains.

Mostly dry elsewhere due to strong high-pressure.

Chance of a cold snap with scattered frosts in the North Great Plains, particularly in the Dakotas at first, then spreading into Iowa and Wisconsin by tomorrow.

SEPTEMBER 24 WEDNESDAY

ISOBARIC

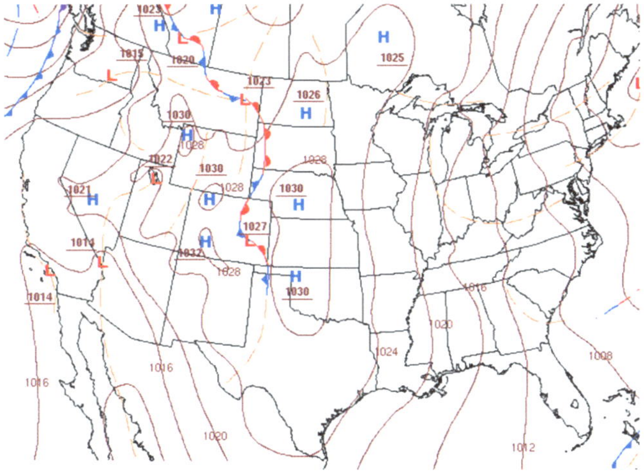

RAIN DISTRIBUTION (Inches)

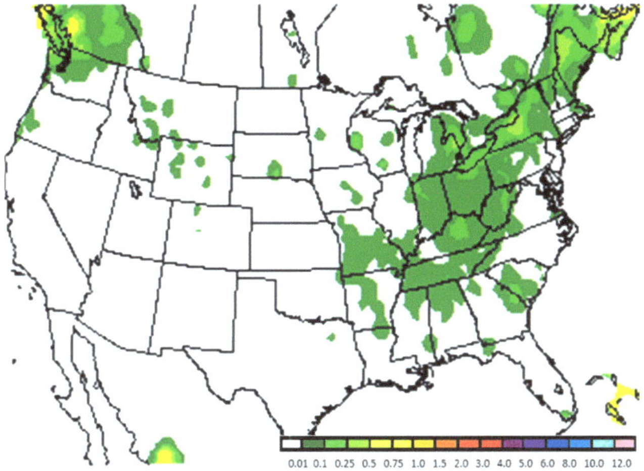

MAXIMUM TEMPERATURE

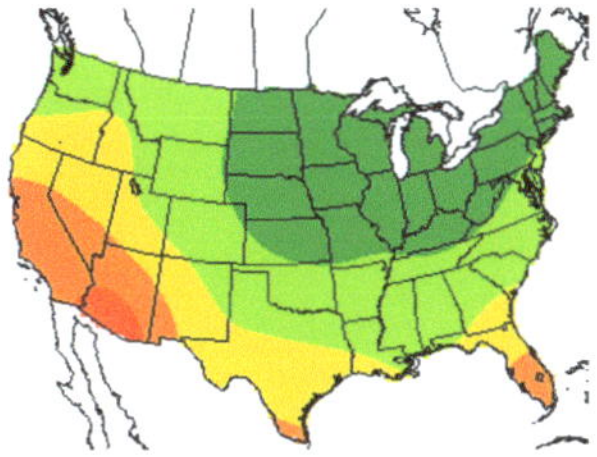

MINIMUM TEMPERATURE

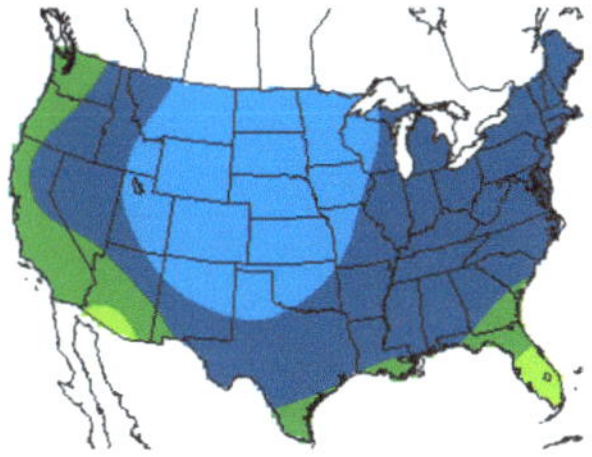

SEPTEMBER 25 THURSDAY

ISOBARIC

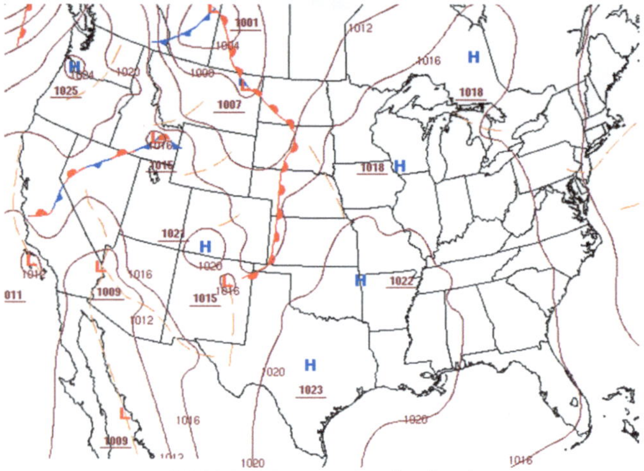

RAIN DISTRIBUTION (Inches)

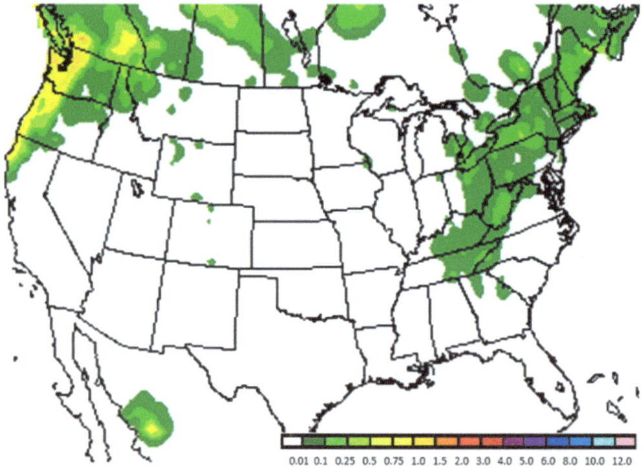

MAXIMUM TEMPERATURE / MINIMUM TEMPERATURE

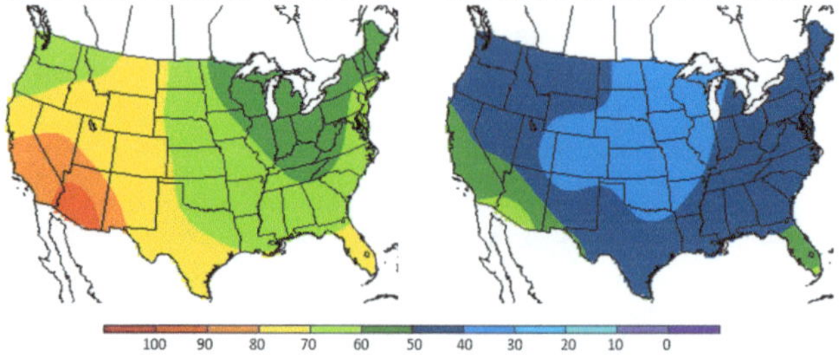

09/25/2025

A depression to the northwest of the country brings rain into the Pacific Northwest and showers into the Inland Northwest and northern California coasts.

Low pressure over the northeast of the country continues to bring widespread rain through the Northeast region and down through the Appalachians, while clearing to mostly dry around the Great Lakes and Inland Midwest.

High pressure and mostly dry weather prevails elsewhere, possibly resulting in a cold snap with scattered frosts over the North Great Plains to the Upper Inland Midwest including in the Dakotas, Iowa, Wisconsin and Illinois.

09/26/2025

A cold front along the US-Canadian border brings some scattered rain activity to the border states from the Inland Midwest to the North Great Plains.

Some outbreaks of rain also around the Pacific Northwest, mostly on the coasts.

Showers develop in the Southwest mostly in Arizona.

Rain clears to the occasional showery pocket in the Northeast to the Great Lakes and isolated showers in the northern Appalachians as high pressure builds over the eastern regions.

Mostly dry elsewhere.

There may be a sharp difference in temperatures with unusually cold temperatures in Ohio, but unusually high temperatures in Kansas.

Chance of heavy precipitation in Alaska.

SEPTEMBER 26 FRIDAY

ISOBARIC

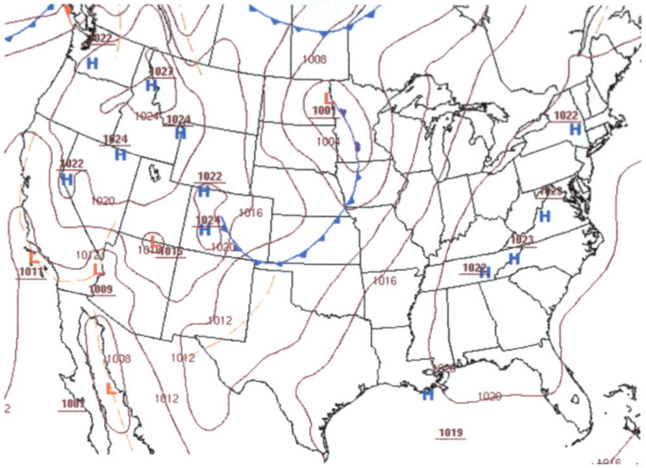

RAIN DISTRIBUTION (Inches)

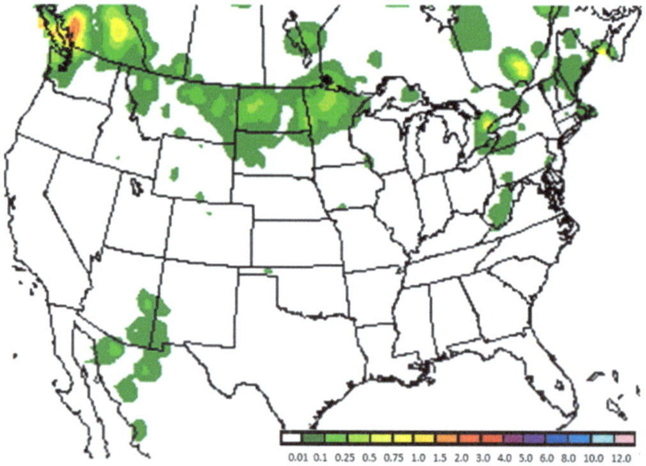

MAXIMUM TEMPERATURE / MINIMUM TEMPERATURE

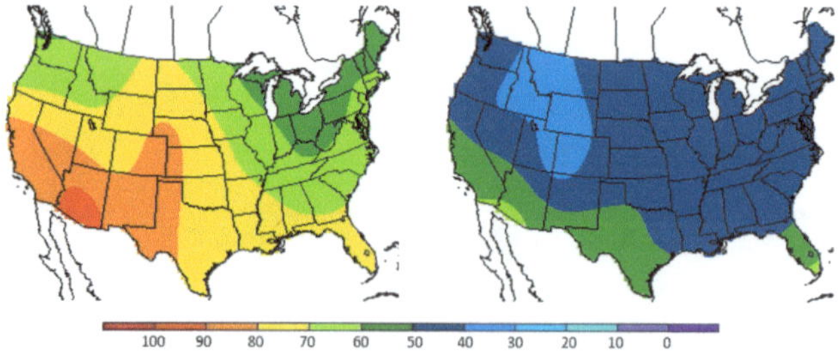

SEPTEMBER 27 SATURDAY

ISOBARIC

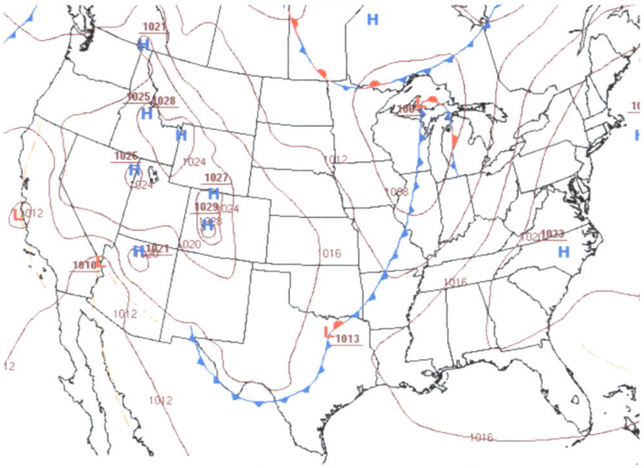

RAIN DISTRIBUTION (Inches)

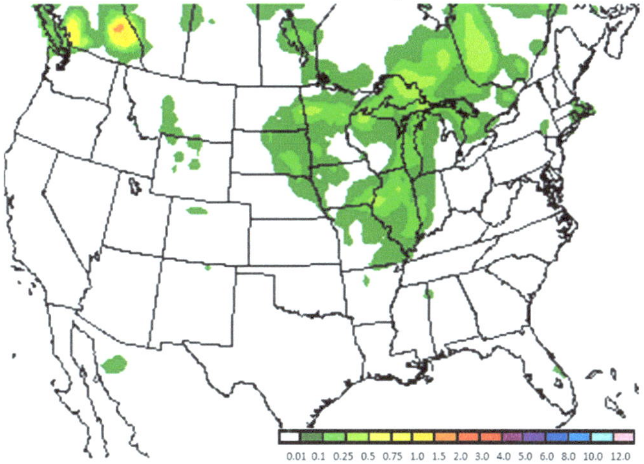

MAXIMUM TEMPERATURE

MINIMUM TEMPERATURE

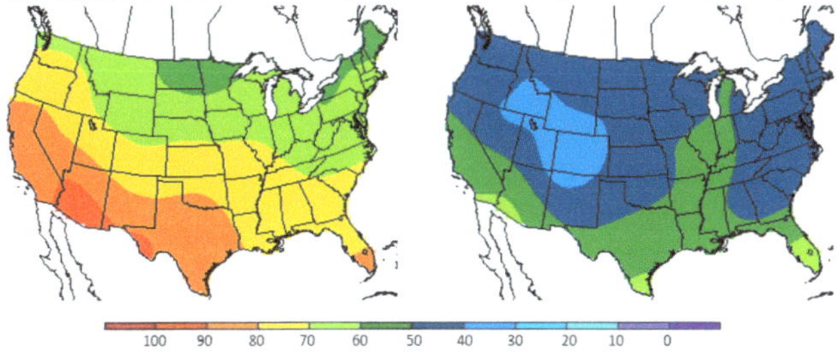

09/27/2025

A weak depression over the Great Lakes extends a cold front down through the country from the Great Lakes to Texas with high pressure prevailing on either side.

Widespread outbreaks of precipitation prevails over the Great Lakes and Inland Midwest regions, including into the eastern half of the Dakotas.

Iisolated showers in the northern Rocky Mountains.

Mostly dry elsewhere.

SEPTEMBER 28 SUNDAY

ISOBARIC

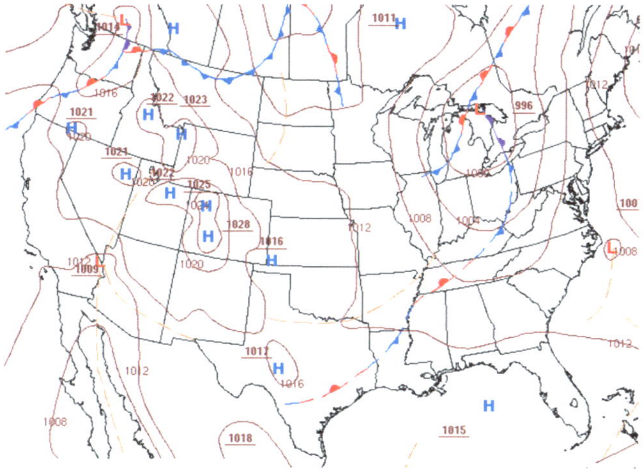

RAIN DISTRIBUTION (Inches)

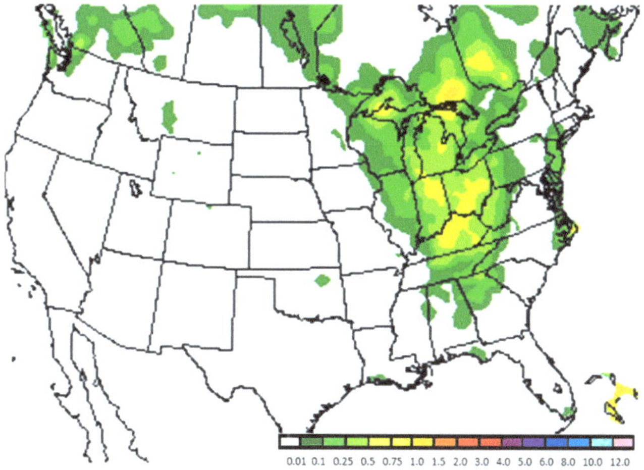

MAXIMUM TEMPERATURE

MINIMUM TEMPERATURE

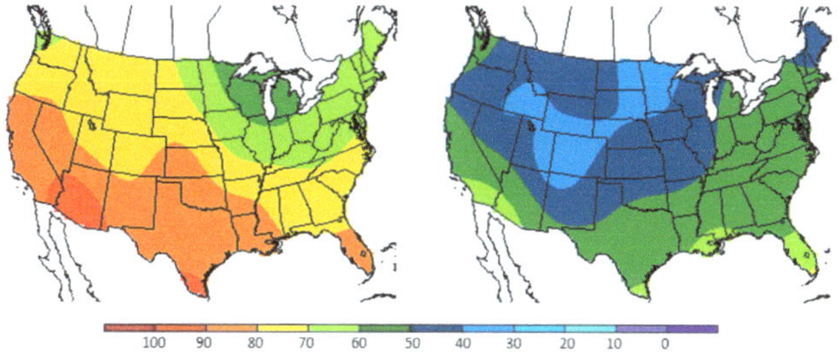

09/28/2025

Low pressure continues to build over the northeast of the country into a large depression, bringing widespread and often heavy rain once more throughout the Great Lakes, Inland Midwest, and down through the Appalachians.

Rain amounts may be particularly heavy in Minnesota over the next few days.

Showers also in the Northeast and in the coastal fringes of the Megalopolis to Chesapeake.

High pressure and mostly dry elsewhere.

Chance of some isolated showers in the coastal fringes of the Pacific Northwest.

Very warm in Illinois and over much of the Midwest.

Chance of a tropical depression forming near the northeast of Florida.

Possible heavy precipitation in Alaska.

SEPTEMBER 29 MONDAY

ISOBARIC

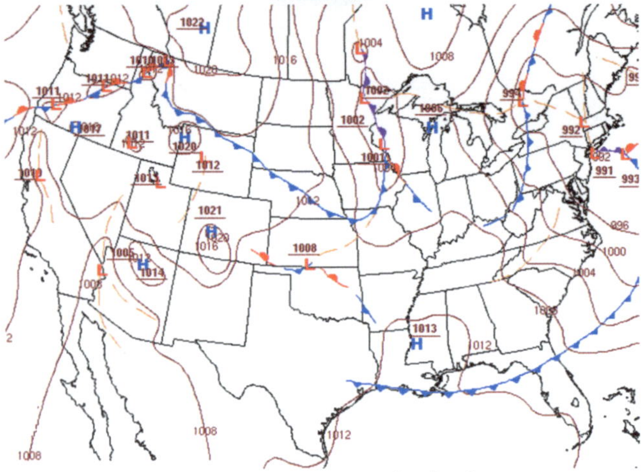

RAIN DISTRIBUTION (Inches)

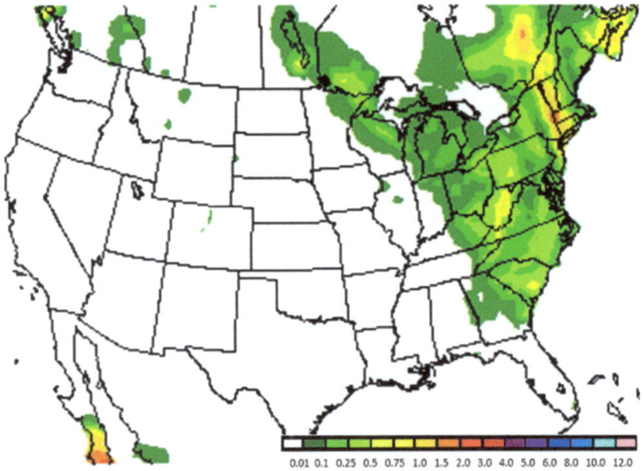

MAXIMUM TEMPERATURE

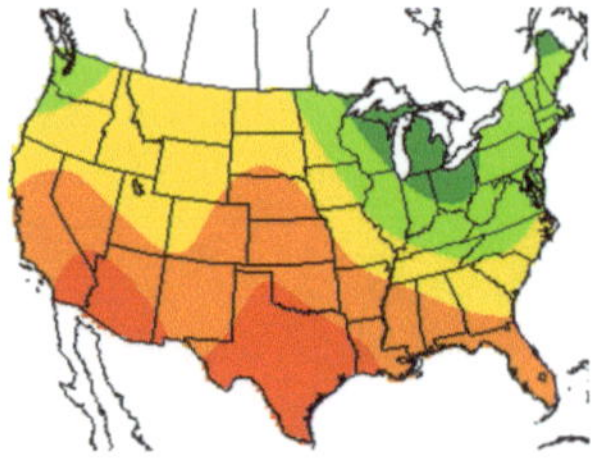

MINIMUM TEMPERATURE

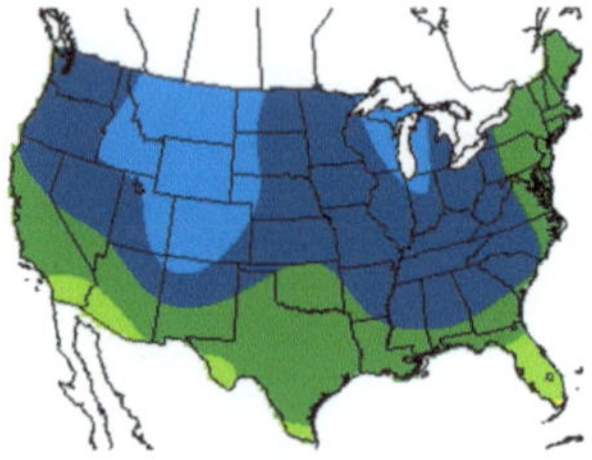

09/29/2025

A large depression over the northeast of the country continues to slowly move eastwards out into the Atlantic Ocean.

Wet fronts and troughs associated with the depression bring stormy conditions of rain and wind, heavy in places, throughout the Great Lakes to the Northeast and down through the east of the Inland Midwest, over the Appalachians and into the Coastal Southeast including Georgia.

Gusty winds associated with the storm may produce some tornado activity in the southeast of the country from Florida to Georgia over the next couple of days.

High pressure and mostly dry conditions elsewhere.

A cold front over the central and North Great Plains may bring some frosts into the regions and up into the Rocky Mountains.

09/30/2025

Stormy conditions associated with a large depression continue over the Northeast to the upper Midwest and around the western fringes of the Great Lakes.

Heavy rain and gusty winds in the Northeast.

Some showers likely along the northern California coast as well as in the northwest of Washington.

Mostly dry elsewhere.

Cold day with frosts likely in Oregon.

SEPTEMBER 30 TUESDAY

ISOBARIC

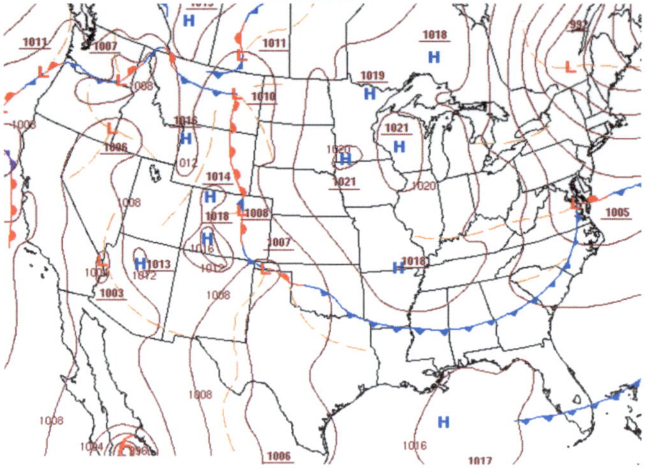

RAIN DISTRIBUTION (Inches)

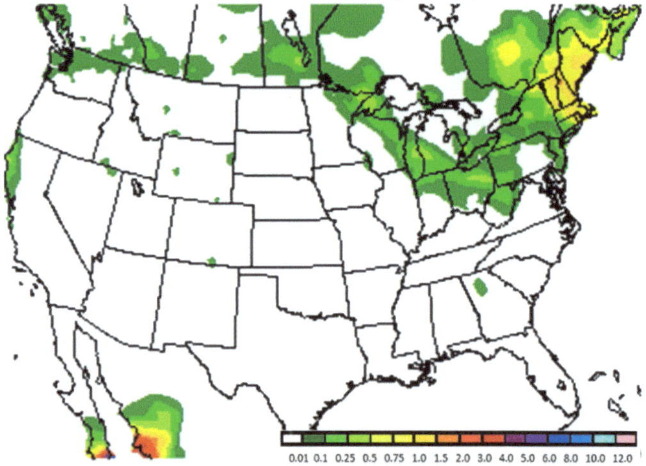

MAXIMUM TEMPERATURE

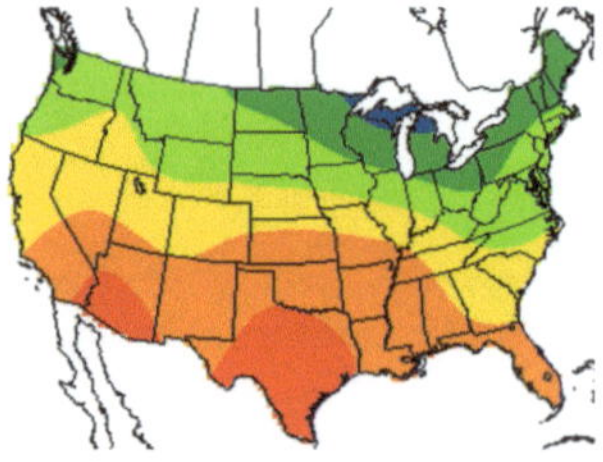

MINIMUM TEMPERATURE

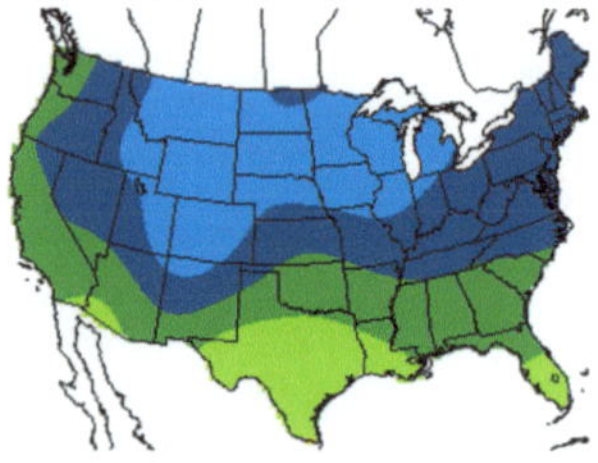

OCTOBER

PHASES OF THE MOON

7th	Full Moon
7th	Crossing Equator
8th	Perigee #6
12th	Northern Declination
13th	Third Quarter
20th	Crossing Equator
21st	New Moon
23rd	Apogee
26th	Southern Declination
29th	First Quarter

1	2	3	4	5	6	7	8	9	10	11	12	13	14	15	16	17	18	19	20	21	22	23	24	25	26	27	28	29	30	31
Wed	Thur	Fri	Sat	Sun	Mon	Tue	Wed	Thur	Fri	Sat	Sun	Mon	Tue	Wed	Thur	Fri	Sat	Sun	Mon	Tue	Wed	Thur	Fri	Sat	Sun	Mon	Tue	Wed	Thur	Fri
						F XhN	P6				^	3Q							XhS	N		A			V			1Q		

ISOBARIC

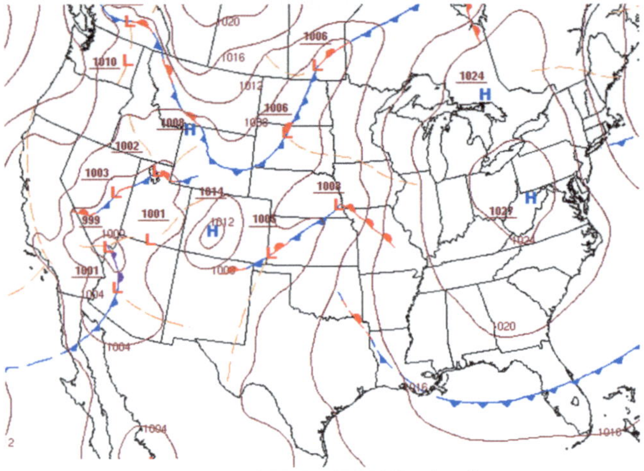

RAIN DISTRIBUTION (Inches)

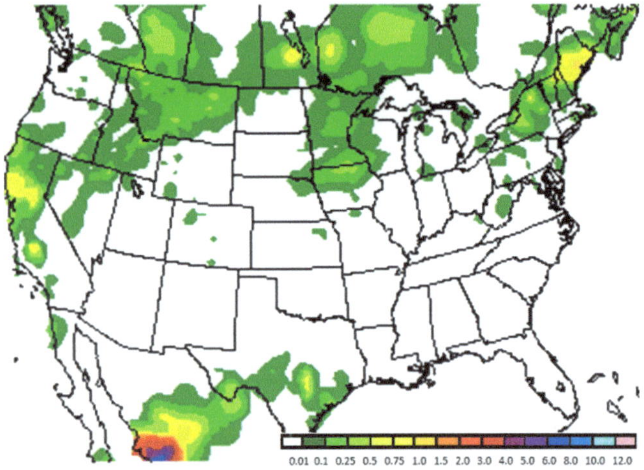

MAXIMUM TEMPERATURE MINIMUM TEMPERATURE

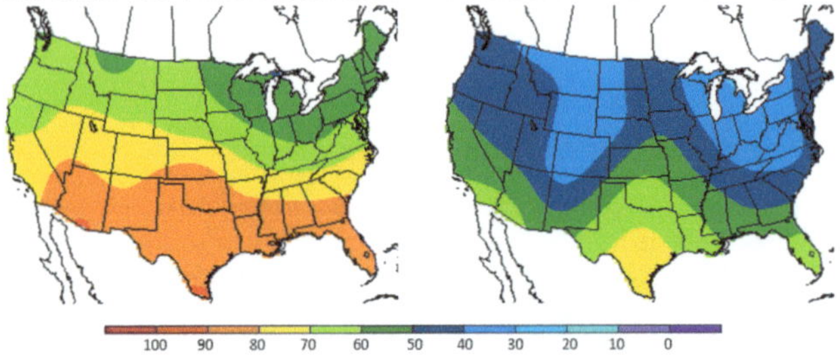

10/01/2025

High pressure over the east. Showers in the Inland Midwest. Rain and gusty winds in the Northeast.

A cold front crosses the North Great Plains bringing rain through the Inland Northwest in the Dakotas.

Showers prevail in the Great Basin.

Rain in northern California, showers in the south.

A rain band crosses central Texas, with possible heavy rain moving into Nevada and Utah.

Mostly dry elsewhere.

A tropical depression in the Gulf of Mexico moves towards Mexico, generating an unstable wind flow in Florida and Georgia, and possibly tornadoes.

Very warm in the Midwest, esp. Illinois and Wisconsin.

Cold with frosts likely in Washington and Alaska.

10/02/2025

Widespread rain from the Inland Northwest over the northern Rockies and North Great Plains, to the Great Lakes. Heavy rain at times in Wisconsin and Minnesota and some patches of showers in the lower Northeast.

A tropical depression in the south of the country brings very heavy rain throughout Texas, Oklahoma, and the Southwest over the next couple of days.

Mostly dry elsewhere.

Temperatures may remain low in the west, but rise on the Plains, possibly up into the 90's in South Dakota.

Chance of snow over the next couple of days in Montana and northern Idaho.

Heavy precipitation in Alaska, esp. the southeast, with freezing temperatures possibly resulting in wintry showers, sleet and snow.

ISOBARIC

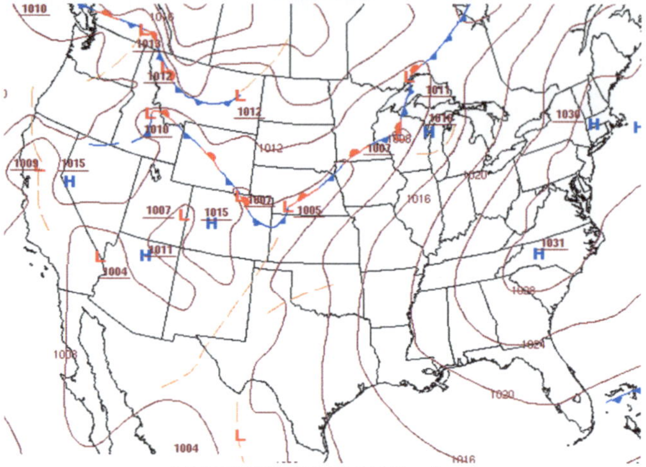

RAIN DISTRIBUTION (Inches)

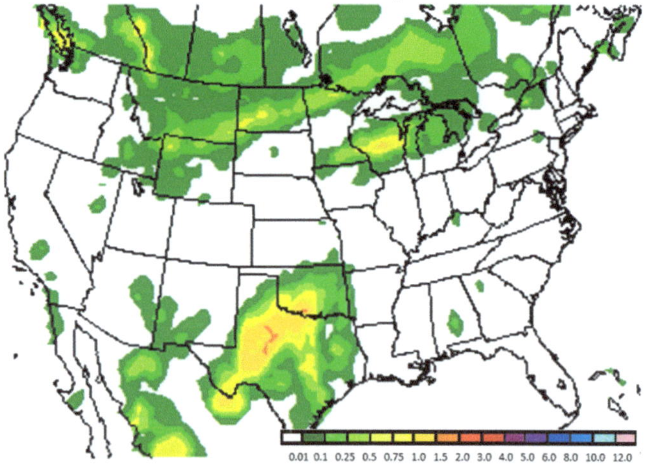

MAXIMUM TEMPERATURE MINIMUM TEMPERATURE

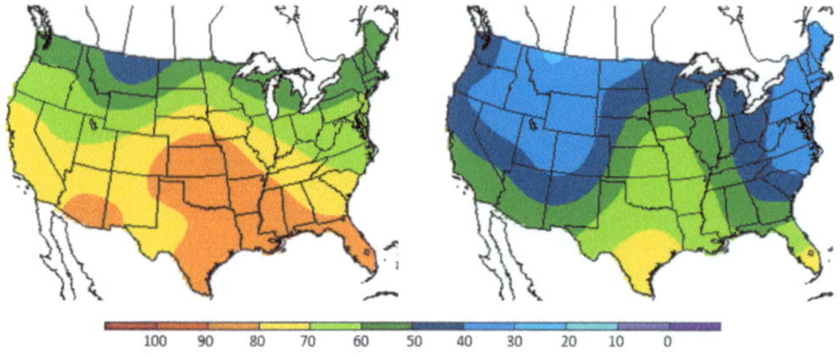

ISOBARIC

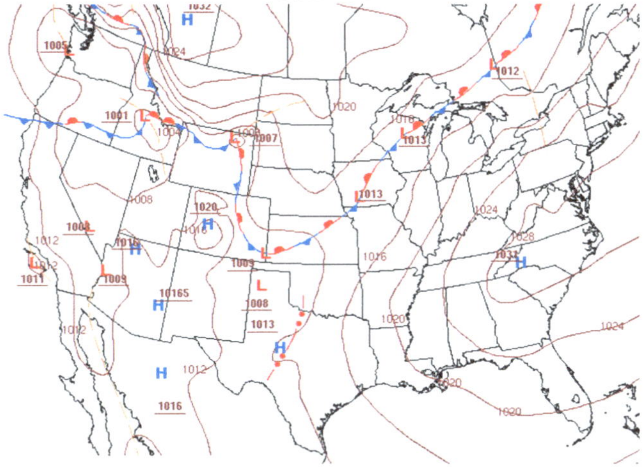

RAIN DISTRIBUTION (Inches)

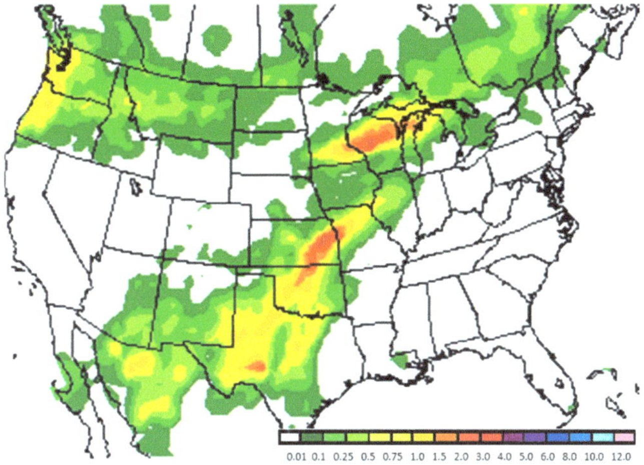

MAXIMUM TEMPERATURE MINIMUM TEMPERATURE

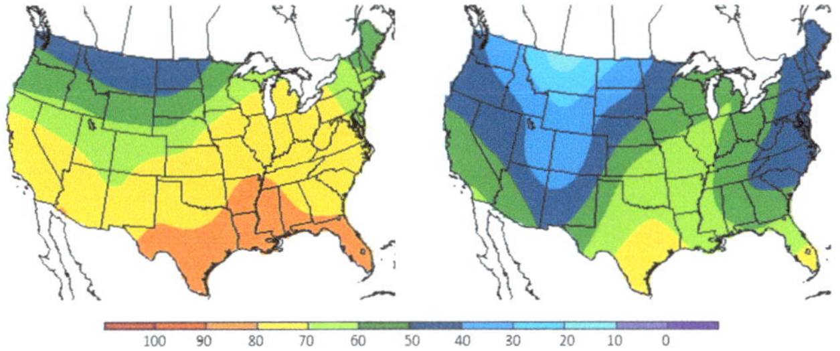

10/03/2025

A cold front brings widespread and heavy rain into the Pacific Northwest, Inland Northwest, and into the North Great Plains, where precipitation may fall as heavy snow in Montana and northern Idaho.

A band of heavy precipitation, possibly as a remnant of an ex-tropical depression, flows through Texas and the Southwest, across the South Great Plains and Ozarks, and over the Inland Midwest to the Great Lakes, with showers possibly reaching into the Northeast.

Mostly dry elsewhere.

It may be unseasonably cold in California over the next few days, while unusually warmer than normal across the Great Plains to the eastern coasts, particularly in South Dakota to Kentucky as well as Indiana.

Chance of snow in Colorado and Utah over the next few days, particularly on the Wasatch Ranges.

10/04/2025

Mostly dry in the east of the country from the Northeast to Deep South, including through the Appalachians, but with some isolated showers likely in Florida to South Carolina.

Similarly, it may be mostly dry in the west from southern California through the southern Rockies and Southwest states.

Widespread precipitation prevails elsewhere, with very heavy downpours and storms in Texas to the Ozarks, and heavy rain at times in the upper Inland Midwest, especially in Minnesota.

Chance of rain spreading into the western half of the Northeast.

It may be warmer than normal in West Virginia over the next couple of days.

Chance of snow in Utah.

ISOBARIC

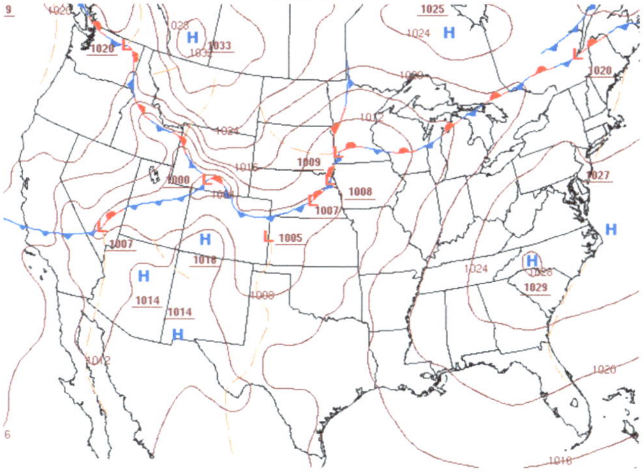

RAIN DISTRIBUTION (Inches)

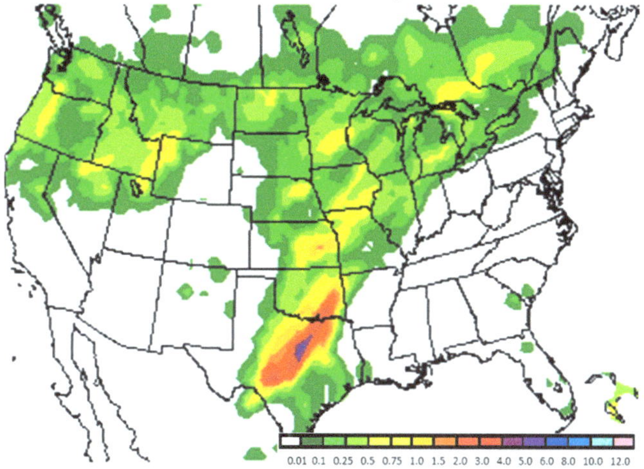

MAXIMUM TEMPERATURE MINIMUM TEMPERATURE

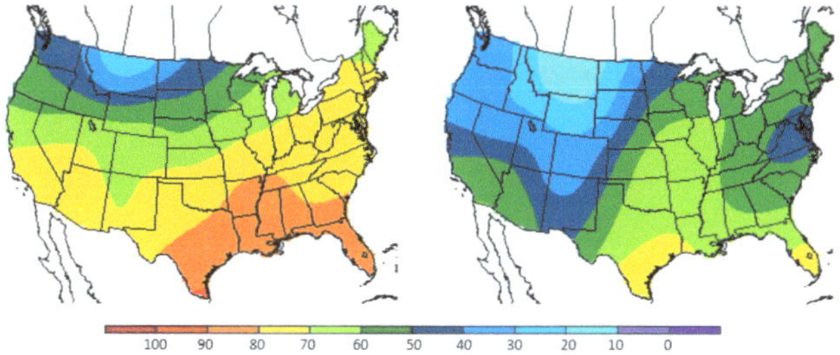

ISOBARIC

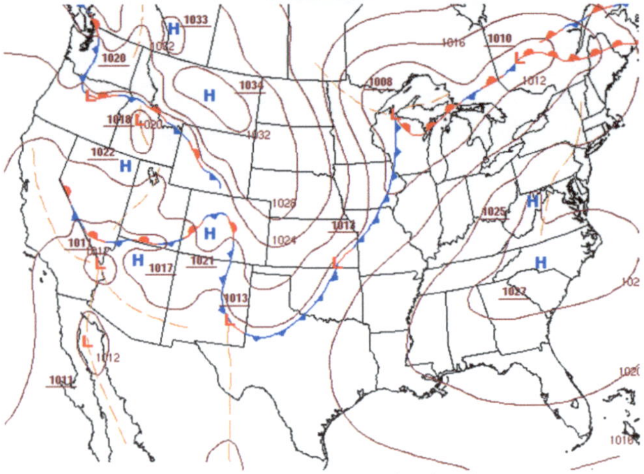

RAIN DISTRIBUTION (Inches)

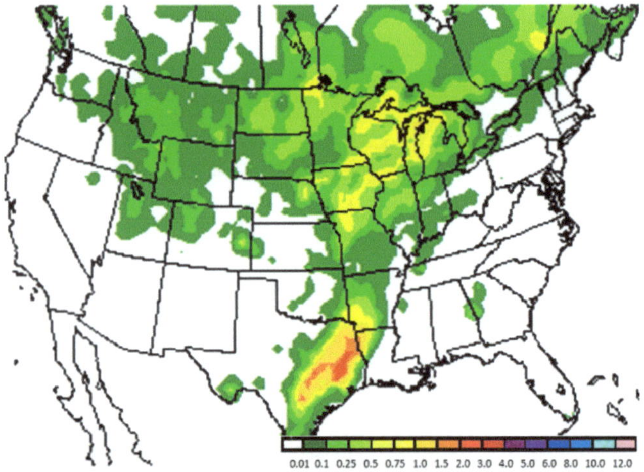

MAXIMUM TEMPERATURE

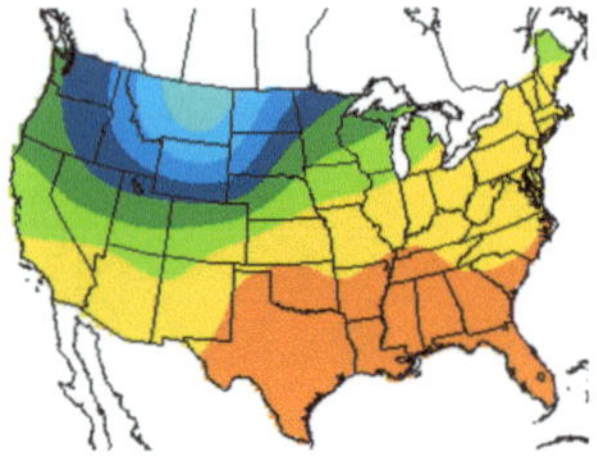

MINIMUM TEMPERATURE

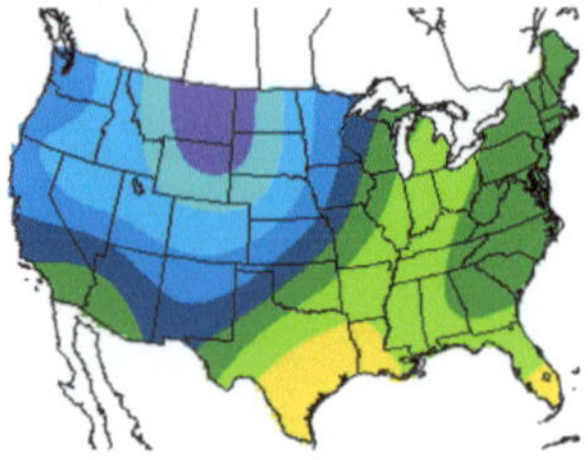

10/05/2025

Widespread precipitation prevails across the northern half of the country around this time from the Inland Northwest to the Great Lakes and up into the Northeast, as well as through the Inland Midwest, the Great Plains and into Texas.

Rain may be particularly heavy in Texas to the Ozarks, as well as very heavy around the Great Lakes.

Chance of the heavy rain briefly spreading into Arizona for a short time over the next couple of days.

Mostly dry elsewhere, although some patches of showers possible in Georgia.

Warmer in Connecticut over the next couple of days, while California may be unusually cooler than normal.

Showery in Alaska, mostly in the southeast.

10/06/2025

A wet front brings widespread rain from Texas in the south to the Inland Midwest and the Great Lakes in the north, with showers into the Northeast and through the western slopes of the Appalachians.

Showers persist in the Inland Northwest to the Great Basin and Rocky Mountains and over the North Great Plains.

Mostly dry elsewhere.

Warmer in Maine and Connecticut, and Kentucky to West Virginia.

Another tropical low may be developing off the coast of Mexico.

ISOBARIC

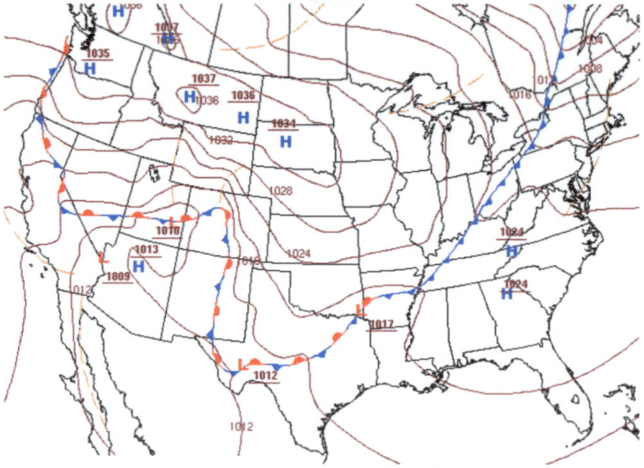

RAIN DISTRIBUTION (Inches)

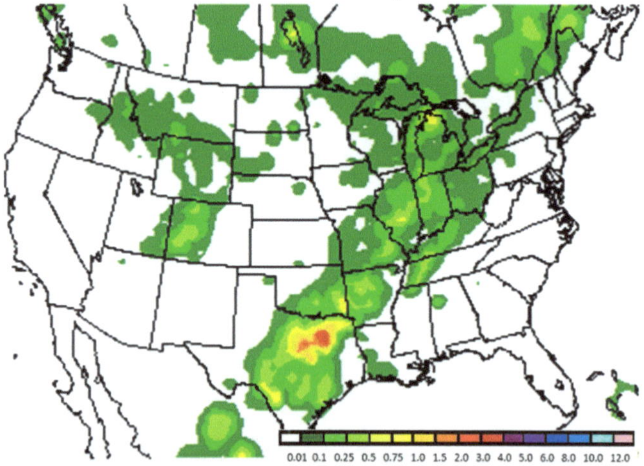

MAXIMUM TEMPERATURE

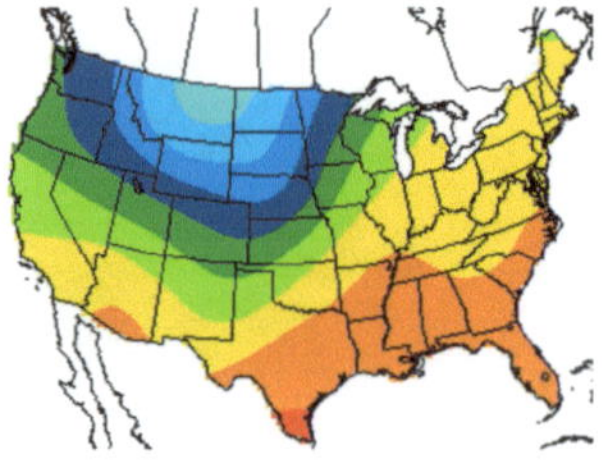

MINIMUM TEMPERATURE

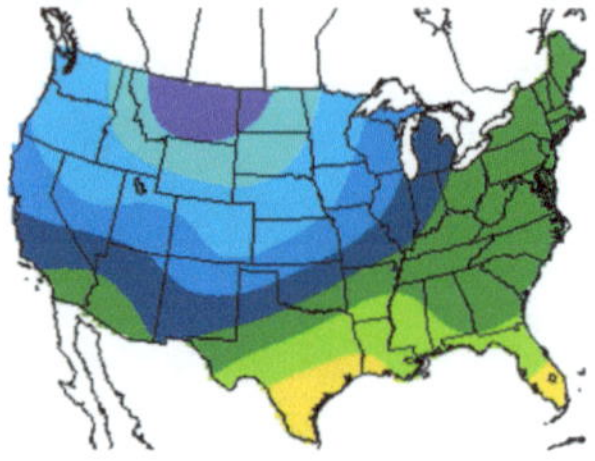

OCTOBER 7 TUESDAY

ISOBARIC

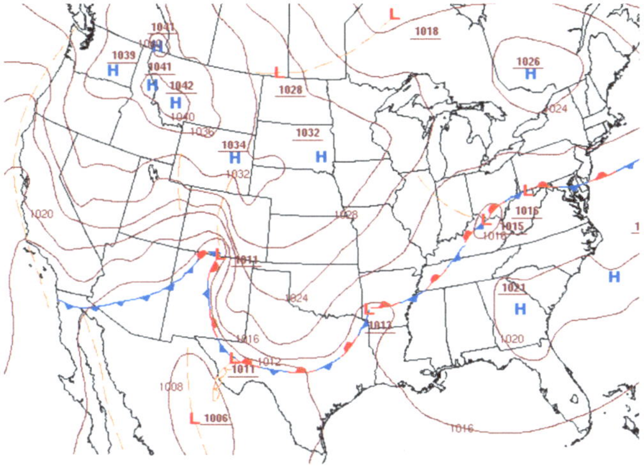

RAIN DISTRIBUTION (Inches)

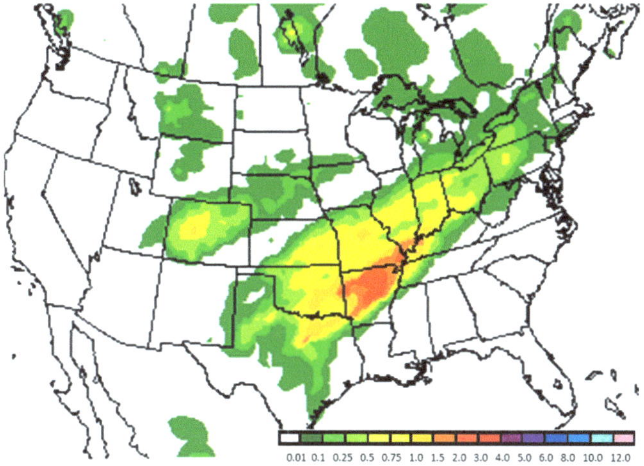

MAXIMUM TEMPERATURE

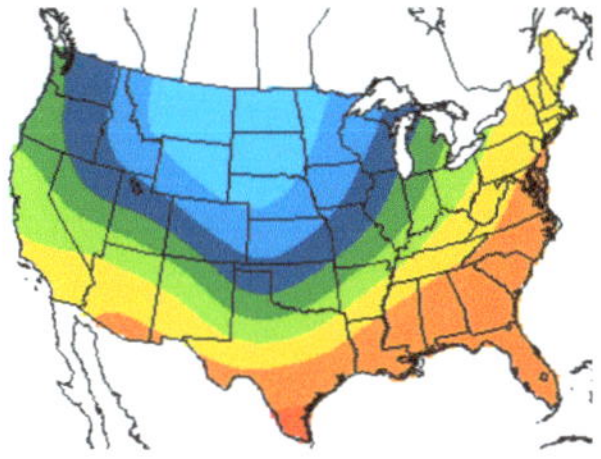

MINIMUM TEMPERATURE

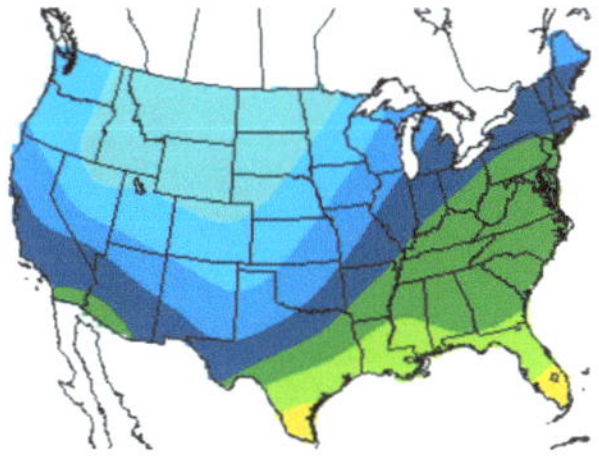

10/07/2025

An almost stationary front brings widespread rain from Texas, through the South Great Plains, across the Inland Midwest, up through the Appalachians and into to the Northeast Megalopolis regions.

Very heavy rain likely from Oklahoma and the Ozarks to the Ohio Valley and stretching into the southern Northeast.

Another band of rain crosses the southern Rockies and onto the central Great Plains.

Some shower in the Inland Northwest to the northern Rockies.

Mostly dry elsewhere.

Chance that a tropical low near Mexico strengthens into a low category hurricane.

OCTOBER 8 WEDNESDAY

10/08/2025

An intense cold front maintains widespread rain in a wide diagonal band from Texas to the Northeast including up through the Appalachians and Inland Midwest and into the Megalopolis, with showers westward around the Great Lakes.

Rain may be very heavy at times along this front, especially in Arkansas to Kentucky.

Some showers up through the Rocky Mountains and possible light snow in the higher reaches of Montana, Wyoming, Idaho and Colorado.

Mostly dry elsewhere.

ISOBARIC

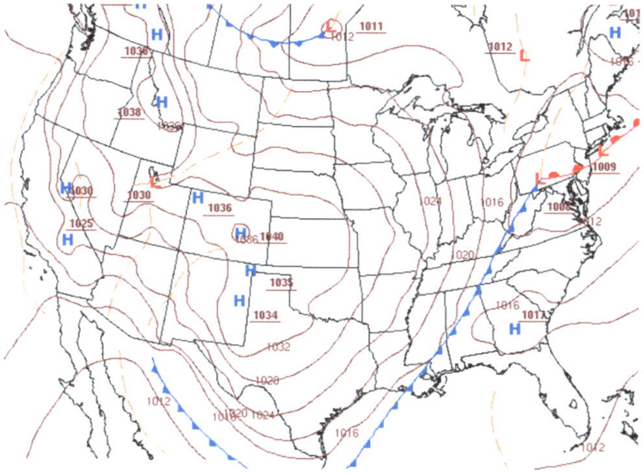

RAIN DISTRIBUTION (Inches)

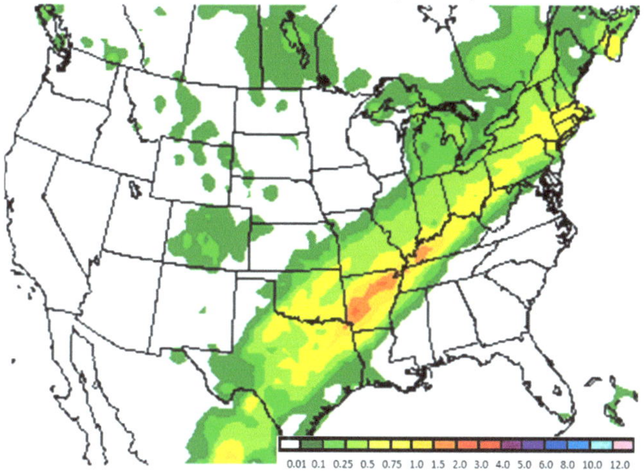

MAXIMUM TEMPERATURE

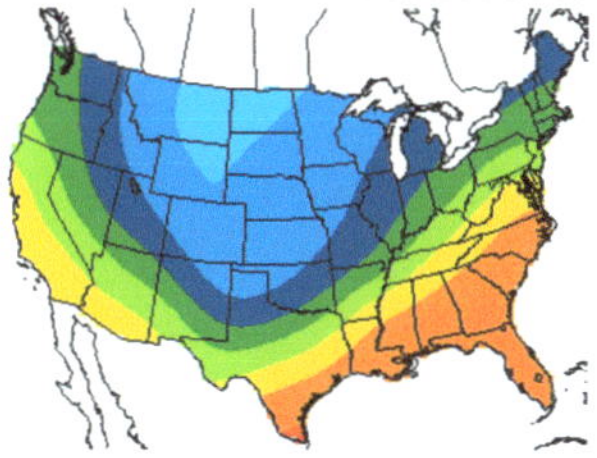

MINIMUM TEMPERATURE

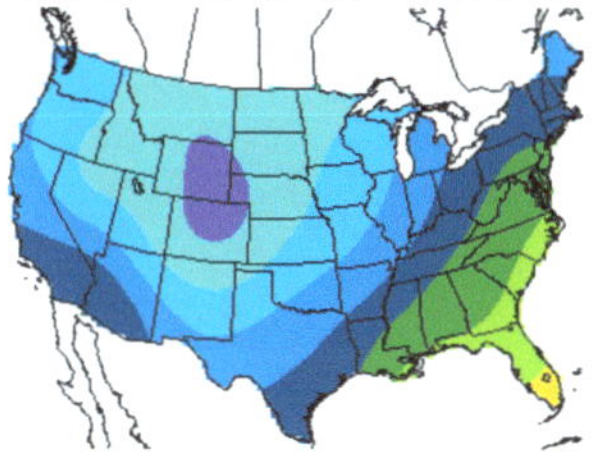

OCTOBER 9 THURSDAY

ISOBARIC

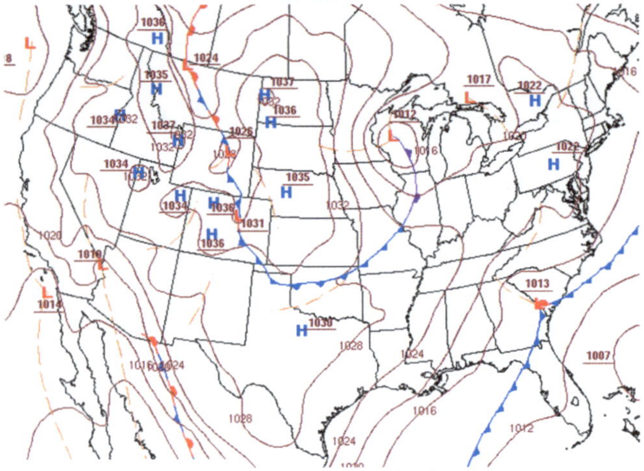

RAIN DISTRIBUTION (Inches)

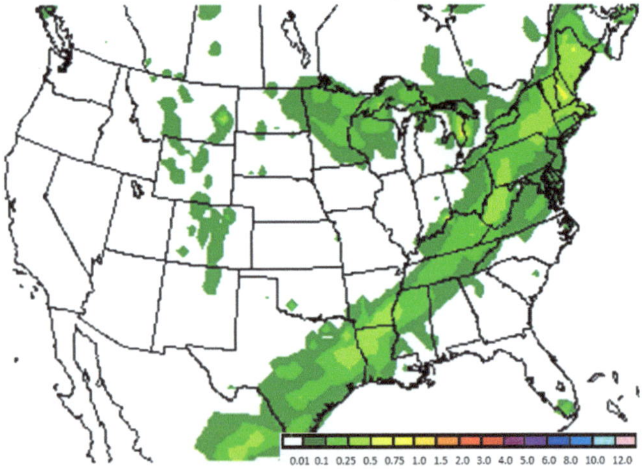

MAXIMUM TEMPERATURE / MINIMUM TEMPERATURE

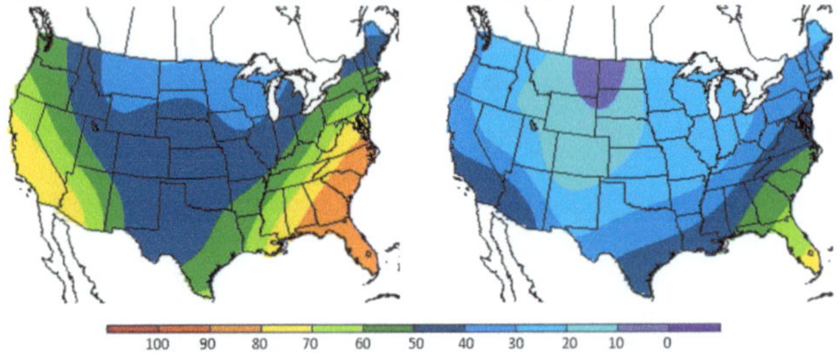

10/09/2025

Low pressure brings rain across the Minnesota to the Great Lakes, while a series of troughs brings brief precipitation and possible light snow and strong winds in Montana, Idaho, Colorado and Wyoming.

Rain prevails in a line from the Northeast region, down through the Appalachians, across the Deep South and into the eastern half of Texas.

Mostly dry elsewhere although chance of showers in the south of Florida.

Unusually high temperatures may return to the North Great Plains. Unusually cool in California.

10/10/2025

High pressure swells over the country clearing rain almost everywhere except for lingering showers around the Great Lakes, in some parts of the upper Midwest, over the north of the Appalachians and into the Coastal Southeast.

Some heavy showers for a time in the west of the Inland Midwest Corn Belt states, including in Iowa.

Some showers also likely in isolated pockets in the Northeast and around the Texas Rio Grande Valley.

Mostly dry elsewhere.

It may be unusually cold in Utah, Idaho, and California.

Rain may be threatening in the Pacific Northwest with heavy rain likely to develop over the next couple of days around the region, particularly in Oregon.

OCTOBER 10 FRIDAY

ISOBARIC

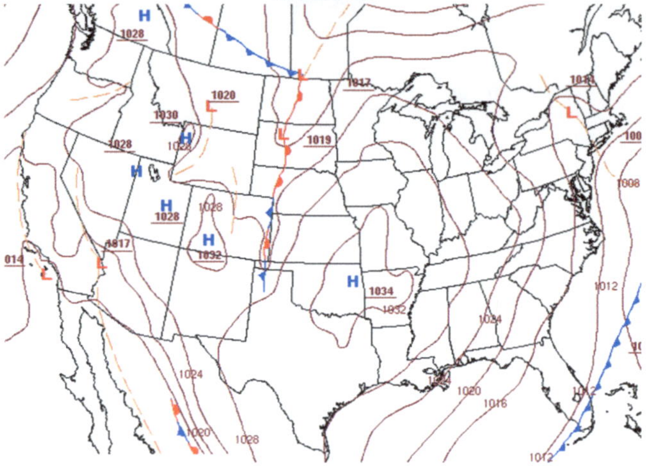

RAIN DISTRIBUTION (Inches)

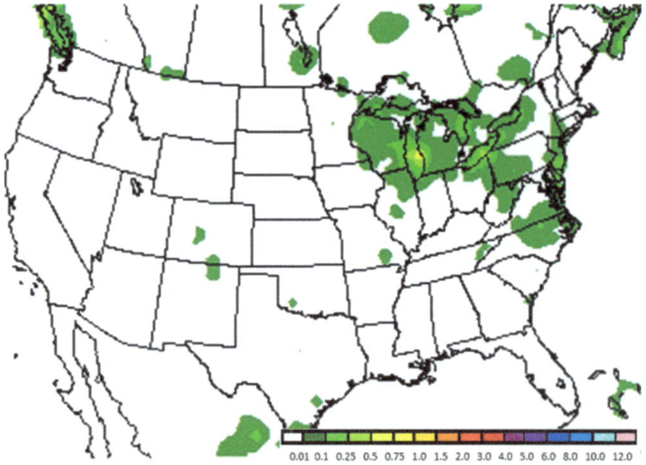

MAXIMUM TEMPERATURE / MINIMUM TEMPERATURE

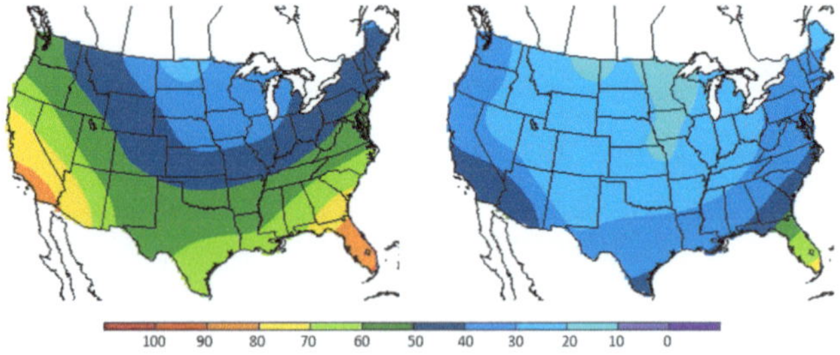

OCTOBER **11** SATURDAY

ISOBARIC

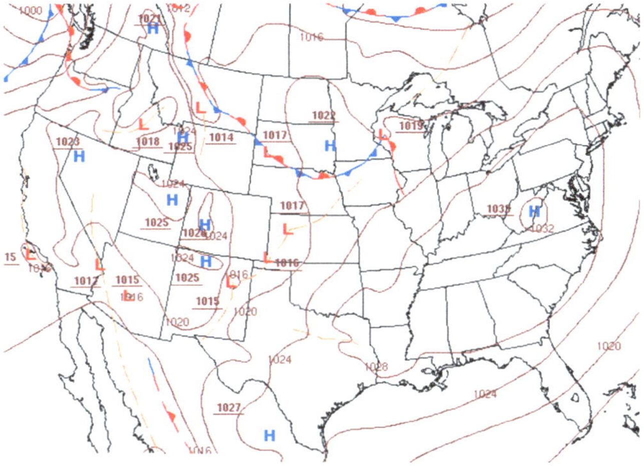

RAIN DISTRIBUTION (Inches)

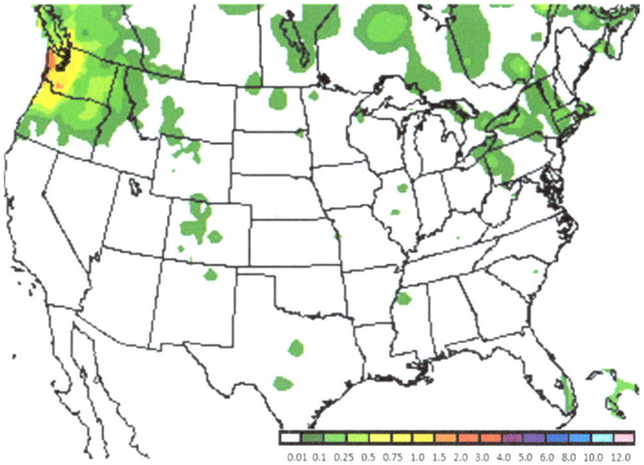

MAXIMUM TEMPERATURE / MINIMUM TEMPERATURE

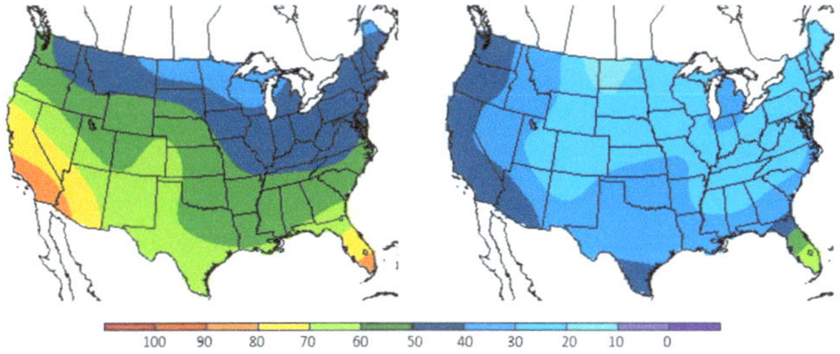

10/11/2025

Rain returns to the Pacific Northwest with very heavy falls in Oregon and isolated showers extending down through the Rocky Mountains.

Isolated spots of precipitation in the North Great Plains and clearing to mostly dry around the Great Lakes.

Some showers linger in the Northeast.

Mostly dry elsewhere, although chance of isolated showers briefly in central Texas and in New Mexico to Colorado.

High pressure dominates the east half of the country.

10/12/2025

Widespread precipitation spreads from the Pacific Northwest into the Inland Northwest and onto the western half of the North Great Plains.

Rain may be heavy for a time in Washington.

Some showers also likely in the northern parts of California and into the north of the Great Basin.

Isolated showers through the Rockies and possible brief showers around the Great Lakes.

Chance of light snow in the higher reaches of Montana.

Mostly dry elsewhere, particularly in the east, which sits under a large high-pressure influence.

OCTOBER **12** SUNDAY

ISOBARIC

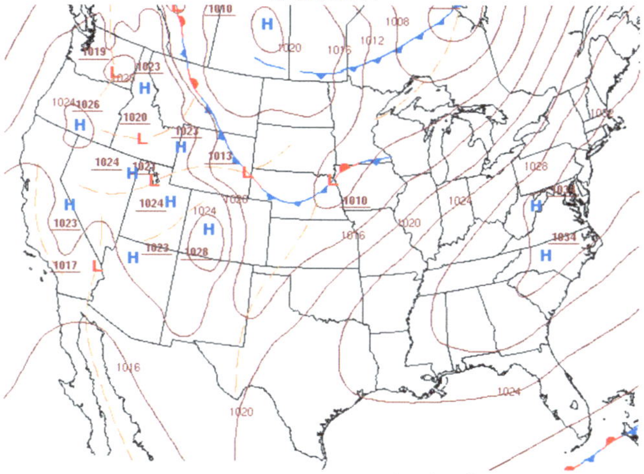

RAIN DISTRIBUTION (Inches)

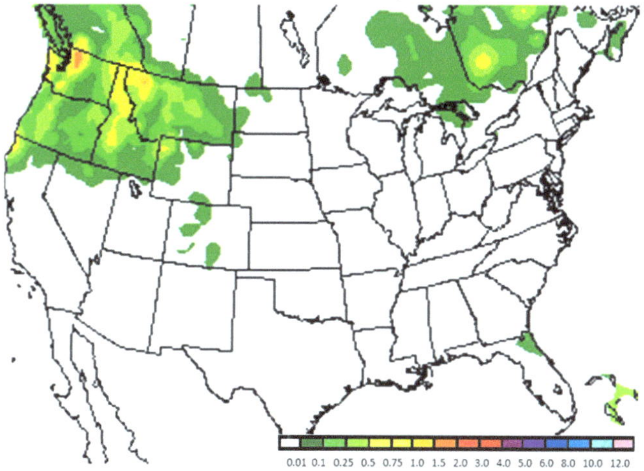

MAXIMUM TEMPERATURE / MINIMUM TEMPERATURE

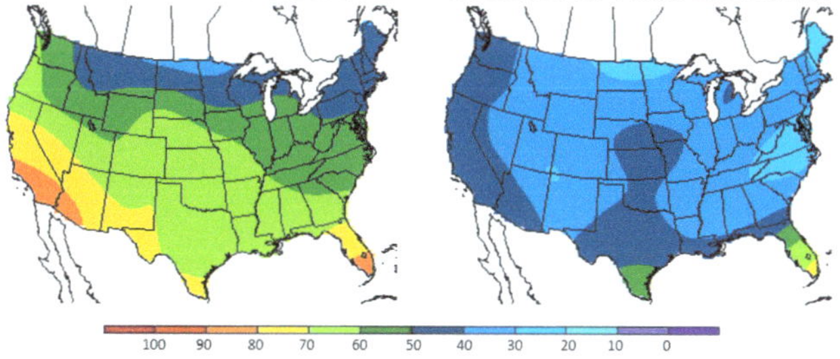

ISOBARIC

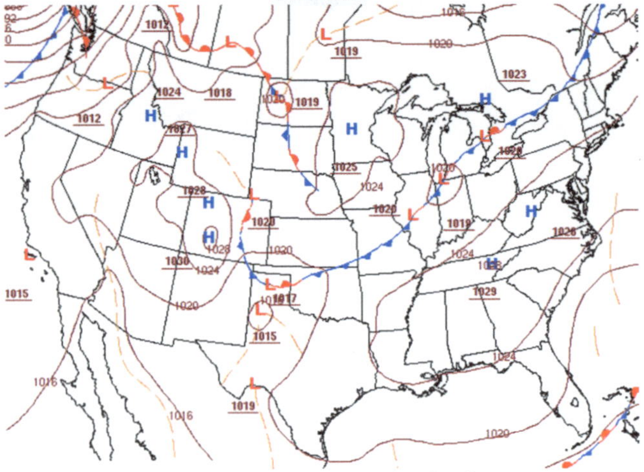

RAIN DISTRIBUTION (Inches)

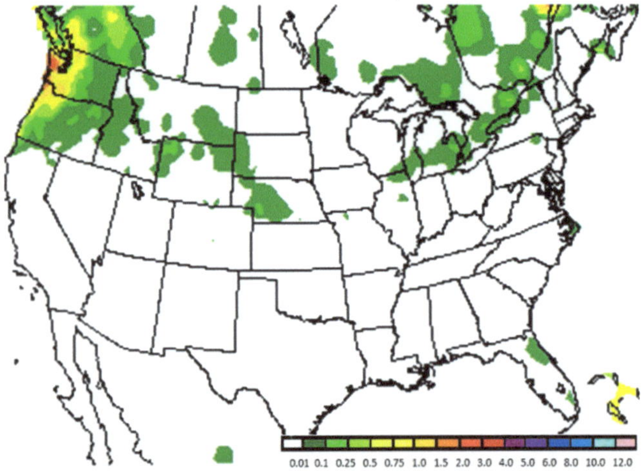

MAXIMUM TEMPERATURE

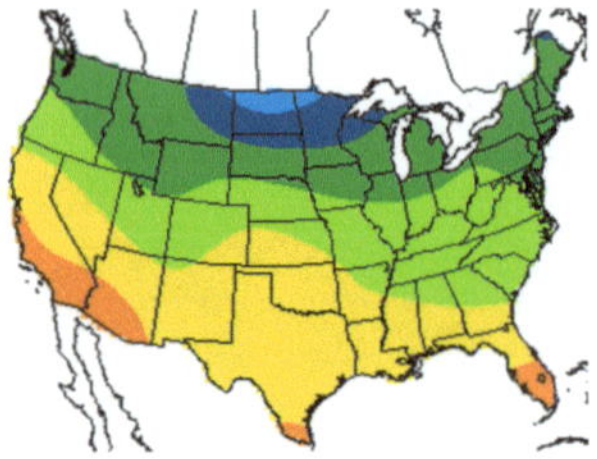

MINIMUM TEMPERATURE

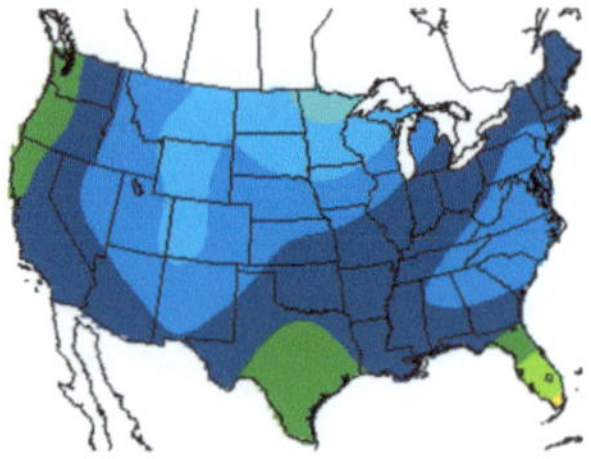

10/13/2025

An intense Pacific Ocean depression near the northwest coasts bears down on the Pacific Northwest states bringing heavy rain and strong winds to the region and showers into the adjacent Rockies.

A cold front pushes a band of showers over the Great Lakes to the western fringes of the Northeast.

Chance of isolated showers in Florida.

Mostly dry elsewhere.

A warmer change may develop in the West, while temperatures may also return to summer-like levels from the Great Plains to the eastern seaboard.

Very warm in Tennessee to West Virginia.

10/14/2025

Heavy rain and strong winds spread in the northwest of the country from the Pacific Northwest coasts to the Inland Northwest region and into the northern parts of the Rocky Mountains, as well as showers spread into the Great Basin and northern California.

Spotty showers also over the South Great Plains and some showers persist around the Great Lakes to the Northeast.

Mostly dry elsewhere.

It may be unusually warm in Texas to Florida as well as in Pennsylvania.

Snow likely in central Alaska over the next couple of days.

ISOBARIC

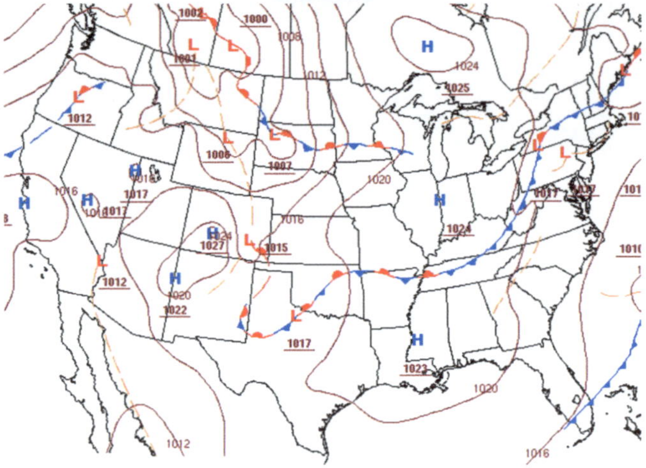

RAIN DISTRIBUTION (Inches)

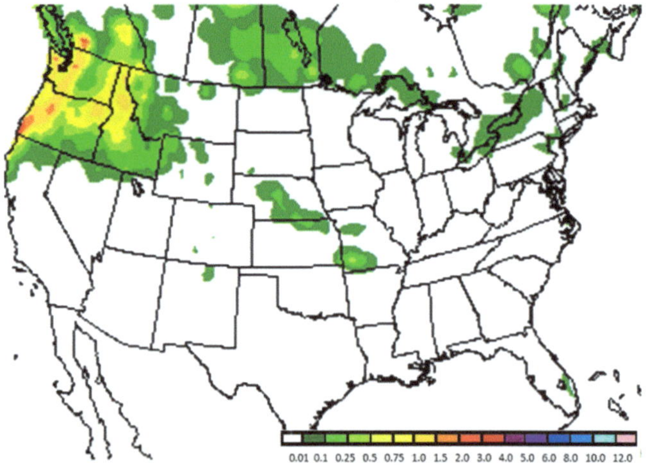

MAXIMUM TEMPERATURE

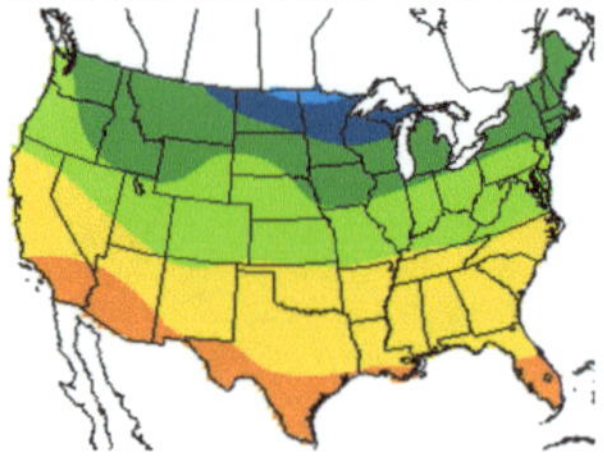

MINIMUM TEMPERATURE

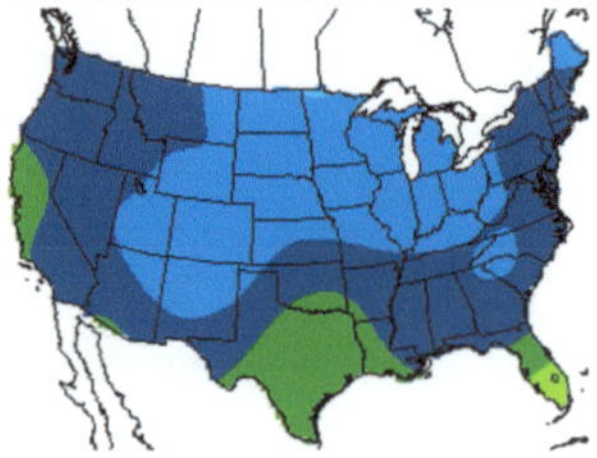

OCTOBER 15 WEDNESDAY

ISOBARIC

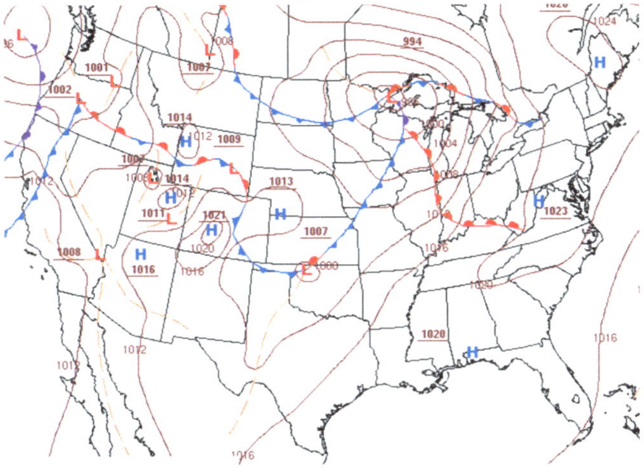

RAIN DISTRIBUTION (Inches)

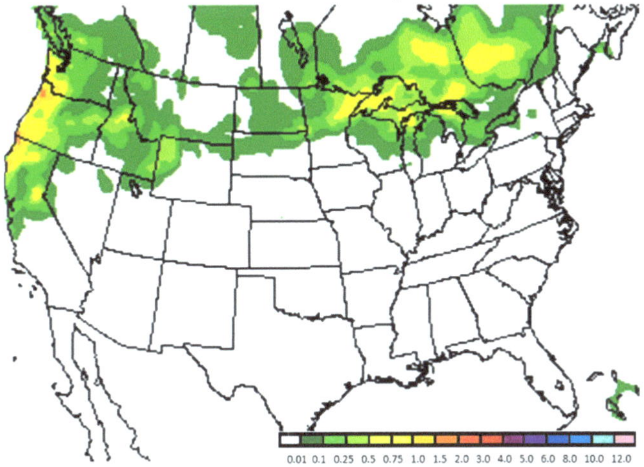

MAXIMUM TEMPERATURE / MINIMUM TEMPERATURE

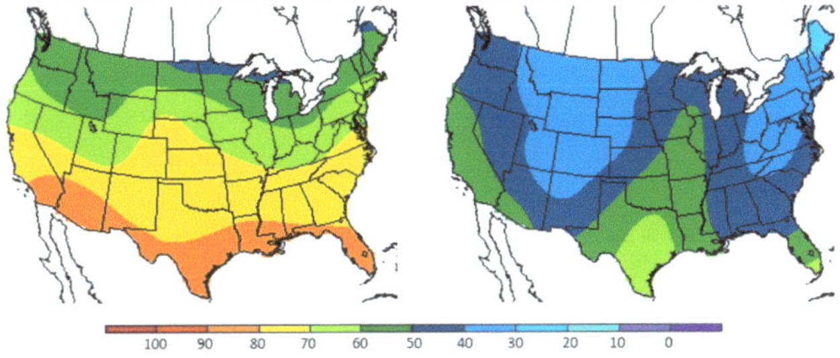

10/15/2025

Low pressure builds into a large depression over the Great Lakes with associated fronts extending rain from the north of the Great Lakes and across the Great North Plains to meet with a complex of fronts over the Pacific and Inland Northwest regions.

The depression off the northwest of the country continues to bring widespread and often heavy rain into the Pacific Northwest and down into the northern half of California and northern parts of the Great Basin.

Mostly dry elsewhere.

A cool spell in the Northwest states and California.

Chance of heavy precipitation developing over the next couple of days in Idaho, Montana, and the Dakotas.

Chance of snow in central parts of Alaska today.

10/16/2025

Widespread precipitation across most of the US-Canada border regions from the Pacific Northwest to the Northeast.

Heavy rain likely in the Pacific coastal states from Washington to the northern half of California.

Heavy rain also over much of the Great Basin, Inland Northwest, the northern half of the Rocky Mountains, and in the Dakotas.

Chance of snowfalls in Nevada, Montana, and Idaho.

Mostly dry elsewhere.

Chance that hotter temperatures continue to build across the Midwest, southern, and eastern states.

Showers in the southeast of Alaska.

OCTOBER 16 THURSDAY

ISOBARIC

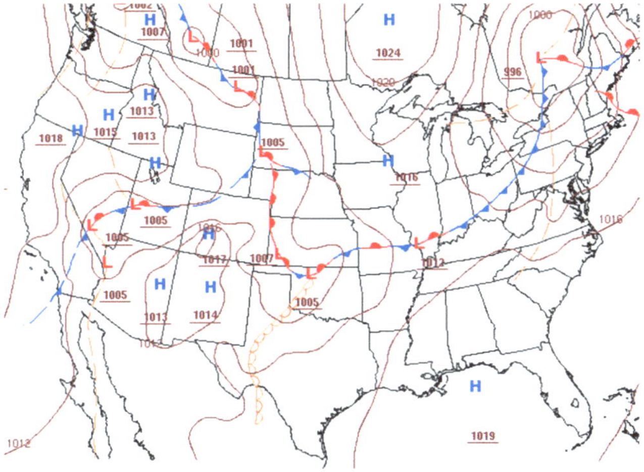

RAIN DISTRIBUTION (Inches)

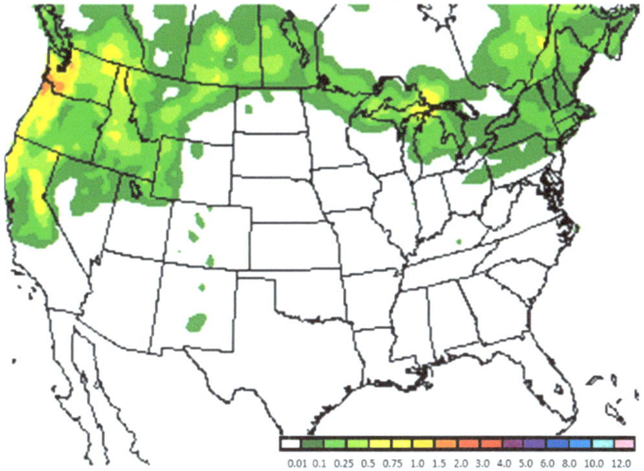

MAXIMUM TEMPERATURE / MINIMUM TEMPERATURE

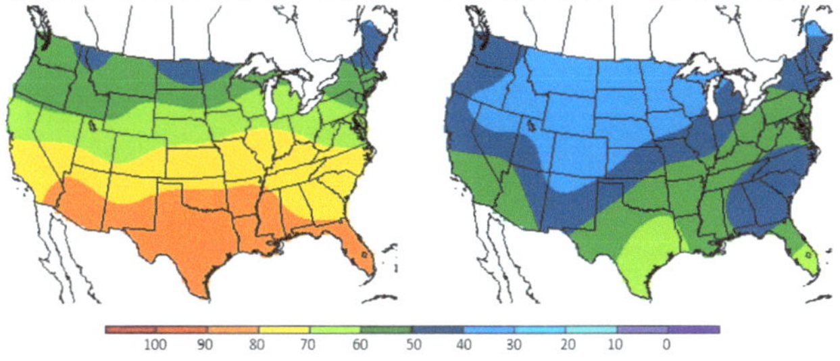

OCTOBER **17** FRIDAY

ISOBARIC

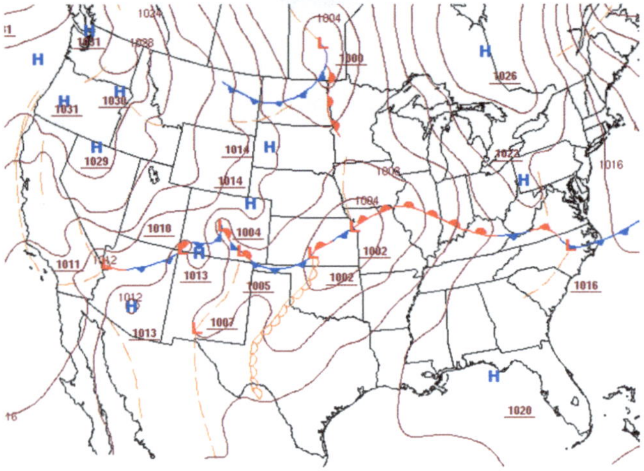

RAIN DISTRIBUTION (Inches)

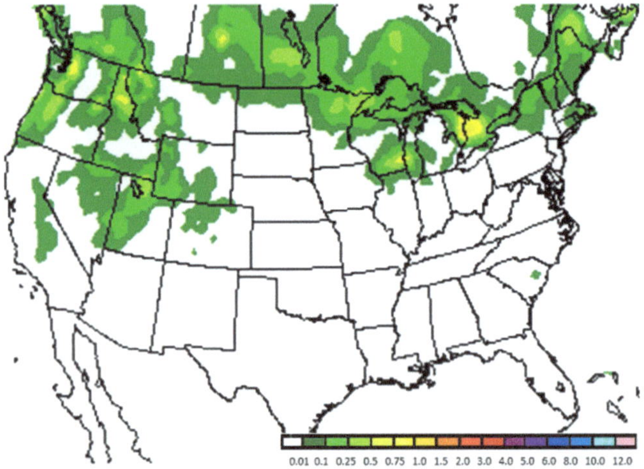

MAXIMUM TEMPERATURE

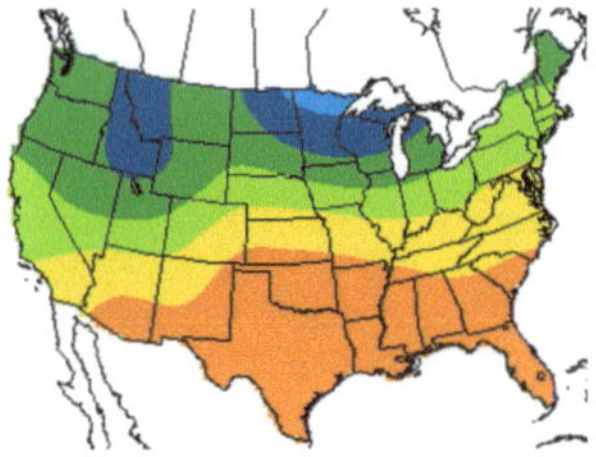

MINIMUM TEMPERATURE

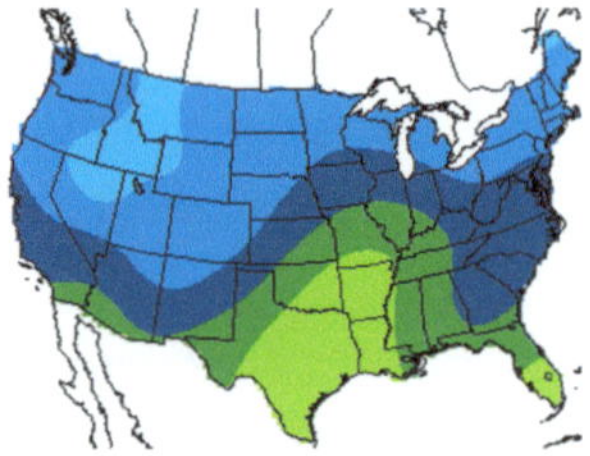

10/17/2025

High pressure over the northwest of the country eases rain to showers in the Pacific and Inland Northwest, the Great Basin and northern half of the Rocky Mountains.

Rain clears to isolated showers in northern California.

Precipitation, some of it wintry, along the US-Canadian border, down the Great Lakes and into the Northeast, with chance of showers into the upper Midwest.

Chance of a storm in Utah with strong winds and snow.

Mostly dry elsewhere. Unsettled winds in California.

It may be very hot over the next few days in the Midwest, and south and eastern parts of the country, in particular in Ohio, Virginia, Kentucky and North Carolina, with temperatures possibly up to 90°F.

Some showers likely in Hawaii over the next couple of days, mostly on the Big Island.

10/18/2025

A weak depression over the northeast brings widespread rain from the Great Lakes to the southern half of the Northeast, with showers extending into the Northeast Highlands and into New England, as well as in the upper Midwest to the northern Appalachians.

It may be unusually warm but very wet in Michigan.

Some showers linger in the Great Basin to the southern Rocky Mountains.

Some precipitation in the west of Texas.

Rain clears to mostly dry in the northwest quarter of the country under a high-pressure influence.

Mostly dry elsewhere. Unusually warm Indiana, Kentucky, Tennessee, and West Virginia.

Some showers likely on the Hawaiian Big Island.

Cold in Alaska with wintry showers including snow.

OCTOBER **18** SATURDAY

ISOBARIC

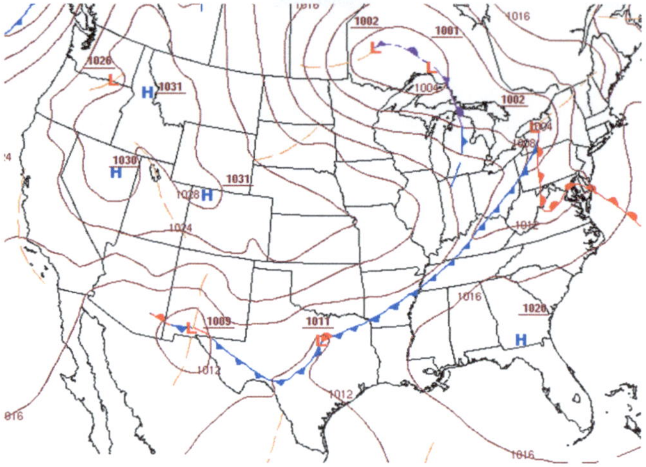

RAIN DISTRIBUTION (Inches)

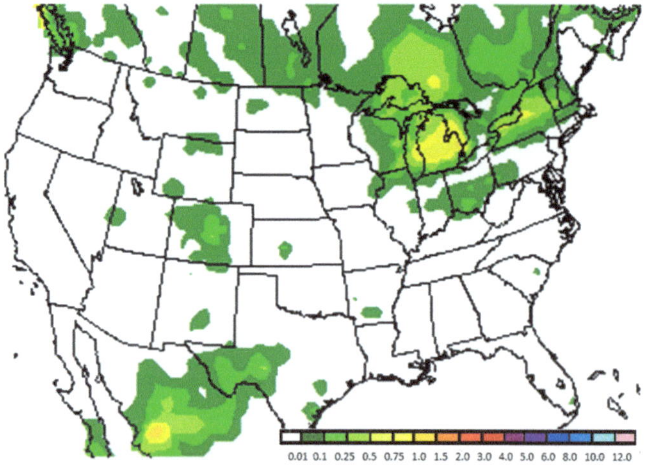

MAXIMUM TEMPERATURE

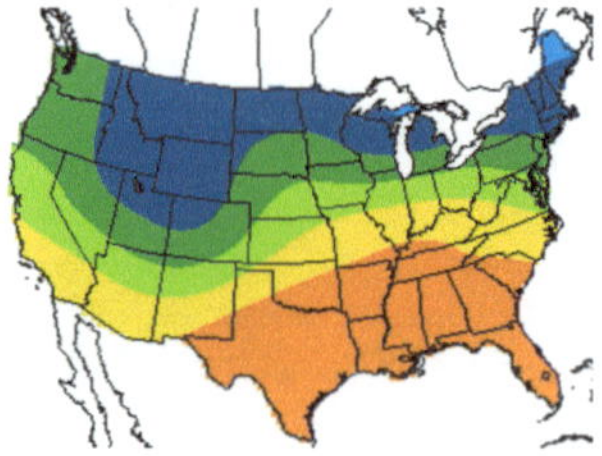

MINIMUM TEMPERATURE

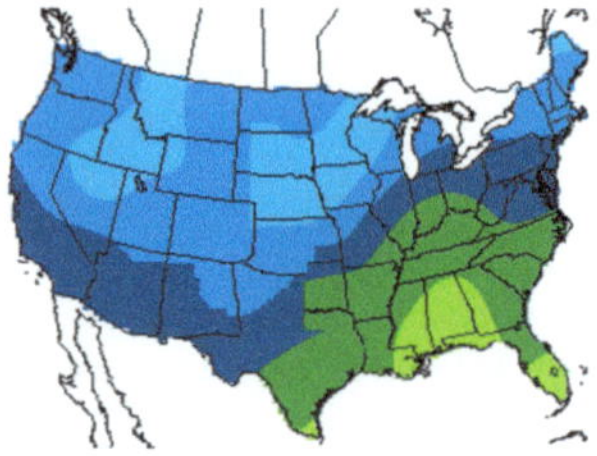

OCTOBER 19 FRIDAY

ISOBARIC

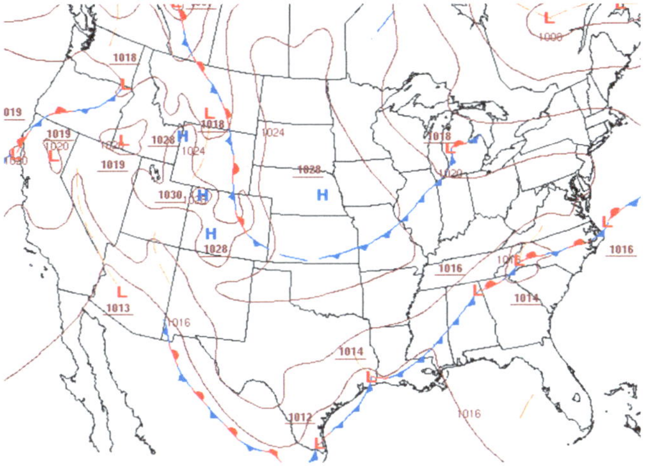

RAIN DISTRIBUTION (Inches)

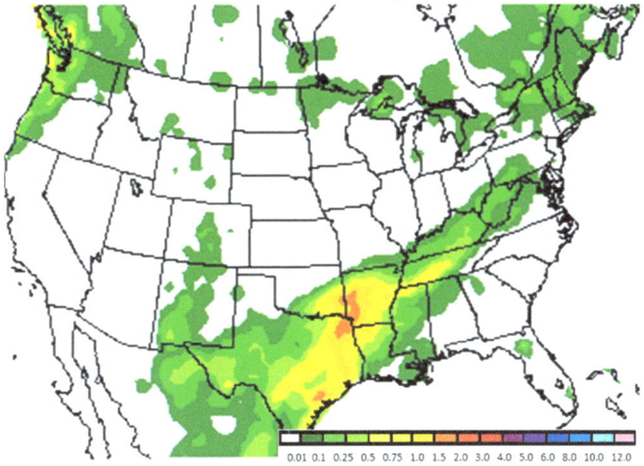

MAXIMUM TEMPERATURE

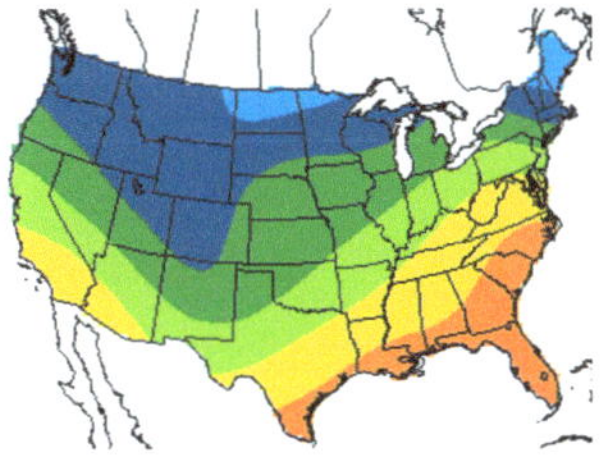

MINIMUM TEMPERATURE

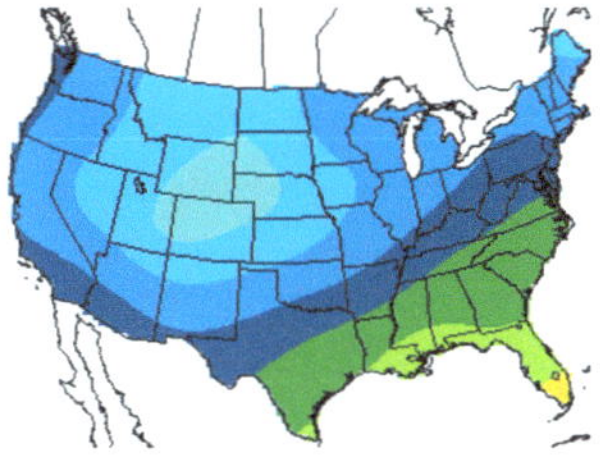

10/19/2025

A wet front crosses the south of the country bringing widespread rain from Texas and the western half of the Deep South and up through the Appalachians, with showers into the Southwest to the southern foothills of the Rocky Mountains.

Some rain returns in the Pacific Northwest region, and isolated showers in the northern half of the Rockies.

Rain eases to occasional showers from Minnesota through the Great Lakes, while showers are more frequent in the Northeast.

Chance of snow in the Great Lakes region, including in Minnesota.

Mostly dry elsewhere.

It may still be very hot in North Carolina.

Snow in Alaska.

OCTOBER 20 SATURDAY

ISOBARIC

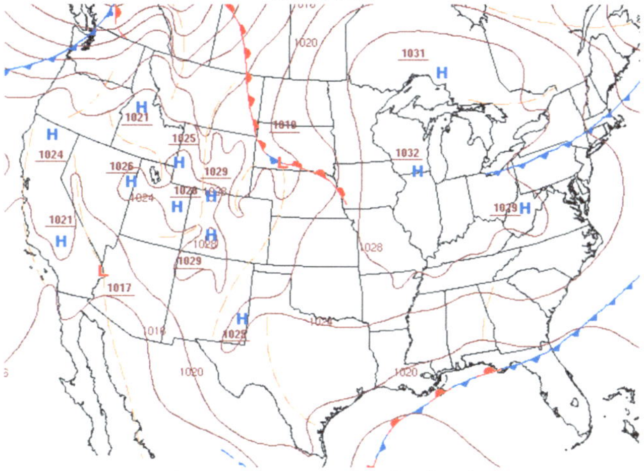

RAIN DISTRIBUTION (Inches)

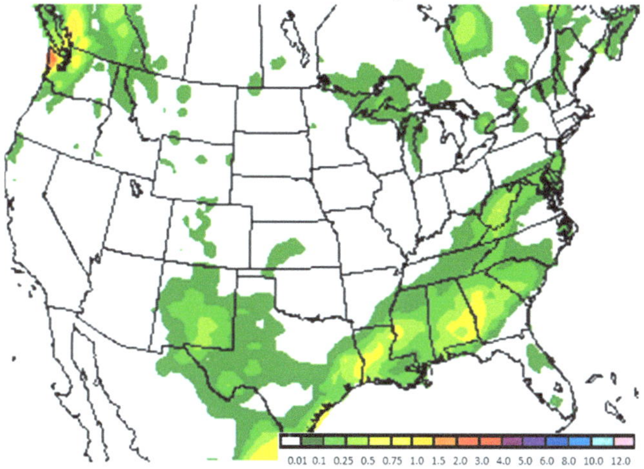

MAXIMUM TEMPERATURE

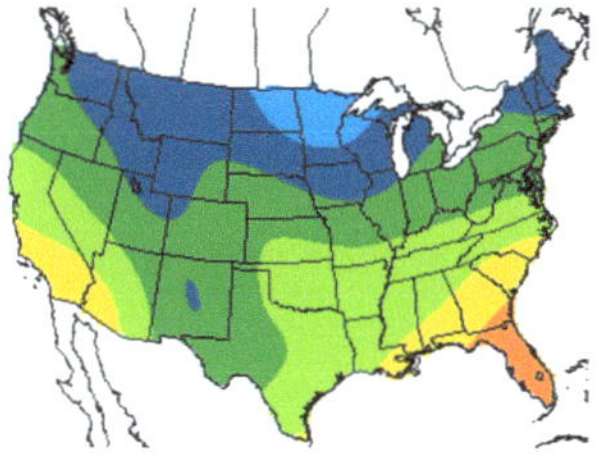

MINIMUM TEMPERATURE

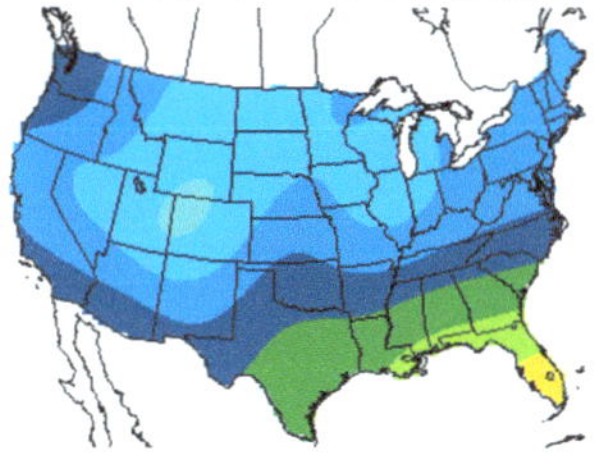

10/20/2025

A large high fills over the northeast quarter of the country with just the odd shower over the Great Lakes to the Northeast.

Some wintry falls around Michigan to the Great Lakes may become snow over the next couple of days.

A cold front brings rain in the Pacific Northwest, with very heavy falls likely around the Washington coasts.

Rain embedded troughs maintain widespread rain over the south of the country from the Southwest, across Texas, throughout most of the Deep South and up along the Appalachians to the Coastal Southeast, with some showers into Florida and in the south of the Great Plains.

Mostly dry elsewhere.

A warm front crosses the Great Plains while it may be unusually cold in California.

OCTOBER 21 TUESDAY

ISOBARIC

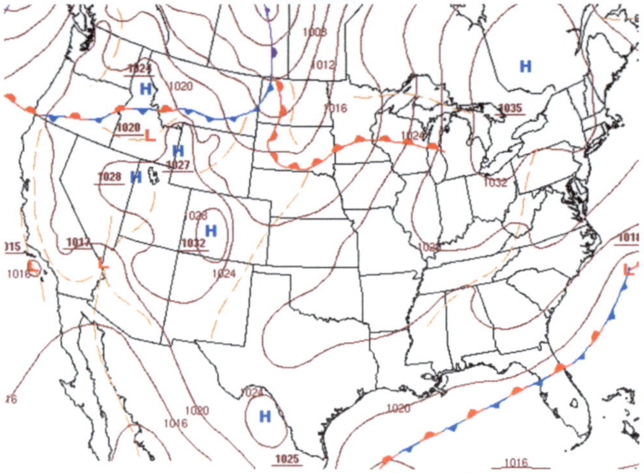

RAIN DISTRIBUTION (Inches)

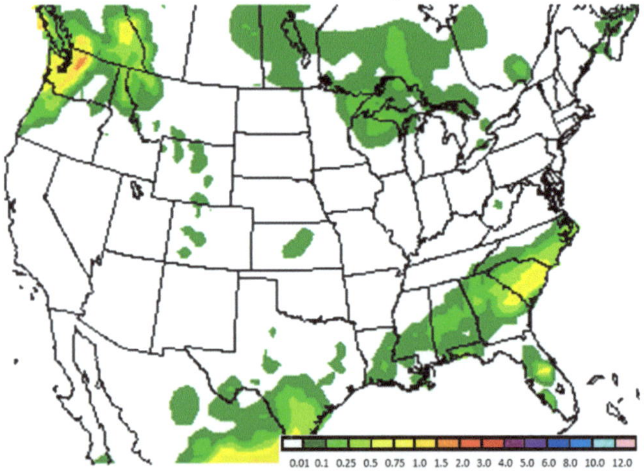

MAXIMUM TEMPERATURE / MINIMUM TEMPERATURE

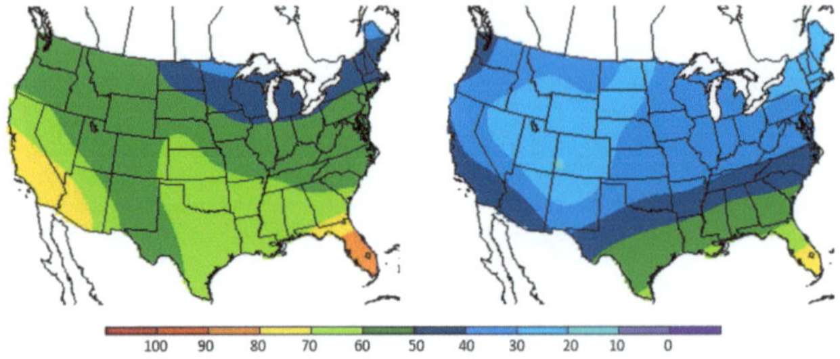

10/21/2025

Wintry precipitation around the Great Lakes continues, while high pressure in the region clears rain throughout the Northeast.

Rain persists in the Pacific Northwest to the Inland Northwest, with isolated showers down through the Rocky Mountains.

The rain-embedded front over the south of the country maintains its slow southerly trek into the Gulf of Mexico, with rain lingering over the southern half of Texas, through the Deep South and into the Coastal Southeast, with some heavy showers likely in Florida.

Mostly dry elsewhere.

10/22/2025

A cold front over the Great Lakes brings some showers mostly over the northern side of the Lakes and some isolated showers into the Northeast.

Rain in the Pacific Northwest mostly along the coastal regions of Washington and Oregon.

Isolated showers in the Rocky Mountains and in the southeastern parts from Florida to the Coastal Southeast.

Heavy rain along the north side of a slow moving front still prevails over the southern half of Texas as the front continues to drift further south into the Gulf of Mexico.

Mostly dry elsewhere.

Cold in Alaska with snow likely.

OCTOBER 22 WEDNESDAY

ISOBARIC

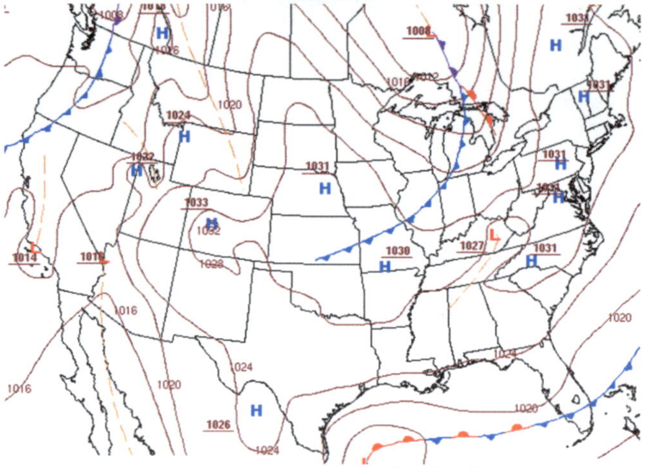

RAIN DISTRIBUTION (Inches)

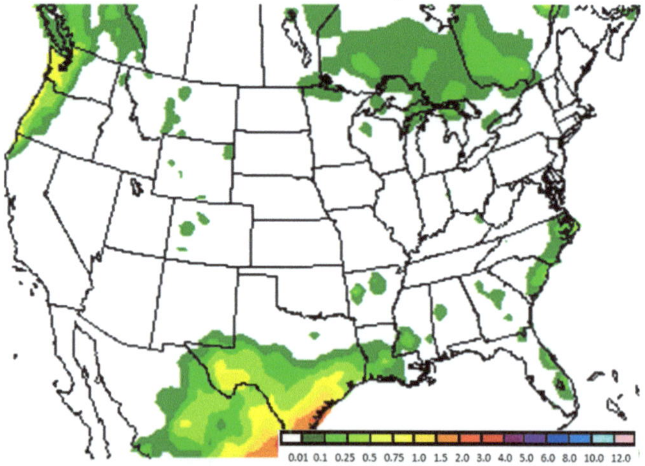

MAXIMUM TEMPERATURE / MINIMUM TEMPERATURE

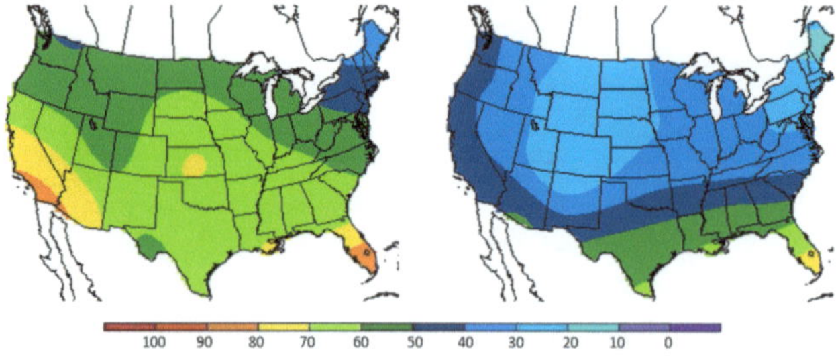

OCTOBER **23** THURSDAY

ISOBARIC

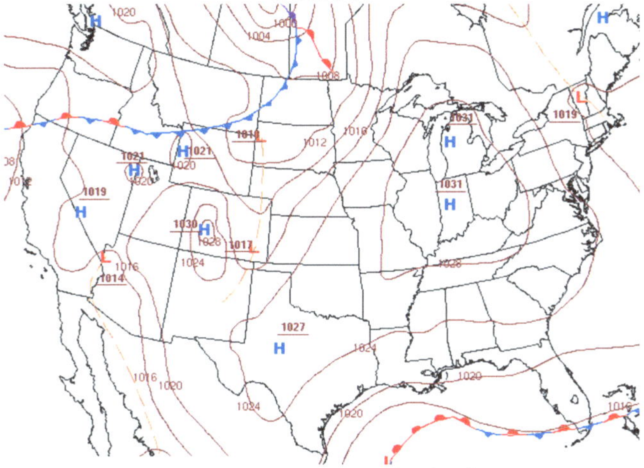

RAIN DISTRIBUTION (Inches)

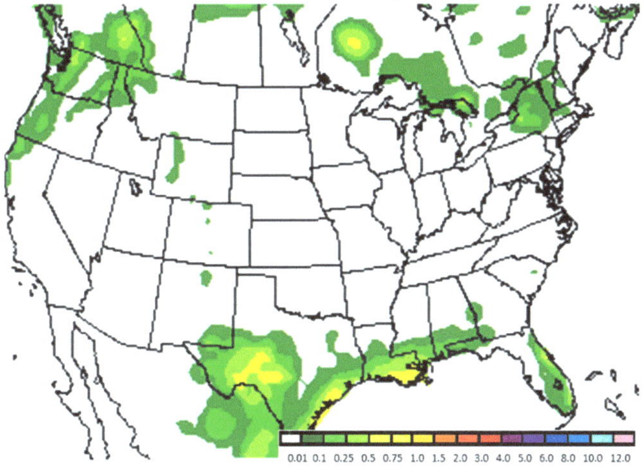

MAXIMUM TEMPERATURE

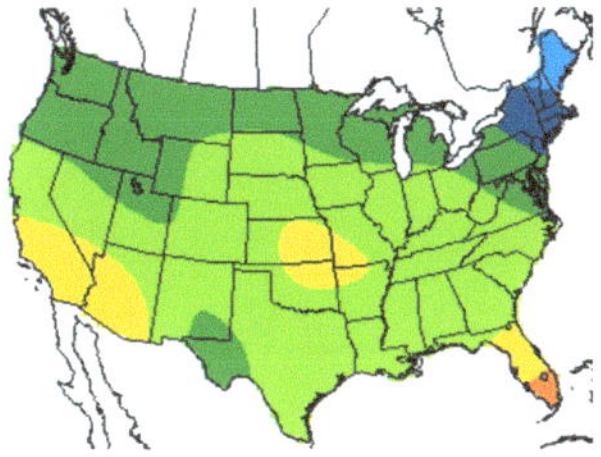

MINIMUM TEMPERATURE

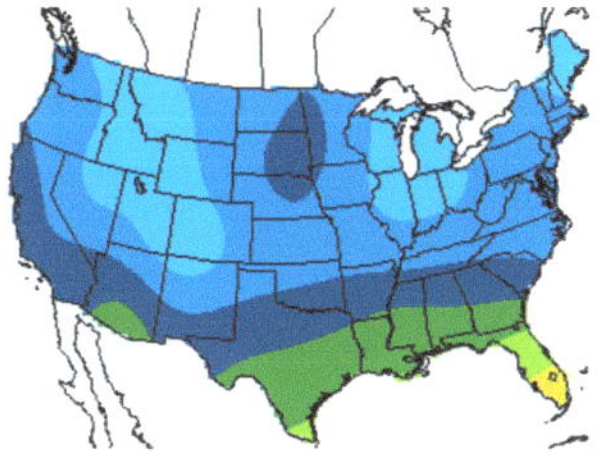

10/23/2025

A stationary front in the Gulf of Mexico still maintains widespread rain activity through the southern half of Texas and along the coastal regions of the Deep South to Florida.

Showers persist in the Pacific Northwest, as well as around the northern side of the Great Lakes to the southern regions of the Northeast.

Chance of isolated showers in the Rocky Mountains.

Mostly dry elsewhere.

Unusually cool day in Hawaii.

Snow in Alaska.

10/24/2025

A slow moving front drifts through the Gulf of Mexico spreads rain throughout most of the Deep South and Florida, while clearing to mostly dry in Texas.

Showers in the Pacific Northwest, mostly in Oregon, as well as into northern California coastal regions.

Some isolated showers in the Southwest and the odd shower may linger along the north of the Great Lakes.

Mostly dry elsewhere.

OCTOBER **24** FRIDAY

ISOBARIC

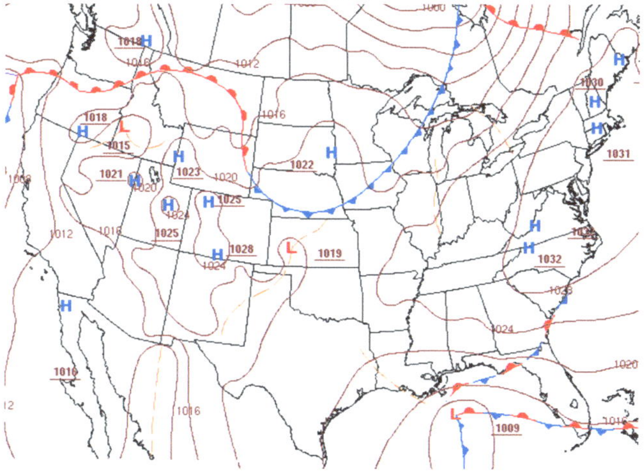

RAIN DISTRIBUTION (Inches)

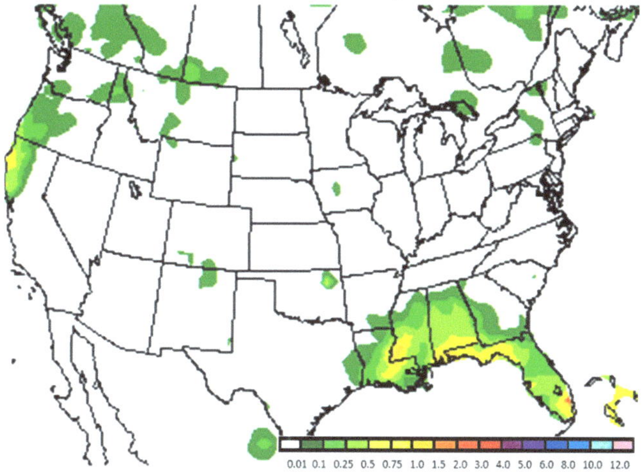

MAXIMUM TEMPERATURE

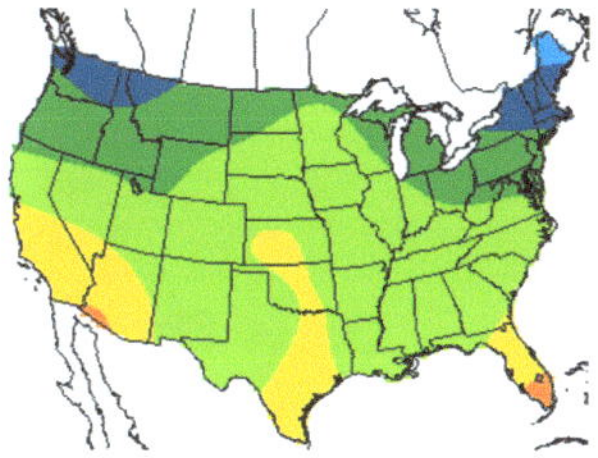

MINIMUM TEMPERATURE

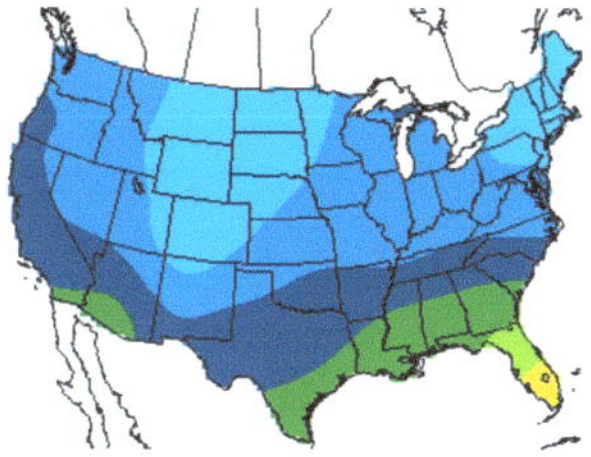

OCTOBER 25 SATURDAY

ISOBARIC

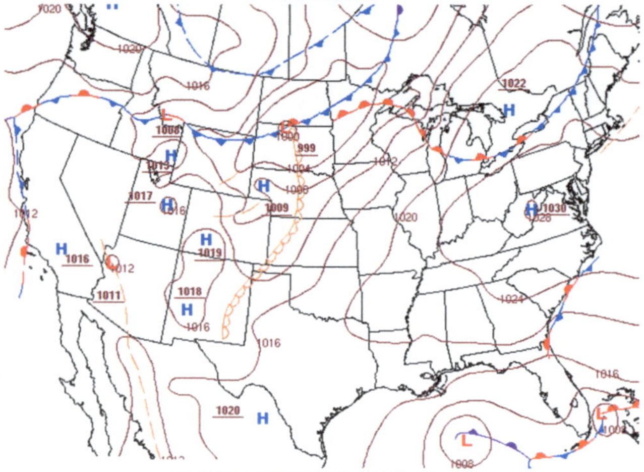

RAIN DISTRIBUTION (Inches)

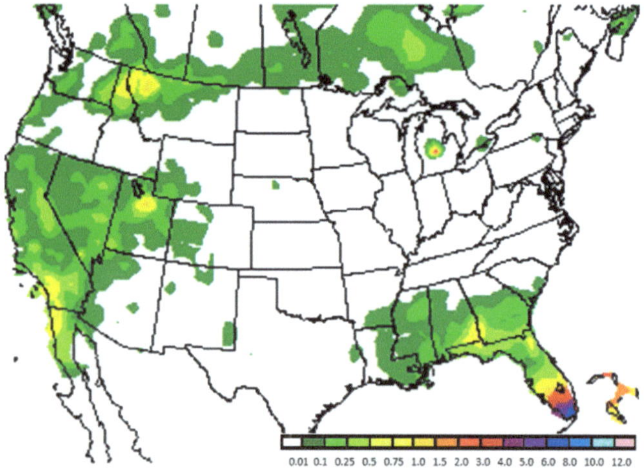

MAXIMUM TEMPERATURE

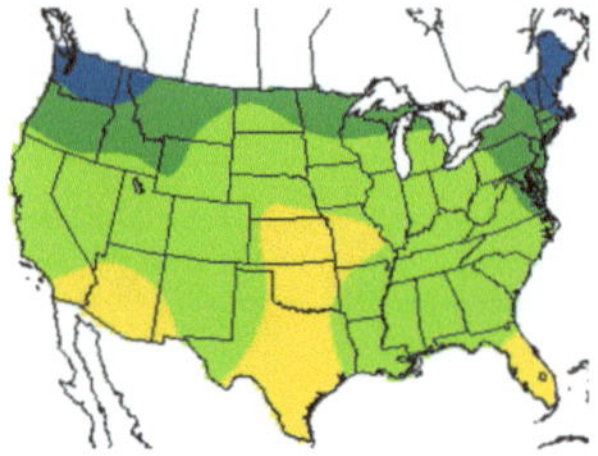

MINIMUM TEMPERATURE

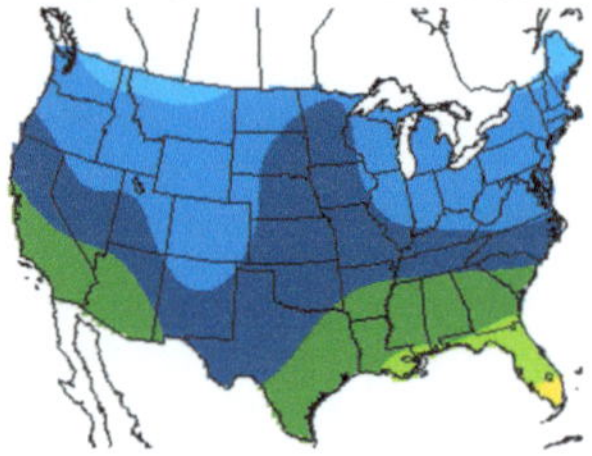

10/25/2025

Widespread rain returns over most of the west of the country with only Oregon and the southern half of Idaho mostly dry.

Elsewhere in the west, widespread rain prevails throughout the Northwest regions, California, the Great Basin and the southern reaches of the Rocky Mountains.

Rain is likely to be very heavy for a time in California, with showers spreading as far east as the central Great Plains by tomorrow, and the western parts of the Corn Belt in a couple of days.

Rain persists throughout the Deep South and into Florida, where heavy rain may cause some localized flooding in the south of the state.

Mostly dry elsewhere.

Chance of an isolated pocket of heavy rain for a short time in Michigan.

10/26/2025

Rain clears to showers over the southwest of the country from the Great Basin to the Southwest region, including in California, while clearing to mostly dry in the Pacific Northwest.

Some heavy rain spells still likely however, in the southern half of the Rocky Mountains.

Precipitation returns over the Great Lakes, with a band of showers following a wet front down through the Inland Midwest, west of the Appalachians and down into the west of the Deep South.

Rain may be particularly heavy in Wisconsin and Michigan.

Chance of torrential rain in Florida, with showers up into the Coastal Southeast Atlantic fringes.

Mostly dry elsewhere.

OCTOBER 26 SUNDAY

ISOBARIC

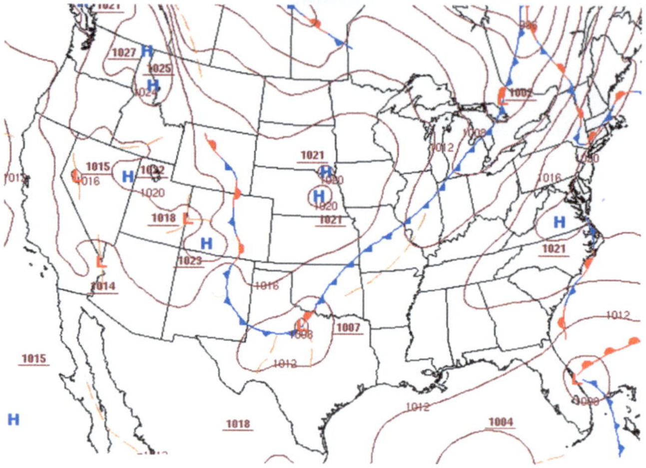

RAIN DISTRIBUTION (Inches)

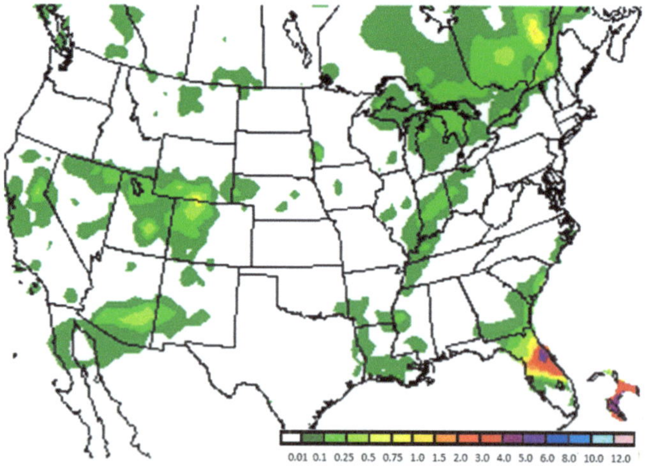

MAXIMUM TEMPERATURE

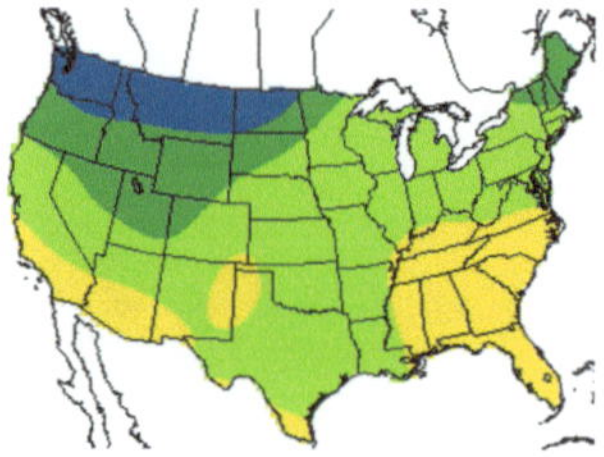

MINIMUM TEMPERATURE

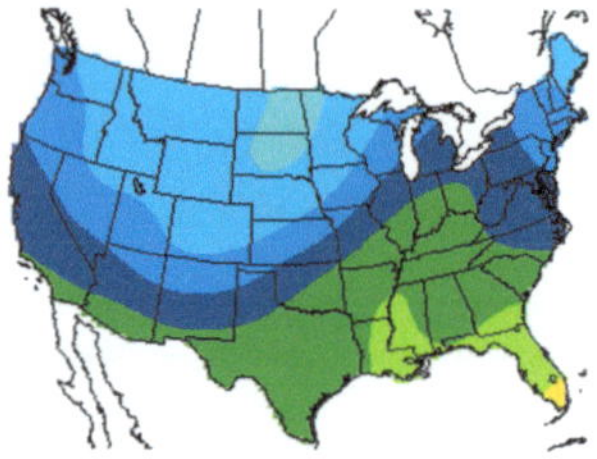

OCTOBER 27 MONDAY

ISOBARIC

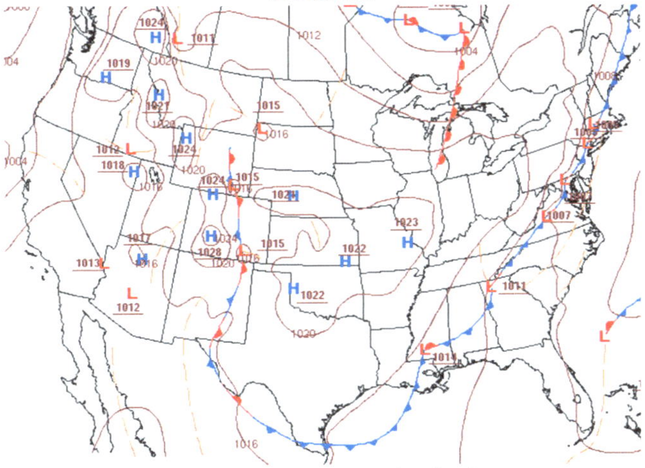

RAIN DISTRIBUTION (Inches)

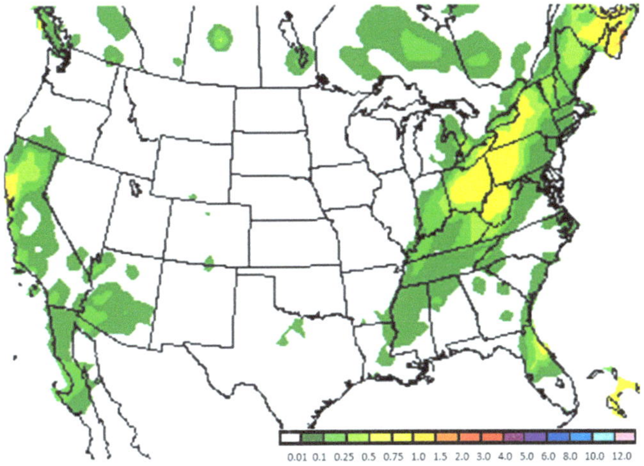

MAXIMUM TEMPERATURE

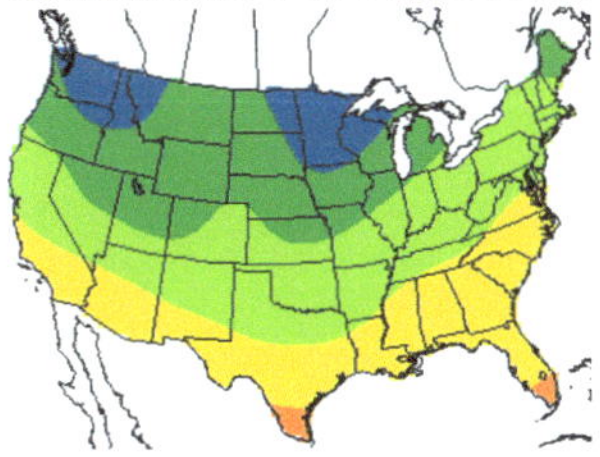

MINIMUM TEMPERATURE

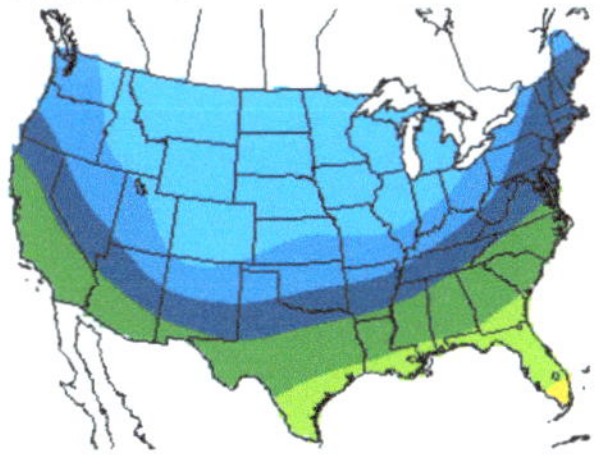

10/27/2025

A cold front crosses the east of the country bringing widespread rain from the Northeast down through the Appalachians and into the Deep South.

A band of thunderstorms is likely within this cold front mostly affecting Tennessee to Kentucky.

Scattered outbreaks of rain persist from Florida to along the coastal fringes of Georgia and the Carolinas.

Rain returns to California and west into Arizona.

Mostly dry elsewhere.

There may be a tornado outbreak over the next couple of days in the east from Michigan to the Gulf Coast regions.

Unusually high temperatures in Alabama.

10/28/2025

A complex of cold fronts and troughs bring widespread storm activity into the western half of the country with heavy rain from the Pacific Northwest to California and east into the Great Basin, Southwest, and southern Rockies.

Some rain also reaches the South Great Plains.

Rain clears in the east except for lingering showers in the Northeast region and around the north of the Great Lakes.

Chance of tornadoes between Michigan and the Gulf Coast continues.

Mostly dry elsewhere.

Temperatures may be unusually high in Tennessee.

Snow likely in Alaska over the next couple of days.

OCTOBER 28 TUESDAY

ISOBARIC

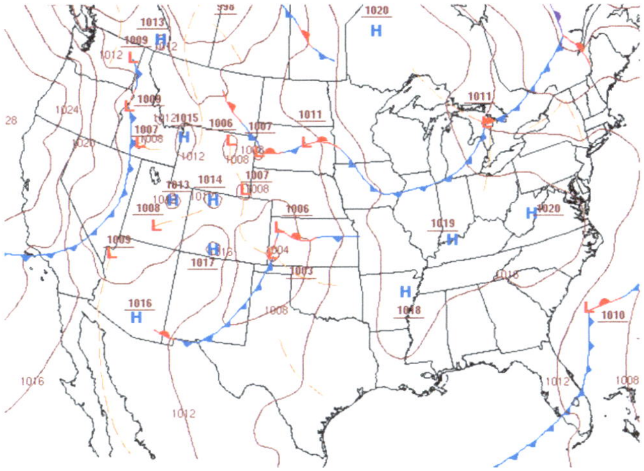

RAIN DISTRIBUTION (Inches)

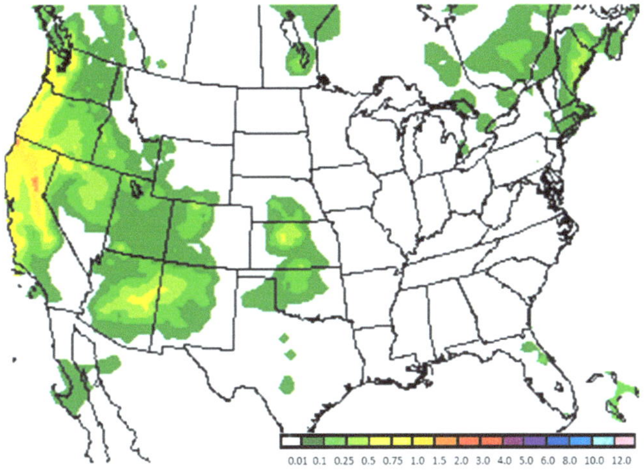

MAXIMUM TEMPERATURE

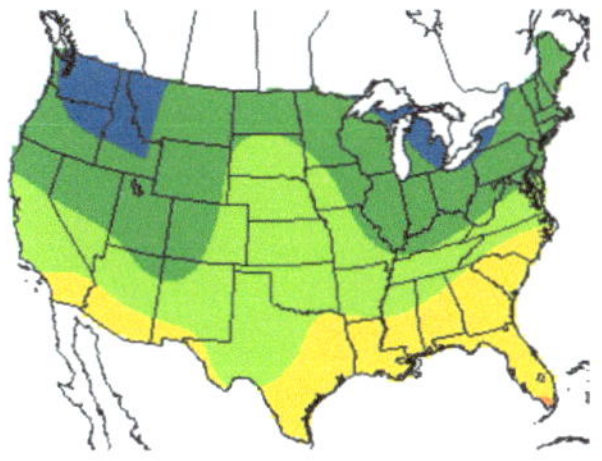

MINIMUM TEMPERATURE

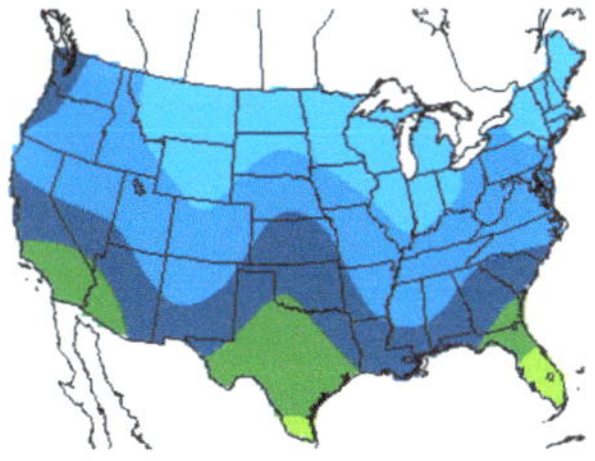

OCTOBER 29 WEDNESDAY

ISOBARIC

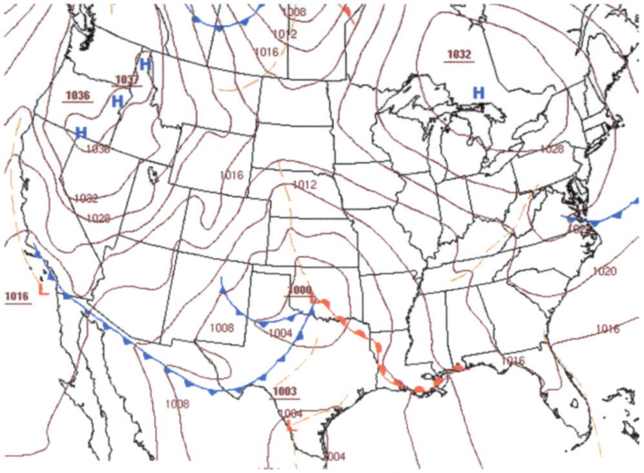

RAIN DISTRIBUTION (Inches)

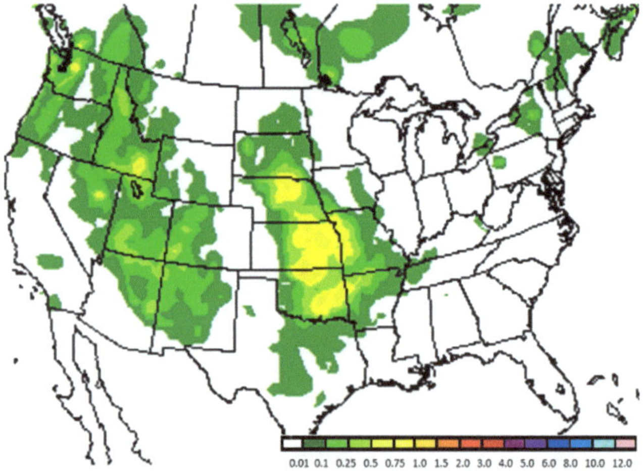

MAXIMUM TEMPERATURE / MINIMUM TEMPERATURE

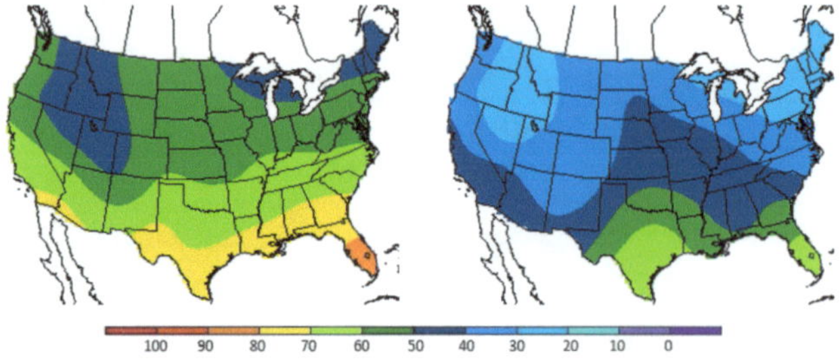

10/29/2025

Storms cross the west of the country from the Pacific Northwest and into Montana to the east, and southwards through the Great Basin and the Rocky Mountains and into the Southwest, while rain clears in California.

Similarly, a trough brings widespread and often heavy rain through the Great Plains to the Ozarks and with showers into the northern inland parts of Texas.

Showers spread into the west of the Inland Midwest, and isolated showers linger in the Northwest.

Mostly dry elsewhere.

Chance of snow in southern Alaska.

10/30/2025

A low-pressure system and its associated fronts bring widespread rain throughout the eastern half of the country, with only the Northeast to the Megalopolis, and Florida regions being mostly dry.

Most regions west of the Rocky Mountains will also be dry, while all other parts from the Rocky Mountains eastward can expect some rain activity over the next few days.

Rain may be very heavy at times in Illinois to Ohio, and throughout the Deep South and into the southern Appalachians.

Washington DC may receive its first significant rainfall ending a prolonged dry spell in the region.

Heavy rain likely to move into the Northeast to the Megalopolis by tomorrow.

Winds may be very strong in Oregon to California.

OCTOBER 30 THURSDAY

ISOBARIC

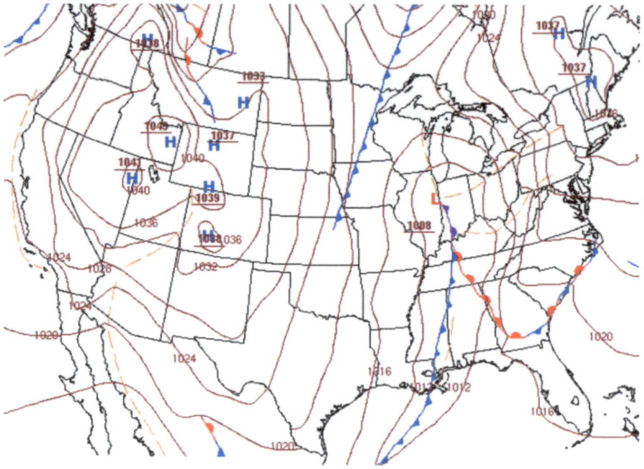

RAIN DISTRIBUTION (Inches)

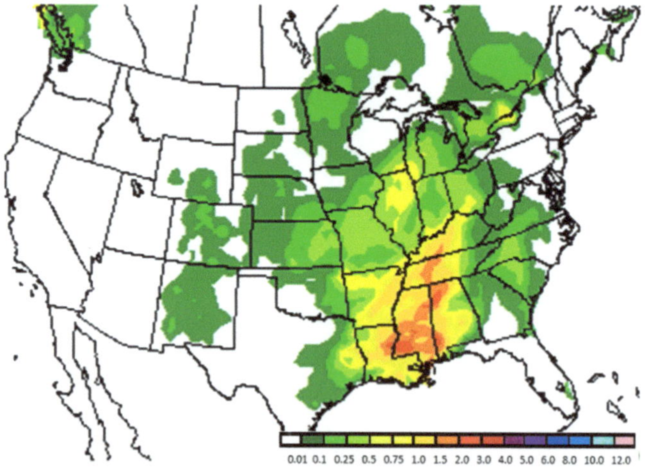

MAXIMUM TEMPERATURE / MINIMUM TEMPERATURE

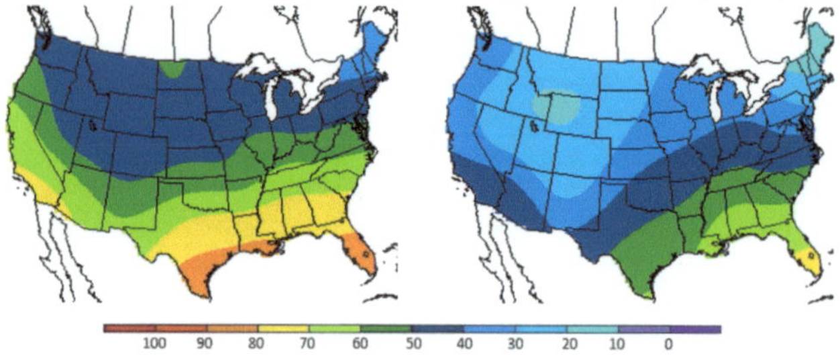

ISOBARIC

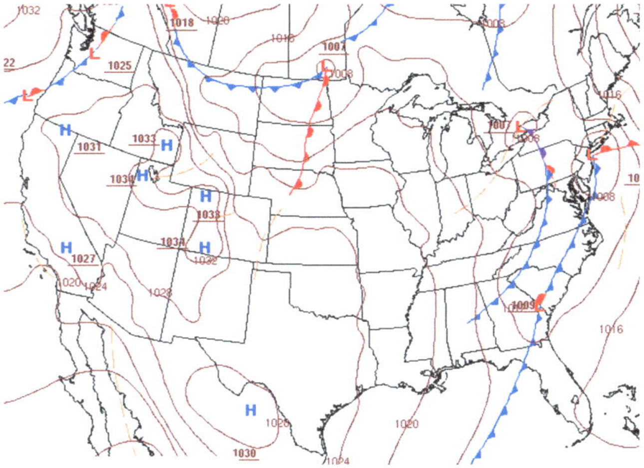

RAIN DISTRIBUTION (Inches)

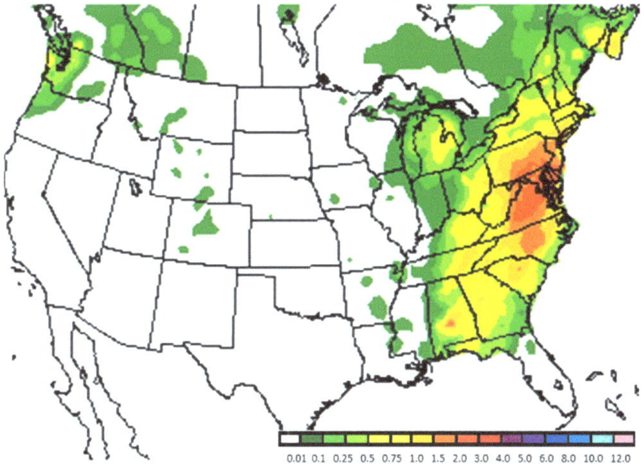

MAXIMUM TEMPERATURE   MINIMUM TEMPERATURE

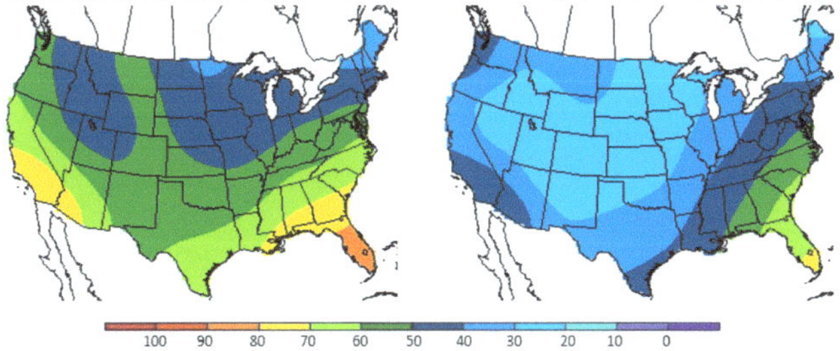

10/31/2025

Cold fronts continue to spread storms over the east from the Great Lakes, Inland Midwest and Deep South regions to the Atlantic Coasts, with very heavy rain from Virginia to the Megalopolis.

Some rain prevails around the Pacific Northwest coasts, and isolated showers in the Rocky Mountains.

Mostly dry elsewhere.

Strong winds over southern California may aggravate any wildfires that may exist.

It may be unusually cool in Utah and Wyoming with patches of frost likely, while being unusually warm in Texas and Michigan.

Minimum temperatures may be very low in Hawaii.

NOVEMBER

PHASES OF THE MOON

3rd	Crossing Equator
5th	Full Moon
5th	Perigee #1
8th	Northern Declination
12th	Third Quarter
16th	Crossing Equator
20th	New Moon
20th	Apogee
22nd	Southern Declination
28th	First Quarter
30th	Crossing Equator

1 Sat	2 Sun	3 Mon	4 Tue	5 Wed	6 Thur	7 Fri	8 Sat	9 Sun	10 Mon	11 Tue	12 Wed	13 Thur	14 Fri	15 Sat	16 Sun	17 Mon	18 Tue	19 Wed	20 Thur	21 Fri	22 Sat	23 Sun	24 Mon	25 Tue	26 Wed	27 Thur	28 Fri	29 Sat	30 Sun
		XhN		F P1			^				3Q				XhS				N A		V						1Q		XhN

NOVEMBER 1 SATURDAY

ISOBARIC

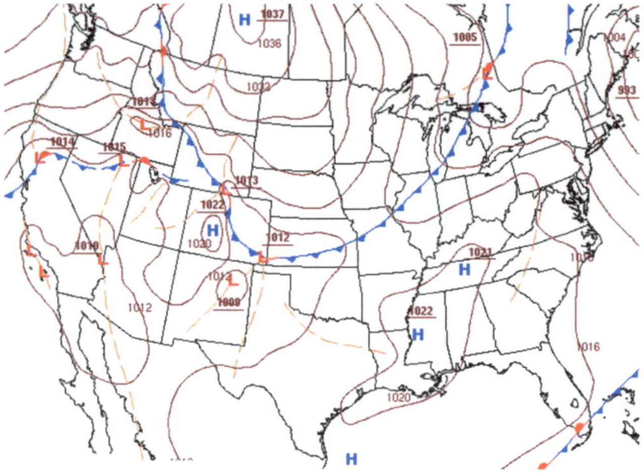

RAIN DISTRIBUTION (Inches)

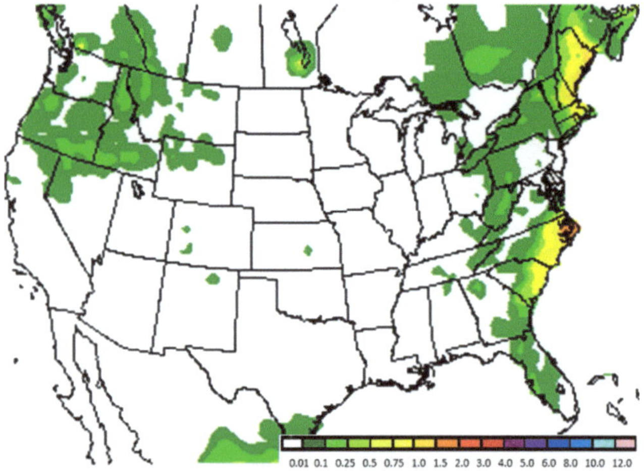

MAXIMUM TEMPERATURE **MINIMUM TEMPERATURE**

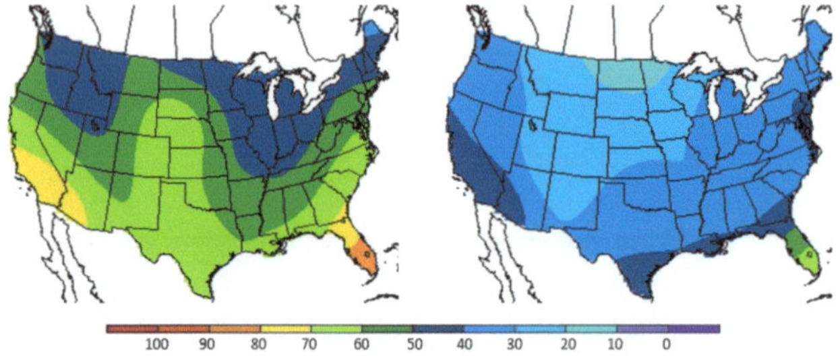

11/01/2025

Scattered outbreaks of rain in the Pacific Northwest to the northern fringes of the Great Basin and isolated showers eastward into the northern Rockies.

An Atlantic depression maintains rain over the Northeast with showers into the Appalachians and the eastern edge of the Inland Midwest.

Heavy rain continues along the Atlantic coastal fringes from the Coastal Southeast to Florida.

Some lingering rain around the southern tip of Texas.

Mostly dry elsewhere.

Unusually warm in the Northeast, including in Maine and Vermont.

There may be a large variation in temperatures, with many places noting unusually low overnight minimums, while daily maximums up to 5°F above the norm for this time of year.

NOVEMBER 2 SUNDAY

ISOBARIC

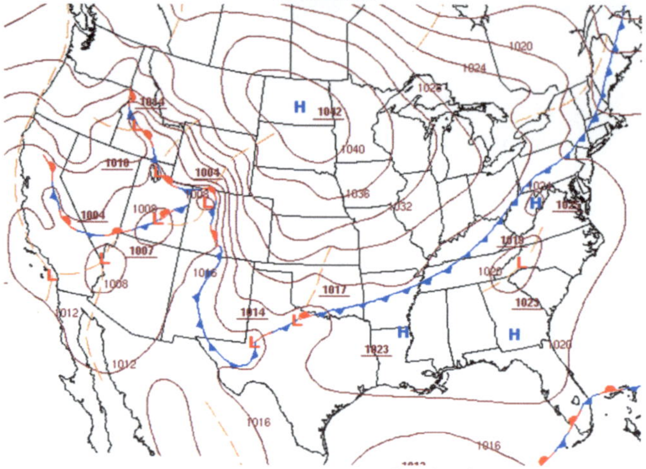

RAIN DISTRIBUTION (Inches)

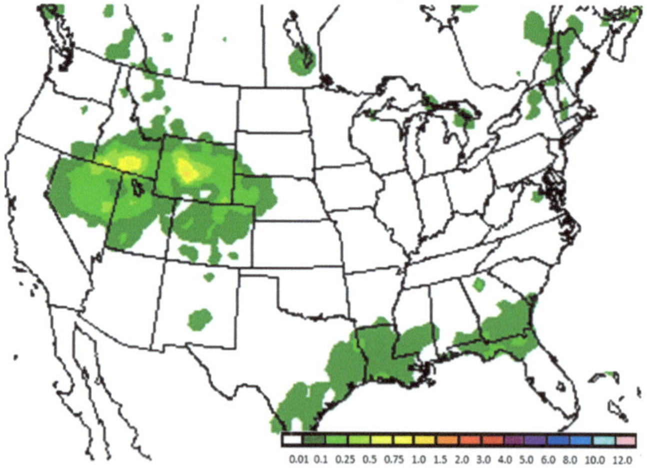

MAXIMUM TEMPERATURE **MINIMUM TEMPERATURE**

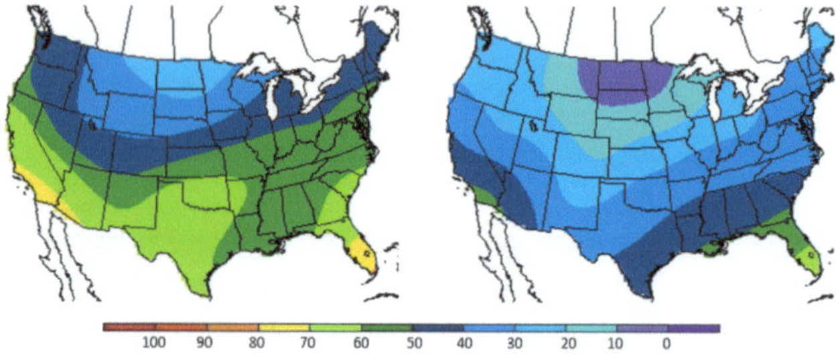

11/02/2025

Rain clears in the northeast, with just a few isolated showers lingering around the Northeast.

Cold fronts bring widespread rain through the Great Basin to the central and southern Rocky Mountains.

Some rain is also likely to prevail from the eastern half of Texas to through the southern half of the Deep South, possibly with showers into northern Florida.

Mostly dry elsewhere.

In California, the heat may be reaching its peak, with maximum temperatures still nearing 100°F the next couple of days.

Further east in places like Georgia and Florida, maximum temperatures may still reach into the 90's, while it may be cooler across the South Great Plains and into Texas with possible patches of frost.

NOVEMBER 3 MONDAY

ISOBARIC

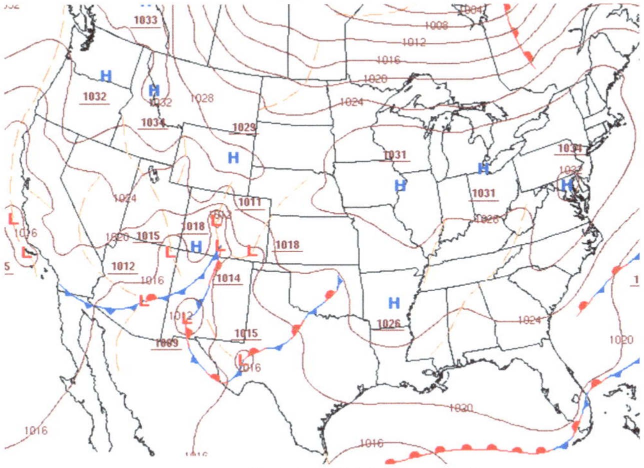

RAIN DISTRIBUTION (Inches)

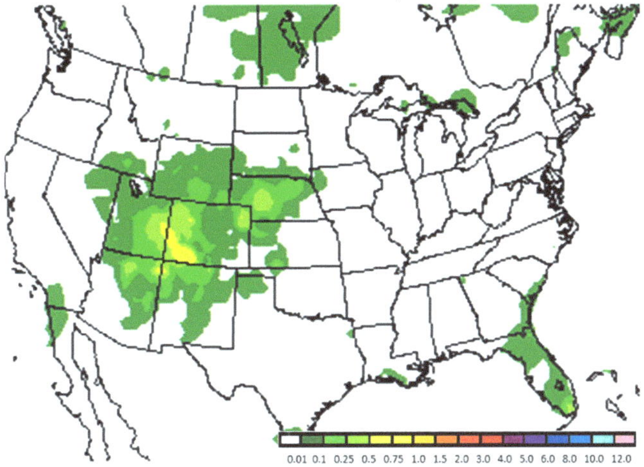

MAXIMUM TEMPERATURE / MINIMUM TEMPERATURE

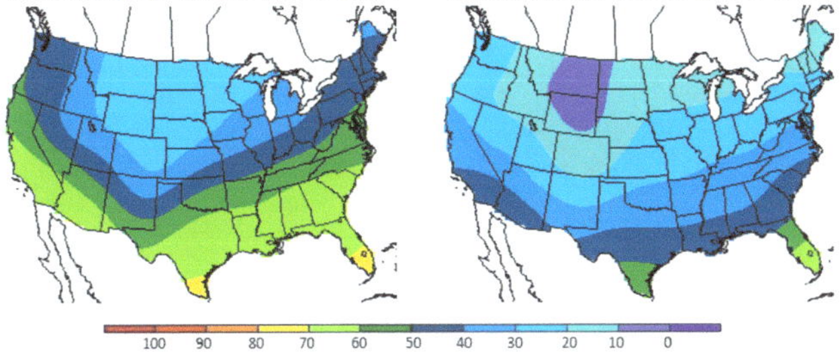

11/03/2025

A complex of low pressure, fronts and troughs bring widespread and occasionally heavy rain in the Great Basin, Southwest, southern Rocky Mountains and east into the South Great Plains.

Some isolated showers around the Northeast, occasional showers in Florida and up through the coasts of South Carolina and some lingering showers in the Rio Grande Valley.

Mostly dry elsewhere.

It may be very warm in Texas to New Mexico ahead of advancing storm, and very hot in California.

11/04/2025

Cold fronts cross the mid-section of the country returning widespread rain throughout the Great Plains, over the Great Lakes and the western half of the Inland Midwest, as well as down through the Ozarks to the Deep South and central Texas.

Rain clears in the Great Basin, but showers persist in the southern Rocky Mountains to the Southwest, as well as spread into the northern Rockies.

Chance of some showers in coastal parts of the Carolinas, southern parts of the Northeast and isolated showers in Florida.

Mostly dry elsewhere

Warm temperatures may return to the central Great Plains with maximums possibly reaching as high as 80°F in some parts of Montana and Colorado.

Chance of heavy rain in the Carolinas to Washington DC developing in the next couple of days.

NOVEMBER 4 TUESDAY

ISOBARIC

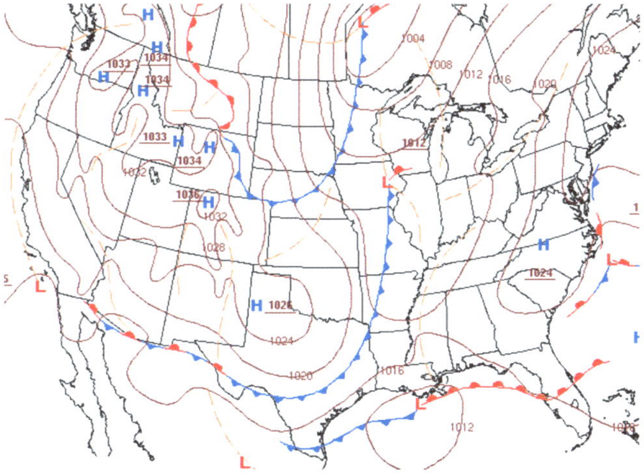

RAIN DISTRIBUTION (Inches)

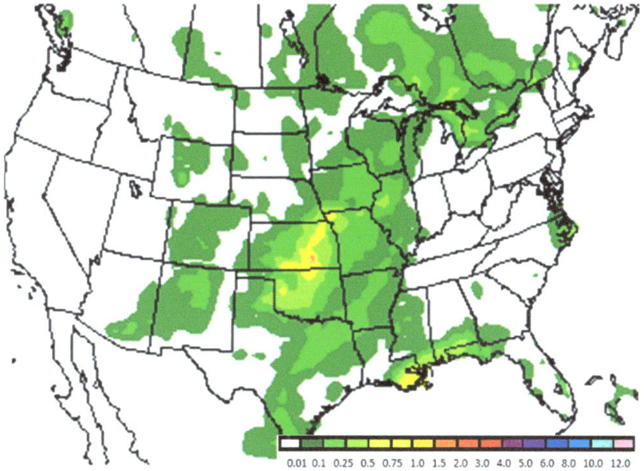

MAXIMUM TEMPERATURE / MINIMUM TEMPERATURE

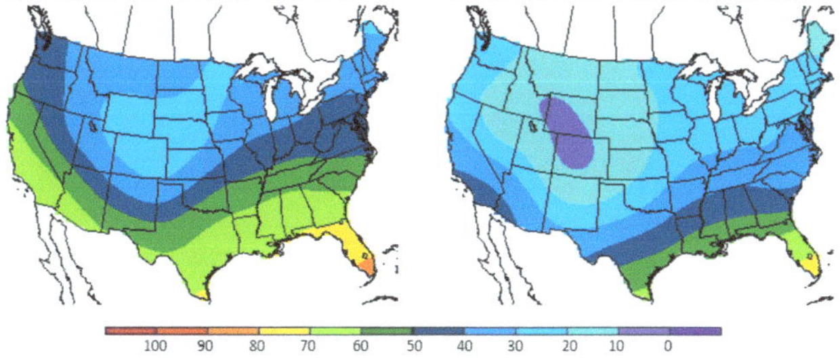

NOVEMBER 5 WEDNESDAY

ISOBARIC

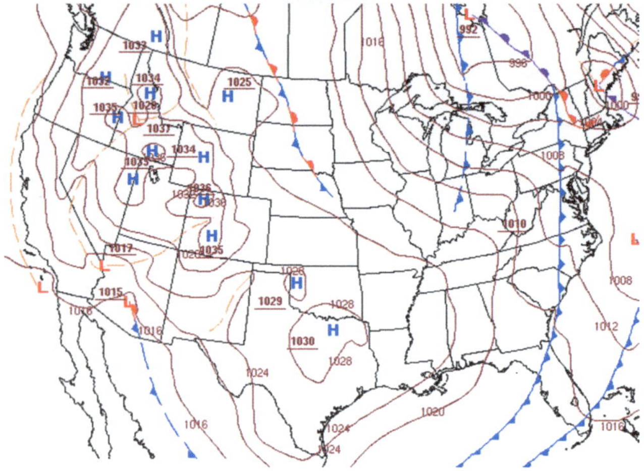

RAIN DISTRIBUTION (Inches)

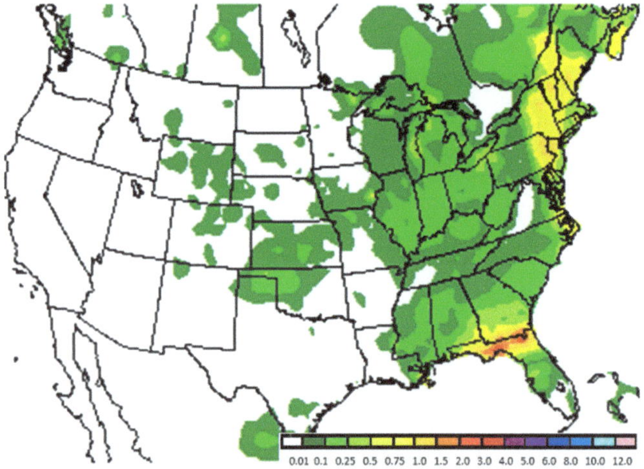

MAXIMUM TEMPERATURE

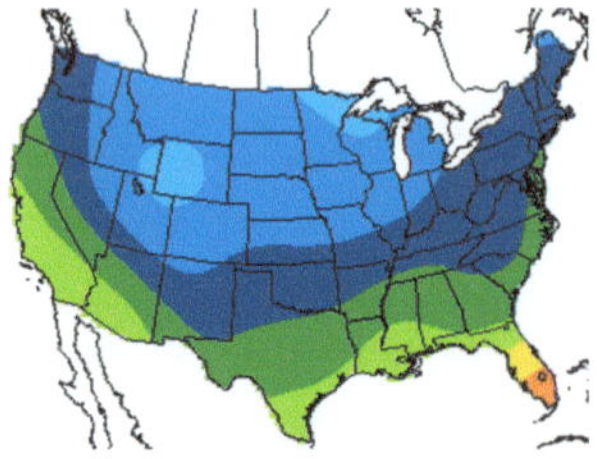

MINIMUM TEMPERATURE

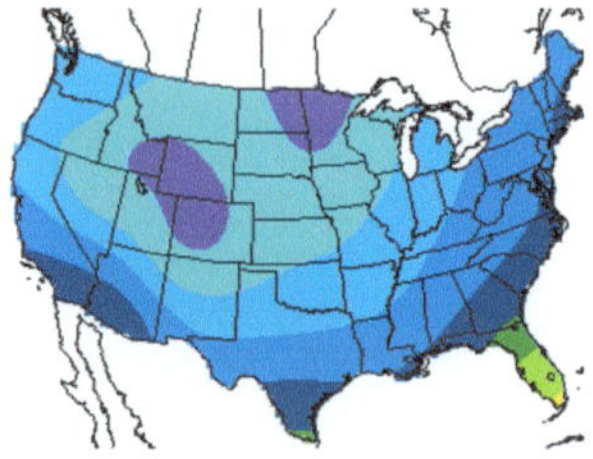

11/05/2025

The passage of wet fronts brings showers over the Great Plains and heavier and widespread rain from the Inland Midwest and Great Lakes to the Atlantic Coast in the east and south into the Deep South to Florida.

Precipitation may be particularly heavy in the Northeast to the Megalopolis regions as well as in Washington DC, the Carolinas, and the northern parts of Florida.

While widespread precipitation prevails in the east of the country, the west will be mostly dry, except for some spotty showers in the Rocky Mountains.

Showers also likely to persist in the Rio Grande Valley.

Frosts likely in the Northwest, including in Oregon, over the next couple of days.

It may be very wet in southern Alaska.

11/06/2025

The wet fronts in the east of the country weaken as they drift away into the Atlantic Ocean, clearing much of the precipitation that had prevailed in the past couple of days.

Precipitation lingers over the Great Lakes to the Northeast, while it clears to mostly dry elsewhere.

Also, mostly dry in the western half of the country, except for a few showery patches mostly in the southern Rocky Mountains to the Southwest and South Great Plains.

Chance of frosts in the Northwest, including in Oregon and Montana.

NOVEMBER 6 THURSDAY

ISOBARIC

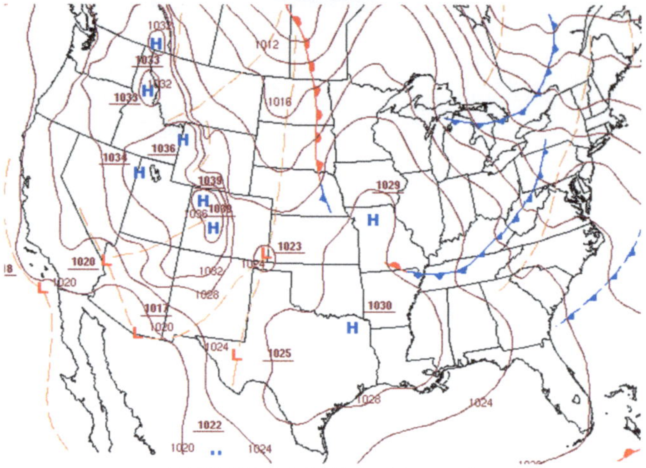

RAIN DISTRIBUTION (Inches)

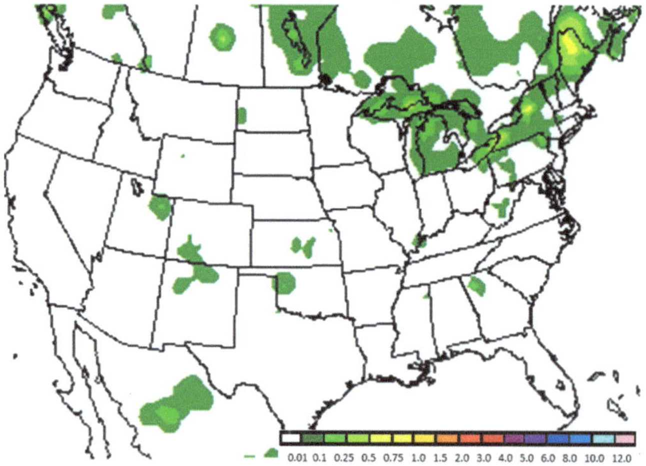

MAXIMUM TEMPERATURE

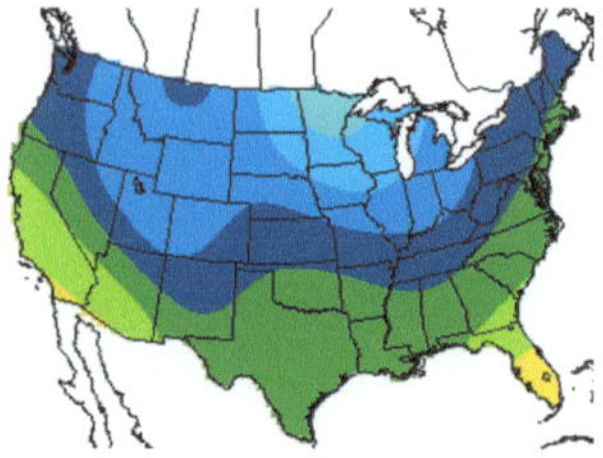

MINIMUM TEMPERATURE

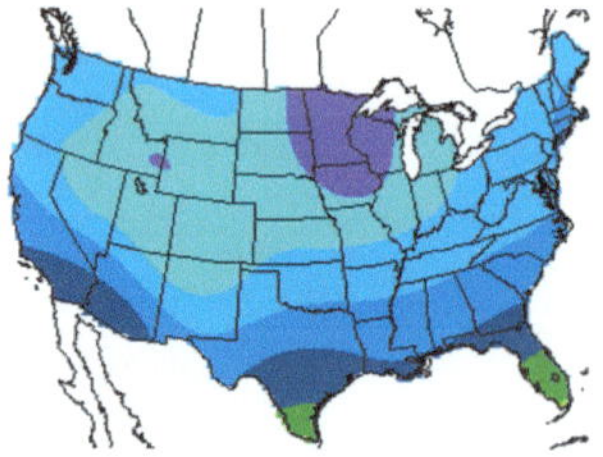

NOVEMBER 7 FRIDAY

ISOBARIC

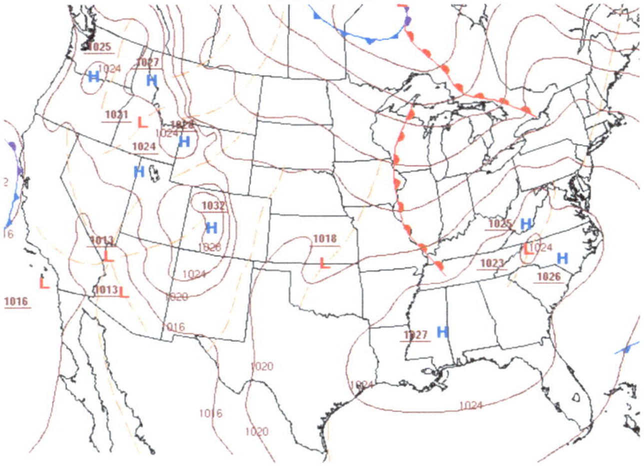

RAIN DISTRIBUTION (Inches)

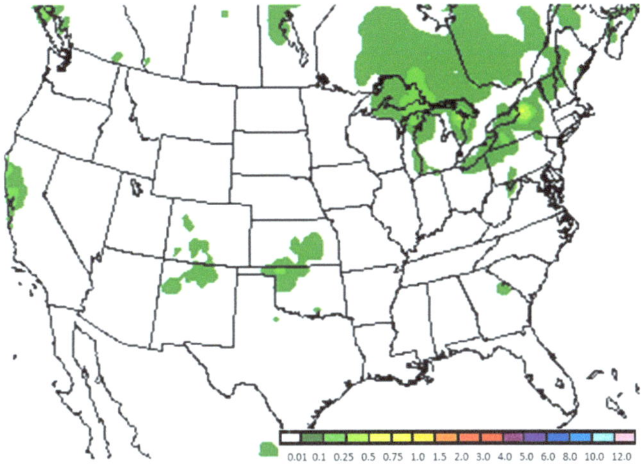

MAXIMUM TEMPERATURE

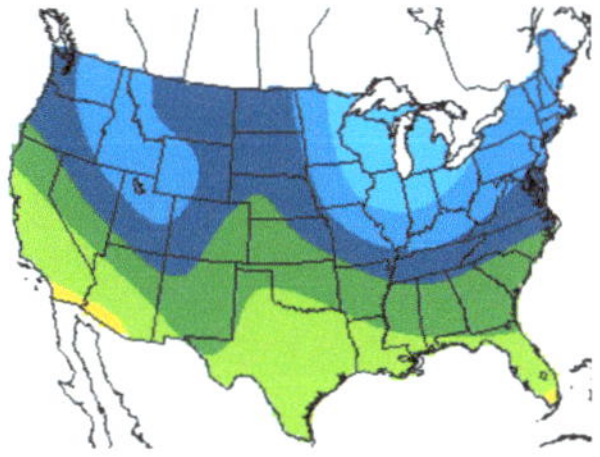

MINIMUM TEMPERATURE

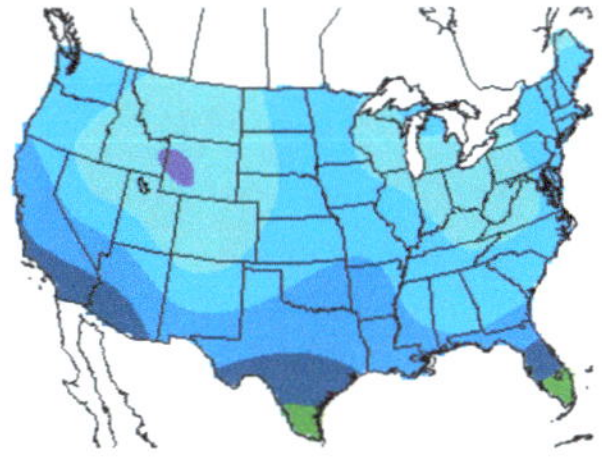

11/07/2025

Some showers persist around the Great Lakes to the Northeast.

Isolated showers in the Southwest to the South Great Plains and briefly in the Coastal Southeast to Georgia.

An occluded front may bring some showers along the coastal fringes of northern California.

Mostly dry elsewhere.

It may become blustery for a time along the Florida coastlines over the next few days with the passage of an Atlantic tropical depression east of the state.

Very cold with frosts in central and eastern parts of the country from Virginia to Ohio, while it may be unusually warm in Arizona.

11/08/2025

A trough crossing Texas brings outbreaks of rain over the southern half of the state and with showers up into the Ozarks and along the coastal fringes of the Deep South.

Showers also in the Southwest, along the Californian and Oregon coasts, and isolated showers in the South Great Plains.

Chance of some isolated showers on the northern side of the Great Lakes to the Northeast.

Mostly dry elsewhere with a slightly warmer airflow, especially over the eastern half of the country.

NOVEMBER 8 SATURDAY

ISOBARIC

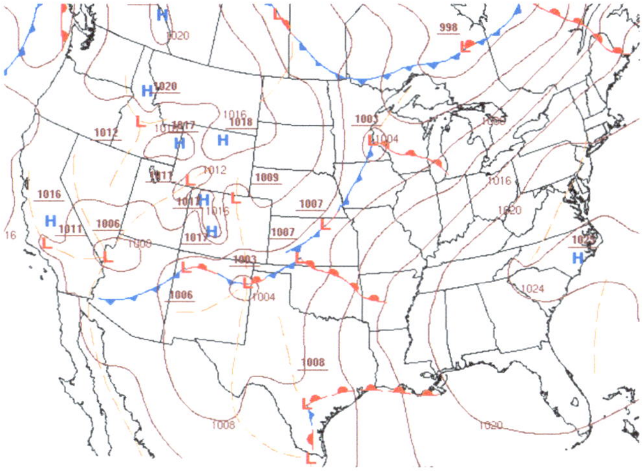

RAIN DISTRIBUTION (Inches)

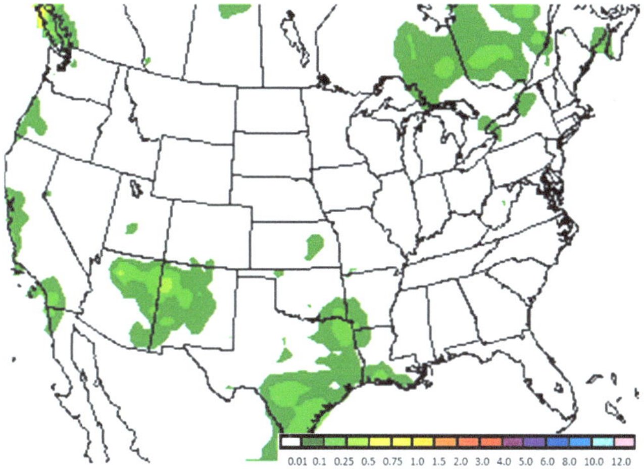

MAXIMUM TEMPERATURE

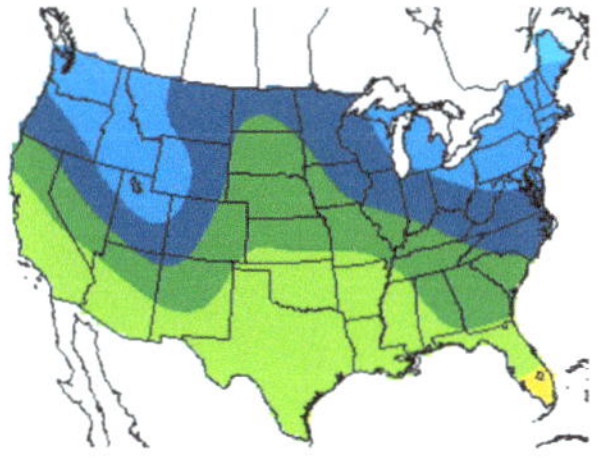

MINIMUM TEMPERATURE

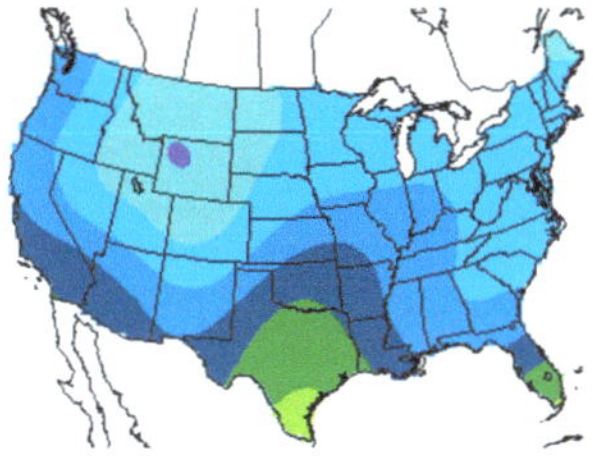

NOVEMBER 9 SUNDAY

ISOBARIC

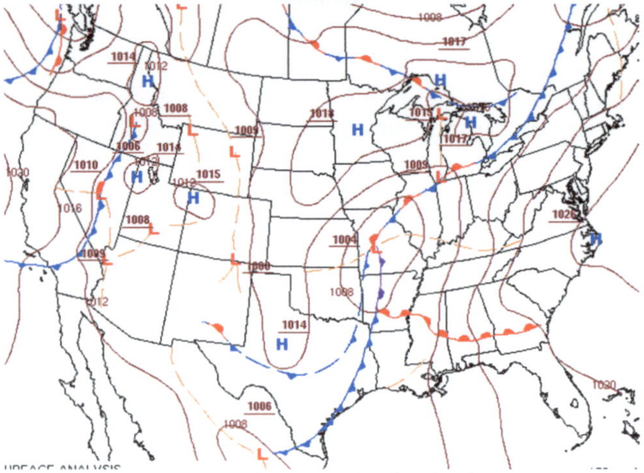

RAIN DISTRIBUTION (Inches)

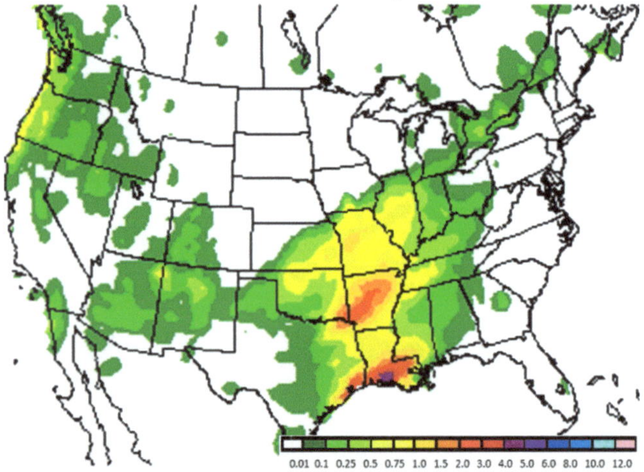

MAXIMUM TEMPERATURE

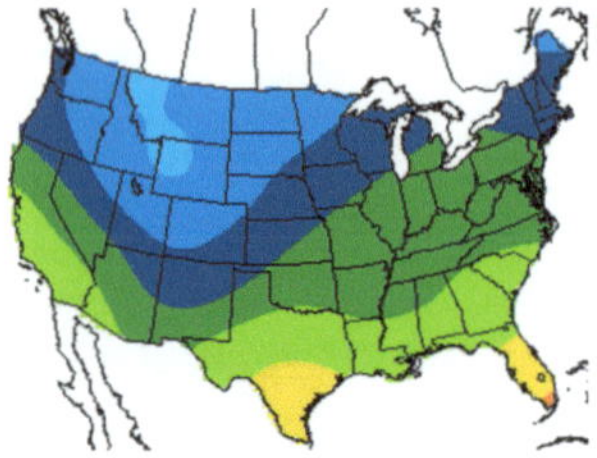

MINIMUM TEMPERATURE

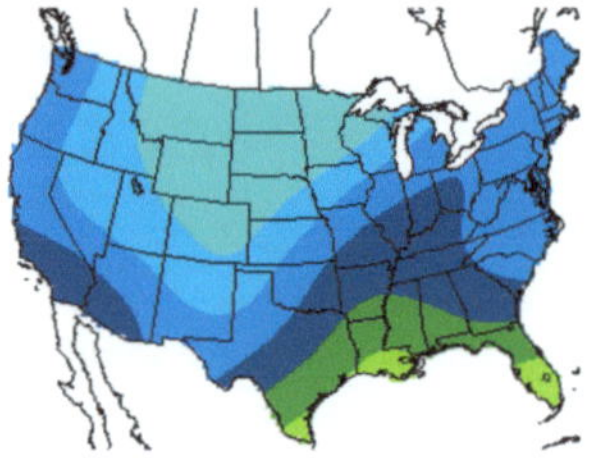

11/09/2025

Cold fronts strengthen as they move out of the mid-section towards the east of the country, returning widespread precipitation from Texas, up through the Appalachians and over the Inland Midwest to the east of the Great Lakes into the lower Northeast.

Rain may be particularly heavy in the Ozarks and in a line from Missouri, through Arkansas and down into Louisiana.

Rain also spreads westward into the Southwest and southern Rocky Mountains.

Rain develops throughout the Pacific Northwest to the Rocky Mountains as well as showers in the Great Basin and California.

Mostly dry over the North Great Plains and through the Atlantic seaboard states from the Northeast to Florida.

11/10/2025

Wet fronts spread rain eastward from the Inland Midwest to the Atlantic coasts, with heavy falls still likely in Arkansas to Louisiana.

Rain also in the eastern coasts of Texas.

I depression north of Washington brings widespread rain into the Pacific and Inland Northwest regions, down through the west of the Rocky Mountains, across the Great Basin and showers into the Southwest and northern California.

Mostly dry through the mid-section states, and although mostly dry, rain may threaten in the Coastal Southeast and Florida, mostly along the coasts.

NOVEMBER 10 MONDAY

ISOBARIC

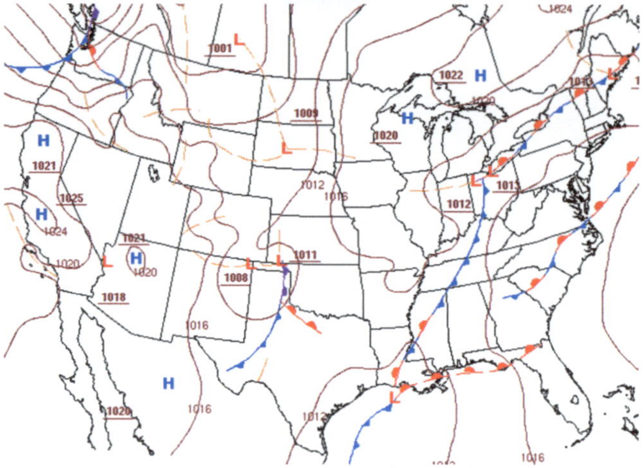

RAIN DISTRIBUTION (Inches)

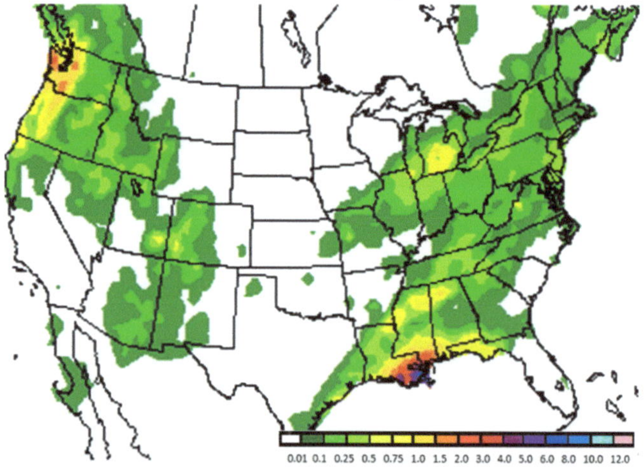

MAXIMUM TEMPERATURE

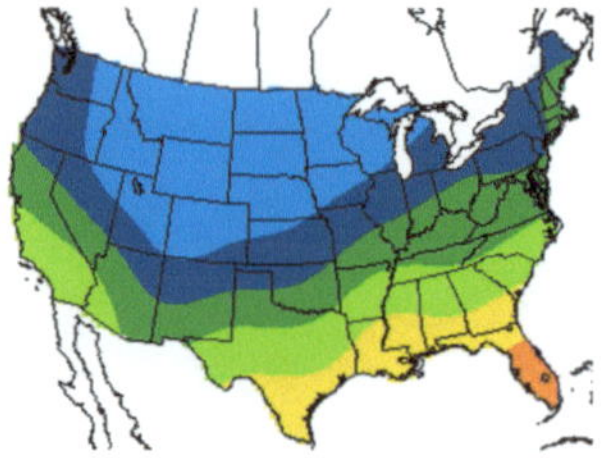

MINIMUM TEMPERATURE

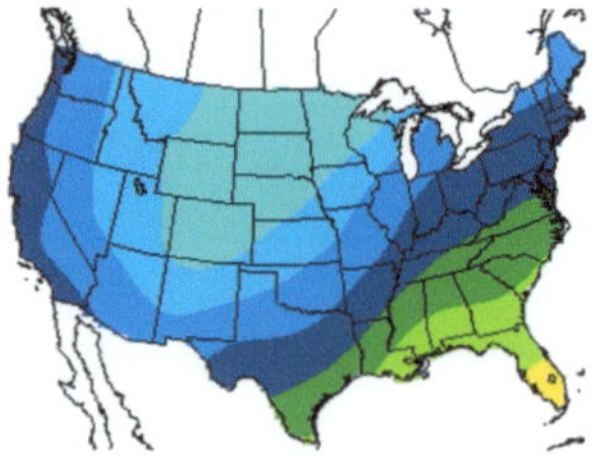

NOVEMBER 11 TUESDAY

ISOBARIC

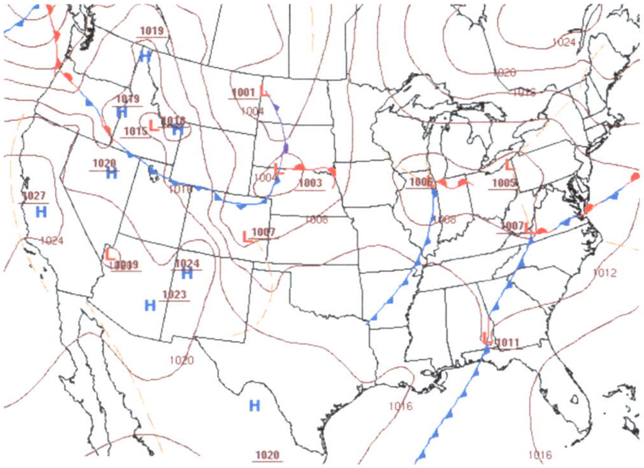

RAIN DISTRIBUTION (Inches)

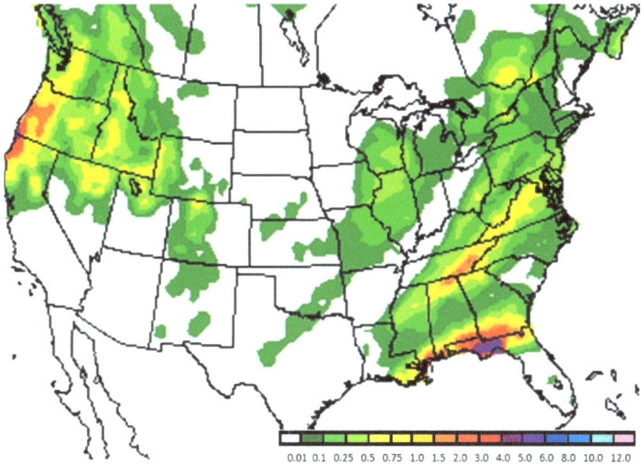

MAXIMUM TEMPERATURE MINIMUM TEMPERATURE

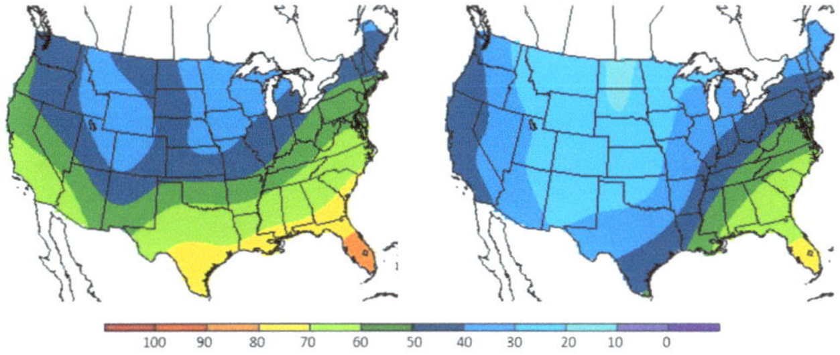

11/11/2025

A depression near the Great Lakes brings widespread rain and strong winds throughout the northeast of the country, including the chance of the strong winds generating large waves on the Great Lakes.

Rain in the Inland Midwest and all areas from the Appalachians to the Atlantic Ocean.

Heavy rain along the Gulf Coast may cause some localized flooding in coastal regions of the Deep South.

In the West, widespread precipitation from the Pacific and Inland Northwest regions, over the Great Basin and down the southern half of the Rocky Mountains.

Heavy rain also in northern California and the Sierra Nevada and Cascade Ranges.

Some showers in the South Great Plains to the central north of Texas. Mostly dry elsewhere.

There is a chance of a late season tropical storm developing to the southeast of Florida bringing unsettled and very strong winds over the State.

11/12/2025

A depression to the northwest of Washington brings stormy conditions into the Inland Northwest, Pacific, northern California, Sierra Nevada and Cascade Ranges, the Great Basin, and the Rocky Mountains.

Some isolated showers in the central Great Plains.

Stormy and wet over the Great Lakes, Inland Midwest and Northeast regions.

Patches of rain remain in the southeast from Georgia to Florida.

Mostly dry elsewhere.

Winds may be unsettled with strong gusts at times along the Atlantic coasts from North Carolina to Maine and Massachusetts over the next couple of days.

NOVEMBER 12 WEDNESDAY

ISOBARIC

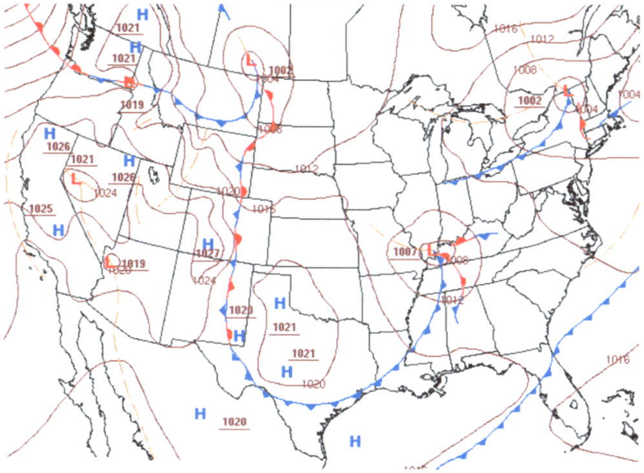

RAIN DISTRIBUTION (Inches)

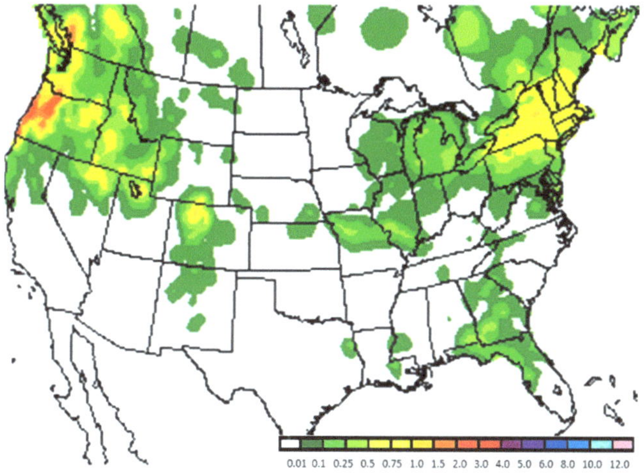

MAXIMUM TEMPERATURE MINIMUM TEMPERATURE

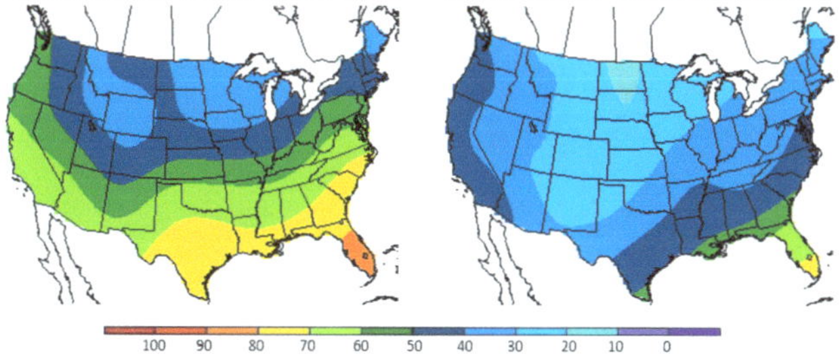

NOVEMBER **13** THURSDAY

ISOBARIC

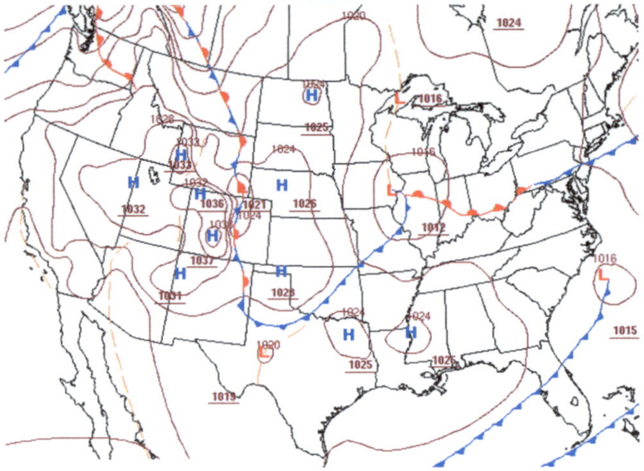

RAIN DISTRIBUTION (Inches)

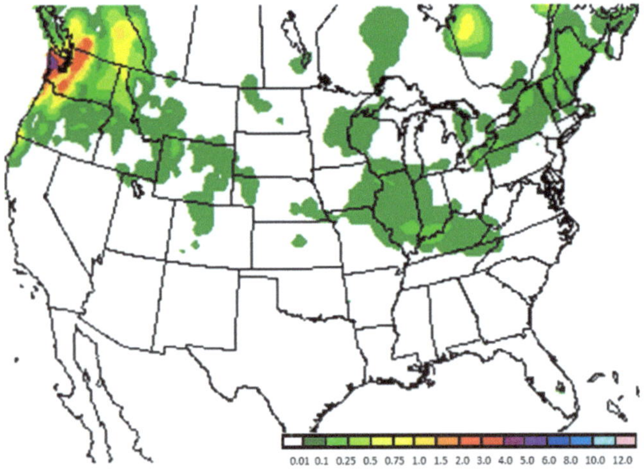

MAXIMUM TEMPERATURE / MINIMUM TEMPERATURE

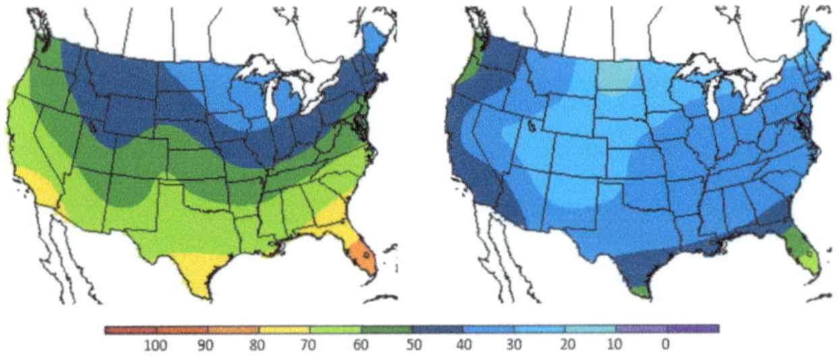

11/13/2025

Rain in the Pacific Northwest with heavy falls in Washington likely, while lighter rain and showers prevail through Oregon and into the northern Rocky Mountains.

Some rain persists in the Inland Midwest to the Great Lakes as well as throughout the Northeast.

Mostly dry throughout the south of the country.

Chance of a tropical storm brushing the Northeast coastline over the next few days.

Snow may also arrive in the Northeast, including in Maine.

Heavy rain in Hawaii over the next few days and heavy snow in the southeast of Alaska.

11/14/2025

Chance of an Atlantic tropical depression building east of the Megalopolis.

Rain in the northeast quarter of the country from the Great Lakes to the Inland Midwest and northern half of the Appalachians to the Coastal Southeast.

Some showers in the far south of the Northeast.

Heavy rain also in the far northwest, through Washington and Oregon, and with the odd patches of heavy precipitation into the west of Montana and isolated showers over the higher reaches of the Rocky Mountains.

Mostly dry elsewhere.

Chance of heavy showers in Hawaii.

Stormy weather bringing snow in Alaska over the next couple of days, particularly in the southwest of the state.

It may be unusually mild in Colorado, Kansas and Washington.

NOVEMBER **14** FRIDAY

ISOBARIC

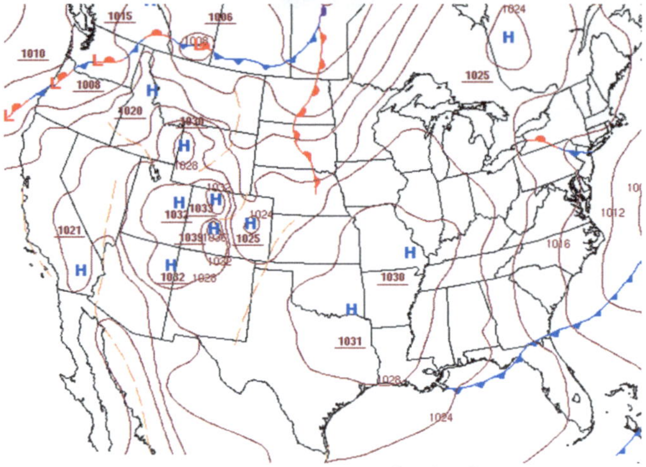

RAIN DISTRIBUTION (Inches)

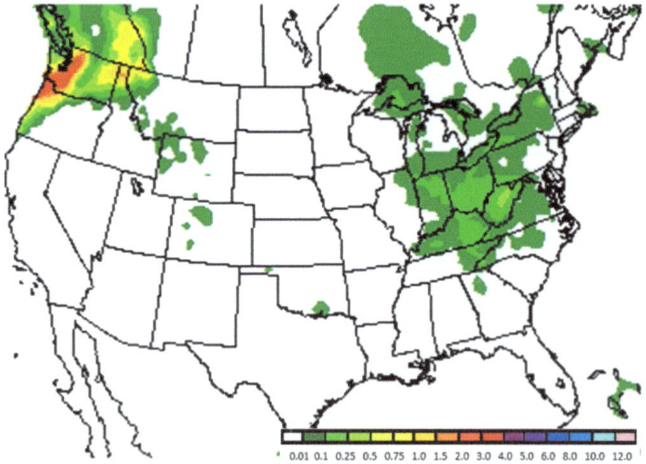

MAXIMUM TEMPERATURE / MINIMUM TEMPERATURE

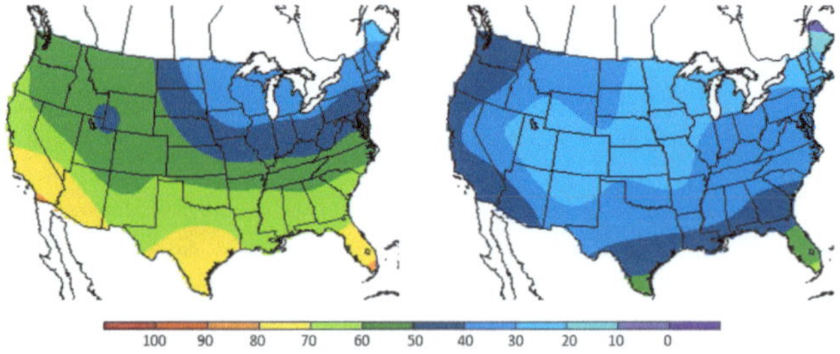

NOVEMBER 15 SATURDAY

ISOBARIC

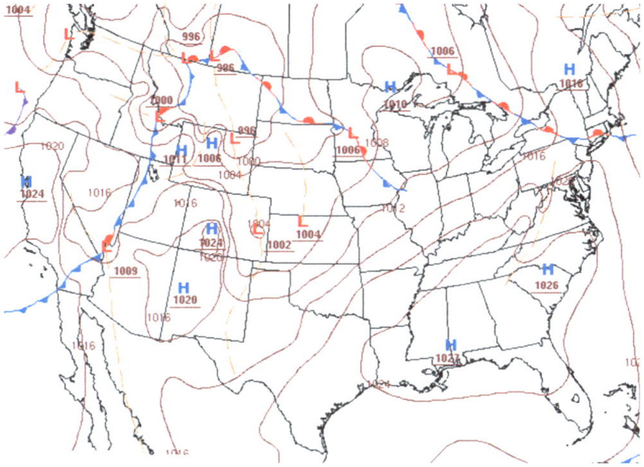

RAIN DISTRIBUTION (Inches)

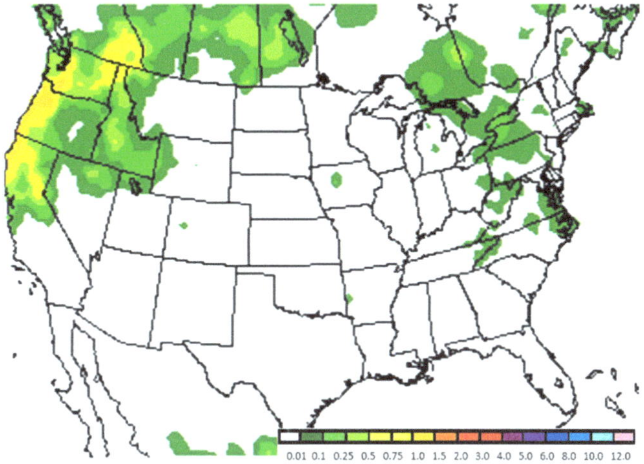

MAXIMUM TEMPERATURE

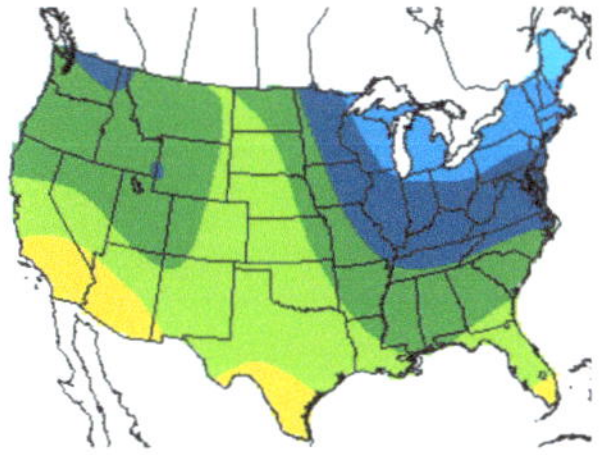

MINIMUM TEMPERATURE

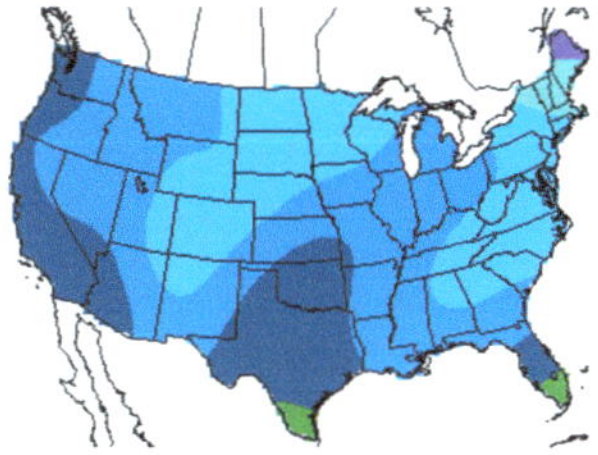

11/15/2025

A cold front brings precipitation into the Pacific and Inland Northwest regions and in the northern Rocky Mountains, as well as down through northern California, with heavy falls likely in the coasts from Oregon to northern California.

Some showers over the Great Lakes to the northern Appalachians.

Mostly dry elsewhere.

It may be unusually warm in Arizona and Texas for this time of year.

A storm brings snow in Alaska.

11/16/2025

A large depression moves into the north of the mid-section extending a cold front from the Northwest, over the North Great Plains and back to the Great Basin, with widespread outbreaks of precipitation in the northwest quarter of the country.

Rain may be very heavy along the Pacific coasts of the Northwest and in northern California.

Wintry precipitation, including the chance of snow, may affect the northern Rocky Mountains, Montana, Idaho, Wyoming, and into the Dakotas.

Chance of some showers in the southern Northeast including in Vermont to New York, where precipitation may fall as snow for a time.

Mostly dry elsewhere.

It may be unusually warm in Arizona over the next few days, possibly with maximum temperatures into the 90's, while very cold in Missouri to Alabama and Florida.

NOVEMBER 16 SUNDAY

ISOBARIC

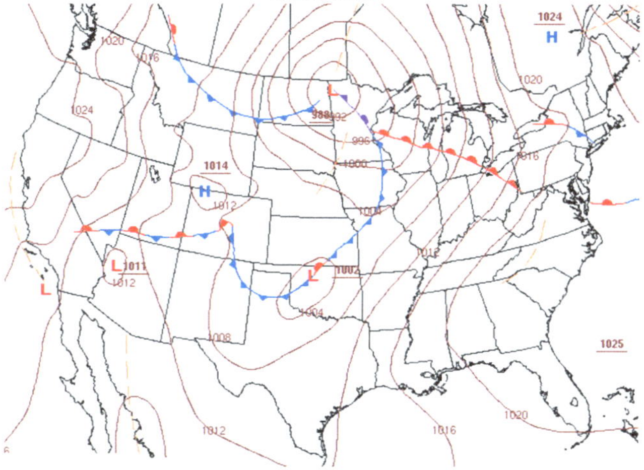

RAIN DISTRIBUTION (Inches)

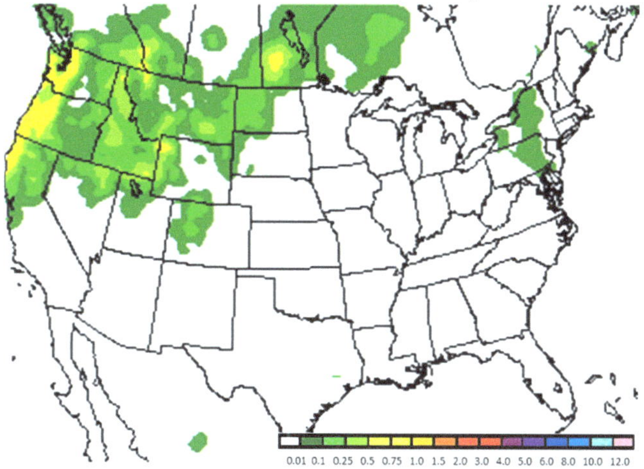

MAXIMUM TEMPERATURE

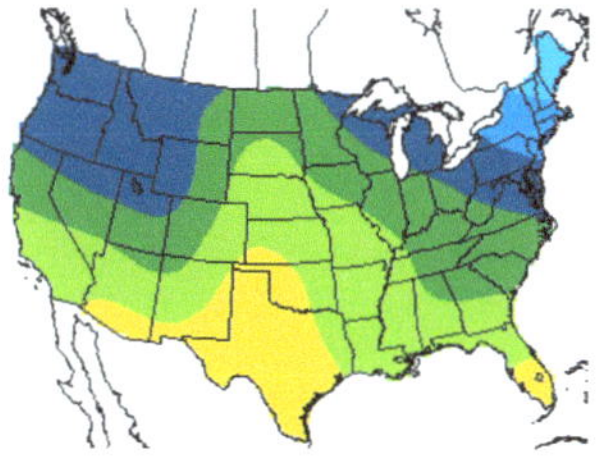

MINIMUM TEMPERATURE

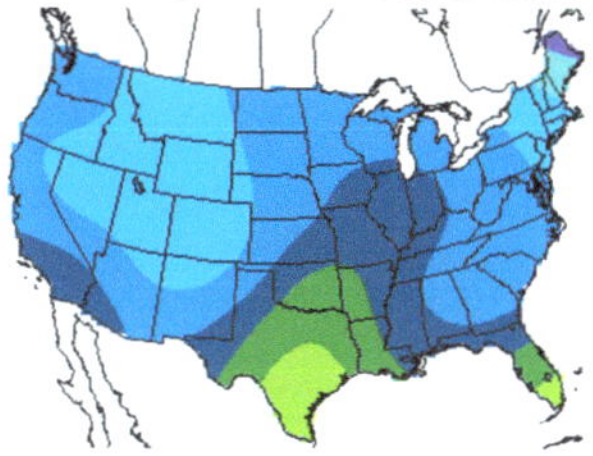

ISOBARIC

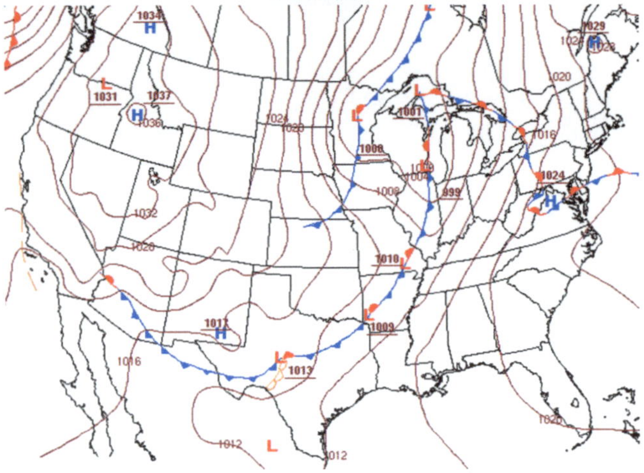

RAIN DISTRIBUTION (Inches)

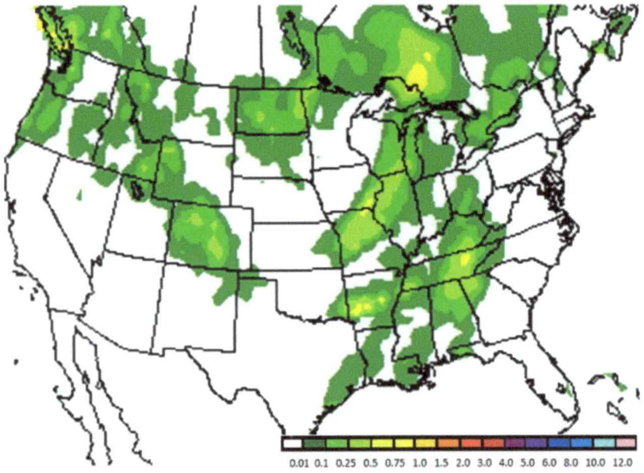

MAXIMUM TEMPERATURE

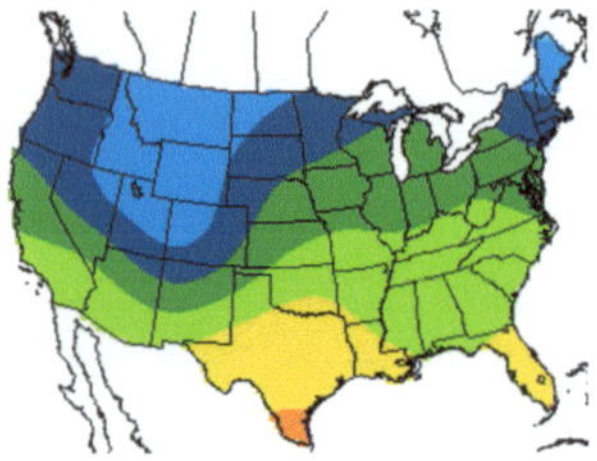

MINIMUM TEMPERATURE

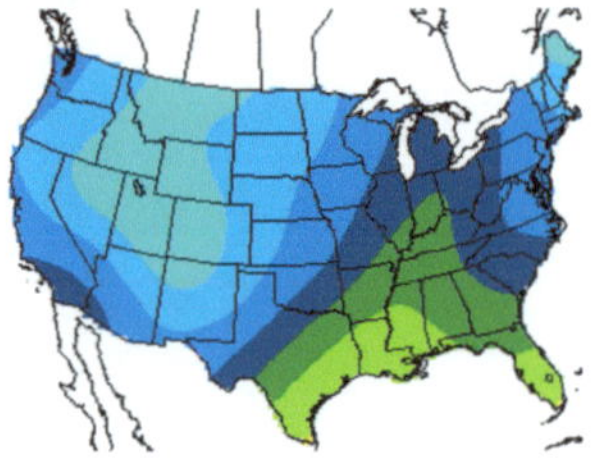

11/17/2025

Cold fronts in the east bring outbreaks of wintry precipitation from the Dakota's to the Great Lakes and south through the Inland Midwest to the Deep South, including across the Appalachians and in the Ozarks.

Some light snow may fall in Illinois, including Chicago.

In the west, rain eases to showers in the Pacific and Inland Northwest, while wintry precipitation extends through the Rocky Mountains to the far north of Texas.

Mostly dry along the Atlantic coastal states from Florida to the Northeast, and California to Texas

Frosts may become widespread in the east, including in the Deep South to the Carolinas, and even into northern parts of Florida over the next couple of days, while it may be very warm in Arizona and California.

A storm brings snow to Alaska.

11/18/2025

Cold fronts bring widespread precipitation through most of the east of the country, except along the coastal fringes from Florida to the Megalopolis.

Precipitation may be heavy from the Inland Midwest, over the Appalachians and into the Deep South, particularly in Tennessee and South Carolina.

Heavy wintry precipitation over the western half of the Northeast to throughout the Great Lakes.

Widespread precipitation also affects the Pacific and Inland Northwest regions, with particularly heavy falls in coastal Washington to the northwest of Oregon.

Some showers down through the Rocky Mountains.

Mostly dry elsewhere.

Unusually cold in the Deep South and the Carolinas to Florida, while remaining unusually warm in Arizona.

Another storm may bring further snow to Alaska.

ISOBARIC

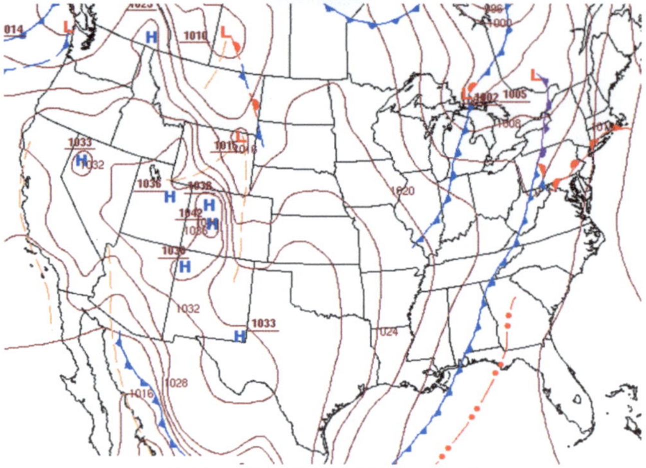

RAIN DISTRIBUTION (Inches)

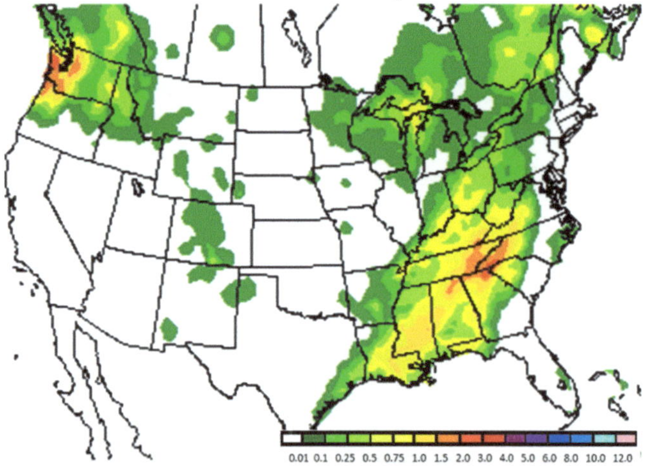

MAXIMUM TEMPERATURE

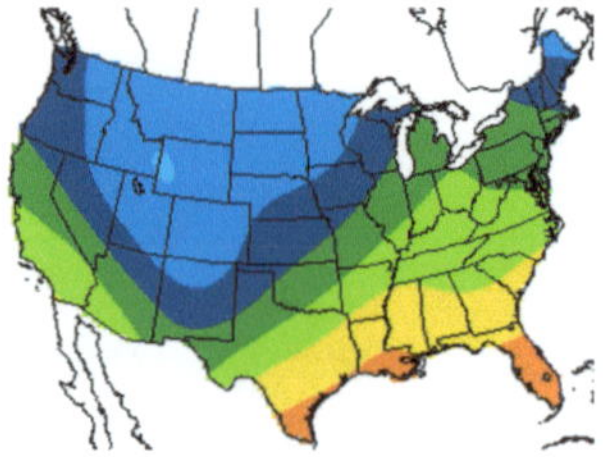

MINIMUM TEMPERATURE

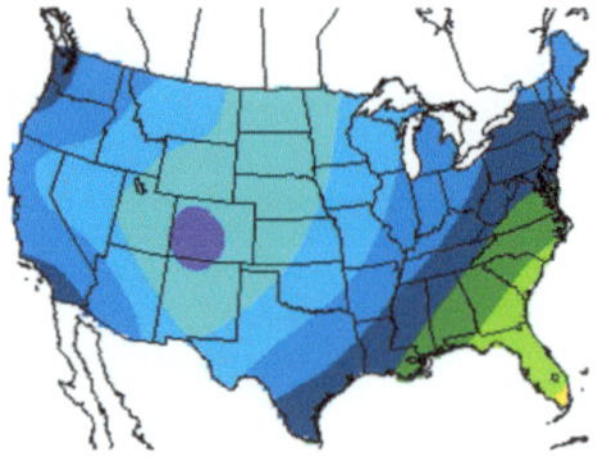

NOVEMBER **19** WEDNESDAY

ISOBARIC

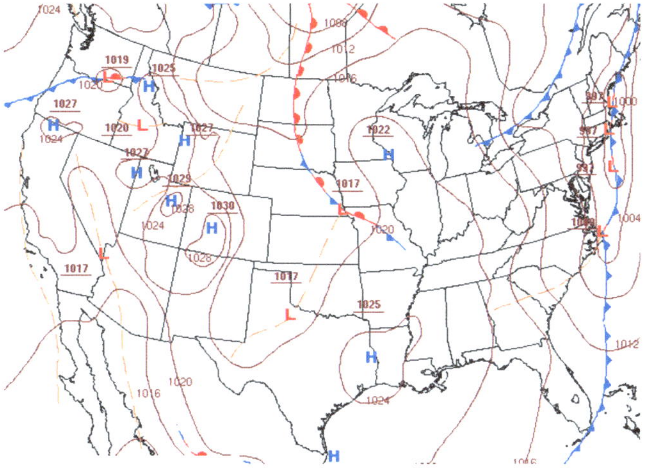

RAIN DISTRIBUTION (Inches)

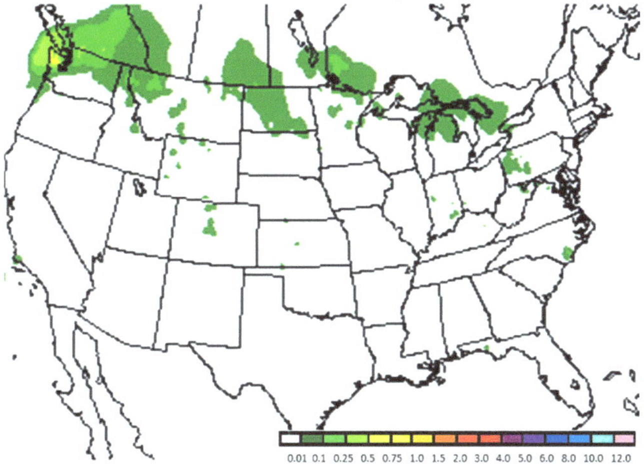

MAXIMUM TEMPERATURE

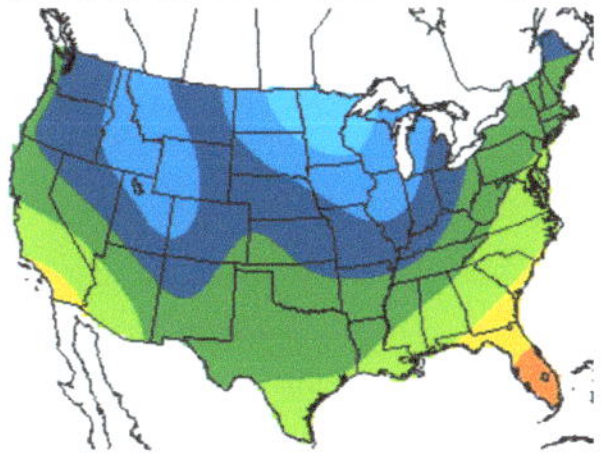

MINIMUM TEMPERATURE

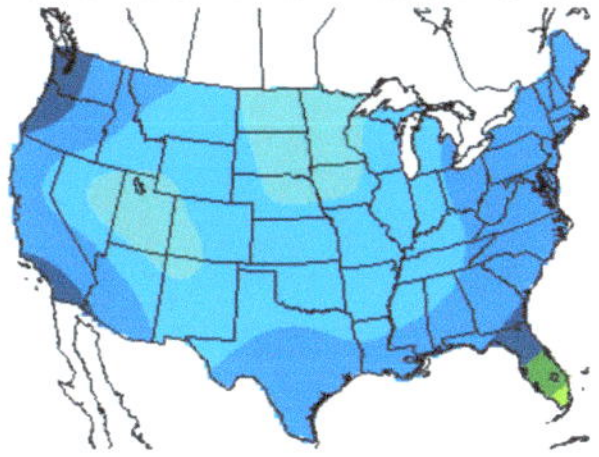

11/19/2025

The cold fronts that delivered plenty of precipitation over the east of the country the past couple of days drift offshore and into the Atlantic Ocean leaving only some lingering wintry showers over the Great Lakes.

Similarly, precipitation contracts to the far northern fringes of the Pacific Northwest and mostly lingers in the North Dakotas.

Mostly dry elsewhere.

It may be continue to be unusually cold in the southeast of the country, especially overnight.

11/20/2025

Outbreaks of rain prevail over the northwestern quarter, in the Pacific and Inland Northwest regions to the northern parts of the Rocky Mountains, although most of Washington state will be dry.

In the east, another front swiftly crosses the Great Lakes and up into the Northeast bringing widespread rain in many northeastern states to the Megalopolis, with showers around the Great Lakes and into the northern fringes of the Appalachians.

Mostly dry elsewhere.

It may be a very warm spell on the Great Plains, including in Texas and Montana.

Another storm crosses Alaska over the next couple of days bringing further snow.

NOVEMBER **20** THURSDAY

ISOBARIC

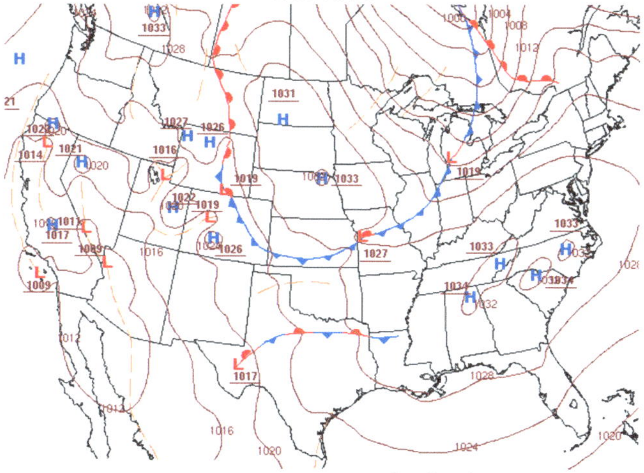

RAIN DISTRIBUTION (Inches)

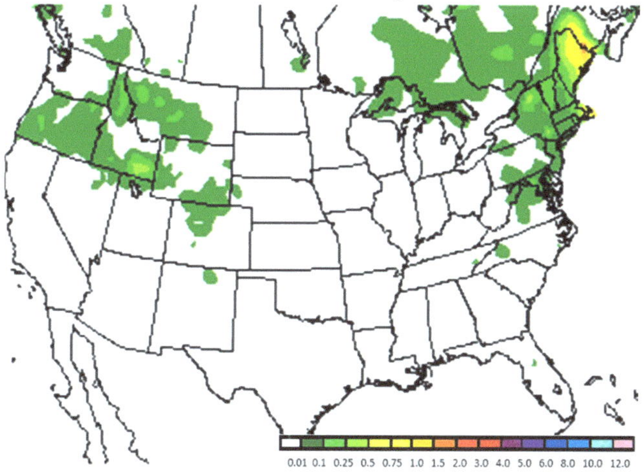

MAXIMUM TEMPERATURE

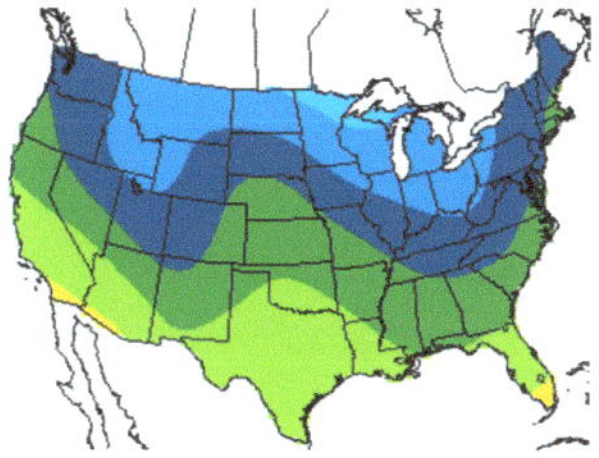

MINIMUM TEMPERATURE

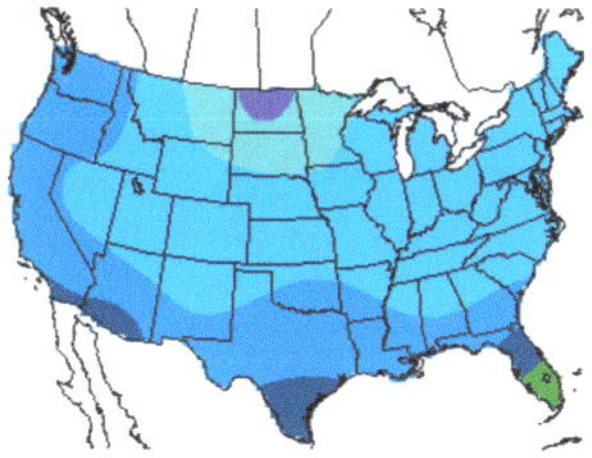

NOVEMBER 21 FRIDAY

ISOBARIC

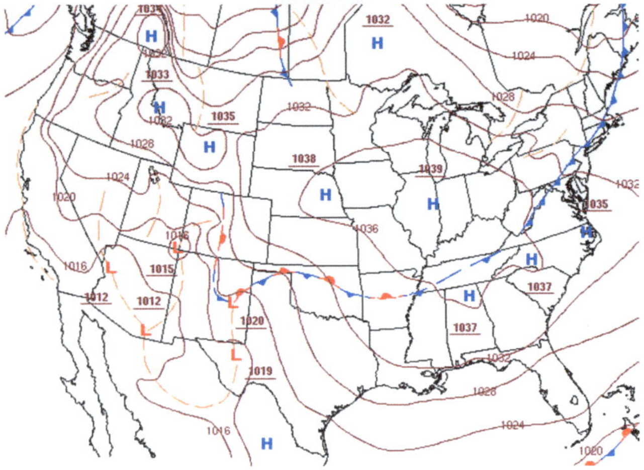

RAIN DISTRIBUTION (Inches)

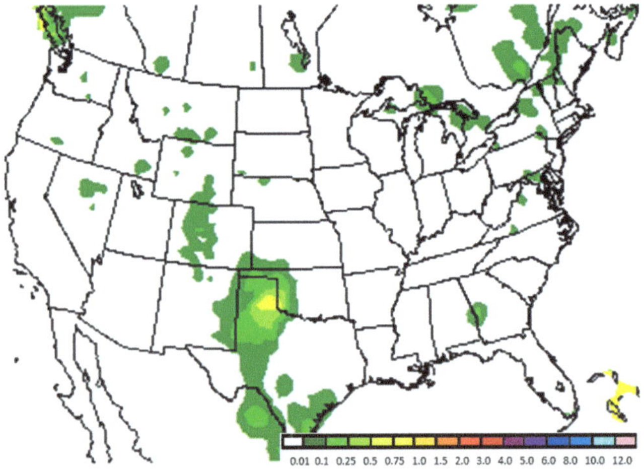

MAXIMUM TEMPERATURE

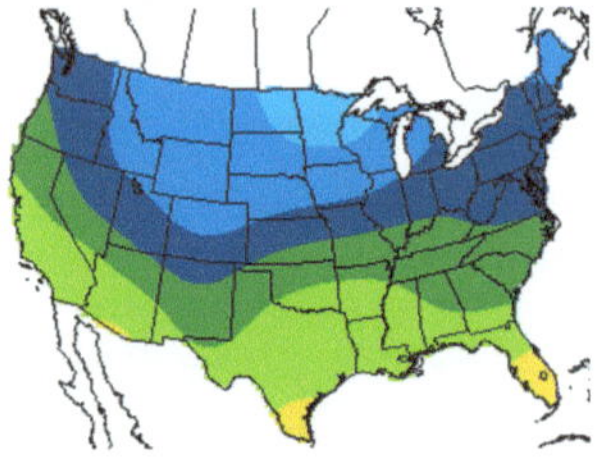

MINIMUM TEMPERATURE

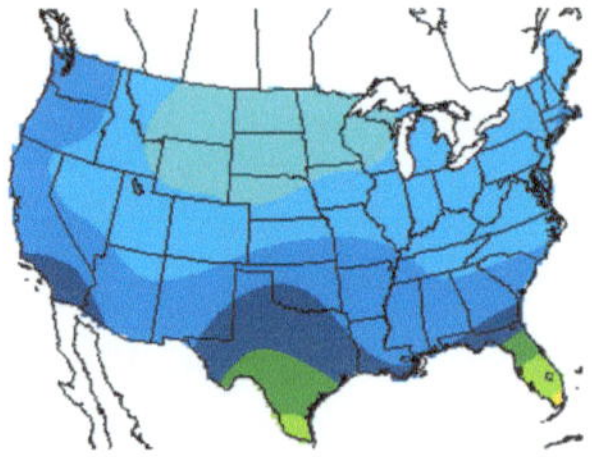

11/21/2025

High pressure prevails across the mid-section and east of the country while low pressure prevails in the west.

Mostly dry everywhere except for some isolated showery pockets in the Rocky Mountains and around the north of the Great Lakes and Northeast, and with the only widespread rain likely in the north of Texas to the Rio Grande Valley.

It may continue to be unusually warm in the central and southern Great Plains, including in Kansas, as well as down into Texas.

A storm brings more snow in Alaska.

11/22/2025

A wet front brings rain from the southern half of the Rocky Mountains and the adjacent South Great Plains, down through the Southwest and into the west and south of Texas, where rain may be heavy for a time in the High Plains.

Some isolated showers around the north of the Great Lakes to the Northeast, as well as in Florida.

Chance of some rain along the Washington coasts.

Mostly dry elsewhere under a strong high-pressure influence.

It may be unusually warm in Texas.

Unsettled winds are likely in the Pacific Northwest.

Chance of snow in the higher reaches of the Rocky Mountains over the next couple of days and gale force winds in Montana.

NOVEMBER 22 SATURDAY

ISOBARIC

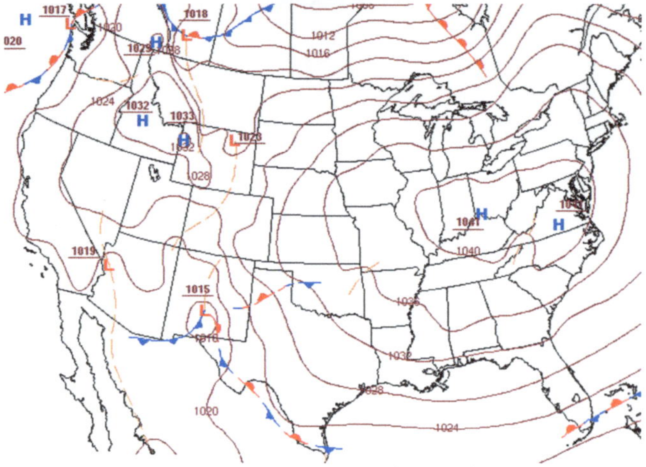

RAIN DISTRIBUTION (Inches)

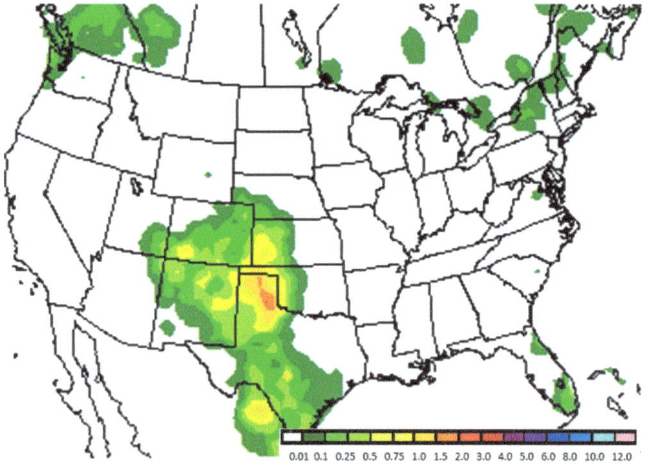

MAXIMUM TEMPERATURE

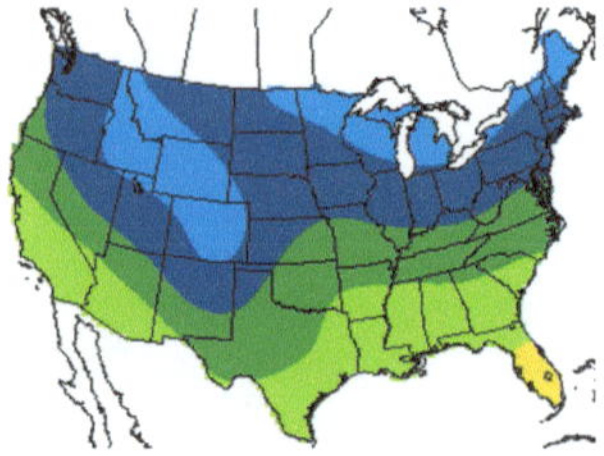

MINIMUM TEMPERATURE

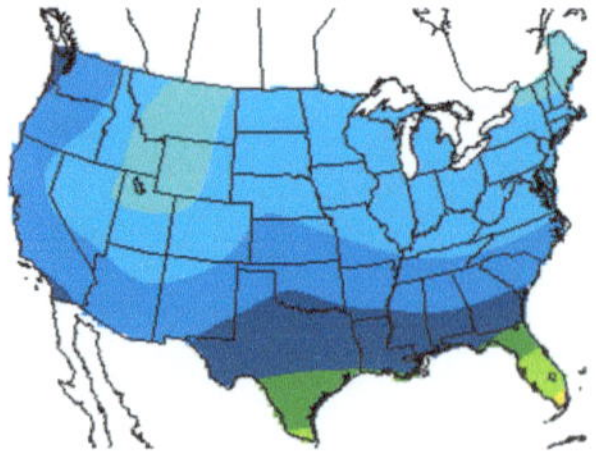

ISOBARIC

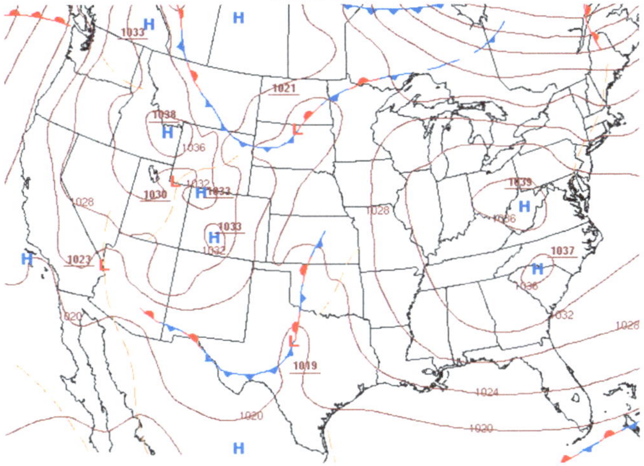

RAIN DISTRIBUTION (Inches)

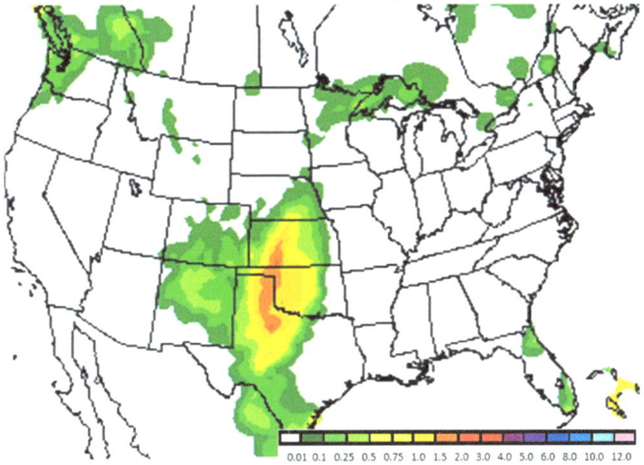

MAXIMUM TEMPERATURE / MINIMUM TEMPERATURE

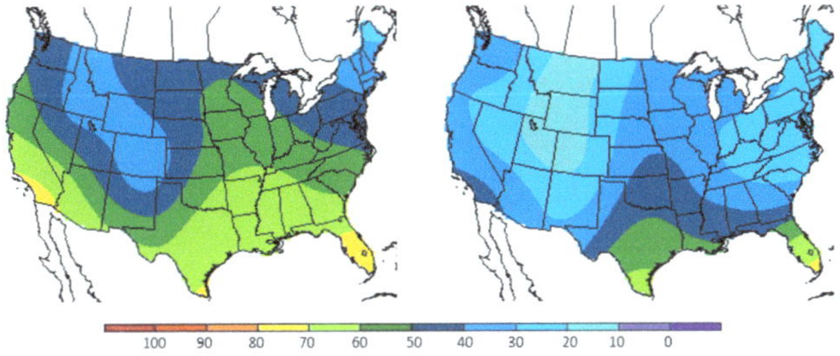

11/23/2025

Widespread precipitation continues throughout the south central areas from the South Great Plains and down through central Texas to the Gulf of Mexico, and west into New Mexico.

Precipitation in the western half of Washington, occasional showers are likely around the northern edges of the Great Lakes, and showers in Florida.

Mostly dry elsewhere.

Chance of snow in the southern Rocky Mountains.

Temperatures may continue to be unusually warm over the Great Plains, including in North Dakota and Kansas, ahead of a cold front, as well as unusually warm in California over the next couple of days.

Snow in southern Alaska over the next few days.

11/24/2025

A cold front extends throughout the mid-section of the country to north of the Great Lakes bringing widespread rain from throughout the eastern half of Texas, up through the Ozarks and the Great Plains and into the Great Lakes, later spreading into the Inland Midwest and Corn Belt over the next couple of days.

Some showers in the south of Florida.

Mostly dry elsewhere.

An intense storm may be building in the Gulf of Mexico with the possibility of it strengthening to hurricane force over the next few days.

Unstable winds in the Deep South may see the generation of tornadoes in the states of Alabama, Kentucky, and Tennessee.

Unusually warm temperatures may continue in southern California.

Snow in Alaska.

ISOBARIC

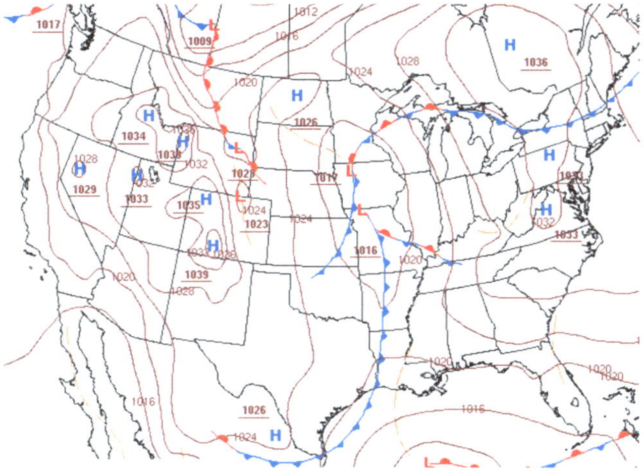

RAIN DISTRIBUTION (Inches)

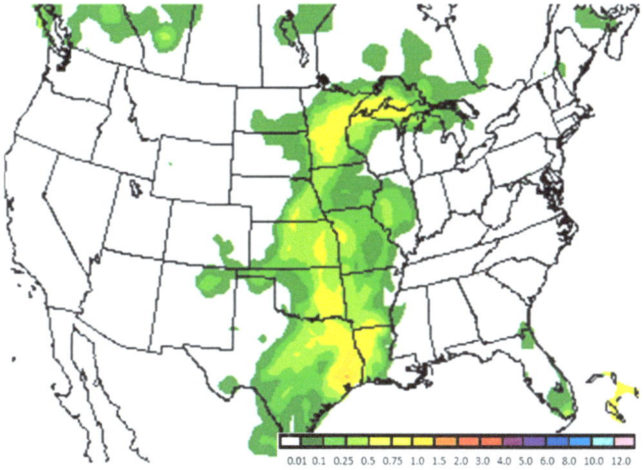

MAXIMUM TEMPERATURE / MINIMUM TEMPERATURE

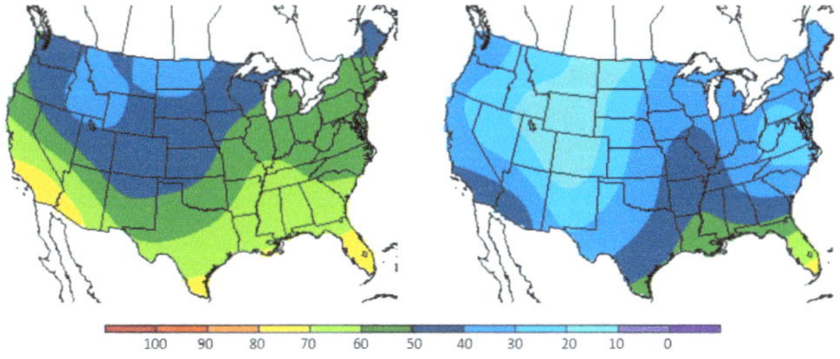

NOVEMBER 25 TUESDAY

ISOBARIC

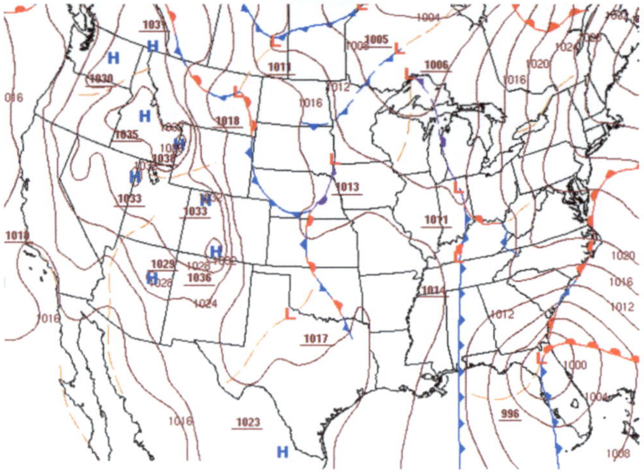

RAIN DISTRIBUTION (Inches)

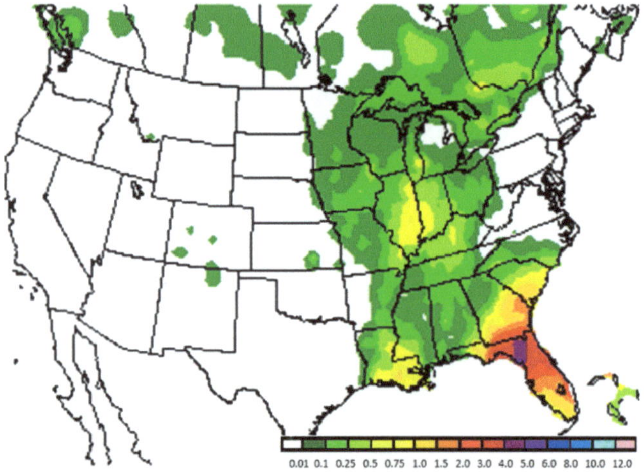

MAXIMUM TEMPERATURE

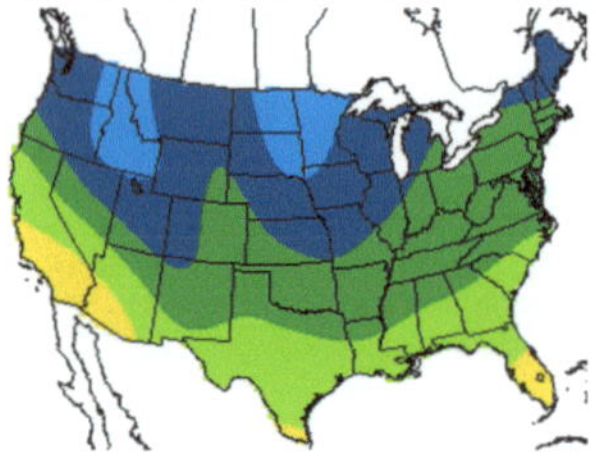

MINIMUM TEMPERATURE

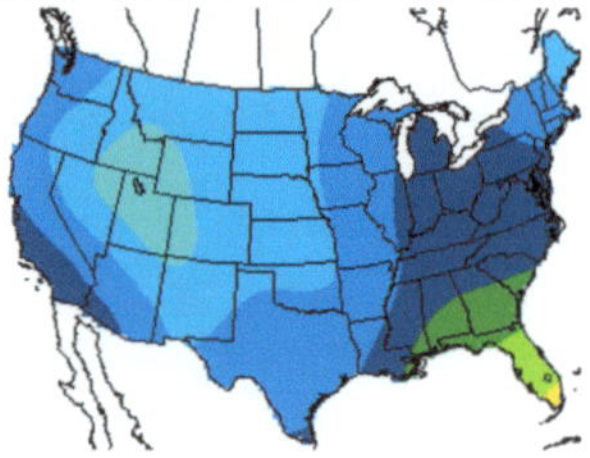

11/25/2025

An intense storm moves up from the Gulf of Mexico and makes landfall in Florida, possibly as a recognized hurricane by this time.

Widespread storm conditions prevail throughout the southeast of the country with wet fronts spreading heavy rain from Florida, through the Deep South and the Inland Midwest to the Great Lakes.

Storms in Florida and up into the Coastal Southeast states may see flooding and damage to infrastructure around this time, and chance of tornado formation winds a threat in the southeast states to the Midwest.

Mostly dry elsewhere. Unusually warm in California.

Chance of heavy rain and snow in the Northeast over the next couple of days, with possible blizzard conditions developing, especially in New England.

Snow in southern Alaska continues.

11/26/2025

The Atlantic storm continues to travel up the eastern seaboard bringing hurricane force conditions of damaging winds and flooding rains from the Florida, up through the Coastal Southeast to the Megalopolis regions and possible blizzard conditions into the Northeast and around the Great Lakes.

Storm conditions also prevail over the Appalachians and into the Inland Midwest but begin to clear in the western half of the Deep South.

Some rain also in northern California.

Mostly dry elsewhere.

Despite the storms in the east, the unusual warmth in the southwest is likely to expand north into many other states over the next couple of days, including in Utah, Arizona, and Oregon. Very cold and frosty in the southeast around the Carolinas, and Missouri.

NOVEMBER 26 WEDNESDAY

ISOBARIC

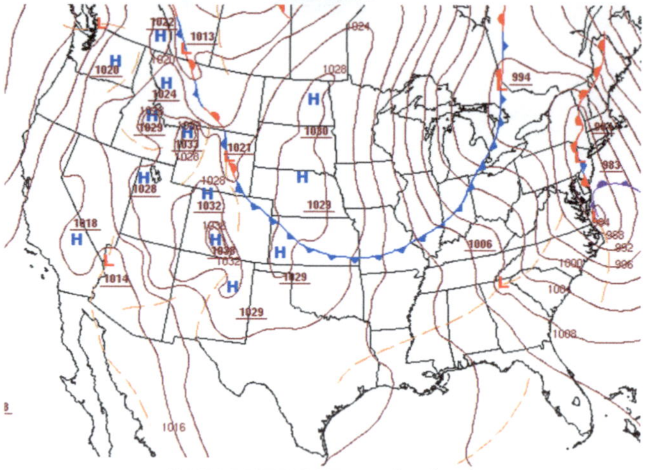

RAIN DISTRIBUTION (Inches)

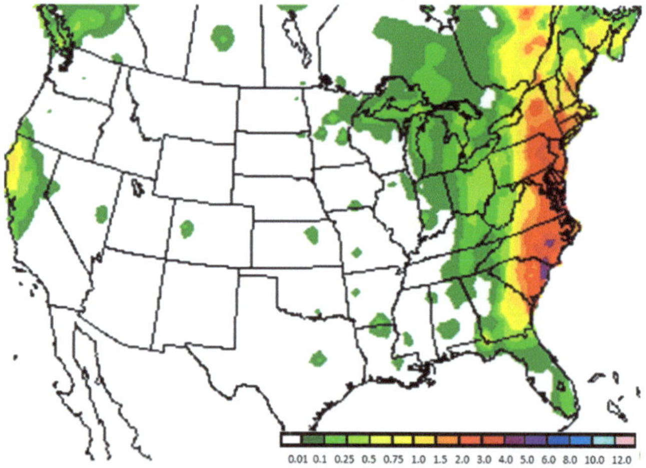

MAXIMUM TEMPERATURE

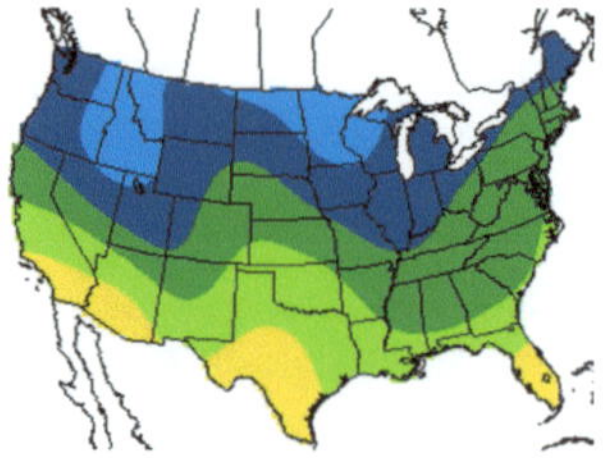

MINIMUM TEMPERATURE

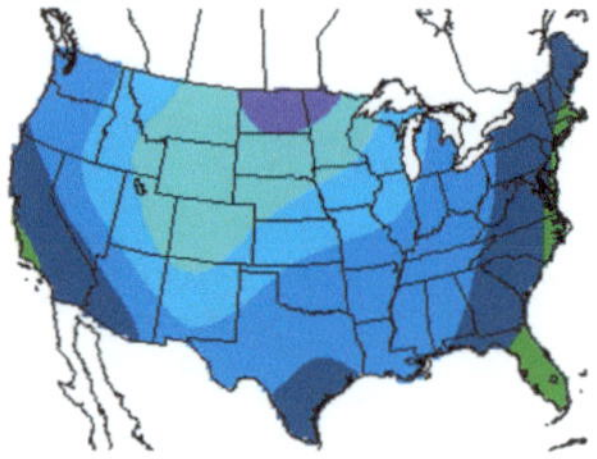

ISOBARIC

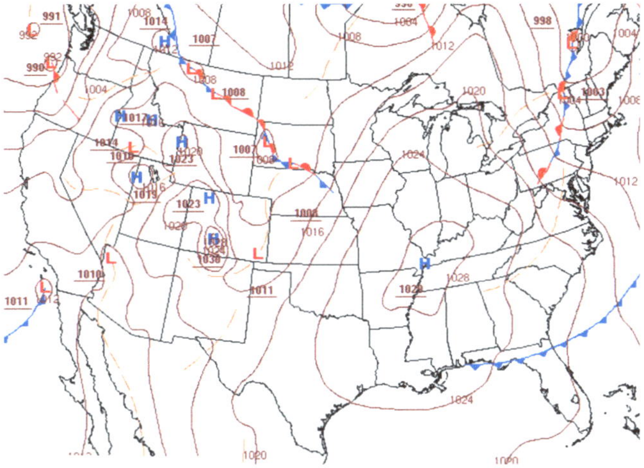

RAIN DISTRIBUTION (Inches)

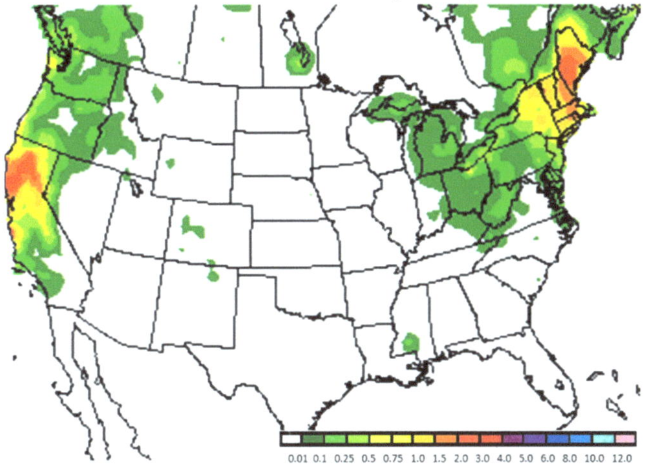

MAXIMUM TEMPERATURE

MINIMUM TEMPERATURE

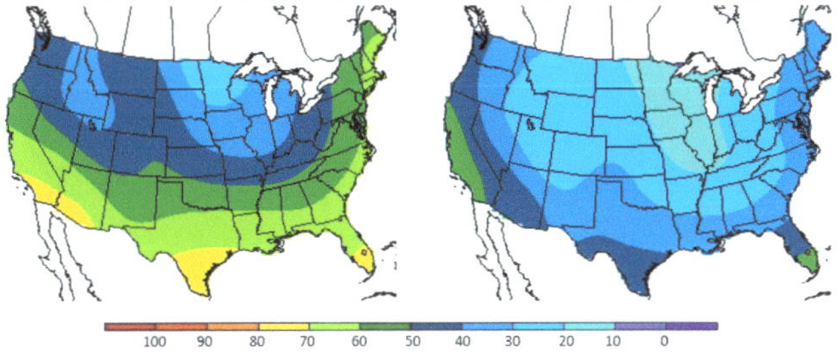

11/27/2025

Storm conditions ease over the southeast with high pressure swiftly moving in behind the storm bringing drier conditions from the Deep South to Florida.

However, storms and heavy precipitation continue in the northeast quarter from the Inland Midwest and Great Lakes to possible blizzard conditions prevailing over the Northeast, and snow likely in Minnesota to Wisconsin.

Rain spreads along the Pacific coasts with heavy falls in California at times, and rain into the Pacific Northwest states over the next couple of days.

Mostly dry elsewhere.

The unusual warmth of recent in California may spread into Utah, Arizona and Oregon.

11/28/2025

High pressure settles over the east of the country for the next few days bringing mostly dry conditions from the Great Plains to the Atlantic coast.

However, some showers may briefly linger around the Great Lakes to the Northeast, including the chance of snow in New Hampshire to Maine.

Rain persists in California to the Pacific Northwest, with an occluded front bringing heavy falls into the northern half of California including in the Sierra Nevada Ranges.

Chance of snow also in the higher reaches of the northern Rocky Mountains in the week ahead.

Mostly dry elsewhere.

Chance of heavy snow in Alaska.

ISOBARIC

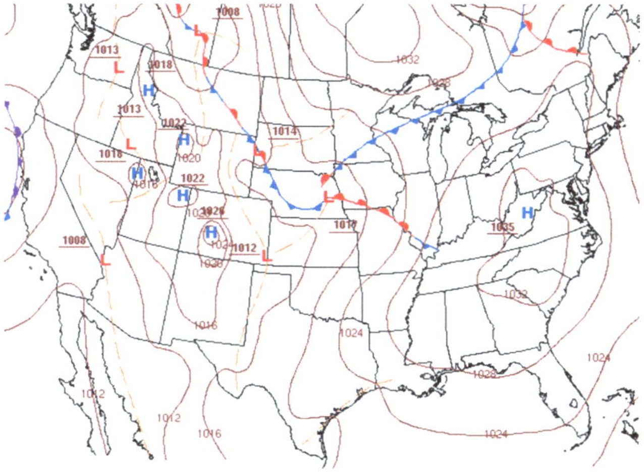

RAIN DISTRIBUTION (Inches)

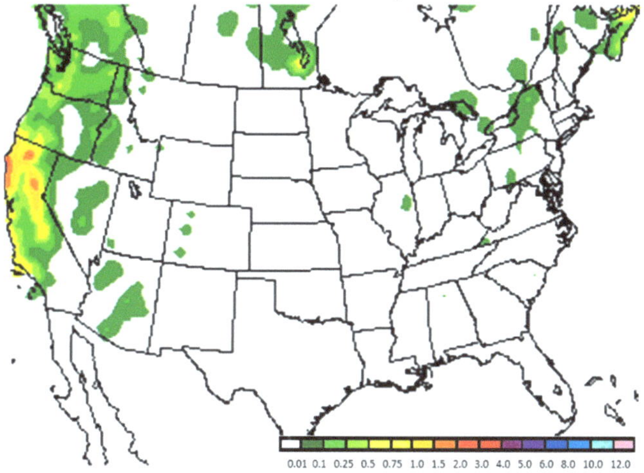

MAXIMUM TEMPERATURE

MINIMUM TEMPERATURE

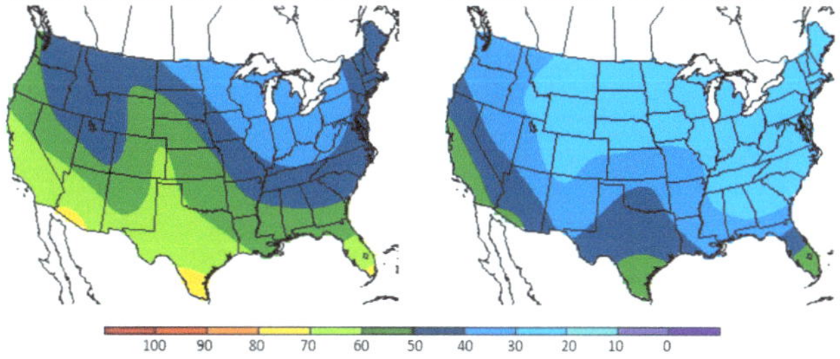

ISOBARIC

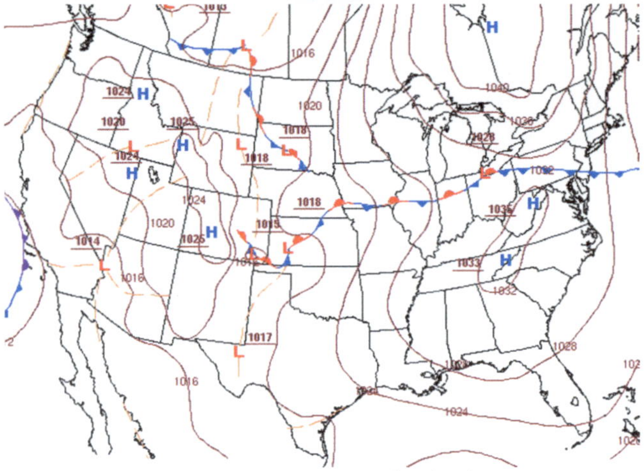

RAIN DISTRIBUTION (Inches)

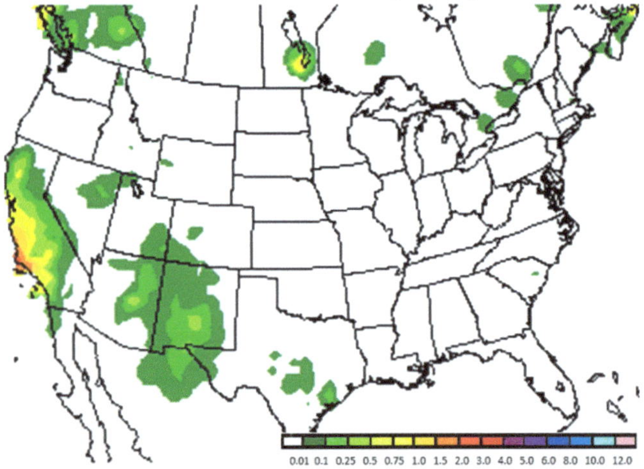

MAXIMUM TEMPERATURE

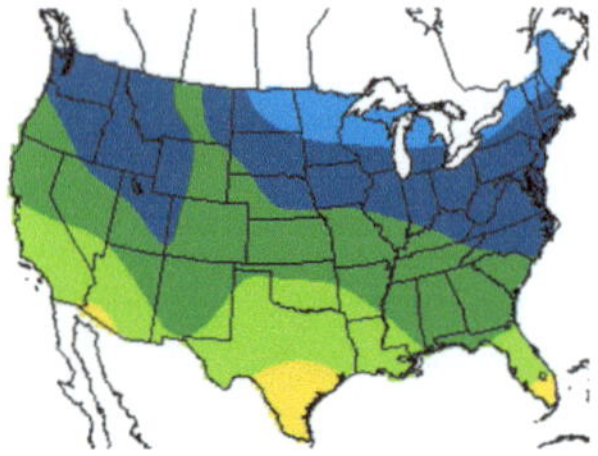

MINIMUM TEMPERATURE

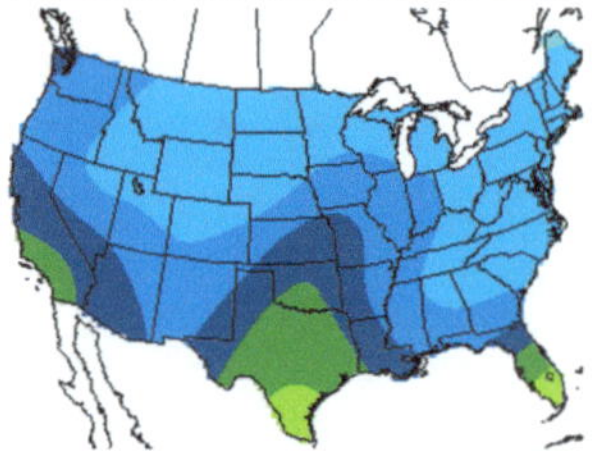

11/29/2025

An occluded front brings heavy rain into California, while a trough of rain crosses the Southwest and another trough moves over the Great Basin.

Some showers likely in the southern half of Texas, and isolated wintry showers may still prevail in the Northeast.

Mostly dry elsewhere.

It may continue to be unusually warm in Texas, Kansas and Nebraska.

Showers in Hawaii over the next couple of days.

11/30/2025

The occluded front near California drifts south along the Pacific coastline, clearing rain over northern California, but persisting in the south with some heavy falls likely mostly around the coasts.

A wet front crosses through the mid-section of the country bringing rain up through the center of Texas and into South Great Plains and Ozarks.

Chance of lingering showers in the far north of the Northeast.

Mostly dry elsewhere.

A warm southerly flow shifts into the south of the country around this time, including down into Texas and Alabama.

Showers in Hawaii.

ISOBARIC

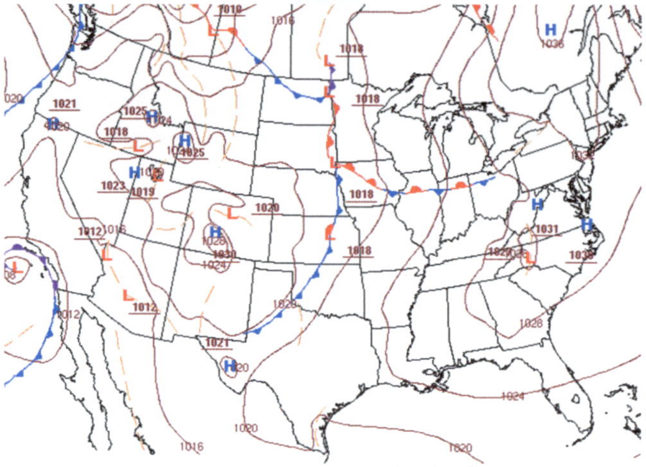

RAIN DISTRIBUTION (Inches)

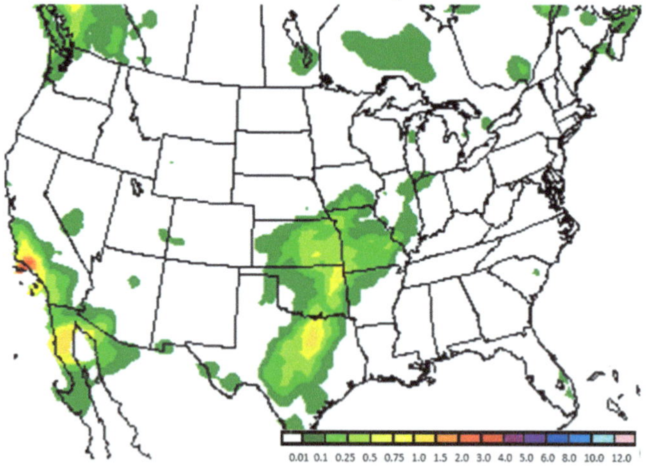

MAXIMUM TEMPERATURE

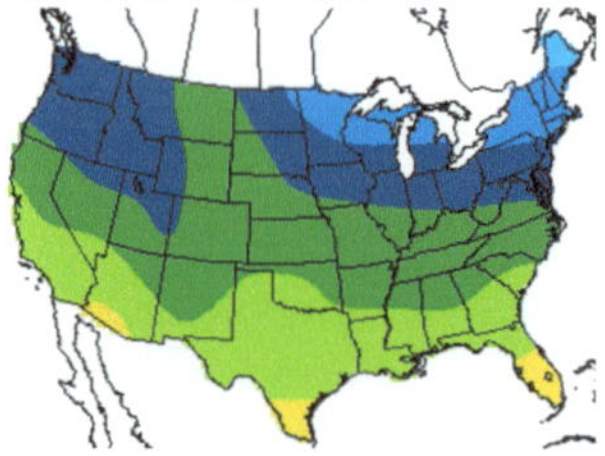

MINIMUM TEMPERATURE

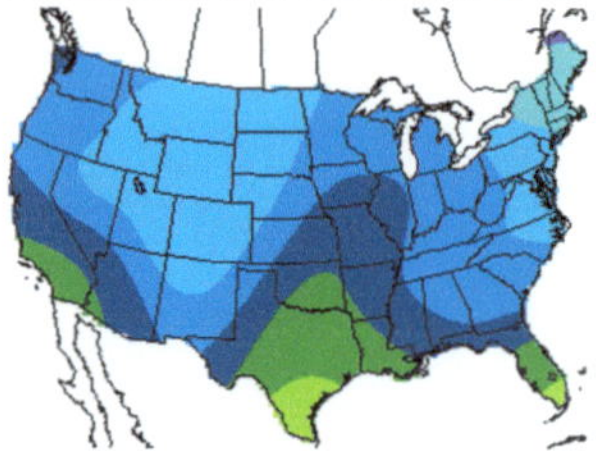

DECEMBER

PHASES OF THE MOON

4th	Full Moon
4th	Perigee #2
5th	Northern Declination
11th	Third Quarter
13th	Crossing Equator
17th	Apogee
19th	Southern Declination
20th	New Moon
27th	First Quarter
27th	Crossing Equator

1 Mon	2 Tue	3 Wed	4 Thur	5 Fri	6 Sat	7 Sun	8 Mon	9 Tue	10 Wed	11 Thur	12 Fri	13 Sat	14 Sun	15 Mon	16 Tue	17 Wed	18 Thur	19 Fri	20 Sat	21 Sun	22 Mon	23 Tue	24 Wed	25 Thur	26 Fri	27 Sat	28 Sun	29 Mon	30 Tue	31 Wed
			F P2	^						3Q		XhS				A		V	N							1Q XhN				

ISOBARIC

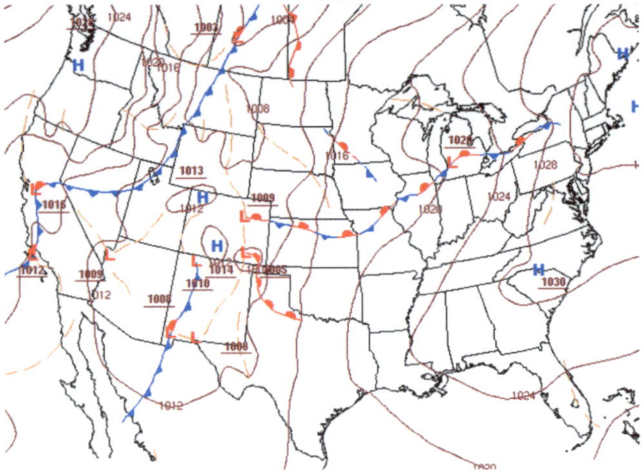

RAIN DISTRIBUTION (Inches)

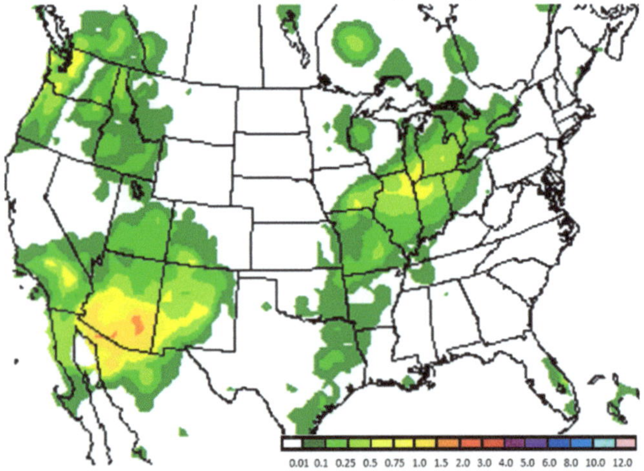

MAXIMUM TEMPERATURE

MINIMUM TEMPERATURE

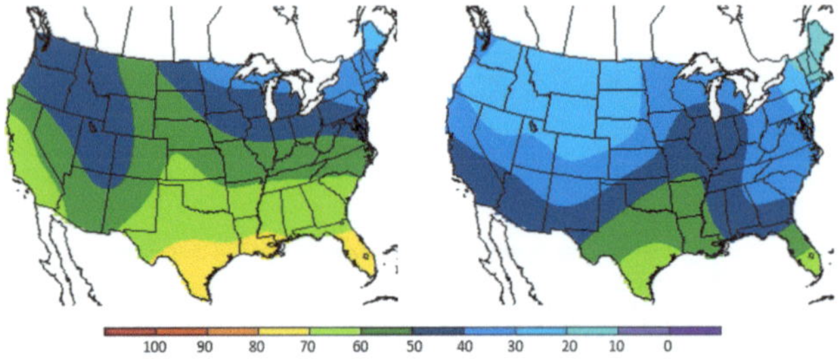

12/01/2025

Cold fronts over the west of the country brings scattered precipitation through the Pacific and Inland Northwest regions into the Rocky Mountains.

Widespread rain spreads from southern California eastward into the Southwest and southern Rockies.

Wintry precipitation in the east from the Great Lakes, through the Inland Midwest and Ozarks, and into the western Deep South states to the eastern half of Texas.

Chance of snow in the Midwest, Iowa and Wisconsin.

Mostly dry elsewhere except for the chance of some showers in Florida.

Chance of heavy snowfall in Montana.

The band of warmth of the past few days may now shift further eastward into the Carolinas to Virginia, as well as continue in Texas.

12/02/2025

Widespread precipitation throughout the midsection of the country from the Rocky Mountains to the Great Plains, the Southwest to Texas, the western Deep South, along the Appalachians and over the east of the Great Lakes to the southern Northeast.

Heavy snow may prevail through Texas to New Mexico over the next few days, as well as snow spreading into the Rocky Mountains and over the South Great Plains in the week ahead.

Snow in Michigan as well as northeast towards Maine.

Mostly dry west of the Rockies and east of the Appalachians from the Coastal Southeast to the Deep South and in the west Inland Midwest to the Ozarks.

Unusually warm in the Coastal Southeast. Widespread sub-zero temperatures in Wyoming, Utah, and Kansas.

Heavy snow in the south of Alaska.

ISOBARIC

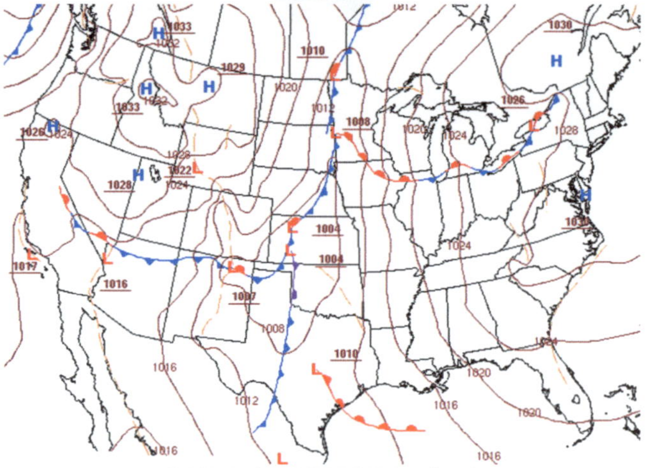

RAIN DISTRIBUTION (Inches)

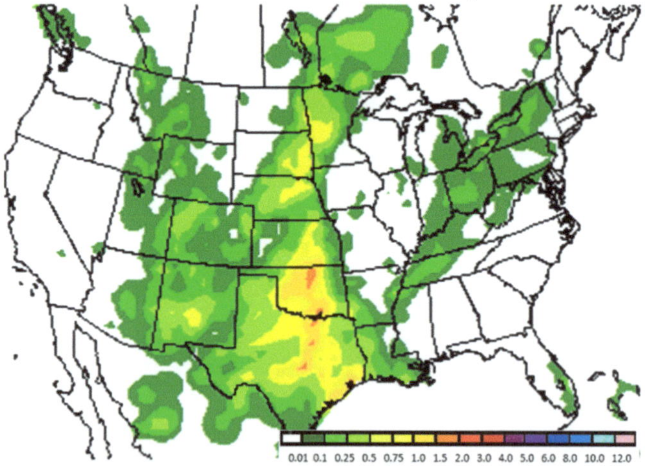

MAXIMUM TEMPERATURE

MINIMUM TEMPERATURE

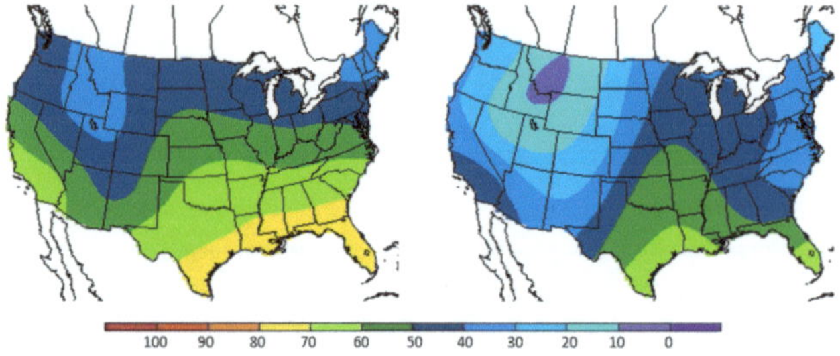

DECEMBER 3 WEDNESDAY

ISOBARIC

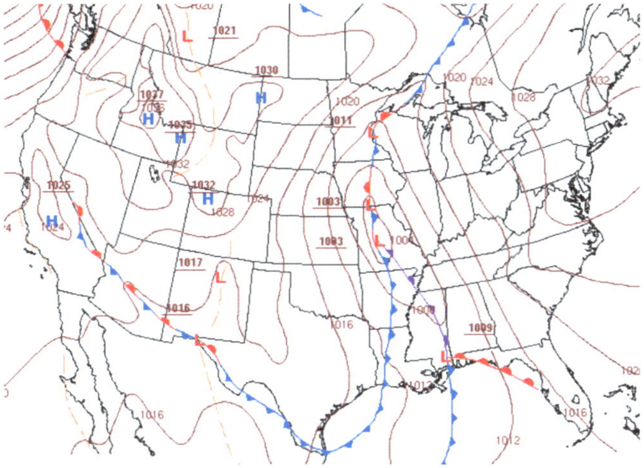

RAIN DISTRIBUTION (Inches)

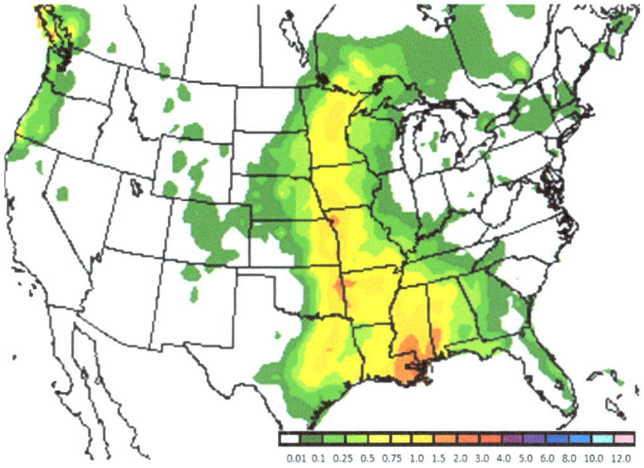

MAXIMUM TEMPERATURE

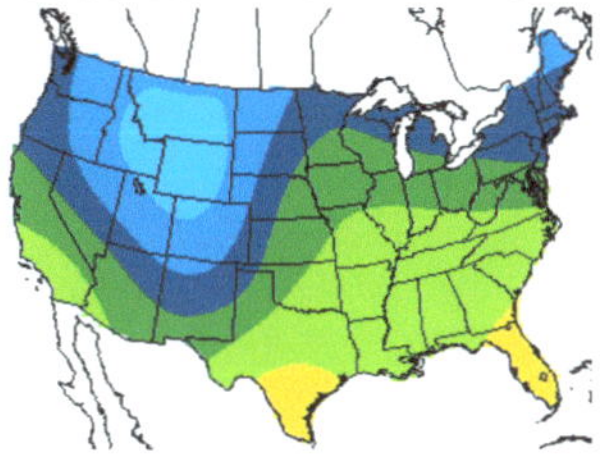

MINIMUM TEMPERATURE

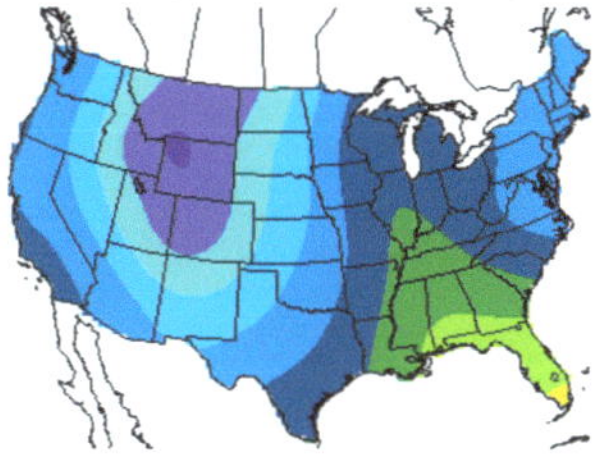

12/03/2025

Widespread storm activity persists throughout the midsection of the country from the north of the Great Plains to Texas, and eastward over the west of the Great Lakes and Inland Midwest, as well as throughout the Deep South to Florida.

Some precipitation in Pacific Northwest coastal fringes.

Wintry showers may linger in the Rocky Mountains and the lower Northeast.

Mostly dry elsewhere.

Chance of heavy snow in Texas to New Mexico, in the southern Rockies, and across the South Great Plains.

It will be very cold in Missouri as well as unusually cold in California.

Snow may be heavy in southern Alaska.

12/04/2025

An intense storm system strengthens over the midsection of the country with a mix of occluded, stationary, and cold fronts prevailing.

Widespread wintry precipitation prevails throughout the Great Plains to the Atlantic Coastal Southeast regions, and from the Great Lakes to Florida, with very heavy precipitation likely around the southern Appalachians to the Carolinas.

Heavy precipitation also moves deeper into the northwest from the Pacific coast to the Inland Northwest and northern fringes of the Great Basin.

Wintry showers through the Rocky Mountains, with chance of snow in the southern ranges, as well as in the higher reaches of New Mexico, Texas, in the South Great Plains, and in Kansas.

Mostly dry elsewhere.

Winds may be unstable and gusty in California.

DECEMBER 4 THURSDAY

ISOBARIC

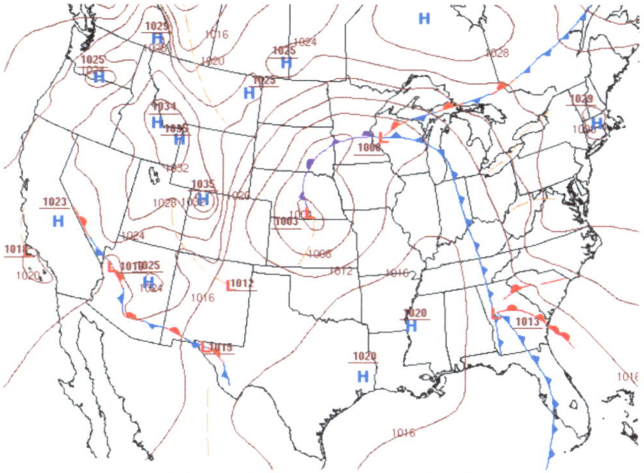

RAIN DISTRIBUTION (Inches)

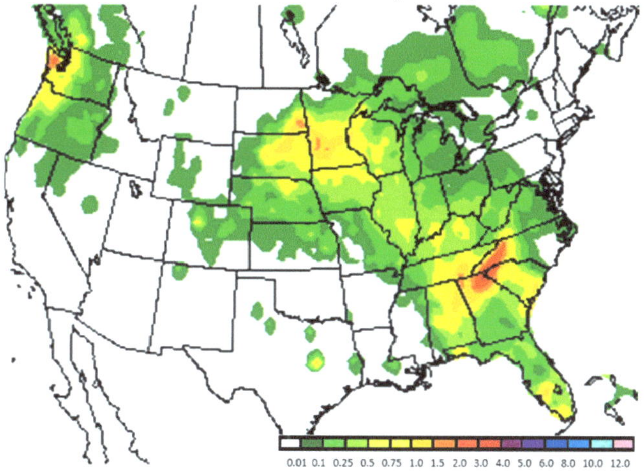

MAXIMUM TEMPERATURE

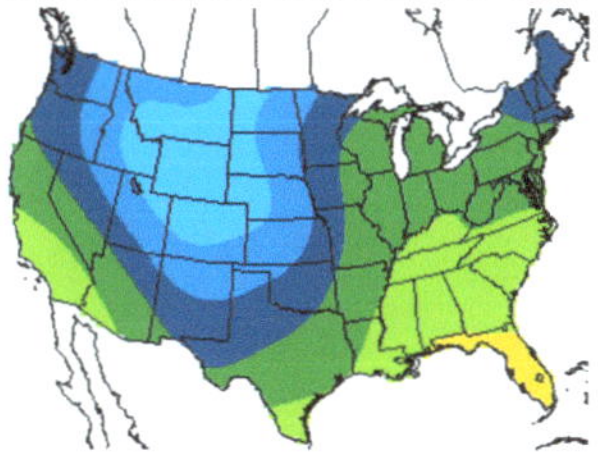

MINIMUM TEMPERATURE

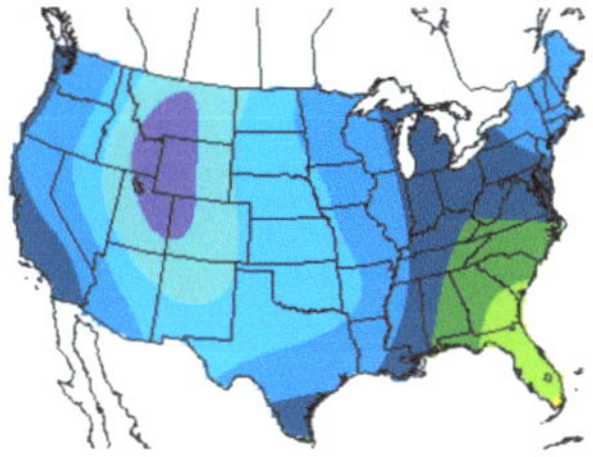

DECEMBER 5 FRIDAY

ISOBARIC

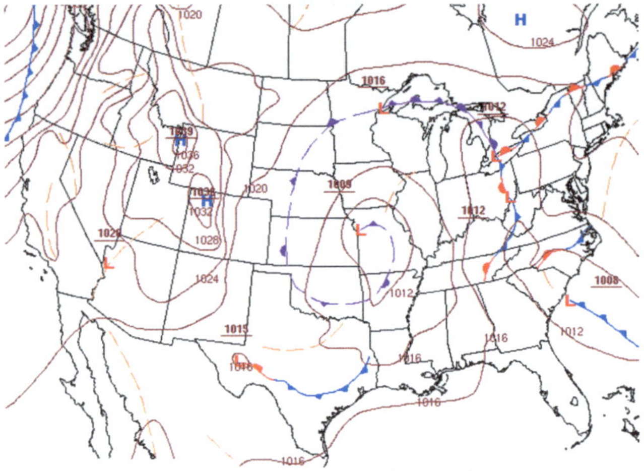

RAIN DISTRIBUTION (Inches)

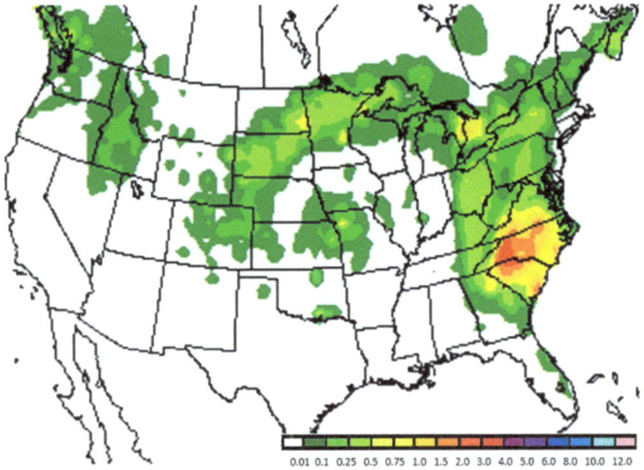

MAXIMUM TEMPERATURE

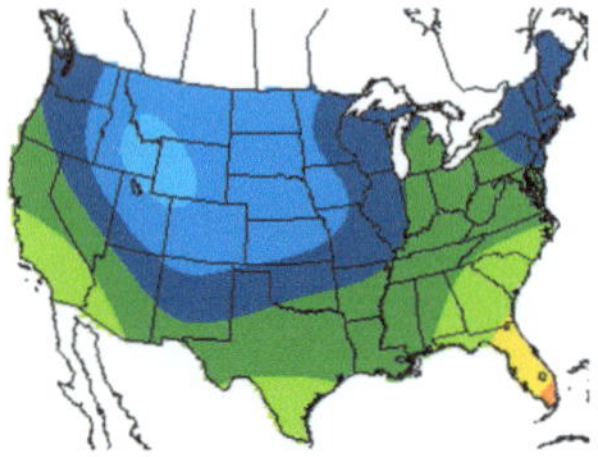

MINIMUM TEMPERATURE

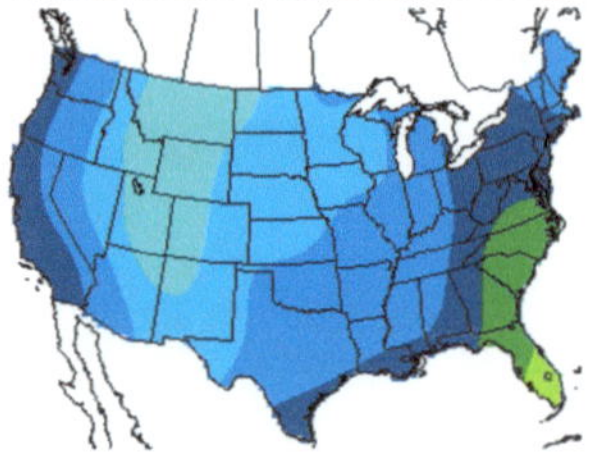

12/05/2025

The storm system over the eastern half of the country eases in the south with a large occluded front prevailing over the Great Plains.

Nevertheless, widespread wintry precipitation strengthens over the Rocky Mountains, and further outbreaks prevail throughout the Great Plains, over the Great Lakes and into the Inland Midwest.

The storms continue to track eastward spreading over the Northeast, the northern half of the Appalachians to the adjacent Atlantic coasts, and in the east of the Deep South to the Coastal Southeast, with very heavy precipitation likely in the Carolinas to Arkansas.

An intense depression to the northwest of Washington also extends outbreaks of rain over the Pacific and Inland Midwest, and into the Rocky Mountains.

Mostly dry in the south from California to Texas.

Snow over the southern Rockies to the Great Plains.

12/06/2025

The storm system over the east of the country continues to trek eastward and offshore out into the Atlantic, with storm activity easing down through the Rocky Mountains and across the south.

Widespread wintry precipitation throughout the Northeast and Megalopolis to the Coastal Southeast, as well as in the Ozarks to the Inland Midwest.

Lingering wintry showers persist over the Great Plains.

Rain may still be heavy in Tennessee to Kentucky, and showers widespread in Florida.

A cold change in Texas likely.

Rain also in the Pacific Northwest and down along the Pacific coasts into the northern half of California.

Mostly dry elsewhere.

DECEMBER 6 SATURDAY

ISOBARIC

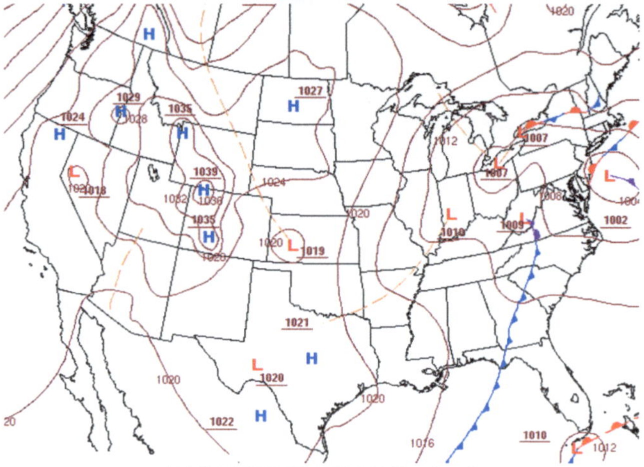

RAIN DISTRIBUTION (Inches)

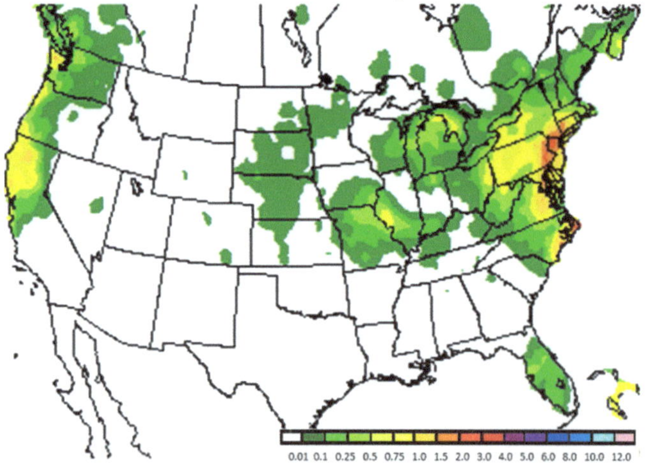

MAXIMUM TEMPERATURE

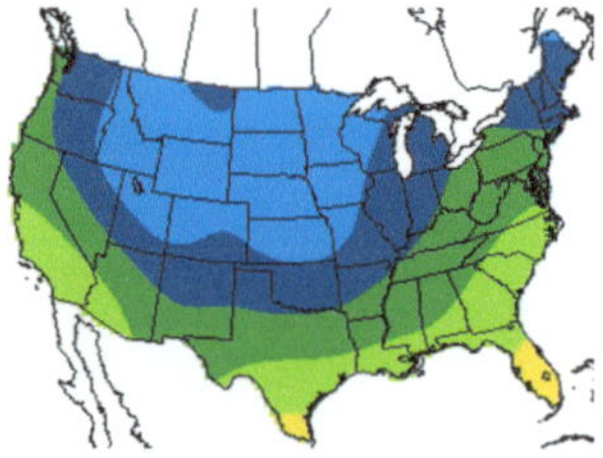

MINIMUM TEMPERATURE

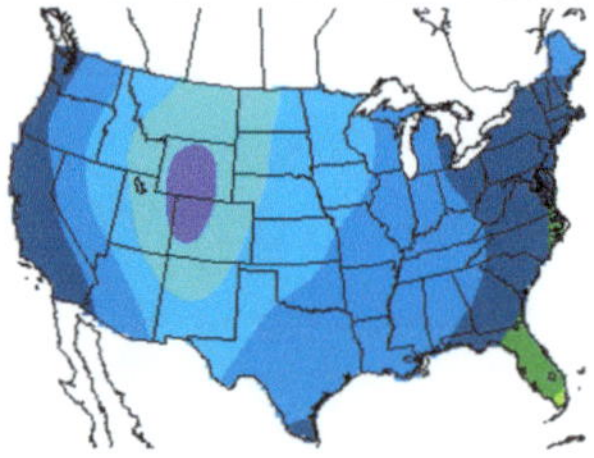

DECEMBER 7 SUNDAY

ISOBARIC

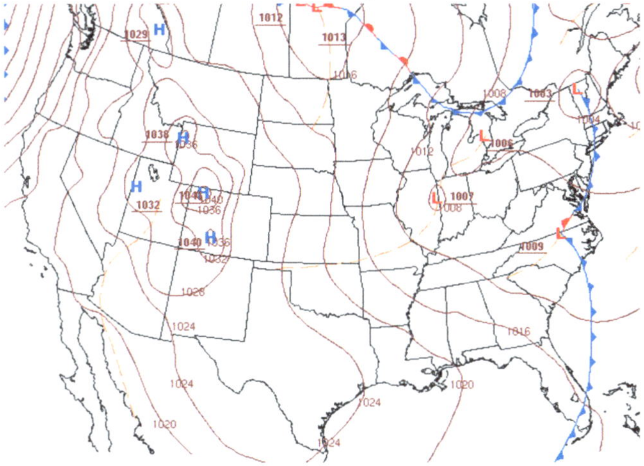

RAIN DISTRIBUTION (Inches)

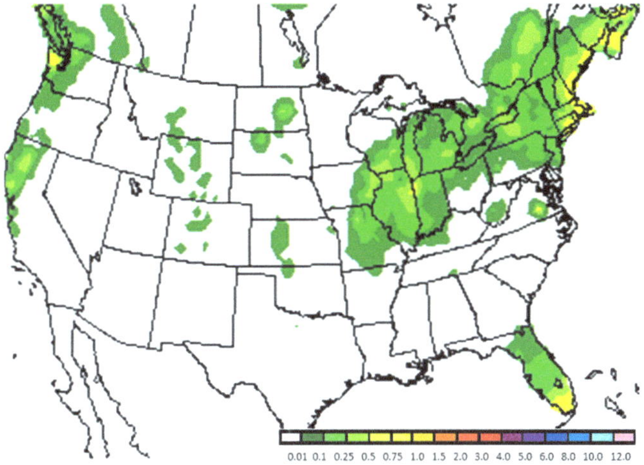

MAXIMUM TEMPERATURE MINIMUM TEMPERATURE

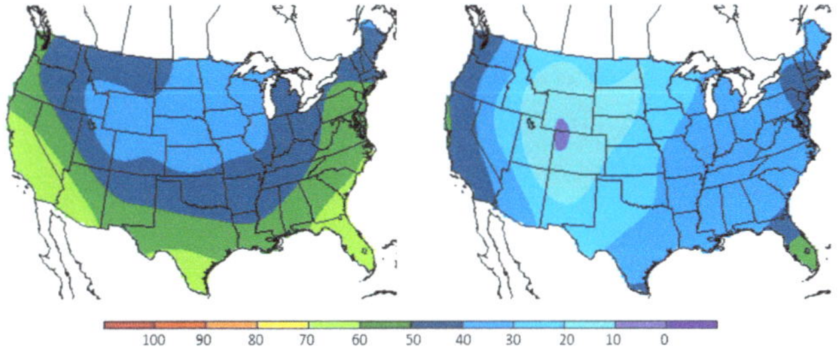

12/07/2025

Wintry precipitation continues throughout the northeast quarter of the country from throughout the Northeast to the Great Lakes and over the Inland Midwest and Corn Belt states.

Rain also in Florida.

Some isolated showers may linger in the Great Plains and up into the Rocky Mountains.

Rain eases in the northwest to lingering showers in northern California to the Pacific Northwest, mostly in the coastal fringes.

Mostly dry elsewhere.

Heavy rain likely in Hawaii with the chance of flash flooding in some places, particularly on the Big Island during the next couple of days.

12/08/2025

Stormy conditions continue over the Northeast, while easing in to wintry showers in the Great Lakes, Inland Midwest and over the Appalachians to the Coastal Southeast.

An intense depression to the northwest of Washington brings widespread rain to the coastal fringes of the Pacific Northwest, through the western half of Oregon and down through the northern half of California.

Mostly dry elsewhere.

Heavy rain may still prevail in Hawaii with the chance of flash flooding in some places, particularly on the Big Island.

DECEMBER 8 MONDAY

ISOBARIC

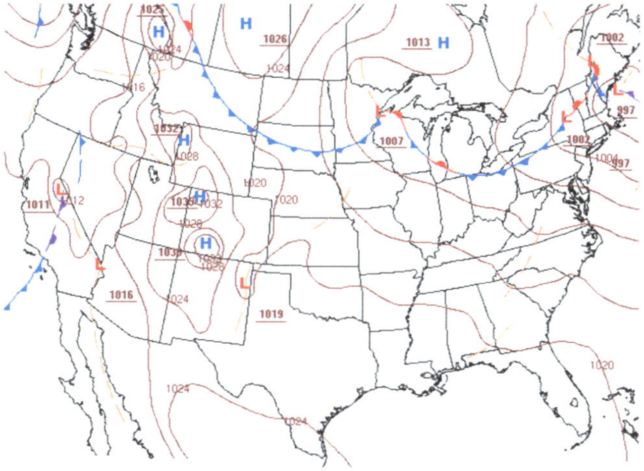

RAIN DISTRIBUTION (Inches)

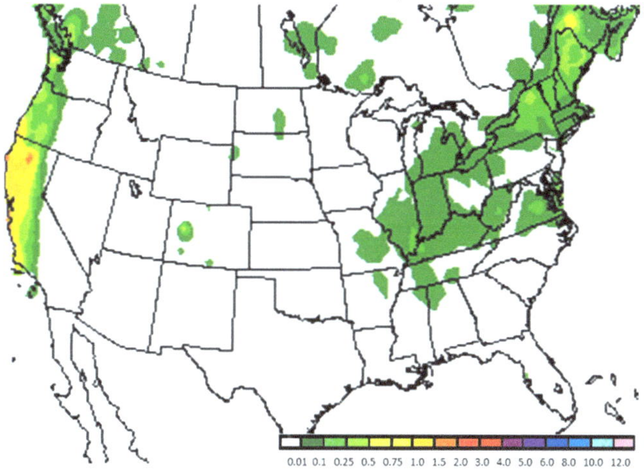

MAXIMUM TEMPERATURE MINIMUM TEMPERATURE

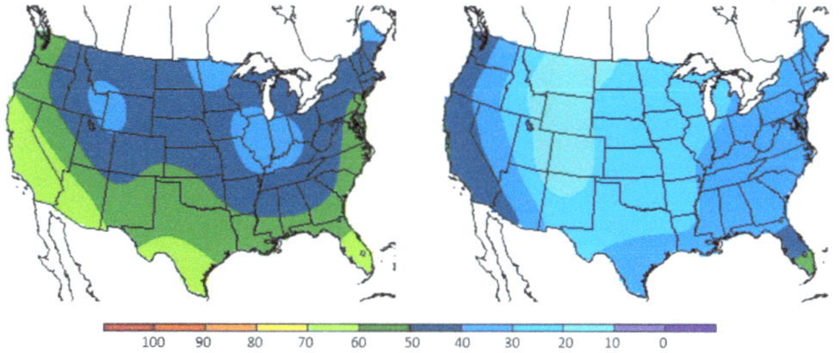

DECEMBER 9 TUESDAY

ISOBARIC

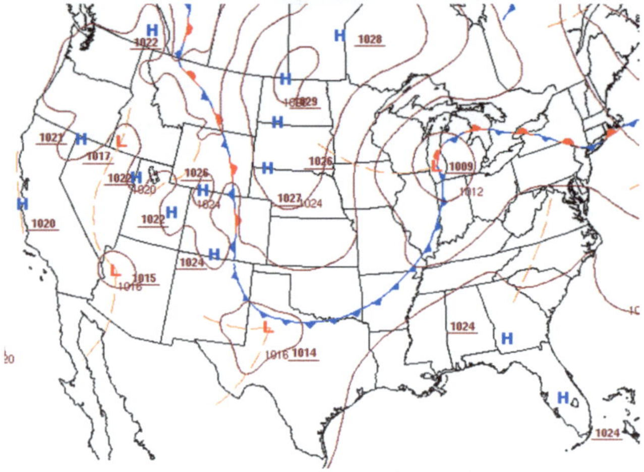

RAIN DISTRIBUTION (Inches)

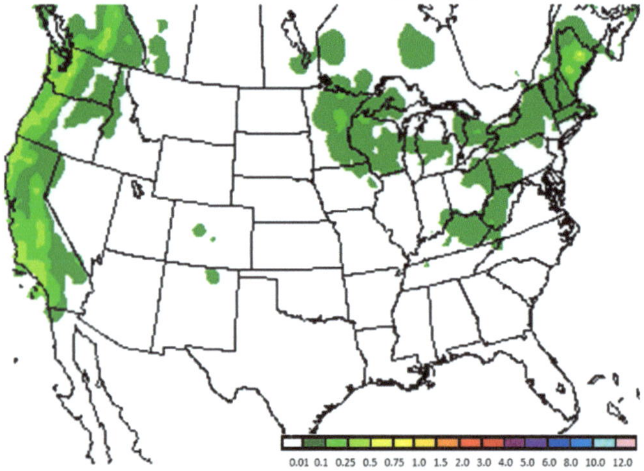

MAXIMUM TEMPERATURE

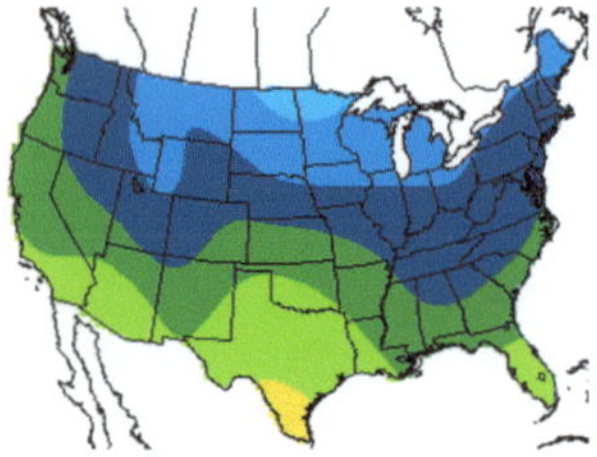

MINIMUM TEMPERATURE

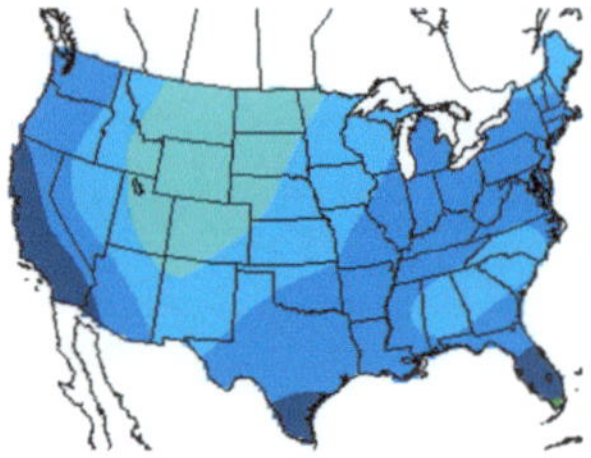

12/09/2025

A weak depression reforms briefly over the northeast bringing wintry showers over the Great Lakes, Inland Midwest, along through the northern half of the Appalachians and into the Northeast.

Widespread outbreaks of rain along the Pacific coastal states from Washington in the north down into southern California, as well as some showers spreading into the Inland Northwest.

Mostly dry elsewhere.

Storm systems may cross both the Northwest and Southwest in the next few days with heavy snow and storm winds in the Inland Pacific Northwest including in Oregon and Washington.

Heavy wintry precipitation in the southwest of Alaska over the next couple of days.

12/10/2025

Wintry precipitation including snow and gales prevails over the Pacific and Inland Northwest, in Montana, Wyoming, Idaho, down through the Rockies and into Utah to Colorado.

Heavy rain in the Southwest including in Arizona, with showers in southern California.

Heavy showers briefly cross Florida, and wintry showers persist over the Great Lakes to the Northeast.

Mostly dry elsewhere.

It may be unusually warm in Alaska, particularly in central parts, despite the chance of heavy wintry precipitation in the southwest.

DECEMBER 10 WEDNESDAY

ISOBARIC

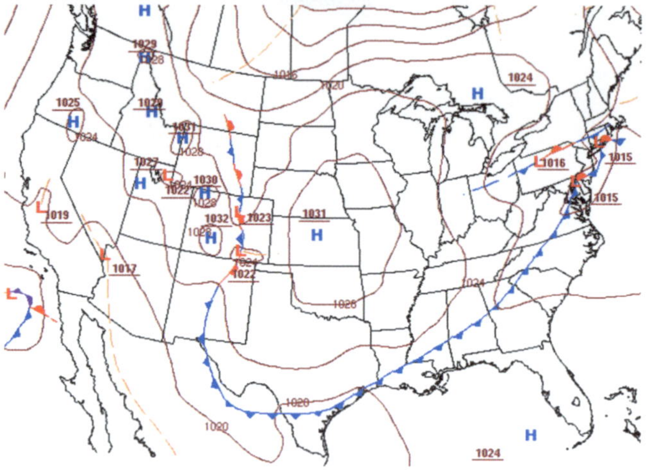

RAIN DISTRIBUTION (Inches)

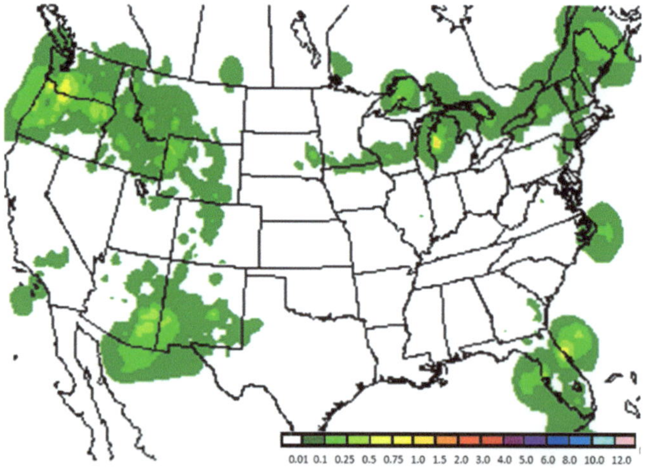

MAXIMUM TEMPERATURE

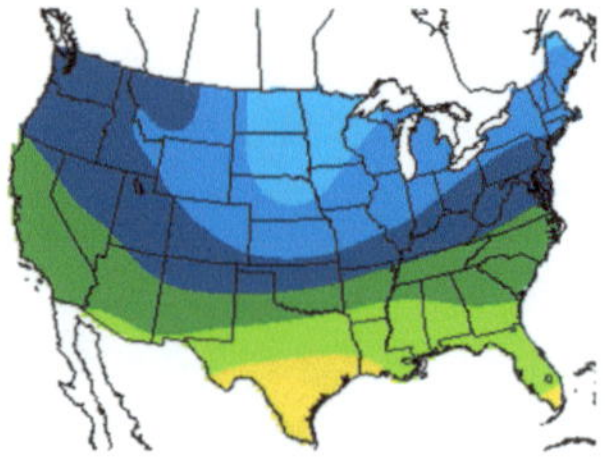

MINIMUM TEMPERATURE

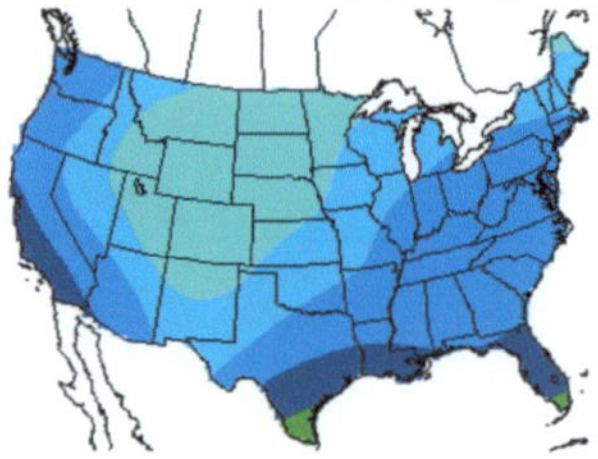

DECEMBER 11 THURSDAY

ISOBARIC

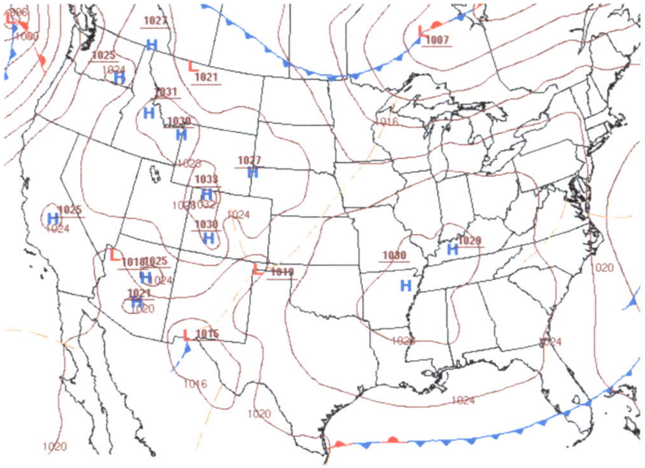

RAIN DISTRIBUTION (Inches)

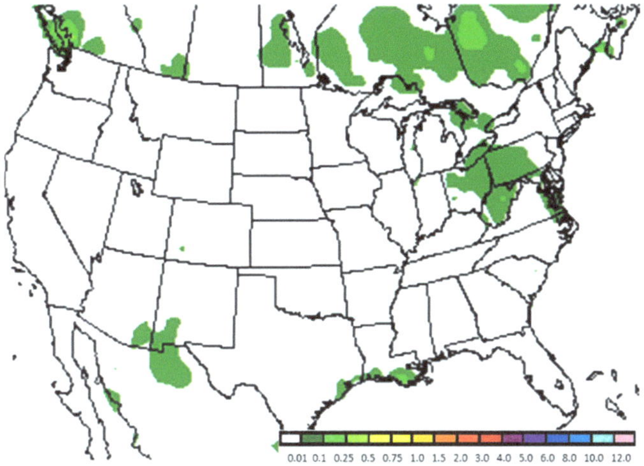

MAXIMUM TEMPERATURE

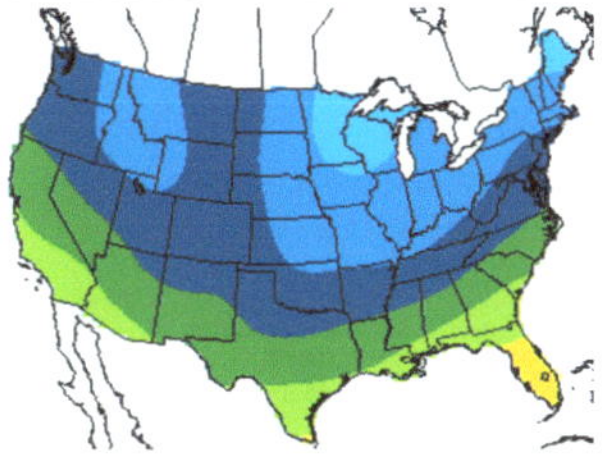

MINIMUM TEMPERATURE

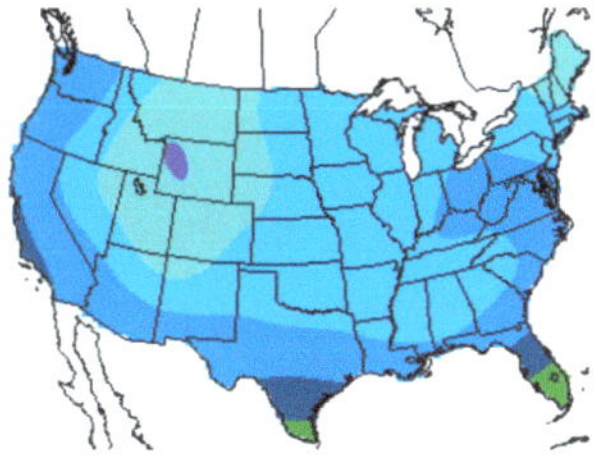

12/11/2025

Some isolated showers around the Southwest, the eastern coastal fringes of Texas and around the Great Lakes to the northern Appalachians.

Mostly dry elsewhere as high pressure settles in the east as well as over the Rocky Mountains to California.

An intense depression northwest of Washington bears down onto the Pacific Northwest in the coming days.

A storm associated with a strong cold front may extend heavy rain and strong winds across Hawaii this week which may also see blizzard warnings issued for the higher reaches on the Big Island.

It will be very unsettled in Alaska despite fairly mild temperatures except in the southeast, where it will be very cold, and a storm may bring high winds and then further snow in the next couple of days.

12/12/2025

Wintry precipitation associated with a cold front lingers over the northern Great Lakes, possibly falling as snow.

Further south, another front brings widespread and often heavy precipitation into the eastern half of Texas and into Louisiana.

In the northwest, a storm makes landfall into the Pacific Northwest and may bring gale winds and flooding into the coastal states from Washington to the northern half of California over the next couple of days.

This storm may also bring snow into the higher reaches of the northwestern states to the northern Rocky Mountains, spreading as far east as Idaho.

Mostly dry elsewhere.

It may be unusually warm in Missouri and Texas around this time.

A storm with high winds prevail in Alaska.

DECEMBER 12 FRIDAY

ISOBARIC

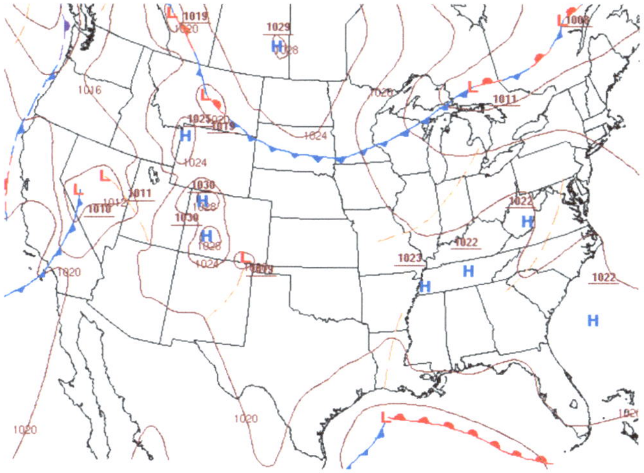

RAIN DISTRIBUTION (Inches)

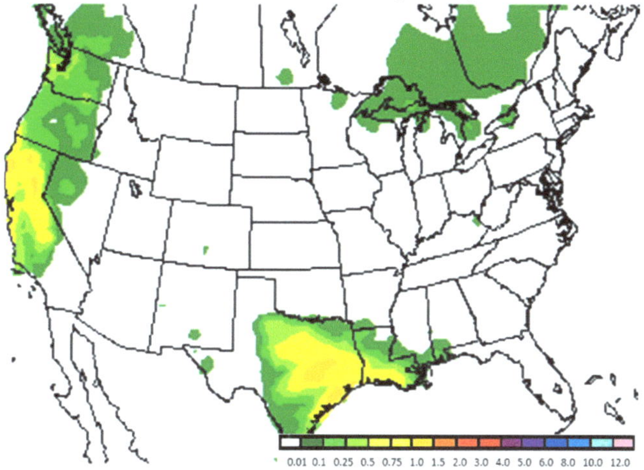

MAXIMUM TEMPERATURE

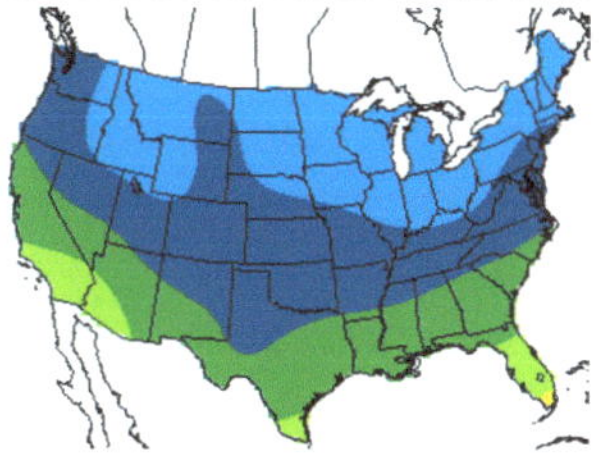

MINIMUM TEMPERATURE

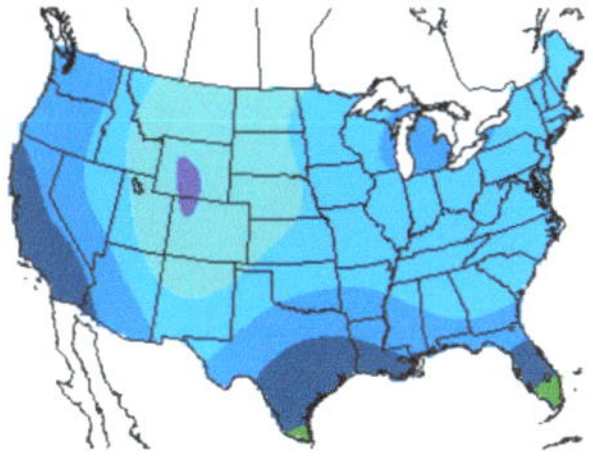

DECEMBER 13 SATURDAY

ISOBARIC

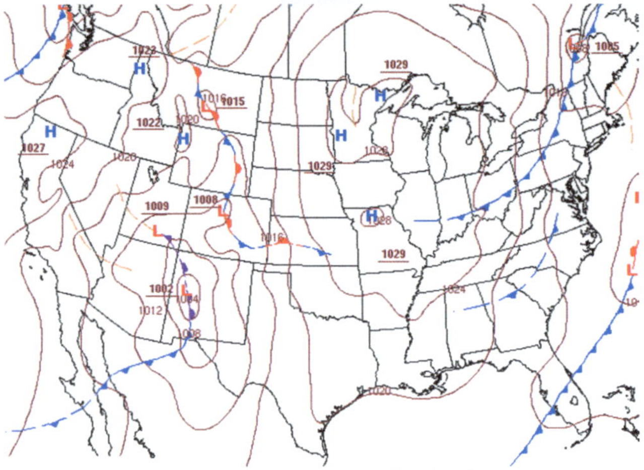

RAIN DISTRIBUTION (Inches)

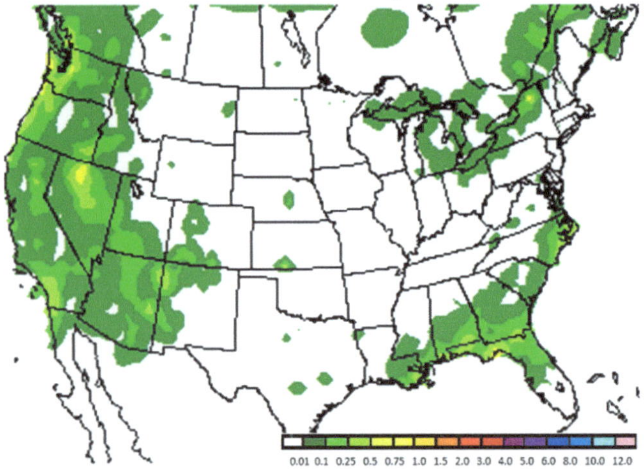

MAXIMUM TEMPERATURE

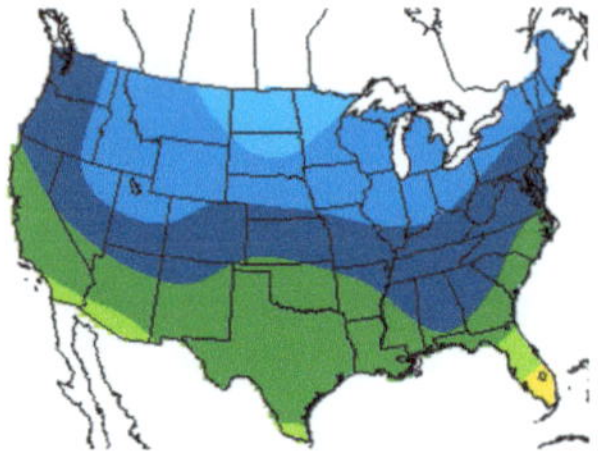

MINIMUM TEMPERATURE

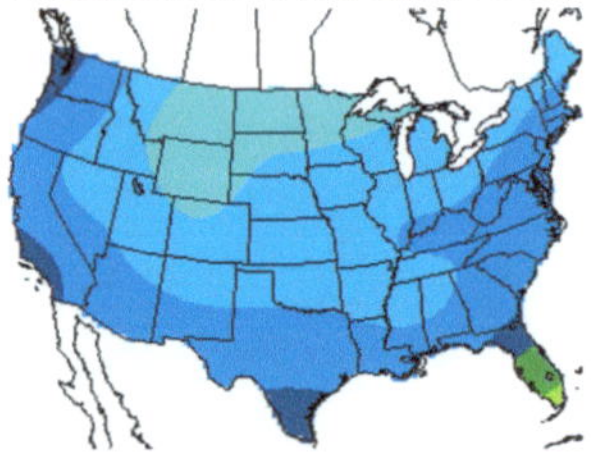

12/13/2025

Storms with gales and heavy rain over the west of the country from the Pacific and Inland Northwest, through the Great Basin, California, and into the Southwest regions.

This storm may bring flooding along the Chehalis River system in Washington as well as snow into the higher reaches of the state, Oregon, and the northern Rocky Mountains, and heavy precipitation into Idaho.

Wintry precipitation including snow also prevails over the Great Lakes to the western half of the Northeast.

Rain in the Deep South to the Coastal Southeast and into northern Florida.

Mostly dry elsewhere.

A storm brings high winds in Alaska.

12/14/2025

Widespread wintry precipitation continues in the Pacific and Inland Northwest, across the Great Basin and spreads southward throughout the Southwest and into Texas.

Snow in the higher reaches of Washington and Oregon, in Idaho and in the southern Rocky Mountains, in parts of New Mexico and over the higher reaches of Texas.

Mostly dry elsewhere except for some lingering snowfalls around the Great Lakes to the Northeast.

Similarly, it may be mostly dry over the North Great Plains and rain may clear throughout California.

Temperatures may be unusually mild in Colorado.

A storm crosses Hawaii over the next couple of days bringing heavy rain and high winds throughout and the chance of snow on the higher reaches of the Big Island.

DECEMBER 14 SUNDAY

ISOBARIC

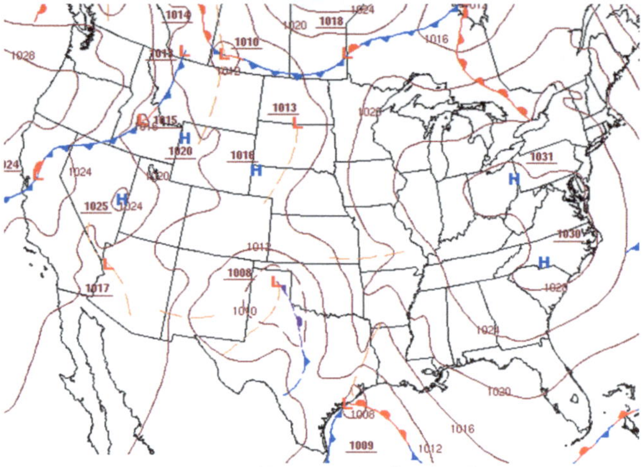

RAIN DISTRIBUTION (Inches)

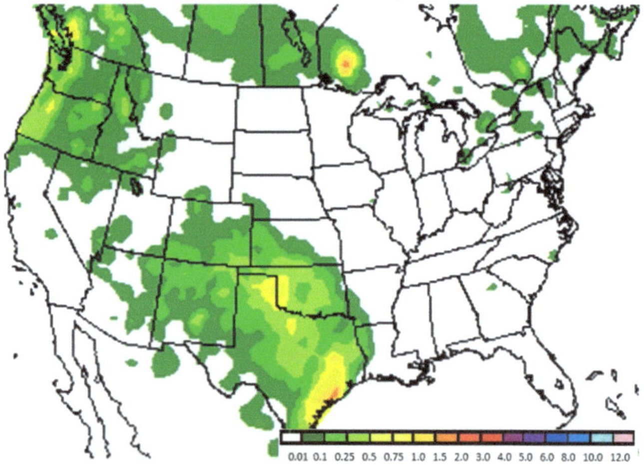

MAXIMUM TEMPERATURE

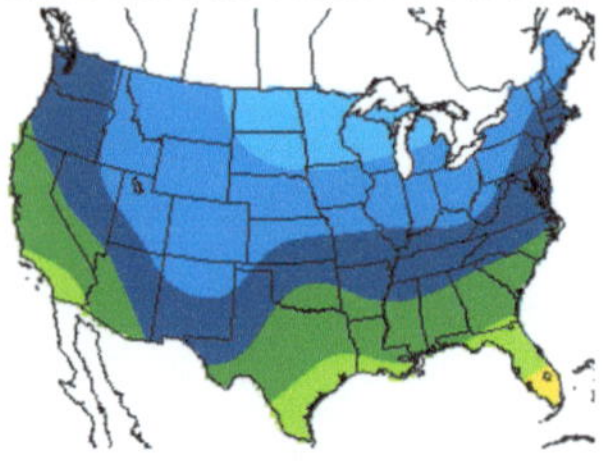

MINIMUM TEMPERATURE

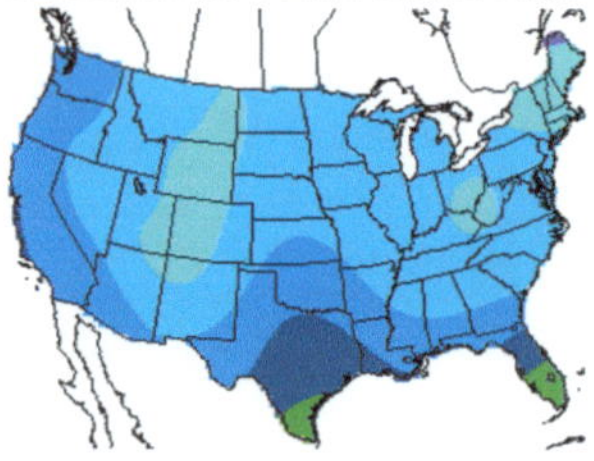

DECEMBER 15 MONDAY

ISOBARIC

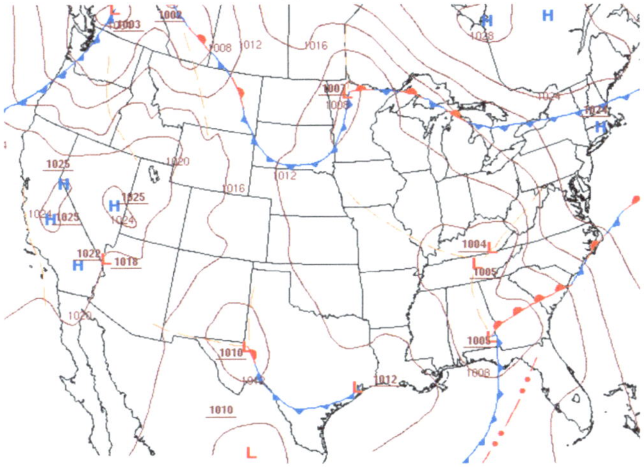

RAIN DISTRIBUTION (Inches)

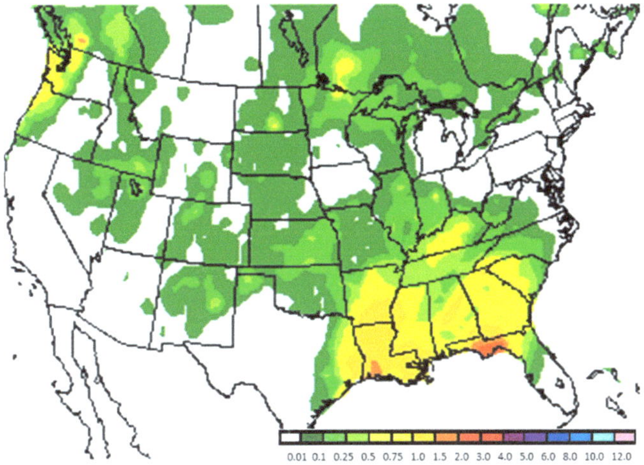

MAXIMUM TEMPERATURE

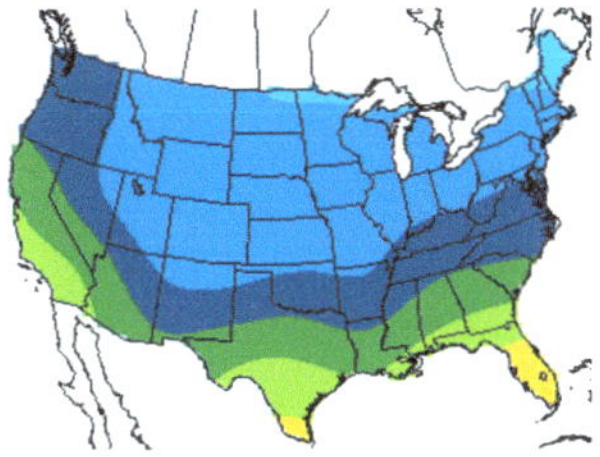

MINIMUM TEMPERATURE

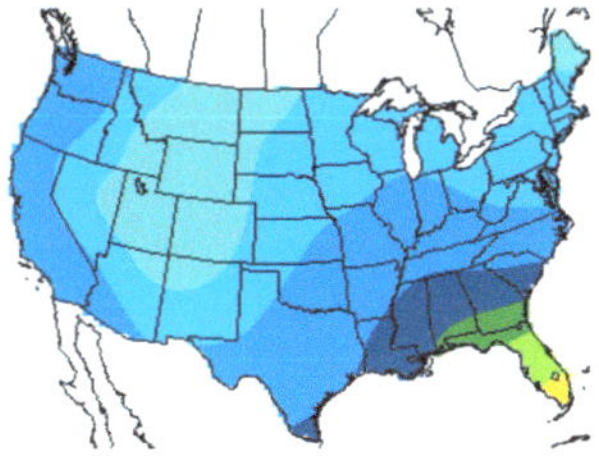

12/15/2025

Wintry precipitation eases in the west. Heavy rain still likely around the Pacific Northwest coast, and wintry showers, including snow into the Inland Northwest to the Great Basin, and east into the Rocky Mountains.

Chance of heavy snow in the Sierra Nevada region.

The storm drifts southeast with wintry precipitation spreading through the Great Plains to the Great Lakes, including snow in the Dakotas and Wisconsin.

Widespread heavy rain in the southeast, in most areas south of the Midwest to the Gulf of Mexico.

Heavy rain in the Deep South may cause flash flooding.

Mostly dry in California, the western half of Texas and in the Megalopolis to the Northeast.

Stormy in Hawaii.

12/16/2025

Storms cover the east of the country from Florida to the Great Lakes and into the Northeast.

Similarly, widespread precipitation prevails in the west from California and the Southwest to the Pacific and Inland Northwest, over the Rocky Mountains.

Snow likely from the Great Lakes to the Megalopolis, as well as heavy snow in the Sierra Nevada ranges and Utah, Nevada, and Colorado.

Chance of flooding in the Carolinas to Virginia.

Most dry through the midsection of the country in the Great Plains to Texas, although some showers may prevail on the South Great Plains.

A frigid airflow may bring severe cold into the Great Lakes region especially Wisconsin, Illinois, and Ohio.

Storms continue across Hawaii.

Snow in Alaska, especially in the southeast.

DECEMBER 16 TUESDAY

ISOBARIC

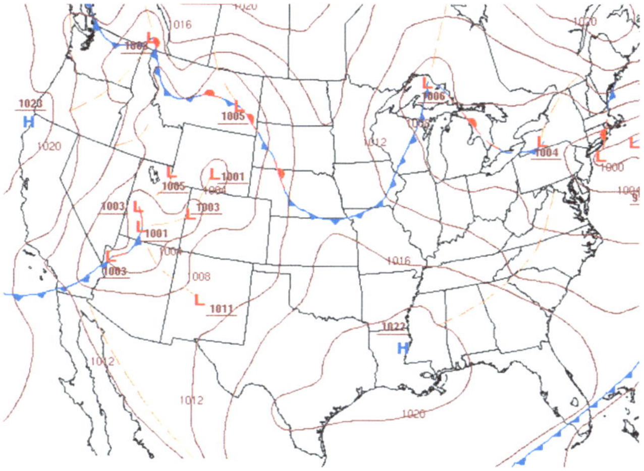

RAIN DISTRIBUTION (Inches)

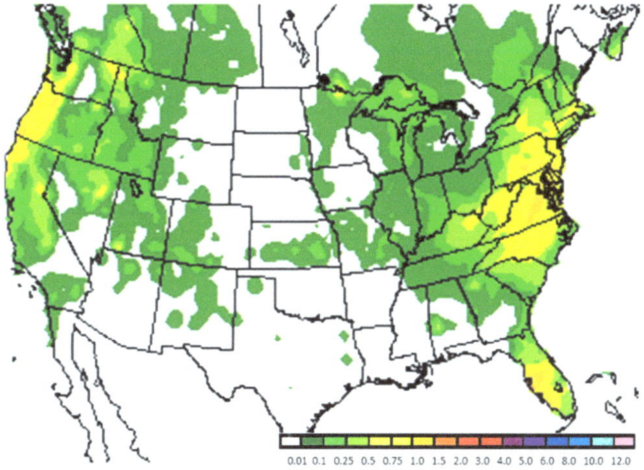

MAXIMUM TEMPERATURE

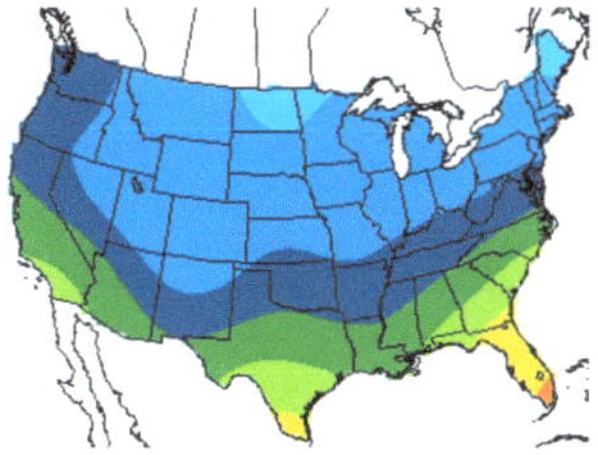

MINIMUM TEMPERATURE

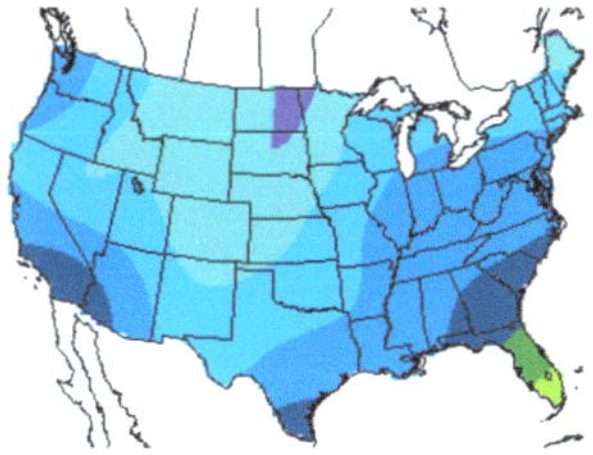

DECEMBER 17 WEDNESDAY

ISOBARIC

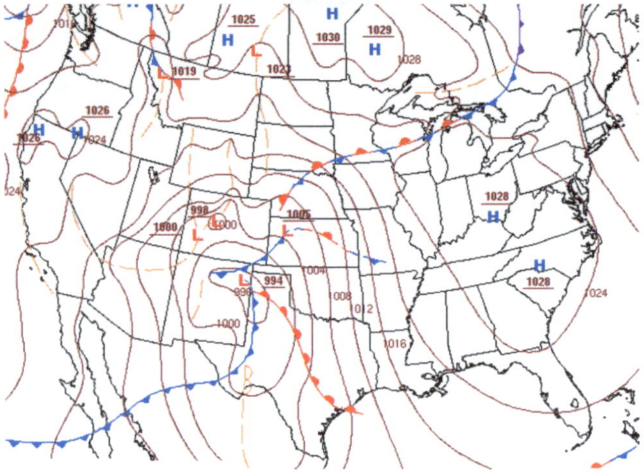

RAIN DISTRIBUTION (Inches)

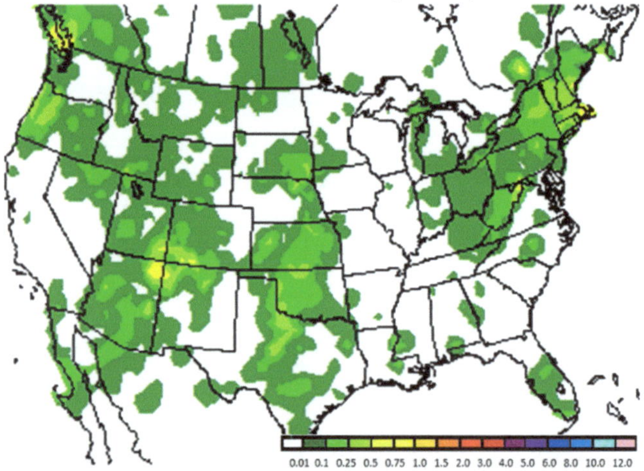

MAXIMUM TEMPERATURE

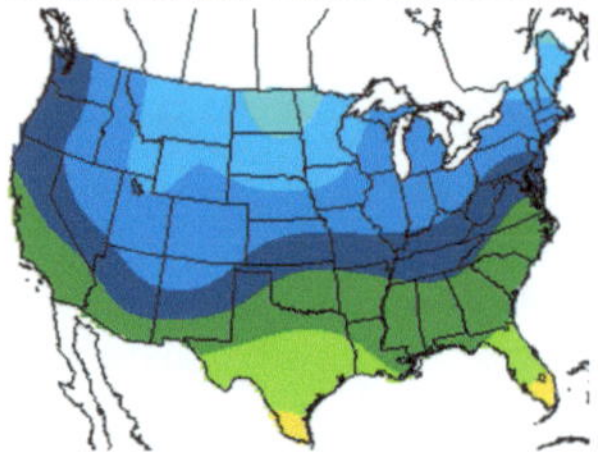

MINIMUM TEMPERATURE

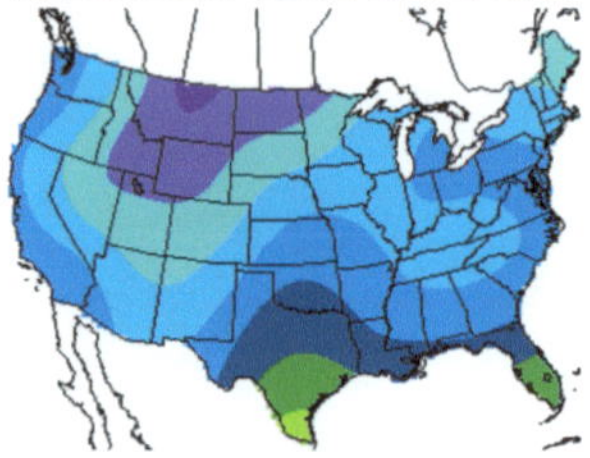

12/17/2025

A depression crosses the south of the country with cold fronts branching to the northeast and southwest.

Patches of rain almost everywhere. Mostly dry in California, west of the Inland Midwest to the Ozarks, and the Deep South to the Coastal Southeast.

Winds very strong and stormy in the Southwest to the Deep South and up through the Great Plains.

Snow in Utah, Nevada, Wyoming, and Colorado, as well as the Rocky Mountains to the central Great Plains.

Snow also likely in North Dakota to the Northeast.

Heavy rain likely in Arizona.

12/18/2025

A large depression that formed in the center of the country earlier intensifies as it slowly moves eastward across the country with widespread storms and heavy precipitation throughout the eastern half of the country from the Great Plains to the Atlantic coasts.

Blizzard conditions prevail throughout many parts of the east from the Great Lakes to the Inland Midwest and southern parts of the Northeast.

Another intense depression to the north of Washington brings storms into the Pacific and Inland Northwest region and into the north of California.

Snow in Utah, Nevada, Colorado, Wyoming, North Dakota, and the Rockies to the central Great Plains.

A storm in Oklahoma may bring some damage to property and infrastructure around this time.

Rain may be very heavy in the Deep South with chance of flooding in low-lying areas.

Mostly dry in California to Arizona.

It may be unusually warm in Louisiana, Georgia, the Carolinas and Texas.

DECEMBER 18 THURSDAY

ISOBARIC

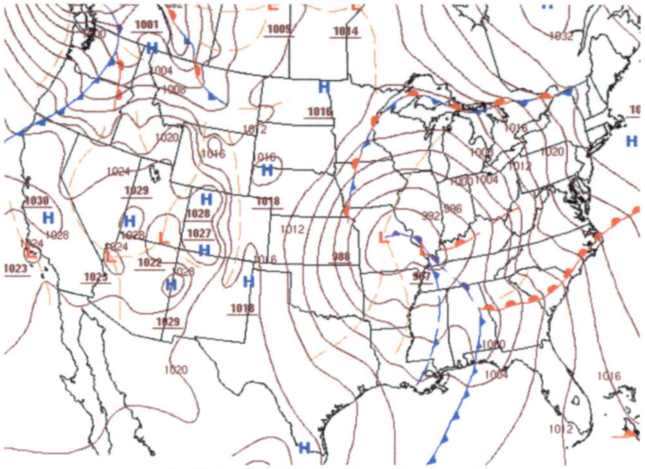

RAIN DISTRIBUTION (Inches)

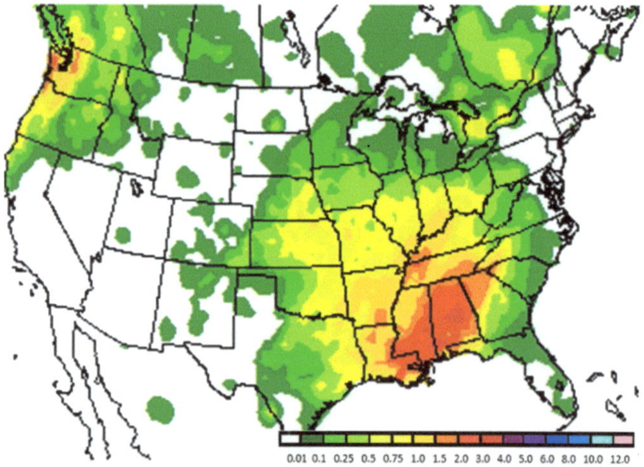

MAXIMUM TEMPERATURE

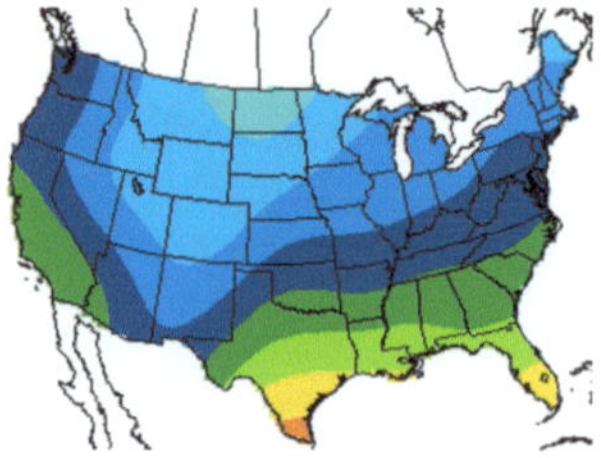

MINIMUM TEMPERATURE

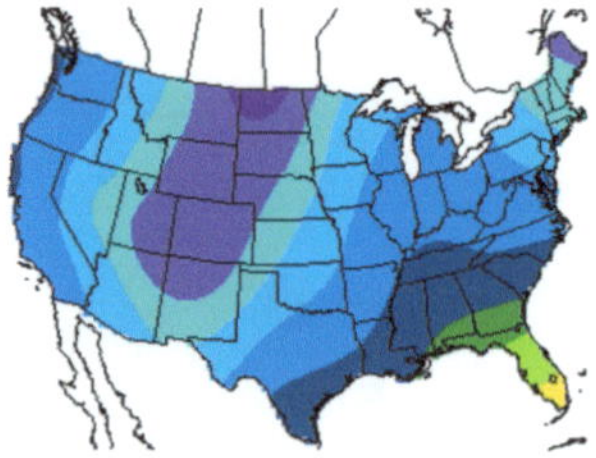

DECEMBER 19 FRIDAY

ISOBARIC

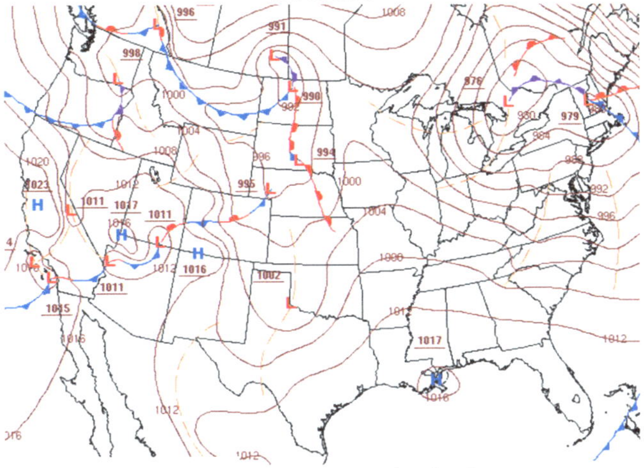

RAIN DISTRIBUTION (Inches)

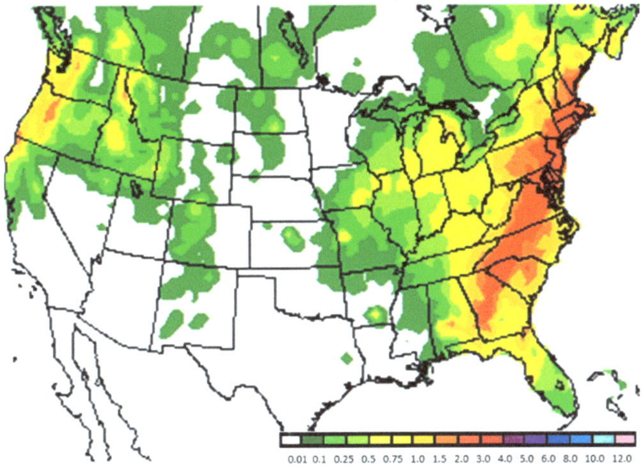

MAXIMUM TEMPERATURE

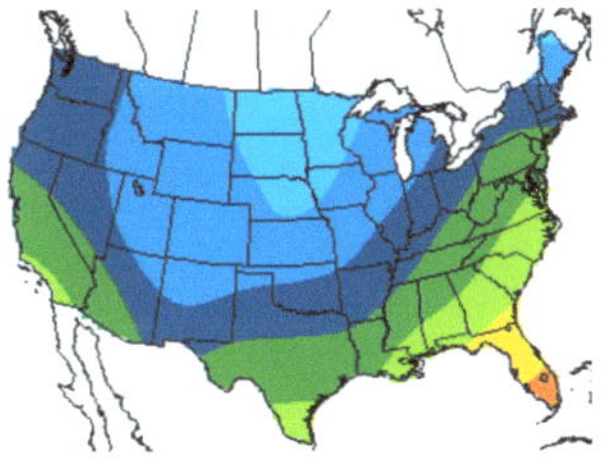

MINIMUM TEMPERATURE

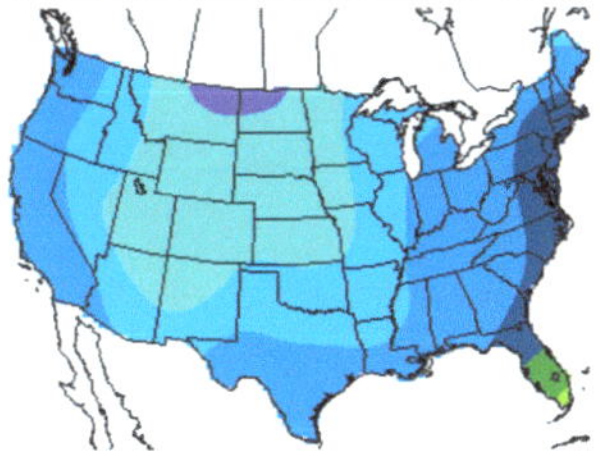

12/19/2025

The storm conditions across the east over the past few days continues its northeasterly trek and may now center over the northeast quarter of the country.

Widespread storms from the east of the Great Plains to the Atlantic coasts and the Northeast to Florida. Chance of damage to local infrastructure.

Blizzards around the Great Lakes to the Northeast. Heavy snow from the Megalopolis to the Appalachians.

Similarly, storms prevail over the northeast quarter of the country from the Pacific Northwest and northern California eastward to the Rocky Mountains, with very heavy and wintry precipitation in the west to the coasts in Washington and Oregon.

Snow in the Rocky Mountains, Colorado, Wyoming, the central Great Plains, and possibly even Arizona.

Mostly dry elsewhere.

12/20/2025

Widespread precipitation and storms prevail over the northern half of the country.

Heavy falls likely on the Pacific coast from Washington to northern California with widespread wintry rain and snow spreading through the Great Basin and over Rocky Mountains, and snow possibly as far south as Arizona and east over the central Great Plains.

Another storm batters the northeast quarter of the country from the Great Lakes to the Northeast, but clears in the southern areas of the Inland Midwest. Heavy rain likely in Missouri for a time.

Mostly dry in the southern states of the country from the Southwest to the Coastal Southeast, with only the odd shower here and there.

Temperatures may be unusually mild in Louisiana and Georgia to the Carolinas.

DECEMBER 20 SATURDAY

ISOBARIC

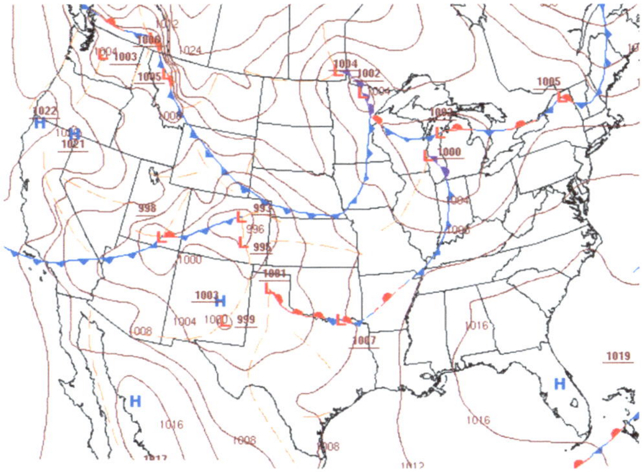

RAIN DISTRIBUTION (Inches)

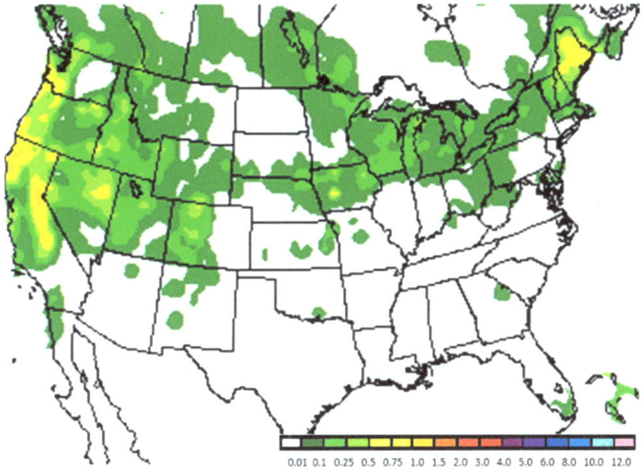

MAXIMUM TEMPERATURE

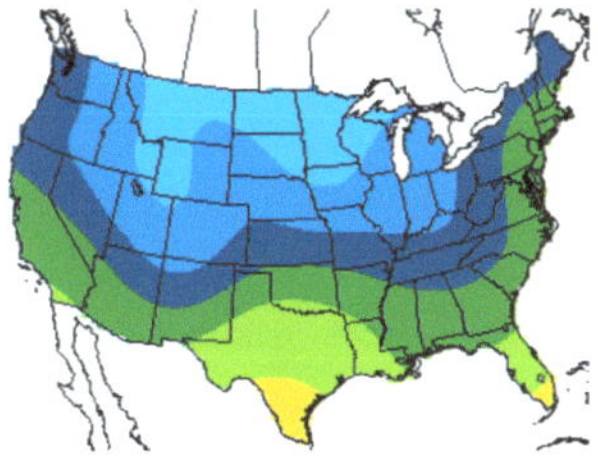

MINIMUM TEMPERATURE

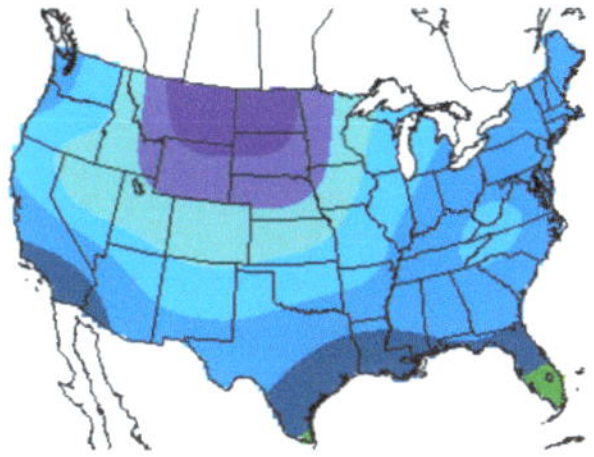

DECEMBER 21 SUNDAY

ISOBARIC

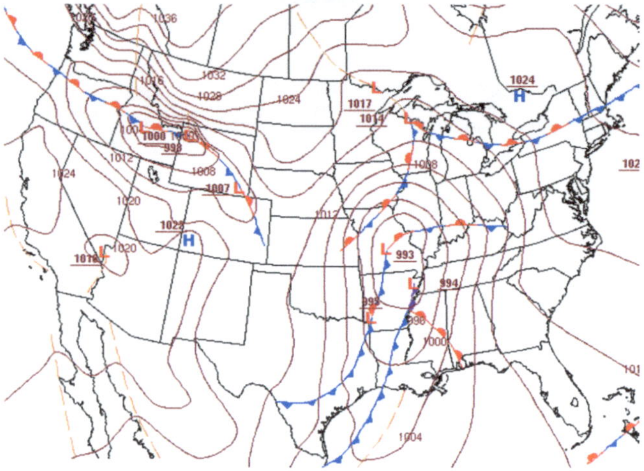

RAIN DISTRIBUTION (Inches)

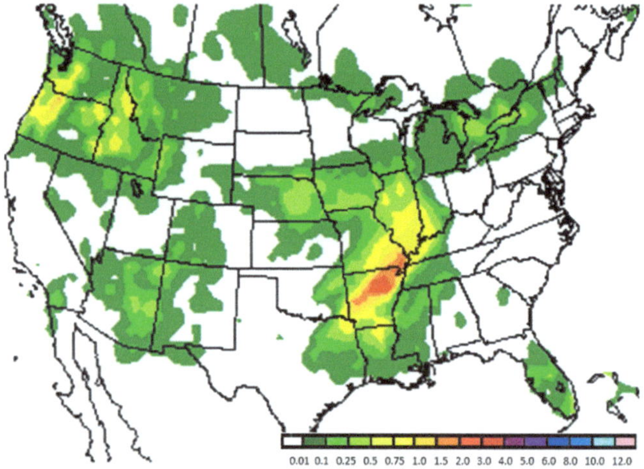

MAXIMUM TEMPERATURE

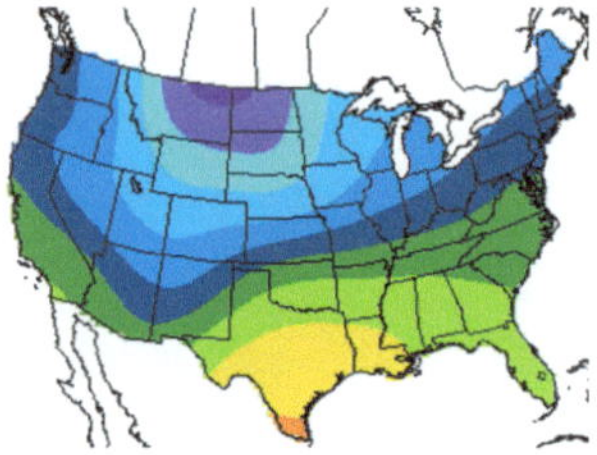

MINIMUM TEMPERATURE

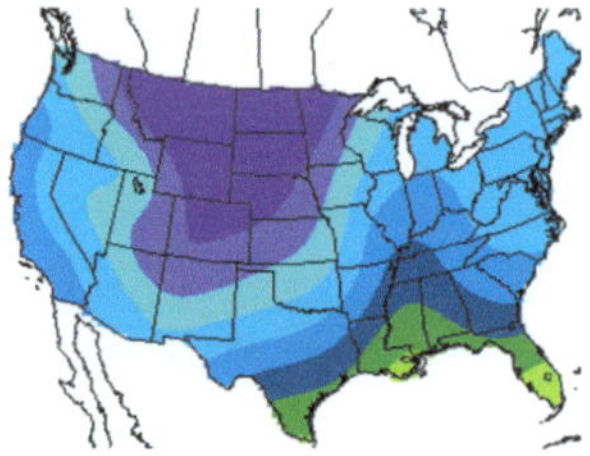

12/21/2025

A depression forms over the east bringing more storms throughout the Inland Midwest to the western half of the Deep South, with particularly heavy precipitation from Missouri, through the Ozarks and into Arkansas.

A stationary front over the Great Lakes and Northeast brings further storms and snow into these regions, as well as across the central and southern Great Plains.

A depression brings intense precipitation and winds from the Pacific Northwest to the northern Rocky Mountains and into the Great Basin, with chance of damage to infrastructure.

Blizzard conditions in the northwestern states from Washington to Montana and Idaho. Snow in Arizona.

Rain through the Southwest. Snow in the Rockies.

Mostly dry in the Dakotas, western Texas and in the Atlantic states from the Northeast to Georgia.

12/22/2025

An intense depression brings storm conditions to the eastern half of the country from the east of the Great Plains to the Atlantic coasts, including in Florida.

Blizzards may cause damage to infrastructure in the Great Lakes, Inland Midwest, Northeast, and Megalopolis regions.

Precipitation may be particularly heavy either side of the Appalachians.

Further west, a stationary front from the Pacific coast to the South Great Plains brings heavy rain from Oregon to northern California, across the Great Basin and into the southern Rocky Mountains.

Mostly dry in the Southwest to Texas, and showery over the North Great Plains.

Temperatures may be unusually mild in Louisiana to North Carolina, and in Georgia.

DECEMBER 22 MONDAY

ISOBARIC

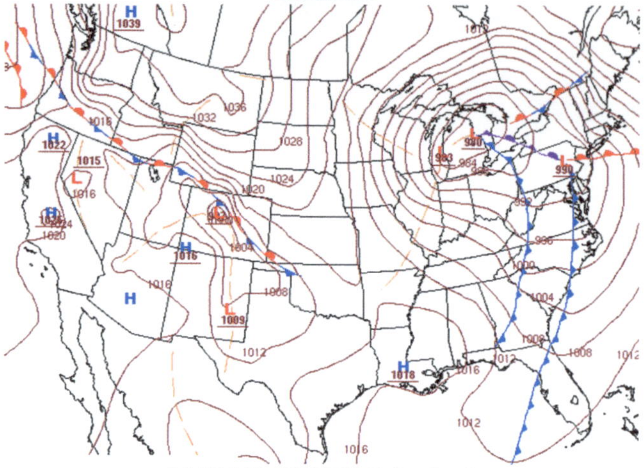

RAIN DISTRIBUTION (Inches)

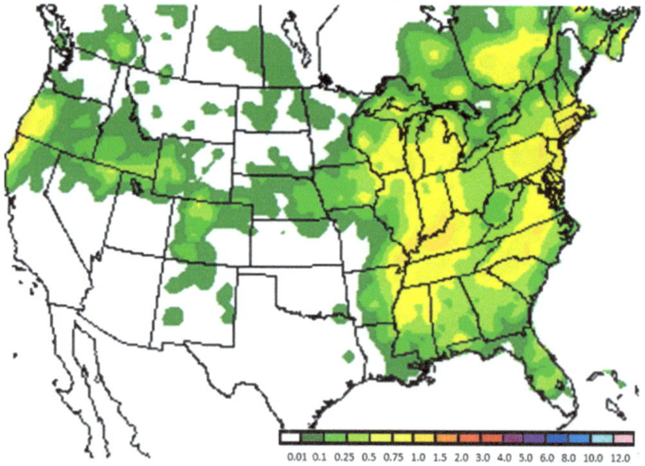

MAXIMUM TEMPERATURE

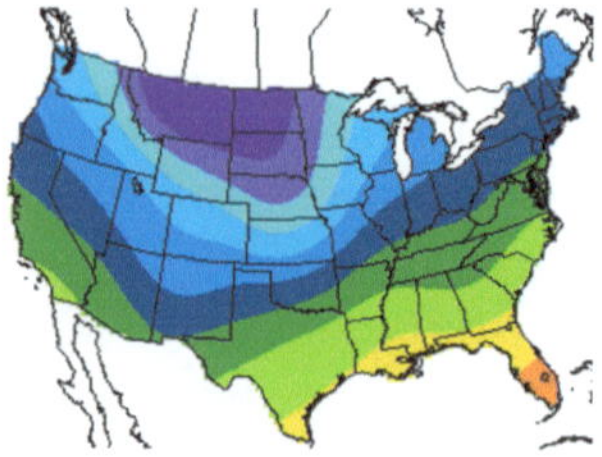

MINIMUM TEMPERATURE

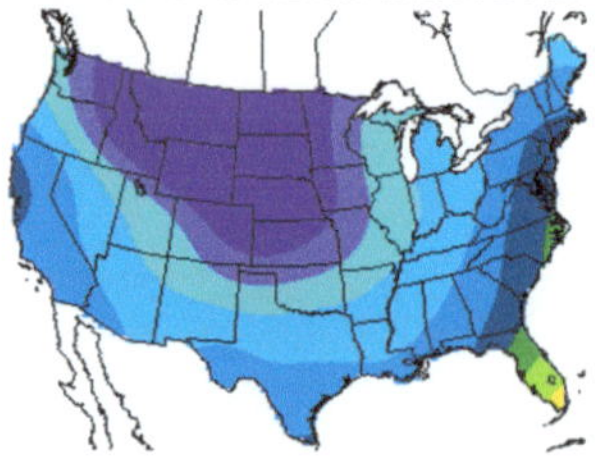

DECEMBER 23 TUESDAY

ISOBARIC

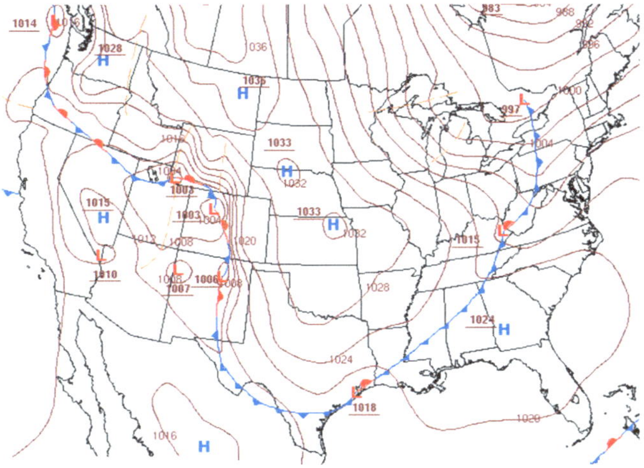

RAIN DISTRIBUTION (Inches)

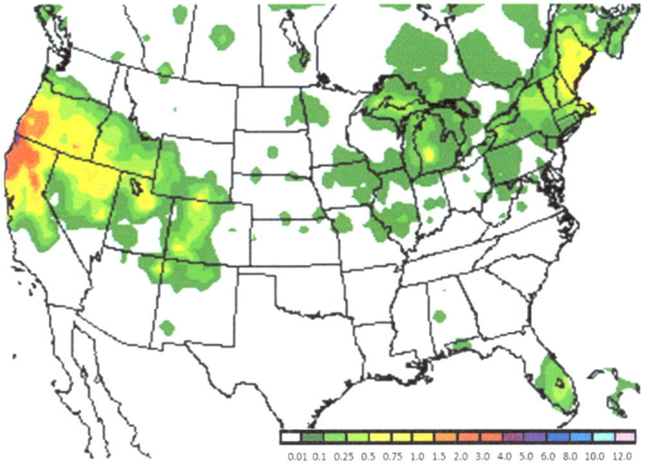

MAXIMUM TEMPERATURE   MINIMUM TEMPERATURE

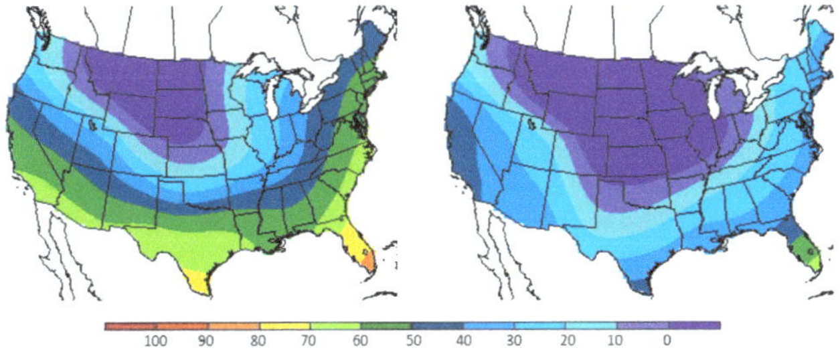

12/23/2025

The stormy conditions move north, easing the rain activity in the southeast, but maintaining some wintry shower activity over the Inland Midwest, and snow over the Great Lakes and Northeast.

Snow may be heavy in Massachusetts to Rhode Island and New York.

Widespread rain over the west from Oregon to the northern half of California and through the Great Basin to the southern Rocky Mountains with very heavy falls at times in the Sierra Nevada and Cascade Ranges.

Heavy rain around the Six Rivers National Forest in northern California may cause localized flooding in the region and up into the southwest of Oregon.

Mostly dry elsewhere with just the odd shower around the Great Plains and in the southern half of Florida.

A cold snap in California over the next couple of days.

DECEMBER 24 WEDNESDAY

ISOBARIC

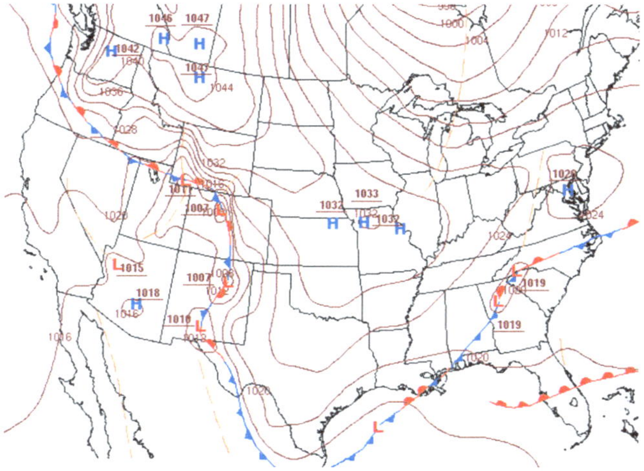

RAIN DISTRIBUTION (Inches)

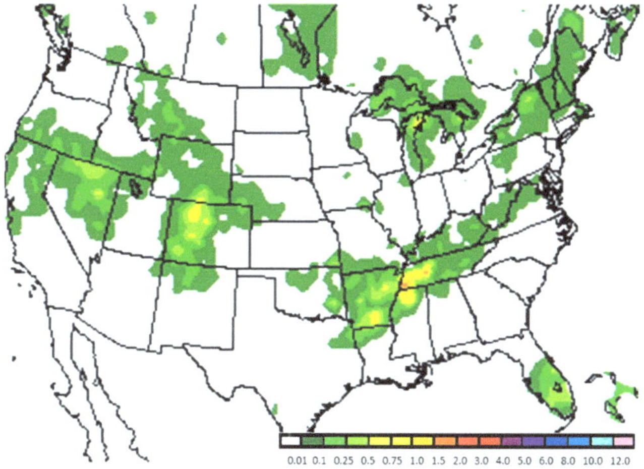

MAXIMUM TEMPERATURE   MINIMUM TEMPERATURE

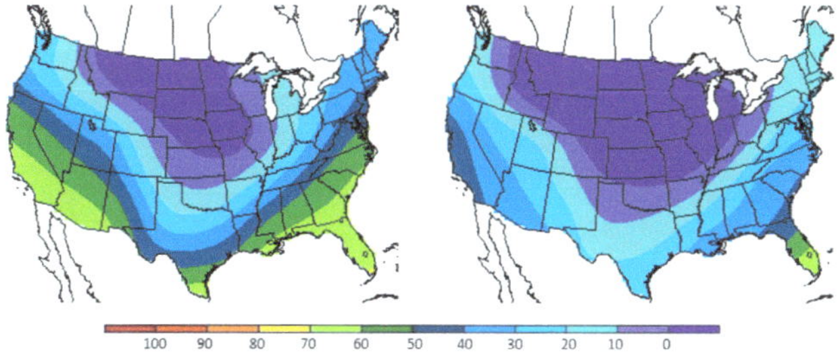

12/24/2025

Storm conditions over the northeast of the country continues to ease as the depression slips further northwards into Canada but some wintry falls continue in the Northeast to the Great Lakes.

Further south, a band of storms cross the northern half of the Deep South and over the southern Appalachians to the Chesapeake coast. Rain in southern Florida.

In the west, rain eases to showers in Oregon to northern California and in the Great Basin, while wintry falls continue along the Rocky Mountains.

Mostly dry elsewhere.

A storm crossing Missouri may bring very heavy wintry falls of sleet, snow, and freezing rain, while a storm crossing the midsection of the country may bring heavy snow in to Kansas, and Illinois.

It may be bitterly cold in Michigan.

ISOBARIC

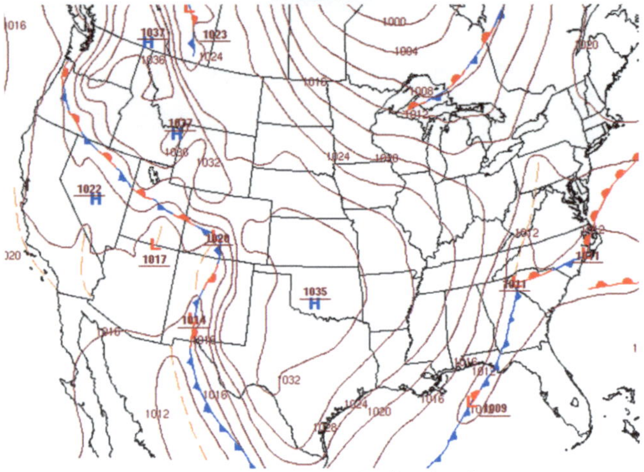

RAIN DISTRIBUTION (Inches)

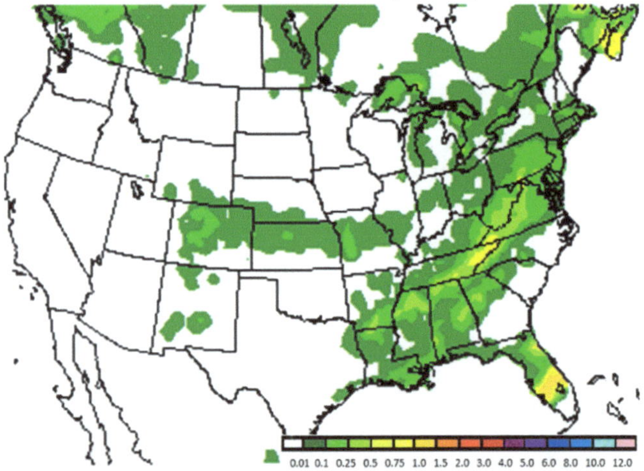

MAXIMUM TEMPERATURE **MINIMUM TEMPERATURE**

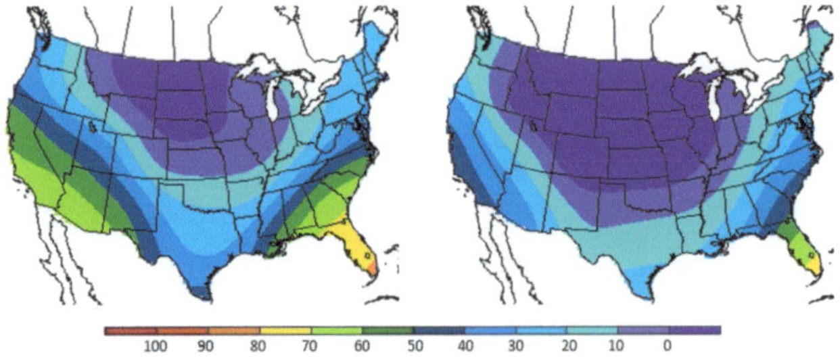

12/25/2025

Storms cross the country from the southern Rocky Mountains, over the central Great Plains and into the Inland Midwest, which may bring some heavy snow into Kansas and Illinois.

Widespread precipitation and storms from the Deep South, along the Appalachians to the Metropolitan with a blast of wintry showers over the Great Lakes and Northeast, and it may be very snowy in Missouri.

Some rain in Florida.

Mostly dry elsewhere.

Despite the high pressure through the midsection of the country, major ice problems are likely in parts of the Great Plains to the Inland Midwest, and it may be bitterly cold in Michigan, Wyoming, Utah, and Arizona.

Chance of severe thunderstorms crossing the Deep South. Potential for the formation of tornadoes in the region, including in Georgia and Florida.

12/26/2025

Widespread precipitation prevails along the Atlantic coasts from the Northeast to Florida and into the eastern Midwest to the Great Lakes.

Snow may be heavy in Massachusetts, while snow may also prevail around the Great Lakes including in Michigan and Indiana and over the Northeast, including in New England and Maine.

Further south along the east coastal fringes, snow turns to heavy rain as it falls in Florida and Delaware.

Widespread wintry precipitation over the northwest of the country as well, throughout the Pacific and Inland Northwest and south into northern California and the Great Basin, with lighter precipitation, including snow, into the Rocky Mountains. Mostly dry elsewhere.

Cold in the Rocky Mountains. Very cold in Utah.

ISOBARIC

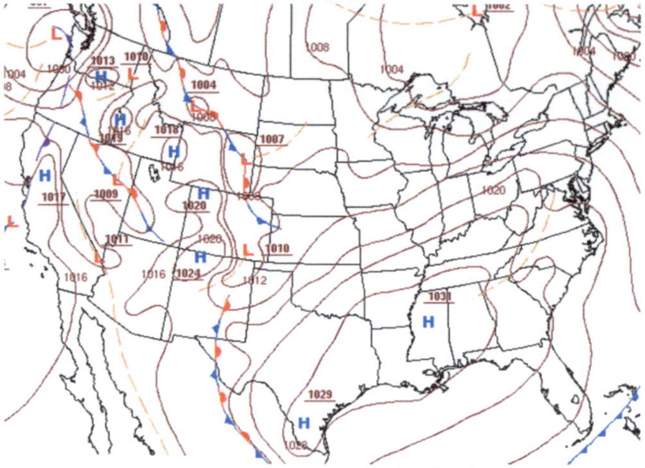

RAIN DISTRIBUTION (Inches)

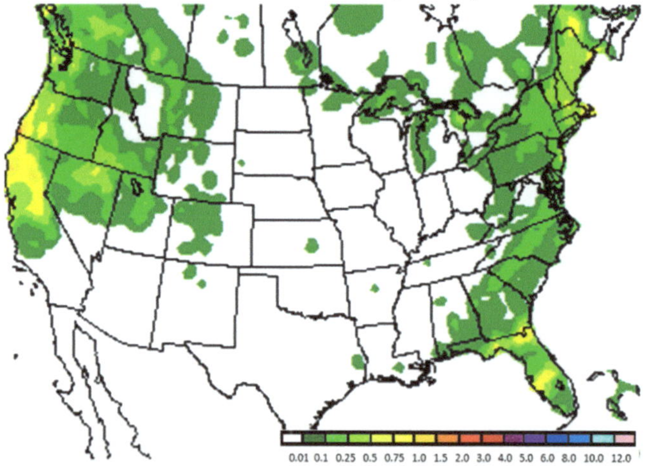

MAXIMUM TEMPERATURE **MINIMUM TEMPERATURE**

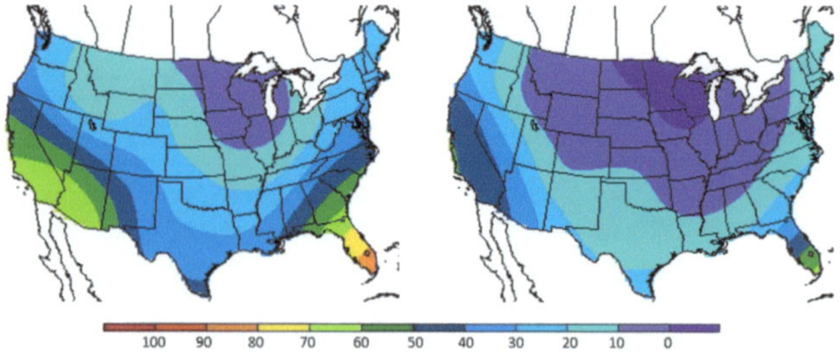

DECEMBER 27 SATURDAY

ISOBARIC

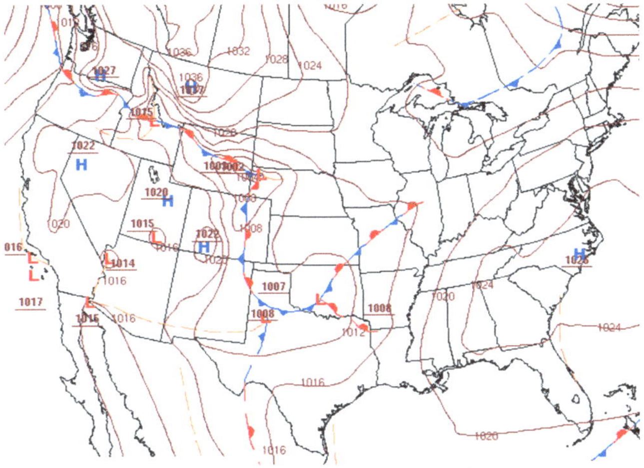

RAIN DISTRIBUTION (Inches)

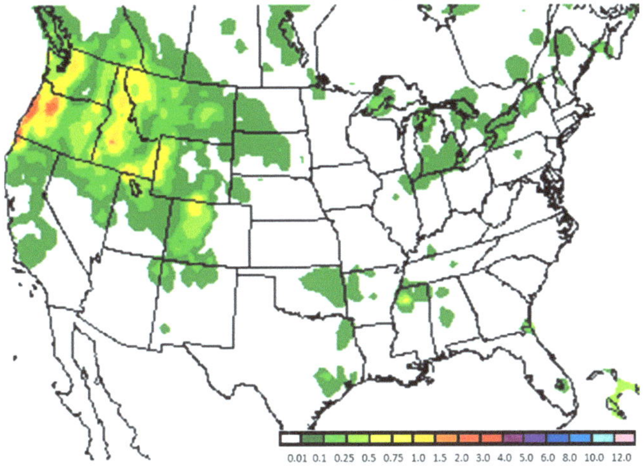

MAXIMUM TEMPERATURE | MINIMUM TEMPERATURE

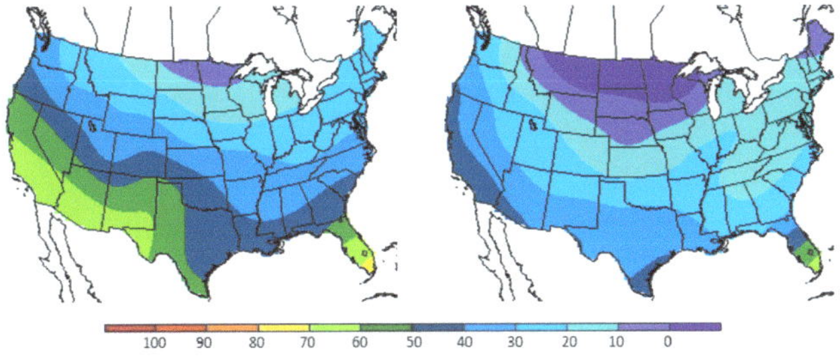

12/27/2025

Widespread precipitation throughout the northwest quarter of the country from northern California to the Pacific and Inland Northwest states, across the Rocky Mountains and into the North Great Plains.

Deep snow is likely in the Sierra Nevada's over the next few days, as well as in Utah.

Mostly dry over the eastern half of the country except for the isolated pockets of showers around the Great Lakes to the Northeast, in Oklahoma and central Deep South states.

It may also be mostly dry from southern California, across the Southwest and over Texas, as well as in most parts of the South Great Plains.

Very cold in Arizona, and the cold air behind the recent storms may bring very cold temperatures to the Carolinas over the next few days.

12/28/2025

An ex-tropical storm may affect the east of the US as a winter storm bringing a wintry blast over the northern and eastern half of the country, and a band of severe thunderstorms over the Deep South to Florida, which may result in the formation of a number of tornadoes.

Wintry precipitation prevails over the northern states from the Metropolitan to the Great Lakes and onto the North Great Plains as well as throughout the Inland Midwest and down through the Appalachians.

Some snow likely in the central Rocky Mountains, while storms prevail in the Pacific and Inland Midwest regions into Idaho.

Snowfalls may still occur in the Sierra Nevada's and Utah.

Mostly dry elsewhere.

DECEMBER 28 SUNDAY

ISOBARIC

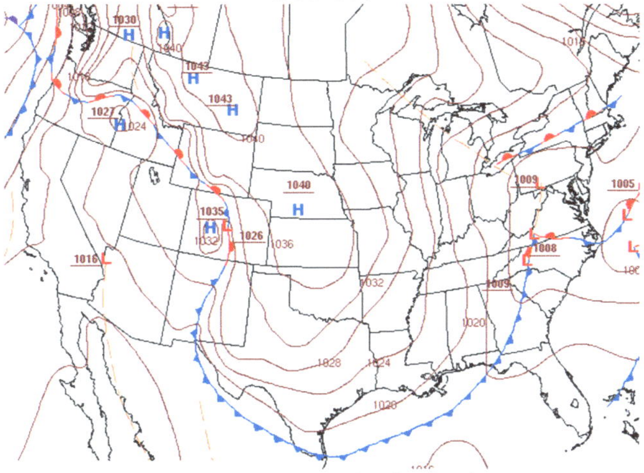

RAIN DISTRIBUTION (Inches)

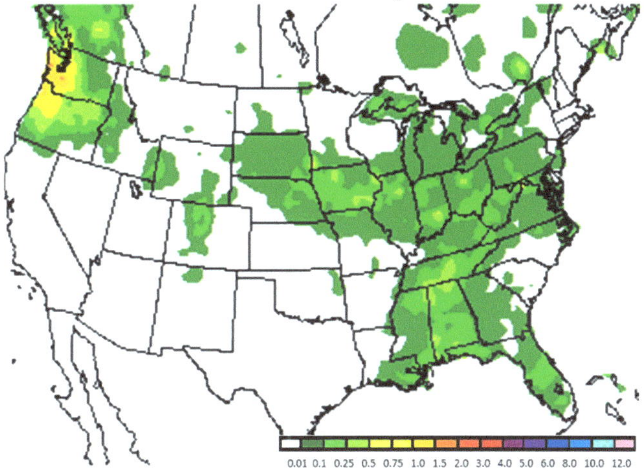

MAXIMUM TEMPERATURE | MINIMUM TEMPERATURE

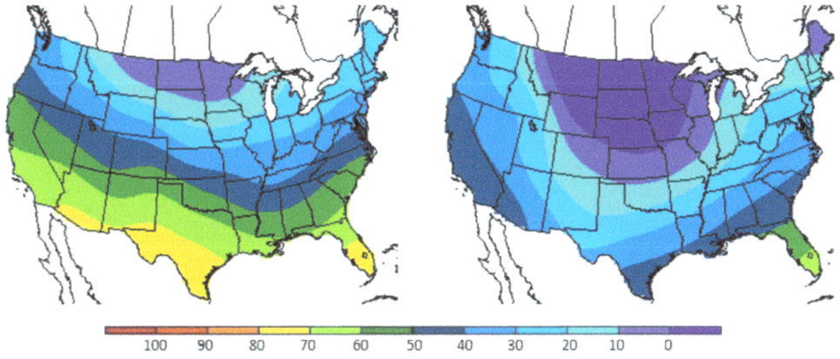

DECEMBER 29 MONDAY

ISOBARIC

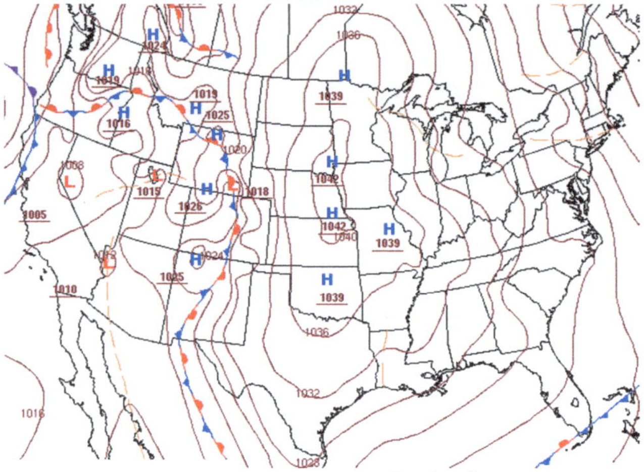

RAIN DISTRIBUTION (Inches)

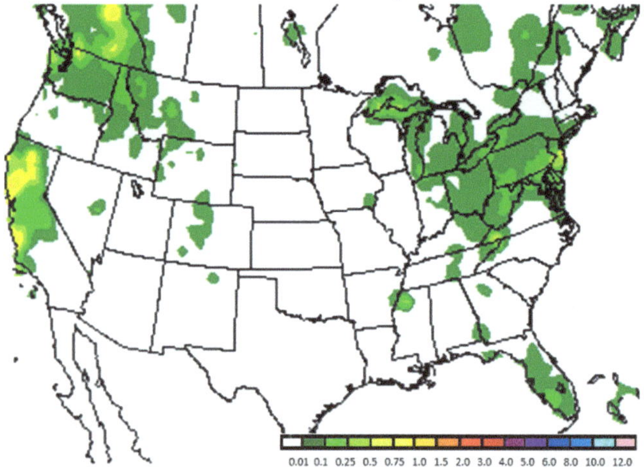

MAXIMUM TEMPERATURE

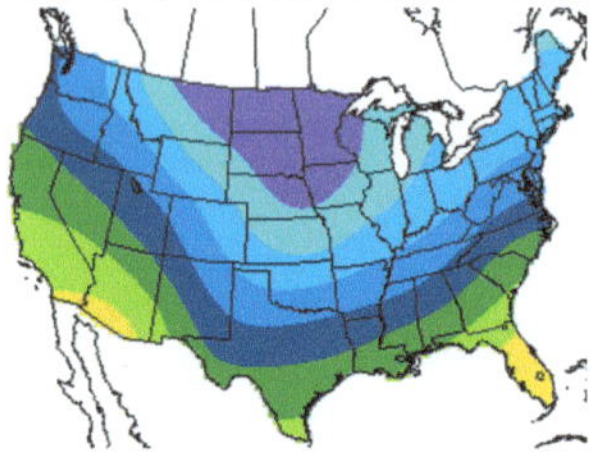

MINIMUM TEMPERATURE

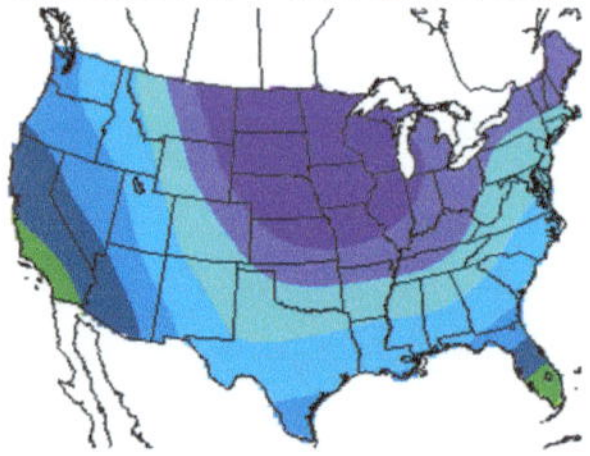

12/29/2025

Wet conditions in the Pacific and Inland Northwest with wintry showers over the Rocky Mountains.

Heavy precipitation in the northern half of California with snow in the Sierra Nevada and Utah still possible.

Widespread precipitation continues over the Great Lakes, through the eastern half of the Inland Midwest and into the Metropolitan and northern Appalachians.

Chance of snow still in the Northeast, particularly in New Hampshire over the next couple of days.

Some showers also persist in Florida.

Mostly dry elsewhere with very cold temperatures throughout the Great Plains and south into Texas.

It will be severely cold in Alaska.

12/30/2025

Very unsettled weather cover the western half of the country from the Rocky Mountains to the Pacific coasts with widespread precipitation throughout the Pacific Northwest to California, and east to the Rocky Mountains.

Snow likely in the Sierra Nevada to Utah in the south and Idaho to the north, as well as in the Rocky Mountains and higher reaches of Oregon and Washington.

Some lingering wintry showers over the Northeast to the Great Lakes and Inland Midwest.

Mostly dry but cold elsewhere.

A warmer spell may develop in the south with the arrival of thunderstorms and scattered showers over the next couple of days, including in Texas and across the Deep South.

Severely cold in Alaska where temperatures may be as low as -40°F in places.

DECEMBER 30 TUESDAY

ISOBARIC

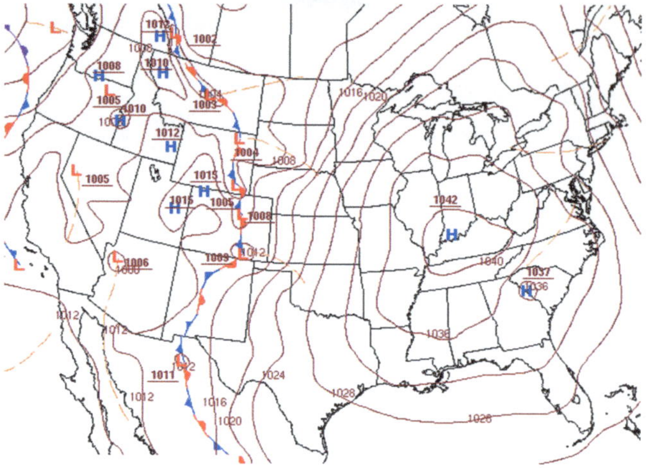

RAIN DISTRIBUTION (Inches)

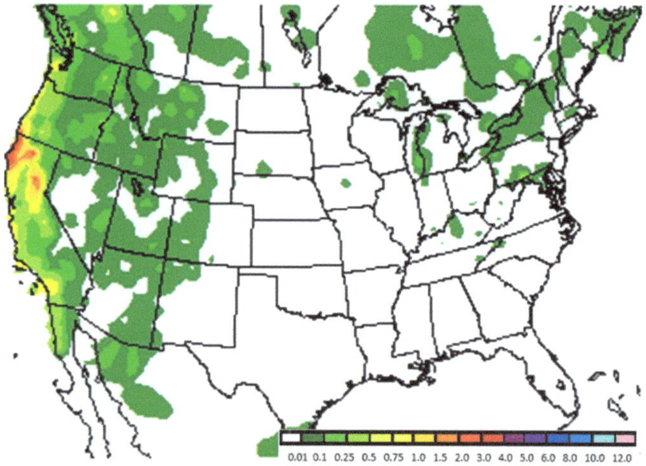

MAXIMUM TEMPERATURE

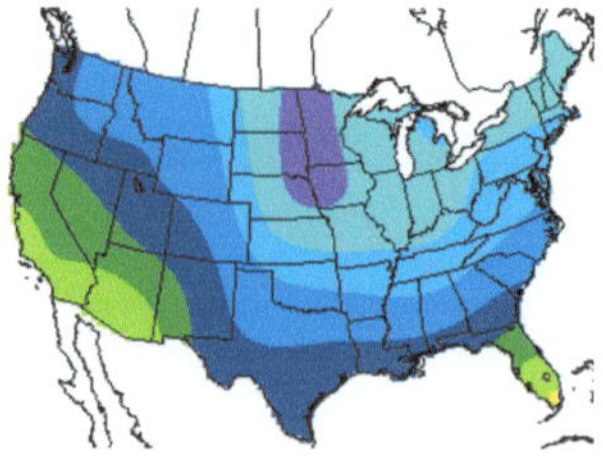

MINIMUM TEMPERATURE

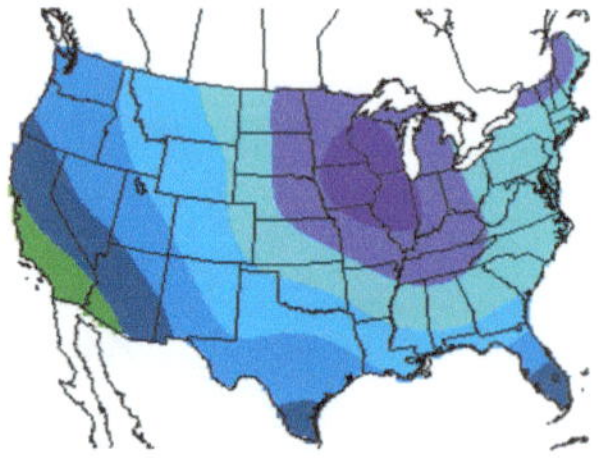

ISOBARIC

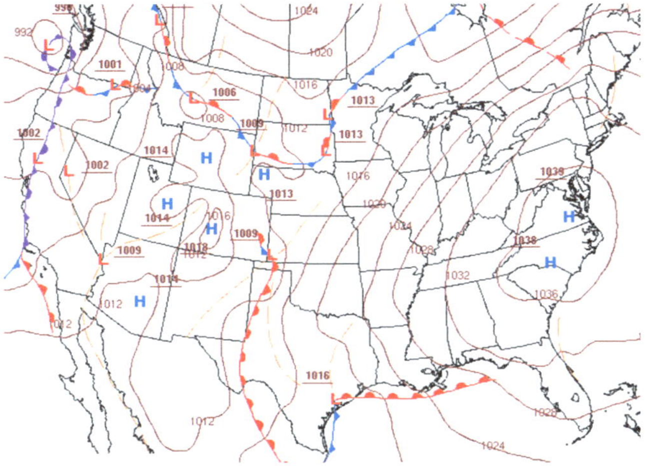

RAIN DISTRIBUTION (Inches)

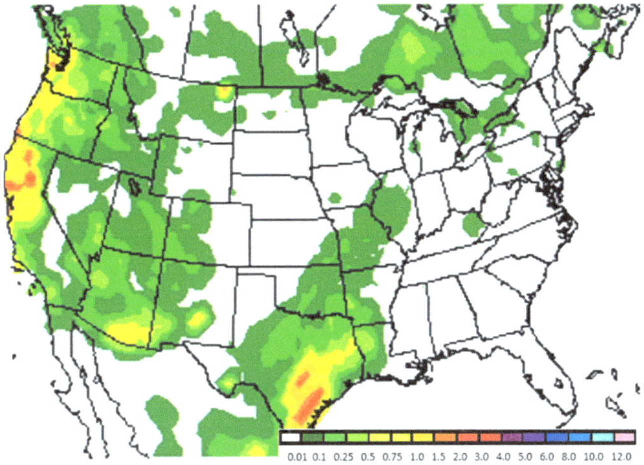

MAXIMUM TEMPERATURE

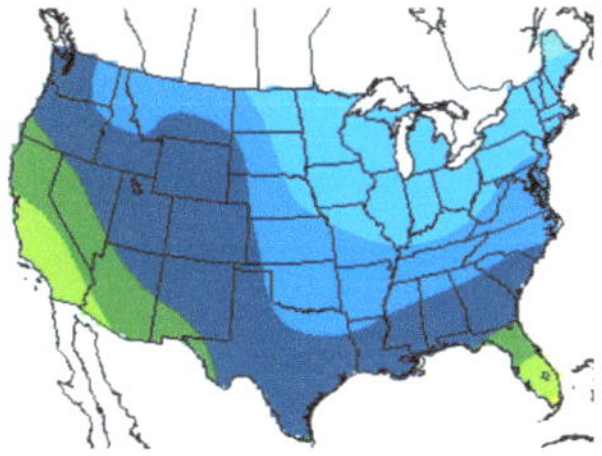

MINIMUM TEMPERATURE

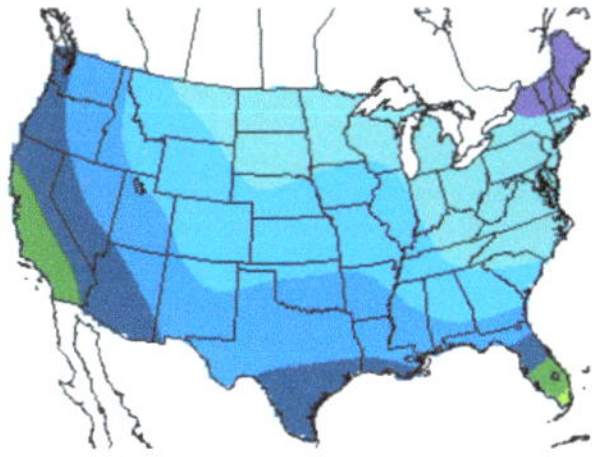

12/31/2025

Widespread precipitation dominates the western half of the country with very heavy falls along the Pacific coasts from Washington to California, heavy at times in the Great Basin and in Arizona to New Mexico.

Further snow in the Sierra Nevada to Utah and Colorado as well as in the western foothills of the Rocky Mountains, Montana and Wyoming.

Moving eastward, a band of intense thunderstorm embedded within a trough spreads rain and storms throughout Texas, Oklahoma and northeastward into Illinois.

Mostly dry elsewhere.

Some snowfalls may still also prevail across the upper Inland Midwest over the next couple of days, accompanied by very strong winds, especially in Minnesota and Wisconsin.

By New Year's Day, a severe storm may affect central and southern parts of the Great Plains, bringing heavy snow to the regions, including in Kansas and Texas.

APPENDIX

SKI SEASON 2025
OVERVIEW FOR ALL

DECEMBER 2024-APRIL 2025:

Prior to Christmas, very good snowfalls are expected in the northwest, although the second week of December may be mild. The third week of December 2024, can expect a large storm. This storm in late December to New Year brings blizzard conditions in the northern Rockies to Colorado.

Another round of blizzard conditions occurs in the second week of January 2025, with most storms concentrated particularly in the Northwest and into western Canada. After this, the next few weeks are drier than the New Year events, mostly due to a large Artic high-pressure system building around the third week of January, which may bring cold temperatures, and although mostly dry, this will allow for the productive use of snowmaking operations.

By the second week of February 2025, snow may fall in the central and southern Rockies, while mostly dry conditions persist in the Sierras and on the west coasts.

Widespread snowfall events prevail, bringing excellent conditions into most western ski resorts, over the second half of February and into early March.

Expect snowstorms in Utah and the northern Rockies regions around mid-March, before most remaining snowfall events contract to western Canada, but also prevail elsewhere in the United States except in the northern Rockies.

The wet and warmer conditions of spring arrive by April in many ski regions, although some parts of Utah and Colorado may enjoy a late burst of snow in the second week of April.

OCTOBER 2025 - APRIL 2026:

Despite a slow start, overall, the season is good and many areas, particularly from Oregon to Colorado, receiving exceptionally high snowfall.

The first significant snow layers in the Northwest are expected around mid to late October, but this may be brief, because a mostly dry November should follow, resulting in a later opening of many ski fields due to an insufficient snow base.

The Whistler ski region may have enough snow to open up around the last week of November, possibly in time for Thanksgiving, but other ski regions like Alta and Kirkwood, may not open until the first week of December 2025. Many western US ski areas that open before December may only be open due to the vigorous use of snowmaking equipment and still only have a fraction of their usual terrain open.

Despite some good storms in December, it may not be until the week before Christmas before the first substantial snow arrives, particularly in the Southeast and Northeast.

After Christmas, the northern Rockies to Utah can expect multiple storms allowing all ski regions to recover from their slow start, and instead, developing above-average snowfield conditions for the New Year 2026.

Only the Sierras may take longer to build a stable snow base, taking a bit longer into the New Year to stabilize.

The second half of January 2026 can expect almost continual snowfall events in the Northwest, Northern Rockies, Utah and part of Colorado.

Significant snow events prevail during February and March 2026, as well as into mid-April before many snow fields close for the season.

BELOW IS AN OVERALL SUMMARY OF EACH MAJOR SKI FIELDS REGIONS ACROSS THE UNITED STATES, WITH SAMPLE GRAPHS OF SELECTED FIELDS SHOWING THE POTENTIALS FOR WHEN SNOW MAY OCCUR FOR THE 2025-2026 SKI SEASON AND FOR THE YEAR OF 2025 OVERALL, WHICH INCLUDES THE JANUARY TO APRIL MONTHS FOR THE END OF THE 2024-2025 SKI SEASON.

SOUTHWEST – CALIFORNIA TO ARIZONA

ARIZONA SNOWBOWL SKI RESORT, KIRKWOOD, MAMMOTH MOUNTAIN, PALISADES TAHOE CITY, SOUTHERN CALIFORNIA (INCL. BIG BEAR LAKE, WRIGHTWOOD, BIG PINES).

DECEMBER 2024 - APRIL 2025:

Despite a couple of storms in November to early December 2024, overall, most precipitation will be too wet to form into snow, leaving the snow base by the turn of the New Year relatively insignificant, and with most ski trails not suitable for use until around the third week of January 2025.

From mid-January to early February, a series of smaller storms bring snow, but no big dumps.

February 2025 sees very little precipitation with low snowfall events over the first couple of weeks of the month, before a storm at the end of the month brings significant snow. During the first half of February snowmaking machines benefit from the cold temperatures.

The best precipitation forming a deep snow base may be over the first half of March after a series of storms.

Most ski resorts may be able to remain in full operation during the second half of March to mid-April, before the rising trend of the temperatures signal the arrival of spring, and the spring, resulting in the closure of most resorts by the end of the month.

OCTOBER 2025 - APRIL 2026:

This season's snowfall may be off to a slow start in not just the California region, but also in Utah, Arizona, and Colorado ski fields, with most places only receiving less than average snowfall.

Much of October and November is snowfree with long dry spells and unusually milder than normal temperatures, and it may not be until the first week of December before temperatures develop suitable for snowmaking machines, and the first arrival of natural snow may occur. Most regions may not open for Thanksgiving, but should be open within a couple of weeks after. Some snow is expected in the California ski regions during the second and third weeks of December, allowing most fields to open some of their slopes, including in areas like Northstar, the Sugar Bowl, and the Sierra-Tahoe regions.

By the end of December, some fields may be mostly open with a reasonable snow base.

The third week of January 2026, sees the passage of a major storm dropping snow into the Southwest regions including on the California to Arizona, and Sierra Mountains ski fields. This dump follows later, in the second week of February, creating a solid snow base.

Mostly dry conditions return, over the latter half of February, before another storm brings further snow in modest quantity early in March.

By mid-March, the first of the spring conditions start to make their appearances on the lower elevations, with increased sunshine and warmer temperatures, and April sees insignificant amounts of fresh snow, with very drier conditions mostly prevailing for prolonged lengths of time.

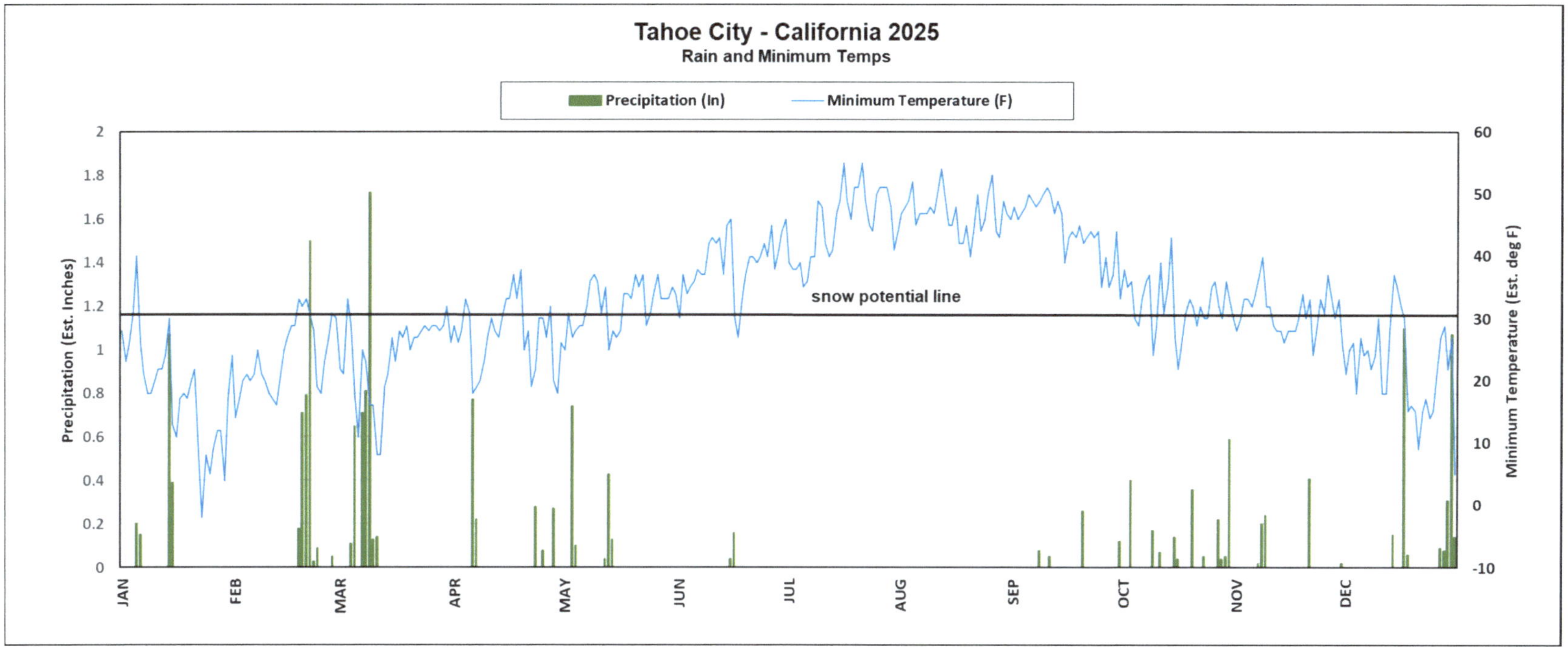

Tahoe City - California 2025
Rain and Minimum Temps
Precipitation (In)
Minimum Temperature (F)
snow potential line
Precipitation (Est. Inches)
Minimum Temperature (Est. deg F)
JAN
FEB
MAR
APR
MAY
JUN
JUL
AUG
SEP
OCT
NOV
DEC

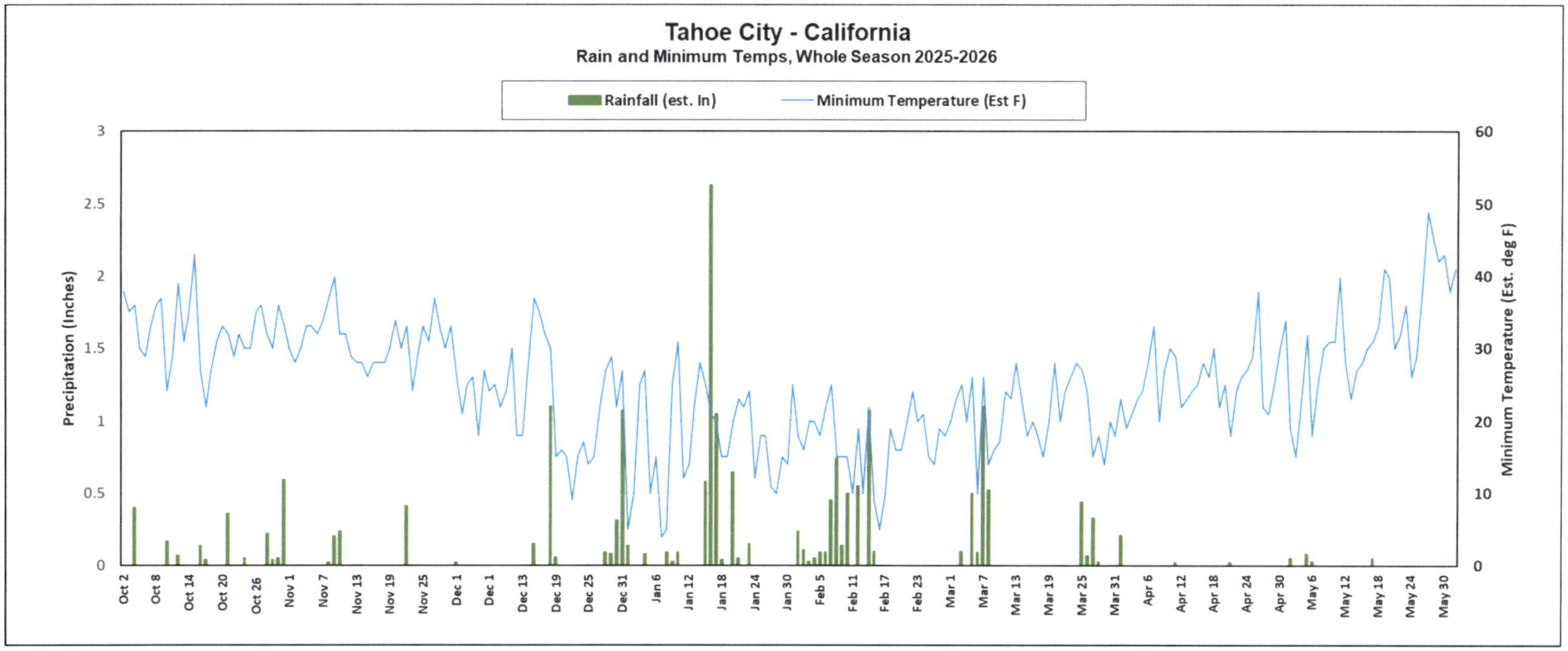

Tahoe City - California
Rain and Minimum Temps, Whole Season 2025-2026
Rainfall (est. In)
Minimum Temperature (Est F)
Precipitation (Inches)
Minimum Temperature (Est. deg F)

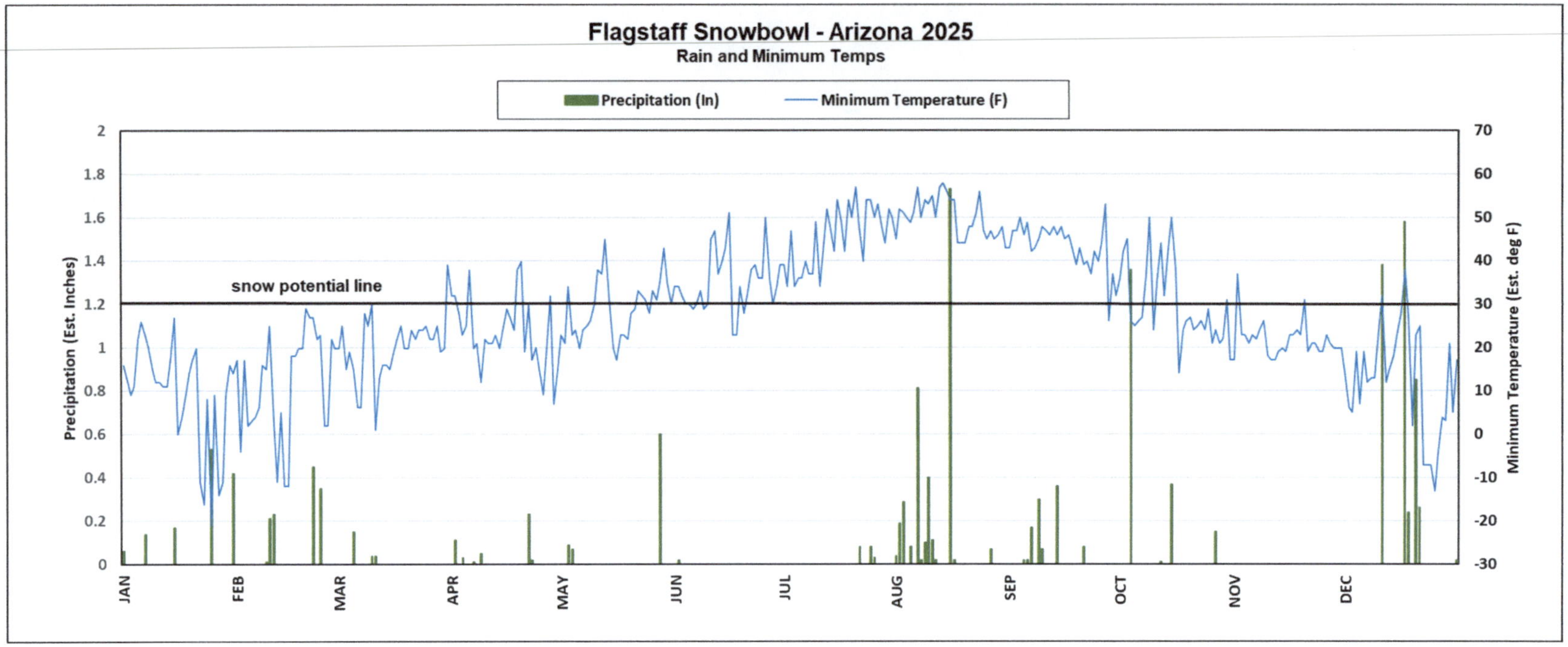
Flagstaff Snowbowl - Arizona 2025
Rain and Minimum Temps
Precipitation (In)
Minimum Temperature (F)
snow potential line
Precipitation (Est. Inches)
Minimum Temperature (Est. deg F)
JAN
FEB
MAR
APR
MAY
JUN
JUL
AUG
SEP
OCT
NOV
DEC

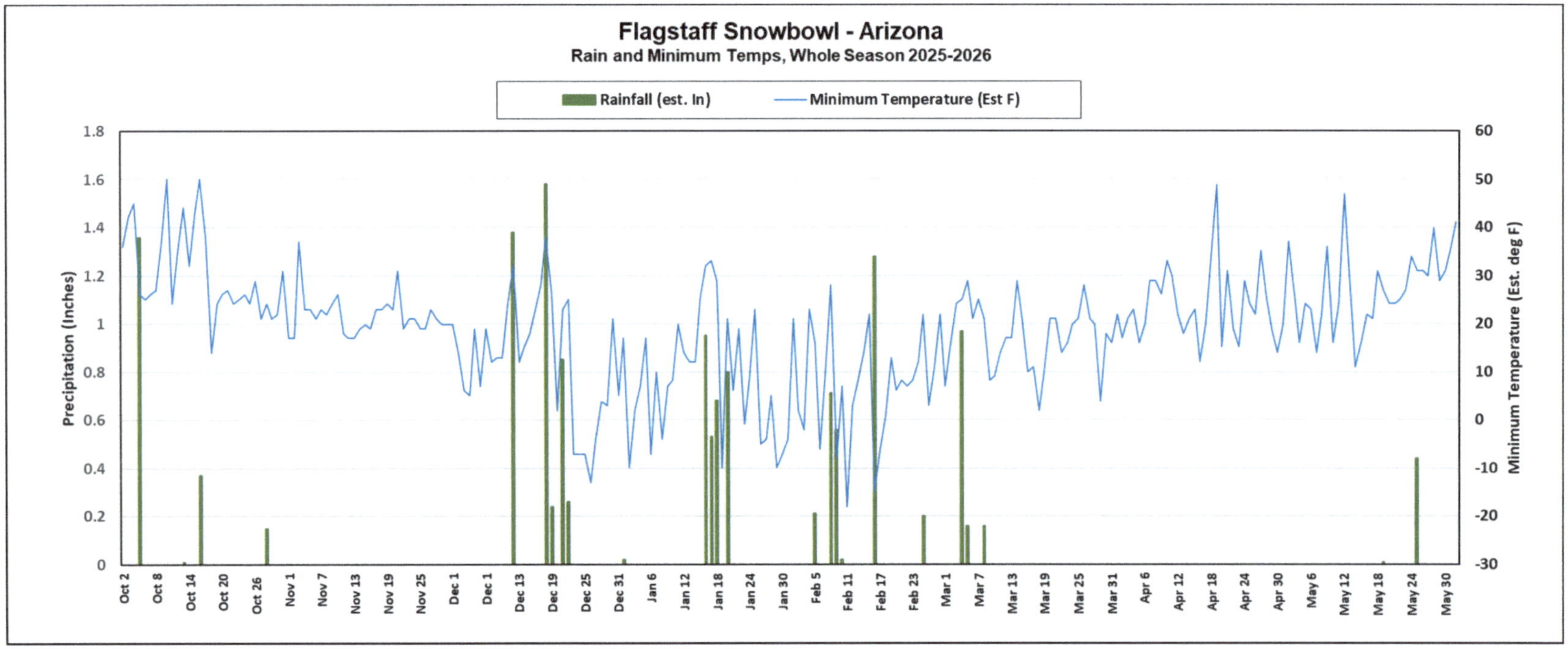
Flagstaff Snowbowl - Arizona
Rain and Minimum Temps, Whole Season 2025-2026
Rainfall (est. In)
Minimum Temperature (Est F)
Precipitation (Inches)
Minimum Temperature (Est. deg F)

PACIFIC NORTHWEST

CRYSTAL MOUNTAIN, STEVENS PASS, MOUNT BACHELOR, WENATCHEE.

DECEMBER 2024 - APRIL 2025:

An early opening to the season, with good snow established by the end of November 2024, as a result of some significant storms during the month. November storms cause outbreaks of localised flooding in the lower reaches and a good amount of snow in the higher levels. By the end of November, the Pacific Northwest should be open.

Around mid-December, further storm systems bring more to the snow base, with further snow falling in the time between Christmas and the New Year 2025.

Snowfall events continue to be productive in January and another couple of feet over the second half of the month.

The next major storm event bringing further snow is in late February 2025, and then again during the second week of March.

Regular storms continue over the second half of March, but due to warmer temperatures becoming more frequent, and the signs of spring increasing, snow levels fluctuate more regularly.

OCTOBER 2025 - APRIL 2026:

October and November see little in the way of opportunities for snow to build in the ski regions.

The second week of December brings snowbearing storms, but mostly in the higher reaches, with rain in the lower areas. By mid-December, a storm event is promising but ends in rain.

Some ski fields may be able to open in the week before Christmas, mostly in Washington, with most other places having a very limited natural snow-base.

Late December to early in the New Year 2026, brings significant snowfall events, building most ski regions to a good snow-base depth.

In mid-January, mostly between the 11th and 24th there may be almost continuous storms in the region, resulting in snow in many ski regions.

In the last week of January to the second week of February, more snowfall is expected. This may be followed in mid-February by more snow, especially in the Washington/Oregon region.

March is milder and mostly dry with very little snowfall except in the highest of reaches.

This mostly dry spell in March is broken near the end of the month into early April with another late but short-lived dump of snow before the season ends with the arrival of spring.

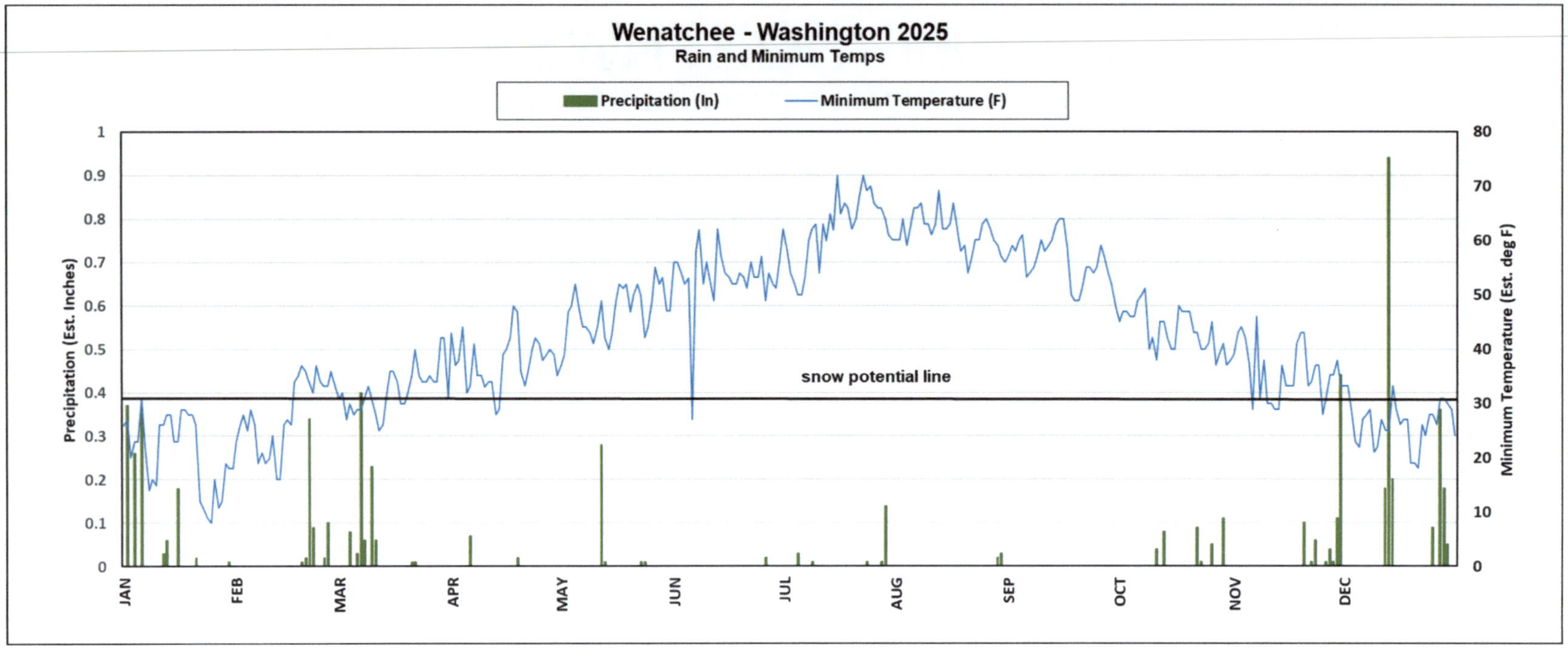

Wenatchee - Washington 2025
Rain and Minimum Temps
Precipitation (In)
Minimum Temperature (F)
Precipitation (Est. Inches)
Minimum Temperature (Est. deg F)
snow potential line
JAN
FEB
MAR
APR
MAY
JUN
JUL
AUG
SEP
OCT
NOV
DEC

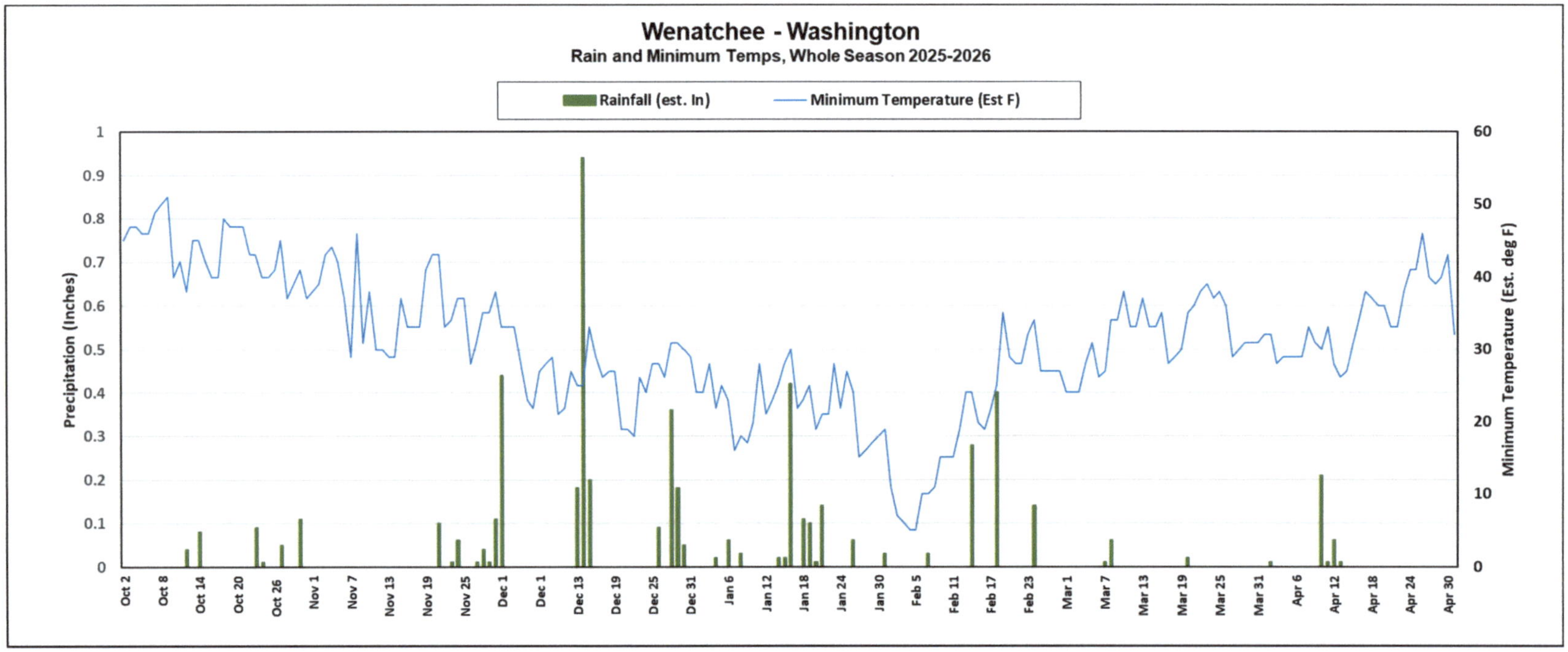

Wenatchee - Washington
Rain and Minimum Temps, Whole Season 2025-2026
Rainfall (est. In)
Minimum Temperature (Est F)
Precipitation (Inches)
Minimum Temperature (Est. deg F)

U.S. NORTHERN ROCKIES

ASPEN, BIG MOUNTAIN – WHITEFISH MOUNTAIN RESORT, BRIDGER BOWL, GRAND TARGHEE, JACKSON HOLE, SCHWEITZER – SELKIRK MOUNTAINS, SUN VALLEY.

DECEMBER 2024 - APRIL 2025:

Good snowfalls occur in Idaho in mid-November, and other States have average to below average rainfall. Even fields like those at Big Sky may only be fractionally open, and some not open until the New Year.

The only exception of significance is expected to be in the Grand Targhee region, with most parts of this terrain receiving more snow and having more fields open than other places.

Most fields have enough snow to open the majority of their fields by the first week of January, but the snow base is still quite shallow.

The Grand Targhee field likely is open by Christmas 2024 and all other fields may be open by the second week of January 2025.

Most fields should receive snow around mid-January 2025, and a further amounts in the last week of January.

February brings moderate dumps of snow in most areas on an almost weekly basis, resulting in peak conditions of the fields for this season.

Snowfall amounts ease back in March, with a small amount arriving around midway through the month. By the end of March, the region transitions towards spring conditions, forcing a number of runs to close early and very few places remaining open into early April.

OCTOBER 2025 - APRIL 2026:

The season has a moderate start, with only occasional light snowfalls during October and November. Overall, snowfalls are around average at the higher elevations but limited in lower levels and more consistent with rain, rather than snow.

December brings the first significant snow dumps of the season in many places. The Grand Targhee runs is mostly open by the second week of December, while many other fields have to wait until closer to the end of the month for a natural snow base to be adequately established.

Most can expect a good snow base by the end of the first week of January, although places like Big Sky are only half open, while others may have received enough snow and cold enough temperatures to operate snowmaking machines during the second half of December 2025.

Over the next couple of weeks until mid-February, further snow may arrive in the Northern Rockies ski regions, particularly accompanying a storm as it crosses north-easterly over the region bringing some good falls.

More snow arrives with another storm over the second half of February in most ski regions, with good base around Big Mountain.

March sees some regular snowfall events, despite some spring conditions peaking in occasionally, before spring becomes firmly established in late March into early in April.

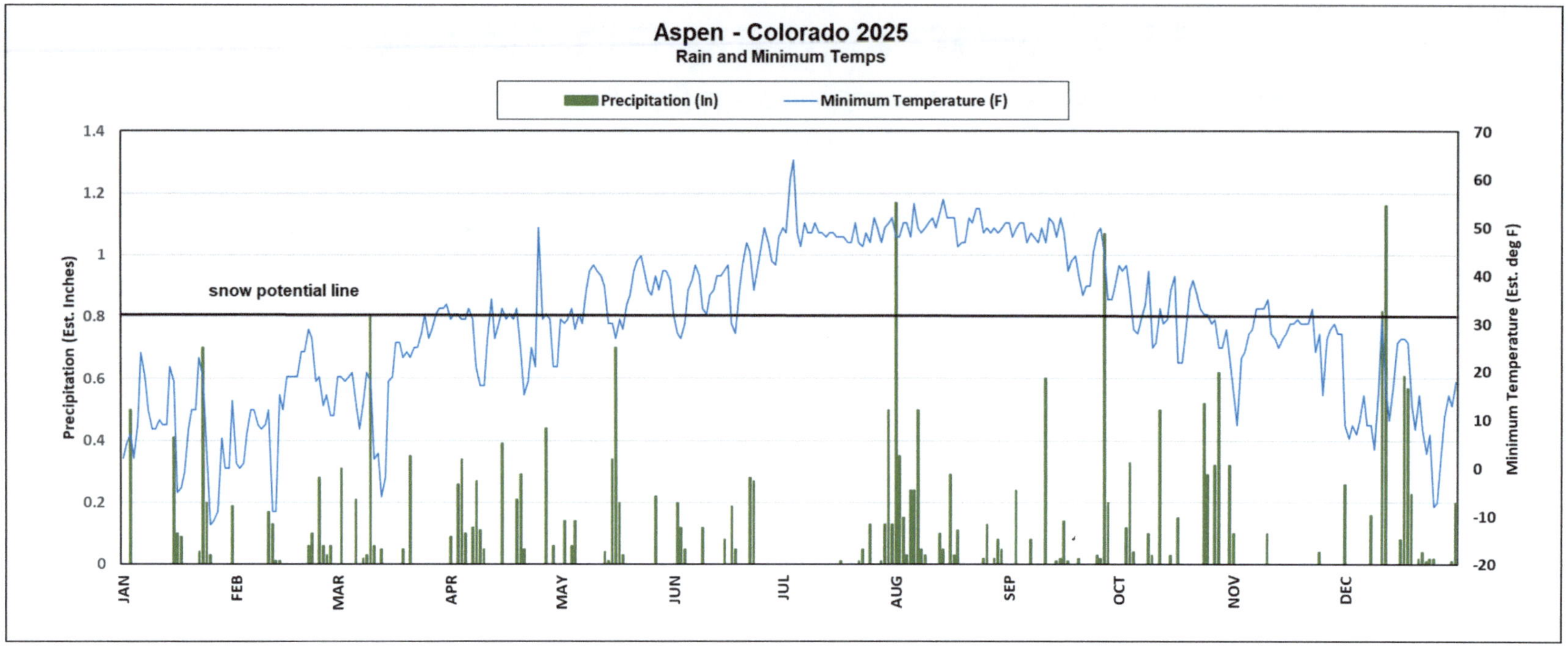
Aspen - Colorado 2025
Rain and Minimum Temps
Precipitation (In)
Minimum Temperature (F)
snow potential line
Precipitation (Est. Inches)
Minimum Temperature (Est. deg F)
1.4
1.2
1
0.8
0.6
0.4
0.2
0
70
60
50
40
30
20
10
0
-10
-20
JAN
FEB
MAR
APR
MAY
JUN
JUL
AUG
SEP
OCT
NOV
DEC

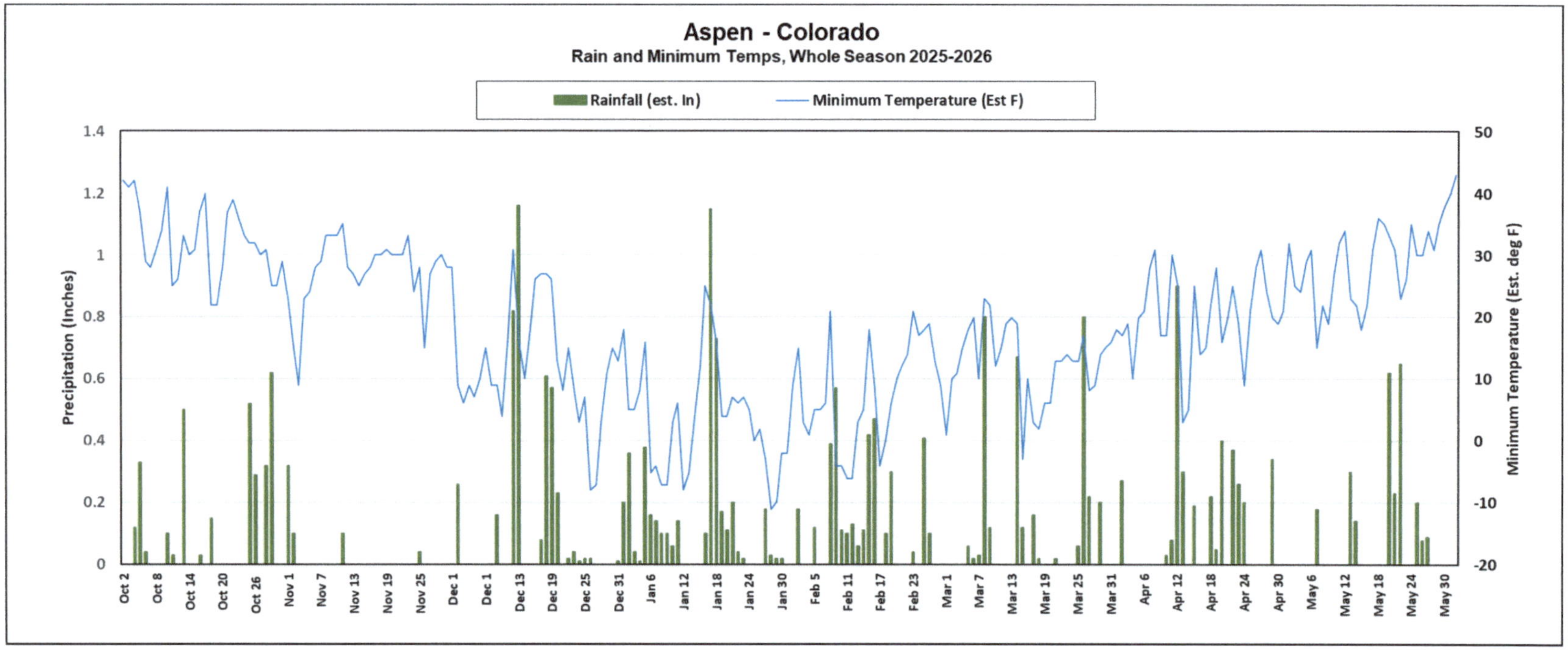
Aspen - Colorado
Rain and Minimum Temps, Whole Season 2025-2026
Rainfall (est. In)
Minimum Temperature (Est F)
Precipitation (Inches)
Minimum Temperature (Est. deg F)
1.4
1.2
1
0.8
0.6
0.4
0.2
0
50
40
30
20
10
0
-10
-20
Oct 2
Oct 8
Oct 14
Oct 20
Oct 26
Nov 1
Nov 7
Nov 13
Nov 19
Nov 25
Dec 1
Dec 1
Dec 13
Dec 19
Dec 25
Dec 31
Jan 6
Jan 12
Jan 18
Jan 24
Jan 30
Feb 5
Feb 11
Feb 17
Feb 23
Mar 1
Mar 7
Mar 13
Mar 19
Mar 25
Mar 31
Apr 6
Apr 12
Apr 18
Apr 24
Apr 30
May 6
May 12
May 18
May 24
May 30

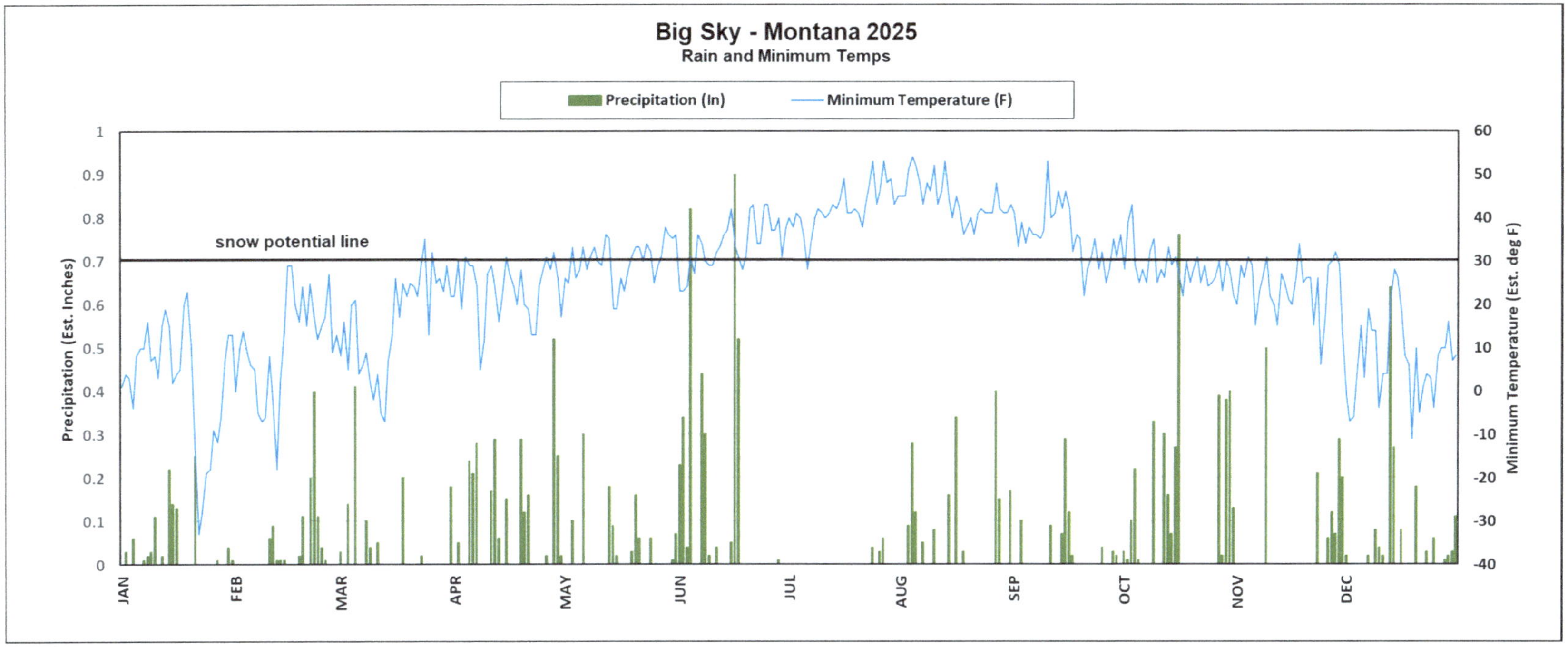
Big Sky - Montana 2025
Rain and Minimum Temps
Precipitation (In)
Minimum Temperature (F)
snow potential line
Precipitation (Est. Inches)
Minimum Temperature (Est. deg F)
JAN
FEB
MAR
APR
MAY
JUN
JUL
AUG
SEP
OCT
NOV
DEC

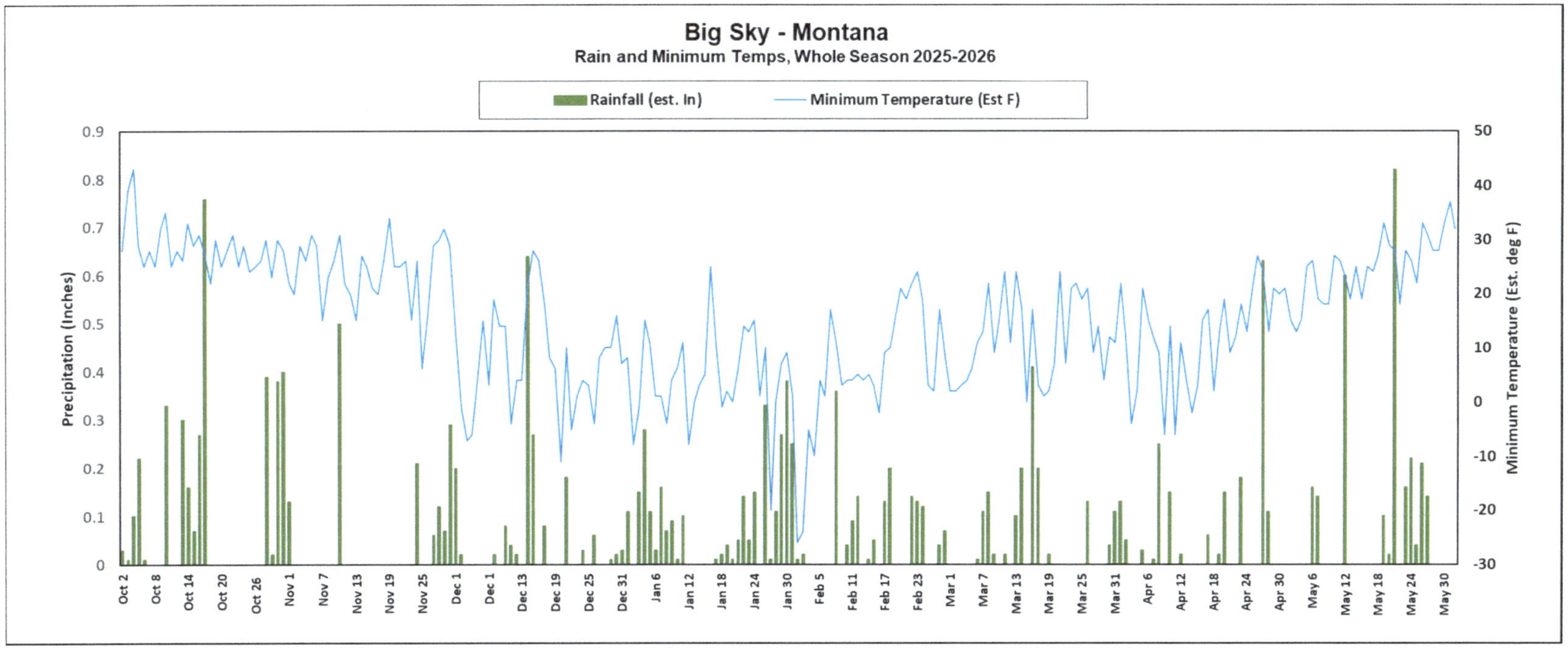
Big Sky - Montana
Rain and Minimum Temps, Whole Season 2025-2026
Rainfall (est. In)
Minimum Temperature (Est F)
Precipitation (Inches)
Minimum Temperature (Est. deg F)

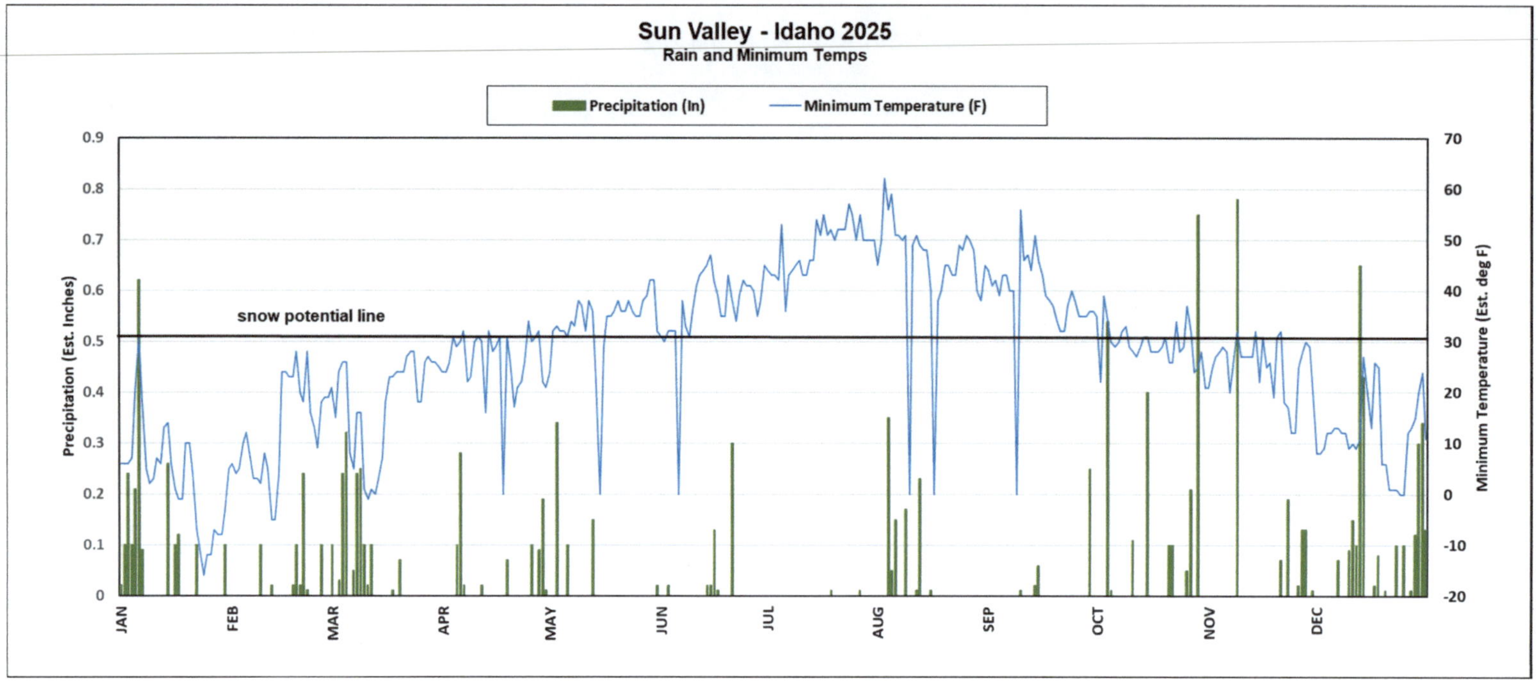
Sun Valley - Idaho 2025
Rain and Minimum Temps
Precipitation (In)
Minimum Temperature (F)
snow potential line
Precipitation (Est. Inches)
Minimum Temperature (Est. deg F)
JAN
FEB
MAR
APR
MAY
JUN
JUL
AUG
SEP
OCT
NOV
DEC

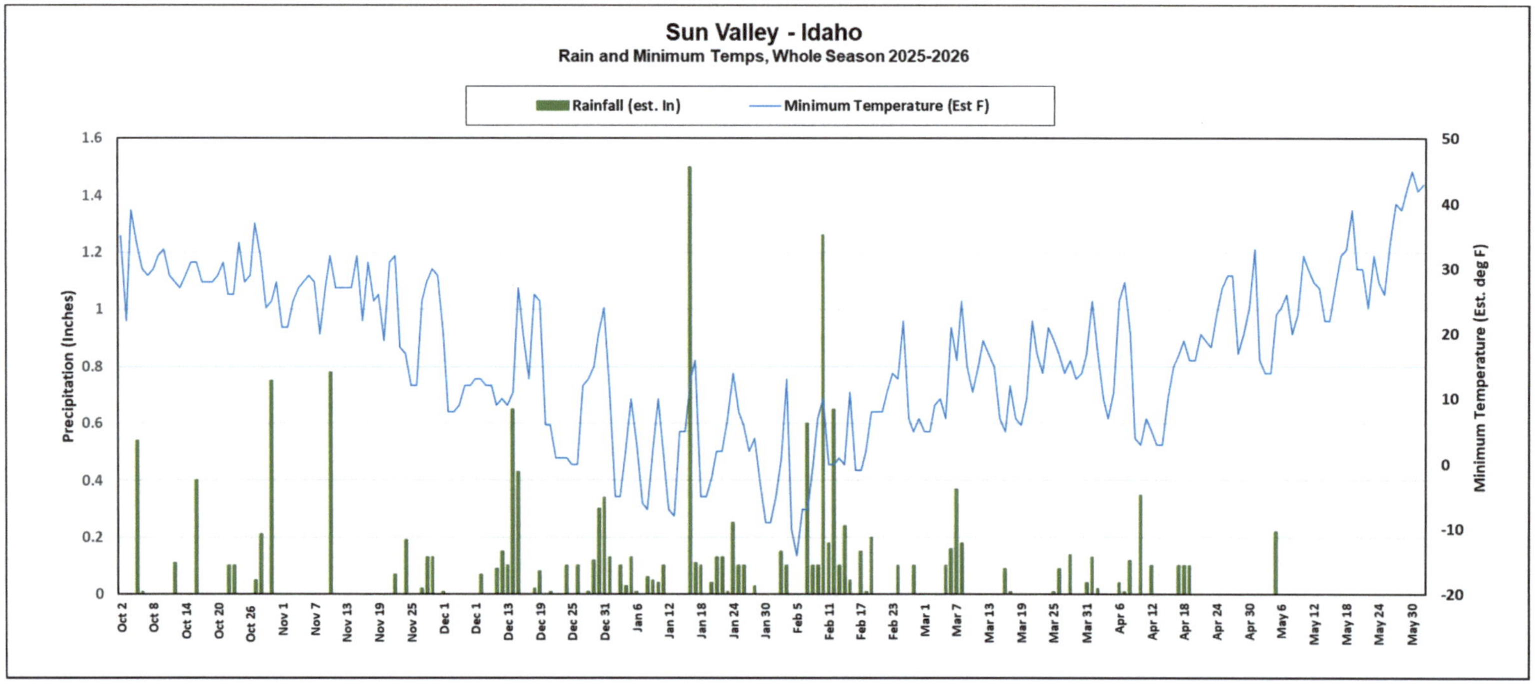
Sun Valley - Idaho
Rain and Minimum Temps, Whole Season 2025-2026
Rainfall (est. In)
Minimum Temperature (Est F)
Precipitation (Inches)
Minimum Temperature (Est. deg F)

UTAH

ALTA – WASATCH MOUNTAINS, SNOWBIRD SKI RESORT, BRIGHTEN – WASATCH MOUNTAINS, SOLITUDE – WASATCH MOUNTAINS, PARK CITY GROUP, SNOWBASIN, BRIAN HEAD.

DECEMBER 2024 - APRIL 2025:

Snowfall is below average in the first part of this season, particularly during October and November 2024. It is not until a large storm moves up from Mexico into the southwest of the country in early to mid-December when the first snowfalls arrive. This event brings good snow base not only in Utah, but also throughout Colorado, New Mexico, and Arizona, although the snow base depths may be small in most places.

The next good snow events arrive during the first couple of weeks of the 2025 New Year, when a solid base is established, although many areas may have less than average.

Mid-January to mid-February is particularly snowy, allowing many places to receive more snowfall accumulation than normal. While it may snow around mid to late February, much of it will be very light and mostly blow away without adding to the snowpack. More snow arrives towards the end of the month.

By the second week of March, conditions on the fields should be excellent. Mid-March may see more, with this reaching the peak of the snow-base for the season.

After mid-March, the spring season makes a steady appearance with steadily warming temperatures over most days, although this is a chance of a later burst of brief snow around mid-April.

OCTOBER 2025 - APRIL 2026:

This season gets off to a relatively good start with conditions steadily improving throughout October and November, before significant falls arrive in December.

A good snow event arrives in late October.

Mid-November to mid-December is unusually drier than average, with most areas only partially operational up to the second week of December, because of snowmaking machine activity maintaining the October base.

The third week of December brings more snow to the fields, with the season in full swing. Very few ski-fields have adequate natural snow coverage, with most assisted by snowmaking machines. By the New Year, only the Alta ski fields may be near full operation standards.

January 2026 sees the arrival of further storms, bringing the snow base up in some places during the first week of the New Year. The end of the second week of January to the end of the month brings further snow, adding to the stability of the snow bases.

A dry spell is likely in the first half of February before more snowfall comes in the second half of the month.

Up to four storms in March to early April, adds further snow mostly in the Cottonwood Canyons region.

Spring conditions arrive in the lower reaches by the end of March, but the higher reaches and snowfields can expect further snow in April and into early May before the season ends around mid-May.

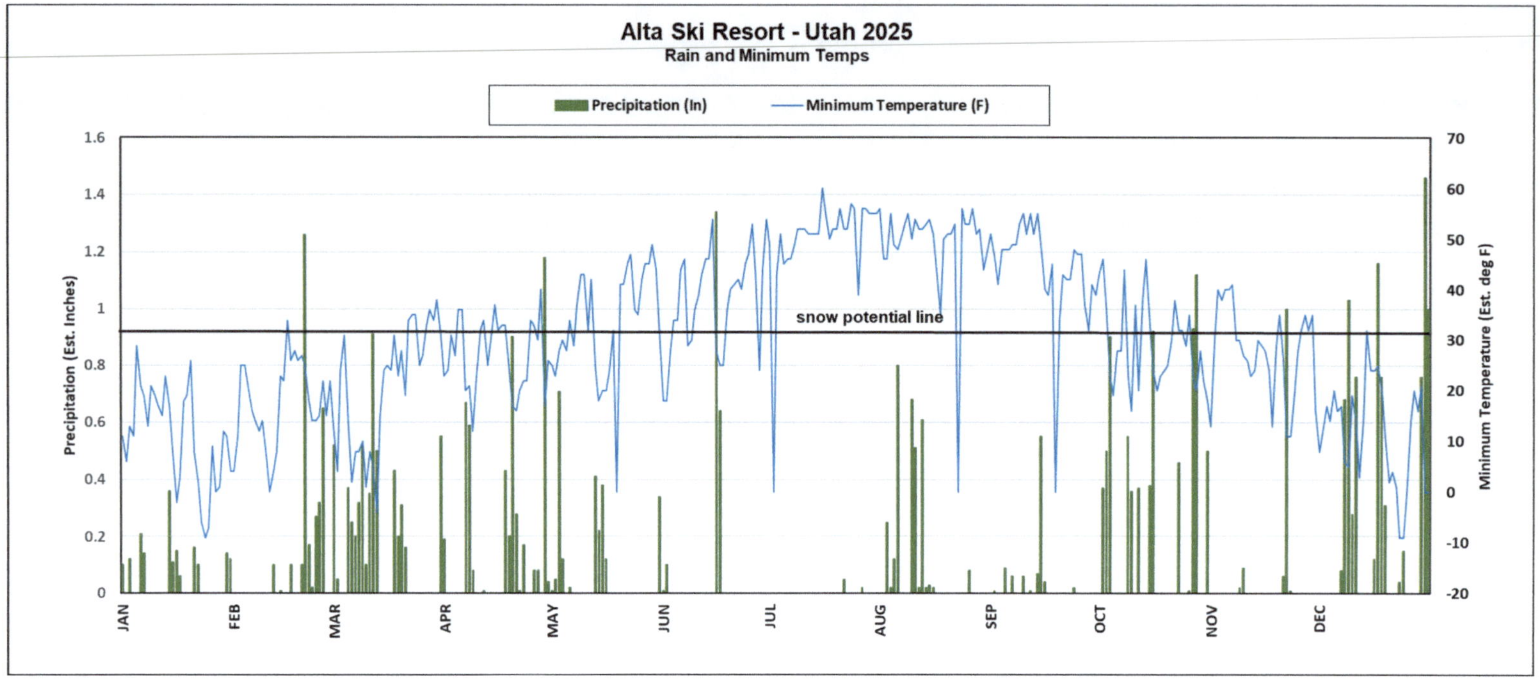

Alta Ski Resort - Utah 2025
Rain and Minimum Temps
Precipitation (In)
Minimum Temperature (F)
Precipitation (Est. Inches)
Minimum Temperature (Est. deg F)
snow potential line
JAN
FEB
MAR
APR
MAY
JUN
JUL
AUG
SEP
OCT
NOV
DEC

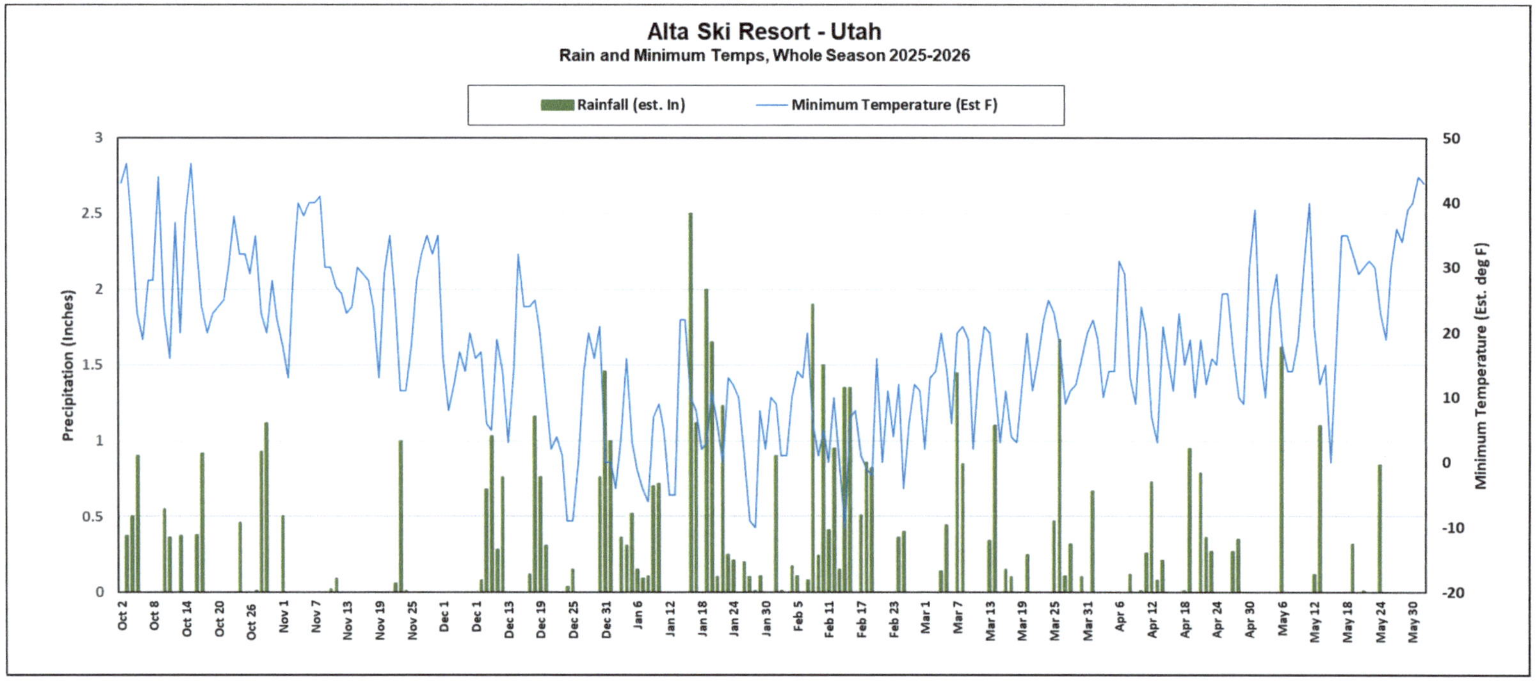

Alta Ski Resort - Utah
Rain and Minimum Temps, Whole Season 2025-2026
Rainfall (est. In)
Minimum Temperature (Est F)
Precipitation (Inches)
Minimum Temperature (Est. deg F)

NORTHERN AND CENTRAL COLORADO

BEAVER CREEK, BRECKENRIDGE, COPPER MOUNTAIN, KEYSTONE, LOVELAND, STEAMBOAT – MT WERNER, VAIL MOUNTAIN, WINTER PARK.

DECEMBER 2024 - APRIL 2025:

October and November sees many places receiving less than half their normal amounts for this time of year. This trend may also continue into the first half of December.

An intense snowstorm in mid-December brings significant falls throughout, developing into severe blizzard conditions over the Christmas and New Year period.

By mid-January 2025, the snow base may be improved. Many places may be twice as much snowpack as average.

Mid-January to early February, snow rates drop to half the average.

In the second and third weeks of February, another burst of wintry precipitation brings more snow to the ski fields, and a further amount as February ends and moves into early March.

The second week of March receives more snow with regular falls between mid-March and early April.

By mid-April, the snow season may be ebbing away with the clear signs of spring showing and bringing snowmelts and slushy conditions in the lower reaches, although the higher reaches may manage to maintain a workable snow-base due to cooler temperatures even into May.

OCTOBER 2025 - APRIL 2026:

The 2025-2026 ski season may start better than the previous year, with good snow falling between mid-October and mid-November, when many fields may consider opening.

Snow events ease up mid-November to mid-December with many places receiving half the average. This situation bounces back quickly over the second half of December, including the possibility of snowfalls on Christmas Day. Most slopes should be open in time for the holidays, despite many areas still having a snow base below their norm.

The first half of January 2026 brings good snow in all the ski regions by mid-month, and a significant dump in the last few days of January 2026.

Early February brings another dump of snow, with more later in the month.

Early March, snow events ease for a time, before the return of new snow.

Generous snow events continue into April, and during spring, cooler temperatures and occasional snowfalls persisting until early June 2026.

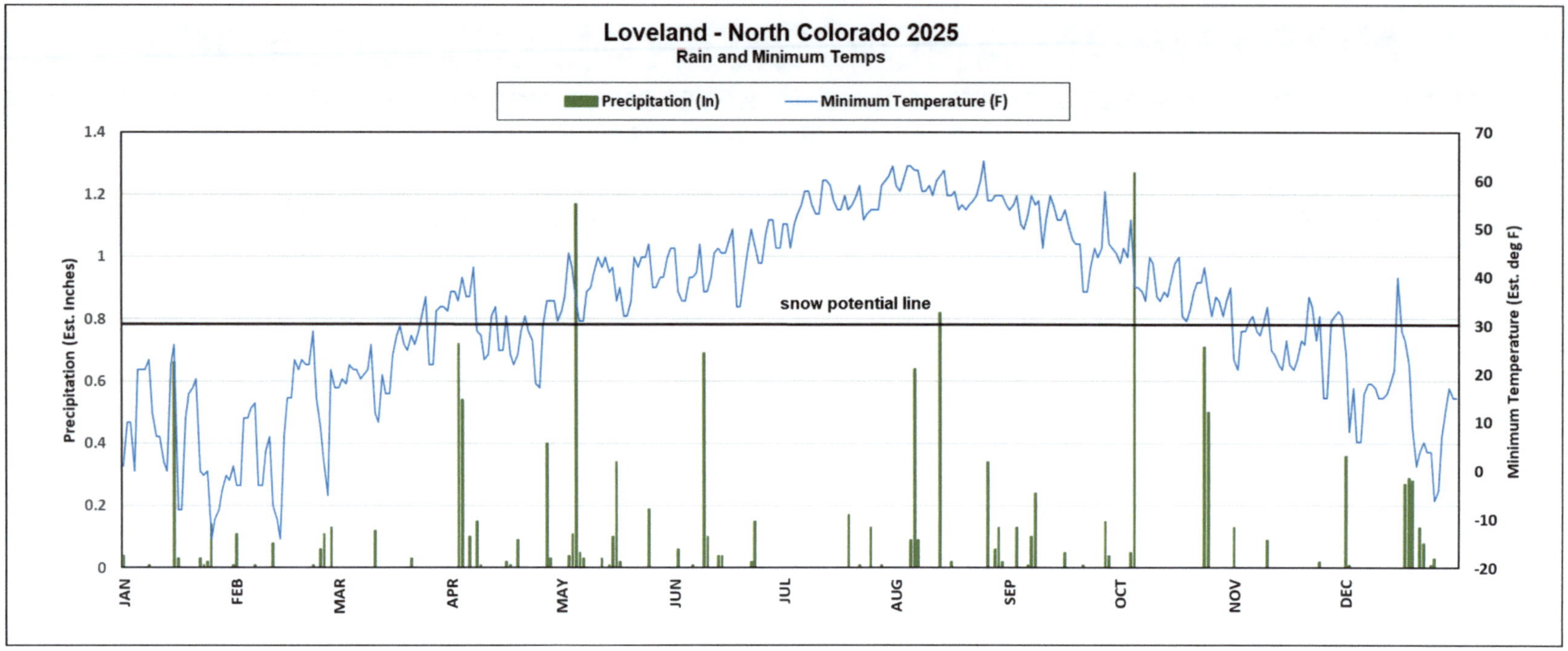

Loveland - North Colorado 2025
Rain and Minimum Temps
Precipitation (In)
Minimum Temperature (F)
Precipitation (Est. Inches)
Minimum Temperature (Est. deg F)
snow potential line
1.4
1.2
1
0.8
0.6
0.4
0.2
0
70
60
50
40
30
20
10
0
-10
-20
JAN
FEB
MAR
APR
MAY
JUN
JUL
AUG
SEP
OCT
NOV
DEC

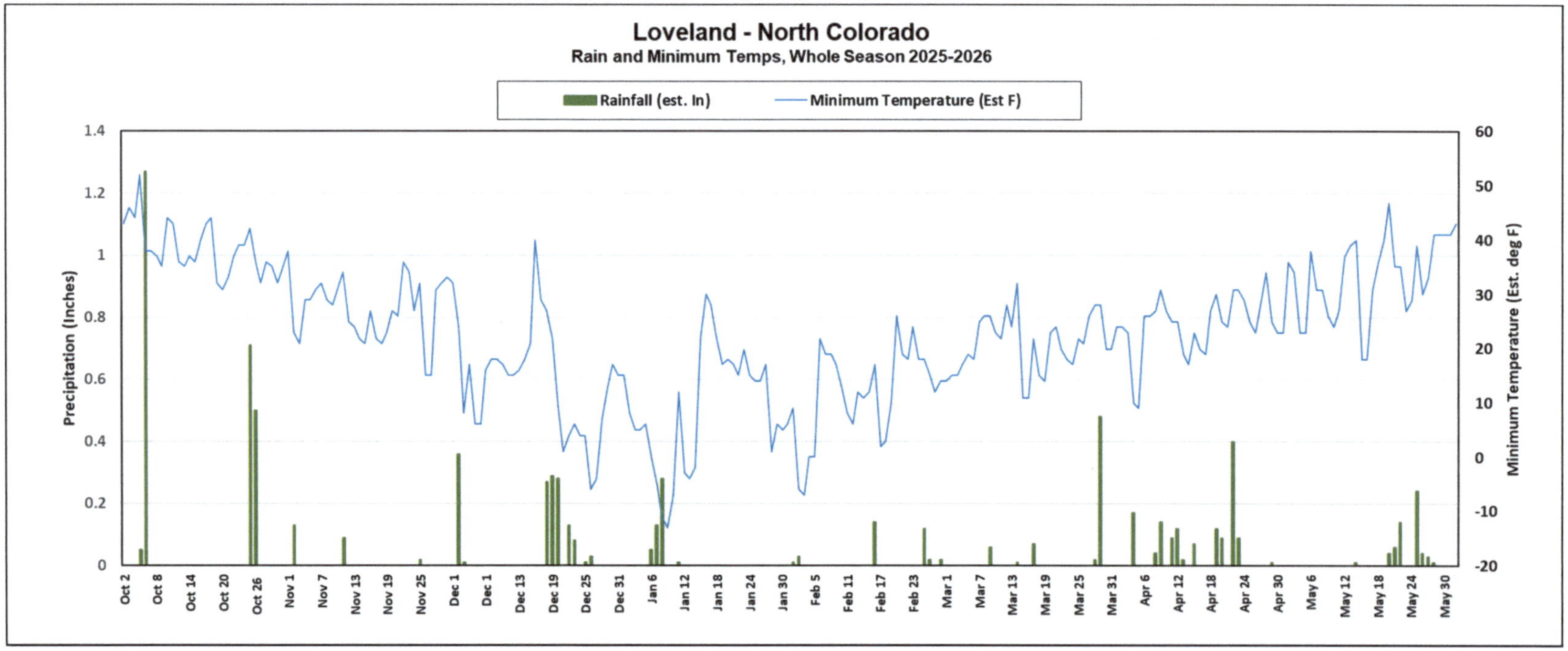

Loveland - North Colorado
Rain and Minimum Temps, Whole Season 2025-2026
Rainfall (est. In)
Minimum Temperature (Est F)
Precipitation (Inches)
Minimum Temperature (Est. deg F)
1.4
1.2
1
0.8
0.6
0.4
0.2
0
60
50
40
30
20
10
0
-10
-20
Oct 2
Oct 8
Oct 14
Oct 20
Oct 26
Nov 1
Nov 7
Nov 13
Nov 19
Nov 25
Dec 1
Dec 1
Dec 13
Dec 19
Dec 25
Dec 31
Jan 6
Jan 12
Jan 18
Jan 24
Jan 30
Feb 5
Feb 11
Feb 17
Feb 23
Mar 1
Mar 7
Mar 13
Mar 19
Mar 25
Mar 31
Apr 6
Apr 12
Apr 18
Apr 24
Apr 30
May 6
May 12
May 18
May 24
May 30

SOUTHERN AND WESTERN COLORADO

ASPEN, CRESTED BUTTE, GOTHIC MOUNTAIN, DURANGO, TELLURIDE, WOLF CREEK.

DECEMBER 2024 - APRIL 2025:

During the months of October and November 2024, enough snowfall allows the opening of a number of fields, including Aspen and Snowmass around the time of the holidays. By the end of December 2024, the snow base should be well above normal in the north of this region, but below normal in the south. This unevenness may equal out at the turn of the New Year, when a large storm with blizzard conditions prevailing, brings a significant dump to most regions. Some Crested Butte's slopes may be unable to open until the third week of January 2025.

Elsewhere, snow falls at the Taos region with a storm around midway through the second week of January 2025, and again in the third week, as well as spreading into the Wolf Creek region. Similarly, good snow in the third week of January falls in the Arizona and New Mexico ski regions. However, the mid-January to mid-February period sees below average snowfall in Aspen and Crested Butte.

Mid-February to mid-March brings further snow, particularly at Wolf Creek.

By mid-March, the snowpack base reduces, but still average, everywhere except in Wolf Creek. This decrease may be due to premature spring conditions developing, resulting in most slopes closing by the end of April.

OCTOBER 2025 - APRIL 2026:

A good start to the season, with snow arriving in mid-October to mid-November, especially in Aspen and Crested Butte.

This generous start to the season may be short-lived, as mid-November to mid-December may see very little in the way of snow, compared to the average.

As a result, the first week of December may only bring limited opening, and some not even open at all until midway through December, after a generous amount of snowfall arrives.

Midway through the second week of December to Christmas Eve, good snow arrives over most areas, especially at Wolf Creek. As a result, a much greater area of slopes and terrain may be open by the New Year and into the first week of January 2026.

Nearly all ski areas in this region receive more snow over the second half of January 2026, creating a strong and stable snow base for the remainder of the season.

February sees a series of snowstorms throughout the month, adding to the snow bases in all regions.

By late February to early March, there may be a brief respite in snowfall volume, particularly in western parts of this region, but regardless, this period may still see another couple of feet added to the snow base, as well as a similar amount on top over the second half of the month.

During the period from mid-March to mid-April, some slopes, including Crested Butte and Aspen, can expect new snow each week, while other regions slightly less.

Finally, the second half of April into early May may continue to bring snow, particularly in the Aspen and Crested Butte slopes again, making this season of 2025-2026 particularly good compared to other years.

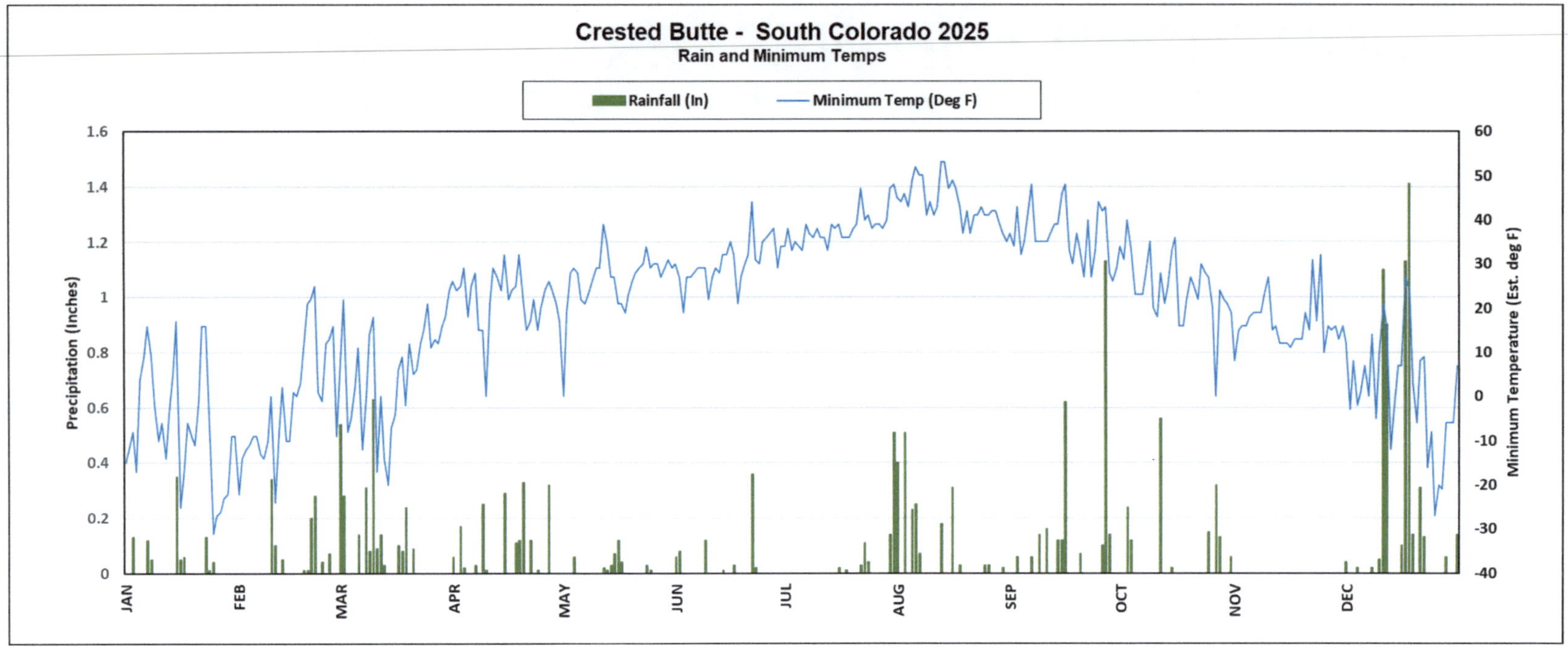
Crested Butte - South Colorado 2025
Rain and Minimum Temps
Rainfall (In)
Minimum Temp (Deg F)
Precipitation (Inches)
Minimum Temperature (Est. deg F)
JAN
FEB
MAR
APR
MAY
JUN
JUL
AUG
SEP
OCT
NOV
DEC

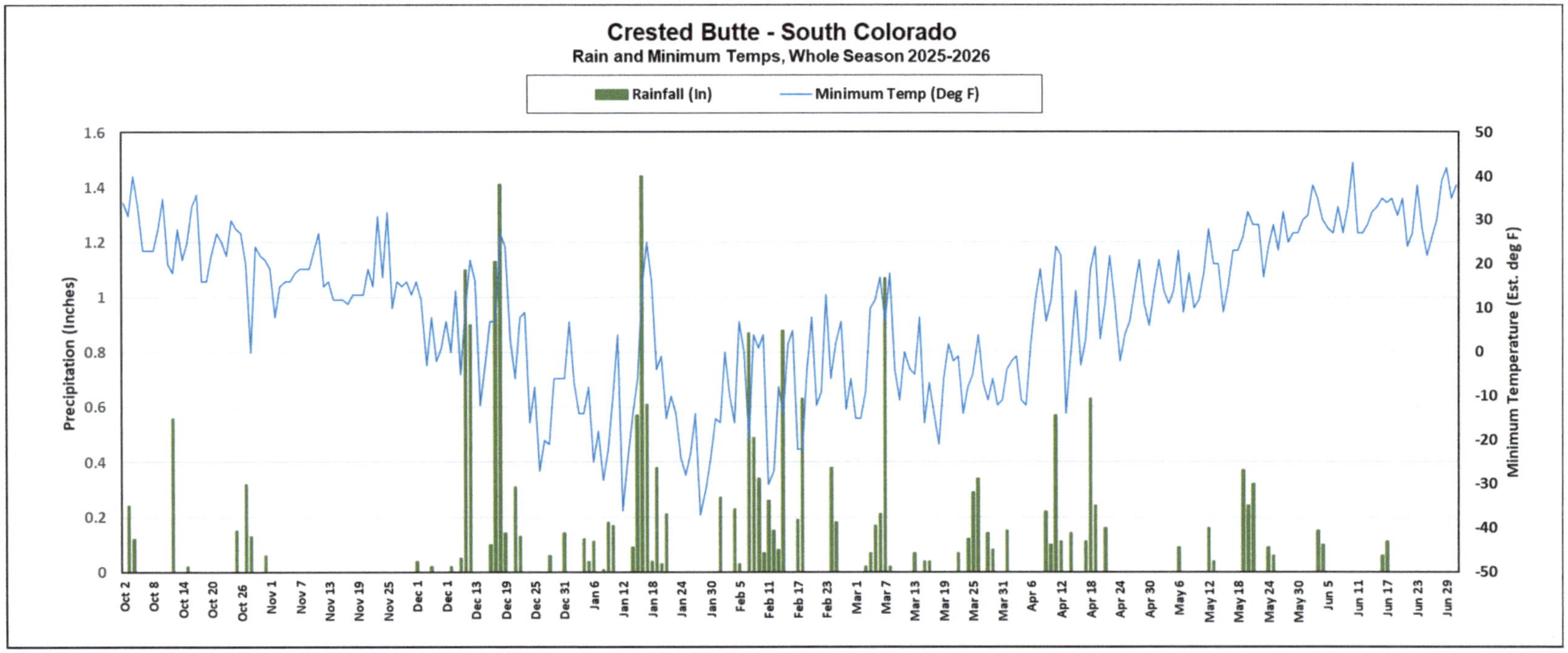
Crested Butte - South Colorado
Rain and Minimum Temps, Whole Season 2025-2026
Rainfall (In)
Minimum Temp (Deg F)
Precipitation (Inches)
Minimum Temperature (Est. deg F)

NORTHEAST

JAY PEAK, SMUGGLER'S NOTCH, STOWE – MOUNT MANSFIELD AND SPRUCE PEAK, SUGARBUSH – MAD RIVER VALLEY, KILLINGTON, CANNON MOUNTAIN, SUGARLOAF – CARRABASSETT VALLEY.

DECEMBER 2024 - APRIL 2025:

October to mid-November 2024 may see some natural snow, but mid-November to mid-December period may be unusually warmer than is typical for this time of year, with nearly all precipitation falling as rain. Temperatures may also be too mild for the use of snowmaking machines to operate until the last few days of November.

There may be a natural snowfall event over the second and third weeks of December, but then temperatures may become too mild again for natural snow to fall, as well as limiting the ability of the use snowmaking machines, with the next natural snowfalls not expected until the first week into the New Year of 2025.

After this brief dusting of snow in early January, the unusual warmth may return for a brief visit for around a week.

It may not be until just after halfway through January 2025 that a few of the ski fields in this region may open for their season, but still with only limited access to many slopes. By the last week of January, only a small amount of natural snow base may coat north-eastern ski areas.

Conditions improve steadily during February, and in particular, a large dump of snow may occur around the end of the month.

Slopes may remain sufficiently stable during most of March due to cold temperatures and a reasonable snow base, assisted by a further couple of feet added to them mid-way through the month. There may be a couple of hiccups of wetter spells of rain, rather than snow during this month however, deteriorating the snow surfaces for a short time, mostly during the third week.

By the second week of April, temperatures may be warming enough to signal the arrival of spring, resulting in the closure of many trails before the end of the month.

OCTOBER 2025 - APRIL 2026:

This season may be off to a slow start with few areas receiving enough snow to open before mid-November, and some opening around late November. Many more slopes may open around the last weekend of November, but the overall amount of slopes opened will be less than average.

Conditions may be average when compared to the previous few years.

Midweek rain during the first week of December may result in setting back the snowpack in many areas, but conditions will swiftly bounce back and develop excellent conditions over the second half of December.

In a similar pattern to December, January may also bring a brief setback with rain in early January 2026, but the snow base depths may be sufficient to minimize any decrease in available trails. This may be followed by more snow over the second week of January, restoring the snow base by mid-month.

A major thaw may occur over the fourth week of January as a result of further rain, resulting in a reduction of available trails being opened. Following the pattern of the past couple of months, this thaw may swiftly be replaced with another dump of snow, restoring the ski fields.

February may follow this snowstorm, then thaw patterns throughout most of the month, resulting in many slopes unable to remain fully operational.

A couple of storms in early March may bring new snow powder, but a mix of rain and snowfalls may bring further instability to the trails during most of this month in the New England snow regions. In contrast, further north in Canada, the storms may be almost all snow and bring outstanding conditions to their slopes.

In early April, a cold snap may bring fresh snowfall in Vermont to Maine, with another foot of snow over the the second week. After this, April may be mostly dry but cold, with most slopes open regularly, as well as a few continuing to open into early May.

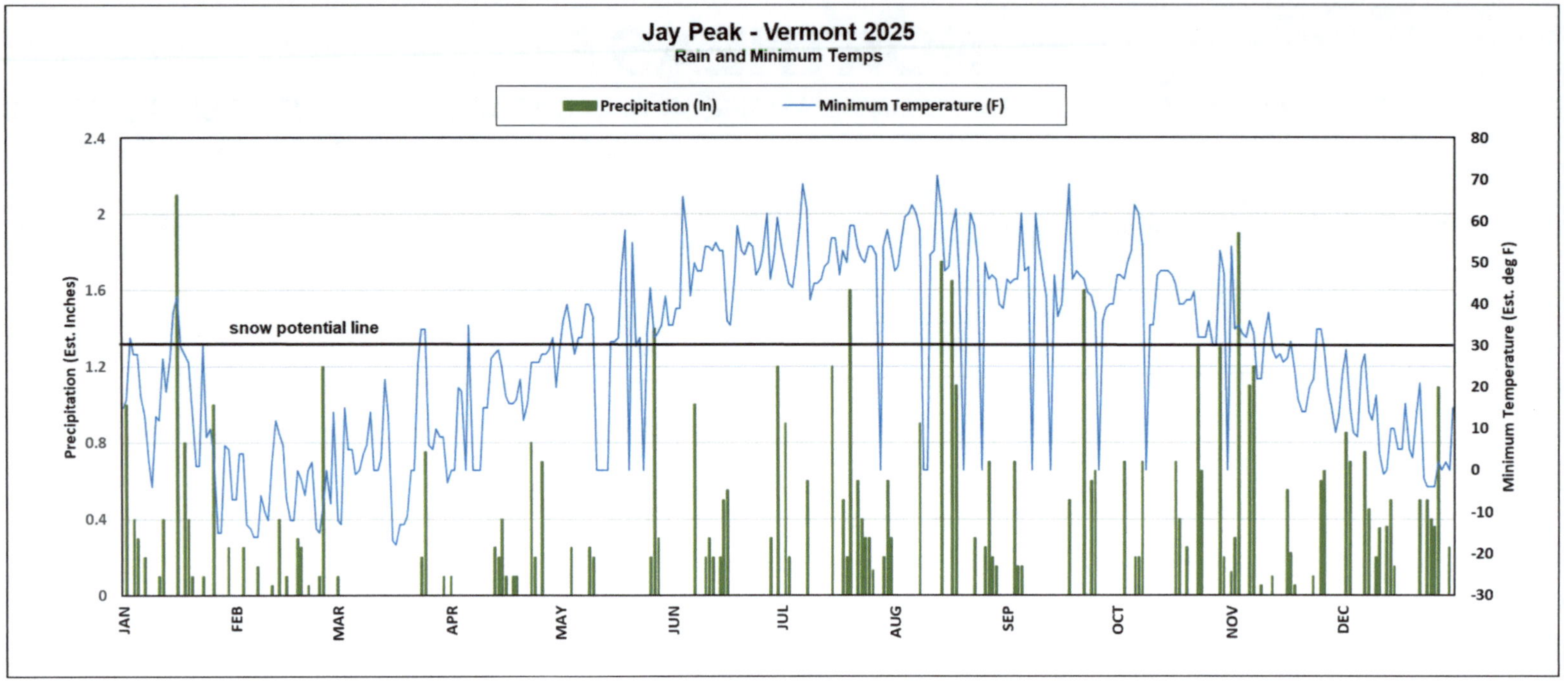

Jay Peak - Vermont 2025
Rain and Minimum Temps
Precipitation (In)
Minimum Temperature (F)
snow potential line
Precipitation (Est. Inches)
Minimum Temperature (Est. deg F)
JAN
FEB
MAR
APR
MAY
JUN
JUL
AUG
SEP
OCT
NOV
DEC

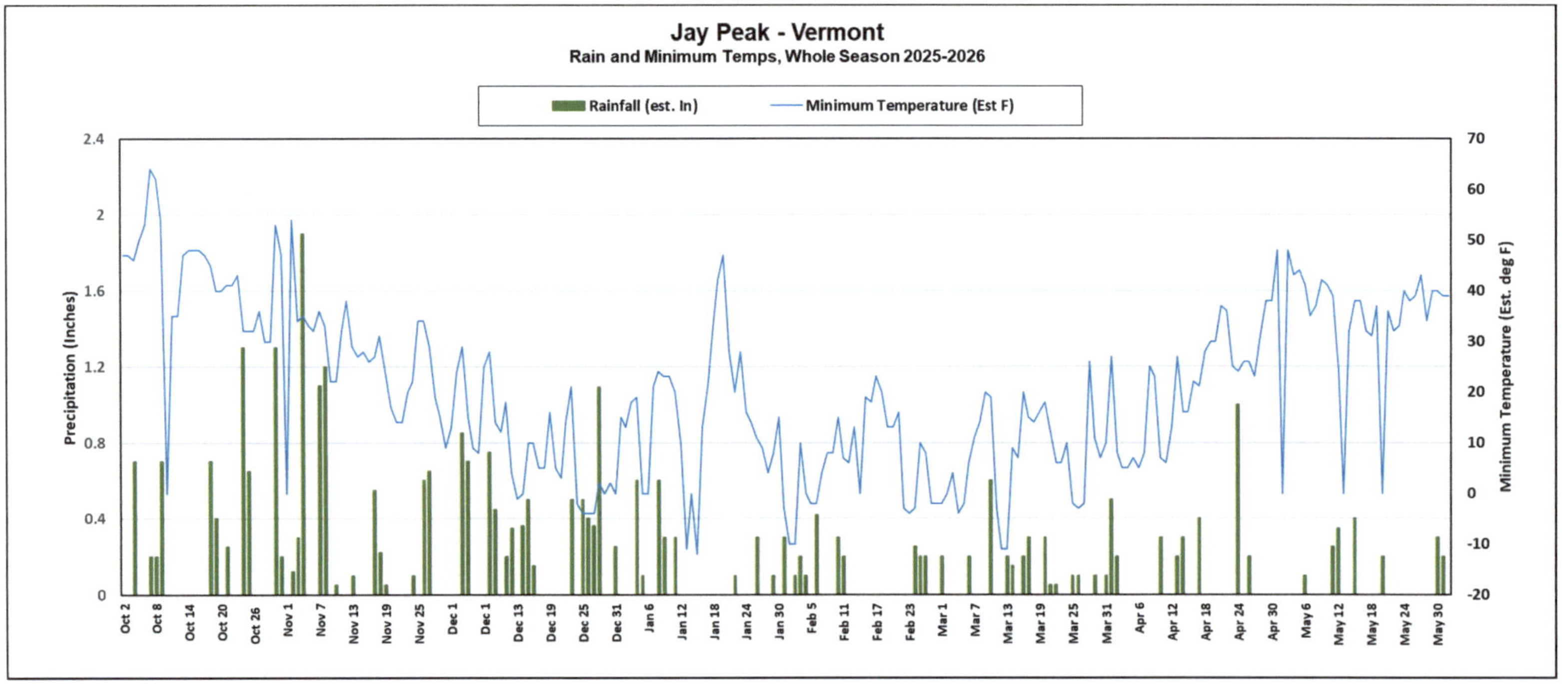

Jay Peak - Vermont
Rain and Minimum Temps, Whole Season 2025-2026
Rainfall (est. In)
Minimum Temperature (Est F)
Precipitation (Inches)
Minimum Temperature (Est. deg F)

HURRICANE POTENTIALS 2025

Dates	Areas Affected
May 19-21	Virginia, North Carolina, South Carolina, Georgia, Florida, The Bahamas
June 12-16	El Salvador, Western Cuba, Florida, East Coast of the United States
August 10-11	Bermuda, Atlantic Canada, Newfoundland
August 21-23	Hawaii
August 26-30	Texas, Oklahoma, Central United States
August 22 - September 2	Lesser Antilles, Puerto Rico, Hispaniola, Jamaica, Cuba, Cayman Islands, Nicaragua, Honduras, Belize, Guatemala, Mexico
September 5	Mexico
September 12	Trinidad and Tobago, Windward Islands, Venezuela, Leeward Antilles, Colombia, Central America, Yucatan Peninsula
September 18-19	North Carolina
September 21	Texas, Kentucky, Tennessee
September 22	Gulf of Mexico
September 23	Texas to Louisiana, Missouri, Florida to North Carolina
September 24	Missouri, Mississippi Delta, Carolina
September 28 – October 1	Gulf of Mexico, Florida, Georgia, Alabama
October 6-7	Mexico
November 11	Cuba
November 7-13	Leeward Islands, Greater Antilles, Lucayan Archipelago, East Coast of the United States, Atlantic Canada, Greenland
November 8-11	Florida
November 12-13	North Carolina to Maine and Massachusetts
November 22	Mexico, Texas, New Mexico
December 19	Puerto Rico, Hispaniola, Yucatán Peninsula, Central Florida
December 25	Deep South, Georgia to Florida

TORNADO POTENTIALS ACROSS THE REGIONS

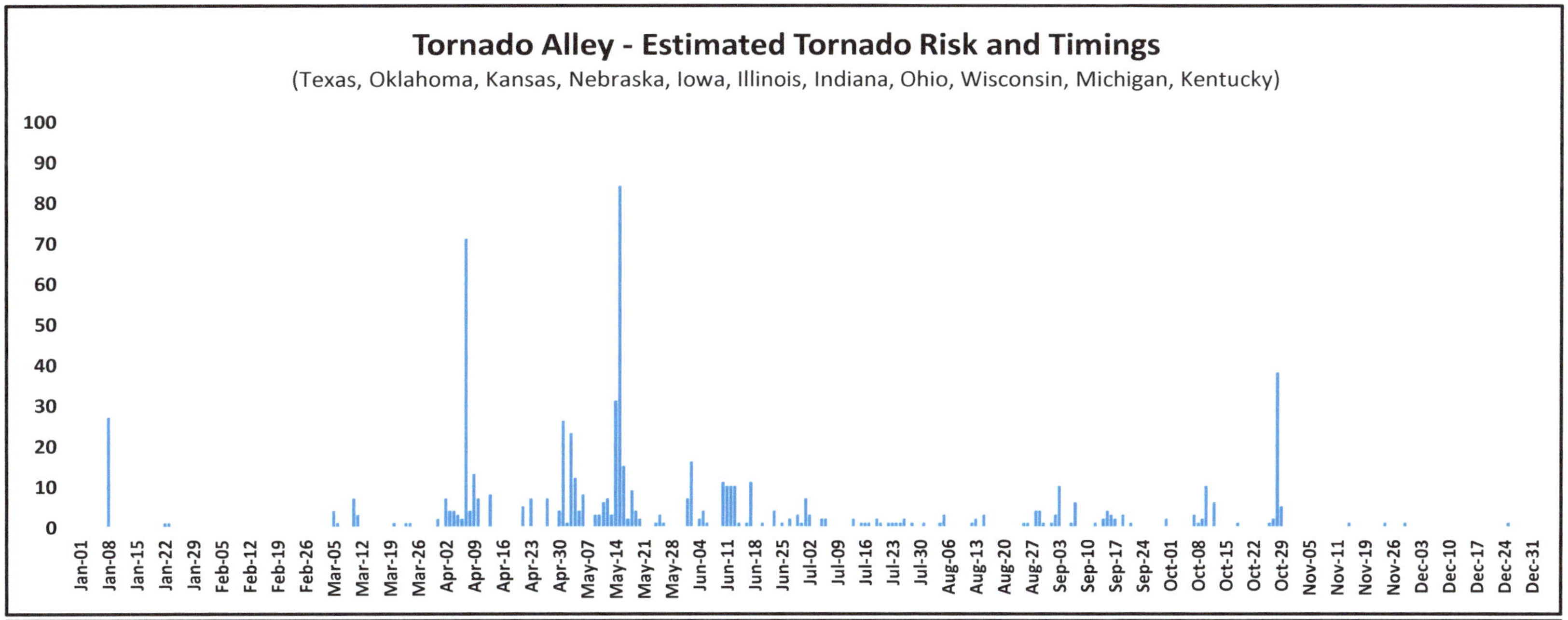

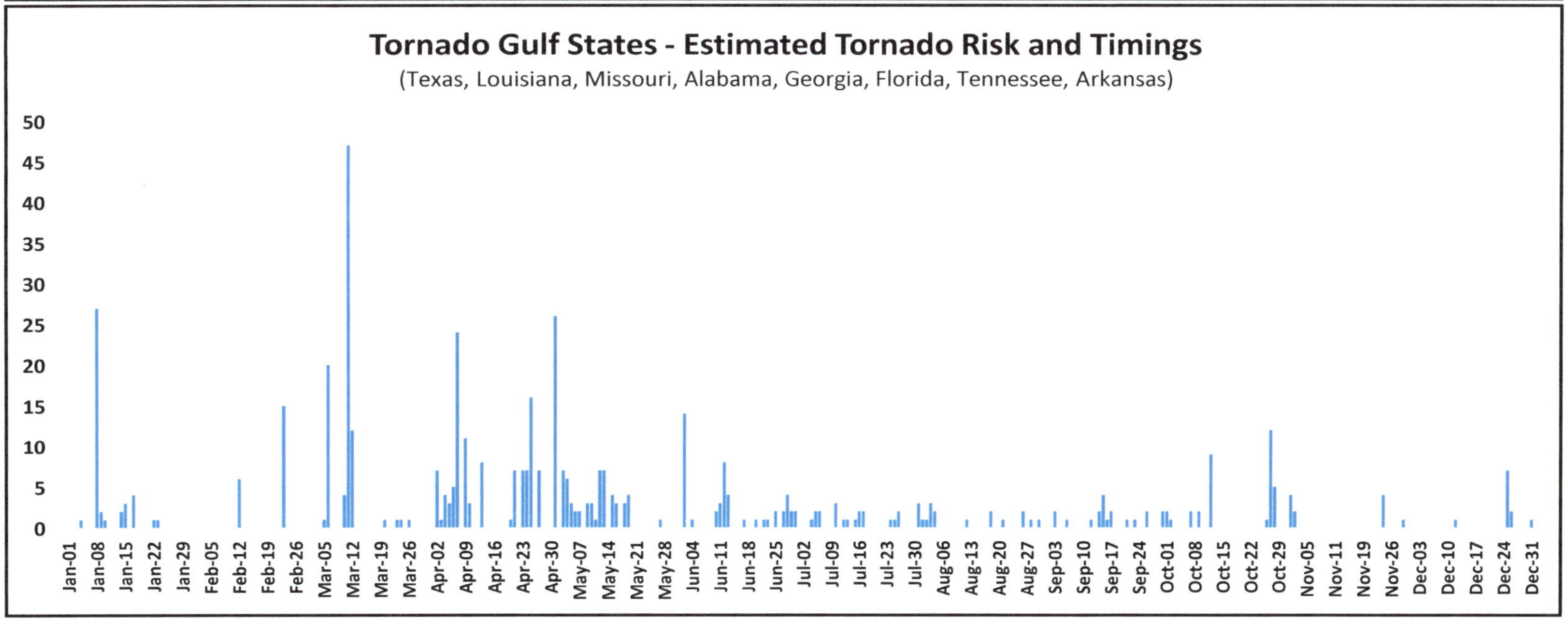

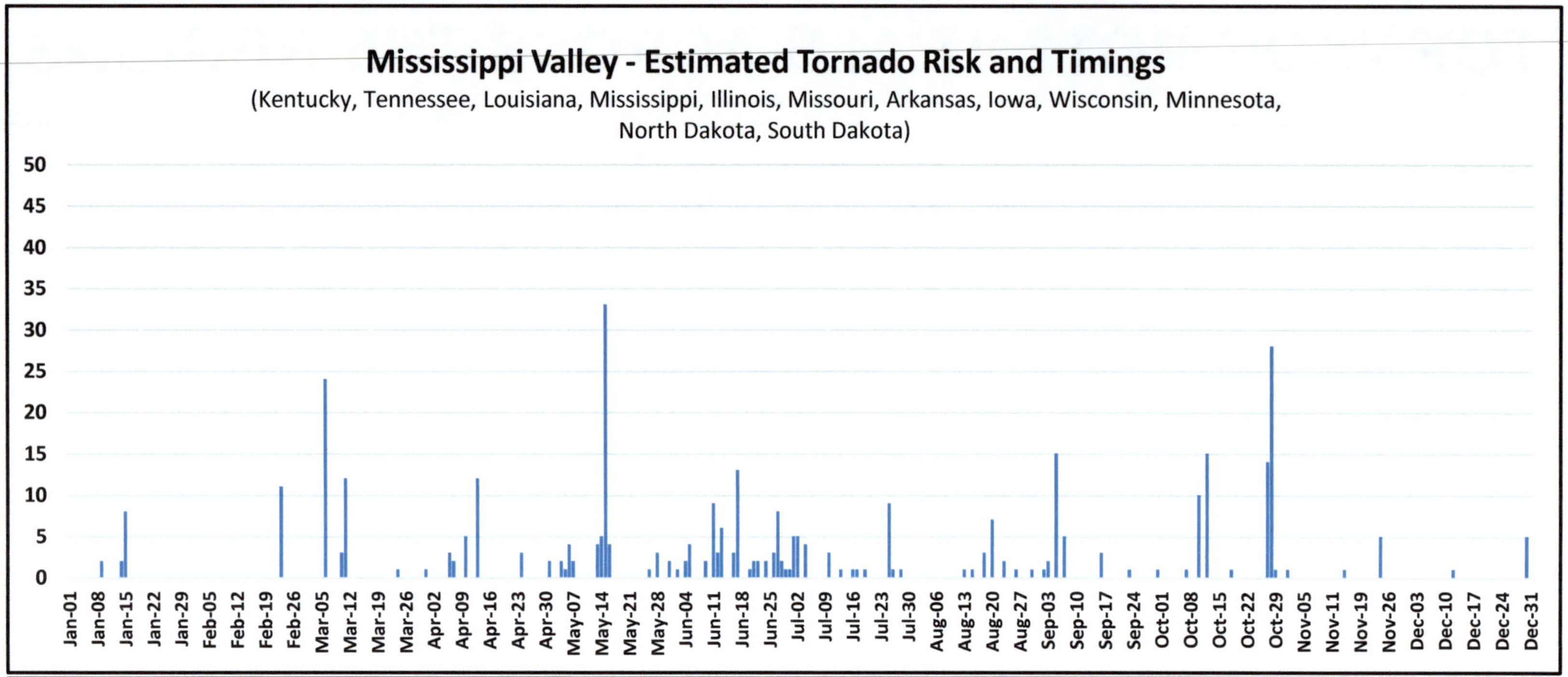
Mississippi Valley - Estimated Tornado Risk and Timings
(Kentucky, Tennessee, Louisiana, Mississippi, Illinois, Missouri, Arkansas, Iowa, Wisconsin, Minnesota, North Dakota, South Dakota)
50
45
40
35
30
25
20
15
10
5
0
Jan-01 Jan-08 Jan-15 Jan-22 Jan-29 Feb-05 Feb-12 Feb-19 Feb-26 Mar-05 Mar-12 Mar-19 Mar-26 Apr-02 Apr-09 Apr-16 Apr-23 Apr-30 May-07 May-14 May-21 May-28 Jun-04 Jun-11 Jun-18 Jun-25 Jul-02 Jul-09 Jul-16 Jul-23 Jul-30 Aug-06 Aug-13 Aug-20 Aug-27 Sep-03 Sep-10 Sep-17 Sep-24 Oct-01 Oct-08 Oct-15 Oct-22 Oct-29 Nov-05 Nov-11 Nov-19 Nov-26 Dec-03 Dec-10 Dec-17 Dec-24 Dec-31

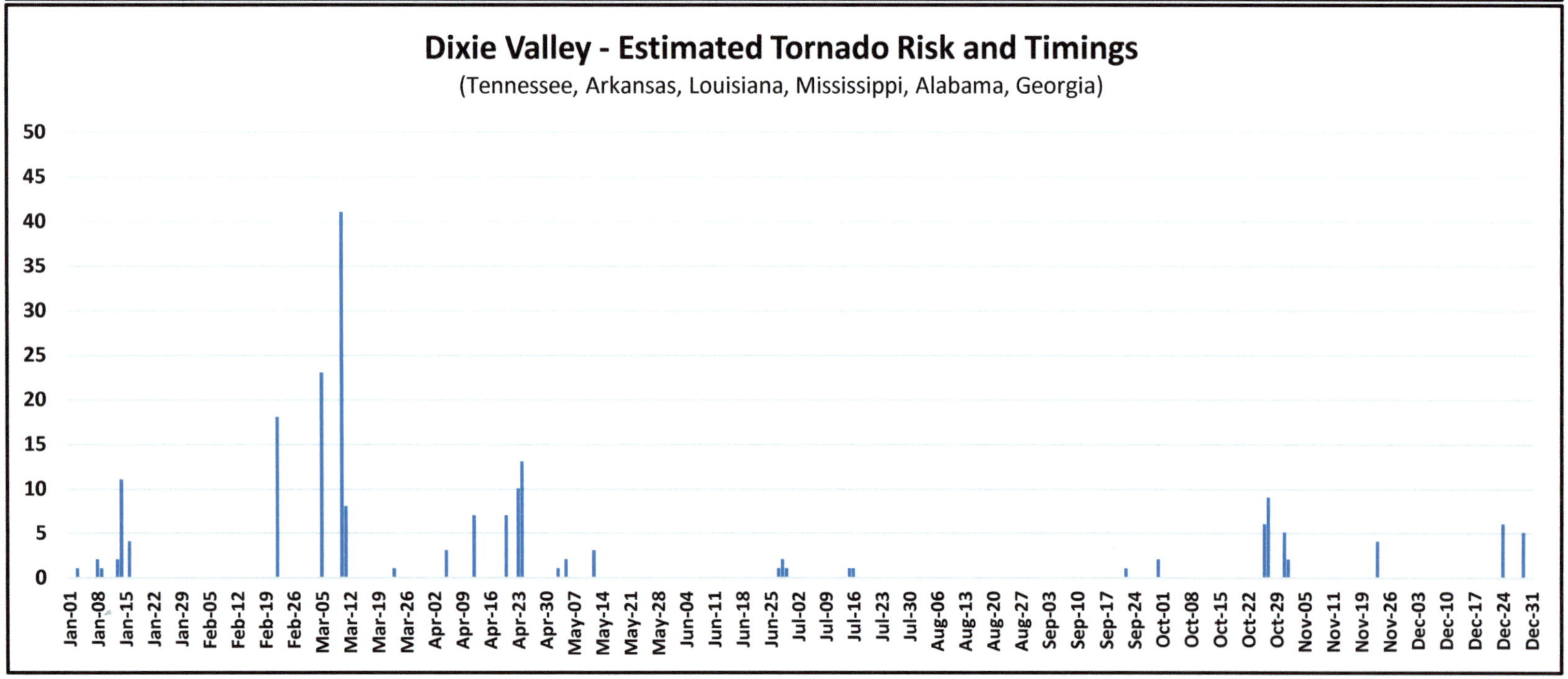
Dixie Valley - Estimated Tornado Risk and Timings
(Tennessee, Arkansas, Louisiana, Mississippi, Alabama, Georgia)
50
45
40
35
30
25
20
15
10
5
0
Jan-01 Jan-08 Jan-15 Jan-22 Jan-29 Feb-05 Feb-12 Feb-19 Feb-26 Mar-05 Mar-12 Mar-19 Mar-26 Apr-02 Apr-09 Apr-16 Apr-23 Apr-30 May-07 May-14 May-21 May-28 Jun-04 Jun-11 Jun-18 Jun-25 Jul-02 Jul-09 Jul-16 Jul-23 Jul-30 Aug-06 Aug-13 Aug-20 Aug-27 Sep-03 Sep-10 Sep-17 Sep-24 Oct-01 Oct-08 Oct-15 Oct-22 Oct-29 Nov-05 Nov-11 Nov-19 Nov-26 Dec-03 Dec-10 Dec-17 Dec-24 Dec-31

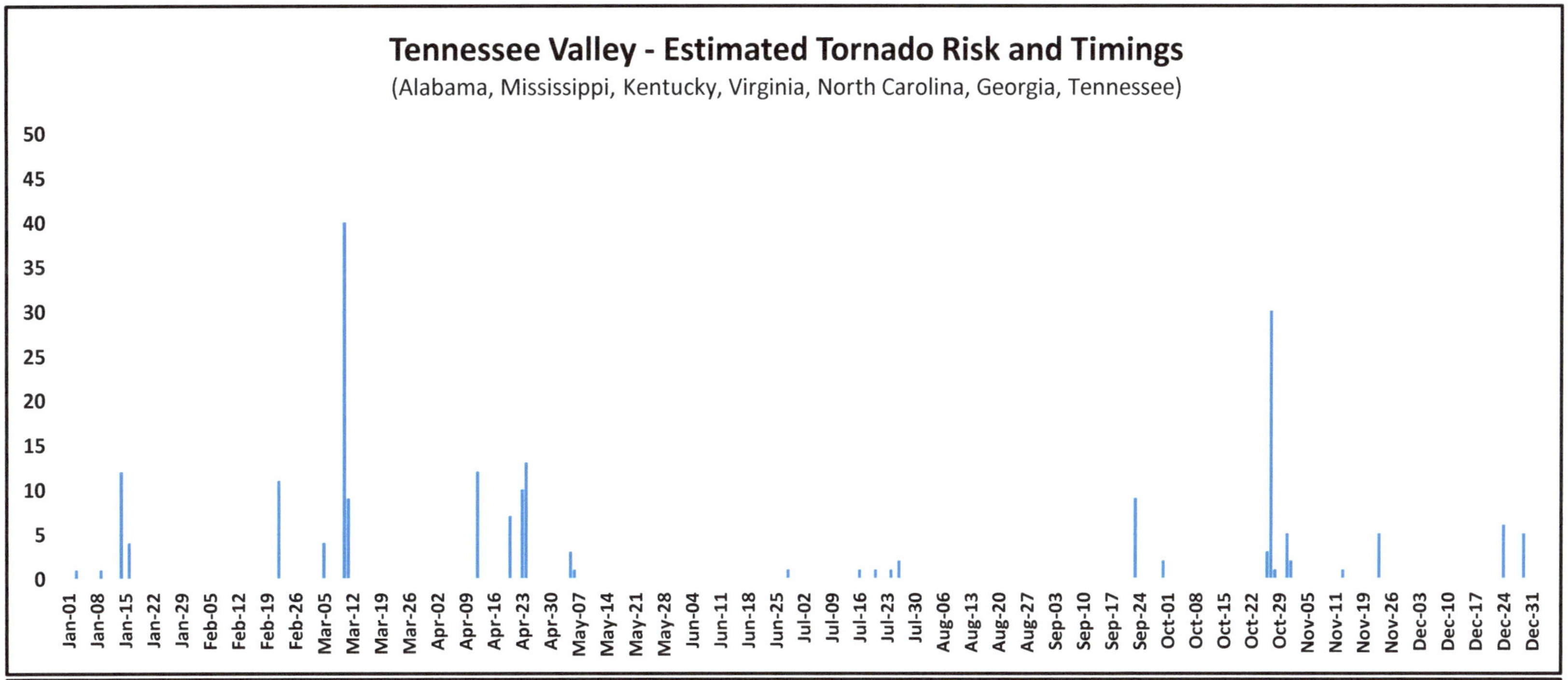
Tennessee Valley - Estimated Tornado Risk and Timings
(Alabama, Mississippi, Kentucky, Virginia, North Carolina, Georgia, Tennessee)
0
5
10
15
20
25
30
35
40
45
50
Jan-01
Jan-08
Jan-15
Jan-22
Jan-29
Feb-05
Feb-12
Feb-19
Feb-26
Mar-05
Mar-12
Mar-19
Mar-26
Apr-02
Apr-09
Apr-16
Apr-23
Apr-30
May-07
May-14
May-21
May-28
Jun-04
Jun-11
Jun-18
Jun-25
Jul-02
Jul-09
Jul-16
Jul-23
Jul-30
Aug-06
Aug-13
Aug-20
Aug-27
Sep-03
Sep-10
Sep-17
Sep-24
Oct-01
Oct-08
Oct-15
Oct-22
Oct-29
Nov-05
Nov-12
Nov-19
Nov-26
Dec-03
Dec-10
Dec-17
Dec-24
Dec-31

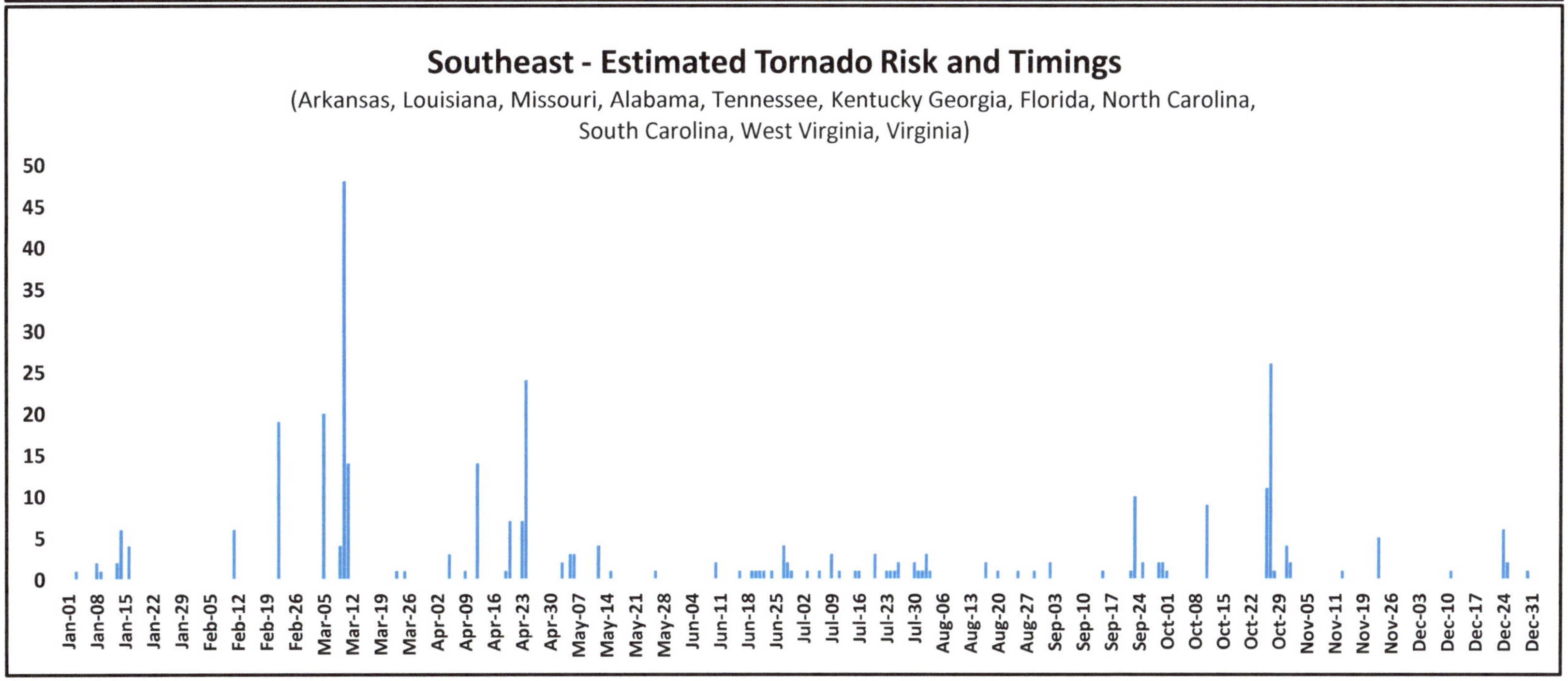
Southeast - Estimated Tornado Risk and Timings
(Arkansas, Louisiana, Missouri, Alabama, Tennessee, Kentucky Georgia, Florida, North Carolina,
South Carolina, West Virginia, Virginia)
0
5
10
15
20
25
30
35
40
45
50
Jan-01
Jan-08
Jan-15
Jan-22
Jan-29
Feb-05
Feb-12
Feb-19
Feb-26
Mar-05
Mar-12
Mar-19
Mar-26
Apr-02
Apr-09
Apr-16
Apr-23
Apr-30
May-07
May-14
May-21
May-28
Jun-04
Jun-11
Jun-18
Jun-25
Jul-02
Jul-09
Jul-16
Jul-23
Jul-30
Aug-06
Aug-13
Aug-20
Aug-27
Sep-03
Sep-10
Sep-17
Sep-24
Oct-01
Oct-08
Oct-15
Oct-22
Oct-29
Nov-05
Nov-12
Nov-19
Nov-26
Dec-03
Dec-10
Dec-17
Dec-24
Dec-31

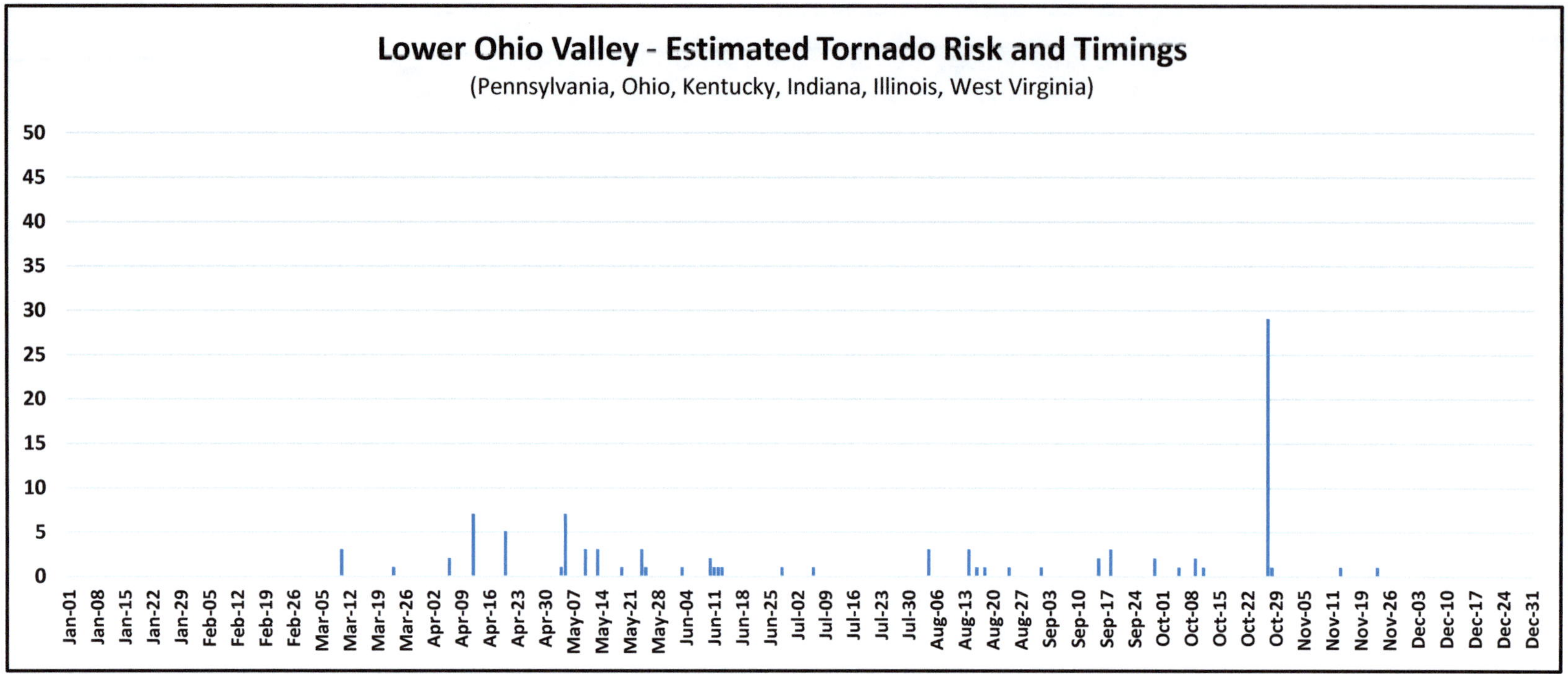

Lower Ohio Valley - Estimated Tornado Risk and Timings
(Pennsylvania, Ohio, Kentucky, Indiana, Illinois, West Virginia)
50
45
40
35
30
25
20
15
10
5
0
Jan-01 Jan-08 Jan-15 Jan-22 Jan-29 Feb-05 Feb-12 Feb-19 Feb-26 Mar-05 Mar-12 Mar-19 Mar-26 Apr-02 Apr-09 Apr-16 Apr-23 Apr-30 May-07 May-14 May-21 May-28 Jun-04 Jun-11 Jun-18 Jun-25 Jul-02 Jul-09 Jul-16 Jul-23 Jul-30 Aug-06 Aug-13 Aug-20 Aug-27 Sep-03 Sep-10 Sep-17 Sep-24 Oct-01 Oct-08 Oct-15 Oct-22 Oct-29 Nov-05 Nov-12 Nov-19 Nov-26 Dec-03 Dec-10 Dec-17 Dec-24 Dec-31

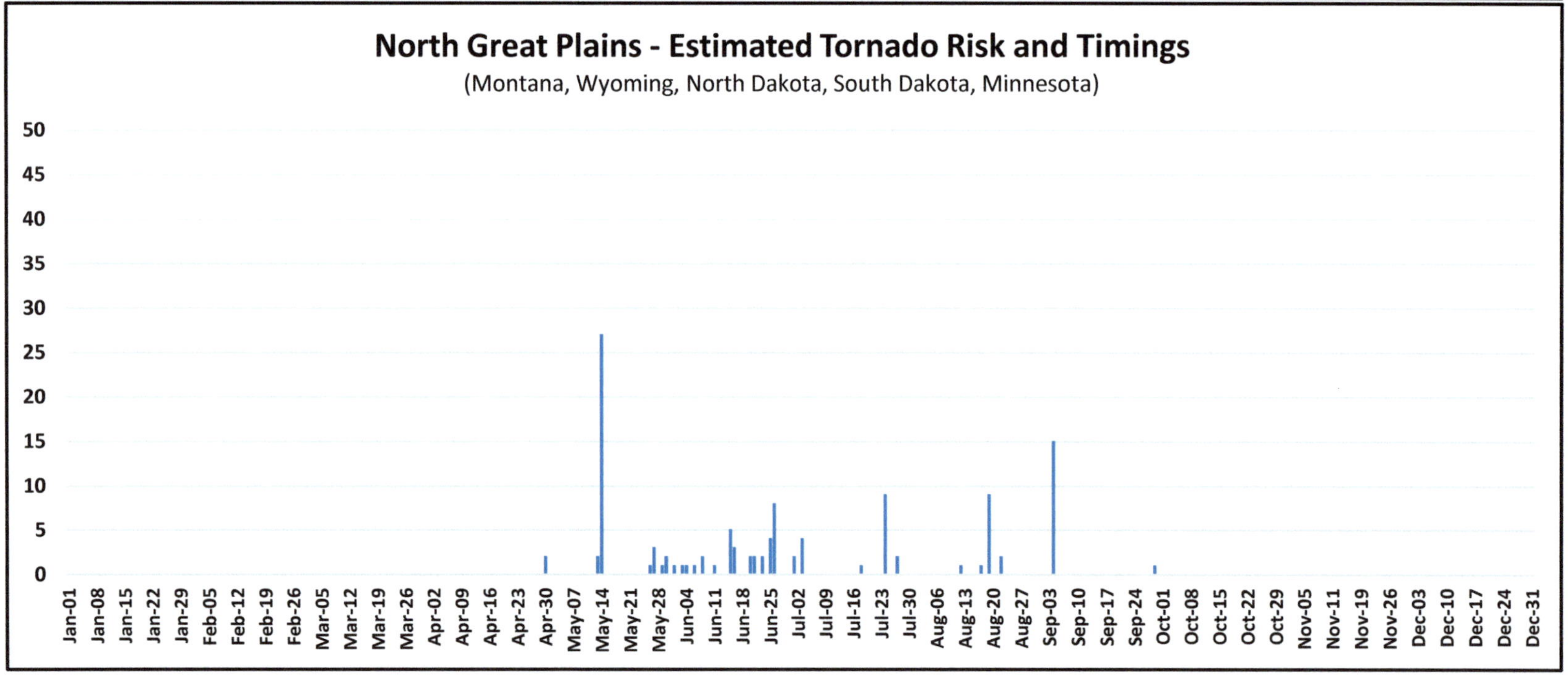

North Great Plains - Estimated Tornado Risk and Timings
(Montana, Wyoming, North Dakota, South Dakota, Minnesota)
50
45
40
35
30
25
20
15
10
5
0
Jan-01 Jan-08 Jan-15 Jan-22 Jan-29 Feb-05 Feb-12 Feb-19 Feb-26 Mar-05 Mar-12 Mar-19 Mar-26 Apr-02 Apr-09 Apr-16 Apr-23 Apr-30 May-07 May-14 May-21 May-28 Jun-04 Jun-11 Jun-18 Jun-25 Jul-02 Jul-09 Jul-16 Jul-23 Jul-30 Aug-06 Aug-13 Aug-20 Aug-27 Sep-03 Sep-10 Sep-17 Sep-24 Oct-01 Oct-08 Oct-15 Oct-22 Oct-29 Nov-05 Nov-12 Nov-19 Nov-26 Dec-03 Dec-10 Dec-17 Dec-24 Dec-31

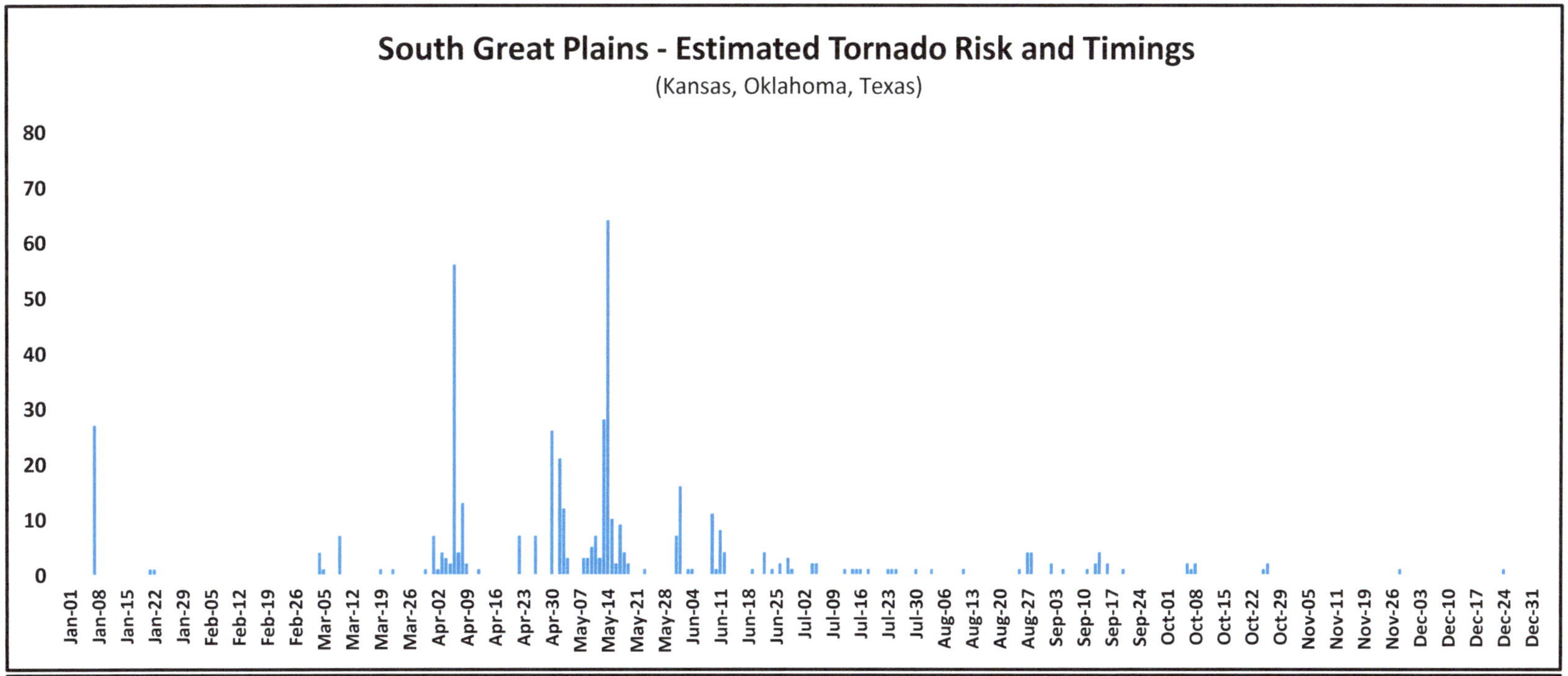
South Great Plains - Estimated Tornado Risk and Timings
(Kansas, Oklahoma, Texas)
80
70
60
50
40
30
20
10
0
Jan-01 Jan-08 Jan-15 Jan-22 Jan-29 Feb-05 Feb-12 Feb-19 Feb-26 Mar-05 Mar-12 Mar-19 Mar-26 Apr-02 Apr-09 Apr-16 Apr-23 Apr-30 May-07 May-14 May-21 May-28 Jun-04 Jun-11 Jun-18 Jun-25 Jul-02 Jul-09 Jul-16 Jul-23 Jul-30 Aug-06 Aug-13 Aug-20 Aug-27 Sep-03 Sep-10 Sep-17 Sep-24 Oct-01 Oct-08 Oct-15 Oct-22 Oct-29 Nov-05 Nov-11 Nov-19 Nov-26 Dec-03 Dec-10 Dec-17 Dec-24 Dec-31

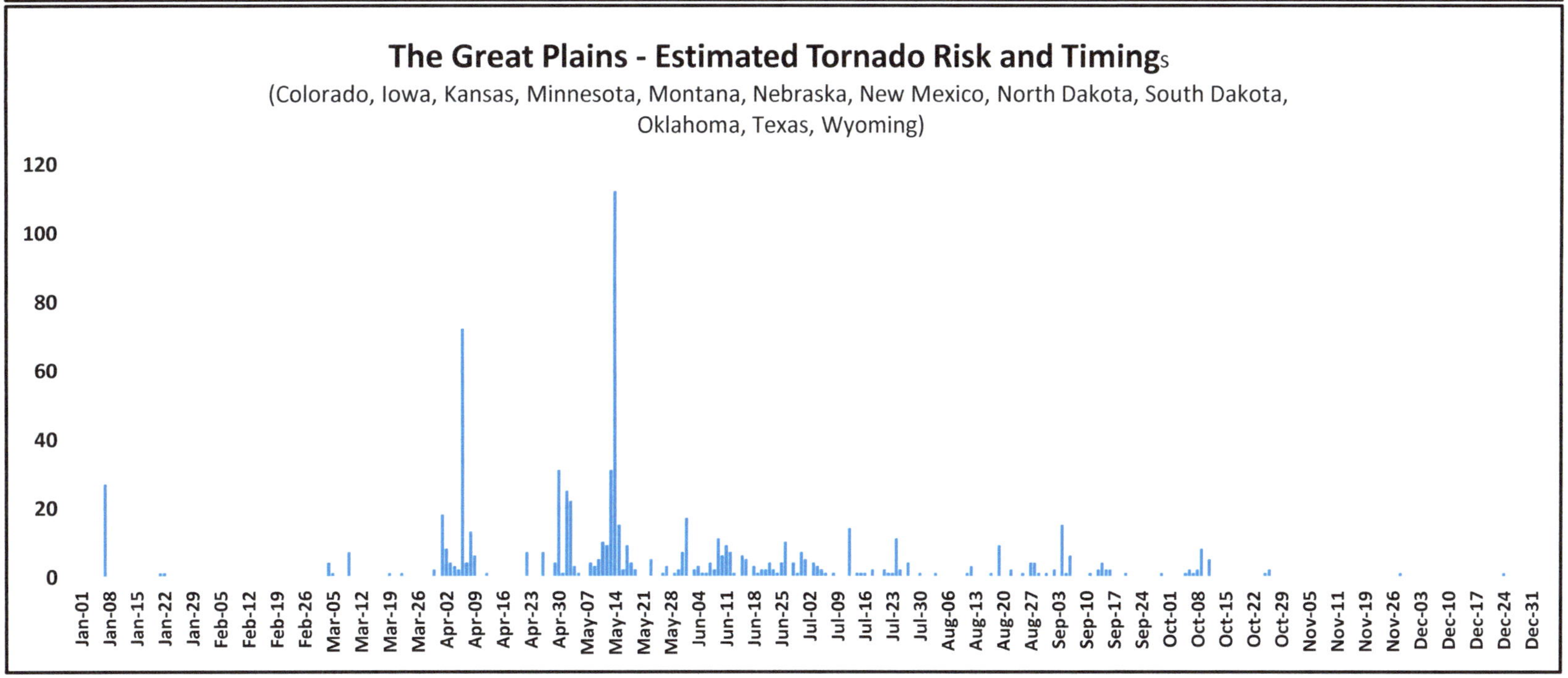
The Great Plains - Estimated Tornado Risk and Timings
(Colorado, Iowa, Kansas, Minnesota, Montana, Nebraska, New Mexico, North Dakota, South Dakota,
Oklahoma, Texas, Wyoming)
120
100
80
60
40
20
0
Jan-01 Jan-08 Jan-15 Jan-22 Jan-29 Feb-05 Feb-12 Feb-19 Feb-26 Mar-05 Mar-12 Mar-19 Mar-26 Apr-02 Apr-09 Apr-16 Apr-23 Apr-30 May-07 May-14 May-21 May-28 Jun-04 Jun-11 Jun-18 Jun-25 Jul-02 Jul-09 Jul-16 Jul-23 Jul-30 Aug-06 Aug-13 Aug-20 Aug-27 Sep-03 Sep-10 Sep-17 Sep-24 Oct-01 Oct-08 Oct-15 Oct-22 Oct-29 Nov-05 Nov-11 Nov-19 Nov-26 Dec-03 Dec-10 Dec-17 Dec-24 Dec-31

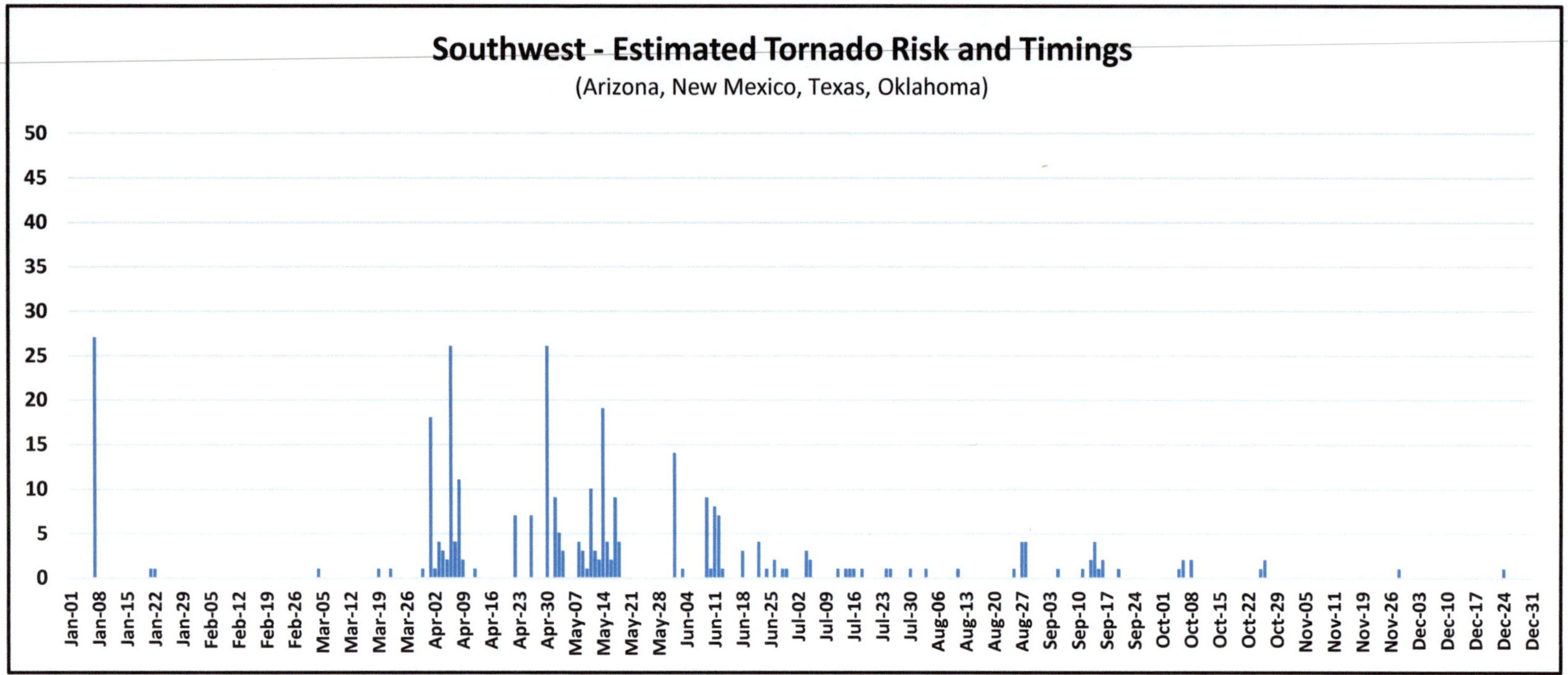
Southwest - Estimated Tornado Risk and Timings
(Arizona, New Mexico, Texas, Oklahoma)
0
5
10
15
20
25
30
35
40
45
50
Jan-01
Jan-08
Jan-15
Jan-22
Jan-29
Feb-05
Feb-12
Feb-19
Feb-26
Mar-05
Mar-12
Mar-19
Mar-26
Apr-02
Apr-09
Apr-16
Apr-23
Apr-30
May-07
May-14
May-21
May-28
Jun-04
Jun-11
Jun-18
Jun-25
Jul-02
Jul-09
Jul-16
Jul-23
Jul-30
Aug-06
Aug-13
Aug-20
Aug-27
Sep-03
Sep-10
Sep-17
Sep-24
Oct-01
Oct-08
Oct-15
Oct-22
Oct-29
Nov-05
Nov-11
Nov-19
Nov-26
Dec-03
Dec-10
Dec-17
Dec-24
Dec-31

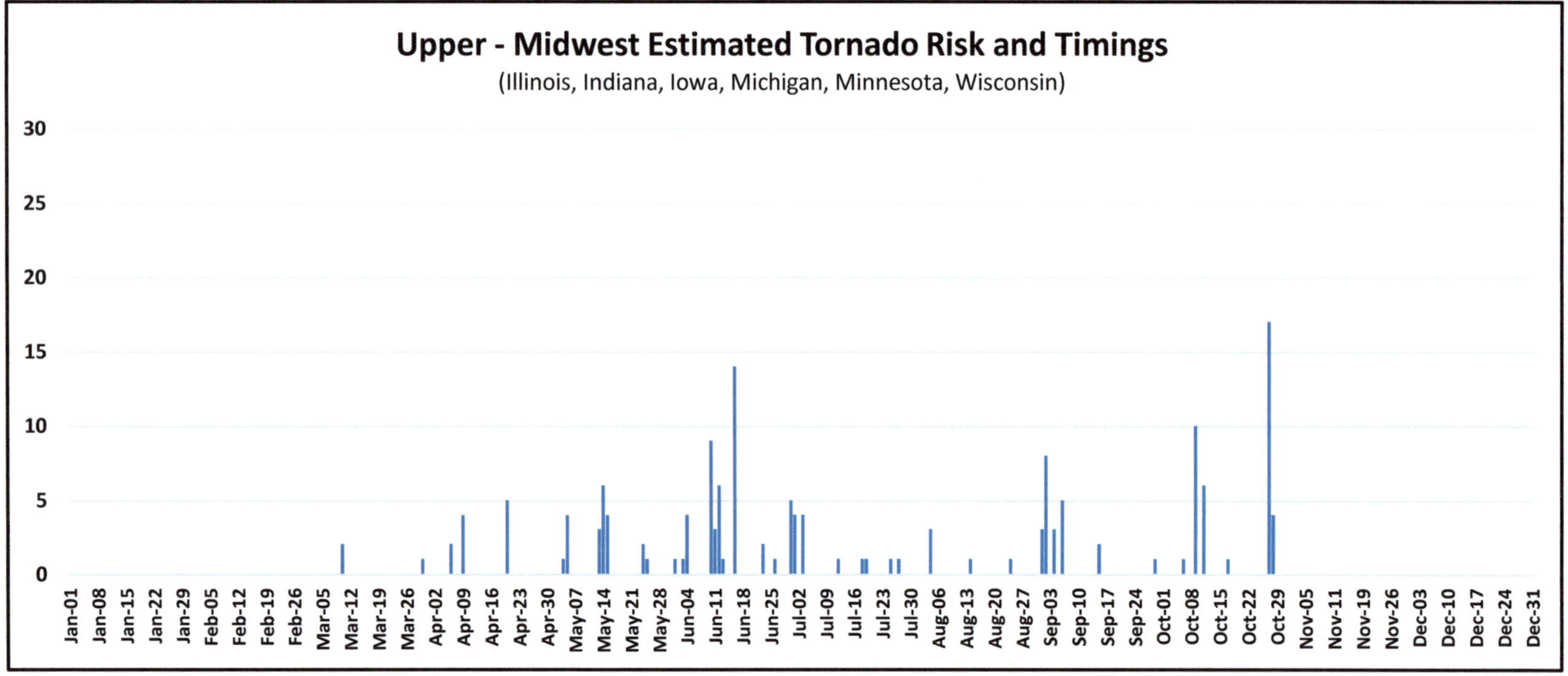
Upper - Midwest Estimated Tornado Risk and Timings
(Illinois, Indiana, Iowa, Michigan, Minnesota, Wisconsin)
0
5
10
15
20
25
30
Jan-01
Jan-08
Jan-15
Jan-22
Jan-29
Feb-05
Feb-12
Feb-19
Feb-26
Mar-05
Mar-12
Mar-19
Mar-26
Apr-02
Apr-09
Apr-16
Apr-23
Apr-30
May-07
May-14
May-21
May-28
Jun-04
Jun-11
Jun-18
Jun-25
Jul-02
Jul-09
Jul-16
Jul-23
Jul-30
Aug-06
Aug-13
Aug-20
Aug-27
Sep-03
Sep-10
Sep-17
Sep-24
Oct-01
Oct-08
Oct-15
Oct-22
Oct-29
Nov-05
Nov-11
Nov-19
Nov-26
Dec-03
Dec-10
Dec-17
Dec-24
Dec-31

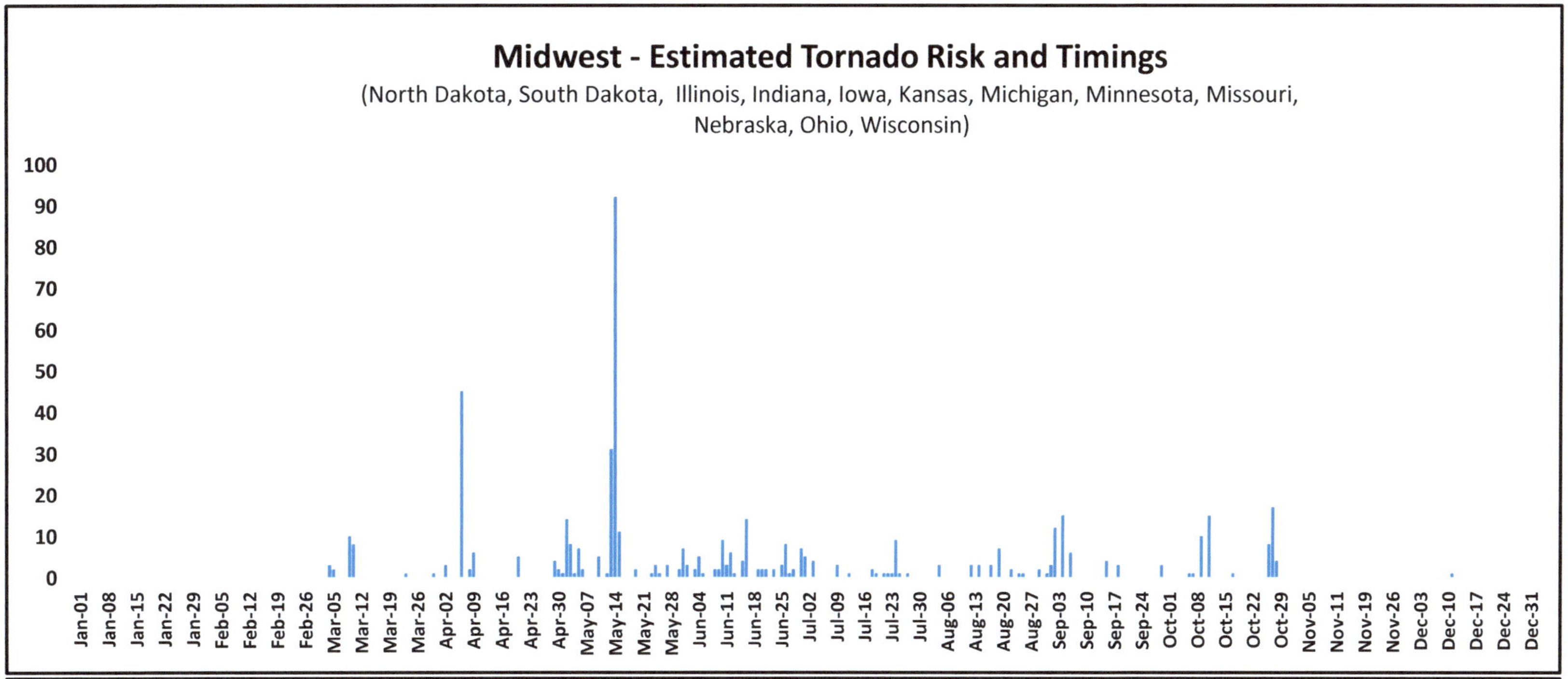
Midwest - Estimated Tornado Risk and Timings
(North Dakota, South Dakota, Illinois, Indiana, Iowa, Kansas, Michigan, Minnesota, Missouri, Nebraska, Ohio, Wisconsin)
100
90
80
70
60
50
40
30
20
10
0
Jan-01 Jan-08 Jan-15 Jan-22 Jan-29 Feb-05 Feb-12 Feb-19 Feb-26 Mar-05 Mar-12 Mar-19 Mar-26 Apr-02 Apr-09 Apr-16 Apr-23 Apr-30 May-07 May-14 May-21 May-28 Jun-04 Jun-11 Jun-18 Jun-25 Jul-02 Jul-09 Jul-16 Jul-23 Jul-30 Aug-06 Aug-13 Aug-20 Aug-27 Sep-03 Sep-10 Sep-17 Sep-24 Oct-01 Oct-08 Oct-15 Oct-22 Oct-29 Nov-05 Nov-11 Nov-19 Nov-26 Dec-03 Dec-10 Dec-17 Dec-24 Dec-31

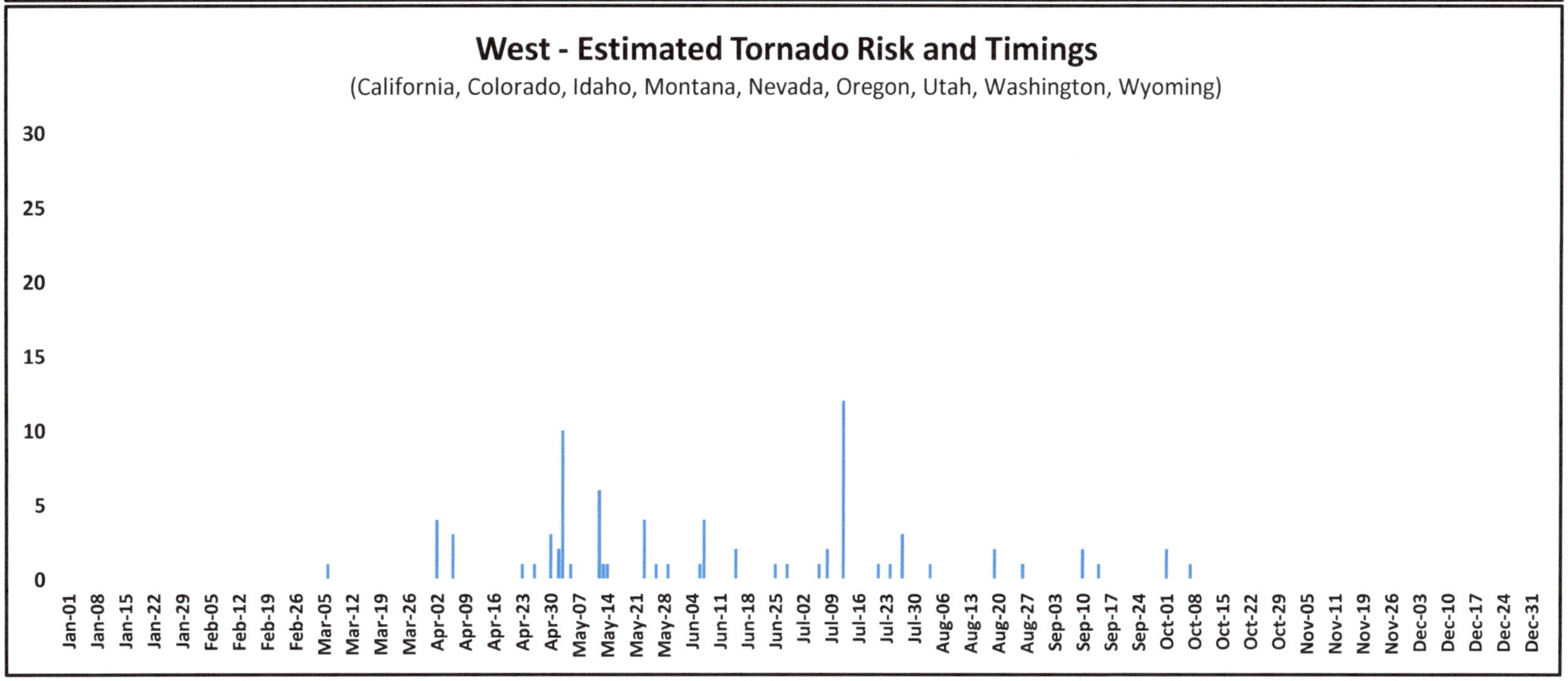
West - Estimated Tornado Risk and Timings
(California, Colorado, Idaho, Montana, Nevada, Oregon, Utah, Washington, Wyoming)
30
25
20
15
10
5
0
Jan-01 Jan-08 Jan-15 Jan-22 Jan-29 Feb-05 Feb-12 Feb-19 Feb-26 Mar-05 Mar-12 Mar-19 Mar-26 Apr-02 Apr-09 Apr-16 Apr-23 Apr-30 May-07 May-14 May-21 May-28 Jun-04 Jun-11 Jun-18 Jun-25 Jul-02 Jul-09 Jul-16 Jul-23 Jul-30 Aug-06 Aug-13 Aug-20 Aug-27 Sep-03 Sep-10 Sep-17 Sep-24 Oct-01 Oct-08 Oct-15 Oct-22 Oct-29 Nov-05 Nov-11 Nov-19 Nov-26 Dec-03 Dec-10 Dec-17 Dec-24 Dec-31

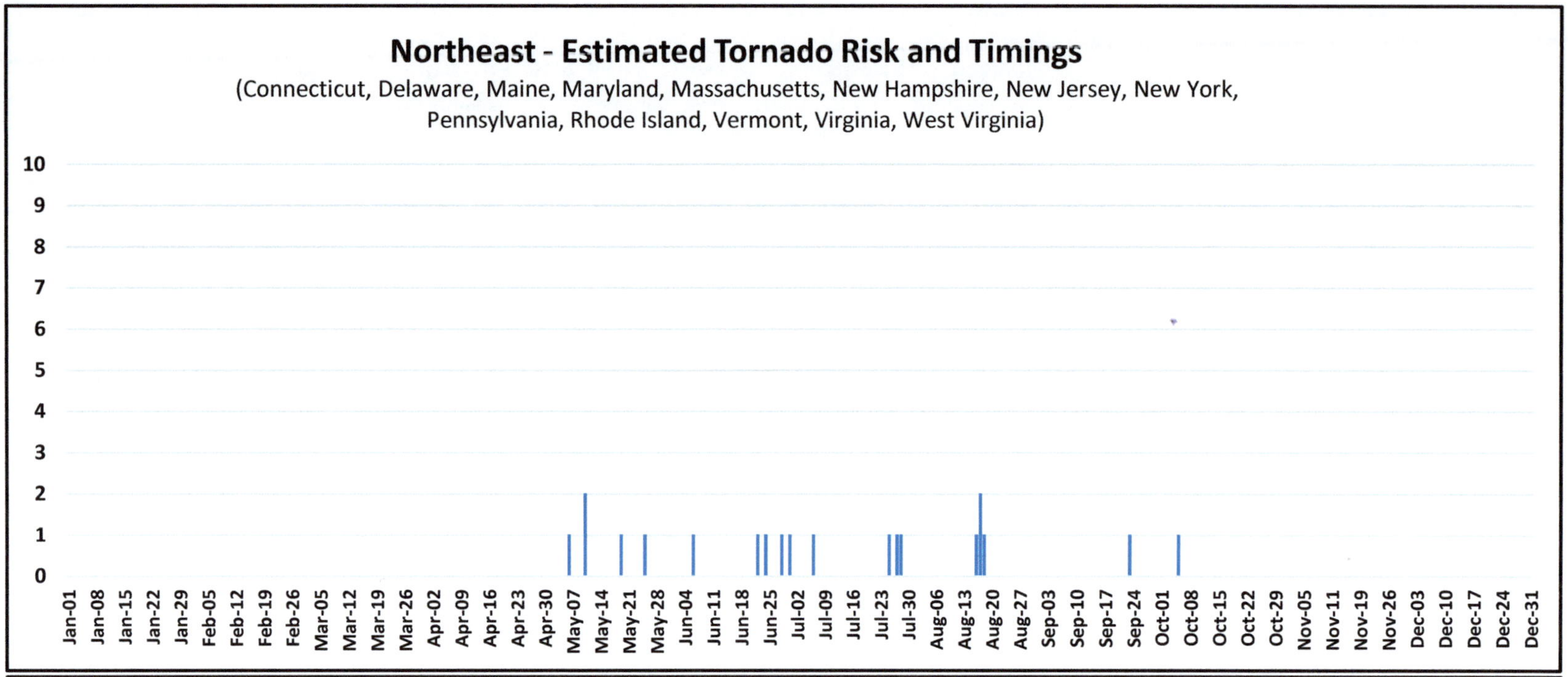
Northeast - Estimated Tornado Risk and Timings
(Connecticut, Delaware, Maine, Maryland, Massachusetts, New Hampshire, New Jersey, New York, Pennsylvania, Rhode Island, Vermont, Virginia, West Virginia)
10
9
8
7
6
5
4
3
2
1
0
Jan-01 Jan-08 Jan-15 Jan-22 Jan-29 Feb-05 Feb-12 Feb-19 Feb-26 Mar-05 Mar-12 Mar-19 Mar-26 Apr-02 Apr-09 Apr-16 Apr-23 Apr-30 May-07 May-14 May-21 May-28 Jun-04 Jun-11 Jun-18 Jun-25 Jul-02 Jul-09 Jul-16 Jul-23 Jul-30 Aug-06 Aug-13 Aug-20 Aug-27 Sep-03 Sep-10 Sep-17 Sep-24 Oct-01 Oct-08 Oct-15 Oct-22 Oct-29 Nov-05 Nov-11 Nov-19 Nov-26 Dec-03 Dec-10 Dec-17 Dec-24 Dec-31

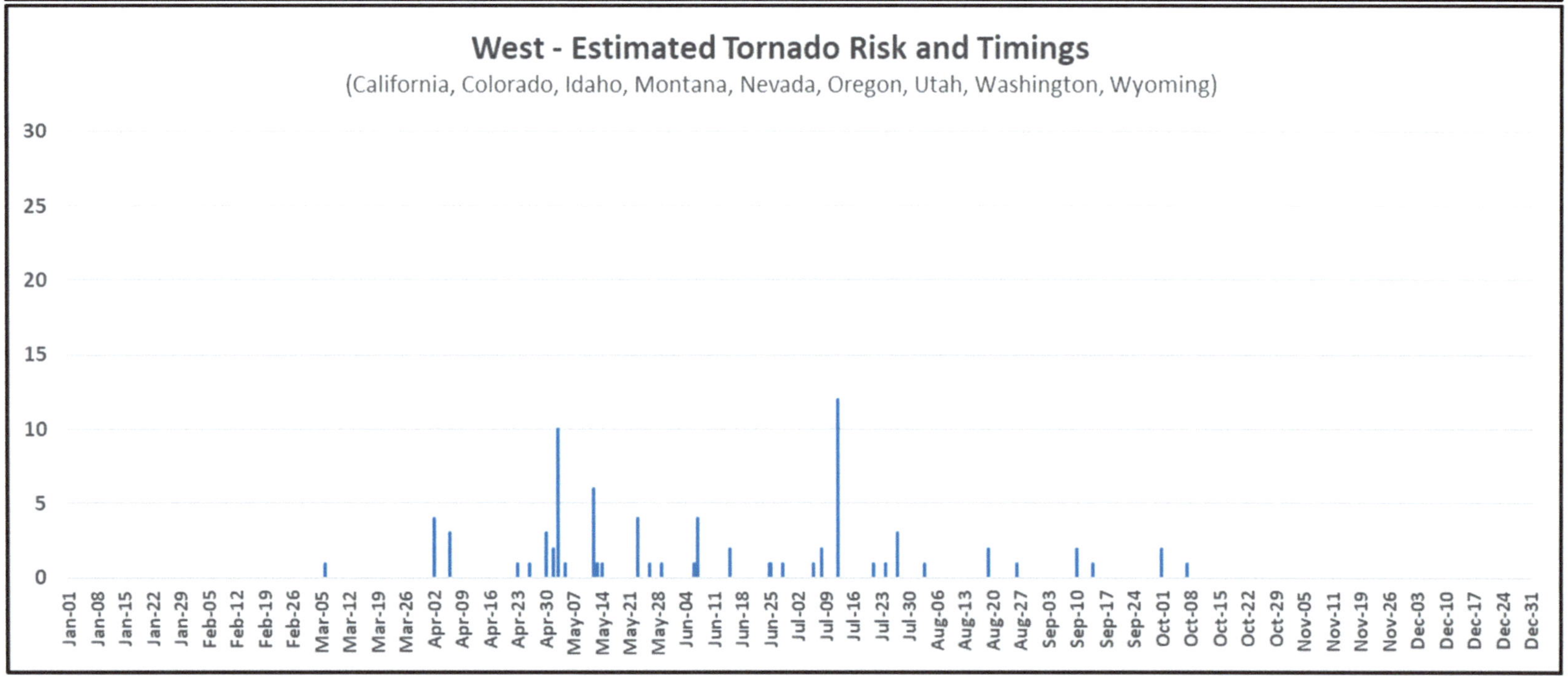
West - Estimated Tornado Risk and Timings
(California, Colorado, Idaho, Montana, Nevada, Oregon, Utah, Washington, Wyoming)
30
25
20
15
10
5
0
Jan-01 Jan-08 Jan-15 Jan-22 Jan-29 Feb-05 Feb-12 Feb-19 Feb-26 Mar-05 Mar-12 Mar-19 Mar-26 Apr-02 Apr-09 Apr-16 Apr-23 Apr-30 May-07 May-14 May-21 May-28 Jun-04 Jun-11 Jun-18 Jun-25 Jul-02 Jul-09 Jul-16 Jul-23 Jul-30 Aug-06 Aug-13 Aug-20 Aug-27 Sep-03 Sep-10 Sep-17 Sep-24 Oct-01 Oct-08 Oct-15 Oct-22 Oct-29 Nov-05 Nov-11 Nov-19 Nov-26 Dec-03 Dec-10 Dec-17 Dec-24 Dec-31

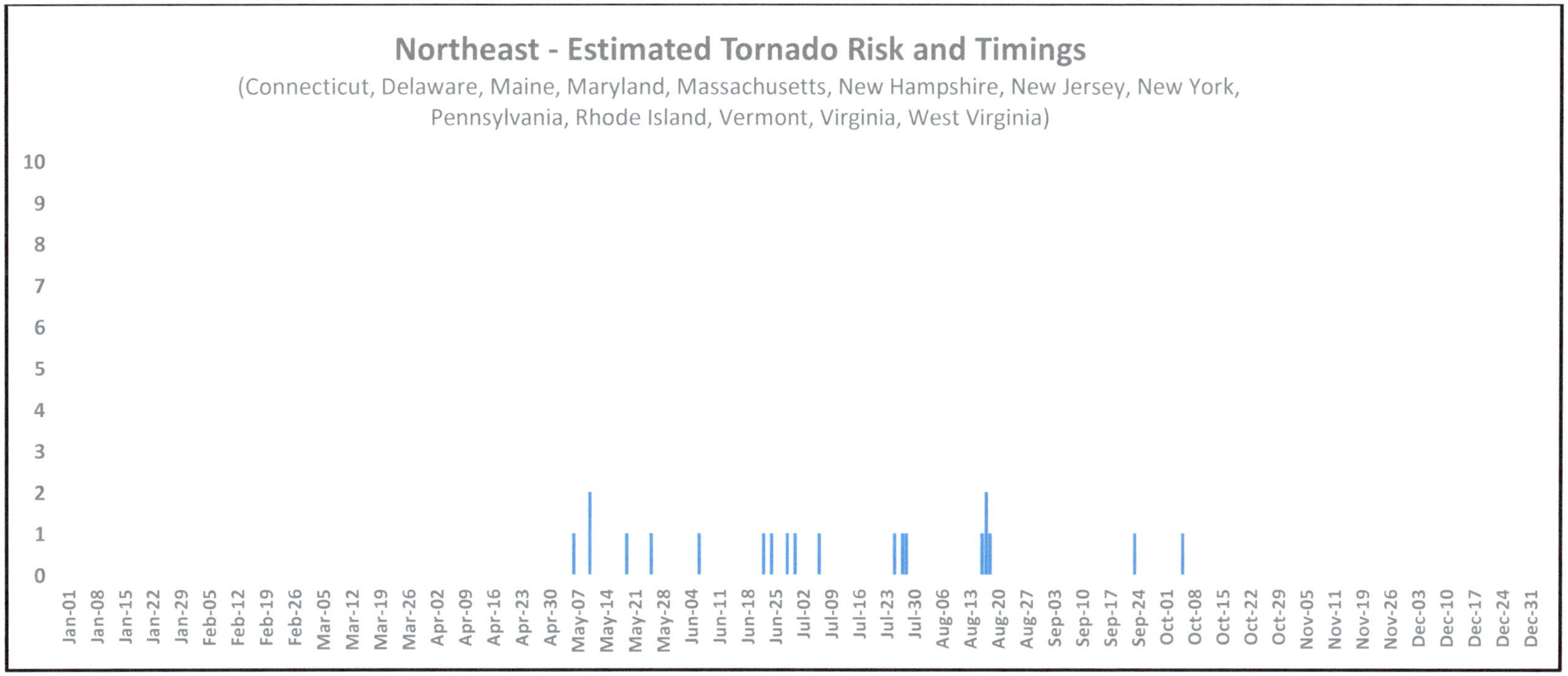
Northeast - Estimated Tornado Risk and Timings
(Connecticut, Delaware, Maine, Maryland, Massachusetts, New Hampshire, New Jersey, New York, Pennsylvania, Rhode Island, Vermont, Virginia, West Virginia)
10
9
8
7
6
5
4
3
2
1
0
Jan-01
Jan-08
Jan-15
Jan-22
Jan-29
Feb-05
Feb-12
Feb-19
Feb-26
Mar-05
Mar-12
Mar-19
Mar-26
Apr-02
Apr-09
Apr-16
Apr-23
Apr-30
May-07
May-14
May-21
May-28
Jun-04
Jun-11
Jun-18
Jun-25
Jul-02
Jul-09
Jul-16
Jul-23
Jul-30
Aug-06
Aug-13
Aug-20
Aug-27
Sep-03
Sep-10
Sep-17
Sep-24
Oct-01
Oct-08
Oct-15
Oct-22
Oct-29
Nov-05
Nov-11
Nov-19
Nov-26
Dec-03
Dec-10
Dec-17
Dec-24
Dec-31

US STATE CAPITALS
RAIN / MINIMUM TEMPERATURE COMPARISON GRAPHS 2025

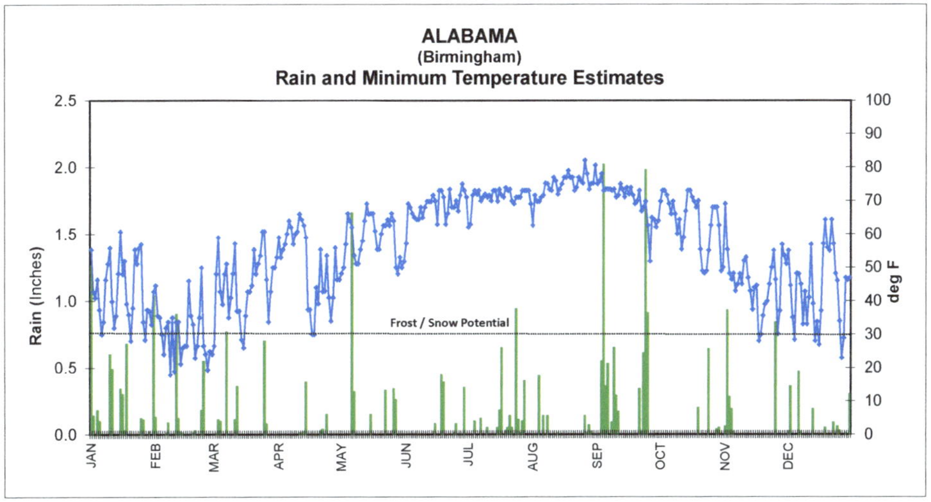

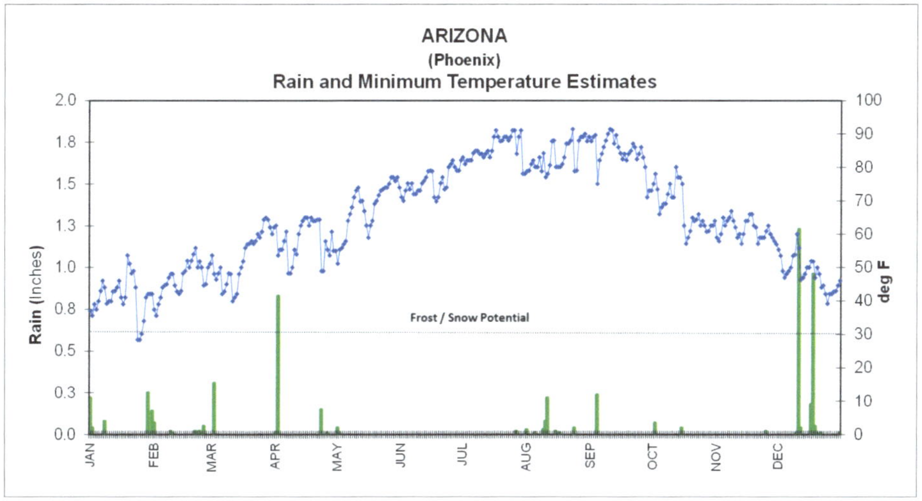

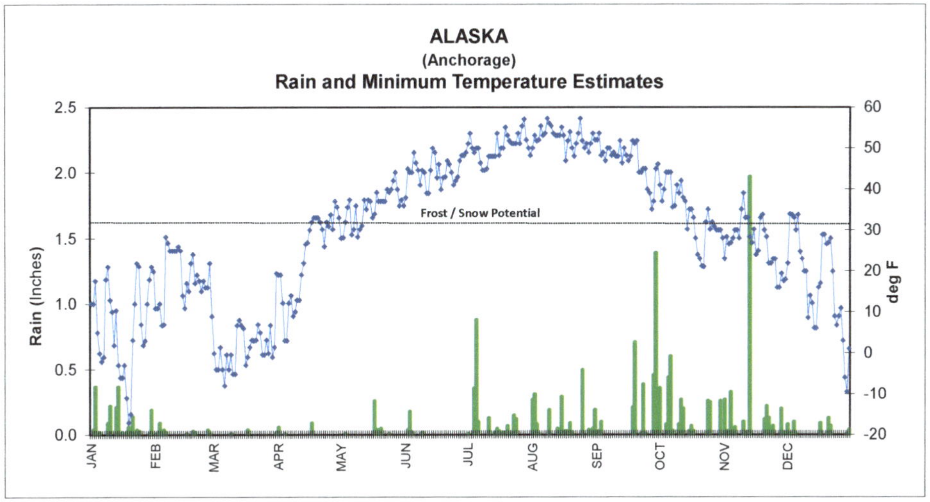

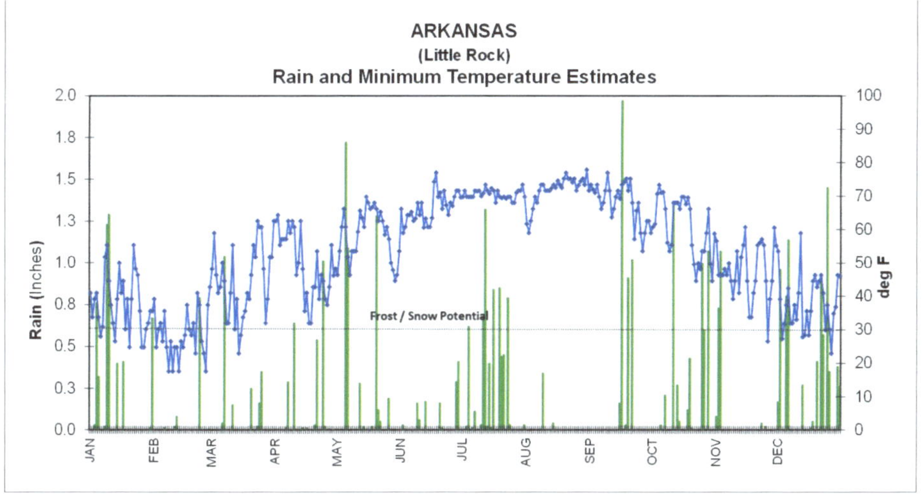

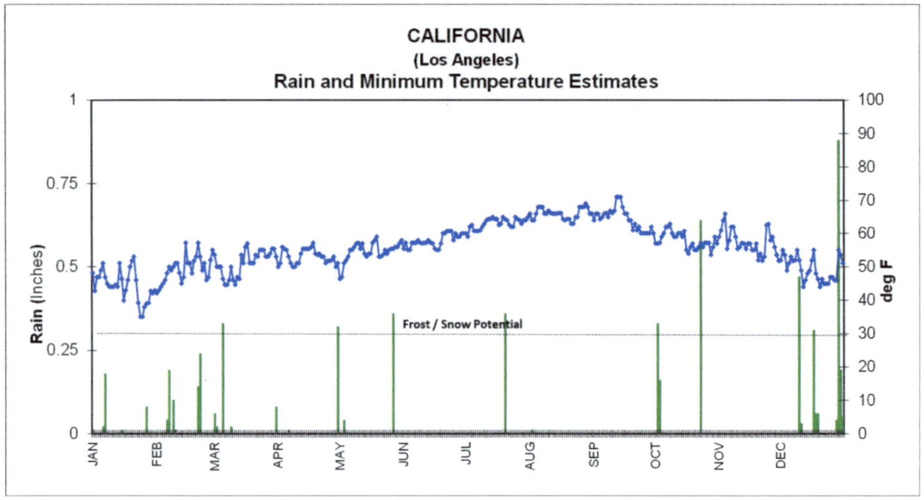
CALIFORNIA
(Los Angeles)
Rain and Minimum Temperature Estimates
Rain (inches)
deg F
Frost / Snow Potential
1
0.75
0.5
0.25
0
100
90
80
70
60
50
40
30
20
10
JAN
FEB
MAR
APR
MAY
JUN
JUL
AUG
SEP
OCT
NOV
DEC

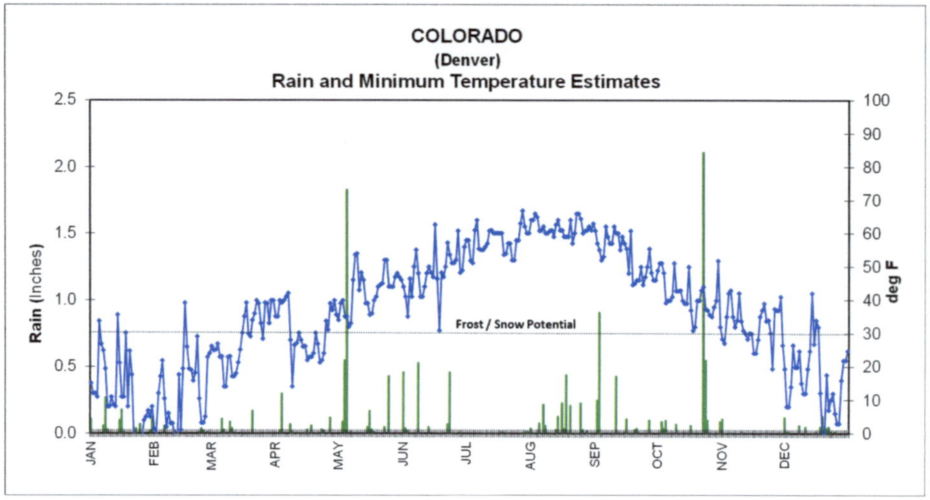
COLORADO
(Denver)
Rain and Minimum Temperature Estimates
Rain (inches)
deg F
Frost / Snow Potential
2.5
2.0
1.5
1.0
0.5
0.0
100
90
80
70
60
50
40
30
20
10
0
JAN
FEB
MAR
APR
MAY
JUN
JUL
AUG
SEP
OCT
NOV
DEC

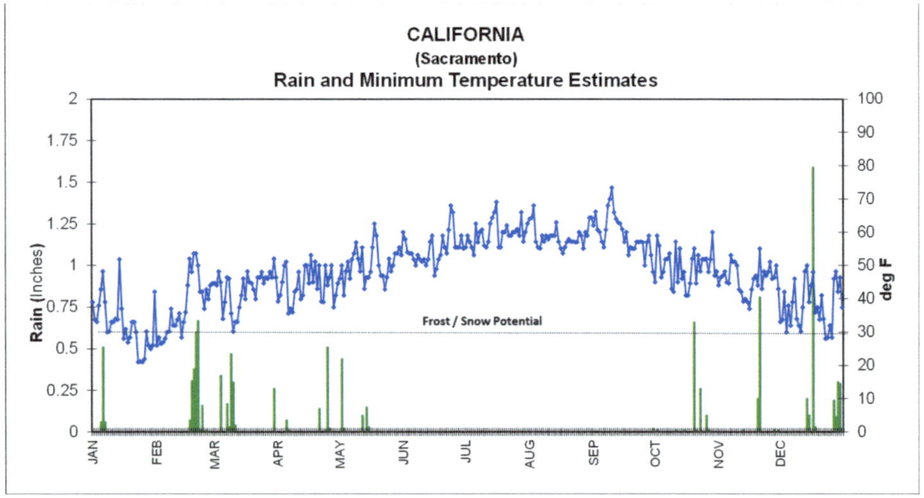
CALIFORNIA
(Sacramento)
Rain and Minimum Temperature Estimates
Rain (inches)
deg F
Frost / Snow Potential
2
1.75
1.5
1.25
1
0.75
0.5
0.25
0
100
90
80
70
60
50
40
30
20
10
JAN
FEB
MAR
APR
MAY
JUN
JUL
AUG
SEP
OCT
NOV
DEC

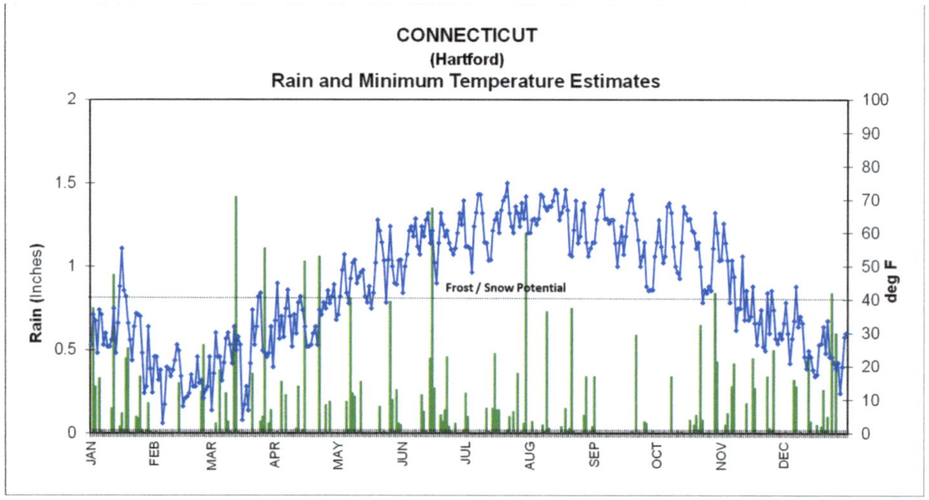
CONNECTICUT
(Hartford)
Rain and Minimum Temperature Estimates
Rain (inches)
deg F
Frost / Snow Potential
2
1.5
1
0.5
0
100
90
80
70
60
50
40
30
20
10
JAN
FEB
MAR
APR
MAY
JUN
JUL
AUG
SEP
OCT
NOV
DEC

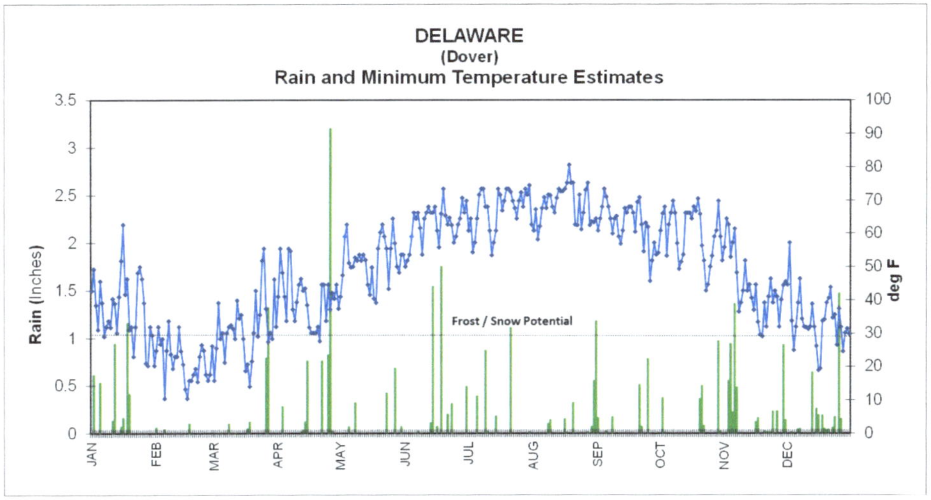
DELAWARE
(Dover)
Rain and Minimum Temperature Estimates
Rain (Inches)
deg F
Frost / Snow Potential
JAN
FEB
MAR
APR
MAY
JUN
JUL
AUG
SEP
OCT
NOV
DEC

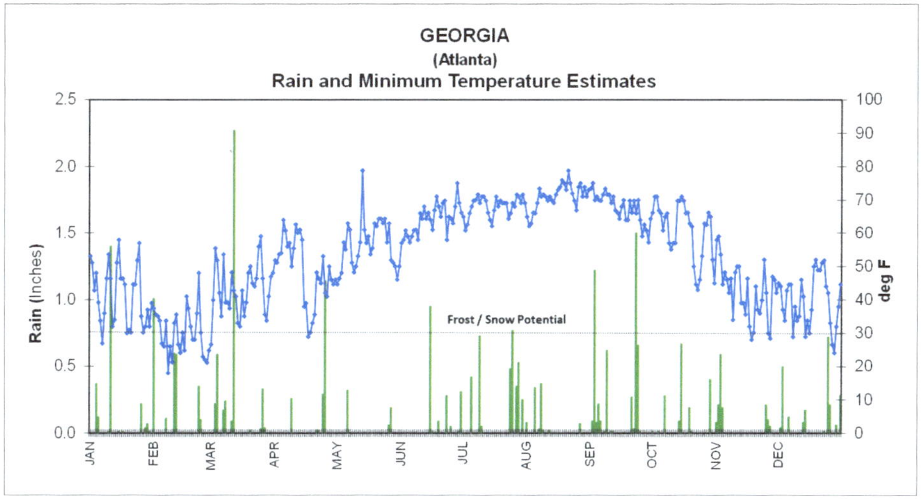
GEORGIA
(Atlanta)
Rain and Minimum Temperature Estimates
Rain (Inches)
deg F
Frost / Snow Potential
JAN
FEB
MAR
APR
MAY
JUN
JUL
AUG
SEP
OCT
NOV
DEC

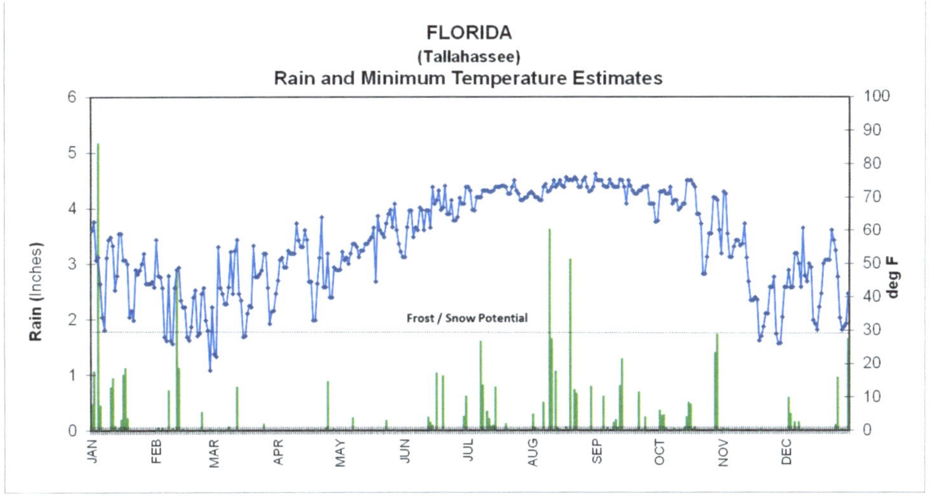
FLORIDA
(Tallahassee)
Rain and Minimum Temperature Estimates
Rain (Inches)
deg F
Frost / Snow Potential
JAN
FEB
MAR
APR
MAY
JUN
JUL
AUG
SEP
OCT
NOV
DEC

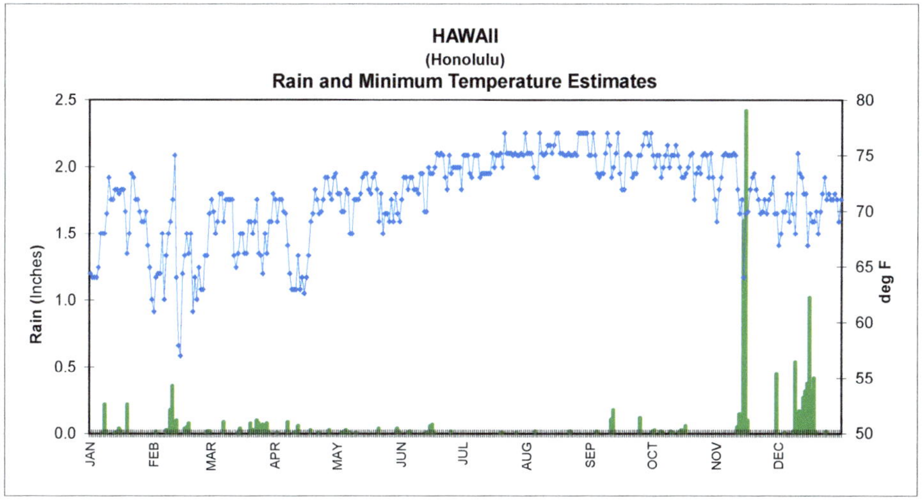
HAWAII
(Honolulu)
Rain and Minimum Temperature Estimates
Rain (Inches)
deg F
JAN
FEB
MAR
APR
MAY
JUN
JUL
AUG
SEP
OCT
NOV
DEC

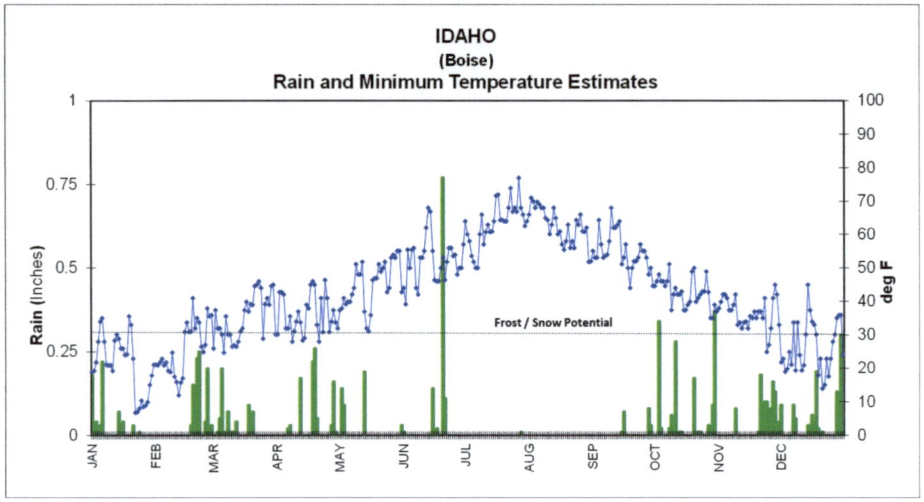
IDAHO
(Boise)
Rain and Minimum Temperature Estimates
Rain (Inches)
deg F
Frost / Snow Potential
JAN FEB MAR APR MAY JUN JUL AUG SEP OCT NOV DEC

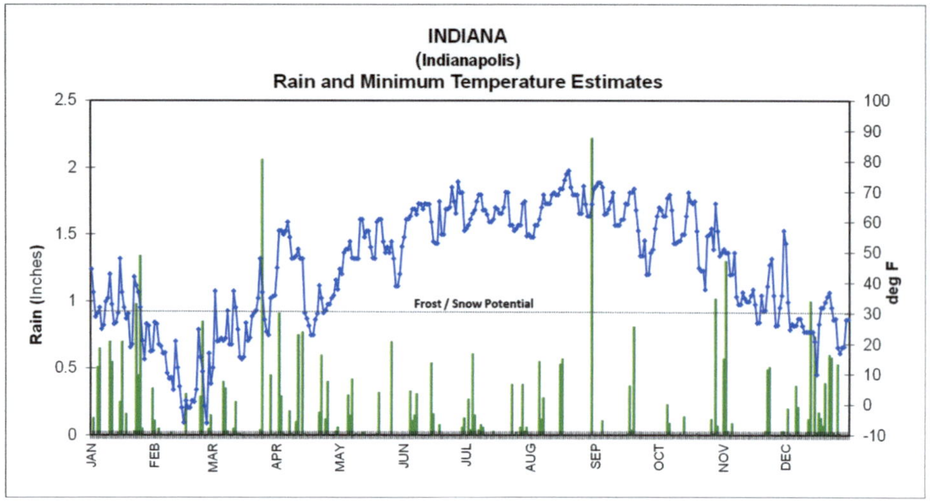
INDIANA
(Indianapolis)
Rain and Minimum Temperature Estimates
Rain (Inches)
deg F
Frost / Snow Potential
JAN FEB MAR APR MAY JUN JUL AUG SEP OCT NOV DEC

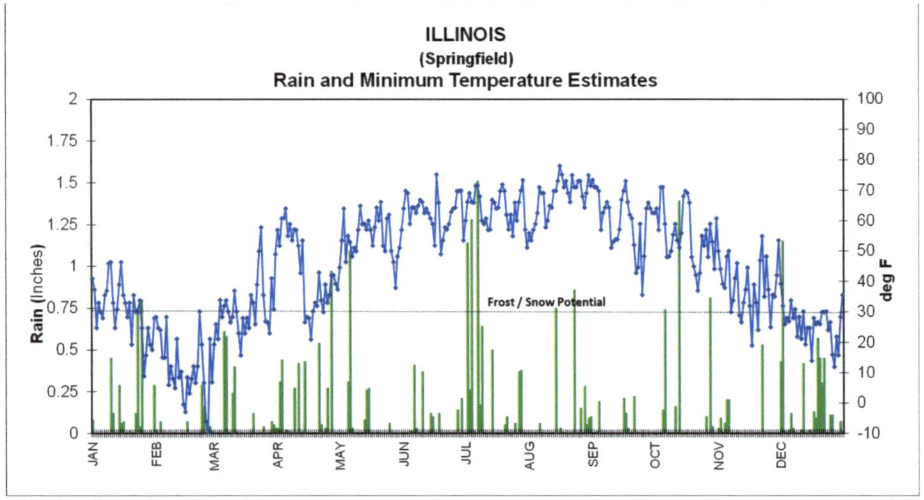
ILLINOIS
(Springfield)
Rain and Minimum Temperature Estimates
Rain (Inches)
deg F
Frost / Snow Potential
JAN FEB MAR APR MAY JUN JUL AUG SEP OCT NOV DEC

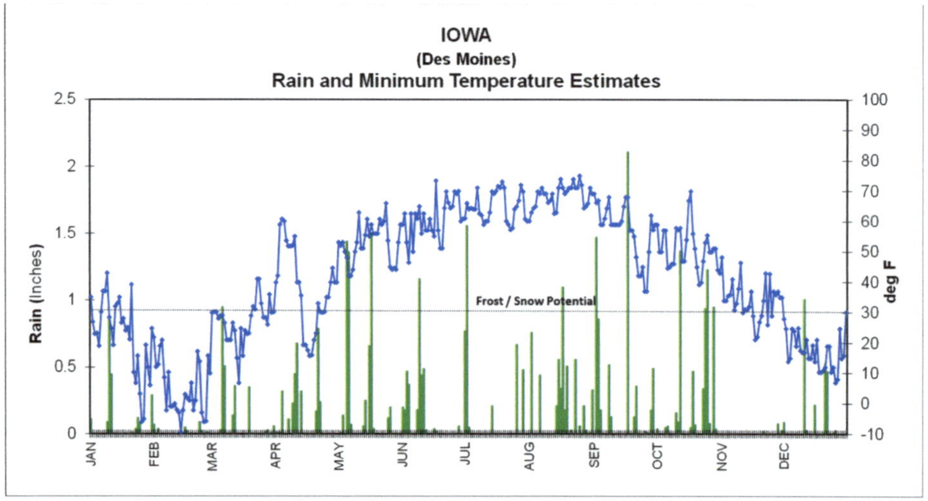
IOWA
(Des Moines)
Rain and Minimum Temperature Estimates
Rain (Inches)
deg F
Frost / Snow Potential
JAN FEB MAR APR MAY JUN JUL AUG SEP OCT NOV DEC

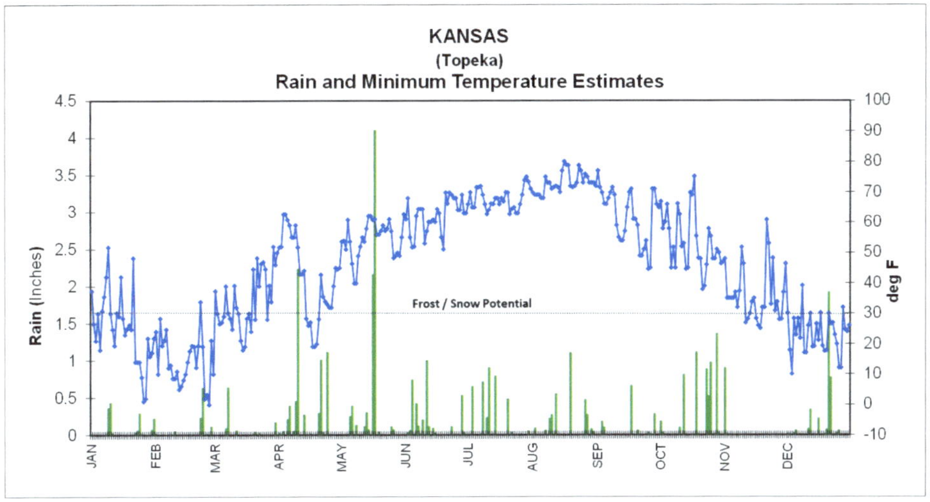
KANSAS
(Topeka)
Rain and Minimum Temperature Estimates
Rain (Inches)
deg F
Frost / Snow Potential
JAN FEB MAR APR MAY JUN JUL AUG SEP OCT NOV DEC

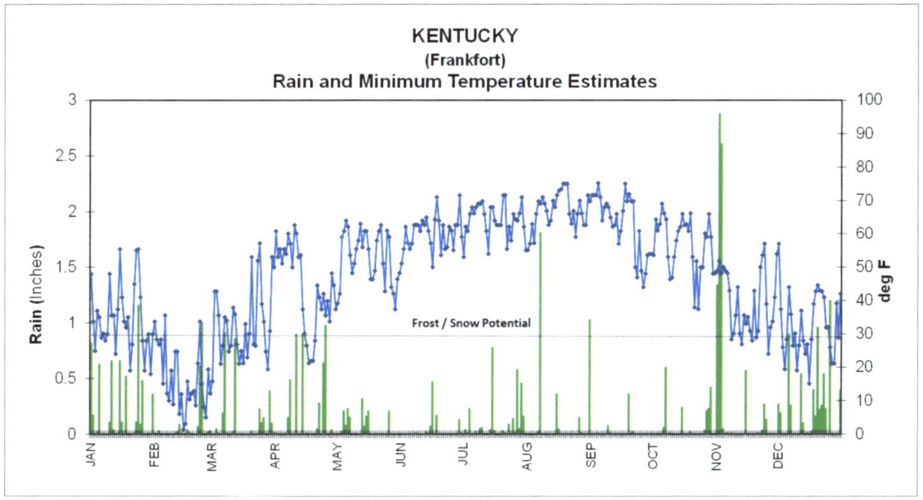
KENTUCKY
(Frankfort)
Rain and Minimum Temperature Estimates
Rain (Inches)
deg F
Frost / Snow Potential
JAN FEB MAR APR MAY JUN JUL AUG SEP OCT NOV DEC

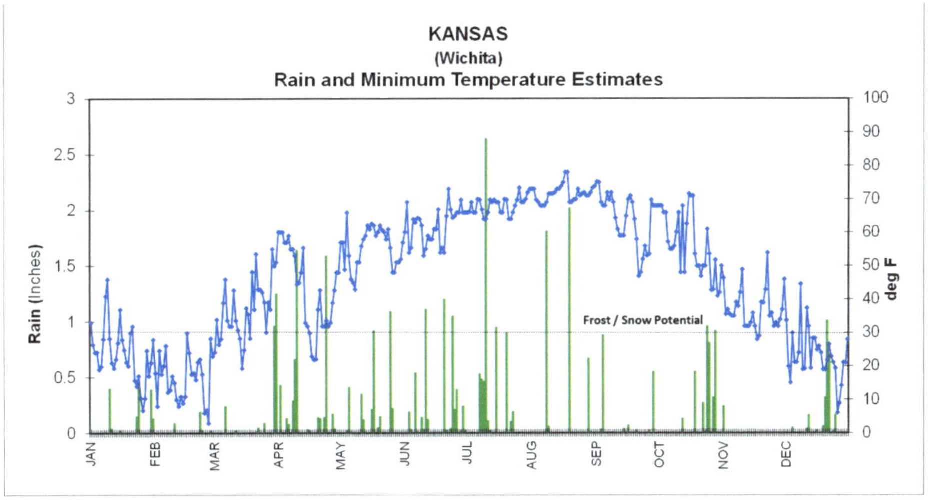
KANSAS
(Wichita)
Rain and Minimum Temperature Estimates
Rain (Inches)
deg F
Frost / Snow Potential
JAN FEB MAR APR MAY JUN JUL AUG SEP OCT NOV DEC

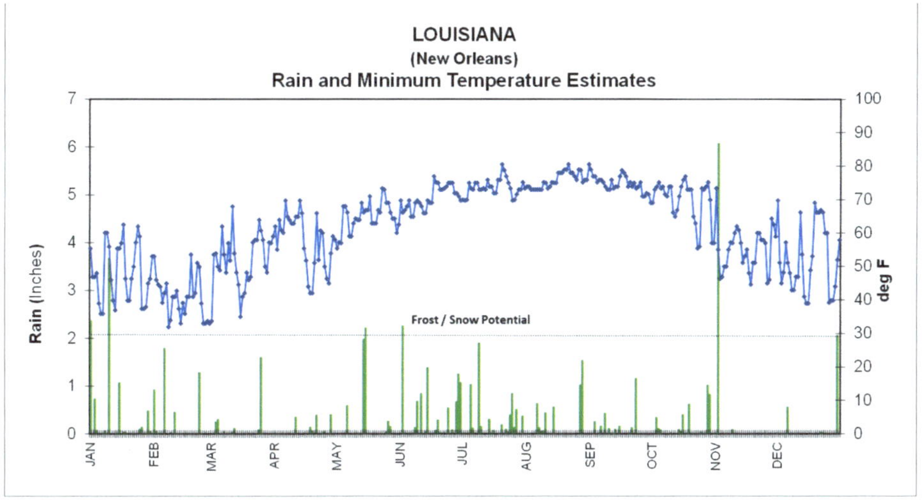
LOUISIANA
(New Orleans)
Rain and Minimum Temperature Estimates
Rain (Inches)
deg F
Frost / Snow Potential
JAN FEB MAR APR MAY JUN JUL AUG SEP OCT NOV DEC

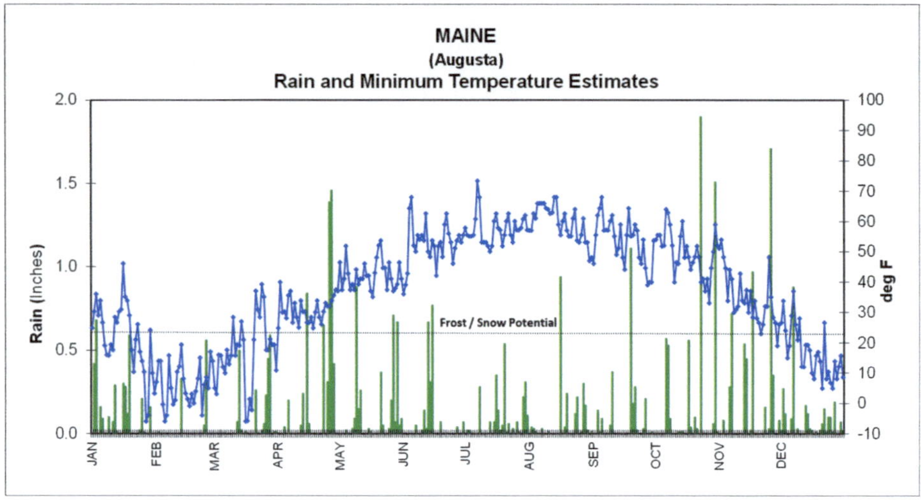
MAINE
(Augusta)
Rain and Minimum Temperature Estimates
Rain (Inches)
deg F
Frost / Snow Potential
JAN FEB MAR APR MAY JUN JUL AUG SEP OCT NOV DEC

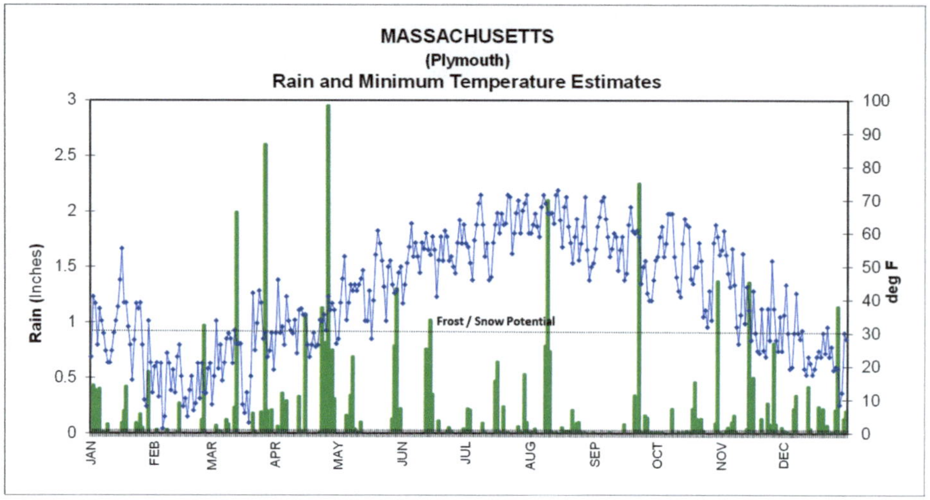
MASSACHUSETTS
(Plymouth)
Rain and Minimum Temperature Estimates
Rain (Inches)
deg F
Frost / Snow Potential
JAN FEB MAR APR MAY JUN JUL AUG SEP OCT NOV DEC

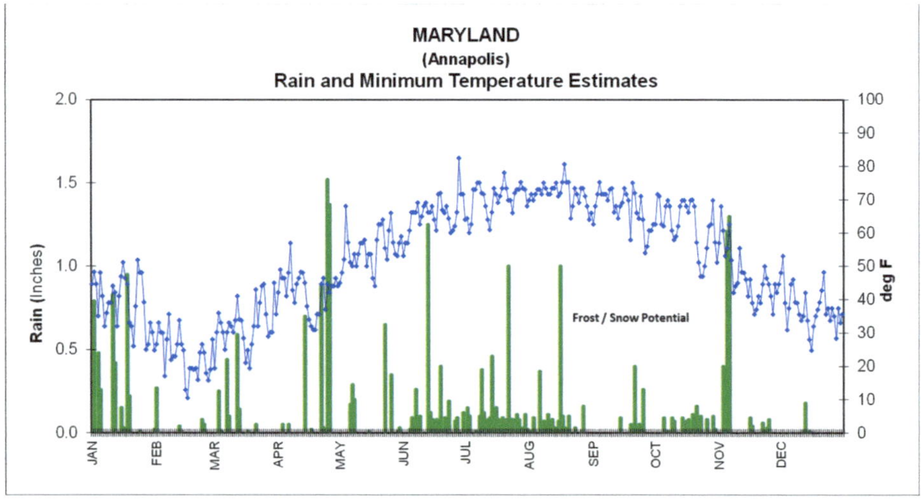
MARYLAND
(Annapolis)
Rain and Minimum Temperature Estimates
Rain (Inches)
deg F
Frost / Snow Potential
JAN FEB MAR APR MAY JUN JUL AUG SEP OCT NOV DEC

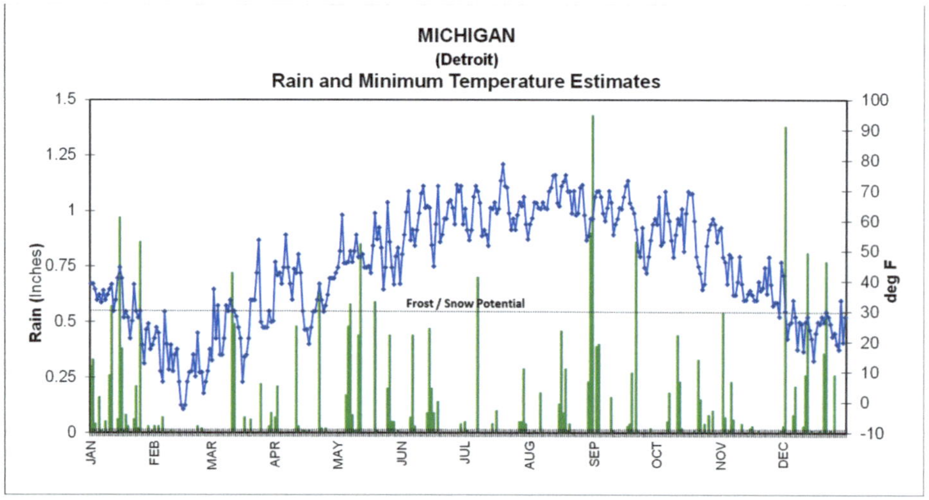
MICHIGAN
(Detroit)
Rain and Minimum Temperature Estimates
Rain (Inches)
deg F
Frost / Snow Potential
JAN FEB MAR APR MAY JUN JUL AUG SEP OCT NOV DEC

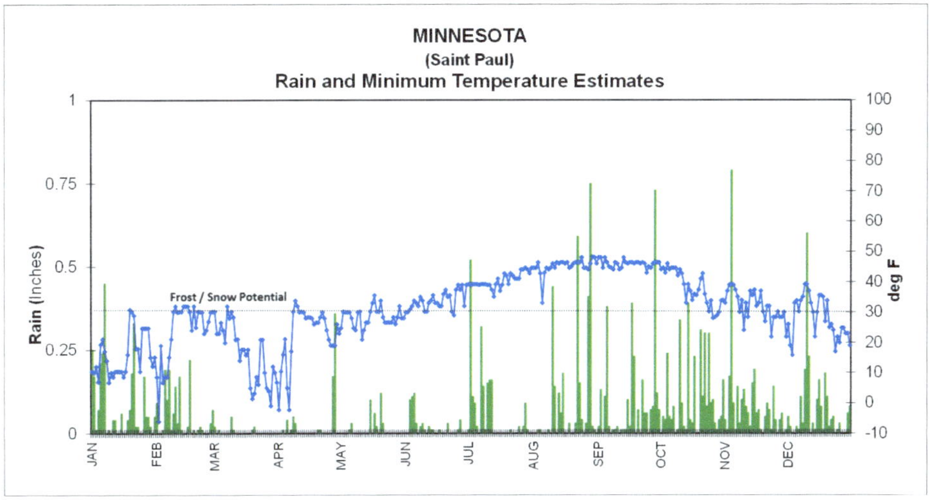
MINNESOTA
(Saint Paul)
Rain and Minimum Temperature Estimates
Rain (Inches)
deg F
Frost / Snow Potential
1
0.75
0.5
0.25
0
100
90
80
70
60
50
40
30
20
10
0
-10
JAN
FEB
MAR
APR
MAY
JUN
JUL
AUG
SEP
OCT
NOV
DEC

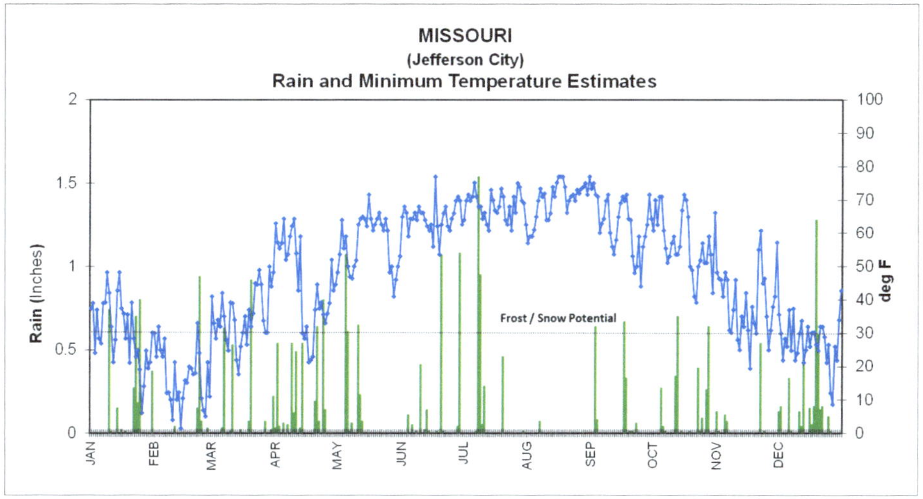
MISSOURI
(Jefferson City)
Rain and Minimum Temperature Estimates
Rain (Inches)
deg F
Frost / Snow Potential
2
1.5
1
0.5
0
100
90
80
70
60
50
40
30
20
10
0
JAN
FEB
MAR
APR
MAY
JUN
JUL
AUG
SEP
OCT
NOV
DEC

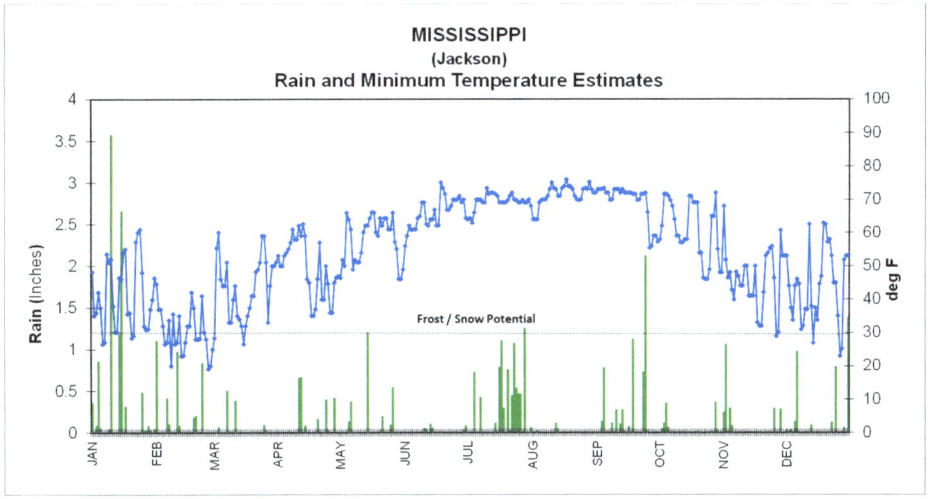
MISSISSIPPI
(Jackson)
Rain and Minimum Temperature Estimates
Rain (Inches)
deg F
Frost / Snow Potential
4
3.5
3
2.5
2
1.5
1
0.5
0
100
90
80
70
60
50
40
30
20
10
0
JAN
FEB
MAR
APR
MAY
JUN
JUL
AUG
SEP
OCT
NOV
DEC

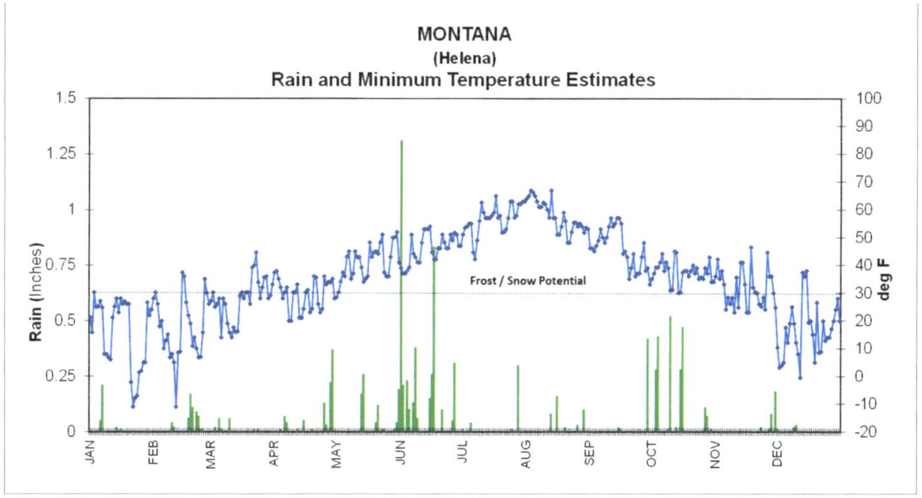
MONTANA
(Helena)
Rain and Minimum Temperature Estimates
Rain (Inches)
deg F
Frost / Snow Potential
1.5
1.25
1
0.75
0.5
0.25
0
100
90
80
70
60
50
40
30
20
10
0
-10
-20
JAN
FEB
MAR
APR
MAY
JUN
JUL
AUG
SEP
OCT
NOV
DEC

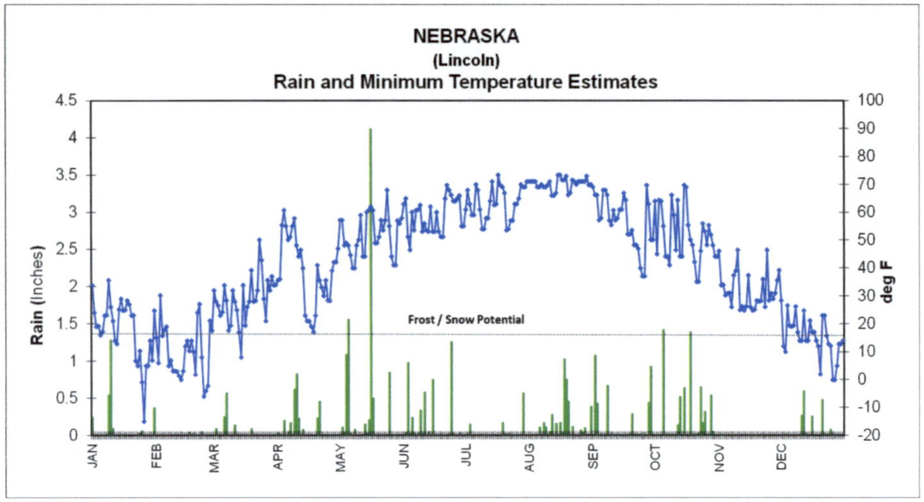
NEBRASKA
(Lincoln)
Rain and Minimum Temperature Estimates
Rain (Inches)
deg F
Frost / Snow Potential
JAN
FEB
MAR
APR
MAY
JUN
JUL
AUG
SEP
OCT
NOV
DEC

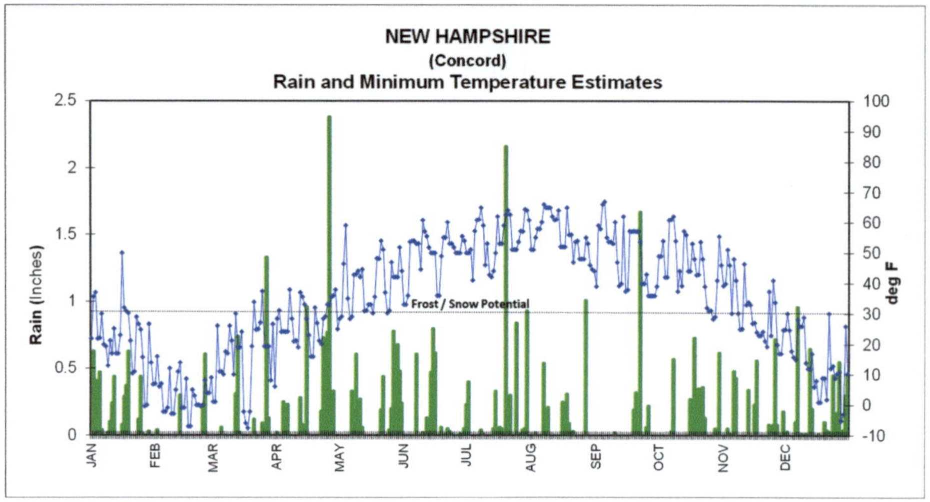
NEW HAMPSHIRE
(Concord)
Rain and Minimum Temperature Estimates
Rain (Inches)
deg F
Frost / Snow Potential
JAN
FEB
MAR
APR
MAY
JUN
JUL
AUG
SEP
OCT
NOV
DEC

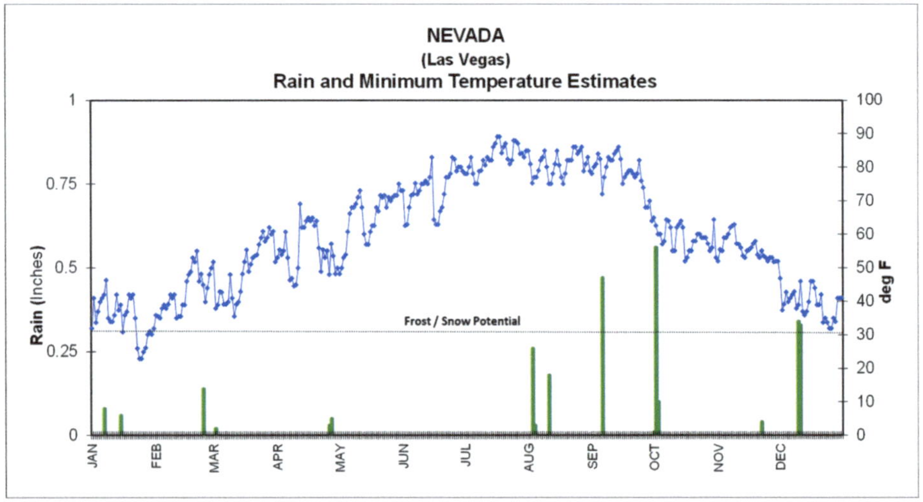
NEVADA
(Las Vegas)
Rain and Minimum Temperature Estimates
Rain (Inches)
deg F
Frost / Snow Potential
JAN
FEB
MAR
APR
MAY
JUN
JUL
AUG
SEP
OCT
NOV
DEC

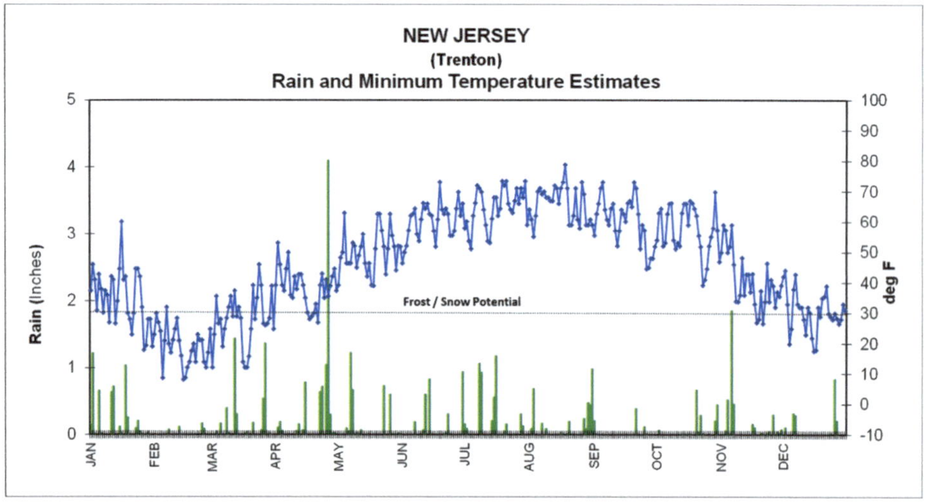
NEW JERSEY
(Trenton)
Rain and Minimum Temperature Estimates
Rain (Inches)
deg F
Frost / Snow Potential
JAN
FEB
MAR
APR
MAY
JUN
JUL
AUG
SEP
OCT
NOV
DEC

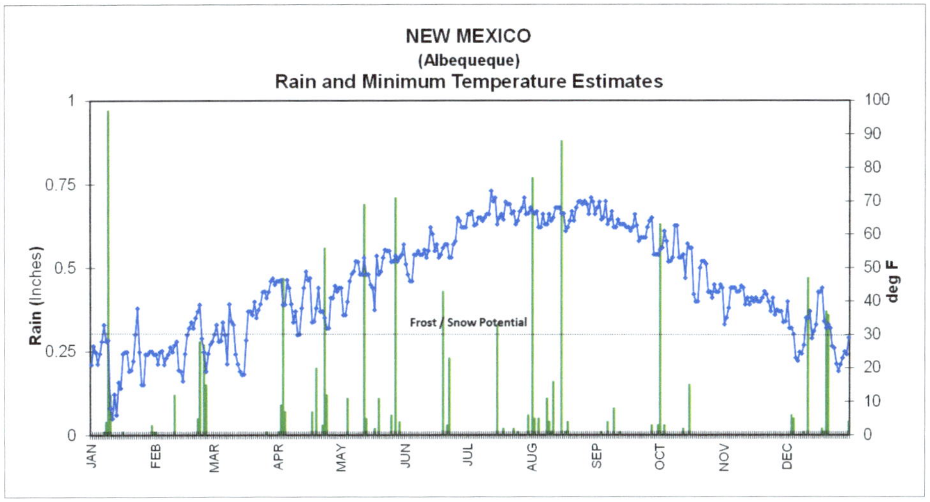
NEW MEXICO
(Albequeque)
Rain and Minimum Temperature Estimates
Rain (Inches)
deg F
Frost / Snow Potential
JAN FEB MAR APR MAY JUN JUL AUG SEP OCT NOV DEC

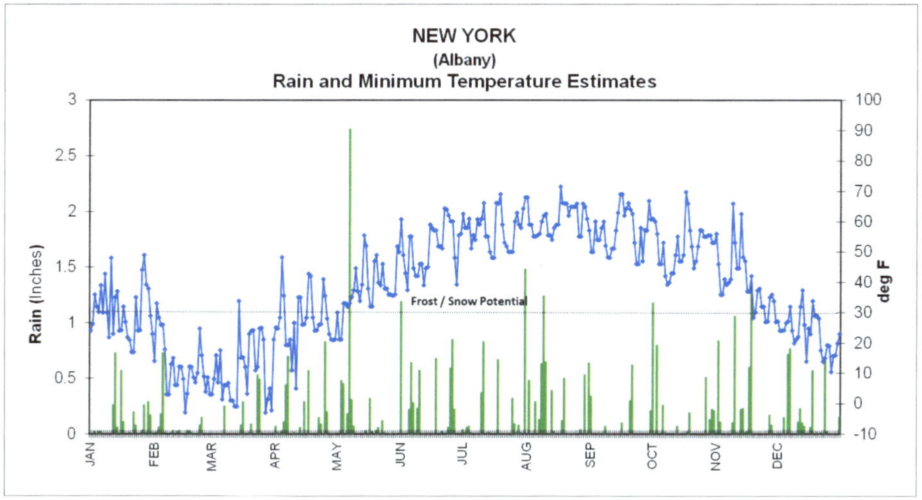
NEW YORK
(Albany)
Rain and Minimum Temperature Estimates
Rain (Inches)
deg F
Frost / Snow Potential
JAN FEB MAR APR MAY JUN JUL AUG SEP OCT NOV DEC

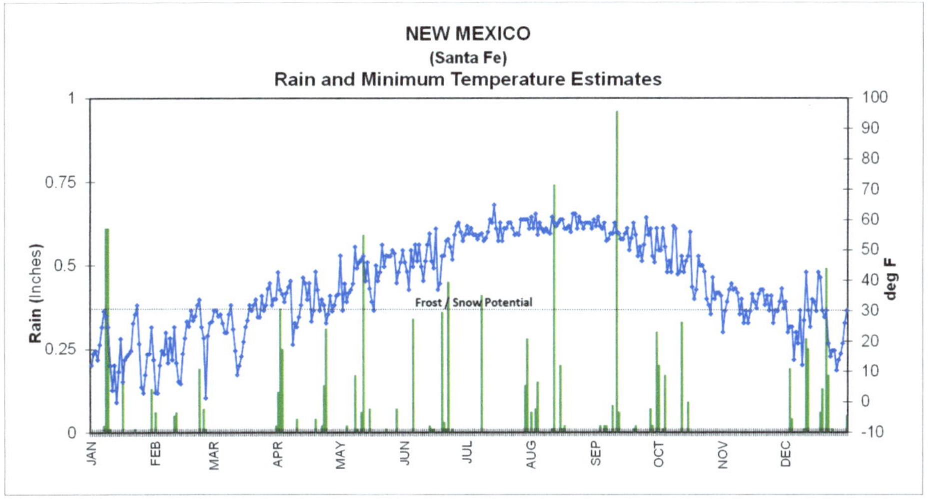
NEW MEXICO
(Santa Fe)
Rain and Minimum Temperature Estimates
Rain (Inches)
deg F
Frost / Snow Potential
JAN FEB MAR APR MAY JUN JUL AUG SEP OCT NOV DEC

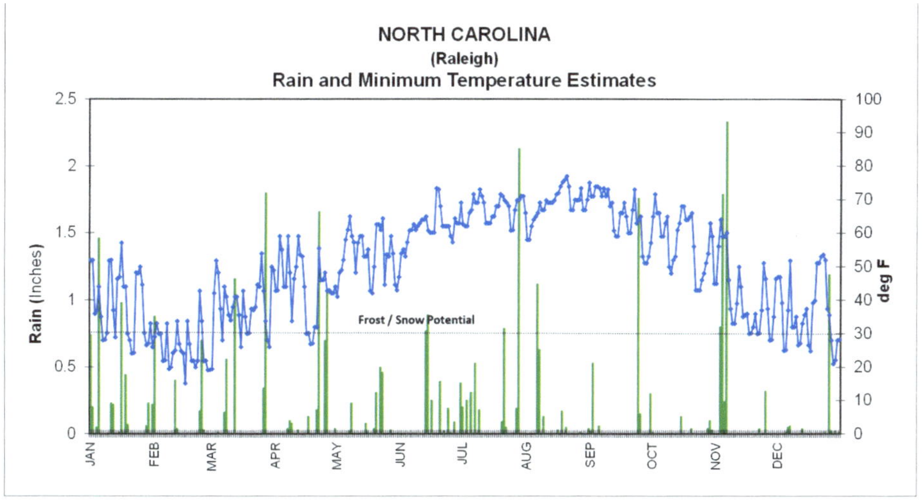
NORTH CAROLINA
(Raleigh)
Rain and Minimum Temperature Estimates
Rain (Inches)
deg F
Frost / Snow Potential
JAN FEB MAR APR MAY JUN JUL AUG SEP OCT NOV DEC

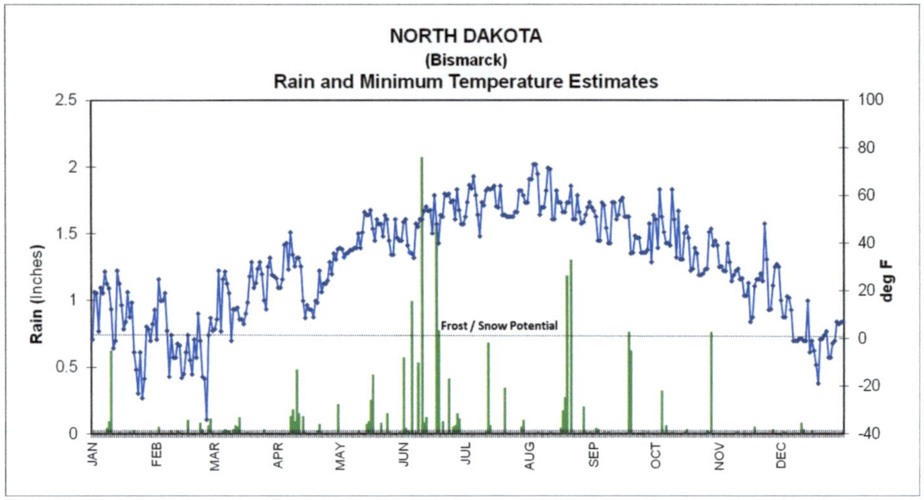
NORTH DAKOTA
(Bismarck)
Rain and Minimum Temperature Estimates
Rain (Inches)
deg F
Frost / Snow Potential
JAN
FEB
MAR
APR
MAY
JUN
JUL
AUG
SEP
OCT
NOV
DEC

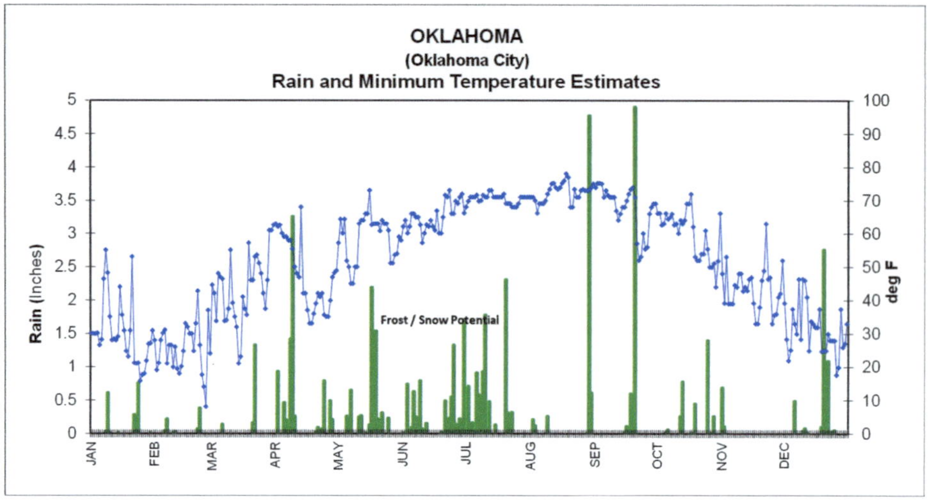
OKLAHOMA
(Oklahoma City)
Rain and Minimum Temperature Estimates
Rain (Inches)
deg F
Frost / Snow Potential
JAN
FEB
MAR
APR
MAY
JUN
JUL
AUG
SEP
OCT
NOV
DEC

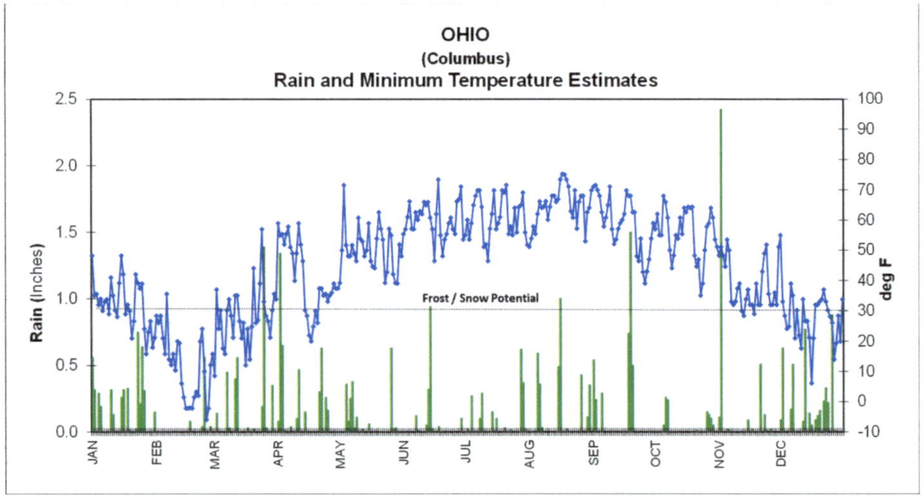
OHIO
(Columbus)
Rain and Minimum Temperature Estimates
Rain (Inches)
deg F
Frost / Snow Potential
JAN
FEB
MAR
APR
MAY
JUN
JUL
AUG
SEP
OCT
NOV
DEC

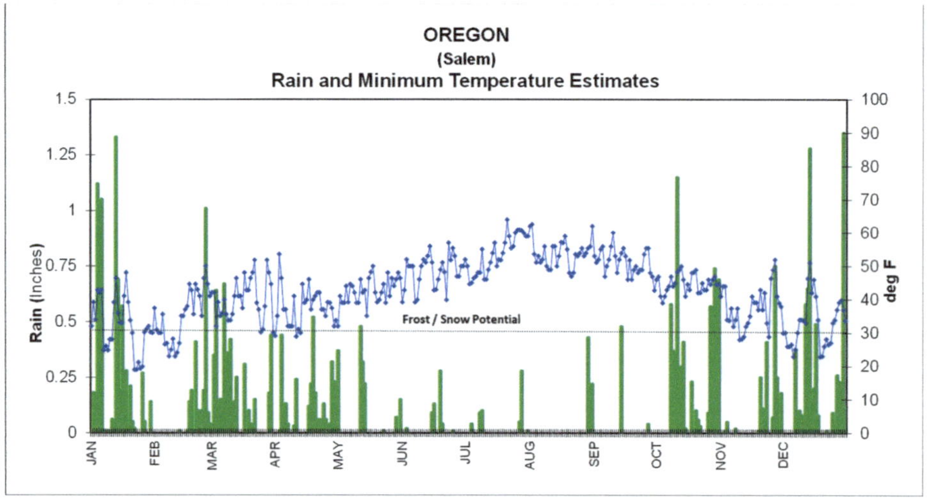
OREGON
(Salem)
Rain and Minimum Temperature Estimates
Rain (Inches)
deg F
Frost / Snow Potential
JAN
FEB
MAR
APR
MAY
JUN
JUL
AUG
SEP
OCT
NOV
DEC

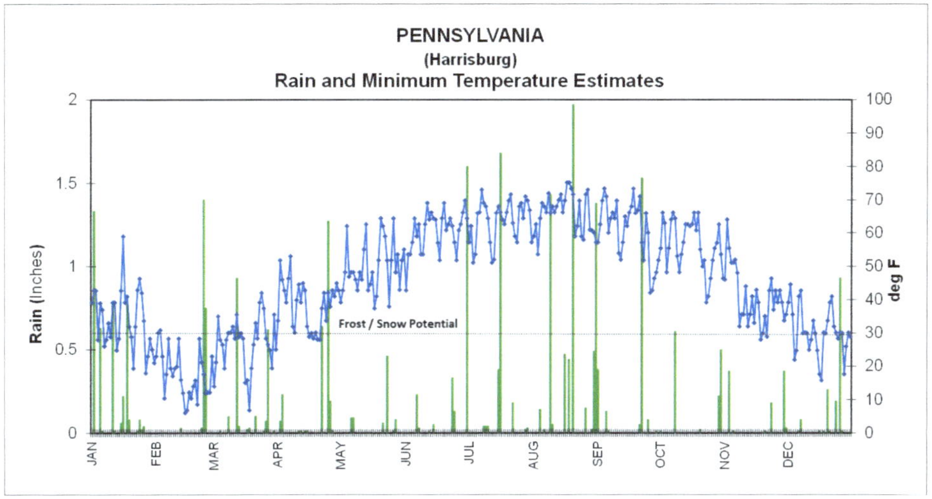
PENNSYLVANIA
(Harrisburg)
Rain and Minimum Temperature Estimates
Rain (Inches)
deg F
Frost / Snow Potential
JAN FEB MAR APR MAY JUN JUL AUG SEP OCT NOV DEC

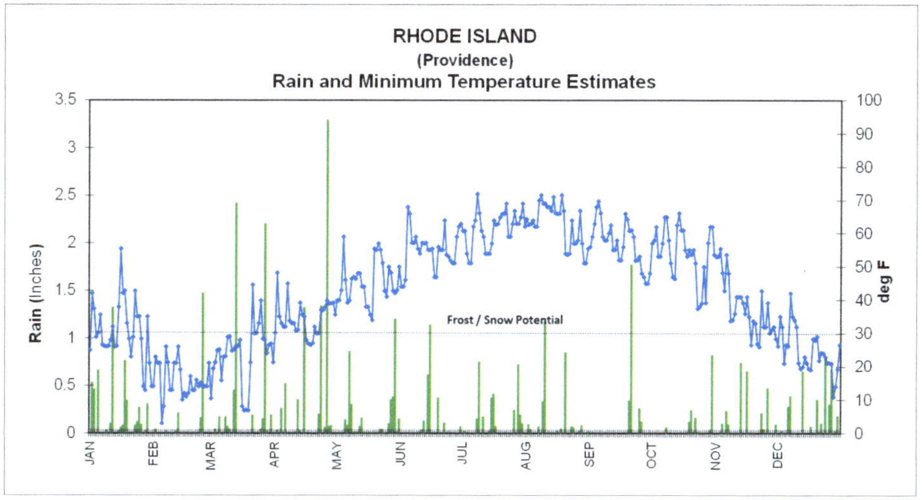
RHODE ISLAND
(Providence)
Rain and Minimum Temperature Estimates
Rain (Inches)
deg F
Frost / Snow Potential
JAN FEB MAR APR MAY JUN JUL AUG SEP OCT NOV DEC

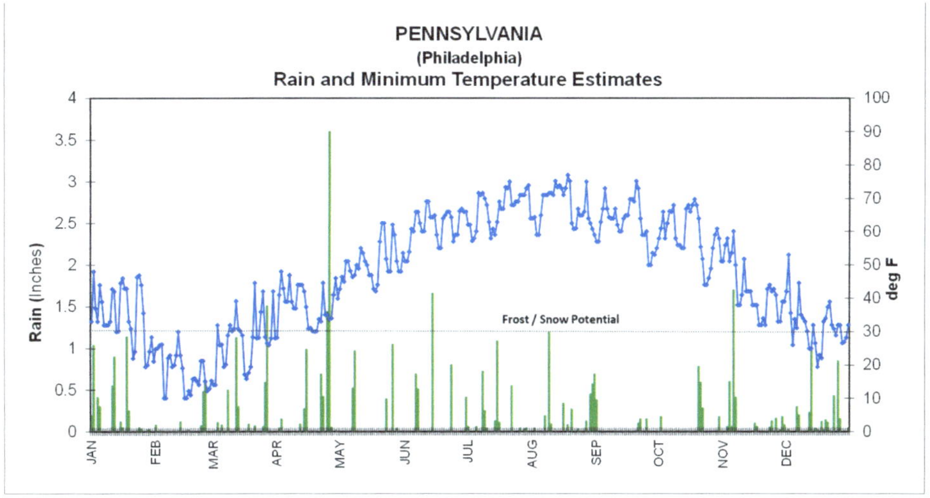
PENNSYLVANIA
(Philadelphia)
Rain and Minimum Temperature Estimates
Rain (Inches)
deg F
Frost / Snow Potential
JAN FEB MAR APR MAY JUN JUL AUG SEP OCT NOV DEC

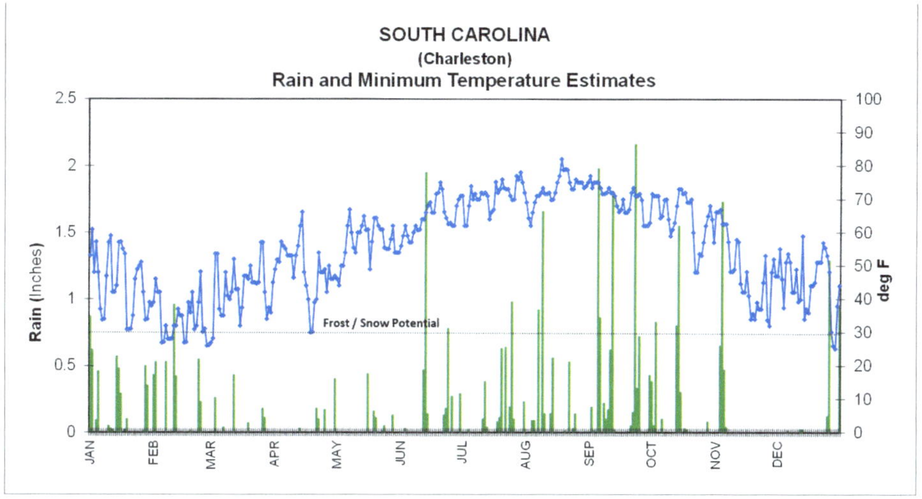
SOUTH CAROLINA
(Charleston)
Rain and Minimum Temperature Estimates
Rain (Inches)
deg F
Frost / Snow Potential
JAN FEB MAR APR MAY JUN JUL AUG SEP OCT NOV DEC

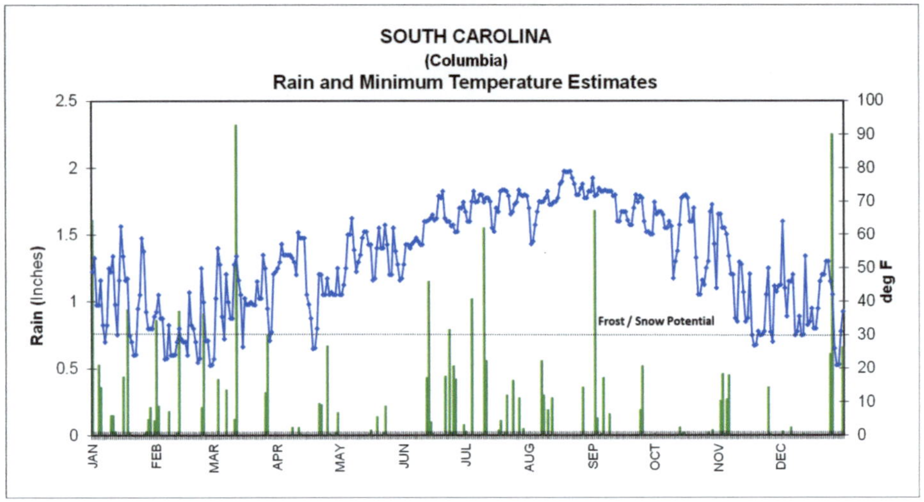
SOUTH CAROLINA
(Columbia)
Rain and Minimum Temperature Estimates
Rain (Inches)
deg F
Frost / Snow Potential
JAN FEB MAR APR MAY JUN JUL AUG SEP OCT NOV DEC

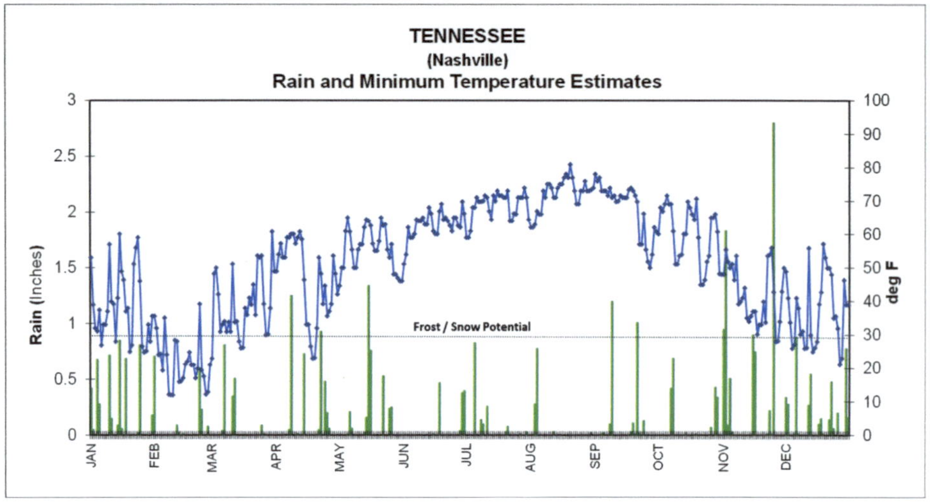
TENNESSEE
(Nashville)
Rain and Minimum Temperature Estimates
Rain (Inches)
deg F
Frost / Snow Potential
JAN FEB MAR APR MAY JUN JUL AUG SEP OCT NOV DEC

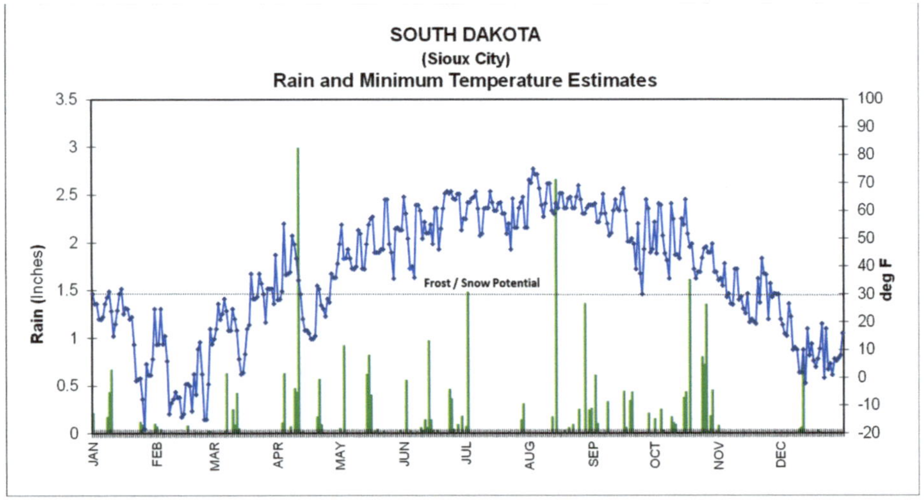
SOUTH DAKOTA
(Sioux City)
Rain and Minimum Temperature Estimates
Rain (Inches)
deg F
Frost / Snow Potential
JAN FEB MAR APR MAY JUN JUL AUG SEP OCT NOV DEC

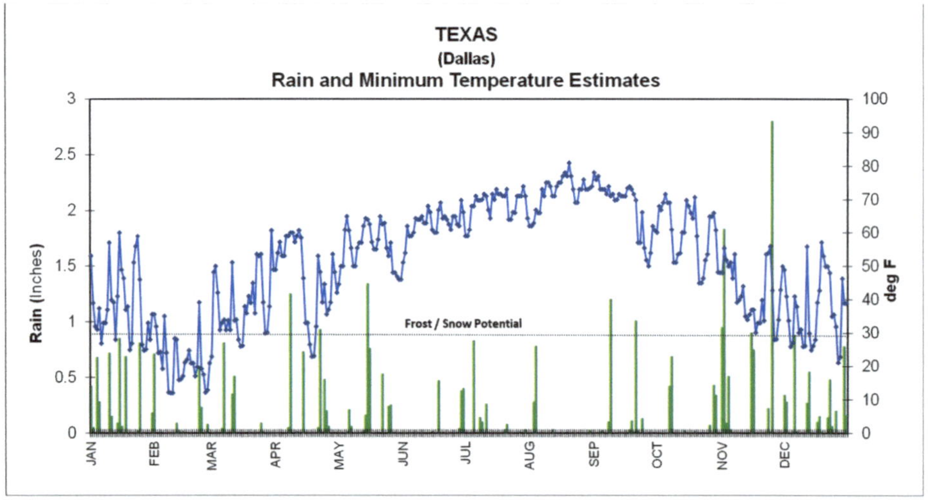
TEXAS
(Dallas)
Rain and Minimum Temperature Estimates
Rain (Inches)
deg F
Frost / Snow Potential
JAN FEB MAR APR MAY JUN JUL AUG SEP OCT NOV DEC

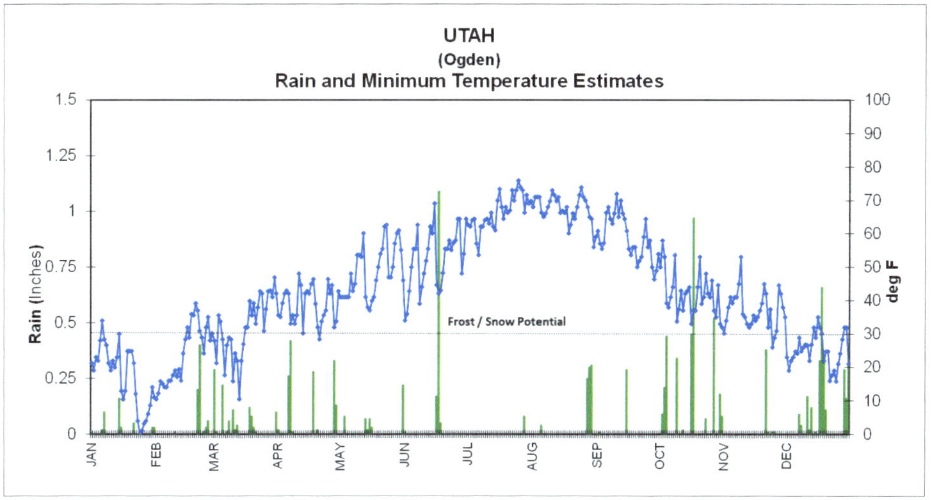

UTAH
(Ogden)
Rain and Minimum Temperature Estimates
Rain (Inches)
deg F
Frost / Snow Potential
JAN FEB MAR APR MAY JUN JUL AUG SEP OCT NOV DEC

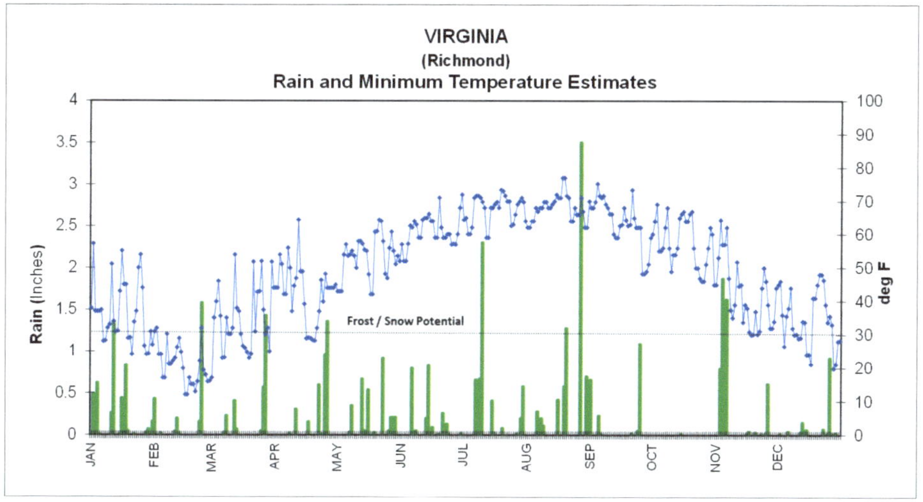

VIRGINIA
(Richmond)
Rain and Minimum Temperature Estimates
Rain (Inches)
deg F
Frost / Snow Potential
JAN FEB MAR APR MAY JUN JUL AUG SEP OCT NOV DEC

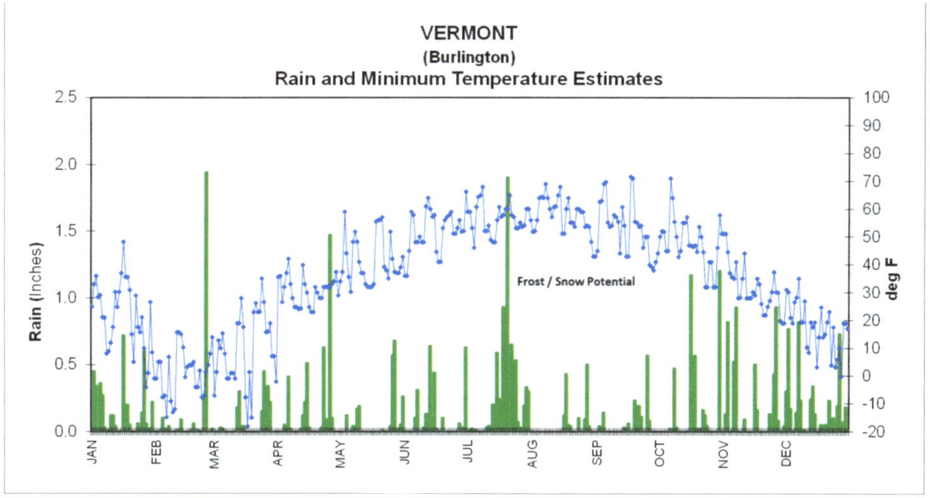

VERMONT
(Burlington)
Rain and Minimum Temperature Estimates
Rain (Inches)
deg F
Frost / Snow Potential
JAN FEB MAR APR MAY JUN JUL AUG SEP OCT NOV DEC

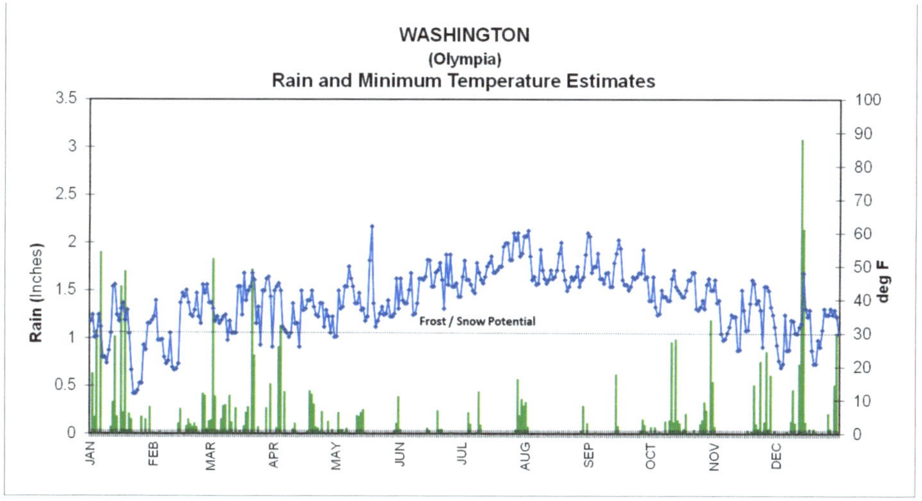

WASHINGTON
(Olympia)
Rain and Minimum Temperature Estimates
Rain (Inches)
deg F
Frost / Snow Potential
JAN FEB MAR APR MAY JUN JUL AUG SEP OCT NOV DEC

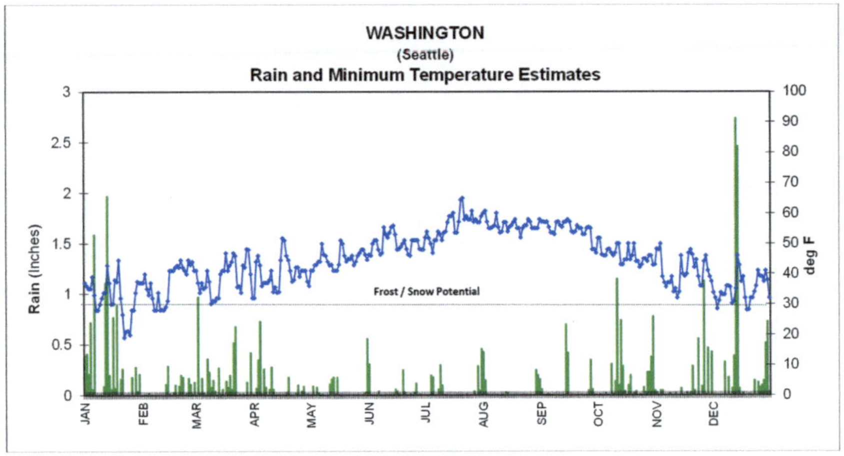
WASHINGTON
(Seattle)
Rain and Minimum Temperature Estimates
Rain (Inches)
deg F
Frost / Snow Potential
JAN
FEB
MAR
APR
MAY
JUN
JUL
AUG
SEP
OCT
NOV
DEC

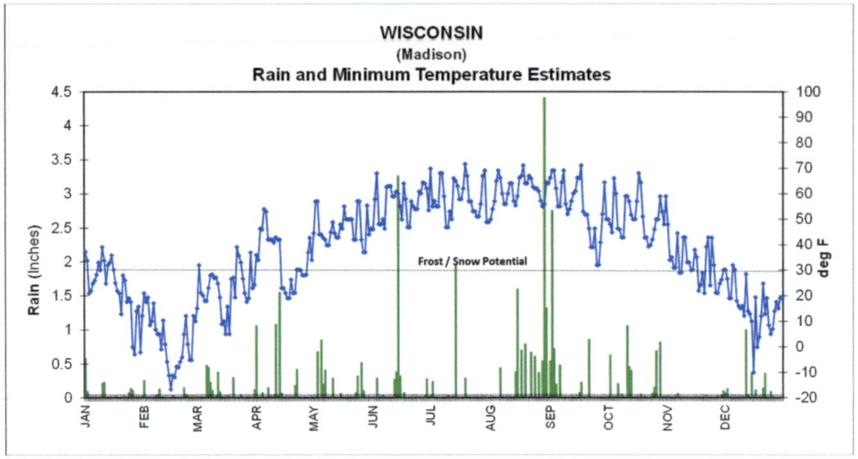
WISCONSIN
(Madison)
Rain and Minimum Temperature Estimates
Rain (Inches)
deg F
Frost / Snow Potential
JAN
FEB
MAR
APR
MAY
JUN
JUL
AUG
SEP
OCT
NOV
DEC

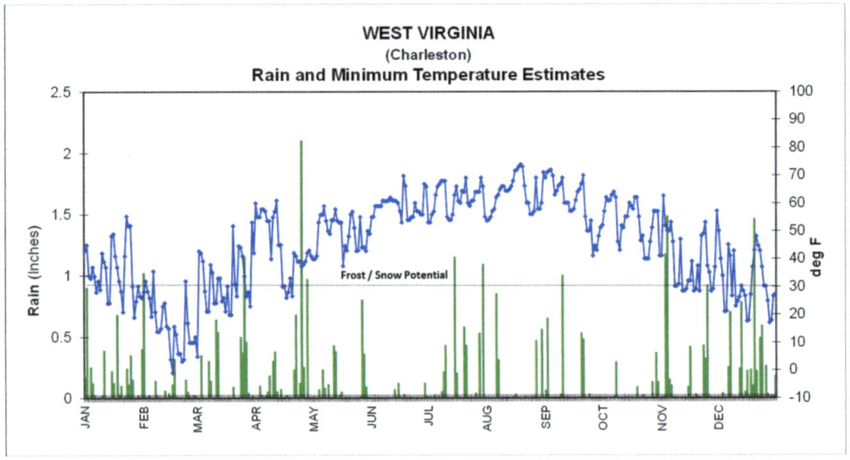
WEST VIRGINIA
(Charleston)
Rain and Minimum Temperature Estimates
Rain (Inches)
deg F
Frost / Snow Potential
JAN
FEB
MAR
APR
MAY
JUN
JUL
AUG
SEP
OCT
NOV
DEC

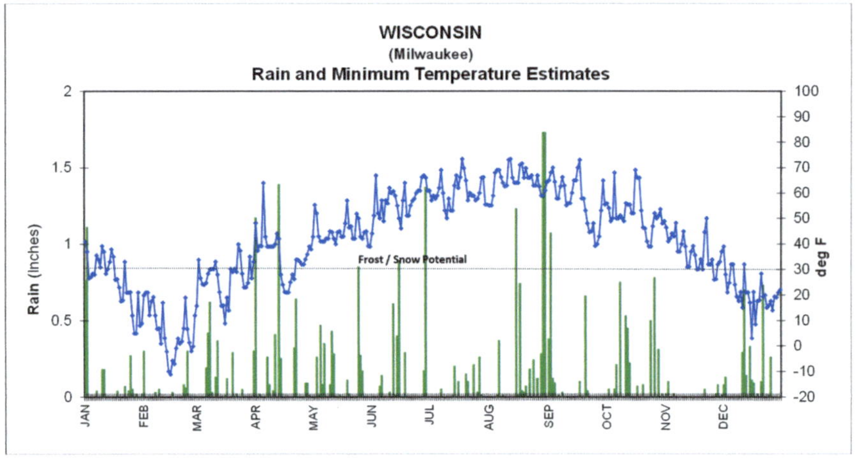
WISCONSIN
(Milwaukee)
Rain and Minimum Temperature Estimates
Rain (Inches)
deg F
Frost / Snow Potential
JAN
FEB
MAR
APR
MAY
JUN
JUL
AUG
SEP
OCT
NOV
DEC

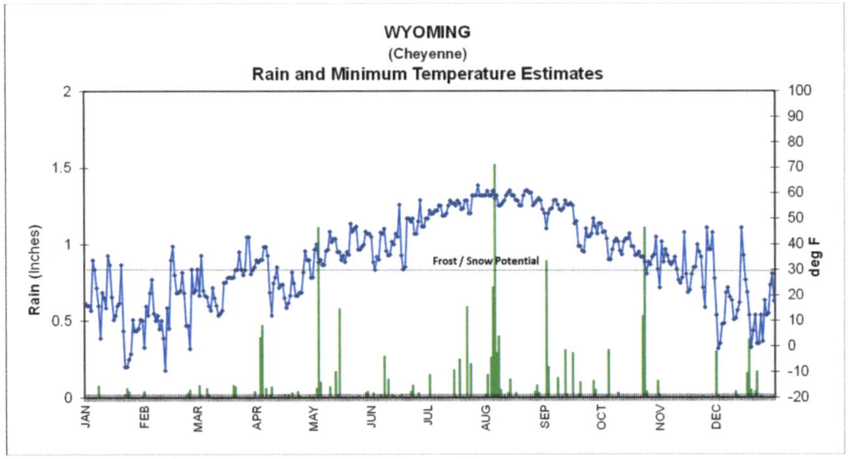
WYOMING
(Cheyenne)
Rain and Minimum Temperature Estimates
Rain (Inches)
deg F
Frost / Snow Potential
JAN
FEB
MAR
APR
MAY
JUN
JUL
AUG
SEP
OCT
NOV
DEC

UNDERSTANDING WEATHER BY THE MOON

WHAT CAUSES WEATHER

Weather is determined by astronomical and physical reasons, which are under relatively fixed control of cycles. It is not determined by any activities by Man. Climate scientists claim that the heavier-than-air "trace" gas, carbon dioxide, which is absent in 99.96% of the atmosphere; and methane, which is absent in 99.9999%, have a controlling effect on the climate. This has not been confirmed scientifically.

Approximately 80-85% of the weather, in this author's opinion, is caused by the moon, changing and repeating its position above the earth. Planning for major droughts or floods is therefore possible, by comparing the position of past lunar cycles, but not by measuring any emissions of a negligible gas.

Here is the weighting system. There is space weather, climatic weather, and everyday weather.

1. CHANGES TO THE WHOLE PLANET: (IRRESPECTIVE OF GEOGRAPHY)

- Deviation of Earth's rotational axis to the vertical.
- Variation of Earth's orbit.
- Magnetosphere
- Heliosphere

2. CLIMATE AT LOCATION: (LONG-TERM TRENDS IN WEATHER PATTERNS)

- Distance from the equator (angle of the Sun's rays = latitude)
- Direction of prevailing winds
- Topographic factors – the shape of the land, proximity to other land masses, elevation, and distance from the nearest coast.

3. WEATHER AT LOCATION: (TRENDS AND DAILY OCCURRENCES, OBSERVABLE AND RECORDABLE)

- Influence of Moon on tides (including air tides), cycle of rain from ocean, and back to ocean. (80-85%), also oceanic cycles due to lunar influence, e.g. El Nino.
- Level and intensity of cloud cover (retains local heat, determines what, where, when, of rain (which in turn is subject to ocean and air tides).
- Solar winds, cycles of Sunspots (contributing to temperature of seasons), hemispherical and seasonal factors.

4. NOT IN THE MIX AT ALL, EITHER FOR CLIMATE OR WEATHER

- Land use changes, deforestation and greenification through agriculture.
- Volcanic activity ongoing, which does not alter climate.
- Ice-reflectivity feedback, constant.
- Industrial and commercial pollution.
- Other activities of Man e.g. diet.

Immediate weather is modified by cycles of solar activity (position and number of occurring Sunspots), and the cyclic position of the Moon relative to the earth. Solar influences can determine floods and drought periods.

The weather has no brain, so it does not know that there is a small percentage of earth that is not covered in the ocean, called the Land. To the moon, it is all oceans. From the height rain falls from, up to 12km, it is immaterial whether there is land below or ocean. That's why rainforests make no difference to rainfall amounts. Forests grow where it rains often, not vice versa.

About 85% of all the weather is controlled, directly or indirectly, by the Moon. If you live by the ocean, then the influence is more obvious and direct. If you don't live on or near the coast, then weather systems take a little longer to reach you. But, you can see, in a broad sense, the influence of the moon. For example, on almost every full moon night, you will see the full moon in all its glory. And that's because, on the night of the full moon, the moon clears the sky of clouds close to midnight, by dragging more air underneath it. This pushes away the cold air of space and prevents condensation of water vapor that makes the clouds visible. The moon controls the weather every night, however, it is just not as obvious.

When the moon is in the sky i.e. above the horizon, it is less likely to rain. This is a Rule. The main times of rain (if rain is about), are when the moon is underfoot (IC), overhead (MH), or on the horizon (ASC=rising or DES=setting). The in-between positions i.e. every three hours, also have rain potential. Sir Isaac Newton identified that when the moon reached a 45° angle to the horizon, a gravitational force corridor was established, and gravity was lessened, i.e. the moon effect was enhanced. Just before it rains, there is a noticeable chill in the air. There are also extra wind gusts in the hours of moonrise and moon set. This was first noticed and recorded by the Greek philosopher Theophrastus in 300 BC. It still applies today. When the trees suddenly rustle for no reason, it may be close to moon-rise.

ORIGINS OF A STORM

All adverse weather comes from *under the boat*. I have never heard a sailor say, 'aw, a violent storm was going on above me, whilst we were sailing *on a flat, calm sea*.' Usually, the flat sea accompanies a high-pressure system. Higher air pressure pushes the sea flat. It is as if your hand was the size of a bath and pushed down on the bath's surface. But the flatter sea arrives first and allows more stable air to gather and consolidate over it, exerting more pressure.

It has been further observed that the deeper a low-pressure system is in the air, the faster the wind blows, imparting more energy, and the longer the wind blows over the same area, the bigger the swells, which are the strong underwater currents. It begins in the sea. If currents increase in the deep of the ocean, then waves will increase at the surface and disturb the air over long distances, enough to generate those faster winds.

Strong winds will always accompany a swell. That's because the waves, which arise from the swell, cause the wind, The etymology says it all. Wind, weather, and wing are words that come from the word *wave*, not the other way around. Fishermen have noticed that the direction in which the wind is blowing matters. The word *west* comes from wind as well. So does the word weather. When there is an onshore wind coming off the sea, underwater waves are being whipped up, which disturbs marine worms out of their burrows, dislodges shellfish such as mussels, limpets, and cockles from their home on rocks, and forces small fish, crabs, and other forms of marine life out of weed beds. The currents change first. How could a surface wind possibly affect a seabed? Surface currents do not transmit downwards to the depths of the ocean.

To prove this theory, put some sand along the bottom of a bath and then splash along the surface, and see if the sand moves. Now place floating things, like corks, on the surface, and generate underwater currents with your hand, and see if the floating things are affected. Which moves more? There is no question that the energy transmission in the ocean is always upward. It means that winds are more likely to be created by currents and swells than vice versa. It would be reasonable to conclude that a clear day indicates little wind BECAUSE the sea is calm, indicating deeper currents have lessened. Deep-sea fish may have a sense of impending stillness in the tides and may plan to be there when small creatures that they prey on emerge from nooks and crannies.

An experienced fisherman can tell when there is a storm coming. The ocean 'speaks' to him and the sound of the swell reaches him first.

Deep ocean currents are a fact, but science does not link them to winds. Seismic activity generates ocean currents, and earthquakes are energy waves up to 20,000 mph through solid rock. It is not likely that ocean currents go downwards to generate earthquakes, and besides we already know that some earthquake epicenters originate about 400 or 500kms below the seafloor.

It is more likely that the wave regimes that earthquakes are part of, transmit through rock to vibrate the ocean floor enough to generate the ocean currents, and on into the air via oscillating waves to generate unstable airflows. The fact that larger earthquakes happen around full moons and what we call perigees, together with stronger currents and king tides, attest to that, also cyclones form around these times. The moon and larger gravitational forces are responsible for all three. So, when it comes to what happens at sea and in the air above the surface of the sea, it is more feasible that ocean currents are in charge, rather than the surface winds.

As soon as the temperatures reverse, the winds reverse. Sailors who monitor Sea Surface Temperatures (SSTs) know this to be more reliable than isobaric maps because sea-surface temperatures change before changes in air pressure, which gives warning of storms before they happen. Just by watching the temperature of the water, you can tell what the weather will do, in the next few hours or longer.

To recap, then, a typical storm begins with the flat sea, the Sun shining above, heating the surface, accompanied by non-mixing of waters at various depths, relatively warm sailing conditions, and SSTs (sea surface temperatures) starting high. The surface of the sea is warm because the layer of water constituting the surface is not being replaced by cooler water. Layers of water closer to the surface are the warmest, being closer to the warmth of the Sun. Then, as mixing begins below the surface, cool water is brought up from below, causing surface cooling. Further mixing brings more cooler waters up from the depths and there is an interchange with the warm waters close to the surface.

Due to the drop in temperature at the surface, clouds may form as cooler air mixes with the remainder of warm air. Because warm air always moves to cold, the air begins to move, and wind "picks up". There is a storm brewing. Warmer air continues to supply the area which has now considerably cooled, and the surface is described as windy, and then as a gale. The clouds increase and darken with the load of water that they gather. Condensation accompanies cooling, supplied by the movement of air. The surface of the sea begins to ripple strongly as the swell increases. Each wave has a convex and concave side. The convex has a higher pressure and the concave has a lower pressure. Waves are created, with higher pressure moving up each crest.

The waves are joined together, and big areas of high pressure are juxtaposed with big areas of low pressure, and strong winds are generated which go from high pressure to low pressure. The storm keeps up if it is fed by the waves and currents, which are the dynamics of the moon, interacting with gravitational factors that produce underwater earthquakes and faster swells, e.g. new moon, full moon, perigees, and when the moon crosses the equator, which is called the lunar equinoxes.

The storm dissipates in reverse order. Winds lessen and the sea surface temperature rises. Mixing below the surface ceases gradually. The Sun comes out as clouds seem to dissipate and the ocean surface flattens. It is all controlled by the size and direction of the swells. Sailors know that to escape a storm, you sail into the incoming wind, which is into the whitecaps. They have known it for thousands of years, and there are ancient accounts in ancient texts, but they may not have realized why.

If we ignore the moon then how could a huge celestial body between a third and a quarter the size of Earth, and only ten Earth circumferences away, not have any effect on Earth, sea, and the air? Astronomers may have let us down over the years. They say that the moon only affects the tides, but nothing else. If it affects something like tides on the earth, then why does it not affect other things that are on the earth too, like humans? And how could all three layers (air, sea, and land) not be interconnected - considering they completely and perfectly interface over most of the surface of the planet? How could it be that only one could get away with having a tide, but not the other two, when all three are joined? Aren't joined things part of a joined system that acts together? Science should answer all questions, for that is the essence of scientific pursuit.

CYCLONES AND HEAVY RAIN

Cyclones are formed in countries when the sun within 6° latitude of the equator warms the waters that must be 26°C -28°C, over a wide ocean area. And that produces massive evaporation. The heat from a full moon day, especially when the moon is close, in certain areas can create a tropical cyclone close to the equator. When the full moon's heating capacity diminishes, then so will the cyclone. The huge accumulation of water in the sky, ready to fall and cause destruction, can only come from the sea.

That is why they immediately begin to fall apart, the minute that full moon moves away. And that's why cyclones don't form south of Fiji or in Europe. The water is just not warm enough That's why you never get cyclones occurring in Ireland.

Cyclones travel from either north or south depending on which hemisphere they are formed in. They will very rarely cross the equator.

Cyclones, hurricanes, and typhoons predominate during equinox months, which are mainly when the atmosphere undergoes a twisting action, through the plane of what we call the ecliptic, due to the increase in speed of the Sun relative to the orbit of the earth. If you can picture the midpoint of a pendulum, that happens in both March and September. Furthermore, with spring and autumn full and new moons, king-tides occur when the moon rises at or near the equator. This represents the midpoint of yet another pendulum, the Sun on the equator being the first, and now the Moon on the equator for the tides. Sun and moon are moving faster relative to earth around equinoxes, so atmospheric disturbance increases.

The end of 2025 is expected to reveal El Nino conditions in the Pacific region, which typically suppresses activity during cyclonic seasons. Currents along the equatorial band, between Tahiti and Darwin stall, allowing for warmth to build up, as it did in 1998. That was an El Nino year, along with 1982/3, 1986/7 and 2016. 1997/8 is still regarded as a powerful El Nino-Southern Oscillation event. Further down the track the Gulf Stream will also slow and stall, allowing for colder Arctic conditions to affect the northern areas of the globe, for a couple of years.

There are twice as many weak La Nina years as there are weak El Nino years. But there are about equal moderate to strong La Nina and El Nino years. La Nina can be viewed as the planks on a deck, with the El Nino akin to the gaps between them. La Nina years are 2x wetter than El Nino years.

It is worth noting that the El Niño/La Nina index has stayed close to the neutral range in 2023 and 2024. In 2025 the index edges up into the El Nino episode range.

PRECIPITATION CYCLES

We return now to what makes more rainfall and what makes less, over a long period. First, we must go back to the cycles of the sun. With each passage of Jupiter, when Jupiter and Saturn sit on one side of the sun in conjunction, there is usually a major drought in the area of Earth that they call, say, eastern Australia. Those years were 1901, 1920, 1940, 1961, 1981, 2020, and they are about 20 years apart. Halfway into this decadal period, Jupiter and Saturn face each other on opposite sides of the sun, and we call that an opposition, and those become flood years. Flood years have higher tides than drought years, and that leads to more rainfall which affect the rivers, and everything falls into place accordingly from the oceans to the farmland.

We can note that the conjunctions i.e. drought years, were 1862, 1901, and every 12, 24, and about 36 years, so the next worst was 1959, and then around 2020 (but more broadly 2016-2023), we call them the solar minimum years, they were years of world-wide droughts in the drier regions because of low rainfall and low agricultural yields.

Solar cycles determine annual temperatures and amounts of annual rain. Moon cycles determine the timing of weather events. Cycle #23 began in 1998, peaked around 2000/1 (solar maximum), and died off (solar minimum) from 2007-2009. After a 2–3-year period of relative inactivity Cycle #24 began around midway through 2010 and peaked in 2014. It "bottomed out" in 2020-21. 2020/2021 saw the end of Sunspot cycle #24. Cycle #25 began in 2022, is expected to peak in 2026, and then bottom out again in 2032.

By combining lunar and solar influences, temperature (solar) may lead to amounts of rainfall (due to evaporation rates), whereas the timing of severe weather events may follow strict lunar cycles.

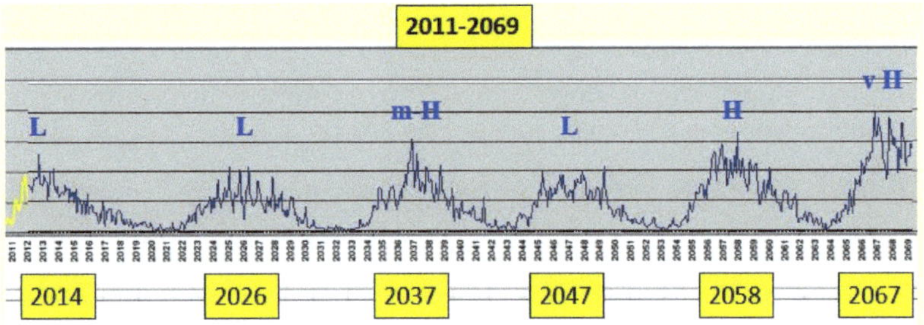

Contrary to what is expected, globally warm periods have fewer and weaker hurricanes while cooler periods have more and stronger hurricanes. 2022 was such a year. It is the Sun's heat that causes evaporation which brings rain.

The amount of heat that the Sun puts out determines the amount of evaporation, which must then fall within about 7 days, near the water source. Coastal locations are generally wetter than places that are some distance inland. Without a source of nearby water, a place quickly dries out. The *amount* of rain falling is due to the amount of solar radiation which causes evaporation. Therefore, rain amounts depend on the sunspot cycle. The amount of rain falling is also dependent on the extent of the air tide, which can be likened to the volume of water in each sea tide. As the solar cycle of radiation is cyclic then so too are rain intensities.

Sea tide timing is predictable, subject to the positions of the Moon and Sun. The volume of water flowing into an estuary depends on wind direction subject to the shape and size of surrounding hills, wind forces increased or confused by surface currents, and less visible factors like the proximity of underwater eruptions, fissures, earthquakes, and time due to temperature gradients. Amounts of rain are influenced by the Sun's warmth for evaporation and then how much cold air comes to the layers of the atmosphere containing clouds.

Sometimes colder air is at a higher or lower altitude than rain-filled clouds, and expected rains must wait for the marrying of the two factors. Until then a cloud may look like rain but heat from the ground is sufficient to buoy up the rain-laden cloud and prevent rain from falling. Analyzing sunspot cycles is at least better than having nothing at all, by looking to see which previous years correspond to the same points of the cycle.

THE JUPITER/SATURN EFFECT

Sunspots have been shown to be electromagnetic disturbances produced by external charges which act on the Sun's internal structure. Just as Sun and Moon, either both on one side of Earth (new Moon) or on opposite sides (full Moon) cause internal stress on Earth to raise tides in land, sea and air, by the same process the largest of our solar system's planetary bodies on one side of the Sun or on opposite sides cause exaggerations in Sun-tides manifesting as sunspots/solar wind.

When the planet Jupiter draws the Sun outwards from its normal line, sunspot maxima occurs, and when the Sun is drawn inwards, the minima are the result. When the gas giants Jupiter and Saturn pass across the line of the Sun's advance (Solar Apex), sunspot minimum is very close to that date. Jupiter, Saturn, Uranus and Neptune, in that order, affect the Sun the most, with a total period of 178 years, roughly 10 lunar cycles. The 11-12-year average sunspot period is roughly the same as the orbit of Jupiter, and any variations are due to the varying positions of the other three great planets.

With each passage of Jupiter, either just before or just after, there is a major drought. In the prolonged drought of 1862, Jupiter and Saturn lined up on one side of the Sun in that year. Both straddled the Sun in 1871, were on one side of the Sun 1900-2, then on opposite sides of the Sun again in 1931. Jupiter/Saturn together go through this cycle every 35-36 years, with the other two planets, Uranus and Neptune, of much slower periodicity. Jup/Sat were together again on one side of the Sun in 2020.

In 1982, Saturn/Jupiter were on the same side of the Sun causing severe drought to flat terrain of the northern hemisphere. In 2001 Jupiter crossed the vertex, bringing further drought. When Jupiter crosses the path of the Sun there is a drought just beforehand.

Jupiter was crossing the Sun in 2007, when world-wide droughts brought a food crisis, sparking riots and e.g in the southern hemisphere the Murray Darling river system in Australia had the lowest inflows on record. No sunspots can result in little or no rain.

Other factors modify air temperatures, such as solar intensity. 1997 was on the rising side of the Sunspot cycle, a quarter of the way in. 1982 was halfway into the decline of cycle #21.

Jupiter has been at high peaks in 1766-69, 1779-81, 1790-93, 1802-4 (drought Jup+Sat), 1814-17 (drought), 1827-30 (drought), 1838-40 (Jup+Sat drought), 1850-52, 1862-64, 1873-76, 1885-88, 1887-1900, 1909-12, 1921-23, 1933-35, 1945-47, 1957-59, 1968-71, 1980-83, 1992-95, 2004-6, 2016-19 (all drought years). Low Jupiter peaks have been 1774, 1786, 1798, 1809, 1821, 1833, 1845, 1857, 1881, 1904, 1928, 1939-40, 1952, 1964, 1976, 1987, 1999, 2011 and 2024.

Jupiter/Saturn peaks and conjunctions have been blamed for low yields and economic panic. Examples have been 1873, 1884, 1893, 1900-2, 1907, 1910, 1958-60, 1987, 1990, 1999, and the recession of 2011-12. The worst have been 1901 and 1959. The next has been 2016-2024, when solar minimum created potential for worldwide droughts through low rainfall and low agricultural yields in 2020. Closer perigees in minimum declination years typically bring longer lasting extreme events.

In 2020 Jupiter and Saturn were in "conjunction". That means they were occupying the same area of sky as seen from earth. When you looked up at them, they appeared very close together, especially in December 2020. Conjunction correlates with drought years. By 2024-5 these two planets will be at right angles to each other, called a "square", and the wetter years will perhaps begin. Finally, in 2030 they will be in "opposition", being 180° apart, and typically this will be a year of floods. Half the cycle is then over. This is a worldwide phenomenon. Each country experiences its own particular cycle of events, according to its unique climate.

The last previous year of equivalent conjunction was 2001. In 2005 was the last square of these planets, and their last equivalent opposition time was 2012. Following that, the last previous square was in 2015. Since 2015 it has been noticeably getting drier.

The order is:

1. J/S conjunction=drought years (2001, 2020),
2. J/S square=coming out of drought years, wetter (2005, 2025)
3. J/S opposition=flood years (2012, 2030)
4. J/S square=getting drier, approaching the next drought (2015, 2036).

Jupiter and Saturn together control the grand cycle of drought/flood. Jupiter orbits the Sun around every 12 years, and Saturn every 29. Jupiter and Saturn are the largest planets in the cosmos, and together have an effect on the Sun, such that every 35-38 years the SSB (solar system barycenter) sees an event-repeat. Thus 2024 may see repeats of events that last occurred in 1987.

THE PLANETS

Jupiter, Saturn, Uranus and Neptune, in that order, affect the Sun, and the sun and moon together affect the oceans. Venus and Mercury affect the tides, but not as much as the sun and moon. Jupiter and Saturn mainly, are responsible for what we call the sunspot cycle. That is when the sun gets overactive and seems to put out visible blotches which some cultures like the Chinese have been watching and tabulating for hundreds of years, for they coincide with extra radiation levels. This in turn translates into increased thunderstorms, and growth, agriculturally.

The sunspot cycle is about 12 years, which is slightly more than the Jupiter cycle of 11.8 years although this can vary between 9 and 14 years. There are variations in the positions of the other planets, and it's the oceans that get affected first. Because the moon is the tiny player in the big team, it gets affected by everything that the big team does, in the same way that a baby is affected by the moods of the whole family. Because the moon is the closest planet to us and so to the oceans, everything in terms of planet aspects involves the moon. There is the timing of the tides, the height of the tides, the speed of the tides, all factors in how the ocean behaves, and all that gets imparted to the weather because all weather begins in the ocean.

That viewpoint is little known, and people still look to the sky for signs of change. But by the time a weather event reaches the sky, where we on the land see it, it's already half-over, it's obsolete, and just the dregs of the system.

THE CASE FOR THE MOON

The bottom of the air and the surface of the sea are invisibly joined. Things that are joined are part of a joined system. Both do indeed comprise one interactive and connected interface, covering a surface area of 73% of the earth. Lunar forces begin with the Earth (or Land) tide, in which the land surface rises vertically around about 20cm per day in NZ and Ireland, 50cm per day in Australia and Spain, and 55cm on the equator, towards the moon, and recedes again in regular daily tidal rhythm, and which in some locations gives rise to earthquakes.

The Land Tide supports the tide of the ocean which then gives rise (and fall) to a daily Air (or atmospheric) Tide. All tides are repeatable, to some extent measurable, and therefore predictable.

Matching the timing of the air-tide to oceanic tide-tables at a location yields a repeating frequency of weather extremes. Nobody questions the predictability of sea tide-tables. The more time spent outdoors, the more it will be realized that nothing in nature is random, but governed by repeating cycles. The rotation of the earth and planets, the recurrence of day and night, summer and winter, animal, bird, insect and fish breeding, feeding and migration habits, and the cycles of the growing seasons can all be predicted, not exactly, but to a degree of usefulness. Weather is the no different.

There are 3 main cycles of the moon that we use for predicting weather, gardening and fishing.

1. (PHASE)

The monthly cycle that we can all see, from New moon to Full and back to New again is called phase. That includes the intermediate phases Waxing Crescent, First Quarter, Waxing Gibbous, Waning Gibbous (Disseminating), Last Quarter, and Waning Crescent (Balsamic). For weather, a change to full or new or quarter moons very often brings a weather pattern change. If raining before a phase change it may stop on the day of changeover, but if dry before and rain is about, it may choose the day of the phase to manifest the new trend.

An old mariners' saying is that the full moon 'eats' clouds, because the night of a full moon is generally clear. That is why you can see the full moon. The full moon actually clears the sky, especially around midnight. Maximum temperature of the day will correspond with tides; and at the beach, temperature is likely to be highest when the tide is lowest.

For gardening, plant above ground with attention to leafy things when the moon is climbing up in the sky in the daytime, which is before and up to full moon. After full moon, the moon is descending in the daytime because it is rising on the other side of the earth, so give attention to root things, e.g. bulbs, carrots, potatoes. For fishing, the new and full moon tides are the best, around 12 noon and midnight, and the quarter moons are good around dawn and dusk.

2. (DECLINATION)

The second cycle to consider is when the moon changes hemispheres because of the tilt of the earth. When it is in the northern hemisphere the tidal variation is greater because of greater gravitational pull, and higher tides mean more water filling a bay which means fish compete for food in the faster currents, so bite-chances may be better, but also the water table rises in the land, and as well the sap tide in the trees. This means more insects appear, also more birds. It is why the Canadian Rockies Hunting calendar is identical to the Maori Fishing Calendar.

When the moon is in the northern hemisphere in winter, around full moon time, Ireland can get very cold winds, dragged southwards over Ireland from the N Pole. Northernmost full moons in summer time bring warm, moist northerly wind systems and warm rain. It is why if the winter full moon rises further in the north it will be a colder season, and if the summer full moon rises far in the south, it will be unusually warm. So, when the winter full moon is also when the moon is furthest north, one may plan for snow. This may only happen in January 2022.

The moon traverses different parts of the sky as it orbits the earth in the course of a month. The "constellation" (or astrological *sign*) is the road map of the declination cycle. In folklore, functions were assigned to signs. For instance, Pisces moon was considered best for fishing, hence the fish zodiac symbol. Cancer full moons were said to bring more snow. In gardening, the constellations favour particular types of plant activity. For example those of a leafy, flowery, fruity, or rooty nature are said to have water (Scorpio, Pisces, Cancer), air (Libra, Aquarius, Gemini), fire (Leo, Sagittarius, Aries) or earth (Capricorn, Taurus, Virgo) signs. Air and fire signs are usually considered barren, and earth and water signs fertile. In weather, the constellations provide the template for changes in barometric pressure.

3. (APSIDAL LINE)

The third lunar cycle is called Perigee-Apogee, or Apsidal Line, which comes about because the moon's orbit around the earth is not a circle but an elliptical or egg shape. The point in the month when the moon is furthest from earth is called the Apogee. At perigee the moon is about 350,000 km from earth and at apogee it is about 400, 000 km from earth. This is a varying difference of about 50,000 km; and even that varies by about 10,000 kms over 12 months of the year.

The gravitational forces that apply between earth and moon are increased at perigee and decrease at apogee and account for extremes in tidal movement, particularly when coinciding with either a new moon or a full moon. In perigee the weather is more turbulent and strong winds are common. These may develop into gales or hurricanes and inflict property damage.

Fishermen use apogee because fish are less skittery at that time. On perigee the tides are bigger but fish don't like the breakers with their bigger swells, so surfcasting isn't the best that day. But estuaries, where fish hide from the turbulence of the shoreline, are excellent for fishing during perigees. Perigee is responsible for the increase in tidal height per month, not the full or new moons.

Just as the moon affects the oceans tides, so it affects the sap 'tide' in a plant's structure. Each plant, seed, and tree, and organism in the garden has a tidal flow or magnetic force that we see daily in the ocean's tidal flow. At perigee (and a couple of days either side) you would avoid planting seeds, as germination can be poor and consequent growth of the seedling may be leggy and weak. But it is an opportune time to maximize on feeding and fertilizing as the nutrient is more easily distributed in the soil. Everything is 'going for it', so to speak.

Perigee favors fungi and they can become prolific over a 3-day period. Perigee is a good time to put fermentation brews to work. Kelp can be gathered because the higher tides bring it onto beaches. At apogee, the magnetic force is weaker and the tidal pull also weaker, gentler and more consistent, so the ground is more settled for putting in things that need a calm environment to get going. While some plants may grow rapidly, others might run to seed quickly if planted in apogee. If the moon is descending and waning, good nourishment is coming to roots, and if an ascending and waxing moon, energies are traveling upwards away from the ground, tending to draw water upwards. Then again, if the apogee, the strongest influence would be lessened, but if perigee, exaggerated.

THE MOON IN PERIGEE

The "perigee" of the Moon is when it comes close to earth once per month. Taking note of averages of perigees over a year yields how in a particular year the Moon is averagely closer to earth. The cycle of perigees is 8.85 years. The closest averagely shorter earth-Moon distance over the past century or so have been 1905, 1922, 1940, 1958, 1976, 1993, 2011, and 2029.

These closest perigee-years happen typically when the Moon is at or near minimum "declination" (also called minor standstill).

This is the slow trekking of the moon, over 18-19 years, between earth latitude lines. It is around minimum declination years that the Moon's inter-latitudinal range is narrower.

In the southern hemisphere, a phenomenon known as the El Niño occurs twice every decade. The ocean currents reverse along the equator, restabilize the difference in single level, and then resume again. The effect on the weather is greatest in the Pacific area, but the Gulf Stream tends to eventually be affected as well. These are the years that the Gulf Stream tends to stall, slow down, and countries susceptible to its usual insulation are less affected by it. This will not happen next year.

The El Niño alternates with La Nina, but the La Nina is twice as long. Mariners have always known about the 'trade' winds, blowing strongly westward along the equatorial band, which is the old name for La Nina. La Nina can be viewed as the planks on a deck, with the El Nino akin to the gaps between them. La Nina is the stronger of the two systems, being the normal situation of strong easterly flows. Although there are twice as many weak La Nina years as there are weak El Nino years, there are about equal moderate to strong La Nina and El Nino years.

On or just after averagely closer perigee years, El Nino has occurred on 1905/6, 1923-6, 1940-2, 1957-8, 1976-7, 1994-5 and 2014-15. It is not expected again until 2032.

As well, there is a monthly N-S declination. In minimum declination years, the speed of changing hemispheres in the course of each monthly declination is less. Such years tend more towards the slackening of sea currents due to a more sluggish lunar lateral pull, and therefore more tendency towards El Nino systems.

The moon in perigee exerts increased gravitational pull on fluids such as air and water (and land, influencing earthquakes), depending on the season, and by generating more turbulence, can cause weather to worsen.

Because the earth rotates beneath the moon once every 24 hours, unsettled conditions can happen in many countries on the same day. Up to three days before or after a perigee, it is common to see the month's most rain, particularly in autumn and winter. Clear skies might return shortly afterwards as the Moon recedes from the earth.

For half the year perigee is accompanied by Full moon, then by New moon during the other half. e.g. in 2022, new moon perigees are in the first two months, then full moon perigees are until October. The last two months see new moon perigees once again. This cycle will repeat in 2031.

As mentioned, perigees vary in closeness. In 2014 the closest perigee was in August, with #2 closest in January. September had the closest perigee in 2015, perigee #2 in February, and perigee #3 in March. Consequently, the closer 2014-15 summer moon was set to fuel increased cyclone and monsoonal activity (in countries prone to cyclones) and some above normal temperatures. But perigees were weak in December 2015 through January 2016, and cyclones were weak.

In 2016 the closest perigee was in November and tied to a full moon, and 2016-17 was a more intense cyclone season for countries not far from the equator. 2016 was also an averagely closer moon across all months, whereas overall, 2017 was an averagely further away moon, but 2018 again started to come closer, with 2019 continuing this trend.

Where 2020 was the 8th closest (in the 125 years from 1905 to 2030), 2021 was not amongst the closest perigee years. Neither was 2022, which was only the 90th closest. 2021 was the 95th closest, and 2024 is the 81st closest. But the moon comes considerably closer again in 2024 (12th) and some intense weather events in Ireland and the UK may be the result.

For most of 2022, perigees were level with the Southern declination. Previous years this has happened have been 2013 and 2004, both of which saw stronger than average cyclones in other countries. But in 2024 the perigees start to trek up towards the equator.

The El Niño/La Nina index stayed close to the neutral range in 2022. During El Nino the moon was often averagely further away in distance, causing a weakening of currents and a slowing of air flows. In 2011, the Moon was averagely much closer to earth, being (since 1900) in the top four closest to earth years. The moon was, in that year, the averagely closest to earth for 19 years.

General trends are:

Closer perigees = more cyclonic activity (1997, 2002, 2007, 2011, 2016, 2024)

Far away perigees = less cyclonic activity (1983, 1991, 2000, 2009, 2013, 2017, 2027)

WHEN IT WILL RAIN THROUGH THE PHASES

The below is an approximation of rain timing, provided rain is about. If no rain, expect a drop in temperatures. Cut and paste this somewhere handy as your guide to year-around weather.

NEW MOON (INVISIBLE)

Most rain, if about, is overnight. If rain occurs during the new moon day, expect rain also the following night. If winter, expect temperature drop. Most new moon days are mainly pleasant and dry; if wet then only isolated dumps, i.e. no long-lasting storm activity.

FIRST-QUARTER MOON (D-SHAPED)

Most rain, if about, is between midnight and noon. If no rain before noon, expect it to stay dry during the afternoon. If wet in the morning, weather should clear in the afternoon.

FULL MOON (FULLY ROUND)

Most rain, if about, is during daylight hours. Otherwise it may have the look of rain all day. If rain does come overnight, which is rare, it may continue the next day. Mostly, clouds will be seen to drift away as the sun sets, and may stay away until dawn.

LAST QUARTER MOON (C-SHAPED)

Most rain, if about, arrives between noon and midnight. If rain arrives before noon, expect more rain between the noon and the evening. If the afternoon is dry, the following morning will usually stay clear.

THE ATMOSPHERE

Temperature, volume, and pressure became the focus of climate science a few centuries ago when the Ideal Gas law was discovered. This states that the three factors are dependent, and if given a closed system, and played around with in the laboratory, then by changing one variable you can make others change as well. These principles were applied to the air in the atmosphere regarded as an enclosed gas. It is supposed to be enclosed because it is held here by gravity. This is where science broke down due to a fundamental misunderstanding.

This is not a laboratory, and the earth does not have a lid by which one can change one factor and watch the changes in the other two. In other words, the ideal gas law, which sounds logical, cannot be applied to the atmosphere, which is open to space. Nor can it be shown to work in the laboratory without enclosing the gas in some container. Gravity is still working in the laboratory when you remove the lid of your container. Therefore, we could arguably scrap the ideal gas law, as it applies to our environment, and the idea of temperature (which is a feeling) along with it.

What of volume and pressure? Well, except in isolated and controlled cases, we cannot watch the effects of air pressure on the air, because it is generated too locally close to the ocean surface to be of any use to the atmosphere. Besides, any wind is generated by currents in the sea that work their way to the surface, and create high and low-pressure areas, and wind between high and low pressure, which is only measured on the ocean's surface. Yet the surface of the sea is NOT where the air pressure is generated.

Which leaves us with volume. Science has not come up with a way of accurately quantifying the amount of air in the world, and therefore telling us that there is a danger of it running out. If every tree on the land was instantly destroyed, we would still have 4 million years' worth of oxygen in the atmosphere, because air contains 21% oxygen produced by photosynthesis on the biomass in the ocean. That means 99.9% of our vegetation, which produces oxygen for the survival of every species of life, operates mostly in the sea, and away from the air. That means $CO2$, or anything else introduced to the air, can have no effect.

The sun warms the earth, but the oceans cool it. Even on the hottest summer's day, a swim in the ocean will be refreshing because the water is always cooler than the air above it. What's more, it can never get hotter than the air above. Everything is in balance and will remain so, and what we call the weather is what we perceive as the outcome of planetary cycles that began thousands of years ago.

SPEED READING OF CLOUDS

METHOD

For many years, there have been people who have looked at the clouds and determined the coming weather. The techniques are relatively simple to master. Generally, the weather in the early morning does not change much during the day. Whatever reading is gained from looking at the clouds, should usually hold for 24 hours. When we look up and take note of the types of cloud shapes that are forming, and then we look again in an hour, what is developing or fading away is noted. Your guru skills have already noted the sky, at least an hour or two before you announce a prediction.

TYPES OF CLOUD

There are essentially 3 types of clouds, which have Latin names to easily describe their shapes –

1. high (cirrus, literally "mares's tails"),
2. middle layer (cumulus, literally "piles')
3. low cloud (stratus, literally "lines").

HIGH (20,000-40,000 FEET)

Cirrus; the cloud of change

After a run of clear or bad weather, light streaks of very high cirrus clouds are often the first signs of change. Because of their height, they only move with the rotation of the earth. Bad weather is about 36 hours away. Check in about an hour; if cirrus increases across the sky, it indicates an approaching depression, but if curls and wisps dissolve or vanish, fine weather continues. The higher the cloud, the finer the weather. Rapid cirrus, in bars, or at right angles to bars, mean gale or wind. The bands generally precede a storm by 20-48 hours.

When cirrus become lines (cirrostratus), expect a quick change.

If cirrus streaks point up, it means rain. If down, wind and dry weather.

If mare's tails are followed by altocumulus (literally "*high piles*") the next day, it means rain within 12 hours. Cirro-stratus, (literally "spread-out hair"), also means rain.

When cirrostratus and stratocumulus gather over the sea, it means rain and probably wind on the second day. Cirrocumulus (literally "*piles of hair*") is a 24-hour warning. It is also called mackerel sky and means fair weather for that day but rains a day or two after.

Before thunder, cirrocumulus can appear as dense and compact masses, in close contact. If their outline is soft and delicate, if fleeces gently merge, it means a continuance of fine weather. Thunder is usually mainly noise but very little rain. But if the clouds are dense, and associated with cirrus, it means lightning in summer or gales in winter. High cirrocumulus commonly appears a few hours before thunderstorms. The harder and more definite the outline, the more unsettled the coming weather. Cirrocumulus in the early morning, which diminishes means a fine and warm day.

MIDDLE (6,000-20,000 FEET)

Altocumulus (literally "*high piles*") is a sign of coming thunder. It is also called tower clouds, or Cauliflowers. It consists of turret-like upward growths. A thicker cloud generally arrives some hours later. Altocumulus often occurs together with altostratus, (literally "high, spread out"), and is the cloud that carries rain and snow. It can cover the whole sky, and when it increases, rain and snow over a wide area may be the result. It may also bring drizzle.

Cumulonimbus, (literally "*gathered piles of rain*"), is a sign of heavy rains. They are cumulus clouds which are large enough to produce showers. From a distance, they appear as tall piles, but to stand underneath and look up is to see the dark look of rain. They bring thunderstorms, heavy rain, and occasionally hail. They can be anything from 1000 to 60,000 feet high. Towering thunder clouds that look like mountains, early on a summer's day can become intense storm activity during the heat of the afternoon and often become torrents of rain in the evening.

When the cumulus rain clouds look smaller toward sunset than they were at noon, they are diminishing, and you can expect fair weather. The formation of cumulus clouds to leeward during a strong wind indicates the approach of a calm with rain. But cumulonimbus clouds forming rapidly in the mountains on warm afternoons may mean an intense lightning storm is on its way in minutes.

Cumulus (or ice cream clouds"), are the most common clouds and are caused by moderate, generally westerly winds. All have a well-defined flat base, between 1500 and 2500 feet high. If they appear early in the morning they will develop into bigger shower clouds. If they do not form till late afternoon, they will die away to be followed by a clear sky night. If clouds open and close, rain will continue.

If the clouds are like sheep's wool, deep and dense, the edges being very white, the surrounding sky bright and blue; they are of a frosty coldness, and will speedily fall in hail, snow or rain.

Strato-cumulus, (literally "spread-out gatherings"), are the doors to a cold front. 1500-6500 feet. They cover the sky, sometimes in parallel bands but they usually bring little or no rain. When at sea, the strato-cumulus clouds appear on the horizon, it is a sign that the weather will break up. If over the land in the winter, they may produce a light sprinkling of snow.

If long stripes of cloud drive at a slow rate, high in the air, and gradually become larger, the sky having been previously clear, expect rain.

LOW (500-2,000 FEET)

Stratus, or hill cloud. Nimbostratus (literally "rain-layer") is a cloud of light rain. Sometimes simply called 'nimbus', these are low, uniform, basic rainclouds, 300-2000 feet. Thick enough to blot out the sun, they can disgorge continuous rain or snow. If a little cloud appears in a clear sky, especially if it comes from the west or somewhere in the south; there is a storm brewing. If there are long points, the look of tails or feathers hanging from rain clouds, close to the horizon, with little wind in the summer, thunder may be expected but the storm will be short. Resembling fog and often covering hills, stratus may bring drizzle.

Harbingers of fine weather are when the day breaks out from the morning fog. Stratus at night is followed by a fine day if the barometer stays high. When mountains extend north and south, if fog or mist comes from the west, expect fair weather. If mist comes down from the top of the mountains, expect rain in summer, and snow in winter.

GENERAL

If two layers of clouds move in opposite directions, then bad weather is on the way.

Clouds do not indicate the presence of water vapor, which is everywhere at once. Clouds represent masses of air of a different temperature. The wind is the movement of an air mass and is not necessarily the movement of clouds. Clouds do not move, that is an illusion. They are disintegrating and reforming at the same time, but we view them at a distance. A fog is a close cloud, and if we were to stand in the middle of a mist we experience the air as still around us.

The wind blows rain in, but also blows it away again. It is more likely to rain when the wind drops. The times when the moon is on the horizon, either rising or setting, and when it is directly overhead or underfoot, are times of greatest rainfall. Atmospheric pressure is highest when the tide is lowest, and vice versa. Atmospheric turbulence, and therefore wind and rain, is more likely at mid-tide going out or coming in.

GARDENING GUIDE 2025
(TIME ZONE USED: UT +00:00)

REST PERIOD:

The Rest Period in the garden is as important as the activity times. It's a time when the Moon is Void of Course, that is, traveling between one sign and another and during this time the Moon is deemed to have "no energizing power", also Rest Periods on the day of the Full Moon, New Moon and when the moon is at Southern and Northern Declinations. Hence it is a good time to sit and enjoy your garden or to attend to non-plant activities such as tidy your shed, repair or maintain outdoor structures or even to sit and plan your next "to do" list when the Moon is once more energized. Eclipse days and the day after the eclipse are also deemed days to rest in the garden.

ALL ABOVE-GROUND ACTIVITY SHOULD TAKE PLACE when the Moon is Waxing – i.e. building from New Moon through the 1st Quarter and up to the Full Moon. This is a time of increasing light and with the Moon pulling the sap upwards into the leaves and flowers.

ALL BELOW-GROUND ACTIVITY SHOULD TAKE PLACE when the Moon is Waning – i.e. from the Full Moon through the 3rd Quarter and up to the New Moon. This is a phase of decreasing light and the general trend for sap to be pulled downwards and into the root and tuber systems.

THE NEW MOON PHASE:

Activities can begin in the garden once the crescent of the New Moon begins to be visible. At this time:

DO start sowing, transplanting, grafting or strike cuttings of all ABOVE-GROUND producers. Flowering annuals, brassicas (cabbage, broccoli etc.) cover crops (mustard, lupins etc.) and annual fruiting crops (cucumber, pumpkin, tomatoes etc.) and houseplants all benefit from attention during this phase. Also, mow lawns should you wish to encourage growth or sow any large areas such as lawns.

DON'T plant root crops (as they will only produce plenty of leaves and little roots, or go to seed too early), nor prune at this time – your plants, and trees in particular, will be prone to excessive sap loss and suffer from die-back.

THE 1ST QUARTER PHASE:

Continue with the New Moon phase activities and especially sow any seeds two days before the Full Moon to yield optimum results. During drought times, sow during this phase when the Moon is in Capricorn or Taurus for the strongest of survival qualities. At this time:

DO sow and transplant all legume, grain, flowering plants (especially roses) and pot plants and start harvesting any above-ground produce, with 2-3 days before the Full Moon being the most beneficial time (especially for herbs and grape harvesting).

DO give plants a food boost if required, but do water it down and apply as close to the Full Moon as possible, particularly for those plants that require phosphorus.

DON'T harvest with the Moon in a water sign (Cancer, Scorpio or Pisces) unless for immediate use, nor prune or plant out root crops.

THE FULL MOON PHASE:

The Moon will have reached its full potency by the time of the Full Moon, and from this point to the New Moon, there is a gradual decrease in its vitality and the sap flow is now downwards into the roots – it is a time when the plants build up their root systems and store their goodness. However, you can continue to harvest above-ground producers for a few more days after the Full Moon. Leeks, onions, garlic etc. are at their best potency when harvested with the Full Moon in Leo. At this time:

DO sow or plant out any root crops (potatoes, sweet potatoes, carrots etc.) especially during the two days in the middle of the Full Moon phase. Also sow or plant out fruiting plants (e.g. strawberries, passion fruit etc.) as well as perennial flowering plants, bulbs and tubers (e.g. daffodils, dahlias, crocus etc.) and this is also a good time to spray, prune and trim hedges.

DO perform drying activities of herbs, flowers and fruits.

DO dig herb roots or harvest leaves and bark intended for medicinal teas.

DO spray fruit trees, mulch your garden and kill weeds.

THE 3RD QUARTER PHASE:

This is considered to be the resting period of the Lunar cycle, the time when the sap in the plants runs at its slowest.

DO harvest all root crops for long-term storage (e.g. potatoes, garlic, onions and root herbs etc.), especially if the Moon is in Taurus or Capricorn. Also, lift your bulbs and tubers for their seasonal storage. You can also use this time to build up your compost, especially when the Moon is in Scorpio and attend to maintenance chores such as spreading compost, mulching, staking taller plants etc., as well as attend to all pruning and trimming of trees, vines and bushes, as there is less chance of die back and excessive sap bleeding. Cut your lawns during this phase to discourage growth.

DO plant trees and saplings for good strong foundations of root growth and bark development as well as plant strawberries or their runners so their growth is well developed before fruiting.

DO fertilize your crops for potassium if needed, as potassium absorption is at its peak at this phase.

WHEN THE MOON IS IN CANCER, SCORPIO OR PISCES:

This is a beneficial time for attending to vegetables such as lettuce, cabbage, herbs, broccoli, leafy pot plants, etc. Avoid any harvesting except for immediate use during these Moon signs as anything harvested for long-term storage will deteriorate faster and be prone to mildew, except of course, unless you are attempting to build compost – choose when the Moon is in Scorpio for best results, as vegetation decomposes fastest under this influence.

WHEN THE MOON IS IN TAURUS, VIRGO OR CAPRICORN:

This is a beneficial time for attending to vegetables such as potatoes, carrots, leeks etc. as well as to non-flowering shrubs and trees and ground covers such as lawns all respond best attended to under these signs. A Virgo Moon (particularly in the 3rd Quarter phase) is best for harvesting any medicinal leafy herbs due to the usual desire for long-term storage.

WHEN THE MOON IS IN GEMINI, LIBRA, AQUARIUS:

This is a beneficial time to attend to vegetables such as globe artichokes, flowers for medicinal purposes, and flowers in general. (Broccoli and cauliflower etc. are considered "leafy" as if planted during a Flower phase they are more likely to bolt straight to seed). The Moon's focus is more directed towards developing perfume, fragrance, flower vibrancy and beauty as well as medicinal properties with the Full Moon in one of these signs.

WHEN THE MOON IS IN ARIES, LEO OR SAGITTARIUS:

A beneficial time to tend any fruit, berry, seed or grain crops such as tomatoes, peas, beans, peppers, wheat or corn, pumpkin, apples, strawberries, raspberries etc. Any seeds you wish to store for next year's planting, or for long term storage are best attended to during these Moon signs, as well as harvesting crops such as onions and garlic for their most potent flavor.

ABOVE-GROUND (LEAFY, FRUITS AND SEEDS)

Vegetables: beans, broccoli, Brussel sprouts, cabbages, cauliflower, chicory, celery, cress, cucumber, endive, lettuces, marrow, mustard, parsley, rhubarb, sage, silver beet, corn, etc.

Flowers: cornflowers, forget-me-nots, marigolds, pansies, poppies, roses etc.

Fruits and seeds: apples, blue-berries, grapes, passion-fruit, pears, peas, peppers, raspberries, strawberries, tomatoes, wheat

BELOW-GROUND (ROOTS, BULBS AND NON-FLOWERING SHRUBS)

Vegetables: beetroot, carrots, garlic, leeks, onions, parsnips, radish, salsify, shallots, swedes, turnips etc.

Flowers: agapanthus, anemone, crocus, daffodils, dahlias, freesias, gladioli, hyacinth, ranunculus, red-hot pokers, tuberose, tulips

Fruits and Seeds: garlic, onions, peanuts

BASIC MOON IN SIGN GUIDE

Moon in Aries	Harvest root and fruit for storage.
	Good for destroying weeds and pests.
Moon in Taurus	Excellent time for planting and transplanting, particularly root crops.
	Also good for leaf crops.
Moon in Gemini	Good for harvesting root and fruit for storage.
	Good for destroying weeds and pests.
	Melon seeds respond well.
Moon in Cancer	Best time for all planting and transplanting.
	Good for grafting and irrigation.
Moon in Leo	Good for harvesting root and fruit for storage.
	Excellent for destroying weeds and pest.
Moon in Virgo	Good for destroying weeds and pests.
	Vines and some flowers respond well.
Moon in Libra	Best time for planting for fragrance and beauty of flowers, vines and herbs.
Moon in Scorpio	Best time for planting sturdy plants and vines.
	Especially favored by tomatoes, corn and squash.
	Good for irrigation and transplanting.
Moon in Sagittarius	Good for harvesting roots.
	Especially favored by onions and fruit trees.
	Good time to cultivate the soil.
Moon in Capricorn	Plant potatoes and root crops and encouraging hardy growth.
	Good for grafting and pruning to promote healing and applying organic fertilizers.
Moon in Aquarius	Harvest root and fruit for storage.
	Good for destroying weeds and pest.
	Good for planting onions.
Moon in Pisces	Good for planting and transplanting, especially for root growth.
	Good for irrigation.

JANUARY	2025		Hour:	1am · 2am · 3am · 4am · 5am · 6am · 7am · 8am · 9am · 10am · 11am · Noon · 1pm · 2pm · 3pm · 4pm · 5pm · 6pm · 7pm · 8pm · 9pm · 10pm · 11pm · Midnight
	1st	Wed	Above	Capricorn / REST / Aquarius
	2nd	Thur	Above	Aquarius
	3rd	Fri	Above	Aquarius / REST / Pisces
	4th	Sat	Above	Pisces
	5th	Sun	Above	Pisces / REST / Aries
1Q	6th	Mon	Above	Aries
P13 XhN	7th	Tue	Above	Aries / REST / Taurus
	8th	Wed	Above	Taurus
	9th	Thur	Above	Taurus / REST
	10th	Fri	Above	REST / Gemini
	11th	Sat	Above	Gemini
^	12th	Sun	Above	REST DAY - Northern Declination
F	13th	Mon	Below	REST DAY - Full Moon
	14th	Tue	Below	Cancer / REST / Leo
	15th	Wed	Below	Leo
	16th	Thur	Below	Leo / REST / Virgo
	17th	Fri	Below	Virgo
	18th	Sat	Below	Virgo
	19th	Sun	Below	Virgo / REST / Libra
XhS	20th	Mon	Below	Libra
3Q A	21st	Tue	Below	Libra / REST / Scorpio
	22nd	Wed	Below	Scorpio
	23rd	Thur	Below	Scorpio
	24th	Fri	Below	REST / Sagittarius
	25th	Sat	Below	Sagittarius
V	26th	Sun	Below	REST DAY - Southern Declination
	27th	Mon	Below	Capricorn
	28th	Tue	Below	Capricorn / REST / Aquarius
N	29th	Wed	Above	REST DAY - New Moon
	30th	Thur	Above	Aquarius / REST / Pisces
	31st	Fri	Above	Pisces

FEBRUARY	2025		Hour:	1am · 2am · 3am · 4am · 5am · 6am · 7am · 8am · 9am · 10am · 11am · Noon · 1pm · 2pm · 3pm · 4pm · 5pm · 6pm · 7pm · 8pm · 9pm · 10pm · 11pm · Midnight
	1st	Sat	Above	Pisces / REST
P10	2nd	Sun	Above	REST / Aries
XhN	3rd	Mon	Above	Aries / REST
	4th	Tue	Above	REST / Taurus
1Q	5th	Wed	Above	Taurus
	6th	Thur	Above	Taurus / REST / Gemini
	7th	Fri	Above	Gemini
^	8th	Sat	Above	REST DAY - Northern Declination
	9th	Sun	Above	Cancer
	10th	Mon	Above	Cancer / REST / Leo
	11th	Tue	Above	Leo
F	12th	Wed	Below	REST DAY - Full Moon
	13th	Thur	Below	REST / Virgo
	14th	Fri	Below	Virgo
	15th	Sat	Below	Virgo / REST / Libra
XhS	16th	Sun	Below	Libra

FEBRUARY	2025		Hour: 1am – Midnight		
	17th	Mon	Below	Libra	REST	
A	18th	Tue		RE	Scorpio	
	19th	Wed		Scorpio		
3Q	20th	Thur		Scorpio	REST	Sagittarius
	21st	Fri		Sagittarius		
V	22nd	Sat		REST DAY - Southern Declination		
	23rd	Sun		Capricorn		
	24th	Mon		Capricorn		
	25th	Tue		Capricorn	REST	Aquarius
	26th	Wed		Aquarius	REST	
	27th	Thur		REST	Pisces	
N	28th	Fri	A	REST DAY - New Moon		

MARCH	2025		Hour: 1am – Midnight		
P7	1st	Sat	Above	Pisces	REST	Aries
XhN	2nd	Sun		Aries	REST	
	3rd	Mon		REST	Taurus	
	4th	Tue		Taurus		
	5th	Wed		Taurus	REST	Gemini
1Q	6th	Thur		Gemini		
^	7th	Fri		REST DAY - Northern Declination		
	8th	Sat		Cancer		
	9th	Sun		Cancer	REST	Leo
	10th	Mon		Leo		
	11th	Tue		Leo	REST	
	12th	Wed		REST	Virgo	
	13th	Thur		Virgo		
F	14th	Fri	Below	REST DAY - Full Moon - Lunar Total Eclipse		
XhS	15th	Sat		REST DAY - Day after Eclipse Day		
	16th	Sun		Libra	REST	
A	17th	Mon		REST	Scorpio	
	18th	Tue		Scorpio		
	19th	Wed		Scorpio	REST	Sagittarius
	20th	Thur		Sagittarius		
	21st	Fri		Sagittarius		
3Q V	22nd	Sat		REST DAY - Southern Declination		
	23rd	Sun		Capricorn		
	24th	Mon		Capricorn	RE	Aquarius
	25th	Tue		Aquarius		
	26th	Wed		Aquarius	REST	Pisces
	27th	Thur		Pisces		
	28th	Fri		Pisces	RE	Aries
N XhN	29th	Sat	Above	REST DAY - New Moon - Partial Solar Eclipse		
P4	30th	Sun		REST DAY - Day after Eclipse Day (Daylight Savings Starts)		
	31st	Mon		Taurus		

APRIL	2025		Hour:	1am – Midnight
	1st	Tue	Above	Taurus / REST / Gemini
	2nd	Wed		Gemini
^	3rd	Thur		REST DAY - Northern Declination
	4th	Fri		Cancer
1Q	5th	Sat		Cancer / REST
	6th	Sun		REST / Leo
	7th	Mon		Leo
	8th	Tue		Leo / REST / Virgo
	9th	Wed		Virgo
	10th	Thur		Virgo / REST
XhS	11th	Fri		REST / Libra
	12th	Sat		Libra
F A	13th	Sun	Below	REST DAY - Full Moon
	14th	Mon		Scorpio
	15th	Tue		Scorpio
	16th	Wed		Scorpio / RE / Sagittarius
	17th	Thur		Sagittarius
V	18th	Fri		REST DAY - Southern Declination
	19th	Sat		Capricorn
	20th	Sun		Capricorn / REST / Aquar
3Q	21st	Mon		Aquarius
	22nd	Tue		Aquarius / REST
	23rd	Wed		REST / Pisces
	24th	Thur		Pisces
	25th	Fri		Pisces / REST / Aries
XhN	26th	Sat		Aries / REST
N P3	27th	Sun	Above	REST DAY - New Moon
	28th	Mon		Taurus
	29th	Tue		Taurus / REST / Gemini
	30th	Wed		Gemini

MAY	2025		Hour:	1am – Midnight
^	1st	Thur	Above	REST DAY - Northern Declination
	2nd	Fri		Cancer
	3rd	Sat		Cancer / REST / Leo
1Q	4th	Sun		Leo
	5th	Mon		Leo / REST / Virgo
	6th	Tue		Virgo
	7th	Wed		Virgo
	8th	Thur		Virgo / REST / Libra
XhS	9th	Fri		Libra
	10th	Sat		Libra / REST / Scorpio
A	11th	Sun		Scorpio
F	12th	Mon	Below	REST DAY - Full Moon
	13th	Tue		Scorpio / REST / Sagittarius
	14th	Wed		Sagittarius
V	15th	Thur		REST DAY - Southern Declination
	16th	Fri		Capricorn
	17th	Sat		Capricorn

Hour: 1am 2am 3am 4am 5am 6am 7am 8am 9am 10am 11am Noon 1pm 2pm 3pm 4pm 5pm 6pm 7pm 8pm 9pm 10pm 11pm Midnight

MAY		2025	Hour:			
	18th	Sun		Capricorn	REST	Aquarius
	19th	Mon		Aquarius		
3Q	20th	Tue		Aquarius	REST	Pisces
	21st	Wed		Pisces		
	22nd	Thur	Below	Pisces	RE	Aries
XhN	23rd	Fri		Aries		
	24th	Sat		Aries	REST	Taurus
	25th	Sun		Taurus		
P5	26th	Mon		Taurus	REST	Gemini
N	27th	Tue		REST DAY - New Moon		
^	28th	Wed		REST DAY - Northern Declination		
	29th	Thur	Above	Cancer		
	30th	Fri		Cancer	REST	Leo
	31st	Sat		Leo		

Hour: 1am 2am 3am 4am 5am 6am 7am 8am 9am 10am 11am Noon 1pm 2pm 3pm 4pm 5pm 6pm 7pm 8pm 9pm 10pm 11pm Midnight

JUNE		2025	Hour:			
	1st	Sun		Leo	REST	
	2nd	Mon		REST	Virgo	
1Q	3rd	Tue		Virgo		
	4th	Wed		Virgo	REST	Libra
XhS	5th	Thur	Above	Libra		
	6th	Fri		Libra		
A	7th	Sat		Libra	REST	Scorpio
	8th	Sun		Scorpio		
	9th	Mon		Scorpio	REST	Sagittarius
	10th	Tue		Sagittarius		
F V	11th	Wed		REST DAY - Full Moon / Southern Declination		
	12th	Thur		REST	Capricorn	
	13th	Fri		Capricorn		
	14th	Sat		Capricorn	REST	Aquarius
	15th	Sun		Aquarius		
	16th	Mon		Aquarius	RE	Pisces
	17th	Tue		Pisces		
3Q	18th	Wed	Below	Pisces	REST	Aries
XhN	19th	Thur		Aries		
	20th	Fri		Aries		
	21st	Sat		Aries	RE	Taurus
	22nd	Sun		Taurus		
P8	23rd	Mon		Taurus	REST	Gemini
	24th	Tue		Gemini		
N ^	25th	Wed		REST DAY - New Moon / Northern Declination		
	26th	Thur		Cancer		
	27th	Fri		Cancer	REST	Leo
	28th	Sat	Above	Leo		
	29th	Sun		Leo	REST	Virgo
	30th	Mon		Virgo		

JULY		2025	Hour:	1am 2am 3am 4am 5am 6am 7am 8am 9am 10am 11am Noon 1pm 2pm 3pm 4pm 5pm 6pm 7pm 8pm 9pm 10pm 11pm Midnight
	1st	Tue		Virgo / REST / Libra
1Q XhS	2nd	Wed		Libra / REST
	3rd	Thur		REST DAY - Moon Void of Course all day
	4th	Fri		REST / Scorpio
A	5th	Sat	Above	Scorpio
	6th	Sun		Scorpio / RE / Sagittarius
	7th	Mon		Sagittarius / REST
	8th	Tue		REST DAY - Moon Void of Course all day
V	9th	Wed		REST DAY - Southern Declination
F	10th	Thur		REST DAY - Full Moon
	11th	Fri		REST / Aquarius
	12th	Sat		Aquarius / REST
	13th	Sun		REST DAY - Moon Void of Course all day
	14th	Mon		Pisces
	15th	Tue		Pisces / REST
	16th	Wed	Below	REST / Aries
XhN	17th	Thur		Aries
3Q	18th	Fri		Ar / REST / Taurus
	19th	Sat		Taurus
P11	20th	Sun		Taurus / REST / Gemini
	21st	Mon		Gemini / REST
^	22nd	Tue		REST DAY - Northern Declination
	23rd	Wed		Cancer
N	24th	Thur		REST DAY - New Moon
	25th	Fri		Leo
	26th	Sat		Leo / REST / Virgo
	27th	Sun	Above	Virgo
	28th	Mon		Virgo
	29th	Tue		Virgo / REST / Libra
XhS	30th	Wed		Libra / REST
	31st	Thur		REST / Scorpio

AUGUST		2025	Hour:	1am 2am 3am 4am 5am 6am 7am 8am 9am 10am 11am Noon 1pm 2pm 3pm 4pm 5pm 6pm 7pm 8pm 9pm 10pm 11pm Midnight
1Q A	1st	Fri		Scorpio
	2nd	Sat		Scorpio
	3rd	Sun		Scorp / REST / Sagittarius
	4th	Mon	Above	Sagittarius
V	5th	Tue		REST DAY - Southern Declination
	6th	Wed		Capricorn / REST
	7th	Thur		REST DAY - Moon Void of Course all day
	8th	Fri		REST / Aquarius
F	9th	Sat		REST DAY - Full Moon
	10th	Sun		REST / Pisces
	11th	Mon		Pisces / REST
	12th	Tue		REST / Aries
XhN	13th	Wed	Below	Aries / REST
P12	14th	Thur		REST / Taurus
	15th	Fri		Taurus
3Q	16th	Sat		Taurus / REST / Gemini
	17th	Sun		Gemini

AUGUST	2025		Hour:	1am – Midnight
^	18th	Mon	Below	REST DAY - Northern Declination
	19th	Tue		Cancer
	20th	Wed		Cancer; REST; Leo
	21st	Thur		Leo; REST
	22nd	Fri		REST DAY - Moon Void of Course all day
N	23rd	Sat	Above	REST DAY - New Moon
	24th	Sun		Virgo
	25th	Mon		Virgo; RE; Libra
XhS	26th	Tue		Libra
	27th	Wed		Libra; REST
	28th	Thur		REST; Scorpio
A	29th	Fri		Scorpio
	30th	Sat		Sco; REST; Sagittarius
1Q	31st	Sun		Sagittarius

SEPTEMBER	2025		Hour:	1am – Midnight
V	1st	Mon	Above	REST DAY - Southern Declination
	2nd	Tue		REST; Capricorn
	3rd	Wed		Capricorn
	4th	Thur		Capricorn; RE; Aquarius
	5th	Fri		Aquarius; REST
	6th	Sat		REST; Pisces
F	7th	Sun	Below	REST DAY - Full Moon - Lunar Total Eclipse
	8th	Mon		REST DAY - Day after Eclipse Day
XhN	9th	Tue		Aries
P9	10th	Wed		Aries; REST; Taurus
	11th	Thur		Taurus
	12th	Fri		Taurus; REST; Gemini
	13th	Sat		Gemini
3Q ^	14th	Sun		REST DAY - Northern Declination
	15th	Mon		RE; Cancer
	16th	Tue		Cancer
	17th	Wed		Cancer; REST; Leo
	18th	Thur		Leo
	19th	Fri		Leo; RE; Virgo
	20th	Sat		Virgo
N	21st	Sun	Above	REST DAY - New Moon / Partial Solar Eclipse
XhS	22nd	Mon		REST DAY - Day after Eclipse Day
	23rd	Tue		Libra; REST
	24th	Wed		REST; Scorpio
	25th	Thur		Scorpio
A	26th	Fri		Scorpio; REST; Sagittarius
	27th	Sat		Sagittarius
	28th	Sun		Sagittarius
1Q V	29th	Mon		REST DAY - Southern Declination
	30th	Tue		Capricorn

OCTOBER		2025	Hour:	1am · 2am · 3am · 4am · 5am · 6am · 7am · 8am · 9am · 10am · 11am · Noon · 1pm · 2pm · 3pm · 4pm · 5pm · 6pm · 7pm · 8pm · 9pm · 10pm · 11pm · Midnight
	1st	Wed	Above	Capricorn · REST · Aquarius
	2nd	Thur	Above	Aquarius
	3rd	Fri	Above	Aquarius · REST
	4th	Sat	Above	REST · Pisces
	5th	Sun	Above	Pisces
	6th	Mon	Above	Pis · REST · Aries
F XhN	7th	Tue	Below	REST DAY - Full Moon
P6	8th	Wed	Below	REST · Taurus
	9th	Thur	Below	Taurus
	10th	Fri	Below	Tau · REST · Gemini
	11th	Sat	Below	Gemini
^	12th	Sun	Below	REST DAY - Northern Declination
3Q	13th	Mon	Below	Cancer
	14th	Tue	Below	Cancer · REST · Leo
	15th	Wed	Below	Leo
	16th	Thur	Below	Leo · REST · Virgo
	17th	Fri	Below	Virgo
	18th	Sat	Below	Virgo · REST
	19th	Sun	Below	REST · Libra
XhS	20th	Mon	Below	Libra
N	21st	Tue	Above	REST DAY - New Moon
	22nd	Wed	Above	Scorpio
A	23rd	Thur	Above	Scorpio
	24th	Fri	Above	Scorpio · RE · Sagittarius
	25th	Sat	Above	Sagittarius
V	26th	Sun	Above	REST DAY - Southern Declination / Daylight Savings Ends
	27th	Mon	Above	Capricorn
	28th	Tue	Above	Capricorn
1Q	29th	Wed	Above	Capricorn · RE · Aquarius
	30th	Thur	Above	Aquarius
	31st	Fri	Above	Aquarius · REST · Pisces

NOVEMBER		2025	Hour:	1am · 2am · 3am · 4am · 5am · 6am · 7am · 8am · 9am · 10am · 11am · Noon · 1pm · 2pm · 3pm · 4pm · 5pm · 6pm · 7pm · 8pm · 9pm · 10pm · 11pm · Midnight
	1st	Sat	Above	Pisces
	2nd	Sun	Above	Pisces · REST · Aries
XhN	3rd	Mon	Above	Aries
	4th	Tue	Above	Aries · REST · Taurus
F P1	5th	Wed	Below	REST DAY - Full Moon
	6th	Thur	Below	Taurus · REST · Gemini
	7th	Fri	Below	Gemini
^	8th	Sat	Below	REST DAY - Northern Declination
	9th	Sun	Below	Cancer
	10th	Mon	Below	Cancer · RE · Leo
	11th	Tue	Below	Leo
3Q	12th	Wed	Below	Leo · RE · Vir
	13th	Thur	Below	Virgo
	14th	Fri	Below	Virgo
	15th	Sat	Below	Virgo · REST · Libra
XhS	16th	Sun	Below	Libra
	17th	Mon	Below	Libra · REST · Scorpio

NOVEMBER		2025	Hour:	1am 2am 3am 4am 5am 6am 7am 8am 9am 10am 11am Noon 1pm 2pm 3pm 4pm 5pm 6pm 7pm 8pm 9pm 10pm 11pm Midnight
	18th	Tue	Belo	Scorpio
	19th	Wed		Scorpio
N A	20th	Thur	Above	REST DAY - New Moon
	21st	Fri		Sagittarius
V	22nd	Sat		REST DAY - Southern Declination
	23rd	Sun		Capricorn
	24th	Mon		Capricorn
	25th	Tue		Capricorn | REST | Aquarius
	26th	Wed		Aquarius
	27th	Thur		Aquarius | REST | Pisces
1Q	28th	Fri		Pisces
	29th	Sat		Pisces
XhN	30th	Sun		REST | Aries

DECEMBER		2025	Hour:	1am 2am 3am 4am 5am 6am 7am 8am 9am 10am 11am Noon 1pm 2pm 3pm 4pm 5pm 6pm 7pm 8pm 9pm 10pm 11pm Midnight
	1st	Mon	Above	Aries | REST
	2nd	Tue		REST | Taurus
	3rd	Wed		Taurus
F P2	4th	Thur	Below	REST DAY - Full Moon
^	5th	Fri		REST DAY - Northern Declination
	6th	Sat		Gemin | REST | Cancer
	7th	Sun		Cancer
	8th	Mon		Cancer | REST | Leo
	9th	Tue		Leo
	10th	Wed		Leo | REST | Virgo
3Q	11th	Thur		Virgo
	12th	Fri		Virgo | REST | Libr
XhS	13th	Sat		Libra
	14th	Sun		Libra
	15th	Mon		Libra | RE | Scorpio
	16th	Tue		Scorpio
A	17th	Wed		Scorpio | REST | Sagittarius
	18th	Thur		Sagittarius
V	19th	Fri		REST DAY - Southern Declination
N	20th	Sat	Above	REST DAY - New Moon
	21st	Sun		Capricorn
	22nd	Mon		Capricorn | REST | Aquarius
	23rd	Tue		Aquarius
	24th	Wed		Aquarius | REST
	25th	Thur		REST | Pisces
	26th	Fri		Pisces
1Q XhN	27th	Sat		Pisces | REST | Aries
	28th	Sun		Aries
	29th	Mon		Aries | REST | Taurus
	30th	Tue		Taurus
	31st	Wed		Taurus | REST | Gemini

UNIVERSAL FISHING CALENDAR 2025

AV	VG	EX
🐟	🐟🐟	🐟🐟🐟

JANUARY	1st	2nd	3rd	4th	5th	6th	7th	8th	9th	10th	11th	12th	13th	14th	15th	16th	17th	18th	19th	20th	21st	22nd	23rd	24th	25th	26th	27th	28th	29th	30th	31st
MOON						First Q	P13						Full Moon								Third Q, Ap								New Moon		
Best Bite Time am/pm	1.00	2.00	3.00	4.00	5.00	6.00	6.30	7.00	8.00	9.00	10.00	11.00	12.00	12.30	1.00	2.00	3.00	4.00	4.30	5.00	6.00	6.30	7.00	8.00	9.00	10.00	10.30	11.00	12.00	12.30	1.00
Next Best Bite Time am/pm	7.00	8.00	9.00	10.00	11.00	12.00	12.30	1.00	2.00	3.00	4.00	5.00	6.00	6.30	7.00	8.00	9.00	10.00	10.30	11.00	12.00	12.30	1.00	2.00	3.00	4.00	4.30	5.00	6.00	6.30	7.00
RATING	🐟🐟🐟	🐟	🐟	🐟	🐟🐟	🐟🐟	🐟🐟	🐟	🐟	🐟	🐟	🐟🐟🐟	🐟🐟🐟	🐟🐟🐟	🐟	🐟	🐟	🐟	🐟	🐟🐟	🐟🐟	🐟🐟	🐟	🐟	🐟	🐟	🐟	🐟🐟🐟	🐟🐟🐟	🐟🐟🐟	🐟

FEBRUARY	1st	2nd	3rd	4th	5th	6th	7th	8th	9th	10th	11th	12th	13th	14th	15th	16th	17th	18th	19th	20th	21st	22nd	23rd	24th	25th	26th	27th	28th
MOON		P10			First Q							Full Moon						Ap			Third Q							New Moon
Best Bite Time am/pm	2.00	3.00	4.00	5.00	6.00	7.00	8.00	9.00	10.00	10.30	11.00	12.00	12.30	1.00	2.00	3.00	3.30	4.00	4.30	5.00	6.00	7.00	8.00	9.00	10.00	11.00	11.30	12.00
Next Best Bite Time am/pm	8.00	9.00	10.00	11.00	12.00	1.00	2.00	3.00	4.00	4.30	5.00	6.00	6.30	7.00	8.00	9.00	9.30	10.00	10.30	11.00	12.00	1.00	2.00	3.00	4.00	5.00	5.30	6.00
RATING	🐟	🐟	🐟	🐟🐟	🐟🐟	🐟🐟	🐟	🐟	🐟	🐟	🐟🐟🐟	🐟🐟🐟	🐟🐟🐟	🐟	🐟	🐟	🐟	🐟	🐟	🐟🐟	🐟🐟	🐟🐟	🐟	🐟	🐟	🐟	🐟🐟🐟	🐟🐟🐟

MARCH	1st	2nd	3rd	4th	5th	6th	7th	8th	9th	10th	11th	12th	13th	14th	15th	16th	17th	18th	19th	20th	21st	22nd	23rd	24th	25th	26th	27th	28th	29th	30th	31st
MOON	P7						First Q							Full Moon				Ap					Third Q					New Moon		P4	
Best Bite Time am/pm	1.00	2.00	3.00	4.00	5.00	5.30	6.00	7.00	8.00	9.00	10.00	11.00	11.30	12.00	1.00	2.00	3.00	3.30	4.00	4.30	5.00	6.00	7.00	8.00	9.00	10.00	11.00	12.00	1.00	2.00	3.00
Next Best Bite Time am/pm	7.00	8.00	9.00	10.00	11.00	11.30	12.00	1.00	2.00	3.00	4.00	5.00	5.30	6.00	7.00	8.00	9.00	9.30	10.00	10.30	11.00	12.00	1.00	2.00	3.00	4.00	5.00	6.00	7.00	8.00	9.00
RATING	🐟🐟🐟	🐟	🐟	🐟	🐟	🐟🐟	🐟🐟	🐟🐟	🐟	🐟	🐟	🐟	🐟🐟🐟	🐟🐟🐟	🐟🐟🐟	🐟	🐟	🐟	🐟	🐟	🐟🐟	🐟🐟	🐟🐟	🐟	🐟	🐟	🐟🐟🐟	🐟🐟🐟	🐟🐟🐟	🐟	🐟

APRIL	1st	2nd	3rd	4th	5th	6th	7th	8th	9th	10th	11th	12th	13th	14th	15th	16th	17th	18th	19th	20th	21st	22nd	23rd	24th	25th	26th	27th	28th	29th	30th
MOON					First Q								Full Moon	Ap						Third Q							P3 New Moon			
Best Bite Time am/pm	3.30	4.00	5.00	5.30	6.00	7.00	8.00	8.30	9.00	10.00	11.00	11.30	12.00	12.30	1.00	2.00	3.00	4.00	5.00	6.00	6.30	7.00	8.00	9.00	10.00	11.00	12.00	12.00	1.00	2.00
Next Best Bite Time am/pm	9.30	10.00	11.00	11.30	12.00	1.00	2.00	2.30	3.00	4.00	5.00	5.30	6.00	6.30	7.00	8.00	9.00	10.00	11.00	12.00	12.30	1.00	2.00	3.00	4.00	5.00	6.00	6.00	7.00	8.00
RATING	🐟	🐟	🐟	🐟🐟	🐟🐟	🐟🐟	🐟	🐟	🐟	🐟	🐟	🐟🐟🐟	🐟🐟🐟	🐟🐟🐟	🐟	🐟	🐟	🐟	🐟🐟	🐟🐟	🐟🐟	🐟	🐟	🐟	🐟	🐟🐟🐟	🐟🐟🐟	🐟🐟🐟	🐟	🐟

MAY	1st	2nd	3rd	4th	5th	6th	7th	8th	9th	10th	11th	12th	13th	14th	15th	16th	17th	18th	19th	20th	21st	22nd	23rd	24th	25th	26th	27th	28th	29th	30th	31st
MOON					First Q						Ap	Full Moon								Third Q						P5	New Moon				
Best Bite Time am/pm	2.30	3.00	4.00	5.00	6.00	7.00	8.00	8.30	9.00	10.00	11.00	12.00	12.30	1.00	2.00	3.00	4.00	4.30	5.00	6.00	7.00	8.00	9.00	9.30	10.00	11.00	12.00	1.00	2.00	3.00	3.30
Next Best Bite Time am/pm	8.30	9.00	10.00	11.00	12.00	1.00	2.00	2.30	3.00	4.00	5.00	6.00	6.30	7.00	8.00	9.00	10.00	10.30	11.00	12.00	1.00	2.00	3.00	3.30	4.00	5.00	6.00	7.00	8.00	9.00	9.30
RATING	🐟	🐟	🐟	🐟🐟	🐟🐟	🐟🐟	🐟	🐟	🐟	🐟	🐟🐟🐟	🐟🐟🐟	🐟🐟🐟	🐟	🐟	🐟	🐟	🐟	🐟🐟	🐟🐟	🐟🐟	🐟	🐟	🐟	🐟	🐟🐟🐟	🐟🐟🐟	🐟🐟🐟	🐟	🐟	🐟

JUNE	1st	2nd	3rd	4th	5th	6th	7th	8th	9th	10th	11th	12th	13th	14th	15th	16th	17th	18th	19th	20th	21st	22nd	23rd	24th	25th	26th	27th	28th	29th	30th
MOON			First Q				Ap				Full Moon								Third Q				P8		New Moon					
Best Bite Time am/pm	4.30	5.00	6.00	6.30	7.00	8.00	9.00	10.00	11.00	11.30	12.00	1.00	2.00	3.00	3.30	4.30	5.00	5.30	6.00	7.00	8.00	9.00	10.00	11.00	12.00	1.00	2.00	3.00	3.30	4.30
Next Best Bite Time am/pm	10.30	11.00	12.00	12.30	1.00	2.00	3.00	4.00	5.00	5.30	6.00	7.00	8.00	9.00	9.30	10.30	11.00	11.30	12.00	1.00	2.00	3.00	4.00	5.00	6.00	7.00	8.00	9.00	9.30	10.30
RATING	🐟	🐟🐟	🐟🐟	🐟🐟	🐟	🐟	🐟	🐟	🐟	🐟🐟🐟	🐟🐟🐟	🐟🐟🐟	🐟	🐟	🐟	🐟	🐟	🐟🐟	🐟🐟	🐟🐟	🐟	🐟	🐟	🐟🐟🐟	🐟🐟🐟	🐟🐟🐟	🐟	🐟	🐟	🐟

AV	VG	EX
🐟	🐟🐟	🐟🐟🐟

JULY	1st	2nd	3rd	4th	5th	6th	7th	8th	9th	10th	11th	12th	13th	14th	15th	16th	17th	18th	19th	20th	21st	22nd	23rd	24th	25th	26th	27th	28th	29th	30th	31st
MOON			First Q		Ap					Full Moon							Third Q			P11				New Moon							
Best Bite Time am/pm	5.00	5.30	6.00	6.30	7.00	8.00	9.00	10.00	11.00	12.00	1.00	2.00	2.30	3.00	4.00	5.00	6.00	6.30	7.00	8.00	9.00	10.00	11.00	12.00	1.00	2.00	2.30	3.00	3.30	4.00	5.00
Next Best Bite Time am/pm	11.00	11.30	12.00	12.30	1.00	2.00	3.00	4.00	5.30	7.00	7.00	8.00	8.30	9.00	10.00	11.00	12.00	12.30	1.00	2.00	3.00	4.00	5.30	7.00	7.00	8.00	8.30	9.00	9.30	10.00	11.00
RATING	🐟	🐟🐟	🐟🐟	🐟🐟	🐟	🐟	🐟	🐟	🐟🐟🐟	🐟🐟🐟	🐟🐟🐟	🐟	🐟	🐟	🐟	🐟🐟	🐟🐟	🐟🐟	🐟	🐟	🐟	🐟	🐟🐟🐟	🐟🐟🐟	🐟🐟🐟	🐟	🐟	🐟	🐟	🐟	🐟

AUGUST	1st	2nd	3rd	4th	5th	6th	7th	8th	9th	10th	11th	12th	13th	14th	15th	16th	17th	18th	19th	20th	21st	22nd	23rd	24th	25th	26th	27th	28th	29th	30th	31st
MOON		First Q Ap							Full Moon					P12		Third Q							New Moon							Ap	First Q
Best Bite Time am/pm	5.30	6.00	7.00	7.30	8.00	9.00	10.00	11.00	12.00	1.00	2.00	3.00	4.00	4.30	5.00	6.00	6.30	7.00	8.00	9.00	10.00	11.00	12.00	1.00	2.00	3.00	4.00	4.30	5.00	5.30	6.00
Next Best Bite Time am/pm	11.30	12.00	1.00	1.30	2.00	3.00	4.00	5.00	6.00	7.00	8.00	9.00	10.00	10.30	11.00	12.00	12.30	1.00	2.00	3.00	4.00	5.00	6.00	7.00	8.00	9.00	10.00	10.30	11.00	11.30	12.00
RATING	🐟🐟	🐟🐟	🐟🐟	🐟	🐟	🐟	🐟	🐟🐟🐟	🐟🐟🐟	🐟🐟🐟	🐟	🐟	🐟	🐟	🐟🐟	🐟🐟	🐟🐟	🐟	🐟	🐟	🐟	🐟🐟🐟	🐟🐟🐟	🐟🐟🐟	🐟	🐟	🐟	🐟	🐟	🐟🐟	🐟🐟

SEPTEMBER	1st	2nd	3rd	4th	5th	6th	7th	8th	9th	10th	11th	12th	13th	14th	15th	16th	17th	18th	19th	20th	21st	22nd	23rd	24th	25th	26th	27th	28th	29th	30th
MOON							Full Moon			P9				Third Q							New Moon					Ap				First Q
Best Bite Time am/pm	6.30	7.00	8.00	9.00	10.00	11.00	12.00	1.00	2.00	3.00	3.30	4.00	5.00	6.00	6.30	7.00	8.00	9.00	10.00	11.00	12.00	1.00	2.00	3.00	3.30	4.00	4.30	5.00	5.30	6.00
Next Best Bite Time am/pm	12.30	1.00	2.00	3.00	4.00	5.00	6.00	7.00	8.00	9.00	9.30	10.00	11.00	12.00	12.30	1.00	2.00	3.00	4.00	5.00	6.00	7.00	8.00	9.00	9.30	10.00	10.30	11.00	11.30	12.00
RATING	🐟🐟	🐟	🐟	🐟	🐟	🐟🐟🐟	🐟🐟🐟	🐟🐟🐟	🐟	🐟	🐟	🐟	🐟🐟	🐟🐟	🐟🐟	🐟	🐟	🐟	🐟	🐟🐟🐟	🐟🐟🐟	🐟🐟🐟	🐟	🐟	🐟	🐟	🐟	🐟	🐟🐟	🐟🐟

OCTOBER	1st	2nd	3rd	4th	5th	6th	7th	8th	9th	10th	11th	12th	13th	14th	15th	16th	17th	18th	19th	20th	21st	22nd	23rd	24th	25th	26th	27th	28th	29th	30th	31st
MOON							Full Moon	P6						Third Q							New Moon			Ap						First Q	
Best Bite Time am/pm	7.00	7.30	8.00	9.00	10.00	11.00	12.00	1.00	2.00	2.30	3.00	4.00	5.00	6.00	6.30	7.00	8.00	9.00	10.00	11.00	12.00	1.00	1.00	2.00	3.00	4.00	4.30	5.00	6.00	6.30	7.00
Next Best Bite Time am/pm	1.00	1.30	2.00	3.00	4.00	5.00	6.00	7.00	8.00	8.30	9.00	10.00	11.00	12.00	12.30	1.00	2.00	3.00	4.00	5.00	6.00	7.00	7.00	8.00	9.00	10.00	10.30	11.00	12.00	12.30	1.00
RATING	🐟🐟	🐟	🐟	🐟	🐟	🐟🐟🐟	🐟🐟🐟	🐟🐟🐟	🐟	🐟	🐟	🐟	🐟🐟	🐟🐟	🐟🐟	🐟	🐟	🐟	🐟	🐟🐟🐟	🐟🐟🐟	🐟🐟🐟	🐟	🐟	🐟	🐟	🐟	🐟🐟	🐟🐟	🐟🐟	🐟

NOVEMBER	1st	2nd	3rd	4th	5th	6th	7th	8th	9th	10th	11th	12th	13th	14th	15th	16th	17th	18th	19th	20th	21st	22nd	23rd	24th	25th	26th	27th	28th	29th	30th
MOON					P1 Full Moon							Third Q								New Moon Ap								First Q		
Best Bite Time am/pm	8.00	9.00	10.00	11.00	12.00	1.00	1.30	2.00	3.00	4.00	5.00	6.00	6.30	7.00	7.30	8.00	9.00	10.00	11.00	12.00	1.00	2.00	2.30	3.00	4.00	5.00	6.00	7.00	8.00	9.00
Next Best Bite Time am/pm	2.00	3.00	4.00	5.00	6.00	7.00	7.00	8.00	9.00	10.00	11.00	12.00	12.30	1.00	1.30	2.00	3.00	4.00	5.00	6.00	7.00	8.00	8.30	9.00	10.00	11.00	12.00	1.00	2.00	3.00
RATING	🐟	🐟	🐟	🐟🐟🐟	🐟🐟🐟	🐟🐟🐟	🐟	🐟	🐟	🐟	🐟🐟	🐟🐟	🐟🐟	🐟	🐟	🐟	🐟	🐟	🐟🐟🐟	🐟🐟🐟	🐟🐟🐟	🐟	🐟	🐟	🐟	🐟🐟	🐟🐟	🐟🐟	🐟	🐟

DECEMBER	1st	2nd	3rd	4th	5th	6th	7th	8th	9th	10th	11th	12th	13th	14th	15th	16th	17th	18th	19th	20th	21st	22nd	23rd	24th	25th	26th	27th	28th	29th	30th	31st
MOON				Full moon, P2								Third Q					Ap			New Moon							First Q				
Best Bite Time am/pm	10.00	10.30	11.00	12.00	12.30	1.00	2.00	3.00	4.00	4.30	5.00	6.00	6.30	7.00	8.00	9.00	10.00	10.30	11.00	12.00	1.00	2.00	3.00	3.30	4.00	5.30	6.00	7.00	7.30	8.00	9.00
Next Best Bite Time am/pm	4.00	4.30	5.00	6.00	6.30	7.00	8.00	9.00	10.00	10.30	11.00	12.00	12.30	1.00	2.00	3.00	4.00	4.30	5.00	6.00	7.00	8.00	9.00	9.30	10.00	11.30	12.00	1.00	1.30	2.00	3.00
RATING	🐟	🐟	🐟🐟🐟	🐟🐟🐟	🐟🐟🐟	🐟	🐟	🐟	🐟	🐟	🐟🐟	🐟🐟	🐟🐟	🐟	🐟	🐟	🐟	🐟	🐟🐟🐟	🐟🐟🐟	🐟🐟🐟	🐟	🐟	🐟	🐟	🐟🐟	🐟🐟	🐟🐟	🐟	🐟	🐟

ACKNOWLEDGEMENTS

The almanacs are not possible without the considerable help and support of Glenys and Victoria in our central North Island office. My huge thanks, also to Vicki in dispatch.

We also acknowledge the NOAA/NWS for ongoing permission to adapt archival MSL maps.

We hope that the information provided here will help in your planning.

Ken Ring